METALWORKING

METALWORKING

Old-Fashioned Tools, Materials,
and Processes for the Handyman

Paul N. Hasluck

Skyhorse Publishing

Skyhorse Publishing books may be purchased in bulk at special discounts for sales promotion, corporate gifts, fund-raising, or educational purposes. Special editions can also be created to specifications. For details, contact the Special Sales Department, Skyhorse Publishing, 307 West 36th Street, 11th Floor, New York, NY 10018 or
info@skyhorsepublishing.com.

www.skyhorsepublishing.com

10 9 8 7 6 5 4 3 2 1

Library of Congress Cataloging-in-Publication Data

Metalworking : old-fashioned tools, materials, and processes for the handyman / edited by Paul N. Hasluck.
    p. cm.
Includes index.
Originally published: Philadelphia : D. McKay, 1907.
ISBN 978-1-61608-185-0 (pbk. : alk. paper)
1. Metal-work. I. Hasluck, Paul N. (Paul Nooncree), 1854-1931.
TT205.H23 2011
684'.09--dc22

                    2010036330

Printed in the United States of America

PREFACE.

THE scope of this book embraces practically the whole art of working metals with hand-tools and with such simple machine-tools as the small engineering shop usually contains. The tool outfit of the average metalworker does not generally include anything more ambitious than a lathe with or without slide-rest, overhead motion, etc., and it is with this limitation in mind that the whole of the contents of this book have been prepared. Even within such limits, the scope is extensive, and has been made to include a large and pleasing variety of work.

In attempting to present a graded course of instruction, difficulties have been encountered which were absent in the preparation of the companion volume—"The Handyman's Book of Tools, Materials, and Processes Employed in Woodworking." The practice of metalworking has been split up into so many branches, and these have been so sub-divided, that a number of distinct trades has been created; and the tools and processes of one trade often have but slight relation to those of another. In face of this fact, it has been found impossible to arrange the work in a series of exercises gradually and successively increasing in difficulty and in the skill necessary for their performance. The book chiefly conforms to another arrangement—that by which the tools and processes are described in their natural sequence; thus, after a brief introduction giving particulars of all the known metals, the book opens with a section on foundry work, the basis of all modern metalworking. This is followed by a section on the art of the blacksmith, and then come detail descriptions of the tools and processes by which the rough surfaces left by casting and forging are chipped, filed, scraped and polished. Succeeding sections deal with annealing, hardening and tempering, drilling and boring, followed by screw threading with taps, screwplates, and dies. Of the greatest importance is the section on soldering, brazing, and riveting, in which an attempt is made to impart a sound knowledge of the whole of the processes in general use. Three sections on the manufacture and ornamentation of sheet metal ware include instructions on pattern drawing, cutting out and shaping, seaming and jointing, wiring and beading, repoussé and oriental decoration, and it is believed that the section on repoussé work forms the most complete treatise on the subject yet published. The finishing, lacquering, and colouring of brass is a subject that follows in natural sequence.

The second part of the book introduces work of a different kind, and considerably more than a hundred pages is devoted to lathes and lathework; all the necessary appliances for and the general processes of chucking, hand turning, slide-rest turning, using revolving cutters, spinning and knurling are described in plain language and fully illustrated. Engineers' tools for setting out, measuring and testing, are illustrated and described in the next section, and then comes the first of the more important examples—a serviceable 4½-in. centre lathe with slide-rest—in the construction of which much of the information given in previous chapters is applied. Later examples include a skeleton clock, small horizontal and vertical steam engines, boilers, a petrol motor, water motors, a dynamo and electric motor, a microscope and a telescope. These are all the subjects of clear working drawings and minute practical instruction. Among miscellaneous branches of metal-working dealt with at the end of the volume are gold and silver working, electro-plating, wire work, electric bell making, etc., etc.

Actual practice is recorded throughout this book, and the discussion of theory has been allowed only when it is an essential preliminary to understanding the principles underlying a method, a process, or the action of a tool. The examples of work have been adapted from existing articles, and the columns of " Work," one of the journals it is my fortune to edit, have been drawn on freely.

The illustrations have been prepared regardless of trouble and expense, with the full consciousness of their value in showing at a glance what pages of letterpress would fail to convey. Many of the engravings show metalworkers' tools and appliances, and in this connection special acknowledgment must be made to the following firms for their great help in kindly having lent electrotypes illustrating modern tools and appliances of approved design: Messrs. G. Birch & Co., Islington Tool Works, Salford, Manchester; Britannia Engineering Co., Ltd., Britannia Works, Colchester; Messrs. W. Canning & Co., Great Hampton Street, Birmingham; Messrs. Charles Churchill & Co., Ltd., 9, Leonard Street, Finsbury, London, E.C.; Messrs. Fletcher, Russell & Co., Ltd., Warrington; Messrs. J. E. Hartley and Son, 13, St. Paul's Square, Birmingham; Messrs. William Marples & Sons, Ltd., Hibernia Works, Sheffield; Messrs. Richard Melhuish & Sons, 84, Fetter Lane, London, E.C.; Mr. Henry Milnes, Ingleby Works, Bradford; Messrs. T. Morris & Co., 172, Hockley Hill, Birmingham; Messrs. Charles Nurse & Co., 182, Walworth Road, London, S.E.; and Messrs. Henri Picard & Frère, 26 & 27, Bartlett's Buildings, Holborn Circus, London, E.C.

The index, containing upwards of 4,500 entries, is a means of readily finding any item of information contained in the work.

P. N. HASLUCK.

CONTENTS.

METALWORKING.

INTRODUCTION.

THE SCOPE AND OBJECT OF THIS BOOK.

METALWORKING is only limited in its subject matter to the extent that its processes are hand-wrought as distinguished from those for which machinery is employed, and that metal is the material worked upon. An idea of the wide scope of the book may be inferred from an outline of its contents. Metalworking in all branches of practical handicraft will be fully dealt with. First comes a concise though comprehensive table showing at a glance the physical qualities of all metals, their specific weights, their strength, their melting points, etc. This is followed by explanations of the first principles of metalworking in the foundry, at the forge, at the vice, and in the lathe. Then will be described in detail general processes employed in metalworking—such as jointing, soldering, drilling, polishing, and lacquering. Electro-metallurgy is included. Those elements of metalworking that are common to many handicrafts having been disposed of, the tools used in metalworking will be illustrated, and how to make them will be explained. Next will follow a large and varied collection of graded examples of work, each one clearly illustrated and described in minute detail. These examples will be typical of the specialised handicrafts of many widely different trades, including wireworking, lathe-building, gunsmithing. motor-building, electroplating, goldsmithing, art metalworking, cutlery, electric bell making, jewellers' work, etc. The contents of this book range from the rudimentary teaching required by the tyro to the construction of high-class examples that will interest the adept craftsman.

CAST AND WROUGHT METALS.

Metalwork of all kinds is readily divisible into two broad classes—cast and hammered. This distinction is sufficient for general purposes, although it does not entirely cover the field, because nearly all malleable iron made by modern processes, and all mild steel, are cast before they are puddled, hammered, or rolled. Neither does it include the method of electrotyping. Both casting and hammering were employed in prehistoric ages, and both methods have continued in use until the present time. Whilst it is comparatively easy to produce intricate forms by casting, the forging of similar forms taxes the very highest skill and patience of the hammerman. Most of the specimens of prehistoric art in metalworking which have been preserved to us are in the form of castings, but the more delicate hammered works are mostly of historic dates. The work of the blacksmith is of comparatively recent origin. Skill in the working of iron dates only from a few centuries before the Christian era. Previous to the introduction of iron, bronze was the metal employed for weapons of war and defence, and for articles of ornament and domestic service. The ancients had acquired very great skill in the composition and use of this alloy, as is proved by the vast number of cutting tools and utensils that have been brought to light by the researches of archæologists.

THE AGE OF BRONZE.

The origin of the age of bronze is lost in remote antiquity. No hard chronological line separates it from the preceding neolithic or new stone age. But the discovery of the use of copper and tin marked a most distinct advance in the history of civilisation; and in this broad sense the bronze-using period may be regarded as a very important age or era in the history of mankind. It is considered probable, and in some isolated districts it is a fact, that there was also a period when pure copper was employed, unalloyed with tin. But the advantages in increased hardness which were gained by alloying tin with copper were so evident, that in most cases bronze, and not pure copper, was used; and as a matter of fact, nearly all the primitive implements of metal as yet found in the old world are made of alloys of copper and tin. The composition of the prehistoric bronzes varied extremely. A good bronze mixture, as used by modern engineers, contains about 88 or 89 of copper to 12 or 11 of tin respectively. Many of the ancient bronzes contained proportions approximating to these, but some contained a much less, some also a much greater proportion of copper. Very considerable traces of lead, nickel, silver, and iron also occur in the early bronzes, the modern art of separating copper from foreign ingredients present in the ores being unknown to the early smelters. In no essential did the earliest known methods of moulding and casting differ from those carried on at the present day. Yet relics have been found that date from a period long anterior to the Christian era; probably many are from 2,000 to 4,000 years old. It has been thought that the bronze age began in England some 1,200 or 1,400 years B.C., and that it lasted about a thousand years, but the knowledge of copper and tin may have been much earlier.

Importance and Value of Iron.

Iron is the most important of all the metals, though the least costly. It is more valuable than any of the precious metals in its usefulness to us. We are at the present time so dependent on iron that it is really difficult for us to imagine a time without iron. Cast-iron is the crude metal derived from the smelting furnace, and imperfectly freed from impurities. Wrought-iron, with which we are directly concerned, is the pure form of the metal, in colour a

metallic, steely grey, but it rusts very rapidly on exposure to damp. In iron we possess a substance which is at once hard, malleable, able to bear a great strain, and yet can be made very brittle; it is also inflexible, so that the most elastic springs can be formed from it. It can also be made to form the thick, heavy ribs and plating of the vessel of war, the slender blade of the surgeon's knife, or the exquisitely artistic and beautiful scroll and leaf work of the chancel screen, the altar railing, or the grille. Iron possesses, in fact, qualities so varied, vast, and useful as at once to mark it out as the central figure amongst the productions of earth. It appears to be quite certain that so important a metal was known from the very earliest times—at least, as far as regards some of its uses. In the early books of the Bible we read continually of iron in various forms and for various uses—domestic and social; also for weapons of war—as iron axes, iron swords, etc.

The Superiority of Wrought-iron.

There are certain qualities possessed by wrought-iron when hammered or rolled out which give it a great superiority over cast-iron for ornamental work and other work where there is no very considerable bulk. First, then, a fibrous texture, rendering it tough in working and able to be bent about in various shapes without cracking or breaking; then again its malleability, enabling the bar of iron to be drawn out or flattened into the required forms; again, its ductility, enabling a thick bar to be drawn out to the very thinnest of wire, or rolled out to the thinnest of sheet. Then again, and lastly, its most valuable quality for our purpose—the quality which we call welding, or the property of uniting together at a heat below the melting point, thus enabling ornamental effects to be produced with the metal alone without having patterns, etc.

The combination of these various qualities in one substance enables the production of ornamental and other effects in iron which would be impossible in any other metal known to us, and without which the smith's art and work would never have been brought into existence. The separate

METALS.

Metal.	Symbol.	Colour.	Derivation of Name.	Discoverer.	Date.	Atom. Wght. (new s'tem)	Specific Gravity.	Specific Heat at 0 C.	Electrical Conductivity. Hg at 0 C.	Heat Conductivity. silver = 100.
Aluminium ..	Al	Tin-white	Lat'n : alumen (alum)	Wöhler	1828	27 1	2·56	·2253	20·97	31·33
Antimony ..	bb	Silver-white	Latin : stibium	Valentine	1490	120·43	6·697	·0523	2·05	4 03
Arsenic ..	As	Steel-grey				75·01	5·727	·083	2 679	
Barium ..	Ba	Ylw-h.-white	Greek : baros (heavy)	Davy	1808	137·43	3·5–4			
Bismuth ..	Bi	White		Agricola	1530	208·11	9·759	·0305 at 20°	·8676	1 8
Cadmium ..	Cd	White, blue tinge	Greek : cadmia (calamin)	Stromeyer	1817	112·3	8·65–8·8	·0548	13·46	20·06
Cæsium ..	Cs	Silver-white	Latin : cæsius (bluish-grey)	Bunsen and Kischhoff	1860	132·9	1·88			
Calcium ..	Ca	Ylwah.-white	Latin : calx (lime)	Davy	1808	40	1·82	·1686	12·5	25 4
Cerium ..	Ce	Steel-grey	Planet Ceres	Klaproth, Hisinger, and Berzelius	1803	140	5·5	·04479		
Chromium ..	Cr	Greyish-white	Greek : chroma (colour)	Vanquelin	1797	52·45	6·8–7 3	·0098		
Cobalt ..	Co	Steel-grey	German : kobold (goblin)	Brandt	1735	58·8	8·52–8·95	·107	9·685	17·2
Copper ..	Cu	Reddish-ylw.	Latin : Cuprum (Cyprian)			63	8·36–8·95	·0933	52 to 54	73·6
Didymium ..	Di + Pr	White	Greek : didymus (double)	Mosander	1842	142	6·544	·04563		
Erbium ..	Er					166				
Gallium ..	Ga	Silver-white	Latin : Gallia (Gaul)	Lecoq de Boisbaudran	1875	70	5·86	·079		
Germanium ..	Ge	Greyish-white	Latin : Germania (German)	Winkler	1885	72·3	5·469	·0737		
Glucinum * ..	Be or Gl	Steel-colord.	Greek : glukus (sweet)	Wöhler	1827	9·08	2·1	·4702		
Gold	Au	Yellow	Hebrew			196·5	19·3	·0316	43 84	53·2
Indium ..	In	Silver-white	Latin : indicum (indigo)	Reich	1863	113·4	7·4	·05995		
Iridium ..	Ir	Grey	Latin : iris (rainbow)	Tennant	1803	192·5	22·38	·0323		
Iron	Fe	Greyish-white	Latin : ferrum			56	6·95–8·2	·114	9 68	11·9
Lanthanum ..	La	White	Greek : lanthancin (conceal)	Mosander	1839	138·5	6·163	·04485		
Lead	Pb	Blue-grey	Latin : plumbum			206·4	11·4	·03065	4·8	8·5
Lithium ..	Li	Silver-white	Greek : lithos (stone)	Arfvesson	1817	7·03	0·578–0·589	·9408	10·69	
Magnesium ..	Mg	Silver-white	Magnesia in Asia Minor	Davy	1808	24·36	1·75	20°·51°·245°	22·84	34 3
Manganese ..	Mn	White-grey	Magnesia in Asia Minor	Guhn	1740	55·02	8	14°·97°		
Mercury ..	Hg	White	The Deity and planet Mercury			200	13·6	·033	1	5·3
Molybdenum ..	Mo	Dull silver	Greek : molybdos (lead)	Hjelm	1782	96	8·62	·0629		
Neodymium ..	Nd					143 6				
Nickel ..	Ni	White	German : kupfernickel	Cronstedt	1751	58·7	8·3–8·7	·10916	7·374	
Niobium † ..	Nb	Steel-grey	Niobe	H. Rose	1841	94	4·6			
Osmium ..	Os	Blue-white	Greek : osmē (odour)	Tennant	1803	190·8	22·477	·03113		
Palladium ..	Pd	White	Planet Pallas	Wollaston	1803	106·5	11·4	·0582		
Platinum ..	Pt	White	Spanish : platæ (silver)	Woods	1741	194·5	21·5	·0314	8·042 Ag = 100	37·0
Potassium ..	K	Silver-white	Potash	Davy	1807	39·04	0·875	·166	11·23	45
Praseodymium ..	Pr					140·5		·0314		
Rhodium ..	Rh	Bluish-white	Greek : rhodon (a rose)	Wollaston	1803	102·7	12·1	·05803		
Rubidium ..	Rb	White	Ruber (dark red)	Bunsen	1860	85·2	1·52			
Ruthenium ..	Ru	White	Rutheria (Russia)	Claus	1843	101·4	12·261	·0611		
Samarium ..	Sm		Samerskite	Lecoq de Boisbaudran		150				
Scandium ..	Sc			Nilson	1879	44				
Silver	Ag	White				107·66	10·4–10·7	·0557	63 845	100
Sodium ..	Na	Silver-white	Latin : salsola (soda)	Davy	1807	22·995	0·9735	·2734	18·3	36·5
Strontium ..	Sr	Ylwash.-white	Strontian, a village in Argyllshire		1790	87·3	2·542			
Tantalum ..	Ta			Eheberg	1802	183	10·8			
Tellurium ..	Te	White shining semi-metal				127·49	6·255	·0475	·000777 Ag at 0° = 1	
Terbium ..	Tr			Muller	1782	160				
Thallium ..	Tl	White	Greek : thallus (green)	Crookes!	1861	203·64	11·88	·0325	5·225	
Thorium ..	Th	Greyish-white	Scandinavian god Thor	Berzelius	1828	232	11·1–11·23	·02787		
Tin	Sn	Silver-white	Latin : stannum (tin)			119	7·3	·0559	8·726	15·2
Titanium ..	Ti	Dark-grey	God Titan	Klaproth	1795	47·9	3·5888	·1135		
Tungsten ..	W	Steel-grey	Swedish : tungsten (heavy stone)	J. and F. d'Elhujar	1785	184·4	18·77	·035		
Uranium ..	U	Silver-white	The Deity and planet Uranus	Klaproth	1789	240	18·7	·0276		
Vanadium ..	V	Light-grey	Scandinavian : Vanadis, deity	Sefström	1830	51·4	5·5			
Ytterbium ..	Yb		Swedish : Ytterby (town in Sweden)	Marignac	1878	173				
Yttrium ..	Y		Swedish : Ytterby (town in Sweden)			89				
Zinc	Zn	Bluish-white	German : Zink (zinc)	Paracelsus	1541	65·4	6·9–7·15	·0935		
Zirconium ..	Zr	Grey	Persian : Zargun (gold coloured)	Klaproth	1789	90·5	4·15	·066		

Several earth metals are omitted from table, as Samarium, Gadolinium, Neon, Xenon, etc.

* Also called Beryllium.   † Also called Columbium.

pieces, when they leave the forge, may be fastened together with straps of heated iron tried round them, with bolts, screws, and rivets, or, where really required, may be fixed together with brass or silver solder. These little details will show that wrought-iron is a material that can practically be used almost as fancy wills, to make either the strongest and most massive, or the finest and most beautiful, work.

MALLEABILITY, DUCTILITY, AND TENACITY.

At this stage it will be well to note these leading characteristics of the metals:— Malleability is the capability of being extended without cracking or breaking, and for the various metals in general use is shown in the following list in order, the most malleable metal being placed first: Gold, silver, copper, tin, lead, zinc, iron, and nickel. Ductility is the property of being drawn into wire, and the metals are ranged in the following order: Gold, silver, iron, copper, nickel, zinc, tin, and lead. Tenacity, or tensile strength, is resistance to being pulled asunder; it is a variable quantity; crystalline construction is often accompanied by brittleness, and fibrous construction by high tenacity, and these are generally diminished by a rise in the temperature of the metal, while the reverse is often the case with regard to malleability

and ductility. Metals with feeble tenacity are known as brittle; this quality may be due to hardness or to molecular construction. The following list places the metals in order of tensile strength: iron, nickel, copper, silver, gold, zinc, tin, and lead.

WEIGHTS OF METALS PER FOOT AND INCH CUBE.

The following list, which gives the weights of most ordinary metals and alloys, will be useful. Weights per cubic foot: Brass, 520 lb.; copper, 549 lb.; nickel, 518 lb.; zinc, 429 lb.; aluminium, 160 lb.; lead, 710 lb.; antimony, 420 lb.; tin, 456 lb.; gunmetal, 544 lb.; and magnolia, 650 lb. Weights per cubic inch: Brass, ·3 lb.; copper, ·318 lb.; nickel, ·318 lb.; zinc, ·248 lb.; aluminium, ·093 lb.; lead, ·41 lb.; antimony, ·242 lb.; tin, ·264 lb.; gunmetal, ·315 lb.; and magnolia, ·376 lb.

MATERIALS FOR THE METALWORKER.

On p. 3 is given, in tabulated form, the most useful particulars of all known metals. Some of the metals tabulated are known only in the restricted field of scientific research. It is, however, useful for the metalworker to have a complete list for reference, and extended particulars of metals and alloys of general use in the arts are given in later pages of this book.

FOUNDRY WORK.

INTRODUCTION.

The work of the foundry falls naturally under two principal heads—the metal, and the moulds. In large shops these are two distinct classes of work, performed by different sets of men. In small shops the moulder must often mix and melt his own metal. The metal may be bought ready prepared in ingots; or the founder may prefer to make his own mixtures. Ingots are convenient, but mixing is cheaper, and the mixer knows what metal he has, which is more than can be said when ingots are bought. An experienced mixer can obtain any grade of metal needed for any class of work, and can use a good deal of waste stuff, together with runners and risers from previous casts.

ALUMINIUM BRONZE.

Aluminium bronze consists of aluminium and copper in varying proportions. A special hard alloy contains 11 per cent. aluminium, and others contain from 10 to 1¼ per cent. aluminium. The 10 per cent. alloy has a tenacity of about 100,000 lb. per square inch, but shrinks upwards of twice as much as brass. As the metal solidifies rapidly, it is necessary to pour with very large gates, and every precaution must be taken to prevent the metal setting in the gates before the casting has been completely poured. The great difficulty with large cored work is to prevent the metal cracking. This must be done by using baked cores consisting of sand and cinders ground together and just held with a little resin. Aluminium bronze of the above quality oxidises rapidly when poured from the crucible to the mould. If the film or skin of oxide passes into the mould, it will appear on the surface as dirt, and will probably spoil the casting. The scum may generally be kept back by the ordinary skimmer, if the pourer takes special care. As aluminium and copper volatilise only at very great heats, there is no appreciable loss in continued casting. Metal made simply by mixing aluminium and copper will most probably be brittle, and does not acquire its best properties till it has been remelted several times. If the casting is uniformly thin, it will generally come sharp, but, if there are thin and thick parts, it is necessary to put runners or offshoots from the thin parts to enable the metal to fill and soak back into the casting as it cools. This will overcome the difficulty often met of thumbing, or the formation of shallow places at the thick parts of the castings. A good bearing brass is an aluminium bronze containing 95 parts of copper and 5 of aluminium, to which is added 8 per cent. of tin. A good method is as follows:—In using a 140-lb. crucible, melt 124 lb. of copper under a good layer of charcoal; then add cautiously, bit by bit, 6½ lb. of ingot aluminium, or fairly heavy aluminium scrap. When the metal is quiet, add 10¼ lb. of ordinary cake tin, thoroughly well stir in, and pour into ingots. When making brasses, take about half the weight of the casting required in scrap metal, and make up with the new ingot metal. This, when melted, must be well stirred and then poured. It is necessary to have large gates to allow proper shrinkage, as there is considerable liability to draw. The mould must be rather loosely rammed, and when poured the metal must not be touched for some time, as it remains liquid in the mould longer than the usual alloys.

Anti-friction Alloys.

The metals that form anti-friction or bearing alloys are copper, tin, antimony, zinc, and lead, although more than three rarely enter into any one mixture. To reduce the friction at a journal, a bearing of brass, bronze, or gunmetal is made, and the bearing surface is lined with a white alloy of comparatively soft nature. On account of its low melting point, when the white bearing shows wear it may be melted out, and a fresh lining inserted, the bronze or other frame thus lasting a very considerable time. The metals are generally melted in an iron ladle, and if allowed to get too hot will volatilise, the nature of the alloy being altered, or will oxidise and form a scum that must on no account be allowed to run into the bearing. These white-metal bearings are used for very fast-running machinery, to prevent the axles or shafts and the brasses heating and consequently binding. Two good anti-friction metals are Babbit's metal and magnolia. Babbit recommended melting together 4 lb. copper, 8 lb. antimony, and 24 lb. tin. This he named hardening, and to every pound of it he added 2 lb. tin. Numerous varieties have from time to time been introduced. one being antimony 2 parts, tin 2 parts, lead 20 parts. Magnolia is a white metal composed approximately of lead 78 parts, antimony 21 parts, and iron 1 part ; it does not appear to heat when subjected to intense pressure.

Bell Metal.

Bell metal is the material used in the manufacture of bells, and it consists principally of copper and tin, although for the commoner classes other materials are often used, such as iron, lead, zinc, antimony, and manganese. Bell metal should be hard, compact, fine-grained, and strong, and should be cast in a manner similar to bronze. The copper is first melted, the tin is next added, and, after vigorous stirring, the metal is then poured.

Brass.

Brass may be considered the most common of the commercial alloys ; it is a compound of copper and zinc in very variable proportions. The commonly used yellow metal consists principally of 60 parts copper and 40 parts zinc, with such slight additions of other metals as are necessary to serve special purposes. This alloy, known as Muntz metal, makes sound castings, and compared to copper is less liable to discolour, is harder, closer in grain, more workable, and fuses at a much lower temperature. The impurities inherent in ordinary copper and zinc are not troublesome in brass foundry works, because it is a general practice to use common brass pigs. For turning work it is well to use 1 lb. of lead to each 100 lb. of molten metal. The molten metal containing the lead must be thoroughly mixed at each pouring, as the lead will liquate or separate out. Common brass used for ordinary bolts, nuts, and turnery work consists of the following mixture : 60 parts copper, 42 parts zinc, and 1 part in the 100 lead. This will work either hot or cold, and forms the usual French alloy. English naval brass, which is stronger than common brass, consists of two qualities, the one used in rolling sheets and rods, the other for making tubes. In the former the alloy is 62 parts copper, 37 parts zinc, and 1 part tin, while in the latter the alloy is 70 parts copper, 29 parts zinc, and 1 part tin. The metals must be thoroughly mixed and stirred before pouring.

Brass Founder's Metal.

The art of mixing alloys for brass can be learned only by practice, but the following broad principles conduce to success. The mixing of the copper alloys becomes the more difficult with increase in the fusibility of the elements. Thus, it is easier to mix hard gun-metal than soft brass, because the zinc in the latter volatilises so rapidly that much of it is liable to be lost in fumes, with considerable diminution in the weight of the alloy, and resulting uncertainty in regard to the composition of the final product. But tin and copper will alloy with little waste, and yield practically certain results. When the soft brasses are required, it is better to mix a large proportion of brass with copper than to make the mixture directly with copper and zinc. When zinc is mixed with copper, the latter must be melted first and the zinc added in small quantities at a time, the surface of the copper being strewn with charcoal powder. Broken glass may be used instead of charcoal dust on the top of molten brass to prevent oxidation. When adding easily fusible metal in fragments to molten copper, they must be perfectly free from moisture, and hot; otherwise the addition will cause a blow up of the melted metal. When blue fumes are given off, that denotes oxidation of the zinc, and the pouring must not be delayed long after. It is better to melt metal at least once, and ingot it, than pour it directly into the mould after a first mixing. For the same reason it is good practice to use a considerable proportion of old metal with new. A bit of borax put into the crucible on the addition of zinc diminishes the volatilisation of the latter. When brass turnings are melted up, it is essential that all iron turnings which become mixed with them in the machine shop shall be entirely removed. They would make the brass castings hard and pinny. These are therefore removed with a row of magnets passed repeatedly through the mixture until the magnets cease to take up any more iron.

Metals Used in Brass Alloys.

Aluminium, the lightest of the metals, is of white colour, soft, very malleable, ductile, with an elasticity and a tenacity about equal to silver. It is not volatile, even when strongly heated out of contact with air. It is practically unacted upon by acids, is unaltered in air, and is non-poisonous. It is now obtained of 98 to 99¾ per cent. of purity, the impurities being silicon and iron. Antimony in its commercial state is often impure, containing iron, lead, arsenic, and sulphur. It is a bluish-white metal of crystalline structure, with fern-like markings on the surface, and is very brittle. It forms useful alloys in consequence of its hardening properties, but it acts very injuriously on the malleable metals, making them hard and brittle. Bismuth is a hard, greyish-white

metal with a bright metallic lustre; it is very useful in preparing fusible alloys on account of its low melting point. Copper is of a salmon-pink colour, is highly malleable, ductile, tough, and tenacious; it readily unites with oxygen when at a red heat, forming oxides. The cuprous oxide is soluble in the molten copper, and this makes the result brittle and spongy, thus forming the greatest obstacle to sound castings. Impurities, such as arsenic, antimony, nickel, and lead, have a very injurious action on copper, making it brittle and hard. Lead is of a bluish-grey colour. It is heavy, malleable, ductile, and tough, but has only a feeble tenacity. Commercial lead is often nearly pure, though it may contain a slight quantity of silver, copper, etc. Nickel has a brilliant white colour, and is malleable, ductile, weldable, and very tenacious. It is magnetic, and does not readily oxidise in air at ordinary temperatures. Commercial nickel, owing to recent improved methods of refining and toughening, contains but small quantities of impurities, and these are generally iron, sulphur, manganese, and cobalt. Tin is a white metal with a brilliant lustre, very soft and malleable. It is supplied in three qualities—common, refined, and grain. Commercial tin often contains small quantities of impurities. Zinc, also called spelter, is a white metal of a bluish shade and with a bright metallic lustre. Ordinary zinc is hard, brittle, and, when broken, exhibits a highly crystalline structure. The general method of testing zinc is by the fracture; the fracture facets, if large, bright, and well formed, generally denote good quality. A tendency to bending instead of to breaking shows the presence of impurities. The lead present in ordinary commercial zinc varies from ·45 to 1·22 per cent., and generally amounts to about 1 per cent.

Brass for Fine Castings, etc.

Brass that is to be used for fine castings and that is to be easily turned and filed requires the quality of liquidity in a much higher degree than ordinary brass. It must not in any case be pasty, and its texture must be uniform and fine-grained throughout. Again, as these castings must be free from such defects as flaws, pinholes, etc., the brass must be capable of remaining liquid until the solidifying point is nearly reached, so that the metal may be poured at the lowest temperature. Generally, casters never think of using a flux to assist in clearing the metal before pouring, or to reduce the oxides or other impurities in the metal, but for this purpose soda, salt, borax, and charcoal, mixed to a paste with oil and used in small quantities, will be found excellent. For fine mosaic work and for castings of a filigree nature that will subsequently be gilded, it is necessary to use a comparatively rich alloy, because with a high percentage of copper the casting requires less gold in the gilding than would be necessary if the brass were yellower.

FUSIBLE ALLOYS.

Fusible alloys consist principally of tin, bismuth, lead, and cadmium ; they are used where a definite temperature is to be indicated, such as for fusible solders, safety cocks, and plugs for boilers and fire apparatus. In the latter case, the rooms are fitted so that the heat of the fire fuses parts of the apparatus and gives warning. These alloys may have much lower melting points as combinations than the metals singly. One alloy consists of 50 parts bismuth, 25 parts tin, and 25 parts lead, and melts at 202° F. Occasionally cadmium is added to the alloy to counteract the crystalline formation caused by the bismuth ; the alloys are thus more malleable. An alloy, consisting of bismuth 15 parts, cadmium 3 parts, lead 8 parts, and tin 4 parts, is often used for soldering Britannia metal and similar materials, which cannot he highly heated, and which are of a white colour.

GERMAN SILVER.

German silver consists essentially of copper, zinc, and nickel. Various metals have been introduced under fanciful names, such as Brazilian and Nevada silver, Argentan, Potosi silver, silveroid, silvene, arguzoid, etc., but these consist of German silver with perhaps slight additions, as, for instance, of iron, manganese, lead, tin, silver, cobalt, and magnesium. German silver is noted for its whiteness, lustre, brilliance, tenacity, and toughness, and its power of resisting chemical influences. Although it is generally used in the rolled, drawn, or spun way, as for spoon and fork making, electro-plate working, etc., yet a considerable quantity is used in the foundry, in making cocks and other fittings for salt water. It is used also for upholstering, instead of brass, nickel, or silverplated metal. For castings that are to be filed and turned, and where malleability is not of first importance, it is sufficient if the alloy has the requisite tenacity and toughness, and sufficient liquidity when melted to adapt it for ordinary moulding purposes. Where a fair percentage of zinc enters into the composition, it is as well first to melt copper and zinc together in equal proportions, allowing 2 lb. extra of zinc for the loss in volatilisation, and then to use this brass in right proportions instead of zinc. The mixture having been decided on, the metal is weighed out, generally allowing about one-half scrap metal. Put the nickel in the bottom of the crucible, the copper next, and if manganese and iron are being used, these must now be put in. The copper melts first. Keep the metal well covered with a layer of charcoal to prevent oxidation. After about an hour or an hour and ten minutes, the brass must be well melted and stirred in. Now add about 1 oz. of a mixture composed of charcoal, borax, and salt in equal quantities, to flux the metals. Next add the tin, lead, or other metal, and cover with a layer of powdered glass. This will run and effectually preserve the metal in the crucible. When small blue jets are seen emerging through the surface, it is time to pour, and here great care is necessary. As each mould is filled, throw a little charcoal into the crucible. The moulds and cores, if used, must be well dried. The greatest difficulty with casting nickel silver, as it is called, is in getting the metal to the right heat, as moulds are liable to be improperly filled, or, if the metal is too hot, to be porous. A teaspoonful of amorphous phosphorus per 100 lb. of metal will render the alloy more liquid, and will tend to reduce any oxides that may have formed, thus producing sounder castings. When alloys of German silver are required for casting purposes, a little lead is generally added. This causes freedom of working, and has a tendency to whiten the alloy. Iron is added when a harder alloy is required. It also has a whitening effect, since from 1 to 1½ per cent. will have the same effect as 4 per cent. of nickel on the colour. Manganese, which is used by some makers, has a very similar effect to that of iron. As German silver becomes altered in composition by repeated melting owing to the volatilisation of a larger percentage of zinc than of copper, it is absolutely necessary to add a quantity of metallic zinc, generally about 2 per cent. All windows, doors, and ventilations must be closed when pouring, as the metal chills very rapidly.

GUNMETAL.

Gunmetal is a term used in a general sense, and may be said to apply to alloys of copper and other metals which are hard and tough. It is generally used in the manufacture of bearings for machine construction, and for this purpose must be capable of resisting hard wear, must be easy to turn, file, and work, and must possess great strength. An excellent plan, in making the various brasses or bronzes, is to add, when just about to pour, a very small quantity of amorphous phosphorus—about a teaspoonful to every 100 lb. of molten metal. This increases the fluidity of the metal, and while it acts as a flux, if not used in larger quantities it has no detrimental effect on the work.

IRON USED FOR CASTING.

Some particulars as to the iron used in foundry work will be useful. It is mostly pig iron, so called on account of its shape and method of pouring; this is an impure iron, made in a blast furnace and containing more than 1½ per cent. of carbon and large quantities of other impurities, such as silicon, sulphur phosphorus, and manganese. The smallest quantity of carbon in pig iron is 1½ per cent., while the largest amount the iron can take up without the aid of other substances is 4½ per cent. Car-

bon is usually found either as a chemical combination or as flaked graphite. Flaked carbon weakens the metal, but otherwise it has little effect. The carbon in combination hardens the metal, and to some extent increases the strength of the iron. In amount it varies from about ¾ per cent. in what is known as No. 1 pig to as much as 1½ per cent., or even more in a white iron. Silicon has an effect in iron closely approaching carbon, and there appears to be no limit to the amount that iron can take up. If it is in excess, the iron becomes hard and brittle. The following amounts of silicon have been given as being the most beneficial for the particular results aimed at. For maximum hardness less than ·8 per cent., for crushing strength about ·8 per cent., for maximum modulus of elasticity about 1 per cent., for maximum density in mass 1 per cent., for maximum tensile strength 1·8 per cent., and for maximum softness and working qualities 2½ per cent. Another ever-present element is phosphorus, which is found in quantities varying from a mere trace to as much as 3 per cent. It always increases the fluidity of the molten metal, but in quantities it has a hardening effect on the iron. Manganese, also always present in pig iron, has a double effect: on the one hand it hardens the iron directly; on the other, it has an indirect softening action by which the sulphur is eliminated. Silicon has the same effect, and iron containing 2½ per cent. of silicon will be found to be hard to some extent, while the carbon will be thrown from the combined condition into that of the graphitic condition. Grey pig iron is usually high in silicon, while white iron is usually low in that element.

MANGANESE BRONZE.

Manganese bronze consists of copper, tin, zinc, manganese, and iron, in varying proportions. Manganese added to an alloy of copper and zinc increases the hardness. Ferro-manganese is used for additional hardness. An ordinary quality of manganese bronze consists of copper, 84 parts; tin, 10 parts; and ferro-manganese, 6 parts. The ferro-manganese can be used up to 12 per cent. by varying the qualities of the tin and copper. For special work, make a trial lot of about 5 lb., casting it under ordinary conditions and treating it as to turning, filing, etc., as the work may require ; then vary the constituents according to the result. It is an excellent plan, for future reference, to enter the results of all casting operations under a distinctive head, with any details which may be useful. This frequently saves much trouble. The difficulty in using ferro-manganese or cupro-manganese is the great heat necessary to melt this alloy. One plan is to place the scrap-metal and the manganese in the crucible, the manganese being lower, and to cover with a layer of charcoal. Then add the other ingredients, except the zinc, if any is to be used, and place in the furnace. The scrap will first melt, and this will assist in carrying down the manganese.

PHOSPHOR BRONZE.

Phosphor bronze consists of copper and tin, with a small amount of phosphorus. It is made by mixing and melting in a clean plumbago crucible copper and tin in varying proportions, according to requirements, and then adding a certain proportion of phosphorus in the form of phosphor tin or phosphor copper, or both. An old crucible, well scraped out and sweated over the furnace to remove all oxides, etc., might answer. Flour charcoal must be put on the top of the metal to prevent undue oxidation. The molten metal must be well stirred previous to pouring. In casting large brasses or bearings the moulds must be thoroughly dried, and covered with a mixture of charcoal and water and again dried, but for ordinary small castings the moulds may be used as in ordinary green sand. This alloy in solidifying passes directly from the liquid to the solid. If constantly remelted it will lose a small part of its phosphorus through volatilisation ; this loss must be made up, however, by the addition of phosphorus as at first. As there is always a tendency in bronzes to separate out when cooling, it is highly advisable to pour the molten metal just before solidification sets in. To determine the point, add the gates or runners or even small ingots to the molten metal ; if the metal begins to adhere to these pieces it is ready for pouring. A good general alloy for phosphor-bronze boiler fittings, pumps, and ornamental castings, consists of about 95 or 96 copper, 4 or 5 tin, with about $\frac{1}{16}$ per cent. phosphorus. As phosphor copper or tin containing guaranteed qualities of phosphorus may be obtained, it is an easy matter to add the desired amount of phosphorus. For phosphor bronze to resist much wearing stress, or friction, a different alloy is generally used ; and one commonly employed for axle bearings, bushes, cogs, etc., consists of about 90 to 91 per cent. copper, 8 to 9 per cent. tin, and $\frac{3}{4}$ to 1 per cent. phosphorus. As the phosphorus increases, the alloys become harder and less malleable, and more than $3\frac{1}{4}$ per cent. of phosphorus renders the bronze useless. The percentage composition of a phosphor bronze used in American locomotive building is copper, 79'70 ; tin, 10 ; lead, $9\frac{1}{2}$; phosphorus, '80.

EXPERIMENTAL FURNACES.

The practical study of metallurgy, either in assaying or the preparation of alloys, involves the use of a furnace that will give a fairly high temperature, sufficient at least to melt copper or cast-iron. In metallurgical laboratories it is usual to have several kinds of furnaces, both for melting and for assay work, and it may be found that for a special purpose one form of furnace is more suitable than another. When, however, it is desired to study the metals, want of space or monetary reasons may compel the worker to limit himself to one, or at most two furnaces, which have to do all the work required. For the ordinary class of metallurgical work—such as the preparation and examination of the alloys—a melting furnace is required ; but for assaying, which is a distinct branch of metallurgical work, both a melting furnace and a muffle furnace are needed. The melting and muffle furnaces may be combined, so as to form only one furnace, but it is not advisable to do this, because the two functions cannot be carried on at the same time, and, also, combined furnaces seldom work as well as separate furnaces.

ADVANTAGES OF HAVING SEVERAL FURNACES.

A choice of furnaces is desirable, and most small foundries have two of different sizes. In large foundries there may be half a dozen, and often a reverberatory furnace in addition, for dealing with extra large quantities. The advantage of having more than one furnace, considered apart from the weight of work done and the number of hands employed, is that fuel is economised by using the small furnaces for light casts, as the smaller the furnace the less the fuel required. In melting there is always a certain amount of fuel consumed in merely heating up the furnace. Each

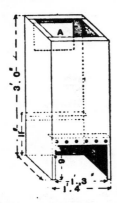

Fig. 1.—Home-made Small Furnace.

furnace is intended to hold but one pot at a time, and the capacities of pots range from a few pounds to a hundredweight. The furnaces for the smallest pots measure 9 in. or 10 in. square inside, and those for the large ones up to 18 in. In laying down a foundry, it would be well to arrange a furnace of the smallest and one of the largest dimensions, even though the latter might be required for occasional use only. It is advantageous to have more than one furnace, even in a small shop. When heavy castings are to be made, the product of one furnace is insufficient, and then the melting must be done in two or more, because brass will not stand hot long in a crucible, though iron will do so in a ladle. Another advantage is that, when

one furnace gets out of repair, another can be used, instead of work being stopped while repairing. In a regular foundry the furnaces are sunk below the floor level, and the top of the masonry on which the covers are laid is only from 9 in. to 12 in. above the floor level. A man standing on the floor can then look into the furnace, and he can stand over it and lower and lift the pots of metal in and out. Furnaces for occasional use are frequently constructed in a different way, with a sheet-iron casing, and portable, but they are inconvenient for foundry work.

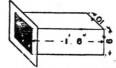

Fig. 2.—Chimney-box for Furnace.

HOME-MADE SMALL FURNACE.

THE small furnace shown by Fig. 1 is suited for analytical and other metallurgical work; it is sufficiently large to melt 7 lb. of metal. The outer casing or frame is made of iron boiler plate about $\frac{1}{16}$ in. thick,

Fig. 3.—Frame for Firebars.

and it carries at the back a cast-iron chimney box (Fig. 2). A rectangular hole A, equal to the inside dimensions of the box end—that is, 10 in. by 6 in.—is cut in the frame 3 in. from the top. The front of the frame is cut away at the bottom to admit air. Two pieces of angle iron, 2 in. by $\frac{1}{16}$ in., as shown in Fig. 3, or, preferably, cast-iron frames, are riveted, one on the front and another on the back, to carry the fire bars. At the top a 2-in. angle iron is riveted all round on the inside to carry the covering bricks. The framing is lined all round with slips of 2-in. firebrick well set in fireclay, sufficient room, say 1¼ in., being left above the inner angle bars to allow of the insertion and easy removal of the four or five wrought-iron firebars 1 in.

square. This space is easily formed by laying flat pieces of wrought iron at a distance of 1¼ in. from the angle irons on the brickwork lining of the other two sides, and building on these with the firebrick. The cast-iron chimney piece has a flange by which it may be riveted to the frame. This casting must be lined with firebrick slips 1 in. thick, so that no part of the ironwork, except the firebars, comes in contact with the fire. This furnace requires a chimney about 20 ft. high, and usually the casting may be built into an existing chimney. Ordinary coke is used as fuel, and all descriptions of alloys, even bronzes and nickel, can be melted in this furnace, which would require to be relined occasionally according to the work done.

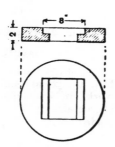

Fig. 4.—Portable   Fig. 5.—Ring for
Coke Furnace.   Firebars.

FIRECLAY.

The fireclay required for lining furnaces of all kinds may usually be purchased at the clay-pits or from builders: about 1½ cwt. or 2 cwt. will be required. It should be mixed with just sufficient water to render it plastic, kneaded and beaten into a square block with a shovel, and then allowed to stand for about a week to temper.

PORTABLE COKE FURNACE.

A furnace to burn coke can be made very readily with a piece of cast-iron piping A (Fig. 4), preferably about 14 in. in diameter by 2 ft. 6 in. long. This forms the casing proper. Two portions on opposite sides and ends of this pipe are cut away to form the air inlet and outlet respectively. The hole B at the back is 6 in. long by 4½ in. wide, and cut at a distance of 4 in. from

the top, acts as the outlet. At the bottom a piece 10 in. long by 8 in. high is cut away, and forms the ash pit, allowing air to pass through the furnace. The pieces may be cut out by drilling a number of holes along the outlines, and then with a thin chisel severing the pieces, afterwards filing the edge to a finish. A cast-iron ring (Fig. 5), about 2 in. thick and of a diameter to fit the pipe, having in the centre a square hole, with two of its sides recessed 1 in. deep and 1 in. broad, to carry the small firebars, must be securely fastened to the inside of the pipe immediately above the bottom hole, and can be

Fig. 6.—Faraday's Blast Furnace.

kept in place by about eight ½-in. bolts passed through the pipe, and screwed about 1 in. into the ring casting. For the flue, a piece of wrought-iron rectangular tubing, say 6 in. by 4½ in., and of sufficient length to connect to any ordinary chimney not less than 20 ft. high, must be stoutly built and flanged, and riveted to the top hole in the cast-iron pipe. This tube must be lined with firebrick slips about 1 in. thick, to prevent the fire burning the iron; the inside of the cast-iron pipe also being lined with firebricks about 7 in. long by 2 in. square. The spaces between the bricks and iron should be well luted with fireclay. This furnace will readily melt any quantity up to 12 lb., and any metal having a melting point not exceeding that of cast-iron. The firebars may consist of four wrought-iron bars about 1 in. square by 7¾ in. long. They rest in the recess of the ring casting. A cover must be used, and may consist either of two or three firebricks or of a piece of cast-iron, and the ends of the flue must be well fixed and mortared into the chimney to cause a good draught.

FARADAY'S BLAST FURNACE.

Fig. 6 shows a section of a very useful furnace that may be constructed without

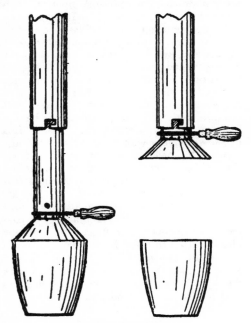

Fig. 7.—Hood for Furnace, Closed.

Fig. 8.—Hood for Furnace, Open.

difficulty by anyone. It is made from two large plumbago melting pots; partly used pots will do quite as well for the purpose. In the outer pot a hole of about 1 in. diameter is drilled in the side as near the bottom as possible. There will be found no difficulty in doing this with the ordinary brace and bit. The second pot is turned upside down and the bottom carefully removed with a hammer and chisel. If a hole is first made in the centre the remainder

of the bottom may easily be chipped out in bits. An ordinary circular iron gulley trap grid (Fig. 9) is placed in this pot; the pot is put into the larger one, and the space between the two is filled with fine sand. The fuel used in this furnace is coke broken into pieces of about the size of a walnut. The fumes from a coke furnace are very unhealthy, and they should be prevented from coming into the room. An open shed would be a suitable place for such a furnace, or it could be placed in an open

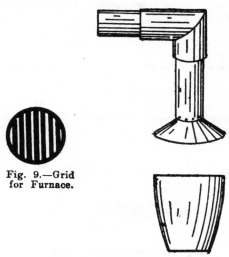

Fig. 9.—Grid for Furnace.

Fig. 10.—Swinging Hood.

fireplace; or a proper hood could be fitted which, being connected with a length of sheet-iron stove-pipe running through an opening in the roof or into a chimney, would carry the fumes away.

HOOD FOR FURNACE.

The hood for a furnace, such as is shown by Fig. 7, may be made adjustable if required, so that it can be raised or pushed aside when a crucible has to be inserted or removed, and its efficiency thereby increased. The hood is made from a circular piece of sheet-iron, cut out as for a lamp shade, and it is fixed firmly upon the end of a loose piece of sheet iron, which should be about 2 ft. 6 in. long. Near to the hood a rivet is fixed in the stove-pipe. The stove-pipe running into the chimney is a trifle larger than the movable piece, so that the latter may slide up or down inside it. At the bottom of the fixed pipe is cut a notch in the shape of a bayonet catch (Fig. 8), and by lifting the movable pipe the rivet may be made to slide into the bayonet catch and the hood held about 2 ft. above the furnace until the crucible is removed, when the movable piece may be gently replaced. The handle for moving it is of wood, and it is fixed to the pipe by iron wire. Another method of moving the hood is to have a piece of stove-pipe about 1 ft. long fitted into the chimney, and into this a smaller elbow-piece is fitted (Fig. 10). The tube from the furnace is fitted to this elbow, and may easily be moved aside. The stove-pipe should be held by a hook and iron wire. The air supply for this furnace may be obtained from a double bellows or fan, and should be sent through a conical pipe placed an inch or two back from the opening.

Fig. 11.—Gas Injector Furnace.

INJECTOR FURNACES.

The melting furnaces known as the "Injector" furnaces are extremely simple in their construction, consisting of a cylindrical body closed at the bottom, made of refractory fireclay; and of a disc-shaped lid, also of fireclay, bound with iron, and provided with a handle for moving it. The heat is produced by a large blowpipe provided with legs, which support it in a horizontal position. The blowpipe is brought against a small hole in the side of the furnace through which the flame passes to the interior. The flames plays around the crucible, and then escapes through a hole in the lid. The blowpipes used with these furnaces are very simple, consisting of a large cast-iron tube provided with a wire gauze at the front and partly open at the back.

The gas tube runs in at the side and the air tube at the back. The blast of air is furnished by foot-bellows, and is regulated by a screw on the burner. The amount of gas required for Injector furnaces varies with the size; a furnace melting 6 oz. of metal uses 7 ft. to 30 ft. of gas per hour, and one for 28 lb. of metal 100 ft. to 300 ft. of gas. The hottest part of the furnace being directly in front of the burner, it is noticeable that the crucibles are rapidly eaten away at the side nearest the blowpipe, and therefore care should be taken that they are put down in a different position each time they are replaced. Injector furnaces are excellent for all classes of metallurgical work, but require constant work at the bellows, and for that reason working for a lengthened period with them may be found rather tedious, which is not the case when employing a draught furnace. There are several useful Injector furnaces that are intended either for petroleum or benzoline, or for carburetted air. These furnaces work very well with carburetted air, formed by passing a portion of the air through a generator containing benzoline or gasoline; the air in passing through this generator becomes saturated with the hydrocarbon and forms an inflammable gas, which may be burnt in a similar way to coal gas. An ordinary paraffin oil burner may also be fitted to these furnaces; it consists of a small reservoir for holding the oil, and a wick tube. Being only an ordinary wick flame, it is not likely to be as efficient as coal gas or carburetted air.

GAS INJECTOR FURNACE.

At Fig. 11 is shown Fletcher's injector gas furnace. This is capable of melting iron, steel, and nickel; the gas being supplied by ½-in. barrel, and the air blast furnished by the smallest foot blower. The furnace consists of a specially prepared casing, to which is attached a specially constructed Bunsen burner. To prepare the furnace for work, the nozzle of the burner is placed tightly against the hole in the casing, and the gas supply is turned on full. The gas being lighted the air is blown with the airway full open before putting on the cover.

of the furnace. When the cover is replaced, the flame should stand out of the hole in the cover about 2 in.; the adjustment then is correct. If the flame is longer, the hole in the air jet should be enlarged, or the gas supply reduced. Before stopping the blower, draw the burner from the hole. To get full duty from this kind of furnace, a boy should be employed to work the foot bellows.

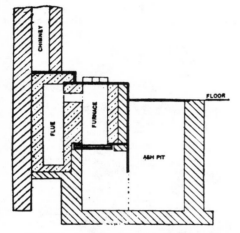

Fig. 12.— Section of Brick-built Furnace.

BRICK-BUILT FURNACES.

In regularly constructed foundries, the furnaces are, as a rule, sunk below the shop floor level, as will be seen from Fig. 12, which shows the ash-pit in the front covered by iron gratings. Usually the furnaces are simple in construction, and range from 9 in. to 16 in. square by about 3 ft. to 3 ft. 9 in. deep. The internal lining of the furnace is of firebrick, well grouted in with fireclay. The firebrick should be composed of a high proportion of silica and a fair proportion of alumina, and should be nearly free from alkalies or other impurities. The analysis of a good Stourbridge firebrick shows : Silica 77˙63 per cent., alumina 19˙48 per cent., lime ˙18 per cent., ferric oxide 1˙29 per cent., magnesia ˙31 per cent., soda, etc., ˙91 per cent. The construction of the furnace flue is most important, as a slight difference in the arrangement of it will con-

siderably alter the draught, and probably give a different result in the working of the metal. Two forms of flues in general use are shown in sectional plants (Figs. 13 and 14). In Fig. 13, H is the firebrick lining ; the arrangement of Fig. 14 is considered preferable ; in this S F shows each separate flue. The top and bottom of each

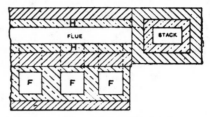

Fig. 13.—Plan of Combined Flues.

furnace F is formed of an iron casting, with movable firebars. The furnace is covered with firebricks 12 in. to 16 in. long, according to size of furnace, three bricks being

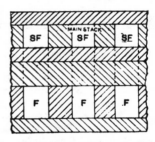

Fig. 14.—Plan of Separate Flues.

used to a furnace ; these are clipped with iron bands to strengthen them, and to enable them more easily to be lifted on or off the furnace.

OIL FURNACE.

Fig. 15 shows an oil furnace consisting of a specially constructed casing with a cast-iron foundation plate. The burner box is an iron casting with holes on one side for two pipes, and a nozzle on the other for insertion in the casing. The oil from the cylinder A flows by gravitation through the tube and cock till it falls into the combustion chamber B containing the burner ; the oil then drips on to the air nozzle C

It flows round the nozzle till it comes to a slot in the bottom, through which the air blast blows it to the floor. When the oil is alight, the blast forces the flame and heat into the chamber D, where it melts the metal in the crucible. To start the furnace, remove the lid E of the combustion chamber, throw in a bit of lighted waste, turn on the oil and air blast, and, as soon as the oil catches light, replace the lid. Then adjust the oil supply until the flame issues about 4 in. above the cover of the casing, or until the furnace smokes. A small foot blower or bellows is required for producing the blast. Ganister, or the very best

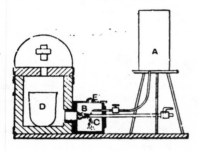

Fig. 15.—Oil Injector Furnace.

fireclay, must be used for lining the combustion chamber to a thickness of about ⅜ in. or ½ in. The oil may be any waste oil, tar oil, blast furnace oil, etc. Possibly it may be necessary to filter the waste oil through a cotton-wool filter before using.

BRICK-BUILT FURNACE.

Fig. 16 shows a side section of what is generally known amongst metal-founders as the *air* furnace. It is of the type used at the present time in a brass-foundry. This kind of furnace is also applicable for the melting of metals in general, unless they are of a very refractory nature. There are times when such a furnace is necessary, and it may be economical to have a furnace for the purpose of making one's own castings. A furnace like the one here illustrated can be built in an outhouse on either a small or large scale. The accompanying illustration being drawn to scale, the reader can decide for himself what size

the furnace shall be built to meet his requirements. In some cases, perhaps a disused flue or fireplace can be utilised to carry away the smoke from the furnace, and if so, it will materially lessen the cost of construction. The furnace should be built on a bricked or cemented floor, and the best firebricks must be used next the furnace; all the other brickwork can be ordinary stock bricks or stone, as shown at E, to fill up space in place of brickwork. The furnace should be built on a base that is square, or nearly so, and the ash-pit B

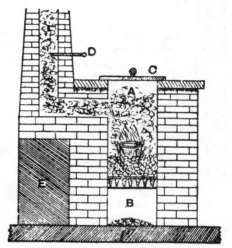

Fig. 16.—Brick-built Air Furnace.

should be open at one side or front to allow a plentiful current of air to the fire. The fire-bars should be of cast-iron, and placed about ½ in. apart, their ends resting on two pieces of iron built in the brickwork to receive them. The iron cover C on the furnace top can be made of ½-in. iron plate, and the damper at D of ¼-in. iron plate. The crucible used in these furnaces should be of plumbago, and if new it must be tempered by heating, when empty, to a red colour, afterwards allowing it to cool gradually as the fire dies out. It is better for the crucible to have a recessed or flanged neck, for the tongs to take a firm hold in lifting in and out, also a spouted mouth to facilitate pouring the melted metal into the mould when casting.

REGULAR BRASS FOUNDER'S FURNACE.

The common type of brass furnace used for the regular work of foundries is shown by Figs. 17 to 21. In these illustrations Fig. 12 shows a plan view above the furnace and damper; Fig. 18 a sectional plan

bars are about 1 in. or 1¼ in. square, resting upon a plate built into the brickwork, and the fire is dropped by pulling out these bars in the pit D. A cast grating E is lifted readily when the attendant desires to descend into D. This grating may be in one or more lengths; or loose single bars fit-

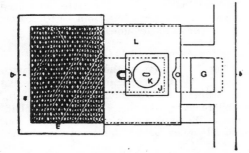

Fig. 17.—Plan of Brass Founder s Furnace.

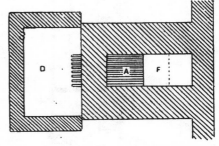

Fig. 18.—Horizontal Section of Brass Founder's Furnace.

through furnace and passage thence into chimney; Fig. 19 a vertical sectional elevation on the section B B in Fig. 17. Two vertical sections through ash-pit and furnace are next shown; Fig. 20 is a section through c c, and Fig. 21 a section through D D in Fig. 19. In these illustrations the furnace is marked A, with

ting into notches may be employed instead. The grating is surrounded by a cast-iron plate A, within the sides of which it fits; F is the opening which connects the furnace to the flue G, and H is the damper. The cover of the furnace is marked J, and K is the tile which covers the opening in J. The

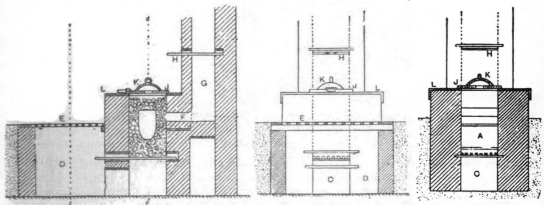

Fig. 19.—Longitudinal Section of Brass Founder's Furnace.

Fig. 20.—Cross Section through Pit of Furnace.

Fig. 21.—Cross Section through Fore of Furnace.

grate bars separating it from the ash-pit below, whence the ashes are raked out through the opening c into the space D, through which also the air is drawn by the chimney draught. The grate

cover J is removed for building the fire, and for lowering or withdrawing the crucibles, but when observing the melting of the metal K only is moved. Both are of cast-iron, and both have eyes or holes, as shown, for effecting their removal. The iron damper H may be of cast or of sheet. Plates of cast-iron or of sheet-iron are built into the brickwork wherever bricks have to be carried over openings. A cast-iron plate

built independent of one another; but if built at once in a group, instead of by additions, they should form members in which the furnaces shall lead into a common flue running horizontally, next the wall; they should have their ash-holes leading into a pit common to all; while the brickwork between adjacent furnaces may be separated by air spaces extending from a little below the cover to the ground.

LARGER BRASS FURNACE.

Fig. 22 shows another foundry furnace suitable for brass or copper. This also can be constructed without great expense. This furnace, like those already described, should be sunk below the floor level, and the chamber below the level for the removal of the ashes is usually covered by iron gratings, which may be made to slide or can be hinged. The ash-pit chamber should be dug about 6 ft. below the floor

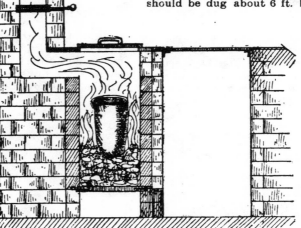

Fig. 22.—Brass Founder's Furnace.

L also covers the top of the furnace, and flanges at the ends help to prevent the sides from bulging. Common bricks are used, except for the furnace and flue linings. A good body of bricks should surround the furnace to confine the heat. As the bricks burn away, the interior must be relined with fireclay or with siliceous sand, which forms a protecting glaze when dried with a slow fire. When two or more furnaces are placed side by side, they may be

level of the foundry, and a brick furnace built up about 1 ft. square (inside) of best bricks, with an inner lining of firebrick, and with furnace bars, on which the fuel is to rest, about 11 in. from the bottom. The firebricks forming the lining must be grouted and set with fireclay, which is a compound of silica, alumina, and water. The furnace should be connected to an upright flue, carried up the wall of the foundry, with a sliding damper as shown to regulate the draught. It is well for commercial reasons to build two or three furnaces in a row ; they can then all be connected to the same central flue, whilst each is regulated by its own damper. This central flue may be further utilised by fixing in it a sheet-iron oven or closet, as shown, and arranging the flue round it. This oven should be fitted with shelves and have a sliding door, and would be used for drying cores used in hollow patterns. The top of the furnace should be covered by a cast-iron plate, 1 in. thick, and the space outside the furnace fire holes is excavated about 3 ft. wide and built round with ordinary bricks to form an ash-pit chamber and admit air to the furnaces. This chamber is covered by a grating, which may slide into place or be hinged, according to convenience. The floor of a foundry may be paved with stone or bricks, or it may be cemented.

CRUCIBLES.

The crucible used is next in importance to the furnace. Crucibles are of two descriptions, clay pots and plumbago pots. The clay pots are composed of fireclay mixed with burnt clay, cement, pipeclay, etc. The crucibles commonly used formerly were made of clay, but plumbago has largely taken its place. These are supplied cheaply by several manufacturers. Clay crucibles absorb moisture, and are affected by frost ; those of plumbago are free from these objections. The crucibles must be capable of resisting a high temperature without softening, must not be friable or crumbling when hot, and should be capable of withstanding sudden changes of temperature, such as occur when the hot crucible is brought out of the furnace for pouring. Some pots are liable to have large pieces break away from the bottom when placed in the furnace, even after annealing ; but the most common fault met with is pinholing, generally due to impurity in the clay. Should the pots pinhole, loss is occasioned by reason of the metal running to waste among the ashes, on the firebars, and about the shop floor. Fireclay crucibles are being very generally superseded by plumbago crucibles : while a clap pot cannot be used

for more than four or five heats, a plumbago crucible may be used for as many as thirty-two heats. Plumbago crucibles are usually made from varying proportions of fireclay mixed with powdered graphite or coke dust. These crucibles, after annealing, will withstand the greatest changes of temperature without cracking, and they very rarely bump. These cost 1d. per pound, calculated on the weight of the metal the pots hold, and may be obtained to hold from 2 lb. upwards, the sizes mostly used being from 60 lb. to 100 lb. each. It is necessary to anneal all crucibles a few hours before melting in them. In annealing, the crucible must be placed mouth downwards in the fire, after which it is ready for use.

PATTERNS FOR CASTINGS.

Success in casting will depend not only on the skill and care of the founder, but on correct and well-made patterns. Patterns are usually made from wood, chiefly boxwood, pine, or mahogany, although metal patterns are used when a large supply of castings is required, iron or brass, or white metal being commonly used for patterns. Core-boxes, as well as patterns, are made in these materials. Iron has to be protected from rusting by varnishing, or beeswaxing ; brass and white metal need no protection. As the thickness of metal is often shaved down as slight as possible, plaster of Paris is frequently used to obtain exact thickness before casting the final pattern or core-box. In moulding, fewer joints between box parts are made in brass founding than an iron moulder would generally make, and fewer core-boxes are employed. The brassfounder makes false cores or drawbacks to fulfil the same purpose. The pattern for a casting must be made larger than the casting required, to allow for shrinkage. The usual allowance is ¼ in. to the foot, and this is sufficient for the shrinkage of metal and the finishing of the casting also. Sharp angles in castings must be avoided wherever possible, and rounded edges with slightly tapering sides must be the rule in all patterns. This will allow them to leave the sand readily without causing defects. Wood patterns should be well varnished or rubbed with powdered blacklead to protect the wood from the moist sand, and also to allow the pattern to leave the sand freely. Should it be required to cast metal patterns for use it will be necessary to make the first wood pattern slightly larger, in addition to the allowance as previously noticed, to allow for shrinkage of metal in pattern.

PATTERNS FOR MOULDING.

For cast work the patterns may be modelled in clay or wax, or made in wood, and the patterns must relieve themselves from the sand when lifted from the moulding boxes. Patterns may also be made of metal, and should be cast in a mixture of lead and tin. These patterns serve to make the mould in sand from which the brass, copper, or iron castings are produced. The patterns should be chased up and finished to the very finest degree, and if the pattern is to be in two or more parts the necessary fitments for fixing them should be soldered on, so that the portions when cast may fit together easily. Exact making and finishing of the patterns will be advantageous, as better castings will result, and the fitting is rendered very much easier. Birds, parts of plants, insects, and such like, may be used as patterns by placing the object in the centre of a box and holding it there by means of pieces of thin string or cotton, and then running in a mixture of plaster-of-Paris and brickdust, about two to one, moistened with water. After the necessary channel for the runner has been made, when set the mould is put in a furnace and heated to burn the contents, which are then carefully blown out, when the mould may be used.

SIMPLE TURNED PATTERNS.

In describing a pattern to be used for making a mould, it will be well to begin with one of the simplest kind. Fig. 23 shows a knob to be turned up in the lathe; it is not given as a model design, because a pattern made to the exact shape shown is almost sure to give trouble. The hollow part A has parallel sides, and lumps of sand will remain in the hollow, with the result that corresponding lumps will be found on

the castings. Sometimes patterns are made of the shape indicated in Fig. 24— that is, with the hollow undercut—and such a pattern will not leave the sand. The proper shape is given in Fig. 25; there it will be seen that all surfaces are sufficiently inclined to allow the pattern to be taken out of the mould without clogging the sand. When about to make such a pattern, first decide as to size, and then as to the quantity of castings likely to be required. If a great number, make the pattern in brass; if only a few the usual method is to make it in boxwood. Beech is much used for large patterns, but for the present requirements is useless. The pattern may be turned in the lathe with ordinary turning tools, and smoothed with fine glasspaper, a good finish being got by taking a handful of clean box

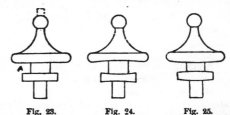

Fig. 23.—Knob Pattern with Chucking Piece.
Fig. 24.—Undercut Knob Pattern.
Fig. 25.—Correct Shape for Pattern.

turnings and holding these against the pattern whilst the lathe is run at a good speed. A boxwood turned pattern finished in this way has a surface that is practically perfect, and will not need varnishing, etc. When making patterns of castings to be finished in the lathe, it must be decided beforehand how they are to be held for turning. The knob here shown will need something by which to hold it, and the plain projecting piece shown at the top of Fig. 23 should be left on for the purpose of chucking it. A screw can then be cut on the foot, or it can be drilled and tapped without removal from the lathe. From such a pattern as the one just described, it will be but a step to the making of spindles and more ornamental work, but if these are of considerable length, as Fig. 26, they must be supported at both ends in turning. This is accomplished by leaving on the dotted portions, and one end is held in a chuck on the lathe, while the other end is supported by the back-centre.

SIMPLE FLAT PATTERNS.

Flat patterns for castings that would be worked in the vice are exemplified by Fig. 27, which shows patterns for a plate and handle to fix on a drawer. The plate portion here illustrated may be made from a piece of boxwood, about ⅛ in. thick, quite smooth on both sides. The outline can then be cut with a fret-saw, but care must be taken when drilling holes that a suitable drill is used, or the wood will split; ordinary metal-workers' drills will be found suitable, and preferably they should be used in the lathe. The plate can now

been brought to a curved shape by chisel and gouge must be glasspapered afterwards to remove the marks of the tool; in planed work it is best dispensed with. In ordinary work No. 2 paper, followed by No. 1½ or No. 1, will be found most useful. All tool work should be done before glasspapering, or the part so treated will probably be soiled by handling before the other part is ready, and the tools will probably be dulled by particles of grit. Sharp edges should be rounded off, and angles filled in with curved fillets. The joint face of each half of a pattern should have screwed upon it a stout rapping plate of wrought- or malleable cast-iron, having two holes, one tapped for a lifting screw, the other clear for a rod by which the moulder loosens the pattern in the sand preparatory to with-

Fig. 26.—Pillar with Chucking Pieces at each end.

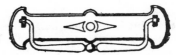

Fig. 27.—Flat Plate with Handle.

be filed to shape, but the pattern should be bevelled uniformly all over, and all holes should be tapered from the front to the back. Circular holes should be left in the plate to receive the lug pieces for the handle to drop in, and these should now be put on. The holes for the handle should not be drilled in the lugs when making the pattern, but they should be made in the castings when finishing. The handle itself may be made from drawn wire, which dispenses with casting; or it may be cast, and if it is to be ornamented, the latter plan will be better. The pattern may be made from a straight length of brass wire of suitable size and turned as required. It should then be annealed by heating to a dull red, and when cold bent to the required shape. Smooth it with fine emery paper, and, if possible, have it finished on a polishing mop. The castings of such a handle may be sprung into place.

FINISHING FOUNDRY PATTERNS.

A few hints on finishing foundry patterns may here be given. A pattern that has

drawing it. The top of these plates should be, say, 1/16 in. lower than the surface of the wood, as the rapping burrs the edges of the hole. If the pattern is small enough to be lifted by one hand, the lifting hole should be vertically above the centre of gravity of the pattern. This point may be found experimentally by driving in a bradawl and suspending the pattern in the hand. When screws are used to connect parts of a pattern, means must be used to make the heads flush, particularly where they would slide upon the sand during the lift. If the thickness of the wood through which the shank of the screw passes is not less than ⅜ in., the neatest method is to make a centre-bit hole and sink the head of the screw below the surface. Wooden plugs are then glued in and cut off flush. With thin stuff the holes are countersunk, the screw head being very slightly below the surface; this depression, with the slot in screw head, being filled with ordinary putty. A paste made of whiting and shellac varnish hardens quick, but when mixed it must be used at once.

Varnishing Foundry Patterns.

A coat of varnish prevents a foundry pattern absorbing moisture from the damp moulding sand. The varnish is made by dissolving shellac in methylated spirit, and is applied with a camel-hair brush. When the first coat has dried, the surface will be found to be roughened by the raising of the grain of the wood. It must be rubbed down with a piece of used glasspaper, and another coat of varnish applied. After this, lightly rub down again and apply a third coat. Red-lead is sometimes put in the varnish to give more body to it, and to fill the pores of the wood more thoroughly. Whilst the pattern body is thus coloured yellow or red by the varnish used, core prints and parts on which loose pieces are to be wired are rendered conspicuous by varnish to which lampblack has been added. Further, to inform the moulder of the shape, the section should be hatched on the joint face on one half of the pattern with another coloured varnish. Patterns should be stamped with a number and, if sent to a jobbing foundry, with the name of the owner. All loose parts should be stamped with the number of the pattern to which they belong, and with a distinguishing mark, to be repeated on the pattern near the position of the piece. A number of loose pieces belonging to one pattern may be kept together by threading on wire passed through small holes bored in the pieces.

Moulding Sand.

The sand best suited for moulding is that to be obtained from the beds of rivers in the neighbourhood of slate and granite, or in the coal districts, although in the latter case it often contains so much iron that it is almost useless, since it is thus rendered very liable to melt, though it may be somewhat improved by adding finely powdered coke dust. The worker in a small way may procure the sand he requires for moulding from any brass-founder, but it should be well sifted of any small lumps before being used. It is then slightly damped and pressed into the boxes to mould from the pattern. The more sand used with the loam in forming the moulds, the more easily the gases escape, and the less the likelihood of the casting turning out a failure. Sand is the best material from which to make moulds, on account of its characteristic porosity, adhesiveness, mobility, and infusibility. Owing to its porous nature, the gases generated during the pouring of the metal escape, its adhesiveness enables a perfect mould to be taken from the pattern, and its mobility enables it to give way and fill in the finest markings and lines of the pattern—so much so, that frequently in moulding from figured wood patterns, a faithful reproduction of the grain of the wood of the pattern is seen. Owing to its infusibility and unaltering nature, the heat of the molten metal does not in the least affect it, neither does it alter the chemical constituents of the sand. The best moulding sands have a generally uniform composition, roughly as follows:—Silica, from 93 to 95 parts per cent. ; clay, from 6 to 3 parts per cent. ; iron oxide, from 1 to 2 parts per cent. Sand containing lime, magnesia, or other metallic oxides is unsuited to the moulder, as it is too close and so will not allow the gases generated to get away freely, or it is too weak and will not retain its form. Should the iron oxide exceed from 1 to 2 per cent., it renders the sand more fusible, so that it may be affected by the molten metal with the result that the surface of the casting assumes a blistered appearance, and blowholes are formed, thus causing waste work. Castings do not all require the same kind of sand, so by mixing it must be modified to suit the nature of the work in hand. One casting will require sand of a porous, coarse, yet adhesive kind, while another will require a fine sand, free from grit, and very adhesive, which will be able readily to conform to all the intricacies of the pattern.

Special Sands.

For fine cast work, loam is usually mixed with the sand, especially when moulding very thin work, and the sand, to allow a larger amount of porosity, is not rammed nearly so hard. The metal contracts in cooling, so if the sand will not give way the metal must, and the result is cracked castings, or those which are said to have drawn.

The skilful moulder overcomes this difficulty by damping the thinnest parts of the impression with charcoal and water; this, by tempering the sand in the mould, allows it to yield, and thus saves the castings. When the oddside of the mould is being taken away from the eye-side and peg-side, it will frequently happen that the sand is broken away at the edges of the impressions. This the moulder has to repair, and it frequently happens that the largest portion of the time of the moulder is taken up

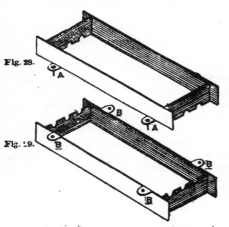

Fig. 28.
Fig. 29.

Fig. 28.—Peg-side of Flask.
Fig. 29.—Eye-side of Flask.

in thus mending. Core sand is for the purpose of making cores, for which it is necessary to have a somewhat coarse, porous, yet very adhesive sand; rock sand, from newly broken felspar rock, is preferred, although the free sand from river banks or from the seashore is often used when mixed with fine strong sand and a little clay to make it adhesive. For ordinary small cores, such as those used in castings for gas fittings and electroliers, ordinary free sand with about one-eighth of clay added is useful. The ordinary new red sandstone is well suited for use in these small castings. Parting sand should be a substance which does not retain moisture, and for this purpose there is no better material than powdered brickdust, and for preference this should be of firebricks, made by powdering the old used bricks, such as are obtained

from the covers and linings of furnaces after the part which generally runs or glazes has been broken off. In the absence of firebrick or brickdust, fine river sand or even ground blast furnace slag or cinder is usable. When molten metal comes into close contact with fresh sand, the surface of the sand fuses, which results in a roughness of the skin of the casting. If the sand be too rough or coarse, the metal will penetrate it somewhat, and a rough casting will result. It becomes necessary, therefore, to use a material that will not be burnt by the metal, and that will give sharp castings as

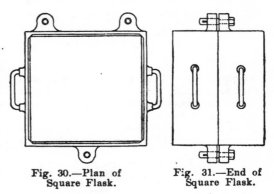

Fig. 30.—Plan of
Square Flask.

Fig. 31.—End of
Square Flask.

well. The general plan is to use a carbonaceous substance, and for this charcoal is used in the form of flour. Charcoal does not of itself readily adhere to the old sand, and it is necessary to dust the mould first with peas flour and then with charcoal. It is generally advisable to blow all excess of this dusting from the moulds by the aid of bellows.

FLASKS OR MOULDING FRAMES.

The moulding frames or flasks consist of a pair of shallow frames, about 3 in. deep, and usually made of cast-iron, having lugs on the one side as at A (Fig. 28), and lugs with holes in on the other side as B (Fig. 29). Wooden flasks are used in brass foundries, though they are seldom seen in English iron foundries. The fitting of the pegs and the eyes of each pair of frames forming a moulding box must be very exact, so that there may be no shifting. The pair are known as flasks in brass foundries, though they are peg-side, while the other is the eye-side. In iron foundries the two parts are called the cope and drag, and these are fitted with lugs having matching pins and holes, the two being either cottered together, or

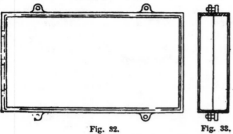

Fig. 32.
Fig. 33.

Fig. 32.—Plan of Oblong Flask.
Fig. 33.—End of Oblong Flask.

weighted and poured while laid horizontal on the sand floor. In brass casting, an extra frame known as the odd-side is nearly always used, but in their main features the flasks used by brass moulders are similar to those employed in the smaller work of the

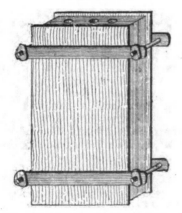

Fig. 34.—Flask ready for Vertical Pouring.

iron founder. There is one type of brass founder's flask, however, which is not employed by the iron founder—that in which the mould is poured while the flask stands vertical, or in a slightly sloping position. Figs. 30 and 31 show a flask used by brass founders and iron founders, and made in sizes from about 9 in. to 18 in. square,

usually without bars or stays. Such flasks are poured while laid horizontal. Figs. 32, 33, and 34 show the type of flask which is used for vertical pouring. These flasks usually are jointed, and have pins in the lugs. The sand is confined by moulding

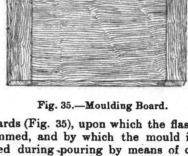

Fig. 35.—Moulding Board.

boards (Fig. 35), upon which the flasks are rammed, and by which the mould is confined during pouring by means of clamps (Fig. 36). The boards are of deal or oak, and the ends are tongued to prevent warping. They are from 1¼ in. to 1½ in. thick. Flasks that are made in wood are strongly dovetailed or tenoned, and clamped at the cor-

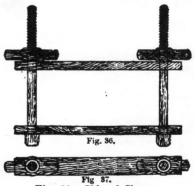

Fig. 36.

Fig 37.

Fig. 36.—Side of Clamp.
Fig. 37.—Edge of Clamp.

ners, and the top end, which is pierced with holes, is made of iron, or protected with iron. A flask made wholly of iron costs rather more in the first place than a wood flask, but it is practically everlasting, besides being more rigid. In Figs. 32 and 33 note the rib cast round the edges of the flasks next the faces, against which the boards are clamped. In some American flasks the section is grooved in order to confine the sand more efficiently. The clamps are of wood, and so are the screw nuts shown in Figs. 36 and 37. Fig. 38 shows the mode generally followed in the moulding of large numbers of small similar parts embedded in one flask, and poured from a large ingate, the pouring taking place through a hole in the end of the flask, out of which end also a riser is brought. Such flasks are sometimes hinged, and, as they cannot be loaded with weights, they are necessarily secured with cotters or bolts, or with wooden screw clamps embracing the outer faces of the flasks.

MOULDING TUBS.

In large foundries the moulding tub to contain the sand usually consists of an iron casting about 6 ft. long, 2 ft. wide at the bottom, and about 2 ft. 3 in. at the top by 15 in. deep. This is placed on brick supports, one at each end, till its top edge is about

the bench. The moulding trough of the brass founder is shown by Figs. 39, 40, and 41, which illustrate in plan, side elevation, and end elevation; this tub affords at once a receptacle for sand and a work bench. It stands against the wall, and is about 2 ft. 6 in. high and wide, and of any length. The tub shown is for one man, the work being done upon the sliding board A, beneath

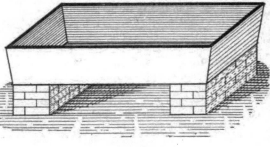

Fig. 38.—Moulding Tub.

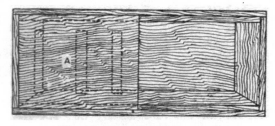

Fig. 39.—Top of Moulding Trough.

Fig. 40.—Side of Moulding Trough.

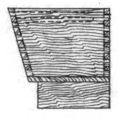

Fig. 41.—End of Moulding Trough.

2 ft. 3 in. from the floor level (see Fig. 38). In smaller foundries very frequently the work bench is made a part of the moulding tub, the advantage being that the sand is handy in the tub beneath the bench, which forms the cover of the tub. When a plain bench is employed, the sand is brought in small quantities from the sand bins or tubs, and placed in a small heap at the back of

which is the trough, which contains the sand. The trough is made of deal of about 1¼ in. thick, dovetailed or nailed at the corners, and tapered at front and sides for convenience of shovelling out the sand.

MAKING AN ODD-SIDE MOULD.

When preparing a mould the sand is first tempered by the addition of water, and passed through a sieve of about five or six meshes to the linear inch. Raw sand must be employed, as used or burnt sand does not possess the adhesiveness necessary for the purpose, since it contains proportions of charcoal and the other substances, such as peaflour, brickdust, etc. The sharp edges of the sand grains too have been rounded by use; they cannot, therefore, wedge together so firmly. The moulding of the odd-side is begun by taking an eye-side frame, and placing it on a board. The inside of this frame is dusted over with parting sand, and then the side is filled with black sand, to which some raw sand has been added. This is rammed down tight, first with the palms of the hands, then with the knuckles, and, lastly, with a wooden mallet; occasionally, the workman will tread it down firmly as the final operation. The sand is then scraped level with a bar

Fig. 42.—Section of Mould.

of iron moulding, a board is laid on the top, and the frame with its entire contents is inverted. The patterns to be moulded are now laid very carefully on the face of the mould, and the dust bag shaken over them; this leaves on the sand a clear outline of the patterns. These are now lifted off, and the sand is carefully cut away, leaving the patterns embedded half-way. A peg-side frame, to fit on the odd-side, is placed on the eye-side, as illustrated by the sectional view (Fig. 42), and then parting sand is dusted over the sand, a mixture of raw and black sand is laid over the half embedded patterns, and the mould filled with black sand. The whole of the sand in the flask is now carefully rammed as tight as possible, as described above. A second moulding board is placed on the top of the flask and well hammered to loosen the patterns, and the whole is turned over and again hammered. The patterns will now be in the eye-side, which is ready for moulding from. Fig. 43 shows a moulding trough in section in which the flask is supported on boards that may be shifted along.

MAKING THE PEG-SIDE MOULD.

The process is now repeated, a peg-side frame is placed on the eye-side, sand pressed in as before, covered by a board, and the whole inverted. The patterns are loosened by hammering on the top board, and the odd peg-side, which now bears perfect impressions of one-half of each pattern in the flask, is taken off, leaving the patterns in the eye-side. The frame called the peg-side is now taken and placed on the eye-side, sand is placed in the mould and well rammed, as in previous operations; the peg-side is removed, leaving the patterns in the eye-side; the odd-side is brought and placed on the eye-side, and the whole inverted. The patterns are loosened by hammering, the eye-side is taken away, and, the patterns being left in the odd peg-side, all is ready for use again, and thus the odd-side can be repeatedly used for moulding purposes.

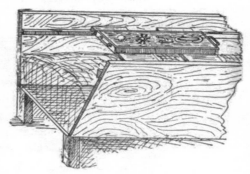

Fig. 43.—Section of Trough.

PLATEWORK AND GATEWORK MOULDING.

In many cases where large quantities of castings of similar work are required, there are two other methods adopted in moulding, namely, platework and gatework. These methods are excellent, and might be adopted with advantage to a much greater extent. They are suitable for use either in hand moulding or in conjunction with a moulding machine. The chief advantage of machine- over hand-moulding is that a perpendicular lift is ensured, thus diminishing the risk of the breaking down of deep face.

In platework the patterns are attached to a match plate or board, while in gatework a number of patterns sufficient to fill the standard flask are riveted or soldered to the gate-piece, practically forming one pattern. Platework and gatework have each their own merits, but generally it is better to put the patterns on plates than to gates, whenever it is possible to do so. Bibs, cocks, plugs, handles, stems, globe, and other valves, the parting lines of which are straight, can all be made on the plate principle, thereby saving the time that would be expended in making odd-sides, trimming the joints of the mould, etc., necessary with gated patterns. There are several kinds of plated patterns, the methods of moulding being different in certain cases. The simplest plated work is that in which two portions of a pattern are put on opposite sides of a plate, wood being employed. The two portions of the pattern are precisely like those which would be used if there were no plate interposed, these portions being those which go into the opposite halves of the flask, the thickness of the plate being immaterial. In moulding, the two sides of a flask are rammed on opposite sides of the plate, being cottered together through it, and the flasks being turned over for the ramming of opposite sides. A superior and more permanent pattern plate is made of iron or brass. In this the opposite portions of the pattern are prepared as though for moulding in the ordinary way, and then they are screwed to the opposite sides of a metal plate about $\frac{3}{8}$ in. or $\frac{1}{2}$ in. in thickness, and the flasks are rammed from the opposite sides of the pattern. In another method the two halves of a pattern are moulded on a separate plate instead of on opposite sides of the same one, and the flasks are rammed apart, and are only brought together when being closed for pouring. In this method also wood and metal plates are both used. Each of these types of plates is used either with flasks rammed in the usual way or upon a moulding machine. Interchangeable flasks are almost a necessity in platework, as the plate must fit on the peg-sides of flasks, which are thus used as dowels.

PLATEWORK AND ODD-SIDE MOULDING COMPARED.

The advantages and disadvantages of platework and odd-side moulding may be summarised. Patterns once put on a plate are only suitable for using in groups, while patterns moulded in an odd-side can be made in any groupings. This latter is a great advantage, when the numbers required off given patterns vary from time to time; work put on plates should be arranged so that sets shall be completed without any parts being in excess on the one hand, or insufficient in numbers on the other. To plate metal patterns properly is costly, and is only economical when large quantities of castings are wanted. The fitting must be done most accurately, otherwise there will be lapping joints and cores out of truth. Small brasswork can only be done cheaply and well by the adoption of either odd-side or the method of plate moulding, and the choice of either must be controlled by circumstances. When the platework or gatework method is adopted, the advantages are that the time of bedding a pattern temporarily into its cope, and of making the joint face and sloping joints on which to ram the drag, is wholly saved. The time in cutting runners is also saved. These occupy a good deal of time when a number of patterns are moulded in one flask, and especially when they are of shapes involving the making of numerous sloping and curved down-joints. These methods of moulding are generally practicable when doing a large run of standard work, but they are seldom available in small foundries, and therefore the odd-side method is generally preferred in these.

BEDDING SAND.

The sand floor usual in a foundry is useful for bedding the heavier work, which is covered by copes. It is also used to lay the flasks upon when moulded and prepared ready for casting. As the sand is receiving constant additions from the new sand used in facings, a depth of from 3 in. to 6 in. will be sufficient to lay down at starting. It is employed over and over again for box filling, reserving the new special sand for facings—that is to say, for the stratum of sand which is immediately next the pattern to a thickness ranging from ½ in. to 1 in., hence called pattern facings. In the smallest moulds facing sand alone would be employed. Partings for brass moulds are made of burnt red sand or of red brickdust. Peameal dusted over is used for facing green sand work, both light and heavy. So are flour, powdered chalk, whiting, and sometimes charcoal. Lime mixed with water is used for facing dry sand moulds and cores for brasswork.

RAMMING BENCH.

In a small foundry it is necessary to have a bench, on which all moulds except those occasionally made in the floor will be rammed. The bench will be of dimensions most suitable to the work required. It may be of iron, but wood is suitable, 3-in. deals being supported on wood brackets, the latter being, if practicable, bolted to the wall with through bolts and wall plates, as shown in Fig. 44. If this cannot be done, the support may be afforded in the manner

moved. These cores are made by ramming raw sand or mixed sand into special core boxes; this sand is then pierced with an iron wire, and allowed to dry over the core stove. Often cores are washed with charcoal and water, and dried, in order to allow the core readily to leave the casting, and also to leave the casting with a good inner surface. This is not done for the common brass castings. In green sand castings, or those in which the moulds are not dried preparatory to pouring, it is usual not to insert the cores till the very last thing previous to pouring, so that the core may not absorb moisture from the mould, and thus prevent the core becoming friable.

THE USE OF CORES.

For castings that are hollow, cores are necessary, the patterns having projections called core prints, which leave hollows in the moulds in which the cores are supported. The core bearing must be of exactly the diameter of the prints in the pattern. A cored pattern and its print are shown at Fig. 29, the print being marked black. The

Fig. 44.—Ramming Bench Bolted to Wall.

Fig. 45.—Ramming Bench Supported by Legs.

illustrated in Fig. 45. Three 11-in. or 9-in. deals will afford sufficient width of bench for the average run of brass founders' work.

CORES.

When an opening is to be left in a casting, a piece of baked sand of the exact size and shape of the opening required is placed in the mould; this occupies the position of the opening required, and when this is in its place the molten metal runs round it, and an opening is left in the casting when the core is re-

white part of the illustration shows the form of the casting, and the black parts show where this core would be placed so as to make a hole through the casting. To

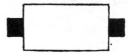

Fig. 46.—Pattern showing Core Prints.

illustrate the use of simple cores, a centre piece and top end suited for a poker may be considered. Such pieces might be made solid and then drilled out, but a sand core laid in the mould will produce hollow castings. Fig. 47 shows the shape of the centre piece, and also the method of putting in the core. The pattern is made of boxwood turned to shape, and the ends have core prints on them the shape of the core.

of suitable size, of either wood or metal. Plaster-of-Paris is then mixed, and run into this box on the sand, and left until it is completely set. The plaster cast, with the pattern, is taken from the sand and trimmed up, grooves or notches are cut in the face to give rigidity to the other half, and the

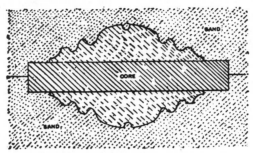

Fig. 47.—Section of Mould showing Core.

The core prints must be of the same diameter as the hole to be cast, and a piece of tube of the correct inside measurement for making the sand core must be sent with the pattern for moulding. The stick of sand which is made in the tube

first half is left to dry. This cast is well oiled, placed on sand, surrounded by a box frame as before, and plaster-of-Paris again poured in. This is allowed to dry thoroughly. The two parts are separated, the core pattern removed, and the core-box is

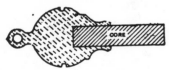

Fig. 48.—Section of Pattern with Balanced Core.

Fig. 49.—Shaped Core in Section.

will then be laid in the core prints, and when the metal is poured this core will leave a hole through the centre. The top piece (Fig. 48), not being cored right through, must have a balanced core. The pattern is made as usual, but the core print projects longer outside the pattern than the hole is to be inside. The core-boxes used to make the cores for small work are often made in plaster-of-Paris, the interior containing the cavity shaped to the cavity desired in the casting.

MAKING CORE-BOXES.

Core-boxes are also known as core-stocks, and are made as follows:—The core pattern is placed halfway in sand, and this is surrounded by a box without lid or bottom,

ready for use. These casts of plaster-of-Paris being somewhat fragile, are used to obtain others either in brass or, principally, in iron, which form permanent core-boxes. Though difficult to make, core-boxes are a necessity for economical work. Often they are not simple straight plugs, but are shaped to follow the outline, more or less, of the main pattern. Suppose, as an example, the pillar shown in Fig. 26, p. 21, is to be made, say, 1 ft. long. This, if cast solid, will be of considerable weight, and the expense of metal will be consequently great. To overcome this, a shaped core (Fig. 49) is put in. It will be noticed that only the general outline of the pattern is followed. First make the main pattern with core prints at each end. This will be made in a similar way to the one with the straight core (Fig. 46). If the prints are to be small, the sand core is strengthened by putting a piece of wire through the centre ; though this has to do with the casting, it is as well to get the cores as strong as possible. Now make a wooden pattern the shape of the core, as shown by the hatching. This must not be too large anywhere, or the castings will come out too thin ; when turned, the wood pattern should be well

Fig. 50.—Half of Core-box.

blackleaded all over, although but little moisture must be used. Bury this pattern half-way in sand, and take care that the surface around it is quite level ; then pro-

Fig. 51.—Whole of Core-box.

ceed to make a plaster cast as already explained. Grooves should then be cut at the sides (see Fig. 50) ; these act as guides for the other half of the core-box. The top surface should next be trimmed quite level, and then well blackleaded. With the plaster mould face upwards, and with the wooden pattern in place, the box is again placed round it, and another lot of plaster poured in. Allow plenty of time for it to set, and then trim the two halves all over until the whole is somewhat of the shape given at Fig. 51. It may then be blackleaded all over, and an iron casting made, this needing a little fitting before use. When the castings are large the core-stocks may be made of wood or iron, according to circumstances.

WIRE CORES.

To illustrate the method of laying in a wire core, the making of a butt-hinge pattern may be described. Half the hinge is shown in Fig. 53, and a wire runs from one end to the other. If this core were not used, it would be necessary to drill through the casting. For this pattern a piece like Fig. 52 is first made, brass cylinders being threaded on the wire, which is about No. 10 B.W.G. The plate, also of brass, is shown

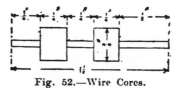

Fig. 52.—Wire Cores.

$\frac{3}{16}$ in. thick, 1 in. long, and $\frac{3}{4}$ in. wide. These dimensions are for a stout hinge, but they can be modified to suit requirements. When filing the hollow, great care must

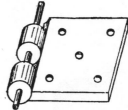

Fig. 53.--Hinged Casting.

be taken that the centre of the wire shall be just level with the top of the plate when soldered in position. The cylinders should next be secured to the plate by soft soldering in the following manner : Moisten the hollow all along with soldering fluid, and heat the plate in a Bunsen or other clear flame ; when it is sufficiently heated, apply a piece of soft solder, and the hollow will immediately flush with the solder. Shake off any excess, moisten the cylinders with soldering fluid, put them in position, and heat again until the solder runs. A neat and firm job will result. The holes for the screws to fix the hinge may next be drilled, and should be well tapered from the top. When castings of hinges are to be made from this pattern, a straight piece of No. 10 B.W.G. wire must be prepared for each hinge. The moulder will coat these with clay wash, and put one in the print of the wire, and when the hinges are cast the wires are knocked out with a mallet. A true hole is thus left, which, as a rule, will not even require clearing out. The two parts of the hinge can then be pinned together with No. 10 B.W.G. brass wire and finished in the usual way.

FALSE CORES OR DRAWBACKS.

Cores which have only one bearing, as those used in casting columns, etc., are called false cores or drawbacks. This false coring is one of the most difficult operations in the moulder's art. In moulding statuary

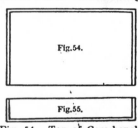

Fig. 54.—Top of Core-bench.
Fig. 55.—Side of Core-bench.

and similar articles, several methods may be adopted, the one chosen usually being similar to the following:—The figure to be cast is laid sufficiently deep in the sand to enable it to leave properly. When the moulder comes to an undercut part, he dusts it well with parting sand and fits in a piece of sand; this is done at each undercut part till the pattern is complete. When he has completed the false cores so that every part will leave without breaking, he completes the mould in the usual way. The top flask is taken off, and each of the cored undercut pieces taken away till the pattern can be removed, when the pieces of core are carefully dried, mended if necessary, and replaced. The mould is dusted, fitted together, and is then ready for the metal.

CORE BENCH AND SPILL TROUGH.

A core bench will be required on which to make the cores. The top of this bench should be a true iron plate—it may be either cast or sheet—with a ledge or flange turned

up round on the top, on the two sides, and at the back (Figs. 54 and 55). An iron plate is employed rather than wood, because it always remains true, though generally covered with damp sand and often made wet. It affords a true, smooth surface for ramming cores upon, the lower faces of which, when rammed on it, will be as true as though rammed on a bottom board. For general use this bench may measure 4 ft. long by 2 ft. 6 in. wide, and the thickness of the cast-iron, being the material generally used, may be ⅜ in. If of wrought-iron, it should be nearly as thick, since it must be rigid enough to withstand the ramming. A spill trough is useful when moulds are poured on end. It is 4 ft. or 5 ft. long, 12 in. to 15 in. wide, and stands 10 in. to 15 in. high, being supported on metal legs, the moulds being leaned against the trough during pouring, and any overflow of metal runs into the trough.

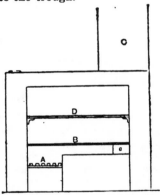

Fig. 56.—Core Drying Stove.

CORE DRYING STOVE.

A small core drying stove is shown by Fig. 56. This is not absolutely necessary, and in some small foundries cores are dried on top of the brass furnace, or in a small cupboard or stove over the furnace. The advantages of a special stove are that the cores are not likely to receive damage and the temperature can be kept nearly uniform. It is the proper appliance to employ for regular and satisfactory work, and small moulds can also be dried in it. An efficient stove is illustrated in Fig. 34; it is cheaply made if built of brick. A fire is lit on the grate below at one end A, and the flame and heat are drawn along under the iron plate B through the opening to the chimney C at the back. No flame or smoke can get into the stove, but it becomes heated thoroughly. The cores are laid upon the iron plates B and D. The heat is regulated by the opening of the fire door, and by a hinged damper on top of the chimney, operated by a chain.

ARRANGEMENT OF RUNNERS.

When moulding a large number of small patterns in a single flask, they, with their runners, must be so arranged that the metal shall not cool before the moulds farthest

during the pouring some amount of pressure is imparted by a head of metal. Generally, this takes the form of a continuation of the head of the ingate a few inches above the front or top mould. The more dense the metal is required the deeper must this head be. Castings for hydraulic work, which have to stand test pressures of 1,000 lb. or more to the inch, have to be very dense. If sufficient head is given above the mould, it is not necessary to cast supplementary heads upon the castings themselves. Such heads are sometimes cast on the ends of pump liners and other cylindrical work. But that is more with a view to take the sullage which gathers in a deep

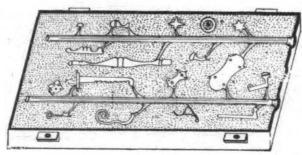

Fig. 57.—Arrangement of Runners in Mould.

from the ingate are reached. To this end it is desirable to maintain something like uniformity in size and mass of the several patterns included in any one flask ; to have the runners of sufficient area, and to pour the metal sufficiently hot to ensure its running to the farther end of the mould. There are two general arrangements of runners. In the one they pass from mould to mould in a flask, the metal running through the successive moulds to the last one. The other is that in which the metal runs down a ridge runner, and passes thence through sprays to the moulds arranged on one side, or on both sides, to right and left. In general the latter is the better plan, and it is the one almost invariably adopted when patterns are of identical forms, or are approximately so. A ridge of large area and small sprays is the ideal arrangement. This is shown in Fig. 57. The patterns are so arranged in a flask that

mould, and many liners are cast without any such addition.

MAKING GATES AND RUNNERS.

Before the moulds are finally closed, channels must be cut in them connecting the various small impressions with the runners and gates, in order to convey the molten metal to the necessary parts of the moulds. A very good way to make the main pouring or git-hole is to procure a thin piece of tin tube and with it cut a hole through the sand in the top box to the parting, afterwards withdrawing the tube and tapering off the mouth or top part of the hole to a trumpet or bell shape, and forming also a small air-hole from the top box, at the other end, through to the parting. This allows the escape of the gases which generate within the mould whilst the metal is being poured, and prevents the casting turning out blown, or damaged by small holes. The channels must be sufficiently large to allow the metal completely to fill the impression made in the mould, but must not be larger, or waste of metal will result. It is usual to put the runners or channels in the eye-side, while if any cores are used they are placed on the peg-side. The moulds, even in common work, are usually dried a little by placing before a stove fire, although this is not absolutely necessary if the impressions have been carefully prepared and dusted with charcoal ; it must not be omitted if fine castings are required. If both sides of the casting be required fine, both sides of the mould must be dried. When sufficient moulds have been made to constitute a heat—say five or six—they are placed against the spilling trough, and are then ready to receive the molten metal.

How Small Bells are Cast.

In making small cycle and other bells, no oddside is used as in ordinary brass casting, the method being to place the patterns on a board, convex side upwards ; then place a casting flask over them. Dust over the patterns a layer of parting sand, then put on a layer of raw sand, and finally fill the mould with ordinary sand. Ram this down exactly as for brass casting, place a board on top, and invert the whole. Take off the board first used, place on a peg-side flask, and dust the inside of the bell pattern, put in sand, and well ram up. Put a board on the top and invert the whole. Well hammer to loosen the pattern, and take off the board and first flask, which leaves the patterns on the inside cores. The patterns are carefully taken off and the mould is dusted with charcoal and peaflour. Place the two parts together, screw up, place against the spilling hearth, and pour. Iron moulds are now used for producing large quantities of small bells.

Common Defects in Castings.

A few of the most common defects may now be mentioned. Sometimes a casting, instead of being of an even, solid surface, is full of minute holes, and no matter how great the care taken a good finish cannot be obtained. These holes are formed through

carelessness in making the moulds, or in handling them after they are made, loose sand being allowed to fall into the pattern-prints, and when the metal is poured in the sand and metal become mixed up. Much of this can be prevented if the halves of the mould are blown clean with a pair of bellows before they are fixed together. Blowing is perhaps the most annoying defect of all, as it often occurs under the surface, and does not show until considerable labour has been expended on the casting, when the tool suddenly drops into a hole, and the article is useless. The moulds, not being properly dried, are unfit for use, and when the metal is poured, steam is generated, and if there is no vent, holes are formed in the castings ; the remedy is to dry the moulds. Clogging is when a casting, instead of coming clear, has lumps of metal in the ridges or recesses of the design. The fault lies in the pattern, and will be dealt with later, when pattern making is described.

Misshapen Castings.

Bad shape arises from two distinct causes, and the first may be styled " ovalness " or " flatness." For example, a simple sphere, instead of coming out a good shape, may be somewhat like Fig. 58, though

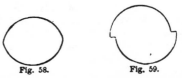

Fig. 58. Fig. 59.

Figs. 58 and 59.—Badly Shaped Castings.

perhaps not so bad. This is caused through the moulds—which were made with the frames not quite close together—being screwed up too tight, the sand being forced out of shape and all the prints flattened. The second cause is that the moulds are out of register, and the result is somewhat like Fig. 59. The moulds do not register if the pins do not fit the holes properly, and when the mould is moved its two parts shift their relative positions. The remedy is to get a new set of pins or a new frame. Sinking is easily detected by a scarcity of metal at some particular place. It comes either through other castings being run from it, or from the shape of the pattern. A slight alteration in the pattern may provide a remedy; but in some patterns the shape cannot be altered, though some larger part is drawing the metal away from a smaller part; in this case, much can be done by pricking the mould with the point of a pen-knife at the place where the defect arises. There are few patterns that will not yield to this treatment, but care should be taken to blow away all loose sand.

DIFFERENCES BETWEEN MOULDING FOR BRASS AND FOR IRON.

The methods of moulding for brass castings do not differ in principle from those adopted in making moulds for iron castings of the same dimensions. With occasional exceptions, castings in brass and gun-metal are always of small size. They rarely exceed a hundredweight, while the greater quantity weigh but a few ounces or pounds each. For iron, as for brass casting, moulds are made in green sand, in dry sand, and in loam, the latter being used for heavy struck-up work only. Green sand, skin dried, is used largely. Speaking generally, finer sand is used for brass than for iron, and the moulds are not poured so damp. Brass moulding differs from iron moulding in two or three respects only. One consists in the occasional use of wooden flasks for brass only, the other in the use of different facings, such as plumbago, pea meal, or other material employed for dusting the surface of the finished mould. In the matter of pouring, too, there are differences. Runners for brass are larger than those for iron, and risers and feeding are less necessary for brass than for iron.

MELTING AND MIXING THE METALS.

The method generally adopted for mixing and pouring the metals is to melt the least volatile metal first, and then with the tongs to plunge the more volatile metals under the surface in small lumps, having heated these lumps first. This method prevents loss from these metals. It is usual to put a portion of the scrap at the bottom of the crucible or pot; then, if any new metal is being used, the copper is placed next, while the zinc is kept on the top of the furnace ready for use. A layer of charcoal, or sometimes of broken glass, is put on the metal to protect it from oxidation. When the metal is melted and of sufficient liquidity, the zinc is added and thoroughly well stirred in bit by bit. It is generally held for a few moments by a pair of tongs in the heat of the furnace to drive off any moisture that may be upon it, and also to warm it slightly. This prevents the spitting or flying of the molten metal. Any other metal that is to be added should be put in immediately before pouring. Add the flux, allow it to clean, and draw the metal out of the furnace and pour. If much

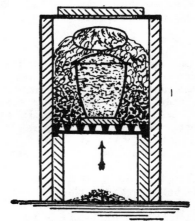

Fig. 60.—Crucible in Furnace.

REMEDY FOR COLD FURNACE.

scrap is being used an additional quantity of zinc, say 2 lb. to the cwt. of metal, must be used to cover the loss by volatilisation: in fact, it is always well to add a small quantity of zinc to cover this loss.

Some furnaces have a bottom grate through which a current of cold air passes, and in these it will be found that when the crucible settles down through the burnt coke on to the fire bars, it may be quite cold at the bottom, while the sides are white hot; this leads to breakage of pots, and prevents proper melting, and a half-fused lump of metal may be found at the bottom of the pot when it is removed. A small disc of fireclay should be placed on the fire-bars before lighting the fire. The crucible will rest upon this (Fig. 60), and the current of air will be prevented from impinging on the bottom of the crucible, thus allowing it to be fully heated throughout.

FUEL FOR MELTING.

The actual melting of metal is not difficult. The fuel used is always coke. Hard coke is the better fuel to employ, as it leaves only a small percentage of ash. Gas coke is frequently used because it is cheap, but its heating power is less than that of the furnace coke, which may cost three times as much, but which goes farther and melts more thoroughly. It is hard, heavy, silvery, and comparatively free from sulphur and other elements which contaminate the gas coke, which is but a by-product; while the furnace coke is a primary product in which the hydrocarbons are retained for heating purposes. Gas coke yields sufficient heat, is easy to obtain, and is cheap. Distinct disadvantages are that it generally contains a large amount of sulphur, and a large percentage of ash. There are patent cokes specially made, which, though costing several shillings per ton more, are in the end more economical, as they last longer in the furnace, contain considerably less sulphur, and produce a smaller quantity of ash.

ARRANGEMENT OF MOULDING AND CASTING SHOP.

In small firms the moulding shop generally includes the casting shop; but when the moulders exceed four, a separate shop is advisable. In the moulding shop are placed the core and drying stoves, and the furnaces also if there is no separate casting shop. In an ordinary small business, the drying stove may consist of a square grate in which a coke fire is kept burning, the sides of this grate being used for drying the moulds, while the top is used for the cores. It is, however, preferable to have proper stoves, which are heated by the hot air from the casting furnaces, the flues from the furnaces being taken under the drying stoves before they pass to the chimney. If a single shop serves for casting and mould-

ing, the furnaces should be placed at one side, while the moulding-tubs may be placed on the other three sides. In Fig. 61 is shown a general arrangement of a moulding and casting shop; D representing a bench to which are fixed two vices for dressing the castings, which are afterwards weighed into the rough warehouse, together with the gates, scrap, and all unused metal. Moulding tubs are lettered A, spilling hearths B, and coke stoves C. The lighting of the shop is from the roof; the door is at one corner, and in each side are openings for two or three hinged doors. Outside this shop is the coke-bin F, from which run two shoots immediately over the furnaces, supplying the caster with the necessary fuel. In front of the furnaces E, E, E are iron gratings, which admit the necessary air to the furnaces, and allow the ashes to be removed, G being the ashpit. One most important point in the construction of the casting shop is good ventilation, in order that fumes, smoke, and steam may

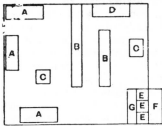

Fig. 61.—Arrangement of Moulding and Casting Shop.

pass off freely. Several openings near the floor, with a louvre opening in the roof, will generally suffice to create an upward current sufficient to remove the fumes. When the business is likely to be large, a separate dressing shop will be needed, although usually dressing is done in the casting shop.

STEEL FOUNDRIES.

Steel foundries differ from iron foundries in the nature of the sand used (which is highly siliceous), in all moulds being dried very hard, and in the flasks or moulding boxes having fewer bars than those used in iron foundries. Many special precautions have to be taken to ensure good castings, and more trouble is experienced in making good castings in steel than in iron. There are several reasons for this— as difficulties connected with temperature, with shrinkage, with composition, with annealing, etc. Most steel castings are made in the open-hearth furnaces, by which it is possible to obtain a more controllable product, and in larger quantities. Steel can be melted in a crucible in an ordinary brass furnace by increasing the height of the chimney to get a sharper draught. To make a large steel casting from crucible metal several furnaces are necessary, so that trouble and expense are saved by the open-hearth furnace. All steel has to be poured at a much higher temperature than iron. It therefore contracts more ; and the hotter it is poured the more it shrinks. Owing to the heat at which it is poured, dried moulds have to be used, and these being also composed largely of the siliceous ganister, bind the shrinking casting, and in the absence of suitable precautions will cause fracture. Hence a casting from the same pattern but made in iron may be accurate, while a similar one made in steel may be either distorted or broken. In any case, the steel will have shrunk twice as much as the iron. Owing to this great shrinkage taking place in a dried mould, so hard that it may be stood upon without damaging it, all weak sections such as the arms of light wheels, and weak parts adjacent to heavy sections, are especially liable to fracture or distortion. Either such weak parts must be thickened, or strengthening brackets must be cast to connect the thick with the thin. It is for this reason that steel castings are often sent home with increased thickness, or added brackets cast in certain sections.

SHRINKAGE IN STEEL CASTINGS.

The shrinkage in steel castings is far from uniform, and it is therefore necessary to allow a larger margin for machining on important surfaces and in cored holes than in the case of iron. Especially should this be done when there are several fixed centres, mutually related, and situated at considerable distances apart. Shrinkage is so uncertain that when a large number of similar castings is required it is safer to have

one or two made first before ordering the entire quantity, lest these should turn out differently from requirements. Faces should have $\frac{1}{2}$ in. or more excess allowance for machining. Many castings not to be machined, can only be poured soundly by having a large self-feeding head of several inches in height, and of large area, cast on. In order to pour an ample volume of hot metal into the mould, the steel founders cut very large runners, and these not infrequently, when there are several on a casting, are a source of distortion. Plenty of metal for feeding is necessary to counteract the great shrinkage of the steel. The greater the variation between the dimensions of adjacent parts, the greater the need for shrinkage heads. For these reasons, it is never well to attempt to get intricate castings in steel, but it is better to build up forms in simple separate castings. These will be much stronger than castings in a condition of severe internal stress. These and many other kindred facts have to be borne in mind when dealing with casting steel. The best mixing of steel for small castings depends on a due balance of chemical elements, chiefly carbon and manganese, and the results are judged by tests. The subject is dealt with in books on the metallurgy of steel and iron. It is usual to assume that small steel castings are required to have a minimum tensile strength of 26 tons to the square inch, with an elongation of at least 10 per cent. in a length of 8 in., the pieces being about 1 in. square. They should also admit of being bent cold in a press, or on a slab or block with an edge of a radius of 1 in., to an angle of 45°. The hammer test is also used to test soundness, and note is taken of surface defects or flaws. Nearly all castings go into annealing ovens, by which they are rendered less brittle than when they come from the moulds.

CLEANING CASTINGS.

When the metal has been poured into the moulds, and sufficient time allowed for it to set and cool sufficiently, the flasks are then opened, the moulds are broken up and the sand is riddled, the castings are taken out and cleaned from core sand, and, if necessary, are dressed at the dressing bench, and finally put into the shaking barrel or tumbling box and barrelled. The barrel generally consists of an octagonal box lined with iron, and rotated by power. The barrels may be of various shapes and sizes, but all are on the same principle, which is that of a revolving churn, either by hand- or machine-power, and with an opening to place the castings in. The castings are tumbled about until all the sand is knocked off them. The continual knocking of the castings one against another cleans off dust, trims bits of brass off, and generally brightens up the castings. They should not be left too long in the tumbling box, or the castings will be knocked out of shape. When working on commercial lines after the barrelling, the castings are taken to the rough warehouse, where they are weighed and booked to the credit of the caster against the metal he has had out. The quantity of work a single caster will do varies, but about six heats of 60 lb. to 80 lb. per day is usual. The number of moulds required for each heat will, of course, depend on the size of the castings and of the crucible used.

Art Casting by the Cire Perdu Process.

The highest refinement in the art of casting in bronze appears to have been attained by the great Italian artists. The method, as described by the Florentine artist Cellini, is as follows: The figure was first roughly modelled in clay to a size very slightly smaller than the casting. Over this was laid a skin or thickness of wax, representing the thickness of metal in the intended casting. A present-day moulder will recognise in this the counterpart of the " thickness " on a loam pattern. The perfect figure of the cast was imparted to this wax with modelling tools, all the fine lines, expressions, etc., being so perfect as to leave little for after touching-up. Then a mixture of clay, pounded brick, and fine ashes, made with water to the consistence of cream, was applied with a brush, completely filling every interstice, and the object was then enclosed in a substantial body of clay, and the whole banded with iron hoops. Then the mass was baked

in an oven, the melted wax escaping through holes left for the purpose. Rods of bronze passing from the outer mass to the inner core maintained constant the thickness of the interspace left by the melted wax, and into this space the metal was poured. After the removal of the baked clay the casting only required touching up.

Aluminium.

Aluminium is the lightest of commercial metals, steel or iron being three times as heavy, copper and nickel three and two-fifths times, silver four times, lead four and two-fifths times, gold seven and a half times, and platinum eight and a quarter times. When pure it is very ductile and malleable ; in fact, it has displaced silver leaf in some instances. When, however, the metal has iron or silicon in solution, it becomes short in character, losing both its ductility and malleability, but increasing in hardness. To be non-corrosive, or subject to the oxidation of the atmosphere, it must be as pure as possible. If, however, hardness and elasticity are of primary importance, then it is an advantage to add slight quantities of other metals, notably copper, nickel, and zinc. Castings require a relatively larger quantity of hardening alloy than work which can be hammered or drop-forged. Aluminium melts at a temperature of about 1,200° F., or between antimony and silver. It does not volatilise, even when raised to a temperature produced by the combustion of carbon, but for good castings it is inadvisable to raise the metal above melting point, or to a point where it loses its silvery appearance, as it has a great tendency to absorb gases during melting and to give spongy castings. Only good blacklead or plumbago crucibles should be used, and in every case it is well to use a plumbago cover or lid to the pot. At an excessive heat the metal will absorb silicon from the crucible, but when raised to its melting point, aluminium, although melted eighteen times, has been found to absorb ·09 per cent. of silicon only. The melting may be done in the ordinary brass furnace. If scrap metal is being used, it must be well pickled and dried, or dipped in turpentine or paraffin, in order to cleanse it from dirt and grease. Small scrap is usually added to a quantity of molten metal, as this prevents loss by drossing. It is well to cover the top of the metal with a thin layer of carbon, but this is not absolutely necessary, as aluminium at ordinary temperatures is not volatile. No flux is necessary, although if the metal is very dirty a small quantity of saltpetre or nitre is used, forced well down to the bottom of the crucible.

ALLOYS OF ALUMINIUM.

In making aluminium alloys, it will be

frames, sewing machines, and the like, consists of aluminium and zinc. Zinc forms a cheap and efficient hardener, and the following quantities are being largely used: 82 parts aluminium, 15 parts zinc, and 3 parts tin. An alloy of 7 parts aluminium and 3 parts zinc has been successfully used. For heavier castings, use aluminium 97 parts, nickel 2 parts, copper $\frac{1}{2}$ part, tungsten or wolfram $\frac{1}{10}$ part, and tin $\frac{1}{8}$ part. For ordinary castings, copper is an excellent hardening medium, an alloy of equal parts of copper and aluminium being

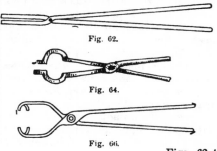

Fig. 62.

Fig. 64.

Fig. 66.

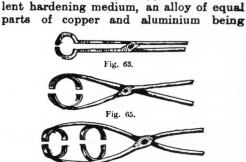

Fig. 63.

Fig. 65.

Fig. 67.

Figs. 62 to 67.—Crucible Tongs.

best first to use the metals in nearly equal proportions; for example, if copper is being used for hardening, make an alloy of 50 parts aluminium and 50 parts copper. Melt the copper first, and add the aluminium bit by bit, keeping the metal to just a little over melting point. Stir continually with a carbon stirrer similar in quality to the plumbago crucible, and, when all is melted, pour into shallow moulds. This alloy will be brittle, readily breaking into fragments; it can easily be alloyed with more aluminium. Aluminium and nickel must be prepared in the same way; such alloys are used principally by jewellers and other special workers, and the following two may be taken as typical: 20 parts nickel and 80 parts aluminium; and 40 parts nickel, 10 parts silver, 30 parts aluminium, and 20 parts tin. A fine alloy, fairly hard and strong, for casting purposes, consists of aluminium 87 parts, tin 10 parts, and nickel 3 parts. Another alloy for casting consists of aluminium 96 parts, copper 1 part, and nickel 3 parts. A useful, cheap alloy utilised in making castings for bicycle

formed, and then the proper alloy made from this. Copper in quantities of from 2 to 10 per cent. may be used, according to the nature of the work required, an average alloy containing 6 per cent. This alloy will be made as follows: To 88 lb. of aluminium when melted add 18 lb. of the cupro-aluminium alloy. The small percentage of copper decreases the shrinkage of the casting, and gives alloys specially useful for art castings.

CASTING ALUMINIUM.

The sand moulds for use in casting aluminium alloys must be made as for brass, but the sand must not be rammed too hard, and large gates and risers are absolutely necessary, particularly at the thickest parts of the castings. The usual shrinkage of aluminium castings is $\frac{1}{4}$ in. per foot. The cores should in no case be hard, as these nearly always cause porous castings or shrinkage cracks; use greensand cores wherever possible. If this is impossible, dry sand must be used, but this must be made of a yielding nature, and yet cohesive. For large cores, it is advisable to use hay rolled round a bar of iron, and to cover this with a layer of greensand, finishing with loam and a coating of powdered whiting and water. For smaller cores the bar may be wrapped with hurds (waste tow). If an extra fine finish is required to the castings, the moulds and cores should be coated with powdered French chalk. The

them from the furnace, tongs of various shapes are used. A number of those that are most generally useful are illustrated by Figs. 62 to 67 on p. 38, and by Figs. 68 to 70 on p. 39. The tongs shown in the first figure are the most simple and are meant to grip the crucible by its edge; these tongs are also the least satisfactory. The second figure shows tongs made so as

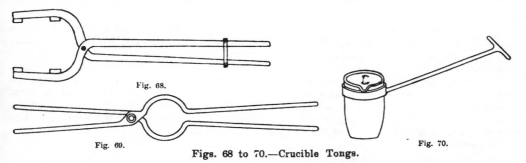

Fig. 68.

Fig. 69.

Fig. 70.

Figs. 68 to 70.—Crucible Tongs.

alloys must be poured at as low a temperature as possible. Pour quickly, keeping the pot or crucible as near to the mould as possible to prevent the running metal carrying in cold air and forming a cold bar. The mould, too, must be tilted so that all the gases confined during pouring may have as much chance as possible of escaping.

CRUCIBLE TONGS.

For the purpose of handling the crucibles, when placing them in and removing

to encircle the crucible and thus afford a much better grip. Some of the later figures illustrate tongs made to hold the crucible with their handles vertical. Fig. 67 is a notable example. Fig. 69 shows two-handled tongs intended for use by two persons, and required when large crucibles heavily charged with metal have to be handled. The last figure shows a crucible holder made with a ring to encircle the crucible, and having a T-handle to facilitate turning out the contents.

SMITHS' WORK.

EARLY MALLEABLE IRON.

Although the art of casting in bronze, as has been already mentioned, was practised in the early ages, yet most of the early work in iron appears to have been wrought under the hammer. The making of huge castings is a comparatively modern branch of metal work. The only distinctions made by the ancient and the early mediæval iron workers were those between malleable iron and steel. Cast-iron, though produced in the furnace, was an abnormal product, for which in its crude state no use could be found. The malleable iron then made was superior to our best, because it was smelted only with charcoal. Very small quantities were produced at one time then, just as is the case now in the Hindoo, Burmese, African, and Catalonian furnaces ; but the quality was admirably adapted for the best and most delicate smiths' work. Probably the early smiths could not have produced such excellent work as they did if they had been compelled to use the inferior bar-iron of our present days.

ANCIENT SMITHS' WORK.

It will be seen at once that the ancient smith had much harder work to accomplish before he could really commence his work than the smith of the present day, who gets his bars of iron rolled to the very section he may require, and produced far straighter and truer than he could possibly make them on his own anvil. This will account in some measure for the absence of straight bars from the oldest smiths' work, as perhaps the most difficult task that could be set was to handle and beat out a long and heavy ingot into a bar with mathematically true angles. In ancient work, in consequence of this fact, we do not find any great variety of section produced, or that the angles of the bars were at all mathematically correct. This must always be borne in mind when estimating ancient work and comparing it with the work of the modern craftsman, and in contrasting the work of different ages.

ANCIENT AND MODERN CRAFTSMEN.

The old craftsman cut from his shingled bar the piece of iron which, with only his eye for guide, he would be able to beat out into the length required to curl up into the scroll form he wanted to make. More or less sufficed for him, and by his method of work he produced a pleasing irregularity even in the most monotonous design. This to us is artistically beautiful, though perhaps to him it was a matter of vexation when he found that his work did not exactly correspond the one part with the other. The modern craftsman, on the other hand, purchases the bars of the required number and shape, cuts them up into pieces of exactly the same length, makes a standard pattern, and when there is much repetition of scrolls of the same size, he uses a tool to gauge the scrolls and ensure their uniformity. If there is any irregularity in the work, it is considered bad smithing ; and if, under the conditions, it is the result of carelessness and want of thought, the result may be inartistic.

HOW ANCIENT IRON WAS PREPARED.

The iron itself reached civilised communities either in the " bloom," direct from the furnace, or more commonly as rudely-shaped ingots small enough to be easy of transport, and in this state it formed, like gold and silver, a current article of barter. The smiths who worked these up were either important citizens or formed separate and honoured communities. The discoveries at Bibracte show an entire town given over to the craft, the members of the guild being buried with their tools and implements around them. Until the tilt or helve hammers formed part of the plant of ironworks, the rude labour of fashioning the object direct from the ingot

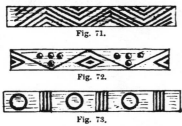

Fig. 71.

Fig. 72.

Fig. 73.

Figs. 71 to 73.—Chisel-worked Bars.

or bloom fell directly on the smith. These hammers weighed from 12 lb. to 1,500 lb., and up to 2,500 lb. Their use was to beat the rough bloom into bars on a slightly tapered anvil, thus relieving the smith of the most laborious part of his task. They were very common in Surrey and Sussex. The furnace-master, who smelted, and the forge-master, who beat the iron out by mechanical means, became distinct callings in England, and had nothing to do with the smiths who produced the finished work. We knew little of the actual manufacture till comparatively recent times.

EARLY EXERCISES IN METALWORKING.

The manipulation of iron in the making of artistic ironwork of various kinds and for varying purposes may well form the subject of early exercises in metalworking. A few of the more often recurring details are here described and illustrated, and these will be of assistance to those who desire to do good artistic work in the best way. Here no pretence is made to give a complete set of working details, those given

being rather in the nature of suggestions, and elaborate designs will be found in later pages. As has been stated, in ancient times the smith had to forge out his bars for himself into the shape and forms required from the iron billets, which were usually from 1 ft. to 1 ft. 6 in. long and from 2 in. to 4 in. square. This was a difficult process, and the bars thus produced were not mathematically true. This work is now not needed to be done by the modern smith, as the bars are rolled at the mills into such shapes as may be wanted, including octagonal, star-shaped, club-shaped,

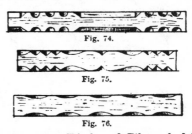

Fig. 74.

Fig. 75.

Fig. 76.

Figs. 74 to 76.—Chisel- and File-worked Bars.

and other ornamental forms, and every bar is true and exact in shape.

CHISEL MARKED ORNAMENTS.

The first method of ornamenting the bar was by chisel marking. Figs. 71, 72, and 73 are taken from ancient examples of this kind of work. Fig. 71 shows perhaps the oldest form of ornamentation. Here the steel chisel, with a thick, short, sharp edge, is driven into the square or flat bars, care being taken to make the junction points clear and sharp. Fig. 72 gives a similar arrangement, but with the addition of a series of holes, which may be drilled with an angle-pointed drill only as deep as the angle of the drill ; or they may be struck with a large centre-punch. Fig. 73 shows an ornament involving a little more work. The triple lines may be struck with the chisel as above described, and then filed to an arched shape with a file. Between these triple lines a circle is placed, which may be struck by a circular punch, or lined out and then struck with a narrow chisel.

Chisel and File Marked Ornaments.

Examples of ornamentation of a different kind are shown by Figs. 74, 75, and 76, though the chisel and file are still used. Fig. 76 is the simplest form; here the sharp edges of the square iron bar are filed across the angle with a half-round file, thus producing a simple, pretty effect. Fig. 74 is a distinct advance on the foregoing; here

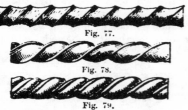

Fig. 77.

Fig. 78.

Fig. 79.

Figs. 77 to 79.—Twisted Bars.

the sharp edges of the iron bar are what is called chamfered, the angle being cut away to a square edge with a chisel, and then filed smooth. The hollows between these chamfers are filed with a small round file. Fig. 75 is similar to this, but the chamfers

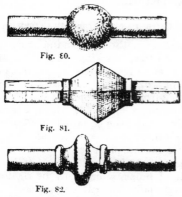

Fig. 80.

Fig. 81.

Fig. 82.

Figs. 80 to 82.—Collars on Bars.

are made ornamental in outline. These are carefully cut with a chisel and then finished with a file.

Twisted Bar Ornaments.

Forms of iron bars twisted are shown by Figs. 77, 78, and 79. This process is accomplished by heating the square iron bar red hot; then fix one end in a vice firmly, holding the other end with tongs if the bar is a light one, or with a wrench if a heavy bar; turn the bar, cooling it with water as the successive twists are made. The iron being cooled retains its shape, and the next turn moves only the heated portion of the bar. Fig. 77 shows the square bar thus plainly twisted. Fig. 78 is a piece of flat iron twisted in the way described above, thus giving a different effect to the iron. Fig. 79 is what is know as a compound twist. In this, a round bar of a small size and a flat bar are twisted together, giving the effect shown in the sketch. This is the most ornamental form of twist, and may be altered in style by using varying shapes of iron to be twisted together. This twist-

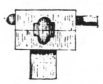

Fig. 83.—Swage for Forging Collars.

ing operation requires care and patience to do the work well.

Collar Ornaments.

Iron bars of various shapes and sizes are ornamented by regular shaped collars at intervals, as shown in Figs. 80, 81, and 82. These may be of various shapes other than the three here illustrated. They are produced by the use of swages (see Fig. 83), which are solid pieces of iron fitting over each other and cut out to the shape of the collar required. One half has a bottom stem which fits in the hole in the anvil, and the top half has affixed to it an iron handle; a collar of iron sufficient in size to make the ornament required is turned round the bar in the position desired. This is then made white hot and placed on the bottom half of the swage, the upper portion of the swage being hammered down on it. Thus the ornament is produced and welded on to the bar.

Open Twist Ornaments.

Another method of ornamenting iron bars is shown by Figs. 84 and 85. This arrangement is very effective as an ornament, and does not detract from the strength of the bar. Fig. 84 shows the open portion twisted, Fig. 85 simply swelled out. Figs. 86, 87, 88, and 89 show the various operations necessary to produce the result. In Fig. 86 the necessary number of round up on an iron plate, when heated red hot regularly all over. Fig. 91, 92, and 93 show differing patterns of collars that may be used. Other shapes will easily suggest themselves.

Ornamenting the Rails.

The rails through which the bars before described have to pass may be dealt with in

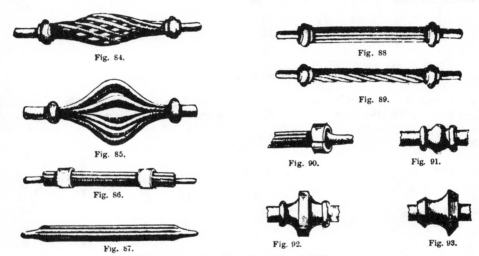

Fig. 84.

Fig. 85.

Fig. 86.

Fig. 87.

Fig. 88

Fig. 89.

Fig. 90.

Fig. 91.

Fig. 92.

Fig. 93.

Figs. 84 to 93.—Open Twists and Collars.

bars are placed round a top and bottom centre short stem and kept in place by welded rings top and bottom; each end is successively placed in the fire and welded up, and the rings are then removed. Fig. 87 shows this operation completed. The collar or ornament at the top and bottom of the twist (see Fig. 88) is produced as described for the iron bar by placing a collar round (see Fig. 90) and welding on in a swage (see Fig. 83). Fig. 89 shows the bars as welded together twisted. They are then made red hot and untwisted to the required shape. This must be carefully done, the portions being cooled with water so that they may retain the shape required whilst the other portions are being worked. Fig. 85 is opened by being hammered end varying ways. Figs. 94 and 95 show a bar of a smaller diameter than the rail passing through, leaving sufficient space on either side to be strong for the work it has to do. The holes may be drilled or punched. Figs. 96, 97, 98, and 99 show the arrangement necessary when the bar has to pass through a rail but little larger than the bar. Then a larger rail is used, and the hole for the bar is forged out as required for a square or round bar (see Figs. 97 and 99) and then welded on to the rail in its proper lengths. Scrolls may be fastened to the upright or horizontal bar by collars fitted on as shown in Fig. 100, in which the collar is made in two portions (see sections, Figs. 101 and 102). The one portion fitting round three sides is left at the ends with shouldered pins; the other side is drilled and riveted on to the side by the pins. Fig. 103 gives four different sections of mouldings suitable for fitted collars. Fig. 104 gives a plain flat collar,

old work of grouped scrolls welded together at the points of junction. The sections (Figs. 110 and 111) show the thinning down of the scroll ends and widening out to form the lap weld. In Fig. 109 a group of single

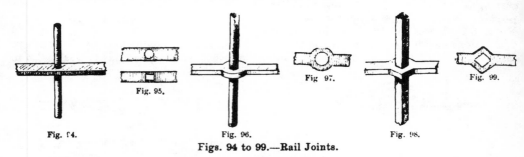

Fig. 95.

Fig. 97.

Fig. 99.

Fig. 94.

Fig. 96.

Fig. 98.

Figs. 94 to 99.—Rail Joints.

the two ends of which are cut slantwise and then bent round the bars. This will require to be pinned to the bars to be kept

leaves formed by hammering out a flat bar is shown welded in with the plain scrolls. In Fig. 104 the portion to be fitted on is

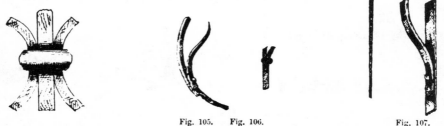

Fig. 105.   Fig. 106.

Fig. 107.

Fig. 100.—Collar Fastening.

Figs. 105 to 107.—Fitting Scrolls to Rails.

in place. Figs. 105 to 107 show other methods of fitting scrolls, etc., to rails and bars.

Scroll Work Ornament.

Much of the beauty of artistic ironwork depends on the beauty of the scroll work,

thinned in the fire as shown, and then riveted to the flat or square bar. Fig. 105 shows a fitted end of scroll; in this a small piece of the size of the end of the iron is filed out of the bar, in which the scroll end is placed and riveted on. Figs. 108 and 109 show a set of scrolls welded together

Figs. 101 and 102.—Sections of Collar.

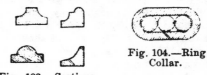

Fig. 103.—Sections of Mouldings.

Fig. 104.—Ring Collar.

which may be varied in many ways according to the taste and skill of the craftsman. Figs. 108 and 109 are specimens from some

at one end so as to be fastened to the bars in one piece. To do this the end of each scroll is thinned down as shown in section in Fig. 107, the thin portions being hammered out wider so as to make a lap joint and then welded together in one.

VOLUTE SCROLL ORNAMENTS.

Fig. 112 is a plain volute scroll thinned down at one end only. Fig. 113 is a similar scroll, but with the end drawn down or

as shown in Figs. 114 and 115. This, when carefully executed, is very effective. Or the end of the scroll may be hammered out wider at the end and tapered down before turning into a scroll, as shown in Fig. 117. If this operation is cleanly and sharply executed, it enhances the value of the scroll. Fig. 118 shows a double-ended scroll of a kind much in vogue in eighteenth-century

Figs. 108 and 109.—Specimens of Grouped Scrolls.   Figs. 110 and 111.—Sections of Scroll Ends.

snubbed. Fig. 116 gives the same volute scroll with the snubbed end drilled through and a rosette of thin iron fastened to the same with a rivet or screw nut for orna-

work, with square welded corners in the centre. Scrolls are made by being turned on a wrought-iron scroll-iron (see Fig. 119) ; this is carefully made of strong flat iron

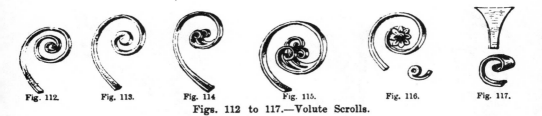

Fig. 112.   Fig. 113.   Fig. 114.   Fig. 115.   Fig. 116.   Fig. 117.

Figs. 112 to 117.—Volute Scrolls.

mentation. In much of the older work this snubbed end is hammered out into leaf shapes, or masks and grotesque faces and heads ; these are cut out with the chisel

1¼ in. or 1½ in. by ½ in., turned in a taper scroll upwards. Round this the heated light iron bar to form the scroll is turned.

Fig. 118.—Double-ended Scrolls.

Fig. 119.   Fig. 120.

Figs. 119 and 120.—Method of Making Scroll.

and filed true and smooth. The ends of the scrolls may be slit up into two or three parts, each of which is scrolled separately

Fig. 120 shows a cast-iron scroll-iron, which answers a similar purpose, but is cast from a wood pattern specially made.

FOLIATED ORNAMENTS.

The scrolls and other portions of artistic ironwork may be, and often are, ornamented by the addition of leaves and husks. In this direction an almost endless variety of pattern is open to the artist craftsman. For the manufacture of leaves the very

generally, especially in cases where a large number were required, the leaves were stamped similar in shape. But even in these cases, whilst using the stamp for the general work, the old craftsman altered the terminations and other portions of his leaves with the hammer and the chisel, so as to change the general effect and relieve it of the sameness that generally belongs to stamped work. In the stamping, the pattern required was cut out in a solid piece

Fig. 121.—Leaf Husk.

Fig. 122.—Plate to form Leaf Husk.

of metal and hardened; and into this the leaf or rose was hammered when hot; or, if the iron used were only thin, it was hammered cold, being heated in the course of

Fig. 123 and 124.—Leaves in Sheet Iron.

Fig. 125.—Group of Leaves.

best iron should be used. Generally Swedish sheet will be found the best to manipulate, though some makes of English iron are exceedingly good, and may be readily used for the heavier work. This class of work was adopted in the olden times for a large amount of the ornamental hinge work for doors and chests and for grilles. The illustrations to this article are all taken from ancient work of various dates. Very

the stamping as the necessity arose for manipulating the metal. A stamp may be made, which would be useful for many shapes of leaves, by having the patterns of various indentations, lines, balls, and other forms cut out, into which the parts of the leaf to be moulded may be hammered as required. By this plan, one stamp may be made to suit many shapes of leaf and flower.

LEAF AND HUSK ORNAMENTS.

Fig. 121 shows a leaf husk that would fit round an iron bar and form a top or bottom termination, or, with the addition of a collar as before described, a centre to a bar standard. Fig. 122 shows the method of

Fig. 126.—Group of Leaves.

its manufacture. A piece of iron is cut out with the chisel as shown, then stamped to give the leaf its indentation, then heated and hollowed, then heated and bent to-

Fig. 127.—Ivy Leaves.

gether so as to give the form shown in Fig. 121. If care is exercised in this operation, the shapes of all the husks may be made to differ slightly. Figs. 123 and 124 show leaves for working on the outer sides of scrolls; these may be riveted to the scroll bars or welded. Fig. 123 is stamped, as before described, in a mould. Figs. 124 and 125 show simply the flat sheet slightly hollowed in the centre, and with the terminations of the leaves turned and twisted according to the taste of the operator and

the effect desired to be produced. Figs. 126 and 127 show groups of leaves with stems that are welded together. Fig. 126 shows a leaf stamped up; this may be accomplished by hammering the leaf into a

Fig. 128.—Simple Leaf and Flower.

plain circular stamp with a round- or oval-headed punch. Fig. 127, ivy leaves, shows each leaf with the terminations simply twisted and curled; this is easily executed

Fig. 129.—Leaf Seen Flatways.

Fig. 130.—Leaf Seen Edgeways.

by heating the iron and twisting with round-nosed pliers or pincers. Welding the leaves to the stems must be very carefully executed, as it is very easy to burn the ends of the small rods of iron.

FLOWER ORNAMENTS.

Flowers may be made in iron as shown in the accompanying illustrations, which give some of the simpler methods. Figs. 128, 129, and 130 show a simple flower and leaves. The flower is formed of round iron rod, bent circularly, each turn being larger than the preceding one as far as the centre, and then smaller to the top. Fig. 128 shows the arrangement at top and bottom of flower. The leaves (Figs. 129 and 130) are cut out of thin iron and bent as shown. Fig. 131 shows a flower made in similar fashion to the last, except that the circular

Fig. 131. Flower and Double Leaf.

Fig. 132.—Plan of Scroll for Flower.

Fig. 133.—Plan of Leaf.

Fig. 134.—Ball Flower with Double Leaves.

Fig. 135.—Ball of Flower.

coils are brought straight up to a point. Fig. 132 shows the bottom of the flower

with the stem screwed on the coil. The rosette under the flower is cut out of charcoal sheet-iron and hollowed in each circular portion. The leaves (Fig. 133) are cut out as shown and hollowed as before described, then welded to the stem as shown in Fig. 131.

Fig. 136.

Fig. 137.

Fig. 136 and 137.—Plans of Section of Rosettes.

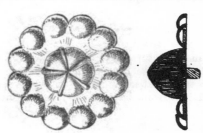

Fig. 138.—Single Rosettes.

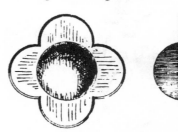

Fig. 139.—Single Rosettes.

ROSETTE ORNAMENTS.

Fig. 134 shows a ball-coil flower with double rosettes underneath. The ball-coil (Fig. 135) is made as described, and screwed on the stem. The two rosettes, which are of different shapes, are cut out of charcoal sheet and hollowed out as shown in section in Figs. 136 and 137. These are then placed on the stem, one over the other, and the ball-coil flower is screwed on. Rosettes are used very much in the ornamentation of ironwork. Figs. 138, 139, and 140, which are taken from ancient examples, show single-sheet rosettes hollowed out as before described and fastened down with a

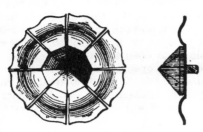

Fig. 140.—Single Rosette.

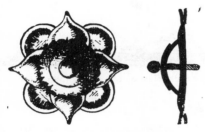

Fig. 141.—Double Rosette.

Fig. 142.—Double Rosette.

Fig. 143.—Double Rosette.

solid-headed screw pin either plain or ornamental, as shown in section in Figs. 141, 142, and 143. Compound rosettes, or rosettes formed of one or more plates, one partially covering the other, are shown in Figs. 141, 142, and 143. The plates are cut out of charcoal iron and hollowed when hot in

stamp irons, as shown in the sectional views on the right. These are fastened with solid-headed screw pins and nuts. The sketches given will doubtless suggest other patterns, and the application of these ornaments will be shown in later pages.

EASILY MADE HANDY FORGE.

A handy forge suitable for light work, and occupying small space when fixed, may be cheaply and easily made in two parts: the pan or hearth, which is fixed to a wall; and the bellows, which are placed beneath on the floor. The pan is made of sheet-iron, 20 gauge, galvanised for preference, so as to avoid rust. A piece of sufficient dimensions for the pan, which is made up

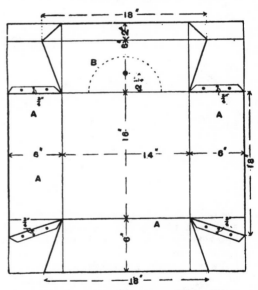

Fig. 144.—Top of Easily Made Handy Forge.

in one piece, should be procured, the corners being secured by two rivets each. The measurements given in Fig. 144 are only intended as a guide, and may be varied to suit requirements. Bend up the sides and front A to the slope shown, and the back B square, fastening the corners as described. In the centre of the back cut a hole sufficiently large to admit easily a piece of ½-in. iron gaspipe. Fig. 145 shows a semicircular piece of thick black sheet-iron, 8 in. in diameter, hollowed slightly, with a hole cut in the centre to admit the pipe. The edge should be dressed back square after hollowing sufficiently wide to take a rivet. This is fixed to the back of the pan inside (see B), to prevent the pan from burning. It also helps to keep the air pipe firm, if well fitted in the hole. Two or three rivets will keep it in position, and care must be taken to keep the two holes perfectly level. Procure a piece of old ½-in. iron gaspipe, and bend both ends at right angles to the required length, which will depend on the height of the pan from the floor. One of the ends is to go through the holes mentioned, and project into the pan about 1½ in.; the other end will be connected with

the bellows to rest on, allowing for a free passage of air. Make a hole in the end of the box to allow the nozzle to pass through, and then secure the under-side of the bellows to the bottom of the box with two screws, so that they are firmly fixed in it. The top of the box may be covered in, but it must be high enough to allow the bellows to open to their full extent. Between the open handles fit a strong wire coil spring, which will keep them in the required position; an old "Moderator" lamp spring will do for this purpose if obtainable.

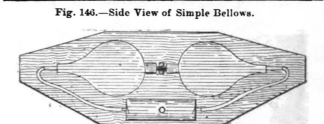

Fig. 146.—Side View of Simple Bellows.

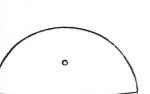

Fig. 145.—Baffle Plate for Forge.

Fig. 147.—Top View of Simple Bellows.

the bellows. The pipe may be fixed to the wall with pipe hooks. A couple of angle brackets fixed to the wall will be found sufficient to carry the pan, and two small bolts, with nuts, fixed through the bottom of the pan and the brackets, will make it quite secure.

BELLOWS FOR THE HANDY FORGE.

Procure an ordinary pair of large size household bellows, and remove the rounded part of the nozzle with a file. If this is done, the nozzle, being tapered, will easily fit into the air pipe. Make a rough wood box, with one end only, large enough for the open bellows to fit in easily, allowing the handles to project to their full length. Fix across the bottom of the box, inside, two pieces of wood, about 1 in. square, for

Fasten the spring with two small staples to prevent moving. Cover the top handle of the bellows with a piece of sheet-iron large enough to place the foot on and fix with small tacks. When this is completed, place the box with bellows on the ground under the forge, and insert the nozzle of the bellows into the air pipe; and, to prevent moving, a weight of some kind should be placed on the top of the box. No other fixing is required. A brick placed in the pan at a convenient distance each side of the air pipe will be found useful in keeping the fire together. The air blast is not continuous, as with a portable forge, but for heating irons, melting lead, small brazing jobs, etc., this will answer very well. The whole cost of material will not exceed five shillings.

Simple Bellows.

A less elaborate blower is shown in elevation and plan by Figs. 146 and 147 respectively. This is made from two pairs of ordinary household bellows; the larger they are the better. The bellows are mounted, valve downwards, on a stout board, and the board must rest on fillets, and have holes cut opposite the valves to supply them with air. The upper boards of the two bellows are connected by a strong leather strap or cord, running over a grooved pulley journalled in a standard mortised into the centre of the bottom board. The nozzles of the bellows are connected by indiarubber tubing to a small valve box at the side, from which the connection to the blast pipe is led. This box (o)

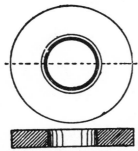

Fig. 148.—Bottom Die for Hollow Boss, Plan and Section.

has two leather clapper valves, opening inwards, and acts to a certain extent as a reservoir. If motion is kept up with the feet, a very powerful blast is obtained, which can be modified instantly at will. This bellows has no troublesome indiarubber diaphragm.

Die Forging or Stamping.

Die forging is for the most part the completion of work which has been previously prepared by the usual methods. All the ordinary tools used on the anvil are required for die forging—not for the actual finish under the hammer, but in the preparatory work. The finishing of work by hand is usually a tedious process, and the time thus taken up is saved by the use of

dies. Absolute accuracy, and uniformity in shape and dimensions of any number of similar pieces, are also secured. In order that dies or stamps shall be real economical factors in the production of work, it is necessary that it be of a repetitive character. The job must not only pay for the cost of the stamp or stamps used, but the stamps must cheapen the job, or be productive of greater uniformity than is practicable with hand work; or fulfil both conditions. Some stamps that are the most expensive to make produce work that is cheaper and better than hand work. The question of the cost of each method is, therefore, relative, each case being settled on its own merits entirely.

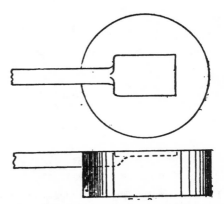

Fig. 149.—Bottom Die for Brake Treadle, Plan and Elevation.

Shapes of Dies or Stamps.

The forms of dies or stamps vary much. The simplest, of which Figs. 148, 149, and 150 are types, consist of a bottom portion only. Fig. 148 is for finishing a shallow boss at the end of a plain lever, Fig. 149 for finishing a brake treadle, Fig. 150 for finishing a boss on the fulcrum of a bell crank lever. These are intended for use under the tup of the steam hammer, and this descends directly on the top flat face of the forging, which, on filling up the die, lies level with the top face of the latter. When the nature of the work is such that the upper face is not flat and level, then top and bottom dies are necessary, and a portion of the forging comes into the upper die. The top and bottom portions are dowelled, or otherwise guided, in order to prevent overlapping of the joints. Fig. 151 shows a pair of dies used for finishing crane hooks ; and, of course, dies of this type would be used when the shapes in top and bottom were dissimilar. A third class of die is used for punching holes. Holes may be punched when double dies are used. For example, if Fig. 151 were made for a

MATERIALS FOR DIES OR STAMPS.

The materials used in stamps are cast-iron, cast steel, and forgings of iron or steel. The first are employed for large blocks, the second for small ones. The first are not very costly, being cast from patterns ; the second often are, because they have to be cut with drills, chisels, files, etc. Cast dies require little tooling. Circular parts are often bored, and dowels

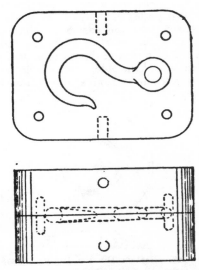

Fig. 151.—Pair of Dies for Crane Hook, Plan and Elevation.

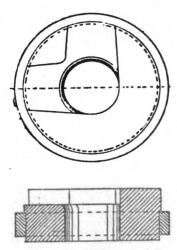

Fig. 150.—Bottom Die for Bell-crank Lever, Plan and Section.

tie-rod end, or for a bossed end, a hole could be punched through the boss centre by forming holes through top and bottom dies central with the boss, and driving a punch, guided by the holes, through the boss enclosed in the dies. The top hole would guide the punch, and the wad would fall out through the bottom hole. When there is no top die, a plate A (Fig. 152) is dowelled on to form a guide for the punch ; or a guide is slipped into the hole, as shown in Fig. 153, where A is the piece being punched, B the loose guide, and C the punch. It is convenient to make the bolster D as a loose piece, though it is often made solid with the body of the die.

have to be fitted into drilled holes. Stamps of cast-iron are usually encircled with wrought-iron bands shrunk on, as in Figs. 150 and 152, to prevent them from bursting. Holes are provided for lifting the heavier blocks (Figs. 151 and 153), wrought-iron bars being thrust into the holes. Sometimes bars are cast on to serve as handles. Small blocks are lifted by means of special tongs, as shown in Fig. 154. Special tools for use with the dies under the steam hammer comprise tools for cutting off and for forming or moulding the hot metal into the outlines preparatory to stamping. Substantially they are anvil tools mounted on long handles for use on work under the hammer. They comprise hot and cold setts, chisels and gouges, swages, plain and of the spring type, blocks for bending work over, mandrels for circular work, V and hexagonal blocks, formers, punches, and bolsters, etc., the

SIMPLE TOP AND BOTTOM STAMPS.

The making of wrought-iron railing tops may illustrate the use of top and bottom stamps. The form of the top is of little

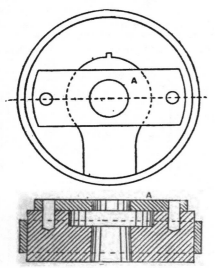

Fig. 152.—Die for Punching Holes, Plan and Section.

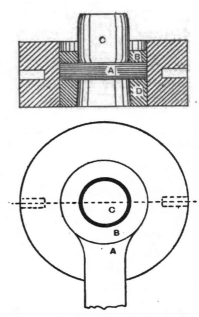

Fig. 153.—Die for Punching Holes, Section and Plan.

description of which would involve a considerable number of illustrations. There is no method of computing the comparative cost of small work except that afforded by experience in a given shop. The me-

consequence, one highly ornamental, ribbed, or veined being stamped nearly as readily as one which is very plain. But

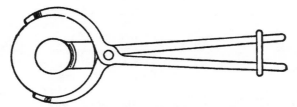

Fig. 154.—Tongs for Lifting Dies.

thods of work differ so much that the practice of one shop is an indifferent guide to that of another. Apart from past experience, the only safe way to price work is to make two or three forgings, time them, and thus fix a price for a quantity.

dimensions will make difference in the forms of the stamps desirable. So also will the nature of the appliances which happen to be available. For the smaller heads, spring stamps are the most suitable; for large heads it is better to have top and bottom stamps separate and dowelled, especially when a steam hammer is available. Mild steel is the best material for

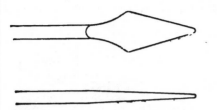

Fig. 155.—Spear-head Railing Top.

small stamps. Cast steel or cast-iron is cheaper for those of larger size. Fig. 155

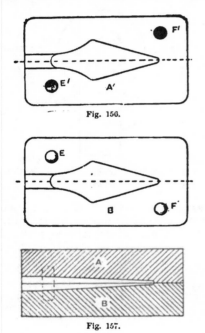

Fig. 156.

Fig. 157.

Figs. 156 and 157.—Dies for Forging Spear-head Railing Top; Top Die, Bottom Die, and Section.

shows a plain railing top. To make this, the bar which is to form the railing is drawn down and flattened at the top at a first heat very roughly to the outline required. Then another heat is taken, and

the top finished in the stamps in two or three blows. Figs. 156 and 157 show the essential construction of the stamps for finishing Fig. 155. Top and bottom blocks each have the form of one-half thickness of the railing top cut or cast in it, according as the material of the stamp is forged or cast. Figs. 156 and 157 are guided into perfect alignment at the time of stamping the head by means of the dowels or pins E and F, which enter into the holes E′, F′.

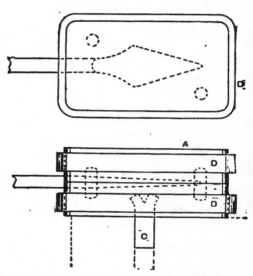

Fig. 158.—Cast-iron Dies or Stamps for Forging Spear-head Railing.

The two half stamps A and B are shown open in their joint faces in the middle and left-hand figures, and closed and in longitudinal section in the lower figure. The stamps shown are intended to be used under the steam hammer. B is laid on the anvil, the rail top laid in it; A placed above, and one blow, or sometimes a couple, causes the roughed-out railing top to fill up the impressions cut in the stamps. A is then removed and the top taken out completed, except for a trifle of fin, which is knocked or ground off subsequently. These stamps are supposed to be of forged steel, and the impressions of the rail top have then to be cut in the steel with cold chisels, gouges, and files. This method would be suitable for rail tops of average dimensions.

mer, of course, the stem is not required. At D D are shown bands of wrought-iron shrunk round the stamps. These are used

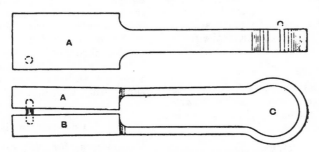

Fig. 159.—Spring Stamps, Plan and Elevation.

CAST DIES.

Fig. 158 shows a pair of dies as made in cast-iron or cast steel, either with tops of large dimensions or for tops of average size, in cases in which so few rails have to be

when cast-iron is the material employed, and are for the purpose of preventing the bursting of the cast-iron under the concussion of repeated hammer blows, which it is not so well able to withstand as cast

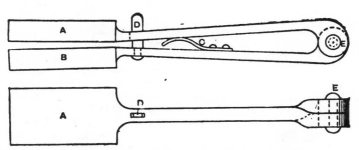

Fig. 160.—Hinged Stamps, Elevation and Plan.

stamped that it is desirable to save the cost of laboriously cutting out the shape of the heads in forged metal with chisels, etc. In the figure the dies are shown closed upon a railing top. For Fig. 158, one half pattern is made in wood, in which the shape of the half thickness of the rail head is cut easily. Two such are cast, and then dowelled together. In Fig. 158 a square stem C of wrought-iron bar is shown cast in the bottom die. This is done when the dies are to be used on an ordinary anvil, the stem fitting one of the holes on the anvil face. It is also cast in if the dies are used on a levelling block. For the steam ham-

and mild steel. If the stamps are very massive, handles should be cast in opposite sides for two men to lift them by; but it is seldom they are required so heavy for railing work.

SPRING STAMPS.

For the smaller stamps, the methods shown in Figs. 159 and 160 are well suited. In Fig. 159 the stamps A B are united in one by spring handles C. The necessary spring is imparted by the forging and setting of the handles, so that the normal condition of the tool is that shown in the figure; B is laid on the forge anvil, or on that of the steam hammer, and A is struck down on the rail top. On the removal of the hammer, A springs up sufficiently high to permit of the withdrawal of the rail top. It is necessary to insert a dowel somewhere to guide A and B into alignment and prevent side-slip. One is shown in the figure.

HINGED STAMPS.

Fig. 160 is a hinged stamp, being pivoted at E; the necessary elasticity is imparted by means of a spring C, which releases A

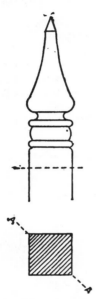

Fig. 161.—Railing of Square Section, showing at A A the Direction of Joint in Stamps.

after the hammer blow. A dowel inserted at D prevents side-slip, the only direction in which sliding can occur. Two or three trials in roughing-out will be necessary before the exact amount of stock is hit upon that will leave the minimum of fin on the one hand, and will not leave deficiency of metal anywhere on the other. The heat for stamping must be nearly that required for welding. In Figs. 156 and 157 it will be observed that the joints of the stamps coincide with the thinner portion, or the breadth of the railing top. This is a

method always followed when practicable in stamped work, because there is less difficulty in releasing the forging from the die than there would be if the joint were in the other direction, and the forging went down deeply into the stamps. In the case of a railing top like Fig. 161, of square section throughout, the stamps should be jointed in the direction A A, so that the forging shall lie diagonally in them, a position most favourable to delivery.

VARIETIES IN HAMMERS.

Hammers for different purposes are made of different materials. The engineer uses hammers faced with hard steel, the stone-breaker or mineralogist hammers faced with soft steel. Again, in another part of his progressive work, the steel hammer with which the engineer commenced his operations gives place to a bronze or a copper one, and this is sometimes displaced by one of lead alloyed with tin, and the usual handle entirely discarded. It then assumes, in all respects save material, the form and mode of handling of the prehistoric stone maul. This leaden cylinder, about 2 in. in diameter and 6 in. long, is the primary form of this second maul, which is about 3½ in. in diameter and 3 in. long, the ends of the cylinder being very much spread out. There is no handle to this tool; occasionally it is made of a mixture of lead and tin, in order to impart a hardness not possessed by lead alone. The plumber dismisses all these, and for direct action upon the material employed in his trade he uses a hammer of wood, discarding not only the material, but also the form of hammers use in allied crafts. One of his hammers serves a double purpose, for if at one moment it is a hammer, at the next it is used as a swage. In some cases, as in the working of copper vessels which have been silver-plated or gilt, the coats of the precious metals are so thin that, although the weight of a hammer-head is required, yet even the wooden hammer, or the still softer leaden hammer, is equally unsuitable, and therefore the workers in these metals cover the face of their hammers at times with one or more layers of cloth held on by a ring which slips round the ordinary hammer-head as the hoop on a drum.

TONGS USED IN SMITHS' WORK.

There are often a dozen tongs to a moderately well-appointed forge, but it is not necessary to get them all at once; a few of the simpler and more useful tongs will now be described. Each of these tools is

elongation of the V-shaped jaws gives a stronger grip, and the rod or bar is less liable to shift sideways. When a bar is so long that it cannot be held with these tongs, a crook-bit tongs (Fig. 165) is used; the jaws being turned aside to allow the bar to pass alongside the handles on one side of the rivet. The lip serves to retain the bar in place, otherwise it would be apt to slip out sideways. With these four

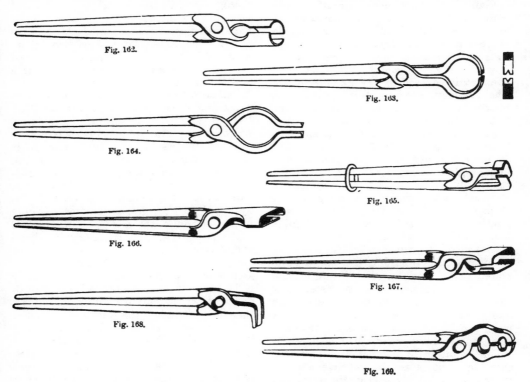

Figs. 162 to 169.—Various Forms of Tongs used in Smiths' Work.

made in several sizes to suit the various kinds of work. Fig. 162 shows the hollow-bit tongs; enclosing and gripping the rod for a length of about 2 in., they take a very firm hold of both rod and bars. When there is a collar or other enlargement at one end of a bar, the pincer tongs (Fig. 163) are sometimes employed to enable a firm grip to be taken; the jaws have V-notches, as shown in the end view. Fig. 164 shows tongs that are more generally useful; the

kinds of tongs work can be commenced on round rods and square bars of iron. The ring encircling the handles or reins of the crook-bit tongs is called a coupler, which is slid over the reins, and tightened by a tap or two with the hammer. The work is thus grasped without the need of any further effort on the part of the smith. For other work there are other forms. For holding flat bars, tongs shaped like those shown at Figs. 166 and 167 are employed. In Fig. 166

the jaws are alike and come into direct apposition. In Fig. 167, a flat jaw falls within the sides of the other. These are made in various widths and proportions, the range of each pair being rather limited. For holding and manipulating rings, hook tongs (Fig. 168) and pick-up tongs (Fig. 169) are employed.

PLANISHING SHEET METAL.

Planishing sheet metal requires great care and attention from the operator, besides considerable skill, which is only developed after ample practice. The art consists in hammering regularly over the surface of a sheet of metal until it has been completely covered by the hammer blows. The hammers used for this purpose are nearly flat, and vary in weight from about 3 lb. to 8 lb. ; the anvils on which the metal is hammered are also nearly flat, and vary in weight from 30 lb. to 80 lb. The surfaces of both hammer and anvil should be of looking-glass brightness, and after use should always be carefully oiled to prevent any rust spots forming. When preparing them for use, a little flour emery or putty powder is sprinkled on an ordinary knife board, and the tools are polished with this after the oil is wiped off. The object of planishing is to harden the metal hammered, and to close the coarse grain usually seen on unhammered sheet metal, this causing the surface to become quite smooth and glossy and consequently rendering it more capable of taking the brilliant polish usually associated with planished work. When metal is tinned before planishing, then, in addition to the improvements in the metal stated above, the tin is worked well into the pores of the planished metal, and consequently the tinned surface remains in good condition much longer than is the case with an article not planished. It is much to be regretted that this is not more generally known, especially in connection with planished tinware, as the extra expense incurred in planishing is more than balanced by the extra wear the article will stand. When planishing large work, it is good practice to commence by working straight across the centre of the sheet, and then from the line of hammer

blows formed, continue working in straight lines backwards and forwards along the length of the sheet until the edge is reached. The metal stretches wherever hammered, and, by working as directed above, the stretching which first occurs along the centre of the sheet causes it to swell there, and this swelling is gradually taken out by working along and out to the edge as stated. In the case of narrow strips of metal, the planishing may be executed across the width or length of the metal, as the stretching of the metal is hardly noticeable in narrow lengths. Sheet copper, instead of being properly planished, is often spotted over with a number of blows, this adding considerably to its stiffness. Hollowed bright work is generally planished on a planing wheel—that is, a machine that has top and bottom wheels of hardened bright steel. The hollowed body is placed between the two wheels, and pressure is then exerted by the wheels on the metal, a number of radial or other strokes, according to the shape of the article, being then made by the backward and forward motion imparted to the wheels. Thus the whole of the surface of the metal, being covered with these strokes, is hardened and brightened. Irregularities in the flatness of sheet metal are generally due to lack of homogeneousness in the metal, or else to faulty rolls. If a sheet of metal has an up-and-down wavy appearance along its edges, then it is tight about the centre of the sheet, and a few blows delivered there will stretch it sufficiently to remove the tightness and allow the sheet to rest flat. If the centre of the sheet has a swelling, then the sheet is loose there and tight at the edges ; a few blows at each corner of the sheet and one or two along the edges will generally stretch the metal enough to remove the hollow from the centre. And in any kind of sheet metal, wherever a loose or wavy place is seen, it is removed by stretching the metal round or opposite the loose place. In planishing, setting, or stiffening, if the metal is required bright, before hammering it must be thoroughly cleaned, otherwise dirt, scale, or anything of a similar nature will work into the metal during the hammering process, and quite spoil the surface.

MAKING A SMALL ANVIL.

The method of making a small anvil as here described is the result of actual practice. A piece about 5½ in. from a wrought-iron bar, 1⅜ in. by 3¾ in., will be wanted; also an odd piece of iron for the horn about 1½ in. square, and a piece of double-sheer steel for the face of the anvil. The first thing is to thicken one of the 5½-in. by 1⅜-in. edges till it becomes 5½ in. by 1½ in. This thickening should extend down for about 1 in. or more, and the sides should

thickened edge should be drawn down a little at one end to form a scarf, and the piece of iron which is to form the beak should also be scarfed. Weld them up, using sand during heating to prevent the iron burning; and let there be plenty of metal around the weld, for this can easily be trimmed off afterwards. Now work up the horn to shape, leaving it about 3 in. long; and apply the flatter all over the body of the anvil, using it also to draw out what might be called the "stern" of the

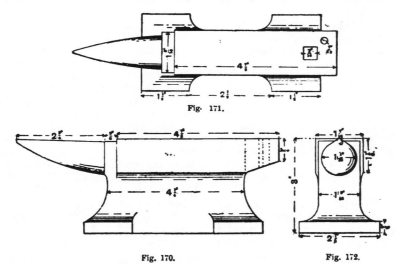

Fig. 171.

Fig. 170.

Fig. 172.

Figs. 170 to 172.—Making a Small Anvil; Side, Top, and End Views.

be worked up during the process with a flatter. When this is done, take a heat on the opposite edge, lay the iron flat on the anvil, and, with a chisel shaped like a very flat U, cut out an arch-shaped piece from the edge, leaving 1½ in. on each end. Now take a good white heat, and, turning the iron up on its thickened edge, hammer down the two projecting pieces left at each end till they spread out and form the feet. This will be plain if the illustrations are referred to. Now cut out the pieces from the ends of the anvil. This is best done on the shaping machine, but if this is not available, they must be cut out with a hot chisel. When this is done, the anvil will be ready to have the beak welded on. The

anvil so that it overhangs the back foot by about ½ in. Use a larger fuller to finish the curves on the tops of the feet. The steel, which should be ¼ in. thick, must now be welded on the top of the anvil. To do this it is advisable to weld or otherwise fasten it to a short rod of iron to form a handle; this dispenses with tongs, which might be in the way. To facilitate welding, one or two grooves should be made with a narrow fuller on the under-side of the steel, parallel to its length. Remember to strike lightly at first when welding, in order to stick the steel, and then hammer up well. This finishes the forging. The beak and its base must be filed up and the face and edges trued on a shaping machine if possible, otherwise with the file. Drill a $\frac{1}{16}$-in. hole through the end, as shown in Fig. 171, and square it with a small file; also drill another hole about $\frac{3}{32}$-in. in the position shown. These holes will be found useful in many ways. Now polish the face; all that then remains is to harden it. This must be done with ferrocyanide of potash if the steel is very mild, and, when quenching, do not dip the whole anvil into water, but let a stream from the tap run on the face so as to drive the heat away from that part first. The body of the anvil may have a coat of dead black. A section of a small tree trunk is the best mount for the anvil, which should be fastened to it by four spikes driven in and bent over the projecting feet.

FLATTENING SHEET METAL.

In giving information as to flattening metal, general principles only can be offered. The best of craftsmen work with nothing but general principles for their guidance; so much depends on the varying circumstances under which the work is carried on. Some metal is most difficult to flatten, while other kinds may be readily levelled; and if machinery is used sometimes instead of manual treatment, it is only a way of causing greater molecular displacement than can be effected by tapping with hammer or mallet. In this book on metalworking we have nothing to do with machinery; therefore it will be assumed that the simplest appliances, and those generally at command, are used—namely, a flat surface of iron or steel, free from indentations, a round-faced hammer, a boxwood mallet, and a piece of wood. A flat surface has been mentioned instead of the regular appliances because it appears this is much more serviceable than the limited dimensions of an anvil. Very often the bed of some machine, which answers the purpose admirably, may be purchased almost at the price of old iron. There are many objections to the ordinary anvil, apart from its expensiveness. The comparatively small surface it presents would be a drawback in the manipulation of a large thin plate, because any metal

extending beyond the surface would ply over the edge, and thus, by affecting the metal on the surface, would create vexatious difficulties.

HOLLOWING SHEET METAL.

For making ordinary stock articles in sheet metal in quantities, the general introduction of stamped and spun work, which will be described in later pages, has rendered skill in hollowing sheet metal of minor importance; but where these articles are made to special sizes, the advantages of the methods here described are obvious. The metal is worked down in concave spaces of suitable curvature, which are cut in the end of a section of an oak or a beech trunk, a convenient size being about 3 ft. high and 2 ft. 6 in. in diameter; the hammers used should also be curved proportionately to the curve required for the work in hand. When cutting out the material, allowances extra to the size of the article must be made, these being absorbed in the curved body. One of the commonest hollowed articles is a circle hollowed to form a cover, such as would be used on a saucepan. The diameter of the circle in the flat can be found by bending a narrow strip of metal to the shape shown by the section of the finished cover, and then straightening this out to show the length. When hollowing a circle, it is usual to commence by working regularly round the edge with a round-faced hammer, the metal being placed over a hollow in the block; the blows are then delivered in a series of concentric circles as far in towards the centre as may be desired. The hollowed circle is smoothed by lighter blows delivered over the surface with regularity, or by a series of radial strokes on a planishing wheel.

METHODS OF PRODUCING PUNCHED HOLES.

Methods of producing punched holes vary according to circumstances. If it is desired to preserve the same amount of metal all around, the hole is partly punched and partly drifted or opened out. Obviously, when a hole is punched entirely with a flat-ended tool, the metal that is removed is equal in area to the area of the point of the punch itself, and the width of the bar is only increased to the slightest perceptible degree (see Fig. 173). Another way of making a hole without weakening the bar to any great extent is by means of a conical punch, which may be inserted, and the hole formed and opened, without the removal of any appreciable quantity of material. Or, if a hot set is driven from one side halfway through the bar, and halfway from the other (Fig. 174), a punch or drift can be driven in afterwards, and the slit opened out into circular or oblong form, as may be required. Long slot-holes are cut in two ways. In one, holes are punched at each end of the intended slot equal to its width, and the metal between is cut out with the hot set, the cuts from opposite sides meeting in the middle (Fig. 175). In the other, holes are punched at each end, a chisel line being made centrally from hole to hole, a drift inserted, and the metal opened out (Figs. 176 and 177). In this way the flanking metal is thrust out sideways, and the bulk of its section retained. Such a slot-hole is finished by inserting a drift or mandrel of the correct size, and hammering the outside edges of the bar upon it. When punching holes, it is necessary to take account of the direction of the fibre. Unless attention is given to this, the iron will become divided, instead of spread out. Punching puts considerable tension on the fibres around the hole, with reduction of area, and it is an operation, therefore, that requires to be done with judgment.

EXAMPLES OF SMITHS' WORK IN THE MIDDLE AGES.

Wrought-iron railings appear not to have been used by the ancient classic races in connection with their buildings, as no reference to such work can be found and no specimens have been preserved. However, from the earliest middle ages to this age, wrought-iron has been largely used in buildings in filling up the windows for protection, in encircling the balconies, in surrounding the open yards of houses and the tombs in cathedrals, and in railing-in the wells in cloisters and courtyards of castles. Many of the railings are very ornamental and rich in design and finish. It is most probable that at first the bars, either square or round, were simply pointed at the top and riveted into a flat bar at the

Fig. 173.—Material Removed by a Flat-ended Punch.

Fig. 174.—Slitting Hole with a Hot Set.

Fig. 175.—Hole Slotted out of the Solid Metal.

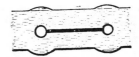

Fig. 176.—Hole Slitted without Removing Metal.

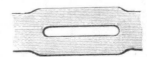

Fig. 177.—Slitted Hole Completed.

foot, with one or more crossbars of flat iron through which the uprights passed. Gradually the bars were forged out into ornamental forms such as fleur-de-lis of varying shapes, as shown by Fig. 178, which is from some railings in a church at Toulouse. The ornamental fleur-de-lis is forged out of square iron flattened down on each side, turned on the anvil to the shape required, and then welded together at the bottom to the upright bar. Fig. 179 illustrates a grotesque head chiselled out of a piece of solid iron and welded to the upright bar. This head is placed alter-

Fig. 178.—Fleur-de-Lis Rail Top.

Figs. 179 and 180.—Grotesque Head Rail Top.

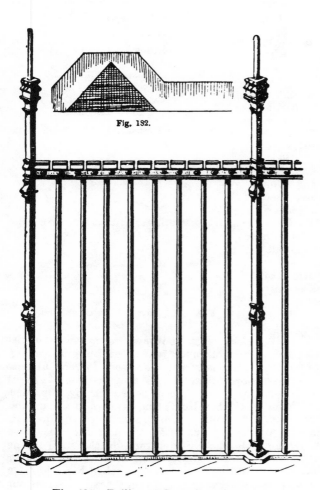

Fig. 182.

Fig. 181.—Railing at Canterbury Cathedral.

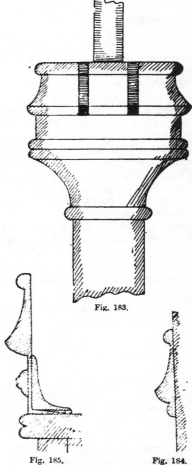

Fig. 183.

Fig. 185.

Fig. 184.

Figs. 182 to 185.—Enlarged Details.

nately with the fleur-de-lis on the same railing. Fig. 180 gives a front view of this
head, which is very vigorous in style, and,
though roughly finished, is very effective.
Many of these heads are found in ironwork
of the Middle Ages, and are very good specimens of chiselled work. In working
these heads, each one would differ from the

above the top rail, are placed at intervals
along the sides as well as at the corners,
and are finished with a welded boss (Fig.
183) which is battlemented—that is, cut out
with a chisel on each of the four sides to
the battlemented forms shown. They are
also ornamented with iron riveted to the
sides of the standards, as shown in Figs.

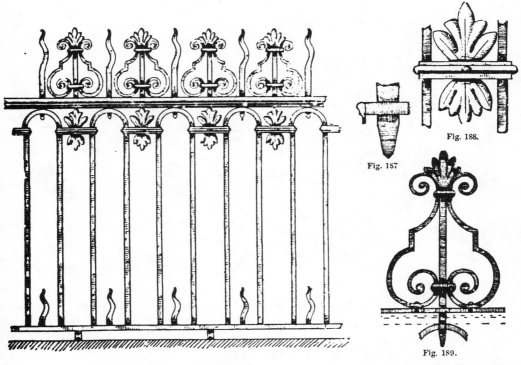

Fig. 186.—Railing at York Minster.

Figs. 187 to 189.—Enlarged
Details.

Fig. 188.

Fig. 187

Fig. 189.

other; thus they give opportunity for the
skill of the blacksmith.

Example from Canterbury Cathedral.

The railing round the tomb of the Black
Prince in Canterbury Cathedral (Fig. 181)
is a typical example of the early railings
of the fourteenth century. This is constructed of square iron bars set anglewise,
and riveted top and bottom into flat iron
cross-bars. Stronger standards, carried

181 and 184. Fig. 182 is a half view of the
bottom bar, showing the welded boss into
which the standards fix. The upper rail
is ornamented on the front edge by being
rounded as shown in Fig. 185, and also by
a piece of flat iron cut out in battlemented
form and fastened to the top rail by an
angle-iron strap with ornamental rosettes
(see Fig. 181 and also Fig. 185, which is an
enlarged section showing the fitting). This
railing is an especially good example of
the work of its time.

EXAMPLE FROM YORK MINSTER.

Fig. 186 is an excellent specimen of a later period from the more highly decorated railing that encloses one of the tombs in York Minster. The upright bars are of square iron and arched at the top, 6 in. from centre to centre. They are riveted to the bottom rail of flat iron and to the flat top rail by a waved bar also of square iron, pointed at top and bottom and form-

decorative ironwork. Quatrefoils with double trefoil leaves are riveted together in the square framework as shown. The trefoil leaves are cut out of flat iron, their stems and the junctions of the quatrefoils being welded together as shown in Fig. 191. The framework, shown enlarged by Fig. 192 and in section by Fig. 193, consists of two bars placed one on the other and riveted or screwed together. This method adds to the decorative appearance of the

Fig. 190.—Railing at Florence.

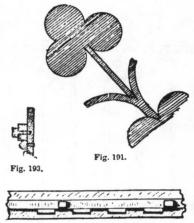

Fig. 193.

Fig. 191.

Fig. 192.

Figs. 191 to 193.—Enlarged Details.

ing a portion of the cresting. Fig. 187 is an enlarged view of this fitting, and also shows the iron moulding riveted to the front of the top rail. A waved spike is riveted to the bottom rail in every other space. At the commencement of the arch in the upright bars a moulding returned on itself is fixed, to which an iron repoussé double leaf is riveted. Fig. 188 is an enlarged view of this. The cresting is formed by fixing a double scroll ornament between the waved bars and connecting to the square central bar by clips of half-round iron and having a single repoussé iron leaf at the top, as shown in Fig. 189. Fig. 190 is a sketch of a railing of the fourteenth century in one of the churches of Florence. In this the framework is filled in with light

framework and at the same time strengthens it.

EXAMPLE FROM SPAIN.

Fig. 194 is a specimen of Spanish railing from a church in Segovia. The upright bars are of round iron with diamond-shaped portions welded at intervals, so as to alternate horizontally on the bars and thus form a pattern. Fig. 195 is an enlarged view of one of the diamond forms. A flat rail (Fig. 196) through which the upright bars pass, and a similar one at the bottom into which the bars are riveted, complete the construction. At the top, the upright bars are flattened out and twisted about two-thirds of their length, as shown in the enlarged sketch (Fig. 197). The cresting of the railing is formed by single S scrolls of flat iron, connected at the top and bottom to the upright bars by a riveted clip, as shown in Fig. 194.

Fig. 194.—Railing at Segovia, Spain.

EXAMPLES FROM GERMANY.

In Germany, a great advance was made in decorative wrought-iron work during the middle ages, and many beautiful works were produced. Fig. 198 is a sketch of a railing at Augsburg, which is typical of this period. The framework is of round iron, to which the scrolls forming the pattern are fixed with ornamental wrought-iron clips (Fig. 199). These clips are also fixed at the junctions of the scrolls and assist in the formation of the general pattern. Fig. 200 is an enlarged view of the ornament at the junction of the scrolls, and is fixed in position by the wrought-iron clip. A flattened leaf is welded on each end of the scrolls, which are of flat iron. The German smiths showed at this period considerable skill in the interlacing of the bars in the design. Fig. 201,

which illustrates a railing from Augsburg, is a typical example of this work, as it shows two methods of simple interlacing. The construction of this railing is similar to those

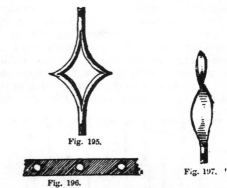

Fig. 195.

Fig. 196.

Fig. 197.

Figs. 195 to 197.—Enlarged Details of Railing at Segovia, Spain.

Fig. 200.

Fig. 198.

Fig. 199.

Fig. 198.—Railing at Augsburg.

Figs. 199 and 200.—Enlarged Details of Railing at Augsburg, Germany.

before described, the upright bars being riveted into flat bars at the top and bottom, while a square bar is fixed at intervals to add strength to the work. The upright bars are round and have an ornament at the bottom made in a pair of moulds and welded on the bar, as shown in the enlarged view, Fig. 202. Fig. 203 shows the upper end of the bars in the top section of the railing; a circular collar is welded to these bars, and a husk, cut out of sheet iron and slightly repoussé before the four leaves are bent over, is fitted at each end of the bar. The special feature of this railing is the interlacing ornaments, the simpler form of which is in the lower section of the railing; it is shown enlarged by Fig. 204.

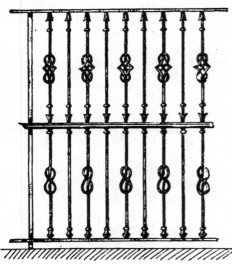

Fig. 201.—Another Railing at Augsburg.

The round bar is bent to the form of a figure 8, being interlaced in the process of turning. The more elaborate pattern in the upper section of the railing is shown enlarged by Fig. 205. The bar is opened in the centre to the form of a curved diamond, and an ornament of the same size iron in the shape of a figure 8 is interlaced with it. This is an operation requiring considerable skill.

EXAMPLE FROM HOLLAND.

Fig. 206 shows a railing in a church at Breda, in the Netherlands, and illustrates a form of interlacing ironwork known as wattle work. In place of the horizontal

flat bars through which the upright bars passed, lengths of round iron are passed in and out of the upright bars in imitation of the basket or wattle work, as shown at A and B (Fig. 206). The railing itself is a typical piece of Flemish work of the stiff Gothic character. The upright bars are of square iron, terminating at the top with an

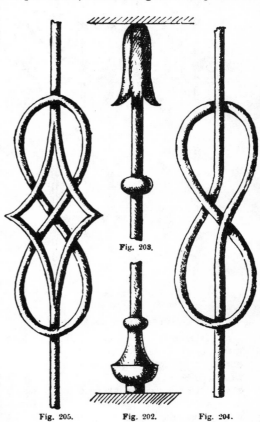

Fig. 203.

Fig. 205.   Fig. 202.   Fig. 204.

Figs. 202 to 205.—Enlarged Details of Railing at Augsburg.

open leaf formed into a fleur-de-lis by the side tongues, which are of flat iron and riveted to the bars. Fleur-de-lis of the same shape but of smaller size occur in the upper section of the railing. The intermediate bars in the cresting, springing from the Gothic arches in the upper section, terminate in a ring and two scrolls. The double Gothic arch panels in the lower portion of the railing, though stiff, are relieved by the hook form of termination at the sides. The iron used in this work is either flat or square, and has the pointed parts carefully welded.

decorative work of this period, and is a piece of work of great beauty. The framework is, as before described, of ordinary square and flat bars. Fig. 208 shows the construction, and also the fitting, of the continuous leaf ornament, hammered out of thin iron and screwed to the front edge of

Fig. 206.—Railing at Breda, Netherlands.

Fig. 207.—Railing at Sheffield.

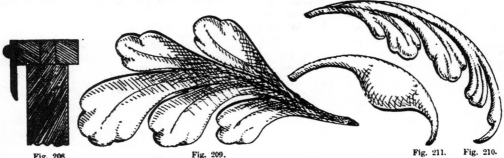

Fig. 208.   Fig. 209.

Fig. 208.—Fitting of Continuous Leaf Ornament on Railing at Sheffield.

Fig. 211.   Fig. 210.

Figs. 209 to 211.—Enlarged Details of Wrought-iron Leaves on Railing at Sheffield.

EXAMPLE FROM SHEFFIELD.

The railing illustrated by Fig. 207, sketched at the Weston Park Museum, Sheffield, is a good example of the most

the flat bars. The cresting shown at Fig. 207 terminates in a rose of repoussé ironwork fixed to the central stem by an oval rosette. The scrollwork is of flat iron bars, to which the ornamental leaves are welded. The scrollwork of the panels is of similar work, and the panels are fixed in pairs facing each other, each pair forming one complete pattern. Figs. 209, 210, and 211

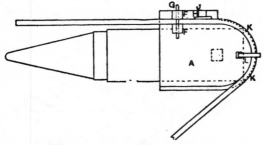

Fig. 212.—Plan of Bending Rig-up.

the radius of the bows, and the section of the edge is such that the $\frac{1}{2}$-in. iron rod will rest in it neatly, while the rod will stand a little way beyond the top face and the plain curved edge of A, so that a ham-

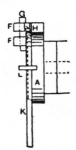

Fig. 214.—End View of Bending Rig-up.

are enlarged sketches of the ironwork leaves, which may be repoussé in the ordinary way or hammered into a mould of the pattern required. Considerable care and skill are needed in welding these leaves to the scrolls, and only the best iron can be used for the making of the leaves, as common iron will not stand the necessary strain. The welding of the various portions of the scrolls exhibits the fullest knowledge of the material and its working. Welding will be treated in detail later.

BENDING WROUGHT-IRON RAILINGS.

Wrought-iron railings can be bent more readily if heated in the fire over the portion to be curved than if bent cold. In either case, the rigs-up shown in

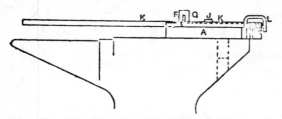

Fig. 213.—Elevation of Bending Rig-up.

Figs. 212 to 217 are suitable for the purpose. In Figs. 212 to 214 a block A of cast-iron is made with a tang cast to fit in the hole in a smith's anvil. The upper edge of the block is cast to

mer or wooden mallet can be used on the rod in order to effect minute corrections and settings of the curvature during the process of bending. The casting is shown by itself in Fig. 215. When bending a rod, it is necessary to confine one side during the bending, otherwise it would spring away from the block. A handy method of securing the rod is shown in Figs. 212 to 214. Two lugs F F confine the rod sideways, and a cottar G driven through them bears down on the rod and prevents it from rising. To secure it further, a wedge H is inserted between the side of the rod and the nib J.

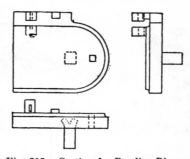

Fig. 215.—Casting for Bending Rig-up.

The lugs F F and the nib J are either cast on the block or made of wrought-iron and pinned or screwed in. If made of cast-iron, they must be stout—say, not less than $1\frac{1}{2}$ in. square in cross section. The cottar G may measure ¾ in. by ¼ in. The reason for using a cottar instead of a screw-bolt is that it is knocked in and back instantly. When the rod κ is secured as shown, it is pulled round the curved end of the block A by hand. But the fitting is facilitated and corrected with

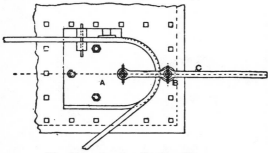

Fig. 216.—Another Bending Rig-up.

blows delivered upon the curved portion with a wooden mallet. When the rod is bent about half-way round, a clip may be inserted at L, in a hole cast in the block. This will prevent the rod from springing away from the block during the completion of the curving, and during the bending of the rod down the straight side of the block.

Another Method of Bending Railings.

Figs. 216 and 217 show an alternative and better method, but more troublesome and costly to rig up. The block is fixed differently, being bolted down upon a levelling and bending block instead of being fitted to the anvil. The block is steadier than the anvil. But the important difference lies in the manner of bending illustrated. The rod is bent round, not by hand, but by means of a roller B, pivoted

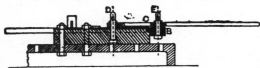

Fig. 217.—Section of Bending Rig-up.

in a radius bar c. If the rod is heated, the roller will turn it against the block so perfectly that little or no subsequent correction will be requisite. The rod c is pivoted

on a pin D, and the roller B on a pin E. This is a method which is used largely in the bending of bars and rods to regular curves. In other respects the block A is similar to that shown in the previous figures.

Making Fireguard.

Fig. 218 shows a fireguard complete as it would stand round the fireplace. It should be of a size to fit against the centre of the mantelpiece jambs, and should stand about 30 in. high, though the height may be varied according to the position. The top rail should be of flat iron ¾ in. wide by ¼ in. thick, and the bottom bar 1 in. by ¼ in. These are bent as shown in Fig. 218, leaving the ends 12 in. long. This size may be either less or more, according to the size

Fig. 218.—Fireguard.

of the room. The rails are drilled to receive the standard bars at intervals, leaving 3 in. space between the bars. The bars of round iron ⅜ in. in diameter must be reduced at each end, and then riveted into the rails (see section, Fig. 219). The back standard bar should be of flat iron 1 in. by ¼ in., with a round hole drilled through at 6 in. from the top to receive the screw on the plate, which is fixed to the mantelpiece, and to which the fireguard is secured by a thumb-nut (see Fig. 220). Another method of securing the guard to the mantelpiece is shown at Figs. 221 and 222. The top rail is turned down to form a hook, which falls into an iron eye on a plate fastened to the mantelpiece. The guard may be made more ornamental by using an angle-iron rail instead of flat iron for the bottom, and fixing on the front a brass ogee moulding (see Fig. 223) and on the top rail a half-round brass moulding (see Figs. 219 and

220). The guard may be painted dead black or any tint of enamel to suit taste.

METAL FENDER.

Of a fender shown by Fig. 224 the base is made of 1½-in. ordinary angle-iron, which may be welded together at the corners by an ordinary blacksmith, or may be mitred at the corners and fixed together by a corner plate riveted on the inside. The former makes the better job. Six pieces of or-

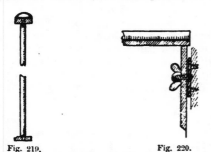

Fig. 219.

Fig. 220.

Fig. 219.—Fireguard Bar.

Fig. 220.— Screwing Fireguard to Mantelpiece.

dinary ¾-in. square iron are cut off to the height required—in this case 5 in.—a hole is tapped top and bottom, and a screw-pin fixed in (see Figs. 225 and 226). The length of fender in front will vary according to the size of the chimney-piece, but it should always be 12 in. wide, as this allows, in case of a tile hearth, sufficient depth for two 6-in. tiles. Holes are drilled in the top flange of the angle-iron base at the corners and ends, and 12 in. from ends at front. The top of the fender is made of iron flat

head screws and nuts. The fender may be completed by the addition of a brass vase or ball knob, as shown in Fig. 226. The ironwork should be well rubbed over and made smooth, then coated with dull black. This makes a very complete and pretty fender kerb.

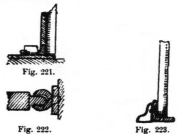

Fig. 221.

Fig. 222.

Fig. 223.

Figs. 221 and 222.—Another Method of Securing Fireguard to Mantelpiece.

Fig. 223.—Brass Moulding on Fireguard Base.

ORNAMENTING THE FENDER.

This fender may be improved very usefully by the addition of a brass or copper plate at the ends, which may be saw-pierced or stamped with a pattern and polished. This plate is supported, as shown in Fig.

Fig. 225.—Fixing Fender Corner Plate.

Fig. 226.—Fender Corner Knob.

225, by two plain iron scrolls screwed to the back side of the pillars and fastened to the

Fig. 224.—Fender with End Stands.

bar ⅞ in. or 1 in. wide by ¼ in. thick (see Fig. 226). The ornament as shown is made of ordinary bent-iron work fixed to the top bar, and the angle-iron base with round-

brass plate with a round-head screw-pin. The brass plates may have the one edge turned up and scalloped, as shown ; or, if preferred, may be quite flat, and in that case the vase or ball knobs at the corners and ends may be retained.

ANOTHER FENDER.

Fig. 227 shows a fender which is constructed, as far as the main portions are concerned, the same as that shown in Fig. 224. The ⅞-in. by ¼-in. flat-iron top bar is

side of corner pillar. This piece of iron has a hole drilled at the end of the proper size to take the pin of the trivet. The trivet may be of brass or copper, of pierced pattern, as shown, or engraved, and with a pin in the centre fixed with a screw to it. This trivet will then revolve in the hole of the bar prepared for it.

Fig. 227.—Fender with Revolving Trivets.

turned over at one end to form a scroll resting on the centre tube. This tube should be ¾-in. brass-cased iron tubing (which may be procured of any ironmonger), and fastened to the two centre front pillars with pin and nut. A brass plate is cut to shape as shown in Fig. 224, and turned over on the bottom edge, see section (Fig. 228), to fix with screw nuts to the iron base. A rose, as shown, may be beaten up in repoussé work, or, if preferred, an engraved pattern may be used. A little piece of ornamental bent ironwork fixed to the

 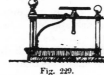

Fig. 228.   Fig. 229.

Fig. 228.—Fixing Brass Plate at End of Trivet.
Fig. 229.—Fitting Revolving Trivet.

base, and with a pin fitted to the brass centre tube, and the brass ball knobs fastening down the top rail, completes the fender in the ordinary way. A useful addition to this fender would be a revolving trivet at each corner.

FENDER TRIVET.

The method of fixing the trivet arrangement is shown in section in Fig. 229. A piece of iron—say, same size as top bar—is fixed on the corner pillar pin, and supported by a plain iron bracket fixed to the back

WROUGHT-IRON UMBRELLA STAND.

The wrought-iron umbrella stand illustrated by Fig. 230 may be made with very little trouble. Fig. 230 shows the umbrella

Fig. 230.—Wrought-iron Umbrella Stand.

stand complete, and Fig. 231 gives an enlarged view of the end, showing its construction. The back pillars are of iron ¾ in. square, hammered flat, slightly wider at the foot, and with the top ends worked into a ball shape. A brass or copper ball screwed on the end of the pillar will give a bright effect. The scrolls which carry the brass rail are of flat iron ¾ in. by 1/16 in., with the upper ends scrolled and the lower ends welded up flat and drilled to allow the rail to pass through. This rail, which may be of brass or copper tubing, is screwed at each end to fit a ball of the same metal. If preferred, a chain hung from these scrolls by rings may be used instead of the tube rail. The front scrolls should also be of flat iron $\frac{3}{4}$ in. by $\frac{3}{16}$ in., curved as shown, and screwed to the back pillars at the bottom and to the upper scrolls with round-headed bolts and nuts. The bottom scroll and the

Fig. 231.—End of Umbrella Stand.

waved ornamental part should be made of flat iron $\frac{1}{2}$ in. by $\frac{3}{16}$ in., and fastened to the back pillars by means of round screws and to the front scroll by bolts and nuts. The back portion of the stand (see Fig. 230) is formed of four rails of flat iron $\frac{3}{4}$ in. wide by $\frac{3}{16}$ in. thick, riveted to the back pillars, the ornamental scrolls and triangular work, of $\frac{1}{2}$-in. by $\frac{3}{16}$-in. flat iron, being screwed into the frames thus made. Fig. 232 shows the upper portion of the back, the scroll-work of which is of $\frac{3}{4}$-in. by $\frac{3}{16}$-in. flat iron.

The panel (see Figs. 230 and 232) is of sheet iron No. 16 Birmingham wire gauge, and cut at the top as shown by Fig. 233, the small portions being let into the top and bottom rails, as shown by Fig. 234.

Fig. 233.

Fig. 232.　　　　　　　　Fig. 234.

Fig. 232.—Top of Umbrella Stand.
Fig. 233.—Top of Lettered Plate.
Fig. 234.—Fitting Plate to Umbrella Stand Rail.

WATER-PAN, ETC., OF UMBRELLA STAND.

Figs. 235 and 236 show the water-pan, which should be made of stout zinc with the top cut out, to give a pleasing effect, and the bead soldered on to give strength, as shown. The water-pan should have two or three iron hooks riveted to the back, as shown in Fig. 236, to clip round the lower flat rail at the back. The stand will look well if painted with dead Berlin black, but any shade of enamel paint may be used, a dark shade being preferred. The framework in a dark colour, with the ornamental scrollwork in a lighter tint, gives a pretty effect. This design might be adapted to fit a corner if necessary.

METAL STAND FOR BRIC-A-BRAC.

The piece of furniture represented by Fig. 237, besides being in itself an effective

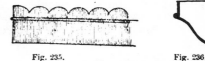

Fig. 235. 　　　　　　　　Fig. 236.

Figs. 235 and 236.—Front and Sectional Views of Umbrella Stand Pan.

ornament, serves as an excellent stand upon which may be displayed the various fancy articles that generally have a place in the modern drawing-room. It is of wrought-iron, with brass or copper trays and ornaments. Fig. 237 shows the complete stand with the three trays in position. The legs are of wrought-iron, $\frac{7}{8}$ in. or $\frac{3}{4}$ in. by $\frac{1}{4}$ in.

Fig. 237.—Metal Stand for Bric-à-brac.

section, and are bent as shown, the upper end being in a single scroll and the lower in a double one, with a third scroll welded on at the foot. Fig. 238 is a plan of the triple connecting stays, or tray supports;

Fig. 238.—Triple Stay for Bric-à-brac Stand.

these should be of iron $\frac{5}{8}$ in. by $\frac{3}{16}$ in., bent at each end into a scroll, as shown in Fig. 237. These triple stays are fixed to the legs in their proper positions by large round-headed screws, each of these carrying a plain, hollowed brass plate, and passing through the plain hole in the leg into the tapped hole in the bent end of the stay. The bottom scrolls are fastened to the bottom triple stay by nuts, thus completing the iron framework.

THE STAND TRAYS.

The tray (Figs. 239 and 240) is of brass or copper, and is turned up round the edge

Fig. 239.—Stay fixed to Leg.

and scalloped out with a file to either of the patterns shown at A and B (Fig. 241). These trays rest on the stays as shown at Fig. 239. It is best to have the brass or copper plates ground and roughly polished in the flat sheet. They then can be finished easily and dollied up when the ornament has been put on. An enlarged view of the ornament for the centre of the trays is given by Fig. 241. The ornamentation may be engraved, with the ground matted. The same pattern could be executed in repoussé work, but an uneven surface would thereby be given to the tray. The advantage of having engraved ornament is that the flat surface of the metal is retained, and there is no difficulty in standing ornaments upon it.

Fig. 240.—Section of Stand Tray.

ORNAMENTING THE STAND.

A better appearance is given to the stand by adopting brass chains, with ball drops hanging from the scroll tops of the legs round the stand. These additions are, of course, not absolutely necessary. The iron legs may be decorated in various ways. If copper is adopted for the trays, a good

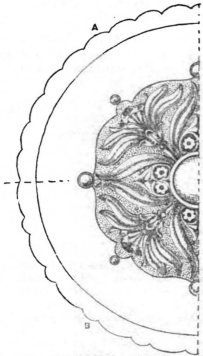

Fig. 241.—Half-plan of Tray.

dead black enamel will be very suitable. If brass is the material used, maroon, plum colour, cream, or French grey enamel paint may be employed.

TEA KETTLE AND STAND.

Fig. 242 gives a design for a tea-kettle and stand with spirit lamp, to be executed in wrought-iron and copper. The kettle is 7 in. in diameter, one spinning forming the top and sides, while the flat disc forms the bottom. The lid, of course, is spun, and has a hinge. (Spinning will be treated fuller later.) The spout may be cut out of flat metal, and either brazed and knocked up to the shape illustrated, or simply made straight and the joint lapped. The three legs of the kettle are wrought from ½-in. round iron rod, and are riveted

and soldered to the body. The handle is also of iron, and should be cut out of No. 16 S.W.G. sheet, and hollowed as shown. It should be split at the point A to allow room for the spout. This method of fixing the handle adds strength, besides giving the kettle a quaint appearance. The tray is of No. 22 S.W.G. copper, worked up with the hammer. Besides being sunk after the ordinary fashion of trays, it has a small well, 2⅝ in. in diameter and ¾ in. deep in the centre, to take the spirit lamp. This should be 2½ in. in diameter and 1 in. deep, and is provided with iron gauze at the top and packed with cotton-wool. The legs of the stand should be cut out of No. 16 S.W.G.

Fig. 242.—Tea Kettle and Stand.

iron beaten up with the hammer; for simplicity they may be made of ⅜-in. by ⅛-in. strap iron, but then would not present such an uncommon appearance as when wrought. Before polishing the copper work, the kettle should be carefully tinned inside and thoroughly cleansed.

FLOWER-STAND.

Fig. 243 illustrates a simple flower-stand

Fig. 243.—Flower-stand.

of iron or brass, which would in itself help to ornament the room. The pots may be of polished brass or copper, or of china or faience ware, and inside these the ordinary flower-pot may be placed. The stand

may be about 3 ft. high to the top of the upper table, which should be about 10 in. in diameter and of cast-brass. For this a wood pattern of the section shown at Fig.

Fig. 244.—Section of Flower-stand Top Table.

244 would be necessary, and should be turned to Fig. 245. A piece of coloured marble, $\frac{7}{8}$ in. or 1 in. thick, may be placed in the top as shown, and on this the ornamental pot would stand. The legs may be of $\frac{5}{8}$-in. or $\frac{3}{4}$-in. brass tubing of strong section. They should be cut to length, and solid pieces of brass, tapped in the centre for $\frac{1}{4}$-in. screws, should be strongly soldered at the top and bottom. The top table is fixed to the legs by three screws with flat heads, as shown at Figs. 244 and 245.

FLOWER-STAND FEET.

For brass feet, a wood pattern must be shaped, as shown in Figs. 243 and 246. Each foot is secured to the leg by a knob with a screw, which enters a tapped hole in the end of the brass tube. The legs are secured in position at the bottom by a

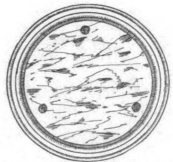

Fig. 245.—Plan of Flower-stand Top.

brass ring with three arms, which fit between the bottoms of the legs and the cast-brass feet; see Figs. 246 and 247. A disc of sheet-brass or iron will form the bottom of this lower table (see Figs. 247 and 248), and over this, if desired, a piece of marble

Fig. 246.—Foot of Flower-stand.

or polished wood may be placed, on which the ornamental pots stand. Steady-pins will be required in the feet to keep them secure in place.

BRACKETS, ETC., OF FLOWER-STAND.

The three brackets to carry the smaller pots on the stand legs are brass castings having sockets fitting over the tube legs.

Fig. 247.—Plan of Flower-stand Bottom Table.

These are secured at the proper height on the legs by set-screws with ball heads, as shown at Fig. 249. The pans for the flower-pots are brass castings (see Fig. 249), and

each is fastened to the bracket by a screwed knob at the bottom. After being fitted together the stand should be taken apart and polished, or bronzed any desired colour and lacquered. Ornaments other than those shown may be used for the brackets and legs, and thus a stand of an entirely different style may be arranged. If faience or china flower-pots are used they should be carefully selected as to colour, so as to harmonise with the other work.

STANDARD OIL-LAMP.

A good design for an oil-lamp floor standard is shown by Fig. 250. It is suitable for making either in wrought-iron or brass, and

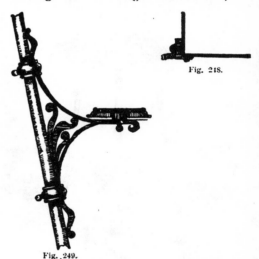

Fig. 248.

Fig. 249.

Fig. 248.—Section of Flower-stand Bottom Table.
Fig. 249.—Bracket on Flower-stand.

the ornamental shield-work, etc., is intended to be made of repoussé sheet metal. Fig. 251 shows the method of fixing the four feet to the main stem of the standard. A turned piece of metal (Fig. 251) is sunk on four sides to a depth corresponding to the thickness of the metal of which the feet are made. To this piece the feet B are screwed, as shown, and it is also tapped 1 in. at the top to take the stem of the standard. To conceal these joints, the turned collar C is slipped over, this being drilled and tapped to take screws D, which hold the shield-work. The standard is intended to be adjustable to various heights, so that the stem is in two parts, the upper sliding into the lower, and being held in

WROUGHT ANDIRONS OR FIREDOGS.

In connection with the designs of andirons or firedogs given by Figs. 252 to 263,

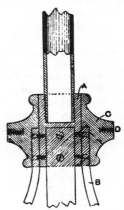

Fig. 251.—Collar for Standard Oil-lamp Feet.

it is interesting to note that the fires of early times were without doubt logs of wood heaped together and lighted upon the

Fig. 250.—Standard Oil-lamp.

position by the thumbscrew E (Fig. 250). The height from the floor to the thumbscrew should be about 3 ft. 3 in., sufficient length of sliding tube being used to allow of raising the lamp another 2 ft. A full-size working drawing should be made before setting to work, keeping all the rest of the fitting in proportion to the above dimensions.

Fig. 252.—Firedog from Museum de Cluny.

hearth. Over this fire probably the pot was affixed which contained the meal to be cooked. As the knowledge of the use of metals increased, it was found that by arranging these logs of wood forming the fire on bars above the ground, a bright fire was the sooner obtained. Hence the arrangement of the andiron, which was originally simply a bar of iron placed across the hearth, supported on blocks of stone or burnt clay, across which and resting on them the logs of wood were placed. Gradually an upright standard in front, and a leg formed out of the cross-bar at the back, removed the necessity for the stone block or clay lump, and the complete andiron became a regular addition to the fireplace. As men grew more dexterous in the work-

Fig. 253.—Firedog from Poictiers.

ing of iron, their skill was applied to the making of domestic articles that were both useful and ornamental.

Various Firedogs.

The illustrations (Figs. 252 to 263) have been selected from a number of sketches to show as far as possible the varying types of these articles. In the home room of ancient times, much of the cooking was doubtless done, and it was in consequence found to be advisable to have some arrangement on the andiron system by which the spits on which the meat was hung could be supported ; such an arrangement is shown in Fig. 252, which is sketched from a pair in the Museum de Cluny. The standard is of square iron, 1¼ in. at the base, and tapering to ⅞ in. ; to this are welded two round twisted rods, which support a circular framework. This framework consists of a

ring of twisted round iron, to which are welded four arms of square iron, twisted in the centre. This forms a kind of basket,

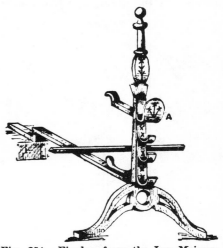

Fig. 254.—Firedog from the Inn Moissac.

which probably carried in winter times the vessel containing the oil or grease and wick which lighted the apartment. The foot is of circular form, of square iron, and having a rude half quatrefoil ornament of flat iron riveted to the inner side. The back leg,

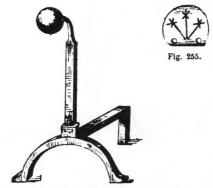

Fig. 255.

Fig. 255.—Firedog Ornament.
Fig. 256.—Firedog from Worcestershire.

which is about 24 in. long, is of iron, 1 in. square, turned down at the back 6 in. ; on these the logs of wood were placed. This is a very early example of this kind of work. Fig. 254, which is the andiron from the Inn

Moissac, is of similar construction, but shows a distinct advance in workmanship. The foot is ornamentally welded of square iron, the upper portion of the standard being filed circular, with ball finial. The hooks for the spits are of flat iron, and ornamented with chisel work, the upper one A having a widened front with three stars and lines converging to a centre. This ornament is shown enlarged at Fig. 255.

SIMPLE AND EFFECTIVE FIREDOGS.

The andirons of the large fireplaces of the Palais de Justice, Poictiers, shown in Fig. 253, are simple but very effective. The foot

as to give a wider standing surface; the standard is of flat iron, $1\frac{1}{4}$ in. by $\frac{1}{2}$ in., at the top of which a round iron ball 2 in. diameter, with a $\frac{1}{2}$-in. stem, has been welded so as to throw the ball forward. Fig. 257, from an old Somersetshire manor house, is a type similar to Fig. 252, but of later work. In this the hooks for the spits are curved over at the front, and are varied in size, the upper one terminating in a figure of a stag's head, with plain horns carefully cut out with a chisel and finished with a file. Fig. 258 gives an enlarged view of this figure. The basket at the top is similar to that described in connection with Fig. 252.

LANCASTRIAN AND BRISTOL FIREDOGS.

Fig. 259, an andiron from the hall of an old Lancastrian mansion, is without the spit hooks. Here the foot and standard are welded in one piece, the bottom portions of the feet being rounded and turned outwards. The basket at the top of

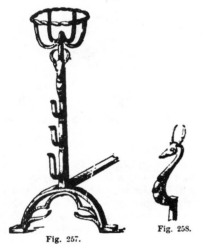

Fig. 257.

Fig. 258.

Fig. 257.—Firedog from Somersetshire.
Fig. 258.—Detail of Somersetshire Firedog.

is of square iron, tapered and turned up into a scroll at each side. The standard is composed of three square rods twisted together to form the pillar, and then widened out at the top to form an oval shape, crossing again above and flattened out as a finial. A rose of hammered sheet iron is affixed at the point of junction of the standard with the foot. The back leg is of square iron as before described. Fig. 256 shows an andiron from an old Worcestershire mansion; this is a type very usual in England. Here the ends of the square iron forming the foot have been upset at the ends and filed out so

Fig. 259.—Firedog from Lancashire.

the standard is more ornate, the upper ring being carried by scrolls of flat iron, which are themselves welded to the upper end of the standard pillar. Fig. 260, a curious and quaint andiron from an old Bristol house, shows workmanship of very high order. It is an entire figure of a cock, but without wings or tail, except at the back where the bar to carry the top is welded on; here a scroll

Fig. 260.—Firedog from Bristol.

finish is given. The head, comb, breast, and legs, with feet, have been chiselled out

Fig. 261.—Firedog from 1889 Paris Exhibition.

of the iron. The whole figure shows great vigour and force in its workmanship.

ORNATE FIREDOGS.

Fig. 261, sketched from the ancient iron-work exhibited at the Paris International Exhibition of 1889, shows an andiron of similar kind, but of much later date and more ornate workmanship. The figure here is a griffin carrying on its head a ball which

Fig. 262.—Firedog from Yorkshire.

is the finial. The wings at the sides of the breast, and the shield in front, are of hammered iron, and riveted to the body. The tail of the animal rests on the back leg, and a claw on each side grasping a small ball is riveted under the wings. The front foot, on which the griffin rests, is welded to

Fig. 263.—Old Foreign Firedog.

the back leg. A very fine specimen of workmanship from a Yorkshire house is shown in Fig. 262. The front foot is of square iron, but welded into shape and finished with the chisel and the file. The standard pillar is of round bars, twisted together and having bosses welded on top and bottom, which are carefully filed up. On the top, as shown at A, are four leaves, worked up to a sharp line on the anvil, and turned over, bringing their ends to the standard pillar. The finial is formed of four similar leaves curved in the opposite

BRASS FIRE-SCREEN.

The screen illustrated by Fig. 264 has been designed for making up in brass. The outer tubes of the frame are ¾ in. external diameter, and are connected horizontally by three cast channel bars (Fig. 265) as shown in Fig. 266. Between the bars and tubes are placed plugs, which are either recessed, as at Fig. 267, or with spigots, as at Fig. 268, for the purpose of keeping the tubes in line and to give a better external finish. A ⅜-in. iron rod, with screwed ends

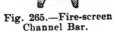

Fig. 265.—Fire-screen Channel Bar.

Fig. 264.—Brass Fire-screen.

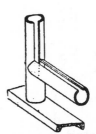

Fig. 266.—Jointing Fire-screen Framework.

direction. The iron shown in Fig. 263 is of foreign workmanship, and differs considerably from those before described. The supporting legs are of round iron, to which the upper portions and back legs are riveted. The ornamental scroll work of flat iron, ⅜ in. by ¼ in., is fastened to the supporting legs with bands of round iron. In the upper portion the spear head forming the central point is open; in the lower portion it is of flat iron hammered out.

and nuts, passes through the tubes and feet, keeping the frame rigid. The tubular frame for the stained glass centre is shown in detail by Figs. 269 to 271. The cross tubes are filed at each end, as indicated in Fig. 270, and when fitted to the vertical tubes the joint should fit as shown by Fig 269. The connecting nuts should fit the tubes snugly, and be soldered or brazed in position. This frame forms a neat and effective setting for the glass.

SLOTTING BRASS TUBES.

The chief difficulty perhaps to the home worker is in cutting the slots, which may be accomplished successfully by either of the following methods: First, by inserting a round iron bar that fits fairly in the bore,

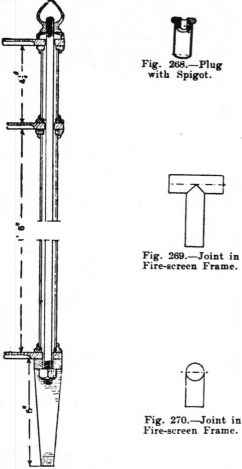

Fig. 267.—Vertical Section of Fire-screen Frame.

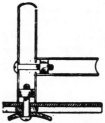

Fig. 268.—Plug with Spigot.

Fig. 269.—Joint in Fire-screen Frame.

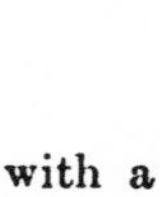

Fig. 270.—Joint in Fire-screen Frame.

placing the tube in a block with a half-round groove of correct size; then with a sharp chisel two parallel cuts may be given about $\frac{1}{32}$ in. apart, the ragged edge being removed by filing. Another way is to melt equal quantities of lead and zinc, and pour this in the tubes; when it has cooled, place

the tubes in the block as before, and split with a chisel, taking only one cut; then melt the lead out and place the tube on end and widen the slot by hammering down on a

Fig. 271.—Connecting Fire-screen Tubes.

piece of sheet iron about $\frac{1}{8}$ in. thick, and finishing to the size required by filing. The scroll ornament is $\frac{1}{32}$ in. by $\frac{7}{8}$ in. wide, and is fixed by small round-head screws. The scrolls that are fastened in the channels of the cross-bars have washers to bring them level with the outer webs (see Fig. 271), and after the screen has been fitted up it should be disconnected and prepared for lacquering. Fig. 272 shows the fire-screen foot.

WROUGHT-IRON FIRE-SCREEN.

The screen shown in Fig. 273 is constructed of wrought-iron, and the principal members of the frame require the use of a forge, although the minor scroll ornament can be bent cold from strips $\frac{1}{8}$ in. thick by $\frac{1}{2}$ in. wide. The section of material shown in Fig. 274 is $\frac{3}{4}$ in. by $\frac{3}{4}$ in. The end of the iron to which the feet are welded is doubled

Fig. 272.—Side View of Fire-screen Foot.

back for about 6 in., and hammered out at welding heat; it is then splayed open and prepared for welding on the feet, which are $\frac{1}{4}$ in. thick, but which vary both in width and thickness towards their ends (see Fig. 275). The two cross-bars and two vertical bars are shouldered down and riveted to the

not bent till the rivet under it is closed home. A section of the leaf is shown in Fig. 278, and Fig. 279 is a detail of E (Fig.

Fig. 273.—Wrought-iron Fire-screen.

standards (see Fig. 276). The scrolls forming the top ornament are welded on, and

273). The centre glass may be cathedral, Muranese, or opaline, and the method of fixing it is shown at Fig. 280, and consists of flat strips in front $\frac{3}{32}$ in. thick by $\frac{7}{8}$ in.

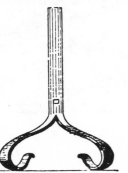

Fig. 274.—Side View of Fire-screen Foot.

Fig. 275.—Plan of Fire-screen Foot.

the handle is riveted in before the adjacent terminals are bent down. The leaf ornament D (Fig. 273) is shown in Fig. 277, and is

wide, fastened with screws; the angles at the back could be of iron or brass $\frac{1}{16}$ in. thick.

The elaborate fire-screen illustrated in Fig. 281 is intended to show the various

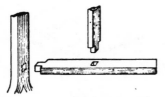

Fig. 276.—Joint in Fire-screen Frame.

ways in which wrought-iron may be manipulated and worked. A screen for use during the summer months when fires are not required should stand within the fender on the hearth, and must therefore be without

Fig. 278.—Section of Wrought-iron Leaf.

Fig. 279.—Another Wrought-iron Leaf.

Fig. 230.—Fixing Fire-screen Glass Centre.

any fixing. Such screens are usually from 18 in. to 24 in. in width, and from 27 in. to

32 in. in height. The framework should be of flat iron 1 in. wide by $\frac{1}{4}$ in. thick, the rails having circular pieces welded on at

Fig. 277.—Wrought-iron Leaf.

each end and drilled through the centre (see Fig. 282) for the screw of the standards. The base of the end standard should be made, as shown in Fig. 281, of flat iron $1\frac{1}{2}$

Fig. 281.—Ornate Wrought-iron Fire-screen.

in. by $\frac{1}{4}$ in., with the side scrolls, 1 in. by $\frac{1}{4}$ in., riveted on the foot and fastened to the centre rod by round-headed screws. The end standards may be of $\frac{5}{8}$-in. or $\frac{3}{4}$-in. round iron, with the bosses welded on. These standards are made in parts and screwed together, and to the framework with pins.

Fig. 282.—End of Fire-screen Framework Rail.

Fig. 283 shows the construction of the upper end, and Fig. 284 illustrates similarly the bottom end with the screw pin fixing the foot to the standard. The centre of the standard is formed of eight pieces of round iron welded together at each end, and then heated and swelled out as shown in Fig. 285, which is a section through the centre, and in Fig. 286, which is the plan of the welding at the top and bottom. The ornamental welded bosses on the standards will need considerable work in execution, and may be dispensed with, the plain iron bar

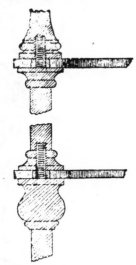

Fig. 283.—Details of Fire-screen Standard.

being substituted for them except at the centre swell ornament, which must be made and welded in. Or the bosses may be cast in brass or iron and fitted on the standard; this would give a similar effect at less cost.

The knobs at the top of the standard may be of brass or copper screwed on, and if

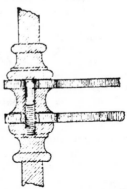

Fig. 284.—Bottom of Fire-screen Standard.

polished and lacquered, will add to the decorative appearance of the screen.

Fig. 285.   Fig. 286.

Figs. 285 and 286.—Section and Plan of Fire-screen Standard Ornament.

SCROLLWORK AND OTHER ORNAMENTS.

The ornamental scrollwork should be made of flat iron $\frac{5}{8}$ in. by $\frac{3}{8}$ in., or $\frac{3}{4}$ in. by

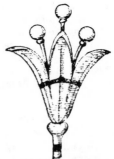

Fig. 287.—Lily in Wrought Iron.

$\frac{7}{16}$ in., and the smaller scroll ornaments of $\frac{1}{2}$ in. by $\frac{1}{4}$ in., or $\frac{1}{2}$ in. by $\frac{3}{16}$ in. flat iron. If the ends of the scrolls are flattened out wider before they are turned a good effect is produced. The scrolls should be fastened together with rivets or ball-headed bolts and nuts. The centre leaf is made of No. 17 B.W.G. iron, cut to shape and hammered up from the back, and Fig. 287 shows enlarged the lily, which has been adopted in this design as the chief ornament. It should be cut from best charcoal iron, No. 16 or No. 17 B.W.G., and when hot, hammered up from the back to shape, as shown by the hatched line across the flower. The three petals with ball ends should be separately forged on to the stem, as shown in Fig. 288. The long leaves of the lily may be cut out of charcoal iron, and hammered up

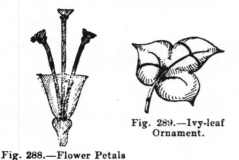

Fig. 289.—Ivy-leaf
Ornament.

Fig. 288.—Flower Petals
Welded to Stem.

as described for the flower, or forged on the anvil from flat iron bars. The ivy leaves (see Fig. 289) in the upper corner ornaments are cut out of charcoal sheet iron, hammered up, and welded on the stems, and the shield towards the bottom of the screen may be made either in brass or copper, and may have a monogram or a date engraved or repoussé on it. The floral portions of the design may be omitted, and a simple scroll introduced if it is desired to make the screen less costly. When finished it may be painted dead black with eggshell enamel.

Wrought-iron Fender Kerbs.

Wrought-iron lends itself easily to the manufacture of fenders, as has already been indicated, and enables a very artistic effect to be produced in an ordinary article at a comparatively small cost. Figs. 290 and 291 give a design of a fender kerb made entirely of wrought-iron, with the exception

of the base. The central portion of the fender has been arranged to give the form

Fig. 291.—End of
Fender Kerb.

Fig. 290.—Front of Wrought-iron Fender Kerb.

of the coat-of-arms of the Prince of Wales, namely, the three feathers. The size of the fender kerb must be regulated by the size of the tile hearth which it is to surround, the inner edge of the kerb standing $\frac{1}{4}$ in. on the tiles. If, as in some cases, the

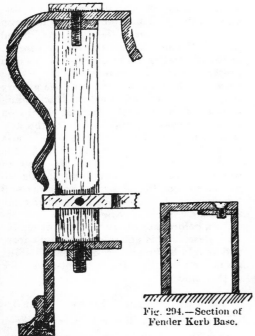

Fig. 292.—Corner Pillar of Fender Kerb.

Fig. 293.—Ornamental Head of Fender Kerb.

tiles are fixed *on* the hearthstone, the inner lining of the base must be cut up the thickness of the tiles.

BASE OF FENDER KERB.

The base is made of 2-in. angle-iron welded together at the corners; or it may be brazed at the corners, though welding makes a better job. To add to the ornamental appearance of the base, a $\frac{3}{4}$-in. ogee moulding may be screwed to the bottom edge of the base, as shown in Figs. 290 and 292, the ornamental head being illustrated in Fig. 293. The inner strip or lining of the base may be of light angle-iron or stout sheet-iron turned over at the top edge and screwed to the angle-iron, as shown at Fig. 294. Fig. 295 gives a part plan of the base, showing the angle moulding with the

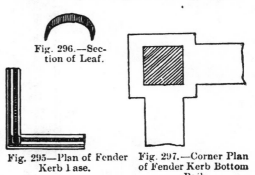

Fig. 296.—Section of Leaf.

Fig. 294.—Section of Fender Kerb Base.

Fig. 295—Plan of Fender Kerb Base.

Fig. 297.—Corner Plan of Fender Kerb Bottom Rail.

ogee moulding on the front edge and the position of the gallery. The corner pillars may be of $\frac{3}{4}$-in. or $\frac{7}{8}$-in. square iron, tapped at the bottom end to receive a $\frac{5}{16}$-in. screw, and at the top to receive a $\frac{5}{16}$-in. screw, as shown in Fig. 292, which also shows the fixing of the top rail of the fender gallery. This may be of flat iron, $\frac{3}{4}$ in. by $\frac{1}{4}$ in., bent to the shape shown in Fig. 290.

ORNAMENTAL HEAD TO FENDER KERB.

The rails of the pillars cross, and are fixed in place by the ornamental head, as shown in Fig. 293. This head is made as follows :—Four leaves are cut out to shape from flat iron, hollowed as shown at Fig. 296; then they are welded to a piece of round iron which has been upset, and the ball at the top is welded on, or, if easier, may be made separately and screwed on. The screw is tapped to fit the square pillar, and fixes the rails at the corner. The rail is secured to a small square pillar at the end of the base. The second rail of the gallery (Fig. 297) is made of flat iron, $\frac{3}{4}$ in. by $\frac{1}{4}$ in., with welded corner to fit over the

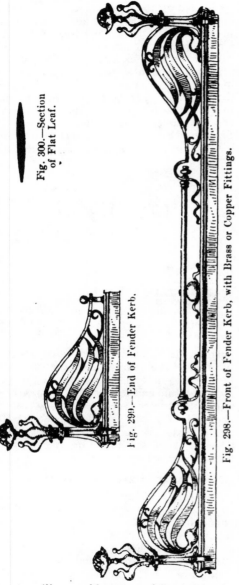

Fig. 300.—Section of Flat Leaf.

Fig. 299.—End of Fender Kerb.

Fig. 298.—Front of Fender Kerb, with Brass or Copper Fittings.

corner pillars and be secured in position by a pin, as shown at Fig. 292. Two short pillars under the centre of the gallery hold this rod in place (see Fig. 290).

ORNAMENTAL FILLING FOR FENDER KERB.

The framework of the fender constructed as above may now be filled in with suitable ornament. This may be made of flat iron $\frac{1}{2}$ in. by $\frac{3}{16}$ in., bent to the design and fitted into the framework with rivets or screws. If round-head rivets or screws are used, a pretty effect is produced. The leaves shown in the central portion of the gallery may be made of bar-iron $\frac{1}{4}$ in. thick, hammered out, or of thick charcoal sheet-iron cut out to shape and welded on to the flat-iron stems. If these leaves are bent and hollowed a little when hot, a better effect is produced. Great care must be used in welding these leaves and light scrolls together so that they may not be burnt and the weld spoilt. To make the scrolls more artistic, the ends may be widened by hammering before they are turned. This fender kerb described is entirely of wrought iron, and should be finished a dull black, which gives the best effect for wrought-iron work of this kind.

IRON AND COPPER FENDER KERB.

Figs. 298 to 305 illustrate a fender kerb with less work in it, and with a little of another metal, say brass or copper, to add brightness. The base of this fender kerb is constructed as described before, but the ogee moulding is fixed to the top edge of the angle-iron, making it a portion of the ornamental gallery. Fig. 301 gives a part plan of the base, which must be of the size of the hearth, usually 4 ft. by 1 ft., and to stand $\frac{1}{4}$ in. on the tiles all round. The corner pillars are of round iron, upset at the bottom end, as shown in Fig. 303, and screwed through the base and fixed with a nut. The upper end is screwed to receive the brass or copper flower. Four pieces of flat iron hollowed out at the tops, as shown in Fig. 305, are fixed to the central stem with round-headed screws, as shown at Fig. 302. The bottom portions of these leaves are bent to fit over the angle base, as shown at Fig. 304. The top rail, of flat iron $\frac{3}{4}$ in. by $\frac{1}{4}$ in., is secured to the centre pillar in the way described before. The central portion of this fender is formed by using a brass or copper tubing with cast turned ends, and fixed to the rail with brass or copper knobs.

ORNAMENTING IRON AND COPPER FENDER KERB.

The brass or copper flower ornament on the corner pillars (Fig. 302) is of cast-copper, for which a pattern must be made fitting into a sheet-metal hollow cup. An ornamental knob may be used if desired. The ornament filling the ends is taken from the leek with its flat leaves, and is made in this way: Flat iron should be hammered out to a sharp edge on both sides, as shown at Fig. 300, and then welded to a ball end and fixed on the little scroll with round-headed screws from the back. The fender kerb

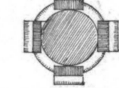

Fig. 305.
Section of Iron.

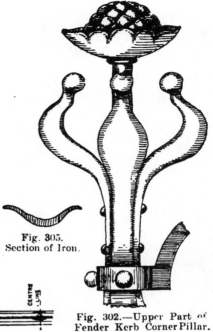

Fig. 302.—Upper Part of Fender Kerb Corner Pillar.

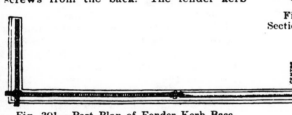

Fig. 301.—Part Plan of Fender Kerb Base.

should be coloured with dull Berlin black, which will give a good contrast with the polished brass or copper central rod and corner knobs. In either of the designs shown here the spaces between the rails or between the rail and the base may be filled in with scrollwork of other designs.

WROUGHT-IRON FIREIRONS.

The wrought-iron fender kerbs described and illustrated above require fireirons to match. These have been designed so as to be easily made by a workman with skill in working wrought-iron. The fireirons should be 25 in. or 27 in. in length for ordinary use, but for grates and fenders of large size the length may be increased to 30 in. or more.

FIRE TONGS.

The tongs (Fig. 306) is the principal article, being the most ornamental, and involving the most fitting, as one part has to work with the other, so that by its action the pieces of coal may be held as required. The pattern has been arranged to be made

Fig. 303.—Section of Fender Kerb Pillar.

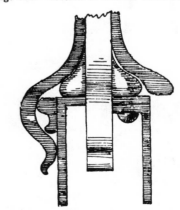

Fig. 304.—Bottom Part of Fender Kerb Pillar.

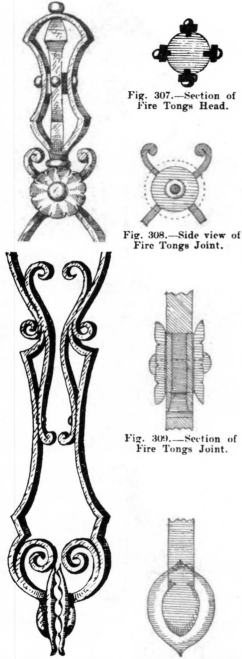

Fig. 307.—Section of Fire Tongs Head.

Fig. 308.—Side view of Fire Tongs Joint.

Fig. 309.—Section of Fire Tongs Joint.

Fig. 306.—Wrought-iron Fire Tongs.

Fig. 310.—Side View of Fire Tongs Claw.

of flat iron, $\frac{1}{2}$ in. by $\frac{1}{4}$ in. for the upper portions, and $\frac{1}{2}$ in. by $\frac{3}{16}$ in. for the lower portions. The head shown in Fig. 306 is used for each of the three articles, tongs, poker, and shovel, and is made of four scrolls of flat iron $\frac{1}{2}$ in. by $\frac{1}{4}$ in., bent as shown, and welded together at the bottom with a rounded collar and screw pin to fix into the central joint. Fig. 307 is a section of the central fitting of the handle, a solid piece of round iron, $\frac{3}{4}$ in. in diameter and $\frac{5}{8}$ in. thick, being tapped, and the scrolls fixed to it with round-headed screws. The scrolls are also fastened at the top with a ball-head screw. The central joint (see Figs. 308 and 309) should be made $1\frac{1}{4}$ in. in diameter, and $\frac{3}{16}$ in. thick, circular in form, and with a neck welded on, into which the head screws. Fig. 308 shows the two arms, which have a circular plate of the same diameter as the central joint, welded to them. They are drilled in the centre to take, as a pivot, a short bolt with a round head and nut. On the outside a rosette made of sheet iron about $\frac{1}{8}$ in. thick, and hollowed up as shown, is fixed under the bolt to cover the circular joint. The arms of the tongs are shown in Fig. 306. The scrolls are separately made, and welded together in the centre and at the ends. Each arm carries a kind of claw to hold the coal; this is arranged by hollowing pieces of sheet-iron $\frac{1}{8}$ in. thick, cutting the edges, and riveting to the swells at the ends of the arms, as shown at Fig. 310.

FIRE SHOVEL.

Fig. 311 shows the ornament on the arm of the shovel; this is made of flat iron as described for the tongs, the portion between the scroll ornaments, as shown in Figs. 311 and 312, being of plain iron. The shovel pan should be of sheet iron, No. 16 B.W.G., hammered out to shape. It should be about 8 in. long, and about $5\frac{1}{2}$ in. to 6 in. wide at the bottom. The ornament may be cut out with the chisel, or be saw-pierced. Or, if preferred, it may be repoussé, that is, hammered up from the back. To fix this pan a flat plate shaped like the shovel pan is welded on the arm, and to it the pan is riveted or bolted.

POKER.

The poker end (Fig. 313) should be made
'om ⅜-in. square iron, tapered down to a
quare point. This poker bit is welded on
he arm to which the scrolls have been

Fig. 311.—Upper Part of Shovel.

Fig. 313.—Lower End of Poker.

affixed, as shown. The work should be
made quite free from oil and dirt, and then
blackened with at least two coats of dead
black, and stoved, if possible ; if this is not
possible, a good self-drying eggshell black
should be used.

TABLE FRUIT AND FLOWER STANDS.

An épergne which may be cheaply made
in two or three ways is illustrated by Fig.
314. It may be constructed in all brass, polished and lacquered, or brass with copper

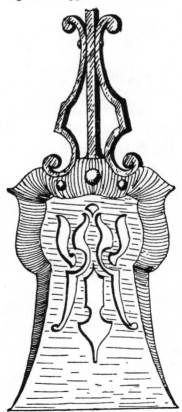

Fig. 312.—Lower End of Shovel.

leaves and ornaments, or brass or copper stands and fittings and wrought-iron scrolls with copper leaves. The construction is as follows : The stand is three-way in shape, standing on three feet, as shown at Fig. 315. For this part a wood pattern should be

Fig. 314.—Elevation of Fruit and Flower Stand.

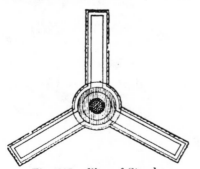

Fig. 315.—Plan of Stand.

made and then cast in brass, the brass feet being fitted on the base with screws. If preferred, the ordinary claw foot may be adopted. The centre stem should be of brass or copper tubing, ½ in. or ⁷⁄₁₆ in. in diameter. This tube pillar is screwed into a

brass base, which itself is screwed firmly to the three-way base. At the top of the tube a cast socket should be turned, as shown in Figs. 316 and 317, and screwed on to the tube pillar. Into this the glass upper vase will fix, having a brass socket to fit into the

Fig. 316.   Fig. 317.   Fig. 318.

Figs. 316 and 317.—Sections of Top Socket.
Fig. 318.—Section of Centre Socket.

upper socket. In the upright of the central pillar at the required height is fitted another socket, shown at Fig. 314, and enlarged at Fig. 318, to which the scrolls are screwed and secured by the double roses and round-headed screws. These rosettes are of No. 17 or No. 18 B.W.G. brass or copper, cut out to the shape and hammered hollow, as shown.

ORNAMENTAL SCROLLS, LEAVES, ETC.

The ornamental scrolls may be of wrought brass, copper, or iron, according to taste, and may be of strip metal, ⅜ in. or ¼ in. wide, and ⅛ in. or ¹⁄₁₆ in. thick. They should be fixed to the pillar and stand as shown

Fig. 319.   Fig. 320.

Fig. 319.—Section of Side Socket.
Fig. 320.—Section of Lower Socket.

with round-headed screws. The ornamental leaves shown should be cut out of brass or copper, repoussé to shape, and fixed to the scrolls with screws.

FIXING GLASS DISHES AND VASES.

The glass dishes are solid at the under side, and drop into a turned brass socket shown enlarged at Fig. 319. This socket is screwed into the upper end of the scroll arm, which has been thickened for the purpose. Another form of fixing would be to file down the end of the scroll to a pin, on which the solid part of the socket would fix. The lower glass vases are fixed to the lower scrolls by brass sockets (see Fig. 319), into which they drop in the same way as the upper vase. These glass dishes may be arranged so as to be interchangeable. When the stand is fitted together, it may be taken apart, and the brass and copper portions polished and lacquered. If the scrolls are in iron, they should be painted a dead black. If preferred, the iron portions may be enamelled any dark colour to taste, but the combination of dead black and polished metals is always better than colour.

A SMALLER STAND.

Fig. 321 illustrates a stand of a smaller size, such as may be used at either end of

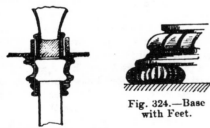

Fig. 323.—Section of Top Socket.

Fig. 324.—Base with Feet.

the table, whilst the larger one occupies the centre. The arrangement here adopted is a central vase for flowers and a glass dish for fruit, etc. (see Fig. 322). The upper vase is of corrugated glass, and fits into a turned brass socket (see Fig. 323) which is screwed to the top of the central tube pillar. The dish fits over the socket, into which the central vase drops. The bottom stand is circular, with an ornamental edge. It may be quite plainly moulded if preferred. This bottom base stands on three fluted, flattened balls (see Fig. 324), which are fixed to the base with screws. For this base a carved wood pattern must be made, and then cast in brass, copper, or iron as preferred. The scroll arms (see Fig. 325) are made as described before, and of material

of the same size, and are affixed to the base and the pillar with round-headed screws.

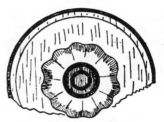

Fig. 322.—Plan of Stand.

Fig. 321.—Smaller Fruit and Flower Stand.

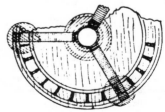

Fig. 325.—Plan of Base.

An alternative pattern of scrollwork and leafwork is shown in the smaller stand, but the patterns are equally suitable.

JARDINIÈRE PLANT-STAND IN WROUGHT IRON.

Jardinières may now be purchased of the most artistic shapes and varying shades of art colours at reasonable prices. Fig. 326 shows such a jardinière fitted into a wrought-iron stand suitable for a plant

Fig. 326.—Elevation of Jardinière and Stand.

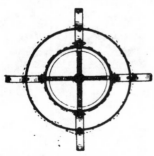

Fig. 327.—Plan of Stand.

which grows to a considerable height, and for which a low stand is therefore required. It should be, say, 18 in. high to top of china pot. The stand on which the jardinière rests is made from angle iron 1 in. by 1 in., the upright edge of which has been cut out as shown. The four supports are made from flat iron 1 in. by $\frac{1}{4}$ in. or $\frac{3}{16}$ in., and

fastened to the central angle rim with rose-head screws, as indicated in Figs. 326 and 327. These supports are kept in place round the china jardinière by a circular

Fig. 328.—Waved Ring.

Fig. 329.—Section of Jardinière Support.

ring, waved, and flattened at each of the supports (see Fig. 328), to which it is secured by a screw bolt having a wrought repoussé leaf as an ornament on the outside (see Fig. 329). This leaf may be of sheet-iron, brass, zinc, or copper. These polished will give a brighter effect without adding much to the cost. The under portions of the supports are kept in place by a scroll secured by a round-head rivet and fastened in the centre with a screw having a ball head, which may be of iron, brass, or sheet-iron, brass, zinc, or copper. These copper (see Fig. 330). The dotted lines on plan (Fig. 327) show the china jardinière in position. If a very heavy plant is used the sizes of the iron to be used in the construction may be increased without altering the design and general effect.

MORE ORNATE JARDINIÈRE STAND.

Fig. 331 is the elevation of a wrought-iron stand to hold a jardinière, constructed on the same lines as the previous one, but taller and more ornate. It should be, say, 30 in. to 33 in. high to the under-side of china jardinière, which should rest in a

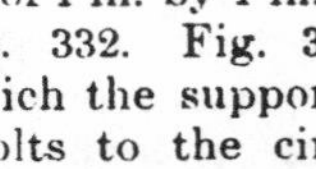

Fig. 330.—Method of Securing Jardinière Supports.

circular frame of 1 in. by 1 in. angle iron, as shown in Fig. 332. Fig. 333 shows the method by which the supports are affixed with screw bolts to the circular frames. The three supports are of flat iron 1 in. by $\frac{3}{16}$ in., and are shown in elevation by Fig.

331. An iron ring fastened to each support by screws having ornamental rosettes on the outer side makes the three supports

Fig. 331.—Elevation of Tall Jardinière and Stand.

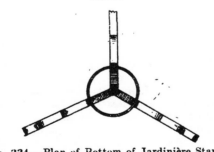

Fig. 334.—Plan of Bottom of Jardinière Stand.

rigid in the centre. Fig. 334 indicates the method of securing the three supports at the foot of the stand so as to give rigidity. The scrolls meet on each side of a three-cornered piece of iron, to which they are affixed by a screw pin tapped into the solid

three-cornered piece, as illustrated in Fig. 334, and at B (Fig. 331). Alternative designs for the groups of leaves in the foot are given at Fig. 335. They should be welded together, flattened out, and riveted into the junction of the scrolls. The scrolls at the upper portion of the support are ornamented with flattened leaves welded together and to the scrolls, as illustrated.

Fig. 332.—Plan of Top of Jardinière Stand.

Fig. 335.—Designs for Leaf Groups.

Fig. 333.—Section of Angle Rim.

Fig. 336.—Elevation and Section of Leaf.

Fig. 336 shows a single leaf and section of same. To increase the decorative effect, a brass ornamental ball A (Fig. 331) may be suspended with chains from each upper scroll. The stands may be finished by giving two or three coats of good paint or enamel of any colour preferred.

ORNAMENTAL IRON GARDEN VASE.

A vase made as illustrated by Fig. 337 would be a pretty and useful addition to a garden, and could be manufactured without serious difficulty. Fig. 337 shows the vase on its stand complete. Fig. 338 is a section of the vase, and shows how it is fixed. Fig. 339 shows the ornamental arm of the vase, Fig. 340 the slotted bar for the top part, Fig. 341 the upper part of stand, Fig. 342 a section of the arm, Fig. 343 a cotter pin, and Fig. 344 shows the profile of the stand. The vase is constructed in the following manner:—An iron stand as shown in Fig. 337 is cast from a wood pattern, which may be turned out from a solid piece of wood having a core print top and bottom, as shown in Fig. 344. This can then be cast hollow. Then turn a collar pattern to fit into the top of this stand (see Fig. 341) grooved out as shown for the lower portion of the vase arms to pass down (see Fig. 338). This upper sec-

wood pattern, of section as shown in Fig. 342, having the back end widened as shown. The flat portion is cut out with a fretsaw to a pattern, as shown in Fig. 339. After it has been sawn, file the edges from the centre outwards, so that the wood pattern will come out of the moulding sand. A flat piece is cast on to the inner side of the arms (see B, Fig. 339). This forms the pattern from which the required number of arms may be cast in iron.

Fig. 337.—Elevation of Garden Vase and Stand.

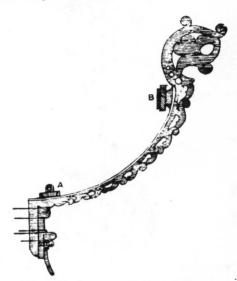

Fig. 339.—Ornamental Arm of Vase.

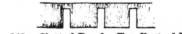

Fig. 340.—Slotted Bar for Top Part of Vase.

SECURING ARMS OF GARDEN VASE.

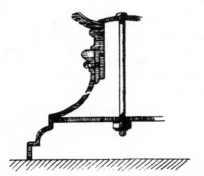

Fig. 338.—Section through Base of Vase Stand.

tion is fastened to the bottom stand with screws (see Figs. 338 and 339). The upper or vase portion is constructed of a number of ornamental arms which slide into the upper section of the bottom stand as before described. These arms are made from a

The arms being placed in position in the bottom stand as described, the bottom arms are secured together by an iron bar with slots in it, to fit over the flat piece at the bottom of the arms (see A, Fig. 339), which is fastened with a cotter pin (Fig. 343). A $\frac{3}{4}$-in. by $\frac{1}{4}$-in. wrought-iron bar, with slots cut out, as shown in Fig. 340, fits over the flat pieces near the top of the arms, and secures the upper portion of the vase. Fig. 338 shows the method of securing the arms and bottom portion of the stand together.

This is done by cutting a circular piece of iron having a long iron bolt through the centre, fastened at the bottom end of the stand with a screw nut passing through a flat iron bar, 1¼ in. by ¾ in. This securely fixes the various portions of the vase together. The circular plate may be perforated to allow for drainage.

Fig. 341.—Upper Part of Stand.

Fig. 342.—Section of Arm.

Finishing Garden Vase.

The sides of the vase may be lined with green moss, and then it may be filled with earth in which the flower roots may be planted. Or a tin or zinc pan may be made to fit the inside, and in this the flower roots may be planted. If the ironwork is painted white or light cream colour, the effect produced by the light colour against the green

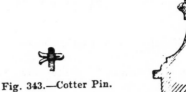

Fig. 343.—Cotter Pin.

Fig. 344.—Profile of Stand.

moss is very pretty and artistic. One form of vase only is shown in the illustration, but with the same arrangement of stand, varying patterns of arms may be used, giving vases of different shapes and form.

SURFACING METALS.

How Castings and Forgings are Treated.

A CASTING as it comes from the foundry, or a forging as the smith supplies it, obviously has yet to receive severe treatment before it is a finished piece of work. A casting has to have its hard skin removed by chipping or by pickling in acid, its new surface filed, and the file marks removed by scraping, the final process being burnishing or polishing. The treatment in the case of a forging is the same, except that there is no necessity, as a rule, for chipping or pickling. This chapter, then, will discuss in detail these various processes (except that of polishing, which will have a later chapter to itself), and the present opportunity will be taken of describing the shapes and functions of the tools used in the work.

The Casting's Skin or Outer Crust.

The crust of the casting, which always retains some sand, is very destructive to the tools unless they can be sent in deep enough to penetrate to the clean metal beneath. Wrought-iron is but seldom pickled previously to being filed, but is either cleaned with an old file or is ground on a stone to remove the outer scale or oxydised surface. The chipping chisel is only required where the work cannot be forged so nearly to the required form as to bring it properly within range of the file.

Pickling Iron Castings.

Iron castings are pickled with a mixture of 1 part of sulphuric acid to 2 parts of water. The castings, if small, are immersed in a trough lined with lead; or else the acid is sprinkled over them. In two or three days a thin crust, like an efflorescence, may be washed off with the aid of water and slight friction.

Pickling Brass and Gunmetal Castings.

Brass and gunmetal when pickled require nitric acid diluted with four to six times as much water. Sulphuric acid is

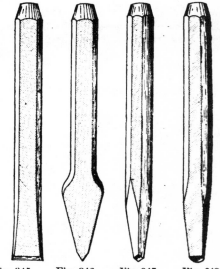

Fig. 345.— Flat Chisel.   Fig. 346.— Cross-cut Chisel.   Fig. 347.— Diamond Chisel.   Fig. 348.— Half-nosed Chisel.

not used in pickling gunmetal and brass, it being used only in the case of iron. The diluted acid should be kept in leaden utensils, or in well-glazed earthenware or glass ones. Brass is commonly hammered all over to increase its density, unless a quantity of tin is added, say a quarter or half an ounce to the pound, which materially stiffens the alloy, so as to render hammering as unnecessary as it is with good gun-metal. Whatever action the metals are subjected to, whether natural in the mould, or artificial under the hammer and tools, etc., it is of primary importance that all parts should be treated as nearly alike as possible.

CHIPPING CHISELS.

There are at least four common kinds of chipping chisels: the flat (Fig. 345), the cross-cut (Fig. 346), the diamond (Fig. 347), and the half-nosed (Fig. 348). The illustrations show clearly the shapes of these tools, of which the functions are about to be described, and later illustrations depicting these chisels will be more in the

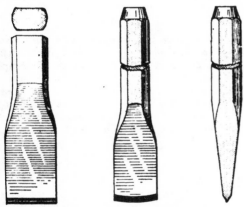

Figs. 349 to 351.—Flat Chipping Chisels.

nature of diagrams, the purpose being to indicate their salient points. Chipping chisels are stocked in eleven lengths between 4 in. and 14 in. inclusive, excluding fractions of inches, and in width they range in seven sizes from $\frac{1}{4}$ in. to 1 in., rising by eighths of an inch.

FLAT CHIPPING CHISELS.

Flat chisels are made as in Figs. 349 to 351. The cutting edge should be parallel to the flats, and this is the more easily discerned in Fig. 349, in which the flat is wide. The broad flat is the best guide in holding the chisel level with the surface to be chipped. Either tool is suitable for

wrought-iron or steel, and in use will take all the power that can be given by an ordinary hammer—that is, one of $1\frac{3}{4}$ lb. weight, and having a 15-in. handle. The

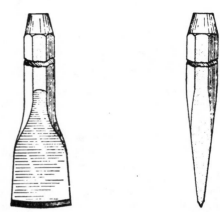

Figs. 352 and 353.—Broad, Flat Chisel.

hammer is held by its end and swung back about vertically over the shoulder. However, if so narrow a chisel were used on cast-iron or brass and driven with full hammer blows, the metal would be broken out instead of cut, and possibly the break would go below the proper depth for chipping, and so leave ugly cavities, the sign of unskilful work. For cast-iron, brass, or similar metals and alloys the broader chisel (Figs. 352 and 353) is used, so as to spread the force of a hammer blow over a

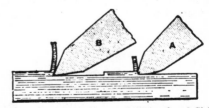

Fig 354.—Diagram Illustrating Angle of Chisel Edge.

greater length of chisel edge, this causing the chisel to move forward through less space at each blow; consequently, the cutting is clean and the metal is not broken out. The broader the chisel, the more easily is its edge held fair with the surface of the work, and the smoother the chipping.

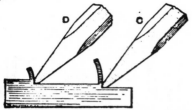

Fig. 355.—Diagram Illustrating Angle of Chisel Edge.

SHAPING CHISEL EDGE.

The blade of the chisel must be made as thin as possible, and Figs. 349 to 353 show the proportions of new chisels. In grinding the two facets, great care must be taken to avoid rounding them, as at A (Fig. 354), and they must be quite flat, as at B. The angle of the two facets must suit to the material; the greater the acuteness the easier is the cutting. The angle shown at C (Fig. 355) is about right for brass, and that shown at D for steel. Naturally, with the hard metal—steel—the more acute angle dulls too quickly. The chisel cutting edge for heavy chipping may be straight (Fig. 349), or preferably curved (Fig. 352); in the latter, the corners are relieved of duty, and therefore are less liable to be broken when the chisel is in use. The curved edge is of still greater advantage in light chipping, because, as shown in Fig. 356, a thin, fine chip can be taken off without the corners of the chisel

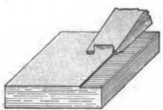

Fig. 356.—Light Chipping.

edge cutting into the metal, and these corners exposed to view aid the eye in keeping the chisel edge level with the surface of the work.

WRONGLY SHAPED CHISEL EDGES.

The edge of a chipping chisel must not be ground hollow as in Figs. 357 and 358,

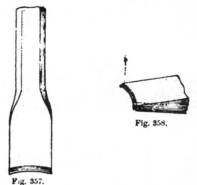

Fig. 358.

Fig. 357.

Figs. 357 and 358 —Hollow-ground Chisel Edges.

because then the corners would dig in and cause the tool to get beyond control; also a force acting on the wedge principle would come into play in the direction of the arrow (Fig. 358), and this would tend to spread the corners and break them off. If the facets of the chisel edge are ground wider on one side than the other (see Fig.

Fig. 359.   Fig. 360.

Figs. 359 and 360.—Badly Ground Chisel Edges.

359), the flat of the chisel will not form a guide as to when the cutting edge is level with the surface of the work; nor must the edge be ground out of square with the chisel body, as in Fig. 360, because then the tool would be apt to jump sideways at each hammer blow.

QUICK METHOD OF CHIPPING.

It is possible to remove a quantity of metal more quickly if first some grooves are cut with a cape chisel (Figs. 361 to 363), as shown at A, B, C (Fig. 364), the space between these grooves being slightly less than the width of the flat chisel, whose corners are thus relieved when it is

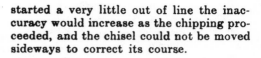

brought into operation. The end of the cape or cross-cut chisel is shaped as at A B (Fig. 362), and not as at C (Fig. 363), because it is desirable to move the chisel sideways to make it travel in a straight

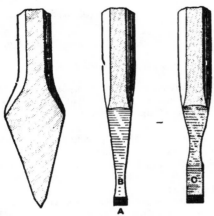

Figs. 361 to 363.—Cape or Cross-cut Chisel.

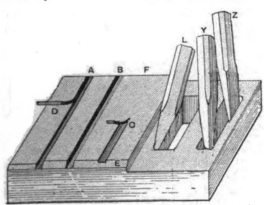

Fig. 364.—Diagram of Chipping Chisels at Work.

line ; the parallel part at C (Fig. 363) would prevent this, so that if the chisel were

started a very little out of line the inaccuracy would increase as the chipping proceeded, and the chisel could not be moved sideways to correct its course.

THE ROUND-NOSE CHISEL.

The round-nose chisel (Figs. 365 and 366) must not be straight on its convex edge. It

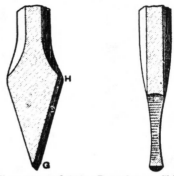

Figs. 365 and 366.—Round-nose Chisel.

may be straight from H to G, but from G to the point it must be bevelled so that by altering the height of the chisel head the depth of the cut can be altered also.

THE COW-MOUTH CHISEL.

The cow-mouth chisel (Figs. 367 and 368) must be bevelled in the same way as the

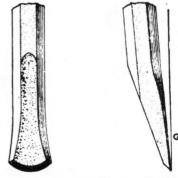

Figs. 367 and 368.—Cow-mouth Chisel.

round-nose, so that when it is required to cut out a round corner as at L (Fig. 364), the head can be moved to the right or to the left and the depth of cut thus governed.

THE OIL-GROOVE CHISEL.

The oil-groove chisel (Figs. 369 and 370) must be made narrower at A than it is across the curve, or it will wedge into the groove it cuts.

THE DIAMOND-POINT CHISEL.

The diamond-point chisel (Figs. 371 to 374) must be shaped to suit the work. If

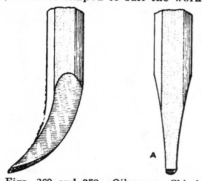

Figs. 369 and 370.—Oil-groove Chisel.

it is not to be used to cut out the corners of very deep holes it can be bevelled at M, and its point x thus brought central to the body of the steel (Q indicates centre line);

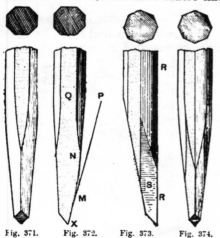

Fig. 371.   Fig. 372.   Fig. 373.   Fig. 374.
Figs. 371—374.—Diamond-point Chisel.

this renders the corner x less liable to break off, which is the great trouble with this shape of tool. However, as the bevel at M necessitates that the chisel be leaned

over as at Y (Fig. 364), the tool could not be kept to its cut in deep holes, so for such work the bevel at M must be omitted, and the edge must be made straight as at R R (Fig. 373).

THE SIDE CHISEL.

The side chisel (Fig. 375) obeys the above rules also, so it may have bevel at w (Fig.

Fig. 375.—Side Chisel.

375), and may be leaned over as at z (Fig. 364), or the side v w (Fig. 375) may be straight along its whole length. In all mortising and slotting chisels, however, when circumstances permit, it is desirable to have no bevel on the side that meets the work, so that the depth can be regulated by moving the chisel head. With the side chisel can be illustrated the points made with reference to the chisels shown in Figs. 365 to 375—that there must be a bevel at the end to enable the depth of cut to be adjusted and governed. For if the straight chisel happens to get too deeply into its cut it cannot be altered, and unless a new cut is begun it will get imbedded deeper and finally will break. But with a slightly bevelled side chisel the depth of cut can be regulated, making it less if it gets too deep, or deepening it if it gets too shallow.

Fig. 376.—Ball Pene Hand Hammer.

HOLDING THE CHIPPING CHISEL.

In all the chipping chisels described above, the chip in the work steadies the cutting end, and obviously the nearer to the head the chisel is grasped by the hand, the steadier is it held, and the less is the liability for the hammer to hit the workmen's fingers. Another advantage is that the chipped surface will be smoother.

THE PROPER POSITION WHEN CHIPPING.

Suppose that a heavy chip is to be taken off a piece of wrought-iron. Stand well

Fig. 377.—Cross Pene Hand Hammer.

away from the vice, and the body will be lithe and supple and will have a slight motion in unison with the hammer. The beginner naturally wishes to get too close, and thus occupies a constrained position, and both feels and looks awkward. For taking a light chip, stand nearer to the work, so as to watch the action of the chisel and keep its depth of cut level. In

Fig. 378.—Straight Pene Hand Hammer.

both cases, push the chisel forward to its cut and hold it near the head as steadily as possible. Many inexperienced persons move it at each blow, but this is a mistake ; because it cannot be maintained so accurately at the proper height.

HAMMERS.

An indispensable tool in chipping is the hammer, and the present is a good oppor-

Fig. 379.—Cross Pene Hand Hammer.

tunity of briefly discussing engineers' hammers in general and chipping hammers in particular. The usual forms of engineers' solid steel hand hammers are shown by

Figs. 376 to 379, and only one of these—Fig. 376—has a ball pene, all the rest having either cross or straight penes. All the shapes illustrated are made in weights

Fig. 380.—Cross Pene Sledge Hammer.

of ¼ lb. and upwards, but there are similar tools made of ½ lb., 3 lb., and 1 lb. weight. Of the sledge hammers (Figs. 380 to 382), the first has a cross pene, the second a straight pene, while the third is double-faced. These are made of 2 lb. weight and upwards.

Fig. 381.—Straight Pene Sledge Hammer.

CHIPPING HAMMERS.

The chipping hammer, or engineer's hand hammer, is made in quite a variety of forms ; thus, in England and in Europe generally the standard form is that shown in Fig. 383, the pene, pane, peen, or pean (these are various spellings of the same

Fig. 382.—Double Faced Sledge Hammer.

word) being at a right angle to the hammer handle, and of the shape shown. In the United States it usually has the pene parallel with the handle, while in all coun-

tries it is sometimes made with a ball
pene.

THE HAMMER PENE.

The pene is mostly used for riveting, and
it is quite a question which is the best

Fig. 383.—Chipping Hammer with Cross Pene.

form. The hammer is one of those tools
which the workman gets used to, or "gets
the hang of"; and there is a good deal in
this term as applied to a hammer, as will
be seen presently. With the pene as in
Fig. 383, and somewhat rounded at the
point A, the hammer handle may be horizontal or not when the pene meets the
work; with the pene as in Fig. 384, if the
handle does not stand horizontal at the
moment of striking, the pene will not meet
the work fair—one corner or the other, B or
C, will make a dent. On the other hand, it
is possible that the form in Fig. 385 is the
best for spreading the metal as required
in riveting, and this is probably the reason

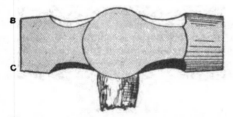

Fig. 384.—Chipping Hammer with Straight
Pene.

why coppersmiths use it, their work consisting largely in stretching or spreading
the metal.

THE EFFECTS OF A HAMMER BLOW.

To understand this matter thoroughly it
is necessary to consider the effects of a
blow with the hammer travelling in different directions. In Fig. 386 are shown the

effects of a blow, the hammer having fallen
vertically at A. The full line denotes the
surface receiving the impact of the blow,
and the dotted ones the direction in which
the effects extend around it. At B are
shown the effects, the hammer having had
while falling a lateral motion from right
to left; hence, as the hammer travelled

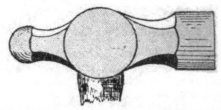

Fig. 385.—Chipping Hammer with Ball Pene.

towards the left both these blows are in
the direction that the pene of the hammer
in Fig. 384 would strike if used naturally.
At C are shown the effects of a blow delivered by the pene of the hammer illustrated by Fig. 383, the hammer head having
moved laterally while descending. The
effects of a ball-pened hammer are shown
in Fig. 387, A representing a vertical blow,
and B representing a blow delivered with a
lateral motion denoted by C. In each case
it will be noted that the direction in which
the effects of the blow extend varies with
any lateral motion of the hammer. The
length of the blow A (Fig. 386) is at a
right angle to the length of C, but the hammer in Figs. 383 and 384 may, by a suitable bending of the arm, be made to cause
the pene blows in either direction, but not,

Fig. 386.—Effects of Hammer Blows.

of course, so handily. As sold, engineers'
hammers and riveting hammers have the
pene as in Fig. 383, and the chipping hammers have the pene in all the three forms
illustrated.

"GETTING THE HANG" OF A HAMMER.

After having become accustomed to one of the shapes shown, it takes about a year

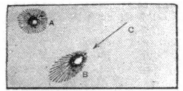

Fig. 387.—Effects of Ball-pened Hammer Blows.

to get thoroughly familiar with the other. Many prefer the ball-pened hammers, and next to them the form in Fig. 383. The form of ball pene best for general use is shown in Fig. 385, and there is a similar shape with the eye much smaller in proportion for pening and for marking work out by lines, as is necessary in very large work. There is a good deal in "getting the hang" of a hammer, and it may be added that a workman often takes a pride in the fact that his hammer shows its wear in the middle of the face and not on one side. It is often thought that that is due to a proper wielding of the hammer, but this is only the case to a limited extent, much depending upon the hammer being properly set upon its handle. The centre line of the handle should be at a right angle to both the length and width of the hammer's head, and the oval of the handle in line

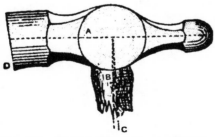

Fig. 388.—Hammer Head Incorrectly Set.

with the length of the head. Suppose, for example, that in Fig. 388 A represents the centre line of the head, and c that of the handle (instead of being at B), the lower part of the face will wear as at D. Or,

suppose the centre line of the hammer to be at c D (Fig. 389), and the major axis of the handle is at A B, then the face will

Fig. 389.—Hammer Head Incorrectly Set.

wear most at E. In getting the hang of a hammer the operator simply accustoms himself to either of these errors, and compensates for it in his method of holding the hammer.

FIXING HAMMER HEADS IN HANDLES.

The eye has to be properly shaped to enable it to hold the hammer firmly and

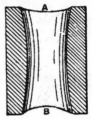

Fig. 390.—Section of Hammer Head.

not in time get loose. One of the best forms is shown in Figs. 390 and 391, the handle end or bottom of the eye being rounded out as at B, and the top A being rounded across the hammer, but not lengthways. The rounding out prevents the handle from going too far through the

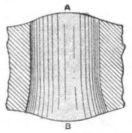

Fig. 391.—Section of Hammer Head.

eye, and when the eye is wedged at the top with a single wedge, as in Fig. 392, the spread across the eye prevents the handle from coming loose. This locks the handle as firmly as can be, and requires one wedge only. A wooden wedge is preferable to an iron one, providing the handle and the wedge are of dry and well seasoned wood.

FILE MANUFACTURE.

In the manufacture of files the pieces of steel, or the blanks intended for files, are forged out of bars of steel that have been either tilted or rolled as nearly as possible to the sections required, so as to leave but little to be done at the forge; the blanks are afterwards annealed with great caution, so that in neither of the processes the temperature known as the blood-red heat may be exceeded. The surfaces of the blanks are now rendered accurate in form and quite clean in surface either by filing or grinding. For smaller files the blanks are mostly filed into shape as the more exact method; for the larger, the blanks

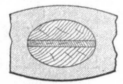

Fig. 392.—Hammer Head Wedged on.

are more commonly ground on large grindstones as the most expeditious method; in some cases the blanks are planed in the planing machine, for those called "dead-parallel files." The blank, before being cut, is slightly greased, that the chisel may slip freely over it. The file cutter is seated before a square stake or anvil, and places the blank straight before him, with the tang towards his person; the ends of the blank are fixed down with two leather straps or loops, one of which is held fast by each foot.

CHISELS FOR FILE CUTTING.

The largest and smallest chisels commonly used in cutting files are shown half size by Figs. 393 and 394. The first is a chisel for large rough files; the length is about 3 in., the width 2½ in., and the angle of the edge about 50°; the edge is per-

fectly straight, but the one bevel is a little more inclined than the other, and the keenness of the edge is rounded off, the object being to indent rather than cut the steel; this chisel requires a hammer weighing about 7 lb. or 8 lb. Fig. 394 is the

Fig. 393.—Chisel for Cutting Rough Files.

chisel used for small superfine files; its length is 2 in., the width ½ in.; it is very thin, and sharpened at about the angle of 35°; the edge is also rounded, but in a smaller degree; it is used with a hammer weighing only 1 oz. or 2 oz.; as will be seen, the weight of the blow mainly determines the distance between the teeth. Other chisels are made in intermediate sizes, but the width of the edge always exceeds that of the file to be cut.

FILE CUTTING.

The first cut is made at the point of the file; the chisel is held in the left hand, at a horizontal angle of about 55° with the central line of the file, as at A (Fig. 395), and with a vertical inclination of about 12° to 4° from the perpendicular, as represented in Fig. 396, supposing the tang

Fig. 394.—Chisel for Cutting Smooth Files.

of the file to be on the left-hand side. The following are nearly the usual angles for the vertical inclination of the chisels, namely: For rough rasps, 15° beyond the perpendicular; rough files, 12°; bastard files, 10°; second-cut files, 5°; and dead-smooth-cut files, 4°. The blow of the hammer upon the chisel causes the latter to indent and slightly to drive forward the steel, thereby throwing up a trifling ridge

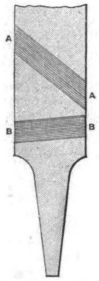

Fig. 395.—Cuts on File Blank.

or burr; the chisel is immediately replaced on the blank, and slid from the operator until it encounters the ridge previously thrown up, which arrests the chisel or prevents it from slipping further back, and thereby determines the succeeding position of the chisel. The chisel having been placed in its second position, is again struck with the hammer, which is made to give the blows as nearly as possible of uniform strength, and the process is repeated with considerable rapidity and regularity, sixty to eighty cuts being made in one minute, until the entire length of the file has been cut with inclined parallel and equi-distant ridges, which are collectively denominated the "first course." So far as this one face is concerned, the file, if intended to be single-cut would be then ready for hardening, and when greatly enlarged its section would be somewhat as in Fig. 396. The teeth of some single-cut files are much less inclined than 58°;

generally those of floats are square across the instrument.

DOUBLE CUTTING.

Most files are double cut, or have two series of courses of chisel cuts, and for these the surface of the file is now smoothed by passing a smooth file once or twice along the face of the teeth, to remove only so much of the roughness as would obstruct the chisel from sliding along the face in receiving its successive positions, and the file is again greased. The second course of teeth is now cut, the chisel being inclined vertically as before, or at about 12°, but horizontally about 5° to 10° from the rectangle as at B (Fig. 395). The blows are now given a little less strongly, so as barely to penetrate to the bottom of the first cuts, and consequently the second course of cuts is somewhat finer than the first. The two series of courses fill the surface of the file with teeth that are inclined toward the point of the file, and that, when highly magnified, much resemble in character the points of cutting tools generally, as seen in Fig. 396.

CUTTING THE OTHER FACE.

If the file is flat and to be cut on two

Fig. 396.—Cutting Files.

faces, it is now turned over, but to protect the teeth from the hard face of the anvil a thin plate of pewter is interposed.

SOME POINTS IN FILE CUTTING.

Triangular and other files require blocks of lead having grooves of the appropriate sections to support the blanks, so that the surface to be cut may be placed horizontally. Taper files require the teeth to be somewhat finer towards the point, to avoid the risk of the blank being weakened or broken in the act of its being cut, which might occur if as much force were used in cutting the teeth at the point of the file as in those at its central and stronger part. Eight courses of cuts are required to complete a double-cut rectangular file that is cut on all faces, but eight, ten, or even more courses are required in cutting only the one rounded face of a half-round file.

CUTTING ROUND-SECTION FILES.

There are various objections to employing chisels with concave edges, and therefore, in cutting round and half-round files, the ordinary straight chisel is used and applied as a tangent to the curve. It will be found that in a smooth, half-round file 1 in. in width, about twenty courses are required for the convex side, and two courses alone serve for the flat side. In some of the double-cut, gullet-tooth saw-files, as many as twenty-three courses are sometimes used for the convex face, and but two for the flat. The same difficulty occurs in a round file, and the surfaces of curvilinear files do not therefore present, under ordinary circumstances, the same uniformity as those of flat files. Hollowed files are rarely used in the arts, and, when they are required, it usually becomes imperative to employ a round-edged chisel, and to cut the file with a single course of teeth.

CUTTING RASPS.

The teeth of rasps are cut with a punch, which is represented in two views (Fig. 397). The punch for a fine cabinet rasp is about 3½ in. long and ⅝ in. square at its widest part. Viewed in front, the two sides of the point meet at an angle of about 60° ; viewed edgewise, or in profile, the edge forms an angle of about 50°, the one face being only a little inclined to the body of the tool. In cutting rasps, the punch is sloped rather more from the operator than the chisel in cutting files, but the distance between the teeth of the rasp

cannot be determined, as in the file, by placing the punch in contact with the burr of the tooth previously made. By dint of habit, the workman moves—or, technically, hops—the punch the required distance ; to facilitate this movement a piece of woollen cloth is placed under his left hand, which cannot then come immediately in contact with, and adhere to, the anvil. The teeth of rasps are cut in rather an arbitrary manner. Thus the lines of teeth in cabinet rasps, wood rasps, and farriers' rasps are cut in lines sloping from the left down to the right-hand side ; the teeth of rasps for boot and shoe makers and saddle-tree makers are cut in circular lines, or crescent form. These directions are quite immaterial, but it is important

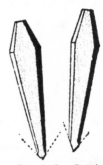

Fig. 397.—Punch for Cutting Rasps.

that every succeeding tooth should cross its predecessor, or be intermediate to the two before it, as, if the teeth followed one another in right lines, they would produce furrows in the work, and not comparatively smooth surfaces.

STRAIGHTENING FILES AND RASPS.

While being cut, files and rasps always become more or less bent, and there would be danger of breaking them if they were set straight while cold ; they are consequently straightened while they are at the red heat, immediately prior to their being hardened and tempered.

HARDENING FILES AND RASPS.

Previous to being hardened the files are drawn through beer grounds, yeast, or other sticky matter, and then through common salt mixed with hoof parings previously roasted and pounded; this compound protects the delicate teeth of the file from the direct action of the fire, and serves as an index of the temperature, as on the fusion of the salt the hardening heat is obtained; also the files are less liable to crack or clink on being immersed in the water. The smeared file is heated gradually to a dull red, and then straightened with a lead hammer on two small lead blocks; then it is heated further until the salt on its surface just fuses, when the file is immediately dipped in water. The file is immersed quickly or slowly, vertically or obliquely, according to its form, the mode adopted being the one best calculated to keep the file straight. The half-round file, owing to its unsymmetrical section, is disposed, on being immersed, to become hollow or bowed on the convex

Fig. 398.—Bastard Grade of File.

side, and this tendency is compensated for by curving the file while soft in a nearly equal degree in the reverse direction.

Straightening Files after Hardening.

It commonly happens that, with every precaution, the file becomes more or less bent in hardening; then it has to be straightened by pressure, either before it is quite cold, or else after it has been partially re-heated. The pressure is variously applied; sometimes by passing one end of the file under a hook, supporting the centre on a prop of lead, and bearing down the opposite end of the file; at other times, by using a support at each end and applying pressure in the middle by means of a lever, the end of which is hooked in the bench. Large files are always straightened before they are quite cooled, after the hardening, and while the central part retains a considerable degree of heat.

When straightened, the file is cooled in oil, which saves the teeth from becoming rusty.

Softening File Tangs.

The tangs are now softened to prevent their fracture; this is done either by

Fig. 399.—Second Cut Grade of File.

grasping the tang in a pair of heated tongs, or by means of a bath of lead contained in an iron vessel with a perforated cover, through the holes in which the tangs are immersed in the melted lead, which is heated to the proper degree. The tang is afterwards cooled in oil, and when the file has been wiped and the teeth brushed clean, it is considered fit for sale.

Qualities of a Good File.

The superiority of the file will be found to depend on four points—the primary excellence of the steel, the proper forging and annealing without excess of heat, the correct formation of the teeth, and the success of the hardening.

Five Grades of File.

The grades of file in general use are classified as follows: Rough, bastard (Fig. 398), second cut (Fig. 399), smooth (Fig. 400), and dead smooth. These five grades are made by giving two separate cuts so as

Fig. 400.—Smooth Grade of File.

to form a diamond-shaped tooth. The first cut put on the file is known as the over-cut and the second as the up-cut. For mill-saw and taper-saw files only, the over-cut is used to form the tooth, the tooth being a sharp ridge of metal extending at a sharp angle across the face of the file, and not intersected by any other stroke of the cutting chisel.

teeth fifty to the inch; "wood rasps" have a punched tooth known as bastard grade, one hundred and five teeth to the square inch; "cabinet rasps" with a punched

Fig. 401.—Flat Bastard File.

This last tooth, as applied to mill and taper files, goes under the two grades of bastard and second cut. The only other tooth which is in use is that applied to horse rasps, wood rasps, cabinet rasps, and shoe rasps. This tooth is raised to a conical form by the use of a punch. On horse rasps it is too large to designate by any grade of cut. On wood and cabinet rasps it is graded under the head of bastard, smooth, and sometimes second-cut.

tooth, and known as smooth, have three hundred teeth to the inch; cabinet rasps with a punched tooth, and known as second cut, have one hundred and eighty-six teeth to the inch.

FLAT FILES.

A flat file is double-cut, and made up of the five grades of cut previously mentioned—namely, rough, bastard, second cut, smooth, and dead smooth. The double cut applies to the faces of the file, the

Fig. 402.—Half-round Bastard File.

As a matter of convenience, the grade of the tooth is given by counting only one of the cuts of the file, and for this is taken the over-cut or first cut.

GRADES OF FILE TEETH.

The exact number of teeth to the linear inch is given below. "Rough" is formed of twenty-two teeth to the running inch; "bastard," thirty-two; "second cut," forty-four; "smooth cut," sixty-eight; "dead smooth," one hundred and twenty; "mill bastard" is formed of a series of

edges being single cut; sometimes the file is made with single cut faces and of the five above-mentioned grades. In shape, this file is generally of parallel sides and thickness for two-thirds its length, and from that distance to the end is drawn to a blunt point, gradually decreasing in width and thickness. Fig. 401 shows a flat bastard file.

HALF-ROUND FILES.

The half-round has the same grade of cut as the flat. It is cut on the flat side the

Fig. 403.—Smooth Hand File.

diagonal teeth thirty-six to the inch; "mill second cut" has a series of diagonal teeth forty-four to the inch; "taper saw single cut," second cut, has a series of diagonal

same as a flat file—that is, it is double cut. On the half-round side is a series of rows of single cuts from shank to point, completely covering it. Some of the rows intersect at different angles; where this occurs it gives the file the appearance of being double cut. It is made in the same

Again, these are the same in cut and grade of cut as the "flat," and the equal-

Fig. 404.—Pillar File.

widths as the various sizes of "flat" files; and is drawn to a point in the same way. A half-round bastard is shown by Fig. 402.

HAND FILE.

This also in cut and grade of cut is the same as the "flat." It is of the same width at the point or end as at the shank

ling file resembles the "hand" file except that it is not only of parallel width, but of parallel thickness also. The cotter or pivot file in shape resembles a very narrow "pillar."

SQUARE FILE.

The square file in cut and grade of cut

Fig. 405.—Square Bastard File.

or tang end. It is gradually drawn down in thickness from about two-thirds its length to the end, leaving it at the end about one-half its original thickness. Sometimes it is drawn very slightly from the centre to the tang as regards its thickness. One edge of the file is left uncut,

is the same as the "flat." It is of equal width and thickness for about two-thirds its length; from there to the end it is gradually drawn to a point, its square shape being retained. Frequently the point or end is left of the same thickness as the tang end. It is then called "paral-

Fig. 406.—Round Bastard File.

and given the name of "safe edge." Sometimes both of the edges are "safe." Fig. 403 shows a smooth cut hand file.

PILLAR FILE.

This is another file the same as the "flat" in cut and grade of cut. In other respects it is the same as a "hand," except that it is only from five-eighths to three-

lel square." A square bastard file is illustrated by Fig. 405.

ROUND FILE.

The round file is in grade of cut the same as the "flat"; the cut itself being made up of a number of rows of single cuts extending from shank to point, the

Fig. 407.—Triangular File.

quarters the width, and in proportion to its width is of greater diameter (see Fig. 404).

rows slightly intersecting. It is gradually drawn to a point, from about two-thirds its length to the end. Sometimes the file is left the same thickness at the point or end as at the tang end, in which case it is called "parallel round." A round bastard file is shown by Fig. 406.

square, and sometimes with both round edges and with one safe edge. Engineers call it a "float file" when single cut. Sometimes it is double cut, as in Fig. 408.

Fig. 408.—Double Cut Mill Saw File.

TRIANGULAR FILE.

The triangular or three-square file (Fig. 407) is still another of those which in cut and grade of cut are the same as "flat." The shape of the file gives it three sharp corners, which vary in sharpness according to grade of cut. It is of uniform

Fig. 409.—Taper Saw File.

dimensions for about two-thirds its length, and is drawn gradually to a point from that distance to the end. There are also "parallel three square" files, these files being the same thickness at the point as at the tang end.

TAPER SAW FILE.

This is known also as taper saw, single cut, slim taper saw single cut, band taper saw single cut, and taper saw double cut. As a rule, it is a file of single cut graded as second cut. It is often cut in all its shapes with a double cut graded as second cut. It is triangular in shape (see Fig. 409), being forged to a point the same as a "three square." Dealing with these in greater particular, the taper saw single cut is a file graded as "second cut," triangular in shape, being cut on the three sides, and on three edges. As a rule, it is cut within a short distance of the end. The slim taper saw single cut is in every particular the same as the preceding, only it is made from stock two-thirds as heavy. The band taper single cut also is much the same;

Fig. 410.—Cabinet Rasp.

MILL SAW FILE, OR FLOAT.

This file of single cut is generally of two grades of cut, bastard and second cut, and sometimes smooth cut. It is flat, slightly drawn at the point, both in thickness and in width, and though of about the same width as a flat file is not so thick. It is

the exceptions are that its edges are ground round and cut with two rows of teeth. In the double cut taper saw, the sides receive a double cut instead of single.

CROSSING FILE.

This is graded the same as the flat. Oval in shape, with two sharp edges, it is

Fig. 411.—Wood Rasp.

generally cut with one round and one square edge, very often with both edges

forged, and drawn gradually from about two-thirds its length to the point. It is cut the same as the back or half-round side of the " half-round " file.

WOOD AND CABINET RASPS.

The cabinet rasp (Fig. 410) is in every respect but two the same as a "wood rasp." Instead of being half-round it is scant half-oval, and in grade of tooth it is

RIFFLER FILES.

The riffler (Fig. 413) is a handy tool for shaping awkward work, and it is formed to go in places where the ordinary file could not go. It has usually a bastard cut.

SELECTING FILES.

To determine a good file, hold it between

Fig. 412.—Warding File.

a smooth, sometimes a second cut. The "wood rasp" (Fig. 411) corresponds in shape to the flat round, and half-round files. The tooth is conical, being raised from the surface of the rasp with a punch, which raises a tooth about equal to the space it removes from the surface. As regards grade of cut, it is generally known as "bastard."

the eye and the light, its point towards the person, so that the cutting edges of every tooth can be seen. These edges should be clean, smooth, and sharp ; in a poor file they may be notched, cracked, uneven, and irregular. Closely examine it for fire-cracks. Hold it up to the light again, but with the ends reversed, and see if the file is all of one colour. A chequered appear-

Fig. 413.—Riffler.

CABINET FILE.

This is like a cabinet rasp. On the flat side it is double cut, and of the grade of "bastard." On the oval side it is of the same grade of cut—that is, bastard, half-round.

WARDING FILE.

This is the same in cut and grade of cut as the "flat," but is seldom more than 8 in. long. It resembles a flat file, but differs in three ways : it is forged more to a point, is not so thick, and is wider. Fig. 412 shows a warding file.

JOINT OR DRILL FILE.

In shape this is a thin "equalling" file, cut on the edges only, and sometimes made with round and sometimes with square edges, according to the wishes of the buyer. As regards thickness it can be ordered by numbers, as shown by some recognised metal gauge.

ance denotes uneven temper, and spots hard and soft. By striking the file with another tool, the ring of the steel will betray a flaw, if there is any.

FILE HANDLES.

File handles should always be propor-

Fig. 414.—Universal File Handle.

tionate to the files to which they are fitted. The hole in the handle should be properly squared out to fit the tang by means of a small "float" ; this may be made from a small bar of steel, and is similar to those used by plane-makers and cabinet-makers. The handles should always have good strong ferrules on them, and the files should be driven home quite straight and firm, so that there is no chance of the tool coming out. Each tool should have its handle permanently fixed; it is false economy to have one handle for many files and to be continually changing, considering the low price of handles, although the universal handle (Fig. 414) is specially adapted for this purpose.

CHEAP METAL HANDLES FOR FILES.

A good temporary, or even permanent, handle for a large file or rasp may be made from a short length of composition pipe. Knock it gently all along until it assumes an elliptical shape—that is, flatten it out slightly for the whole of its length, and then drive it, somewhat gently, over the tang of the file. If the file be first slightly warmed and dipped into powdered resin, it will hold better. Over this metal handle a piece of indiarubber tubing may be drawn, so as to give a better grip, and prevent the hands being blackened. A piece of indiarubber tubing slipped on to the point of a file or rasp will often enable one to get a good firm grasp of what is sometimes an awkward tool to hold.

THE OBJECT OF FILING.

Filing is generally performed with one of four objects in view: first, to remove a quantity of metal; second, to reduce the work to shape; third, to make the work fit; and fourth, to finish the surface. When the object is to remove a mass of metal, the file requires to be as large as can be conveniently handled upon the work, and this for engineers' use need not exceed a 20-in. file for the largest work. To make this bite well and to drive a fair cut will require all the power a man can continuously exert. The cut of the file should be, for roughing wrought-iron, a bastard cut; for cast-steel, a second cut; for soft machine steel a second cut; and for brass, the rough cut of a preferably new file. The hard, outside skin on cast-iron, and the sand adhering to it, make this metal difficult to attack. A new file would be used on such work to no purpose and would be spoiled. One nearly worn out will be nearly as effective, and will not be much injured. As stated many pages back, there is more than one way of removing the "skin," "bark," or "crust." One is to pickle the casting in dilute acid, so as to dissolve the outer crust of the casting, and liberate the sand adhering to the surface. Another is to chip it off with a chipping chisel, and this is a very good plan where much material has to be removed from some particular part of a

Fig. 415.—Vice too Low.

large, unwieldy piece of machinery, though, as has been already indicated, some practice will be required with the hammer and chisel before they can be used satisfactorily. In dealing with a casting straight from the foundry, first thoroughly brush it with a hard brush to remove loose sand. Assuming that neither pickling nor chipping is convenient, take an old file, and work away steadily at the skin till a surface of pure metal is arrived at. Having by then removed the parts that spoil files, the old file, with which but slow progress can be made, is changed for a better one. For the purpose it is best and most economical to use one that has been at work on brass until it has become too much worn for that material. Such a file is in first-class condition for working on cast-iron after the latter's sandy skin has been removed, and when worn out on that it will serve first-rate for steel.

SUBSTITUTES FOR THE FILE.

Filing is an operation that to do well requires very great practice. The use of the file should be avoided when the lathe, the milling machine, or the planer is available. Considerable skill is required to get up surfaces of large area by filing alone, more especially when these surfaces are required to be accurately flat. In large engineering works, filing is superseded by the planing and shaping machines for almost all work of large size. The speed and accuracy of the planing machine cannot be approached by the file when there is a great quantity of material to be removed. Files, then, are used only for the purpose of "fitting," and to smooth up the parts that are inaccessible to the planing tool. A planing machine, however, is a heavy and an expensive piece of machinery, and many are therefore obliged to dispense with its valuable aid.

PROPER HEIGHT OF VICE.

If men were all of the same height there would be no difficulty in regulating and determining the right height of the mechanic's vice, for it would be easy enough in each case to arrive at a standard which would serve equally well for all. There is, however, a considerable difference in the stature of men, some being above and some below the average height. A tall man would be compelled to stoop too much when working at a vice of suitable height for a short man; and a short man would be obliged to stand upright and raise his arms to an inconvenient height if put to work at a vice at which a tall man could work with ease and comfort.

Four sketches that will assist in determining the correct height of vice for every worker are given by Figs. 415 to 418. In Fig. 415 the vice is too low, and the workman has to bend his knees and stoop too much over his work, and thus he loses power. In Fig. 416 the vice is too high, and this compels the workman to assume too erect a position and raise his arms too high, and in this case also power is lost. In Fig. 417 the vice is just at its right height, midway between the too low posi-

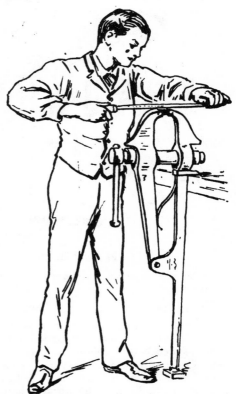

Fig. 416.—Vice too High.

tion in Fig. 415 and the too high position in Fig. 416; and the position assumed by the workman is such as enables him to bring the whole weight of the body, or nearly so, to bear on the stroke, and hence the power exerted is at its maximum. The proper and most convenient height for each workman is obtained as in Fig. 418, when the top of the vice anvil or jaws is just high enough to touch his elbow

Fig. 417.—Vice at Right Height.

when he stands erect and bends his arm as illustrated.

Correct Position for Filing.

The position at the vice to be assumed by the worker is a most important point to be decided before beginning filing proper. The vice should be fixed at the correct height, and so that the work held in its jaws will lie level. The worker should take up his position as follows: the left foot should be about 6 in. to left and 6 in. to "front" of the vice leg; the right foot should be about 30 in. to front—that is to say, 30 in. away from the board in a straight line with the vice post. This position gives command over the work, or, rather, over the tool, and is at once characteristic of a good vice man. The file

must be grasped firmly in the right hand, by the handle. The left hand must just hold the point of the file lightly, so as to guide it, and, when taking the forward cut, a fairly heavy pressure must be applied, proportionate to the size of the tool in use and the work being done. Some workers, for full duty with, say, a 14-in. file, have the left foot near the front of the vice, while the right one stands at least 26 in. behind. On the forward stroke the front foot is relieved almost entirely of the operator's weight, which will fall on the file, while on the back stroke the front foot should take most of the weight, so that the file may be relieved.

How to Use a File.

The art of filing a flat surface is not to be learned without considerable practice, and the following instructions, full as they are, can only serve as a guide to practical experiments. Long and attentive practice is necessary before the novice is able creditably to accomplish one of the most difficult operations that fall to everyday engineering work, and one which even the professional mechanic does not always succeed in. First, the vice must be of the right height, and then the proper position must be assumed, as already described. To obtain the greatest amount of duty, a large file must be applied on the forward stroke with all the power the worker can put on it; or, if the file is small, with as much power as is possible without risk of breaking it. The end of the file handle should abut against the palm of the hand, so that the file is pushed, and not dragged. The file must be relieved of all pressure during the return stroke, or the teeth will be liable to be broken off, just in the same manner that the point of a turning tool would be broken if the lathe were turned the wrong way. It is not necessary to lift the file altogether off the work, but it should have only its bare weight pressing during the back stroke. One of the chief difficulties in filing flat is that the arms have a tendency to move in arcs from the joints, but this will be overcome by practice. The preliminary file strokes should not all be made parallel one to another, but first at one angle and then at another, so that the file marks will cross and recross each other, which enables the file to cut more easily. Work which has been filed up properly will present when finished a flat, even surface, with the file marks running in straight parallel lines. Each stroke of the file will have been made to obtain a like end; whereas work which has been turned out by a careless or inexperienced workman will often bear evidence that each stroke of the file was made without any regard to all others, and the surface will be made up of an unlimited number of facets, varying in size, shape, and position. Amateurs who have never received any practical instruction in the use of files have generally a bad habit of pressing heavily on the tool continuously, during both forward and backward stroke, and at the same time work far too quickly. These habits, combined, will almost invariably spoil good work, producing surfaces more or less rounding, but never flat. The speed of the file may be made as quick as it can be pushed, providing the file is pressed to the work with all the weight possible, or, if a small one, with all the weight its strength will stand.

Filing Work to Shape.

For filing to shape, a smaller file must be used, so that even while removing the mass of the metal, the shape of the work can be readily observed by a slight lateral motion of the file, without entirely removing it from the work, or without stopping the file strokes. When it is necessary to file up a small surface—say, 2 in. or 3 in. square—the file must be applied in continually changing directions, not always at right angles to the jaws of the vice. In that case, though the work might be made perfectly straight in that direction, yet there would not be any means of assuring a like result on the part lying parallel to the jaws. When the surface is fairly flat, the file should be applied diagonally both ways; thus any hollow or high places, otherwise unobservable, will be at once seen, without the aid of straightedges

This method of crossing the file cuts from corner to corner is recommended in all cases. The file invariably should travel right across the work, the whole length of the file, not just an inch or so at some particular part, being brought into play. When in use, the file must be held quite firm, yet not too rigid. The sense of touch plays a big part in successful filing, and a firm grasp of the tool, at the same time preserving a light touch to feel the work, is essential.

Filing Work to Fit.

In filing to fit lies the greatest art of filing, for here it is necessary that the file

Fig. 418.—Testing Correct Height of Vice.

be of true outline, and to be so applied that it touches the work at the required spot only. To accomplish this result, the file should have a curve from end to end, as in Fig. 419 from A to B, in which case it may be held so as to touch the work at the required spot only, as at c. If the

stroke. This screw-like motion, given alternately from right to left, and *vice versâ*, serves to cross the file cuts and keep the hollow true to shape.

Fig. 419.—Filing Work to Shape.

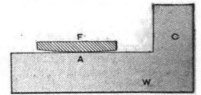

Fig. 421.—Use of Curved File.

file has a hollow place, as at D (Fig. 420), the file will be apt to touch at both E and F, while the operative area of file teeth is increased, and the file does not cut so freely. The file should be slightly curved in its width, if it is to be used on very fine fitting, and more especially if used for draw filing. Thus, in Fig. 421, the file F is so curved, and it is obvious that it can be brought to bear upon a very small area of the work, as at A; but suppose the file to be hollow across its width, as in Fig. 422, or crooked, as in Fig. 423, then it will file two grooves at A B at every cross-filing stroke, while in either case it would be difficult to file flat up in the corner close to the shoulder c of the work w.

FILING CURVED CORNERS.

In curved corners the half-round file should, for cross filing, be swept around so that the file marks appear as in Figs. 424 and 425, for if the file is moved in straight lines it will file a waved surface, such as in Fig. 426. The file should be swept first from left to right, as in Fig. 424, and then

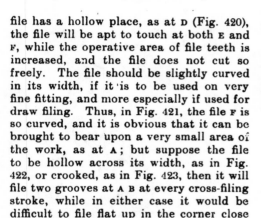

Fig. 422.—Use of Hollow File.

FILING OUT MOULDINGS, GROOVES, ETC.

In filing out mouldings and grooves of curved section, the file is worked in a special way. The files used are generally

from right to left, as in Fig. 425, which will prevent the formation of waves and keep the curve true. If the file is used too

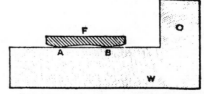

Fig. 423.—Use of Crooked File.

Fig. 420.—Defective Filing with Hollow File.

rats'-tails or half-rounds. The straight-forward stroke so necessary with the ordinary hand files is not employed here; instead a partially rotary motion—a sort of twist axially—is given to the file at each

much in one direction—that is to say, if the file marks are not made to cross and re-cross sufficiently often, or, if it is used with a sweep, more from right to left than from left to right, the defect shown in

Fig. 427 is liable to appear—a depression **running** from A to B.

treated in detail in a later chapter. Nothing can exceed the beauty of well-fin-

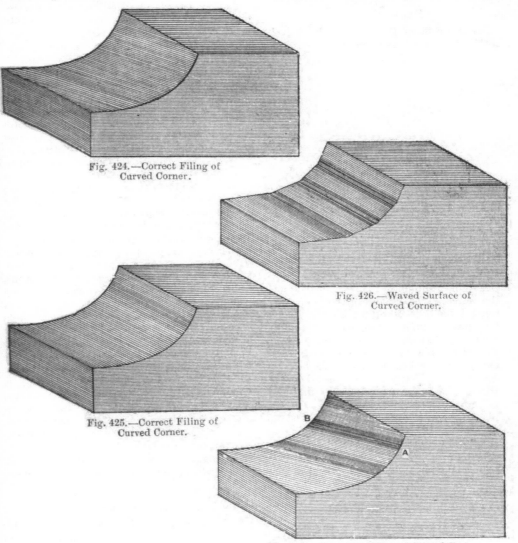

Fig. 424.—Correct Filing of Curved Corner.

Fig. 426.—Waved Surface of Curved Corner.

Fig. 425.—Correct Filing of Curved Corner.

Fig. 427.—Incorrect Filing of Curved Corner.

FINISHING FILED WORK.

With regard to finishing filed work, such as has to be made particularly presentable to the eye, there are many ways of polishing and burnishing, but, properly speaking, these are not filing, and they will be

ished work perfectly square and smooth as left by the file, untouched by any polishing materials. In such work the filing must be got gradually smoother by using successively files of finer cut. When the work is finished sufficiently fine for the purpose the lines should be carefully equalised by "draw filing."

DRAW FILING.

For draw filing the file may be held in both hands, much as a spokeshave is held,

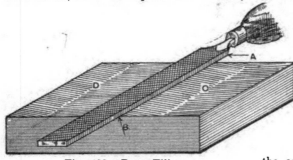

Fig. 428.—Draw Filing.

and drawn over the work so as to produce a series of fine parallel lines, which are considered a beautiful finish for many kinds of high-class engineering work. Draw filing should not be commenced until the work has been smooth cross-filed. To remove the cross-filing marks, the file may be tilted as in Fig. 428, so that it is the edge A B that does the bulk of the work. The draw file marks should cross after every few strokes, as shown in Fig. 428. Short strokes only should be taken, and these under heavy pressure at first until the cross file marks are removed. In draw filing with the half-round file, this should be swept so that its contact during each stroke begins at A (Fig. 429), and ends at B, while its marks cross, as illustrated, owing to a slight end motion that is given to the file. The strokes for all draw filing should be short, because long ones cause the file to pin and cut scratches.

CLEANING CLOGGED FILES.

Files that have become clogged with minute particles of metal, dirt, and grease are not fit to use, and should be vigorously cleaned. For this purpose the scratch brush is used generally; but this is not very efficient in removing those little pieces of metal which get firmly embedded and play havoc with the work. File cards are

also used; these are made by fixing together a quantity of cards—such as a pack of playing cards—by riveting, or by screwing them to a piece of wood. These file cards are used in the same way as the scratch brushes—transversely across the file in the direction of its "cuts"—and though neither tool produces much effect yet they are both often used. Files clogged with oil and grease may be boiled for a few minutes in strong soda water to dissolve the grease and, as a rule, set free most of the dirt and filings. A little scrubbing with an old tooth-brush will be beneficial before rinsing the files in boiling water and drying them before the fire. These methods will remove the ordinary accumulation of dirt, but the "pins," so much to be dreaded when finishing work, can be removed only by picking them out with a scriber point, or preferably by pushing them out with a piece of thin, very hard, sheet brass. To a certain extent, these "pins" may be avoided by using chalk on the file, if it is used dry, or a drop or two of oil will sometimes help matters. Another method of removing grease, old paint, etc., into which particles of metal become embedded, is to give the file a full coat of paraffin oil, wash it in hot soda water with a stiff brush and soap, clean out the cuts with a sharp-pointed awl, apply more paraffin, and brush as before. Then with a hard dry brush and fine emery, or ground cold cinders, brush vigorously across the file in the direction of the cutting. Fragments of hard metal embedded between the teeth must be picked out.

ACID BATHS FOR CLOGGED FILES.

A file choked with such easily soluble metals as copper and brass should be washed in a strong solution of caustic soda, rinsed in water, scrubbed with the file card, washed in hot water, and dried quickly. This removes dirt and grease; for removing the copper or brass immerse in a pickle of 1 pint of nitric acid, 1 gill of muriatic acid, and 1 gallon of water. A smaller quantity of the pickle could be made, keeping to the same proportions. This treatment also serves to roughen or sharpen the worn teeth, and lengthens the tool's effective life. To prevent rapid rusting, transfer the file from the acid to the caustic soda solution, well rinse, and dry quickly. An alternative acid bath is composed of 2 parts of nitric acid, 1 part of sulphuric acid, and 25 parts of water, and its action may be hastened by attaching the file or files to the positive pole of a Daniell battery, and then forming the battery's negative wire into a coil that will pass around but not touch the files. Bend the end of the coil into the solenoid—that is, turn it inwards towards the files, and examine them after the current has been passing for ten minutes. Then rinse and dry.

Effective File-cleaning Process.

Place twenty or thirty clogged files in a strong solution of soda in warm water; if they are not rusty thirty minutes will suffice, and then brush them well with a hard brush to remove all dirt. Submit the files to a wire brushing before placing them edgeways in a wooden bath, leaving a slight space between them; then pour in sufficient water barely to cover them, and add to the water a quarter of its bulk of nitric acid in solution. Fine files should remain thus for fifteen minutes, and rough files thirty minutes. The wooden vessel must be kept at a slow, regular rocking motion, so that the liquor is constantly scouring the files, and should they be laid flat they must be turned over every three or four minutes. At the proper time transfer them to a pail of cold water, and scrub them well with a brush; then replace them in the solution, adding half the previous quantity of nitric acid, and pouring it on the shanks, not on the files. Keep the vessel in motion for fifteen or thirty minutes, according to the character of the files, then take them out and wash again with cold water. Add to the solution in the wooden vessel about 12 per cent. of sulphuric acid, and, when the fumes have subsided a little, replace the files, and continue the rocking motion about one minute

for fine and three minutes for rough ones, again removing them to cold water. Take care to brush away the acid, and then lay the files to dry before a fire. While still warm rub them with oil, and the files are then fit for use. A new file can undergo two such operations. It must then be re-cut, or it may be converted to a screwdriver, or, if large, to a cutting chisel for metals, stone, etc.

Renovating Files by Sand Blast.

Wet sand blast can be used to re-sharpen blunted files; two or three minutes is time enough to put points on the worn teeth of a 14-in. rough file, both sides being done at once. The keenness of a newly cut file is much increased by sand-blast setting. A file with one side so sharpened and the other not was shown by Mr. Holtzapffel at a meeting of the Society of Arts. A block, slowly tilted from the horizontal whilst resting on the file, was shown to slide when an angle of 23° was reached on the unset side. The keenness may be easily increased till the sliding angle rises to 70°, but strength is then lost and the teeth break easily.

Scraping Metal.

The object of scraping is to obtain between two surfaces a closer contact than is possible with planing or filing. Before

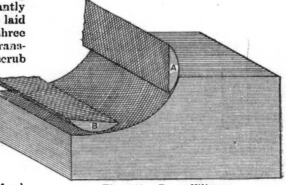

Fig. 429.—Draw Filing.

the introduction of scraping by Sir Joseph Whitworth, emery powder and oil or water were used, the two faces being rubbed together, and the inequalities ground away by the emery. This method had the disadvantage of collecting the powder in the hollow places, and grinding them at the same time as the high ones, and also a further evil in the difficulty of removing

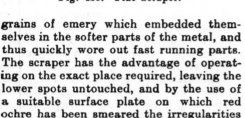

Fig. 430.—Flat Scraper.

grains of emery which embedded themselves in the softer parts of the metal, and thus quickly wore out fast running parts. The scraper has the advantage of operating on the exact place required, leaving the lower spots untouched, and by the use of a suitable surface plate on which red ochre has been smeared the irregularities or high places are made apparent.

Fig. 431.—Round-nosed Scraper.

MAKING SCRAPERS.

The scrapers are formed generally from old files, preferably " smooths," because these have already tangs for the handles, and require but little forging. Fig. 430 is

Fig. 432.—Round-nosed Scraper.

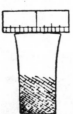

Fig. 433.—Testing Scraper.

a flat scraper, which is used on flat surfaces, and Figs. 431 and 432 are round-nosed scrapers, used lengthways on the bearing of large or small journals. Select files of a size suitable to the work in hand,

grind out the teeth, draw out the point rather thin for 2 in. to 3 in., and do not burn or overheat the steel. Allow them to

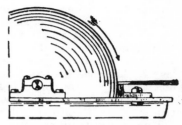

Fig. 434.—Grinding Scraper.

cool slowly, and file or grind to shape. The cutting edge should be square (never chisel shape on flat scrapers), and also not more than $\frac{1}{16}$ in. thick. The scrapers are heated to a clear red, and quenched in clean water, and left quite hard. When

Fig. 435.—Edge of Scraper on Oilstone.

ground, the front edge of a flat scraper should be slightly curved, and this is tested by holding it up to the light with the edge of a steel rule (see Fig. 433). Fig. 434 shows the position of the scraper at the grindstone, and Figs. 435 and 436 show the

Fig. 436.—Side of Scraper on Oilstone.

method of oilstoning the front and sides respectively. The tool in Fig. 435 should be held firmly about 2 in. above the stone and worked to and fro, in a vertical plane.

To test the cutting edge, try the scraper on the thumb-nail, which it should pare easily when properly sharpened.

VARIETIES OF SCRAPERS.

The half-round scraper (Figs. 437 and 439) is used in conjunction with the round-

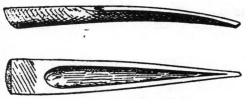

Figs. 437 and 438.—Bent Scraper.

nosed scraper on large and small bearings; the first has a straight motion, the second a circular motion imparted to it, and the cuts should be crossed to prevent scraping the same place too often. The bent scraper (Figs. 437, 438, and 439) is made from a half-round file, and in forging

Fig. 439.—Section of Bent Scraper.

it a flute is made up the centre with a fullering tool, and afterwards trimmed up with a file; the flute lessens the surface to be sharpened, and the edge remains true for a longer period. The three-cornered scraper (Figs. 440 and 441) is made from a file, and when the scraper is a small one the flute can be filed out.

SCRAPING VALVE SEATINGS.

The heel scraper (Fig. 442) has the point turned back at 45°, or to the required

Fig. 440.—Three-corner Scraper.

bevel, and is used on the seatings of screw-down valves, as in Fig. 443. When it is undesirable to remove the box the valve is put in the lathe and skimmed up; then red

ochre mixed with oil is applied to the valve, and the latter is rubbed on the seating a turn or two, when the high places are revealed by the red marks, and may be removed by scraping. When a fair

Fig. 441.—Section of Three-cornered Scraper.

bearing is obtained, the valve and seating is ground with flour emery and oil, the valve being lifted slightly, and dropped in a new position every few seconds to prevent grooving.

OTHER SCRAPERS.

Fig. 444 shows a hook scraper, and Fig. 445 a joggled scraper; these are used on the seatings of sluice boxes or on similar work where direct access is unattainable.

Fig. 442.—Heel Scraper.

Fig. 444 cuts with an inward stroke, and Fig. 445 with a pushing stroke. The cranked file (Fig. 446) is used in confined areas, and Fig. 447 illustrates an iron raised handle provided with a taper groove to fit on the tang of a file; it is useful when large flat surfaces have to be treated (see Fig. 448), and Fig. 449 shows another suitable handle.

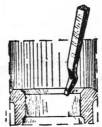

Fig. 443.—Surfacing Valve Seat.

HOW TO USE THE SCRAPER.

Fig. 450 shows the correct way to hold a scraper; the pressure is applied with the left hand, which regulates the amount of cut, and the right hand governs the length of the stroke, which varies from $\frac{3}{8}$ in. when starting to $\frac{1}{4}$ in. in the finishing stages. The rule is not to start scraping till a good

Fig. 444.—Hook Scraper.

face has been obtained with the file. A machine-planed surface, when tested on a surface-plate, will be found slightly out of truth, owing to a change in form after

Fig. 445.—Joggled Scraper.

releasing the chucking bolts, and from internal forces which are set up in cooling, and which to a certain extent are set free when the outer hard skin is removed by the machine tool. If the discrepancy warrants

Fig. 446.—Cranked File.

it, files are used till a fairly dappled bearing is shown by the red ochre. On starting to scrape, begin with the largest and hardest places, keeping the scraper low at the back; then wipe off quite clean, and

Fig. 447.—Raised Handle.

test again with the standard plate, the "marking," or red ochre, being thin and evenly distributed over the plate. Now begin to scrape the work in diagonal rows,

and at the next operation work on the opposite diagonal (see Fig. 451), the object being to break up the large and widely spaced patches into smaller ones, evenly spaced and close together.

Fig. 448.—Raised Handle on File.

COMPLETING THE SCRAPING.

As the work approaches this condition, it is especially necessary that the tools be in first-class cutting order, and the cut must be lighter and shorter in stroke;

Fig. 449.—Another Raised Handle.

great care must be exercised, or the work will appear to get worse instead of improving. Another point is the use of rouge instead of ochre, as the former is ground much finer and is therefore better adapted for the last stages. At the end, wet the tips of the fingers in the thick oil, which grinds up in sharpening the tools, and dab it on the plate; then give the work a good

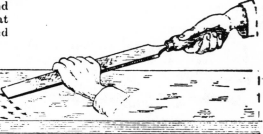

Fig. 450.—Holding Scraper.

rub, remove the plate, and wipe off clean and dry, taking care that not a particle of grit is on the work. Then rub the two surfaces together dry, and on examination it will be found that most of the spots are almost, if not quite, joined together, and when this stage of the work has been

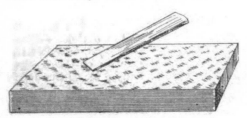

Fig. 451.—Scraping in Diagonal Rows.

reached the job may be considered complete. Often the tang of a converted file is joggled, as in Fig. 452, and this forms a most useful tool.

ARBORING.

Arboring implies the facing-off of bosses through which bolts pass, in order that the heads, or nuts, as the case may be, shall take a perfect bearing on the bosses. This facing-off is performed in several ways. The bosses may be planed or shaped, but that is not termed arboring. They may be faced in a lathe when the holes are drilled in the lathe, but that would be called facing. They may be arbored on the table of a drilling machine, the arbor being driven from the drill spindle, and that is machine arboring. Or they may be done by operating the arbor and cutters with a hand wrench, and that is the kind of arboring here meant. Facing or arboring of bosses by some means or another is necessary in all good work. If boss faces are left black, whether on castings or forgings, there is more or less inequality of surface, and the

Fig. 452.—File with Joggled Tang.

bolt heads, or nuts, do not take a uniform and perfect bearing all over. A certain amount of cross strain is then thrown on

the head and nut in tightening up, a strain which it is desirable to avoid. More than

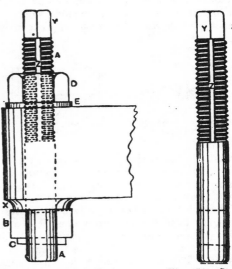

Fig. 453.—Arbor in Position for Facing Boss.

Fig. 454.—Stem of Arbor.

that, when bright heads and nuts are used, good appearance and finish demand that the bosses shall be bright too.

ADVANTAGE OF ARBORING.

One advantage of facing bosses with an arbor is that the holes being drilled first, and the arbor inserted and revolved in them, the cutters are bound to face the bosses exactly at right angles with the drilled holes. Should the holes be a trifle out of truth, the boss faces, being square with them, ensure a perfect bedding of the bolts.

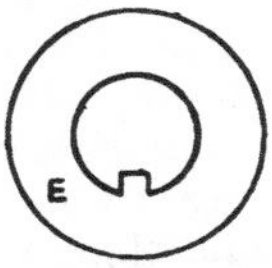

Fig. 455.—Washer.

Fig. 456.—Nib for Washer.

GOOD TYPE OF ARBOR.

A good type of arbor is shown in Fig. 453, in position for facing-off the boss x,

and details are illustrated to scale in Figs. 454 to 458, the arbor being of ¾ in. diameter.   For arbors of smaller or larger diameters the other dimensions would be made proportionate, with the exception

Fig. 457.—Washer   Fig. 458.—Arbor
  with Dovetail.   Cutter.

that the cutter should not be made thinner in any case.   The tool comprises the arbor proper A, the facing cutter B, secured with wedge C, the washer E, and nut D. The arbor itself (A) is preferably made of steel.   It is turned throughout its entire length, and then screwed with a ¾-in. Whitworth thread to the distance shown in Fig. 454.   The square end Y is made small enough to permit the nut and washer to come off the end, and it should be made of a size to suit one of the standard tap wrenches.   A narrow keyway Z (Fig. 454) has to be cut in the arbor continuous with the thread.   It will be from $\frac{1}{16}$ in. to ¼ in. wide, and a trifle deeper than the bottom of the thread.   It should be cut in a slot-drilling machine, but may be chipped out with a narrow chisel, and finished by filing. The nut D (Fig. 453) is a common ¾-in. nut, case-hardened.   The washer E (Fig. 455) is a trifle larger than the nut across corners, and from ⅛ in. to $\frac{3}{16}$ in. thick.   A nib is formed in the bore of the washer to enter into the groove or keyway Z in the arbor. This is made as a separate bit (Fig. 456), and fitted into a dovetail formed in the washer, as shown at Fig. 457.   The dovetail will taper from $\frac{3}{16}$ in. to ¼ in.   The nib will be filed so that the part below the dotted line in Fig. 456 fits tightly into the dovetail by driving.   In order that the nib shall not become loose, the edges of the dovetail are filed off slightly, and the nib, after being driven in, is riveted over these edges, being made just a trifle thicker than the ring to permit of this.   After riveting, the slight excess of metal is filed off level

with the faces of the ring.   The slot for the cutter B is made about $\frac{1}{16}$ in. wide by about 1 in. long, starting ⅝ in. or ½ in. from the bottom end.

ARBOR CUTTER.

The arbor cutter (Fig. 458) is made of tool steel, and makes a close sliding fit sideways in the slot.   The width is less, to permit of the insertion of the wedge C. The length is usually ¼ in. or thereabouts larger than the standard nut across corners.   It is filed to fit against the sides of the arbor, as shown in Fig. 458, so that no end movement can occur, the cutter always remaining central.   The edges are filed off on opposite sides to form suitable cutting angles.   The backing off is slight, as shown.   If it is excessive the edges will not stand much usage.

ANOTHER ARBOR.

Another form of arbor is sometimes used, in which the place of the nut D and washer E is taken by two common nuts, or a nut and lock nut, and no key groove is cut in the arbor.   The advantage of this is that it is more easily made than the form in Fig. 453.   But it is not so efficient, because there is no key groove to prevent possible slacking of the nuts.   In Fig. 453, when the arbor is turned with a wrench, the

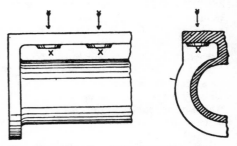

Fig. 459.—Inserting Arbor from Outer Face.

washer must also turn, and the friction of the nut and washer prevents any slackening back from occurring.

USE OF ARBOR.

To use the arbor shown by Fig. 453, the nut and washer are taken off, the arbor,

with the cutter properly wedged in, is
thrust up through the hole, the washer and
nut run on again, and the nut tightened
slightly. Then a wrench (double-handed if

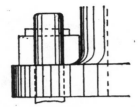

Fig. 460.—Arboring next to Radius.

space will allow) is put on the neck of the
arbor and two or three turns are given to it,
thus removing a little metal from the face
of the boss. Then the arbor necessarily
slackens and the cutter ceases to operate.
So a slight turn is again given to the nut,
tightening the arbor once more, and the
wrench is brought into requisition again, a
little more metal being removed from the
boss face, with consequent slackening of
the arbor. Thus these operations alternate until sufficient has been faced off the
boss. In some instances the arbor cannot
be inserted from that side of the hole which
has to be faced, as for example in Fig. 459
from the side x. In such cases the cutter
and wedge have to be taken out and the
arbor dropped in from the opposite face, in
the direction shown by the arrows. Then
the cutter and wedge are re-inserted, the

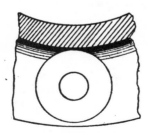

Fig. 461.—Arbored Face next to Radius.

nut and washer run on, and the operation
of facing proceeds in the way already
explained.

ARBORING CLOSE UP TO CYLINDRICAL BODIES.

It very often happens that arboring has
to be done in situations close up to the
body of cylindrical work, in which the

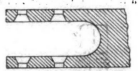

Fig. 462.—Arboring Countersunk Holes.

cylinder and flange merge into one another
with a good radius, as in Fig. 460. In such
cases there may be thin boss facings, or
only the plain flange. When there is only
the plain flange, it is still necessary to
arbor a circular face a little larger in diameter than the dimension of the nut or
bolt-head across corners, and the arboring
often extends into the hollow or radius, as
seen in Fig. 460. Then it saves time to
chip out the larger portion of the hollow
before commencing to arbor, otherwise for
several times the arbor will merely remove
a little bit of the hollow at each turn, and
be cutting wind during nearly half the revolution. The arbored face is shown in
plan in Fig. 461.

ARBORING COUNTERSUNK HOLES.

It sometimes happens that countersunk
holes have to be arbored by hand, as in
Fig. 462. Then the cutter is made like
Fig. 463, in which x y are the cutting

Fig. 463.—Cutter for Arboring Countersunk Holes.

edges. The directions previously given
hold good in these cases. Such cutters
operate much more rapidly than the flat
facing cutters, the angle being more favourable to cutting than the transverse

edges, and the cutter being of a smaller radius is stiffer and therefore more easy to operate.

ARBORING SHALLOW WORK.

In Fig. 453 the arbor is shown for facing through a hole the depth of which sensibly

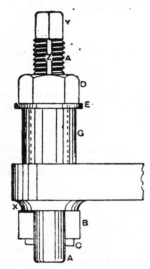

Fig. 464.—Arbor with Tubular Washer.

approximates to the length of the arbor itself—that is, the depth is within the range of the screwed portion, and therefore within the range of adjustment of the nut. But it frequently occurs that the arbor has to be passed through a hole so shallow that the nut cannot be screwed down on the top face of the work, because the screwed part does not extend sufficiently far to permit of that. It is not desirable to screw a greater length of the arbor than that shown, because a good

Fig. 465.—Flat Scraper.

length of plain-turned section is required for guidance and to preserve the points of the screw-thread from becoming damaged

by friction against the sides of the hole. For shallow work, therefore, tubular washers are inserted between the washer E and the face of the work, as shown at

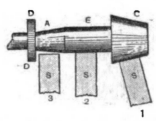

Fig. 466.—Flat Scraper in Use.

G in Fig. 464. These are simply plain lengths of iron tube. Neither exact length nor exact diameter is important. Instead of a bit of tube a large-faced nut is often substituted.

LUBRICANTS FOR ARBORING.

No lubricant is necessary in arboring cast iron and brass, but the arboring of wrought iron and steel is facilitated by the application of a little oil.

SCRAPERS FOR TURNED WORK.

The scraper is used by the engineer as a finishing tool for lathe as well as vice work. The forms used for lathe work are the flat, the bevelled, and the round-nosed scraper, shown respectively in Figs. 465, 467, and 469. The flat scraper, having a straight cutting edge at its end A, can be applied to all surfaces having a straight

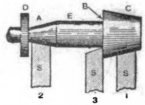

Fig. 467.—Bevelled Scraper in Use.

outline, whether the work is parallel or taper, providing there is no obstruction to prevent their proper application to the work. Thus, in Fig. 466 is shown a piece of work taper at A and C, parallel at E, and having a collar at D, the scraper S being shown presented to each of these sections.

Fig. 468.—Using Bevelled Scraper close to Arm.

The scraper is brought fair with the surface of C, by inclining to one side, as at position 1. It is held straight, as in position 2, for section E; but when it is presented to the taper at A, it will be found that its edge cannot be brought fair with the outline of taper A because collar D is in the way. But the bevelled scraper may be used throughout the whole piece, as shown in Fig. 467, turning it over when moving it from position 1 to position 2, while its edge may be used to face the shoulder B.

Using Bevelled Scrapers on Turned Work.

There are other cases in which the bevelled scraper must be used, as, for example, when it is required to be used close up to the work driver, or to an arm such as H in Fig. 468. As the scraper requires to be turned over, it is obvious that its end

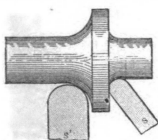

Fig. 469.—Round-nosed Scraper in Use.

should be ground straight across as denoted by the dotted line B in Fig. 465, so that both edges may be equally keen. The

round-nosed scraper is used for rounding and hollow corners, or may be made to conform to any required curve or shape. It is limited in capacity, however, by an

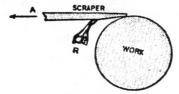

Fig. 470.—Scraper on Small Lathework.

element that affects all scraping tools—namely, if too great a length of cutting edge is brought into action at one time, chattering will ensue; and to prevent this the scraper is only made of the exact curvature of the work when it is very narrow, as at S in Fig. 469. For broad curves it is made of more curvature, so as to limit the length of cutting edge, as is shown at S', and is swept around the work so as to carry the cut around the curve. There are, however, other means employed to prevent chattering, and as these affect the flat as well as the round-nosed scraper, they may as well be explained with reference to the flat one.

Preventing Chattering of Lathework Scrapers.

First, then, by keeping the end A (Fig. 465) of the scraper thin, there is less surface to oilstone when sharpening it; this is important, because scrapers are sharpened several times on the oilstone before

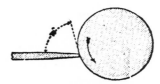

Fig. 471.—Scraper on Large Lathework.

being re-ground. But too thin an end A will cause chattering, it being found that it should not be thinner than about one-twentieth of an inch. Now, if the thickness of the scraper were no greater than one-twentieth of an inch throughout its length, it would vibrate very rapidly and therefore cause chattering marks on the work ; so it is made thick along the body, and is bevelled off just at the end as shown in Fig. 465 ; this gives a strong body and a small area to the oilstone.

Fig. 472.—Holding Scraper at Angle.

Scraping Small Lathework.

In small work it is brought to bear upon the top of the work, as in Fig. 470, in which case the strain falls endwise ot the scraper, as denoted by arrow A ; it is obvious that the scraper offers more resistance to vibration when the strain falls endwise than when it is applied to bend it. But if the work is of large diameter, the distance between the point of support of the scraper and the tool rest R and the cutting edge is too great, and the scraper is not sufficiently steadied ; so it is applied as in Fig. 471, the elevation of the scraer body lessening the angle between the front face of the work, the amount of elevation shown making this angle 75°. When the scraper is used upon the top of the work it is often presented at an angle to the work, as in

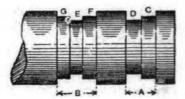

Fig. 473.—Turned Piece with Groove.

Fig. 472, and moved laterally to carry the cut along. This reduces the length of cutting edge to avoid chattering, and enables the surface already scraped to afford a partial guide to the tool in carrying the cut along. Fig. 473 represents a case in which two grooves of the respective widths at A and B are to be cut, and in this case chattering is prevented by using a narrow tool. The groove A is cut in two steps, D and C, while groove B, being wider, is cut in three strips, G, E, and F.

POLISHING METALS: THE MACHINES AND PROCESSES.

INTRODUCTION.

MANY articles of metal require a polished surface in order to give effect to the workmanship or ornament, or to enhance their value. Formerly this polish had doubtless to be obtained entirely by hand labour, and, considering the difficulties, it is surprising that such good effects were produced. However, equally good, and in some cases better, results are now obtained in much quicker time and with far less expenditure of energy ; although, unless great care is exercised, the beauty of the workmanship is damaged, the finer lines and marks being obliterated in the polishing process. Of course, many articles that are required to be highly polished are quite plain in shape, or have only large plain surfaces, and these may be polished much more easily than ornamental surfaces.

THE OBJECT OF POLISHING.

When metal articles are made by casting or forging, their surfaces are left in a more or less rough condition, according to the tools employed and the skill of the worker. The surface lacks lustre, because it is broken up into numerous little eminences and depressions of various shapes and sizes, which absorb the rays of light instead of reflecting them, and the business of the metal polisher is to produce a uniform reflecting surface. If the metal is clean and malleable, the eminences may be reduced by being pressed down with a burnisher, which is a tool having its surface harder and more highly polished than that to be reduced. Burnishing may be done after the metal is polished, to get a higher lustre on the surface, or it may be the process solely employed to polish the surface after its roughness has been reduced by filing, grinding, turning, planishing, or spinning. It is described later in the chapter.

IMPERFECTIONS OF WROUGHT SURFACES.

The rough condition of a metal surface may be first reduced by filing, thus changing it from a matt to a surface which will be found on close examination to be full of small grooves and scratches, the sides of which reflect the light in proportion to their regularity. A similar condition is obtained by turning in a lathe, and by planing. The rough surface may also be removed by grinding on an emery wheel, and the scratches will be finer still if the surface has been ground on a fine grit grindstone kept wet with water. But the smoothness of a metal surface got by filing, turning, and grinding is affected by the closeness and hardness of the grain of the metal. If this is soft, the particles will break away from each other, and will thus leave the surface in a rough, matt condition. If the grain is coarse and spongy, it will be impossible to get a smooth surface by any means, the grains showing numerous tiny pin-holes after the rough surface has been removed. Coarsely cast bars of copper and silver intended for wire will show the coarse nature of the metal even when drawn to a very fine size, and this defect will frequently cause the wire to break before it can be drawn fine. To get a highly polished surface on metal, therefore, it must have a close, fine grain and be moderately hard.

METHODS OF POLISHING.

The business of the metal polisher is to reduce a matt surface to a grooved surface, and to reduce the size of the grooves to the very smallest obtainable. To do this,

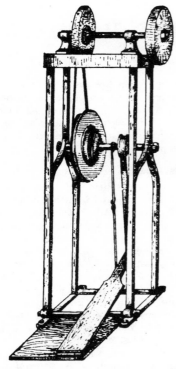

Fig. 474.—Treadle Polishing Machine.

a portion of the surface must be removed by means of suitable tools, or by an abrading material selected, made to suit the metal. For iron and steel, the various grades of emery are employed, starting with a coarse grain and finishing with fine flour emery. If a finer polish is desired, steel is rubbed with rouge or putty powder. Bronze, brass, German silver, and similar hard alloys, may be first rubbed with emery, then with the coarser grades of tripoli, then with finer tripoli, and finally with rouge. Silver, gold, copper, and the softer metals may be reduced by the finer grades of tripoli and of rouge.

HISTORY OF METAL POLISHING.

The history of metal polishing may be divided into three distinct eras. The first embraces all the processes for obtaining a polished surface by rubbing with tools held in the hand, and by means of loose, gritty powders, such as silver sand, Trent sand, Bath brick, pumice-stone, whiting, rotten stone, lime, puttypowder, and rouge. These powders were applied on sticks covered with basil, buff, or chamois leather, or with felt. In the next era were

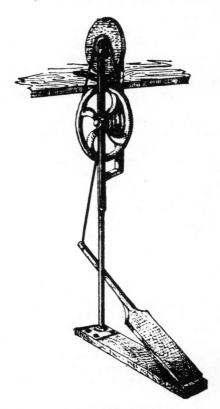

Fig. 475.—Bench Polishing Machine.

introduced revolving discs of various materials, charged with one or more of the loose powders above enumerated, which were made to press against the surfaces to be polished ; and the third and modern era has seen polishing machinery and materials greatly improved, resulting in better workmanship, more rapid processes, and reduced cost.

METAL POLISHING MACHINES.

In order to produce a polish, the first process is by file, scraper, or some similar

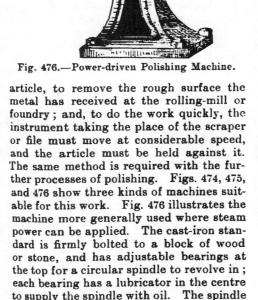

Fig. 476.—Power-driven Polishing Machine.

article, to remove the rough surface the metal has received at the rolling-mill or foundry ; and, to do the work quickly, the instrument taking the place of the scraper or file must move at considerable speed, and the article must be held against it. The same method is required with the further processes of polishing. Figs. 474, 475, and 476 show three kinds of machines suitable for this work. Fig. 476 illustrates the machine more generally used where steam power can be applied. The cast-iron standard is firmly bolted to a block of wood or stone, and has adjustable bearings at the top for a circular spindle to revolve in ; each bearing has a lubricator in the centre to supply the spindle with oil. The spindle has fast and loose pulleys in the centre, and at the ends devices to hold the emery-wheel, grindstone, buff, or dolly, as may be required. Fig. 475 illustrates a simple machine that may be fixed to an ordinary bench and worked by foot. A treadle device is fixed to the floor and carried by a standard having a plate to secure it to the bench and an upper pillar to take the wheel, buff, or dolly. This is the most simple construction that can be adopted for the work. The wearing portions are

few in number, and can easily be replaced, but considerable speed, say 3,600 revolutions per minute, is easily obtained. Fig. 474 illustrates a grinding or polishing machine, having the same treadle movement fixed between standards, and with a small bench at the top, on which two bearings for the spindle are fixed.

EMERY BUFFS.

For the first grinding process emery discs are often used for iron or steel, especially when very rough surfaces have to be dealt with. For brass, copper, or the softer metals, buffs, made in various ways, are better. Fig. 477 illustrates the more usual plan. A piece of wood is cut out in circular form, or, better, four pieces of wood are glued together with their grains crossing and then cut out. Round the circular edge is securely fixed a thick piece of buff leather. The centre hole carries a brass boss that is screwed to fit the end of the spindle, or has a plain hole to fit on the spindle between the collars, as shown in Fig. 476. This buff is prepared for use in the following way: It is painted on the leather edge with hot glue and made to revolve in a tray containing the required emery until it has received an emery surface ; then, when cool, it is ready to be placed in the machine. The article to be faced and polished is then held against it as it is made to revolve swiftly. Fig. 478 shows the

Fig. 477.—Leather Buff.

construction of a leather-covered buff of large size. The wood is of wedge-shaped sections, securely fitted and fixed together and held by brass or iron rings ; the outer edge is covered with leather as before described. Fig. 479 shows a solid leather buff with a rim of thick leather, either bullneck or sea-horse, which may be turned in the lathe to suit various shapes of mouldings and ornaments (see Fig. 480). The central holes in these buffs screw on the point of the spindles shown in Fig. 476. The article to be polished is held against them, and the emery powder held at the

Fig. 478.—Large Leather Buff.

point of contact. Fig. 481 shows a pressed felt buff, which is used in place of leather buffs in many cases, and more often for the further polishing with crocus or rottenstone. It is fixed to the spindle in the same way.

MATERIALS FOR POLISHING.

The materials employed with buffs and dollies (dollies are described on p. 139) may be now referred to. Emery powder of various qualities is used to get the first polish face on the article. Rottenstone and flour emery together make a good material for polishing iron and steel. The materials should be formed into a cake or slab by mixing with soiling suet, and then running into square or oblong moulds, and using cold with the buffs. Crocus and rottenstone mixed in the same way with a little rouge form the best combination for brass or copper and kindred materials. Calico dollies and mops are used with the powders dry to give the finishing effects.

WHAT EMERY IS.

Emery is a well-known compound, which sharpens tools, polishes plate glass or lenses, bright steel, and other metals. As a polishing agent, it is no novelty, for the ancient Greeks used and valued it and called it *smyris*, from which term most European nations have derived their name for the material. Naxos, one of the isles

of the Grecian Archipelago, was for long the sole source of emery, where it was worked by a lessee under contract with the Government of Greece. Other sources have been found. In Asia Minor there are abounding deposits of emery. The Naxos island lacks good harbourage, and the mineral has to be shipped from Syra and Smyrna. The distribution of emery in nature is peculiar. It is found on mountain tops in the vicinity of Ephesus and in the valleys.

WORKING EMERY.

Emery exists in nuggets, great and small, loosely embedded in reddish soil, and also in the form of nodules in crystalline limestone or marble. Masses weighing 30 tons to 40 tons are found, and pieces very much smaller, which are worked by simple exposure of the surface, or by means of shafts and galleries. The blocks are broken by sledge hammers, but sometimes they must first be roasted and cooled to make disintegration possible. The pieces then are carefully hand-picked and sent from Asia Minor by camel and donkey to the railroad, and thence to Smyrna, to be carried as ballast in the sailing ships light-loaded with cargoes of liquorice root.

QUALITIES OF EMERY.

Emery contains 70 per cent. of alumina and from 8 to 25 per cent. of oxide of iron, and has a specific gravity or density of 3·75 and 3·98 for the Naxos and Samos qualities respectively. The native alumina, Al_2O_3, is corundum or adamantine spar, a substance second only to the diamond in hardness. Corundum is a term applied in its

Fig. 479.—Sea-horse or Bull-neck Buff.

wide sense to embrace adamantine spar, sapphire, and emery. In emery the alumina does not crystallise into the rhombohedral shapes typical of its other forms.

Again, emery presents the dull or opaque variety. When blue and transparent corundum is sapphire, when red it is ruby or oriental ruby, and in other tints the same mineral becomes oriental topaz, amethyst, or emerald. The natural colour of emery varies from bluish-grey to brownish. The rich red chocolate tone of much of the emery of commerce is due to artificial colouring. Other hard substances are employed as adulterants, garnet and iron slag with others. It is noteworthy that French manufacturers employ always the original Naxos emery for preference. The blocks as imported are stamped by ore-crushing batteries, and if needful they are first made hot in the fire. To be useful the emery must be reduced to powder, and of this there are many grades, ranging from grains as large as mustard seed down to the superflour emery, which is deposited last in elutriation with water. The finest emery is naturally destined for the most delicate purposes. The makers of optical glasses are most exacting in their requirements.

Testing Emery.

The utility of emery rests in its abrasive power, which is tested in a practical way by its action on plate glass. The pulverised emery is rubbed on a weighed plate of glass by means of an agate muller, and by comparing weights at various stages a very direct guide as to the hardness of the material is obtained. Taking blue sapphire as the standard unit of hardness, it is easily possible to denote in terms of figures the merits of any given sample. Pure corundum shows a durability of from

Fig. 480.—Sections of Solid Leather Buffs.

90 to 97 per cent., while that of emery, according to its sources, may vary between 40 and 60 per cent.

Utilising Emery.

Hones and wheels of emery present emery in a form suitable for use. They are made by incorporating the powder with some plaster material, and moulding and hardening it. Shellac, paper pulp, artificial stone, or loam and water, so baked

Fig. 481.—Solid Felt Buff.

as to constitute a rough terracotta, are all means that are employed. Emery-cake, for use on buffing wheels, may be made by simple kneading in warm beeswax which has been chilled by a plunge into cold water. Emery sticks having a core of wood are dipped first into glue and then into emery powder. Repeated coatings improve the sticks. Emery paper or cloth is fabric coated with glue and dusted with the powder from a sieve. The grains of emery for this purpose are in some six sizes, varying from 30 to 90 mesh in an inch, to suit the requirements of different purposes.

Obtaining Fine Emery by Elutriation.

It will have been noted that emery is the chief of the polishing media. Lens workers use the best emery, and there are many jobs in metalworking also where none but the best should be used. A little of the finest emery on the oilstone would assist in quickly putting a keen edge on woodworking tools. The coarser grades of emery can be bought in bags of 7 lb. weight, with the size of hole marked—that is, the mesh of the sieve through which it is put—but the finer grades are washed from the residue from the roughing processes in large firms. The best sources to get the slush required for the emeries are bottle-stopper grinders; they have no further use for the emeries when once used, which can always be bought for a few shillings per hundredweight. Put the slush in a large pan half full of water, break up all the lumps that may have formed, and put it through a wire sieve of about 126 holes to the inch. The wire gauze can be stretched over a wooden ring, and tied round. Sift the emery and part it—that is. stir it well up—and pour it almost immediately into another pan. By this means all the coarsest emery will remain in the first pan; this can be put aside for grinding tools, etc. Now part again, first with seven seconds between stirring and pouring, and then with fourteen seconds. These two emeries will be fine. Among the finer emery in the pan will be a quantity of "mud" containing powdered glass, emery too fine to cut, etc. This must be washed away, so as to leave the fine emery as pure as possible. This can be done by stirring up the emery and water, and vibrating the pan by knocking it with the closed fist, as in Fig. 482, and pouring off every ten minutes or so, using fresh water every time. Do not pour down too low, or the finest emery may go with the mud.

Fig. 482.—" Knocking Down " Emery.

BARREL METHOD OF PARTING EMERY.

Another way is to get a barrel, into which the emery is put after it has been passed through the sieve and given a good stir up, having previously made a series of holes down the barrel (as Fig. 485). It is obvious that opening the holes, beginning from the top, produces a variety of grades, but this method is not so trustworthy as pouring each one separately. The emery when placed in pots should be numbered according to its fineness, and kept covered.

OTHER WAYS OF PARTING EMERY.

Another way is to let the water dribble on the pan, as in Fig. 483, stirring occasionally, and allowing the mud to float off. When the water is pretty clear after being allowed to settle a few minutes, a finer

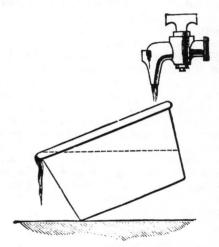

Fig. 483.—Dribbling Mud from Emery.

sieve can be used. Fine muslin, nainsook, or cambric should be tied round the sieve that was used before. Be sure that it is quite clean, and that all the emery in the pan is put through. Do not force it, but knock it gently through, as in Fig. 484. Now it can be parted again three or four times, allowing a longer time, from one to ten minutes, between each pour.

THE BEST WORKING EMERY.

It is not the finest-looking or finest-feeling emery that puts the smoothest surface on the work; a well parted one—that is, one with a homogeneous grain, will be better than many emeries that look and feel finer; having a more perfect grain, it will work down cleaner and farther than a finer emery that is not so well parted. which will drag, owing to its grains being of different size.

SPEED OF BUFFS, DOLLIES, ETC.

The mops, dollies, buffs, and bobs should be run at a surface speed of from 4,000 ft. to 6,000 ft. per minute. Thus, if a bob 12 in. in diameter be running at 2,000 revolutions per minute, as its rim measures quite 3 ft. its speed will be about 6,000 ft. per minute. An 8-in. bob has a rim about 2 ft. long, and its speed on the same spindle would be about 4,000 ft. per minute. Therefore, a speed of 2,000 revolutions per minute meets all ordinary requirements with bobs, mops, buffs, and dollies of from 8 in. to 12 in. in diameter, but for smaller diameters the spindle should be run at a higher speed. It has been found that less pressure is required on bobs of large diameter run at the above speeds than on smaller appliances at low

Fig. 484.—Sifting Emery.

speeds, and, as both pressure and speed are necessary to get best results, high-speed bobs of medium diameter are the most economical. These polishing appliances are run with their upper surface coming toward the operator, who presses the work against the under side. A hood or guard should therefore be placed over the running bob to prevent the abrading material being thrown in the face of the worker, and a tray should be placed beneath to catch the waste particles.

ANOTHER METHOD OF MAKING BUFFS AND BOBS.

Another method of making a bob may be here described. Two discs of well-seasoned beech, of the required diameter and half the thickness of the desired disc,

are glued together side by side, with their grains opposite, and then left under pressure until quite firm. The disc is then turned true, and a number of small grooves are cut in its face. A strip of thick felt

Fig. 485.—Barrel Method of Parting Emery.

of the required width and length is laid out on a bench, the rim of the wood disc is run in some good hot glue until well coated and warm, and is then run along the felt under pressure, the felt being made to cling close to the wood, and afterwards secured to it with long tacks. When the glue has become firm and hard, the tacks are withdrawn, and wooden pegs, dipped in glue, driven into the holes; then

Fig. 486.—Calico Dolly or Mop.

Fig. 487.—Circular Brush.

the whole is trimmed true with a sharp knife. The felt-covered disc is rolled in hot glue, and then over a layer of emery spread on a board or held in a tray, until well coated with emery. The emery employed varies with the work to be done. A bob for the first roughing down is usually coated with No. 60 emery. Then follow bobs coated with Nos. 80, 120, and 140, the last being used to get a finishing polish on steel before employing a mop or dolly. See that the glue penetrates the felt all over before it is rolled in the emery, then apply pressure by means of a stick in the centre

a box, tray, or bench is placed on the other side. As each article gets too hot to be held it is thrown into the box, and another article is reached from the bench, so treating all in succession, and going over them again after they have got cool in the box.

Fig. 488.—Spindle for Mops and Dollies.

Fig. 489.—Spindle for Mops and Dollies.

hole whilst rolling the bob, to get the emery well into the glue. When the glue and emery is quite hard and firm, the bobs are ready for use. [Emery-coated bobs thus prepared cut iron and steel with great rapidity. The friction generates heat, and the article becomes polished, this action being intensified by the pressure.]

EMERY TAPE MACHINE.

For cycle frames and handle-bars, an emery tape machine is preferable to a polishing lathe and bob. The emery-coated tape or band is run over pulleys, and the bars are held to it whilst running; by this means every part of the bar can be reached easily and be speedily polished.

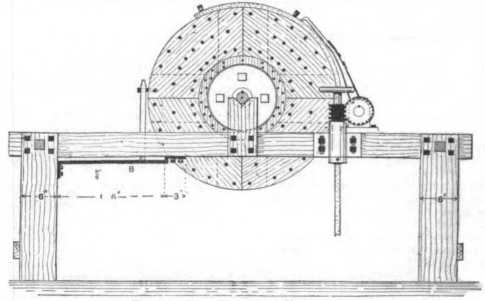

Fig. 490.—Front Elevation of Glazing Buffer.

POLISHING WITH BUFFS.

To polish a number of articles quickly, arrange them on a bench at hand, whilst

MOPS AND DOLLIES.

To eliminate the scratches left by emery bobs and tapes, another class of revolving discs, named mops or dollies (see Fig. 486), are employed. These are made of several circular pieces of calico, clipped and securely sewn together. The pieces of calico are clipped together by brass flanges having a threaded centre to fix on the spindle. Where the article to be polished is very ornamental and the work very fine, a circular brush (Fig. 487) may be used. This brush may be of fibre or mixed fibre and bristle. Dollies are charged with fine flour emery in the form of a composition— that is, emery mixed with tallow, or fine flour emery and oil. Figs. 488 and 489 give alternative forms of spindle, and show the various ways in which the wheels, buffs, etc., may be fixed to suit the work required to be done.

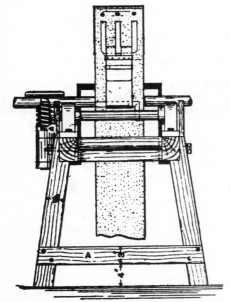

Fig. 491.—End Elevation of Glazing Buffers.

FINAL POLISHING.

The final polish to brass and similar soft metals is given with tripoli in its various grades, and with rouge on calico mops, these also varying in coarseness or fineness with the required finish of the work. Articles to be nickel-plated should be finished off smooth, so as to leave little polishing

afterward, because deposited nickel is very hard and not easily rubbed down with mops. The final polishing is done with fine grade tripoli and with Sheffield lime.

POLISHING PLATED METALS.

Silver deposits leave the vat in a matt condition, resembling white unglazed porcelain. The rough surface is first reduced

Fig. 492.—Marking out Wheel Body.

with brass wire brushes kept wet with stale beer, then washed and dried. It is then polished with mops charged with rouge compo., and a higher polish is got with the hand charged with fine rouge. Gold deposits are also brushed with brass wire scratch-brushes, then dried, and polished with mops charged with rouge.

MAKING AND MOUNTING A GLAZING BUFFER.

The buffer or buff, as has been explained, is a round wooden wheel built up of pine and covered on the edge with walrus hide and emery; its use is to give a fine polish to iron. A mounted buffer is illustrated in side view and end view by Figs. 490 and 491. A very useful size of buffer is 2 ft. 7 in. diameter and 8 in.

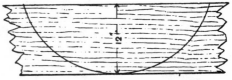

Fig. 493.—Marking out Wheel Body.

broad, and for this the materials needed will be one board 9 ft. by 8 in. by 1¼ in., two boards 9 ft. by 12 in. by 1¼ in., and five boards 8 ft. by 11 in. by 1⅜ in., and plates, bolts, etc., to be described later. From the 8-in. board set out three pieces as in Fig. 492, and from the 12-in. board set out six like Fig. 493. Then saw these boards out roughly to the lines. For setting out the board, draw a 2 ft. 8 in. circle on a sheet of brown paper, and 4 in. on

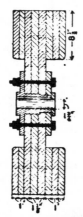

Fig. 494.—Section of Glazing Wheel.

each side of the centre line draw straight lines. When the timber is sawn, plane both sides down to 1⅛ in. thick, and join one of Fig. 492 to two of Fig. 493. Shoot the joining edges and mark them for reference. By using all the prepared boards three complete discs will be formed. Take one of these discs and glue the edges where joined, rubbing the joints to bring them close. Then cramp them together and knock three iron dogs across each joint, one leg in one board and one in the other, taking off the cramp when the glue is set. Then remove the dogs and lay another circle, placing it across the grain. Fasten these together with screws, which must not come through. Place dogs all round the

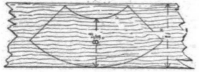

Fig. 496.—Pattern for Segment of Wheel.

edge about 1 ft. apart : then turn over and fix the other disc in the same way, and, when dry, remove all dogs and screws. The grain is crossed so that the boards

will not spring when dry. Next mark the centre of the disc. On circles of 2 ft. and 1 ft. diameter, respectively, bore ¾-in. holes

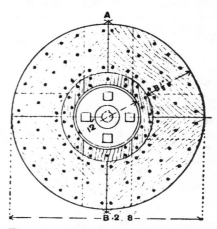

Fig. 495.—Elevation of Glazing Wheel.

right through, about 6 in. apart. Then glue some pine pegs and drive them into these holes, and cut them off level when the glue is set, as shown in Fig. 494 ; an elevation of the buffer wheel is shown by Fig. 495.

THICKENING BUFFER.

Now on a sheet of brown paper describe a circle 2 ft. 8 in. diameter, and another 1 ft. 3 in. diameter. Then draw a diameter, and, from its ends as centres, strike intersecting arcs (A B, Fig. 495) having a radius rather larger than that of the circle first drawn, and join the two intersecting

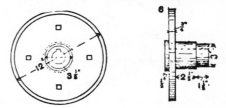

Figs. 497 and 498.—Wheel End Plate.

points by a line so as to divide the circle into four equal parts. Then cut the paper into segments (Fig. 496), only one of which will be needed. From the five boards

1⅛ in. thick set out twenty of these quarter circles, and saw them out and plane them to 1 in. thick. Shoot their ends true and mark each for ready reference. Then take four of these quarters, glue them on the large disc, and screw them down, knocking dogs around the edge. Do the same on the other side of the disc, being careful not to place one joint over the other. When dry, remove all screws and dogs, and fasten four more quarters, keeping the joints separate. In this way fasten two layers on one side and three on the other, so that the two nuts on the shaft that hold the buffer will not project too far. The glue having set, remove all screws and dogs and bore two rows of ¾-in. holes about 6 in. apart on circles, as shown by Figs. 494 and 495, and drive in pegs.

COMPLETING BUFFER WHEEL.

Bore a hole through the centre 3⅜ in.

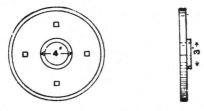

Figs. 499 and 500.—Wheel End Plate.

diameter to receive the boss of the cast-iron plate (Figs. 497 and 498). This plate should be bored for a 1¾-in. shaft, slotted right through for a ½-in. key, and turned on the boss as shown. The other plate (Figs. 499 and 500) must be bored 3 in. to slide on the other easy. Place the plate with boss through the holes in the buffer, and mark the positions of the bolt holes. In the wood, bore ½-in. holes for bolts, and fasten the whole with lock-nuts and washers, as shown in Fig. 494. Four ½-in. bolts, 6 in. long, will be required for this. Now drive a mandrel into the hole, and, in a lathe, turn the sides first to a width of 8 in. If the boss projects through the loose plate, turn it level, and then turn the edge of the disc, making it 2 ft. 7 in. diameter. The wooden buffer is next

covered with walrus hide. This varies in thickness, but the best for the purpose is ¾ in. thick. One hide turns the scale at 150 lb., and is about 11 ft. long by 5 ft. broad, the price being 6s. per lb., or £45

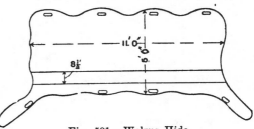

Fig. 501.—Walrus Hide.

per hide. A strip 8½ in. broad must be cut with a saw as in Fig. 501.

TRESTLE FOR GLAZING BUFFER.

A trestle will be required. To make this trestle, take two pieces of deal, 11 ft. long, 4 in. broad, and 3 in. thick ; saw 6 ft. 6 in. off each length, and make the remaining pieces 1 ft. 4 in. long. Then on the 4-in. sides of the long pieces mark at both ends lines 6 in. from the ends, and then other lines 3 in. from these ; saw down these lines for 1 in. and cut out, so that the short lengths will go in 1 in. Then glue and nail up as in Fig. 502. For legs, use four pieces 2 ft. 2 in. long by 6 in. broad and 3 in. thick. Mark a line 4 in. from one end, and saw the legs so that they will splay out when fastened in place, as shown in Fig. 491. In the 6-in. side bore a hole 2 in. from the top for a ⅝-in. bolt ;

Fig. 502.—Plan of Trestle.

then nail the legs in place and bore through the side pieces so that the bolts may go close against the cross-pieces on the outside of the frame. When this is bolted up tight, nail across the legs pieces of timber A about 1 in. thick and 3 in. broad (see Fig. 491) to strengthen the legs. Two wrought-iron slides (see Fig. 490) will

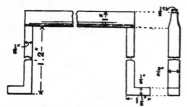

Fig. 503.—Iron Frame.

be required, with four holes for 1¼-in. screws, so tnat they can be fastened as shown. These support another wrought-iron piece (Fig. 503), which holds the buffer so that the stretcher can pull at the other side. The top part must be twisted over, so that the flat will come against the buffer and the steel pegs will go under the edge.

LEATHER HOLDER AND STRETCHER.

The other end of the trestle will need a contrivance to hold and stretch the leather. This is a shaft 2 ft. long and 1½ in. diameter; a hole must be drilled, ¾ in. tapping, 6½ in. from one end. In this hole tap a piece of wrought iron, ¾ in. diameter and 9 in. long, threaded and riveted in place at one end, and bent as shown at c in Fig. 504. Two collars will be needed, as shown, and must be drilled for ½-in. set-screws; where the screws pinch the shaft it must be countersunk, so that the ends of the screws will enter it. This must be done after the shaft has been put into place. A

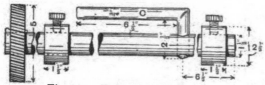

Fig. 504.—Tightening Shaft.

worm-wheel about 5 in. diameter must be keyed on the shaft.

SHAFT OF GLAZING BUFFER.

Two cast-iron pedestals (see Fig. 505) are needed, with a hole on each side for

coach screws. With 4-in. coach screws, fasten one of these on the trestle about 1 ft. 6 in. from one end, as shown in Fig. 490. Then take the shaft, put on one

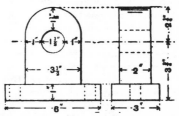

Fig. 505.—Bracket for Tightening Shaft.

collar, place one end into the fixed pedestal, put on the other collar, put the other pedestal on the shaft and then on the trestle, and fasten it down. Put the shaft at the right-hand level with the outside of the pedestal, push up a collar, and fasten it; then push the other collar against the other pedestal and fasten it there, leaving just enough play for the shaft to turn easily by hand. The long end of the shaft will be at the left hand when standing at the end of the trestle. A cast-iron bracket with two long slots for ½-in. bolts, as in Fig. 506, will be wanted. The 1½-in. shaft passing through this must be long enough —namely, 12 in.—to take a collar fastened on the bottom side of the bracket with a

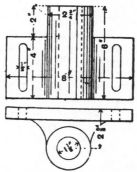

Fig. 506.—Bracket for Vertical Axle.

grub screw; a grub screw has a slot or nick but no head, so that it can be driven in level with the outside of the washer. Resting on the top side of the bracket, a worm must be keyed on the shaft to gear into the worm wheel, and so high that the fingers, when turning the hand-wheel, miss the teeth. The hand-wheel, of about 6 in.

drive through the leather into the wood for 2½ in. The thin pegs are for the outer edge of the leather and the thick for the centre.

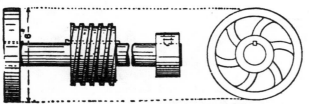

Figs. 507 and 508.—Axle with Driving Wheel.

diameter, must be keyed on the shaft. The axle with driving wheel is shown by Figs. 507 and 508.

STEEL CLAW AND PEGS.

A steel claw like Fig. 509 pulls the leather tight when the shaft is turned. A piece of leather 4 in. broad and about 4 ft. long is required; it should have a large buckle, so that if the claw is too far on the top of the buffer it can be shortened or lengthened. In the centre of the strap and 5 in. from the buckle, a hole must be made for the middle prong. Two wood blocks with **V**'s cut out at top (see Fig. 490), and standing 6 in. above the top, must be fastened on the middle of the trestle with four coach screws in each. The shaft of the wooden buffer runs in these while the leather is put on. For a buffer 2 ft.

FITTING LEATHER OF GLAZING BUFFER.

Soak the leather in clean water and, to see which way the grain runs, cut a thin bit of the outside and rub it down with the back of the knife; the right way will show smooth. Be careful to mark which way is right, as it will tear if put on the wrong way. The wooden buffers run in pairs, one on each end of the shaft, with the driving pulleys between. The buffers have the deeper parts of their hollow sides outwards. They run so that anything held on the top is pulled away from the operator; therefore the leather should smooth down towards him. Get a shaft long enough to go through the buffer and over the **V** pieces; it need not fit the hole in the glazer if it is strong enough. Standing at the end carrying the tightening motion, knock three pegs into the edge of the buffer in a row; turning it round, put the three pegs under the sliding piece; glue the leather for a foot, having the

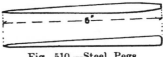

Fig. 510.—Steel Pegs.

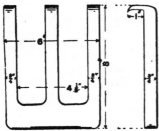

Fig. 509.—Steel Claw.

7 in. diameter, 600 or 700 steel pegs (Fig. 510) will be wanted; 420 will be $\frac{3}{16}$ in. thick at top under flat, and the rest will be $\frac{3}{16}$ in. thick and made sharp, so that they will

smooth way towards you when at the head of the trestle. Knock five pegs through the leather about 2 in. into the wood, using two of the $\frac{5}{16}$-in. pegs and three of the $\frac{3}{16}$-in. pegs. The thin pegs must be driven so as to slant towards the centre, one on each side $\frac{3}{4}$ in. from the edge, and the other three at equal distances, making five in a row. Then, seeing that the three pegs are against the sliding piece, put the strap on the ¾-in. staple, take the claw with the strap on the staple, and pull tight; it

Fig. 511.—Wooden Peg.

should reach about a foot from the five pegs. Drive the ends into the leather with a hammer; then begin to turn the hand-wheel. This will draw the staple round and will pull the strap with the claw on tight. Then, if the three pegs are against the slide, the buffer cannot turn and the leather will be stretched. Knock five more pegs in, being careful to place them where the leather is glued. Now slacken the claw, and take it off; turn up the hide, glue another foot of leather, putting plenty of hot thick glue on. Then turn over the leather again, turn the glazer around until the five pegs go under the sliding piece, and knock the claw in the leather again and tighten up; knock five pegs in, pull the claw off, and proceed in the same manner all round until covered. Bring the ends of the leather as close as possible, then take off the claw and get on the top of the trestle and commence to drive pegs in, five in a row, thin ones on the outside and thick ones in the centre. Any pegs broken in the buffer must be removed. Drive the pegs 2 in. into the wood, with about 1½ in. between each row, until this is done all over. Then place the buffer away for three or four days in a warm room to allow the glue to set.

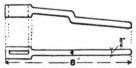

Fig. 512.—Handle for Steel Pegs.

Completing Glazing Buffer.

From ½-in. square soft timber with straight grain, cut 600 or 700 pegs (Fig. 511). Next make a steel handle (Fig. 512), with a slot in one end to fit on the flat part

at the tops of the steel pegs. Stand on top of the trestle, and put this key on the steel pegs; turn them round, at the same time giving them a pulling-up motion. This will release the pegs; do not pull out more than a foot at once, or the leather will spring up. Then dip the wooden pegs in thick hot glue for about 1 in. at the sharp end, and, with a hammer, drive them in the holes where the steel pegs have been. When this is done all over, leave the buffer to dry for three or four days. Then turn the surface of the leather in the lathe, truing the edges so that they project ¼ in. on each side it will then be ready to put on the glazer stands. To balance the buffer, fasten a thin strip of lead in the hollow where it is lightest. Then run it on its own shaft in place, and hold a piece of chalk against the leather;

Fig. 513.—Tumbling Drum.

where the chalk marks, it must be scraped with a knife made out of a half-round file ground to a sharp edge. Keep chalking and scraping until it chalks all round. Then take off the shaft, and place the buffer on a bar of iron driven into the wall. Damp the leather all round with lukewarm water, and then rub thick hot glue all over the leather. The buffer must be rolled in a box about 8 ft. long and 1 ft. wide and 1 ft. high, filled about a quarter of the way up with emery. This covers the buffer with emery where glued. Next place the buffer on the bar to dry; it will then be ready to use.

Polishing Metals in the Tumbling Drum.

The abrasive substances for this purpose are: river sand, emery in various degrees of fineness, crocus, putty powder (oxide of tin), Vienna lime, etc. When a coarse powder has done its work, it must be entirely removed from the article before a finer powder is applied. The reason for this is evident when it is remembered

Fig. 514.—Straight Burnisher.

that polishing means substituting fine scratches for coarser ones, until the fine scratches become so fine as to be imperceptible. Therefore, if all the grades of emery are used, every bit of the coarse quality should be washed away before the finer is used; and the finer, after it has done its work, is to be removed before crocus, putty powder, or lime is applied. Each part of the article must be subject to the application of the powders under friction, until a perfectly regular surface is obtained. If it is required to polish chains of a simple pattern, such as plain, oval, or round links, and with no long, flat, rod-like parts to them, one of many methods may be selected. The revolving brush, dolly, or bob, made of either cloth or leather, can be charged with the polishing powders, as described previously in this chapter, and the work can be held in the hand, so as to present each part in turn to the action of the polishing material.

THE TUMBLING DRUM.

The most usual way, however, is by using what is called a tumbling drum. This varies very much in form and in interior arrangement, but the diagrammatic sketch

Fig. 515.—Curved Burnisher.

shown by Fig. 513 will give the idea. It is a barrel, with a door in it, which runs eccentrically on a spindle, the spindle being driven either by hand or steam. Into this drum or barrel is placed some river sand, moistened by dilute sulphuric acid, and then the articles, leaving about one-quarter of the space empty. When the

drum is revolved the articles are continually being shifted, so that every part of the articles becomes polished by the friction they undergo as they are tumbled about. When they have been carried as far forward as possible in this coarse polishing material, they are next placed in another similar drum, with leather-waste, charcoal, emery, and oil, and are finished in a third drum with Vienna lime.

Fig. 516.—Curved Burnisher.

OTHER METHODS OF POLISHING.

Another way of attaining the same result is by using a long bag, like the leg of a stocking, in which the chains and emery, etc., are put, and which the ends tied up. The friction is obtained by shaking the bag in such a way that the chains are thrown rapidly from one end to the other.

BURNISHING.

There is still a process that has been merely referred to but not described, and that is burnishing. A burnisher is a piece of hardened polished steel, and is applied by friction and pressure to the work, which must be very smooth. Burnishing is a method employed chiefly with metals on which a coat of a superior metal, gold or silver, has been deposited electrically. An electro-deposited coat of silver of moderate thickness presents a dead white surface as it leaves the plating solution, unless this has been made to deposit bright silver by using a brightening solution. This surface is entirely devoid of lustre, and appears, when viewed through a powerful magnifying-glass, to be made up of a fine

Fig. 517.—Curved Burnisher.

network of silver grains cemented together. These grains absorb the light, and cause the deposit to assume a dull surface, known as "matt." It closely resembles frosted silver, and has a pleasing appearance when freshly deposited, rinsed in clean hot water, and quickly dried in a room free from dust. When this "matt" surface is scratch-brushed, it loses, to a certain extent, its peculiar dead whiteness, but no amount of scratch-brushing and polishing will give it the mirror-like surface so much desired in silver plate. To get a highly polished, reflecting surface on

Fig. 518.—Round Burnisher.

electro-plated goods, they are burnished after they have been well washed, scratch-brushed, and dried. Though the following instructions will have especial reference to the treatment of plated goods, they are also applicable to any metal work sufficiently soft to be affected by a hard burnisher.

Fig. 519.—Hooked Burnisher.

BURNISHERS.

Burnishers in their least expensive form are made of steel blades varying in shape, running into the wood. Straight burnishers, shaped as shown at Fig. 514, and in section at A, are used for burnishing stems of spoons and forks, and plane surfaces generally. Curved burnishers, such as those shown in Figs. 515 to 519, and in

Fig. 520.—Curved Burnisher.

section at B, C, D, E, and F, are used for burnishing the insides of the bowls of spoons and for hollow curves. Burnishers made of chips of agate, and of bloodstone or hæmatite, set in brass ferrules and mounted on wood, are more costly than those made of steel, and they also impart to the goods a more finished surface. Some

further forms of burnishers in everyday use are shown in Figs. 520 to 523. Special agate burnishers, as supplied by Reeves, are illustrated by Figs. 524 to 527.

PRELIMINARY PRECAUTIONS.

It is most important that the work to be burnished should have been prepared properly for plating, as on this depends the

Fig. 521.—Burnisher for Corners, etc.

perfection of the finished surface when burnished. In the first place, all scratches, lines, indentations, and corroded pits must be removed by filing, rubbing down with water of Ayr stone, polishing, and burnishing, before the article is pickled and quicked with mercury preparatory to being placed in the plating vat. The slight roughness imparted to the surface by the action of the acid pickle is not in any way

Fig. 522.—Hooked Burnisher.

detrimental, but should a stain be left on the brass or German silver surface, or should the operator leave his finger-marks thereon, these will be distinctly traceable in the surface of the finished article if the spot does not strip under the burnisher. The utmost cleanness must be observed in the preparation of the articles to be burnished, and care must be taken to put the quicking coat of mercury on evenly, or the

Fig. 523.—Pointed Burnisher.

silver will be apt to strip from slightly soiled spots, as also from those where thick blotches of mercury have been left on the surface. Articles made of pewter, Britannia metal, and similar soft alloys are usually difficult to burnish because they are softer than the overlying coat of silver, but they are made worse by lack of care in their preparation; they should be transferred at once from a clean potash dip to the plating solution (after being

Fig. 524.—Agate Burnisher.

properly cleaned and prepared) without any intervening rinsing, because such alloys are readily tarnished when exposed to the air whilst wet. Potash dissolves the tarnish on such metals.

METAL-STRIPPING UNDER THE BURNISHER.

Silver will strip under the burnisher when it is deposited too fast or too slow, since its hardness is greatly affected by its rate of deposition. The plater should, therefore, find out by trial the best rate at which to deposit a coat for burnishing on the several metals, or alloys likely to be employed. Silver will also strip when a plating bath has been made up by dissolving chloride of silver in a solution of cyanide of potassium, or when chloride of silver has been used in building or faking up a plating solution. A similar result will follow on the use of too much brightening solution in the plating bath. Plating solutions thus ruined should be set aside

Fig. 525.—Agate Burnisher.

for the most common work, and allowed to work out, then treated for recovery of silver.

PREPARING FOR BURNISHING.

When the requisite amount of silver has been deposited on the article to be burnished, detach it from the battery or machine, and swill to and fro in the plating solution for a few moments to dissolve off any sub-cyanide of silver adhering to the deposit, which, if left on, might discolour the pure white coat on exposure to daylight. Next rinse in clean hot water to remove all traces of the plating solution, and dry off quickly in hot boxwood sawdust. This sawdust is obtainable from boxwood block makers and engravers on wood, and it should be perfectly clean, else it will soil the silver. For the same reason the dust of other woods is unsuitable, because in all, or nearly all, except box, there is found some trace of acetic or of tannic

Fig. 526.—Agate Burnisher.

acid or of some resinous substance. The dust is best made and kept hot over a water bath or in a steam heated chamber, to prevent charring. The dried articles are scratch-brushed, or else scoured with very fine sand in soap-suds applied with a soft cotton brush, then rinsed in hot water, and transferred to the burnishing bench, where they should be laid on a clean soft pad of rag, on which they will be held whilst being burnished.

LUBRICANT FOR BURNISHING.

If an attempt were made to burnish articles whilst surface and burnisher are dry, the tool would heat and drag off the silver in the form of fine dust. A lubricant is, therefore, essential; linseed tea or a decoction of marsh mallows answer this purpose, as both of these are of a slippery nature, and are harmless when applied to silver. Soap-suds are sometimes used, and these form a fairly good substitute when

Fig. 527.—Agate Burnisher.

freshly made, but they should never be set aside for use a second time, as they are apt to undergo changes whilst standing exposed to the air which result in the formation of acids injurious to the silver coating.

PREPARING THE BURNISHER.

The burnisher must first be polished to a dead black lustre, by rubbing its edge or face briskly along a groove worn in a polishing buff charged with jeweller's rouge. A thin burnisher is first selected to ground the work, and this is afterwards gone over with one having a broader surface, finally finishing off with a broad bloodstone burnisher. The tool is held in the right hand, the lower part of the handle resting on the outside of the little finger, and the upper part resting against the inside of the three other fingers, with the ball of the thumb on the top of the handle. In this position great pressure can be brought to bear upon the tool, if required.

HOW TO USE THE BURNISHER.

The strokes of the burnisher always should be given in one direction, since cross strokes will spoil the appearance of the burnished surface. Each stroke must be applied with some pressure, and the burnisher must be kept supplied freely with the lubricant, to prevent heating. Each succeeding stroke should slightly overlap that of its predecessor, so as not to leave unburnished metal between the strokes, and a clear, mirror-like surface behind. As the surface or the edge of the burnisher gets dull by use, polish it up on a polishing pad, made for the purpose out of buff hide, mounted on a piece of wood like a large razor strop, and charged with rouge or with the finest putty powder. Very pleasing effects on ornamental goods are sometimes obtained by burnishing certain parts, such as bands and raised parts, whilst others are left matt. Gold-plated articles are treated in a similar manner to those of silver, but it is not usual to burnish nickel, since this metal is somewhat hard and intractable under the burnisher. When articles have been burnished, the finishing polish is put on by hand with soft rags, charged with a suitable plate powder, or by a dolly of soft linen revolving in a lathe.

ANNEALING, HARDENING, AND TEMPERING.

ANNEALING.

ANNEALING is a process of rendering metals less brittle. It is performed by allowing them to cool very gradually from a high heat, a sudden reduction of temperature rendering them hard and brittle. Annealing is also a necessary process in the manufacture, by drawing, of wire and small tubing, as well as in making brass, copper, or sheet-iron vessels by hammering and rolling; the metal, by compression, becomes too hard and brittle for further reduction until it is annealed, when it recovers its former softness and pliability. Annealing very greatly increases the malleability of cast iron, and so annealed castings are known as "malleable."

How COOLING AFFECTS HEATED METALS.

Tempering and annealing are nearly allied, but the processes are not confounded in the arts, owing to their different technical applications. The heating is a necessary preliminary whether to withdraw the hardness incident to hammering and rolling of malleable metals, or the hardness incident to the rapid cooling of a casting in its mould. The protraction of the process of cooling the casting has a favourable effect upon its toughness and comparative softness. This is plainly seen by comparing them with chill-hardened articles, which are rendered hard and brittle by the sudden cooling. Exposure of the hot steel to a cold surface renders it hard. This is usually done by dipping the red-hot metal in water, but other cold surfaces which are rapid conductors will answer the same purpose. A thin heated blade placed between the cold hammer and anvil is hardened by rapid cooling. Thicker pieces, in the same circumstances, are somewhat hardened, but may be filed. Placed on cold cinders, or other bad conductor, the steel cools more slowly, and becomes softer. Placed in hot cinders, and allowed to cool by their gradual extinction, it becomes still softer. Encased in a close box with charcoal-powder raised to a red heat, and allowed to cool very slowly, it reaches its softest state, except by a partial decomposition; this last-named process is annealing.

THEORY OF ANNEALING.

Generally, in the annealing of metals, cast iron for instance, the metal is brought to a red heat, and then allowed to cool slowly. It is argued that the molecules of metal take a different arrangement from that assumed when the cooling is rapid. In the latter case, the exterior portion of the metal contracts first, and presses upon the interior portion, and the particles of the latter may thereby be compelled to take an arrangement which they would not were the cooling to take place at an equal rate in every part, while the process of cooling was long protracted. In annealing —that is, making cast iron malleable—the metal is kept for several hours at a temperature a little below the fusing point, and then allowed to cool slowly. The perfect result is best attained by giving the particles time to expand and to contract, and not violently changing their structural relation, unless it be held that chemical changes in the furnace (such as parting with a portion of the carbon) have to be taken into consideration, and that the change is not all in the mechanical disposition of the particles.

CASTINGS FOR ANNEALING.

With regard to the kind of iron castings suitable for malleable work, one essential is that they be white, or nearly so. The

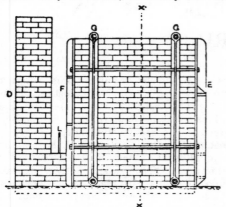

Fig. 528.—Side Elevation of Annealing Furnace.

grey irons are unsuitable, the carbon in them being chiefly in the graphitic or uncombined condition. Charcoal iron is preferable, but iron smelted from coke is also used successfully. There is some difficulty in casting, because the white iron sets so rapidly that it is difficult to fill the mould quickly with small runners, which are desirable, large runners being inconvenient because of their liability to pull from the castings during shrinkage. Many of these castings, being of small size, are melted in crucibles in a brass furnace. But a reverberatory furnace, and also the common cupola, are as frequently made use of. Trial bars are cast and broken to test the quality of the metal previous to pouring the moulds.

PREPARING CASTINGS FOR ANNEALING.

Before annealing, the castings are cleaned thoroughly by grinding and brushing, or by the use of a tumbling drum (p. 144). They are then washed in dilute sulphuric acid and cleansed with water. After annealing, the castings are again cleaned in the tumbling drum and in a current of water.

THE PROCESS OF ANNEALING.

The general rules governing annealing may be summed up as follows: Heat slowly and uniformly the entire piece to a temperature higher than that at which the metal was last worked—if hammered or straightened cold, to a bright red. Allow cooling to take place as slowly as possible, but excluding the air. If the steel has not been heated above dark cherry red, and is low in carbon, annealing in boiling water will give excellent results. The kind of packing usually employed is clean forge scale, and this is considered better than other material. But iron oxide in the form of red hematite ore is also often made use of. The forge scales are cleaner than the iron ore, but they part with some of their oxidising properties at every heat. These are restored by washing them in a dilute solution of sal-ammoniac, which produces a layer of rust on the surfaces. It is not necessary that the substance used should be rich in oxide, since limestone rock ground to powder has been employed with success.

BOXES FOR ANNEALING.

The use of boxes is necessary. If cast-

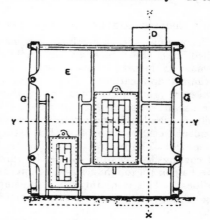

Fig. 529.—Front Elevation of Annealing Furnace.

ings are packed directly on the floor of the oven instead of being put into pots, the annealing occupies longer by a day or two, and the castings become more distorted.

The annealing boxes used are made of cast iron, and usually, though not always, of cylindrical form, and they are arranged in tiers. As the lower ones become packed full, the next one above is put on and packed, and so on. They soon burn out.

CARE IN PACKING CASTINGS.

Care has to be exercised in packing, since, however carefully the scale or powdered ore is packed, the mass settles a little when subjected to the heat of the furnace. But apart from this, if castings are packed loosely they lose their shape, and come out distorted and often useless; hence the need to fill all spaces completely with the powder. Generally, too, the castings should be laid in the way in which they would be least subject to distortion if the packing shifted.

POINTS IN ANNEALING METALS.

In annealing steel, the risk of overheating it is present. Some workers approve of heating the steel and burying it in lime, others of heating it and burying it in cast-iron borings, while others approve of heating it and burying it in sawdust. A far

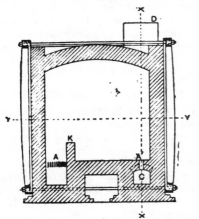

Fig. 530.—Cross Section of Annealing Furnace.

better plan is to put the steel into a box made for the purpose, filling the box with charcoal dust, and plugging up the ends so that the air is kept from the steel, then to

put the box and its contents into the fire till it is heated thoroughly, and the steel is at a low red heat; it must then be taken from the fire and allowed to remain in the box, without opening the box till the steel is cold. Then the steel is clean and soft, and without those bright spots (" pins ")

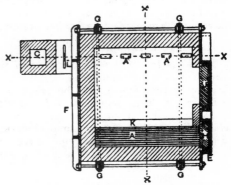

Fig. 531.—Horizontal Section of Annealing Furnace.

which impede the filing and working. A piece of stout gas-pipe, with a bottom welded in, and a plug made for the other end, makes a very good box for a small quantity of steel; but for a large quantity, the box must be large in proportion. If the steel is very large, it is as well to make a charcoal fire to heat it in, and then let the steel and the fire get cold together before it is taken out, and it will be equally soft. When a piece of steel has to be annealed hurriedly, it may be heated in an open fire, and buried in charcoal dust till it is cold, or if it be heated to a red heat sufficient to be seen in a dark place, and then plunged into cold water, it will work more pleasantly, but not so softly as if it were heated in a box with charcoal.

FURNACES FOR ANNEALING.

The annealing, cementation, or, more precisely it may be, the decarbonisation of brittle castings to produce tough, malleable cast iron is not necessarily done in any special type of furnace. The Siemens regenerative gas furnace, reverberatory furnaces (see p. 156), and steel cementation furnaces, are, with various modifications, all used. Any furnace or oven in which a uniformly high temperature can be maintained during several successive days is adapted for the production of malleable cast iron. A description of a typical annealing furnace may be instructive, though it is hardly supposed that the worker in anything but a large way will be able to have such a construction on his premises.

Typical Annealing Furnace.

Figs. 528 to 532 illustrate a type of furnace suitable for annealing. Dimensions are not given, but it may be made from 3 ft. or 3 ft. 6 in. square to 6 ft. or 7 ft. square. It is built of firebricks set in fireclay. Fig. 528 is a longitudinal elevation, Fig. 529 a front elevation, Fig. 530 a transverse section in the plane x x (Fig. 528),

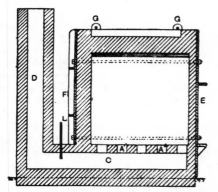

Fig. 532.—Longitudinal Section of Annealing Furnace.

Fig. 531 a horizontal section in the plane y y (Fig. 529), and Fig. 532 a longitudinal section in the plane x x (Fig. 530). The brickwork is bound together by means of front and back castings E and F respectively, and side binders G G G G, all tied together with bolts as shown. The bolts which connect the front and back plates pass through holes cast in the ribs of the binders. The bottom bolts which connect the binders pass through the brickwork— the top bolts pass over it. H and J (Figs. 529 and 531) are the fire and oven doors respectively, sliding in guides cast on the

front plate E, and lifted by chains and levers (not shown), and counterweighted. The doors are made as open frames, either of cast iron or of angle iron bent round into rectangular form and welded at the corner. They are fitted with firebrick, both to preserve them from warping by the heat of the furnace and to prevent the heat from becoming dissipated. The interior of the plate E is also brick-lined, except at the doorways, for the same reasons. K is a low bridge of brickwork separating the firegrate from the oven. Sometimes this is brought higher up, sometimes dispensed with entirely. L is a damper for regulating the draught, and is raised and lowered with a chain (not shown) passing over a pulley, and actuated by a lever. The brickwork chimney D can be carried up to the full height required, or its height can be completed with a wrought- or cast-iron pipe. Provision must be made at one end or one side of the culvert C for cleaning it out periodically.

An Older Type of Furnace.

Fig. 533 is a section taken across an older type of annealing furnace of simpler construction. The firegrate is built beneath the furnace, and the products of combustion pass up through sloping flues into the chamber, whence they escape into the chimney placed directly on the roof of the chamber. This is a wasteful furnace, but readily and cheaply built.

A Modern Annealing Furnace.

The time of lighting up and of cooling down is a considerable fraction of the total time occupied in annealing, varying from two to three days. To save this time the furnace in Fig. 534 was designed. In this there are two firegrates, placed one on each side of the oven. The boxes are carried on a trolley, which runs on rails between the walls which form one side of the firegrates. The top of the carriage is made of firebrick, and the bricks of the furnace floor are made to overlap the edges of the carriage in order to prevent the escape of heat down the sides. The advantage of this form is that, as the charging door for the carriage and the firing doors are situated at opposite ends of the furnace, there is no need to draw the fires and let the furnace cool down before withdrawing the boxes. A tier of boxes can be drawn out and another run in without extinguishing the fires.

STANFORD'S FURNACE.

A modern type of annealing furnace which has given good results is that of Mr. Stanford. It is in principle a reverberatory furnace, but less wasteful of heat than

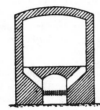

Fig. 533.—Simple Annealing Furnace.

the common types. The products of combustion pass from the fire-hole into the furnace and heat the annealing boxes, thence they escape slowly through small holes in the opposite wall, and pass down into bottom flues, winding underneath the floor area, round which they circulate, still giving up heat to the brickwork floor. Finally they pass out through the main flue to the chimney. The heat is prevented from becoming dissipated through the walls of the furnace by the narrow downtake through which the gases pass into the flues beneath the floor on the one side, and by air spaces formed in the side opposite. The annealing boxes are introduced through brick-lined doors situated at each end.

VARIOUS KINDS OF IRON AND STEEL.

Of iron and steel, only the latter can be hardened and tempered, though the former can be "case-hardened" by a special process. In physical structure steel is midway between malleable iron and pig iron. Malleable iron may contain from the merest trace up to ·3 per cent. of carbon, but in the latter case the metal will have decidedly steely characteristics, and will be classed as iron or steel according to the presence of other materials which influence the one which predominates. Pig iron contains 2·0 per cent. of carbon; intermediate proportions give steel proper, which partakes more or less of the characteristics of pig or malleable iron as the content of carbon decreases. The precise point at which metal ceases to be steel and becomes pig or malleable iron, as the case may be, is hard to determine, but a generally accepted distinction is that, when heated to a full blood-red heat and plunged into cold water, steel becomes hardened, whereas the same process has not that effect on iron. This facility of hardening steel places this metal foremost amongst those used for cutting tools and general implements.

STEEL AND IRON DEFINED.

Dr. Percy defines steel as iron containing a small percentage of carbon, the alloy having the property of taking a temper; but other investigators have defined as steel all alloys of iron cast in malleable masses. Sir Joseph Whitworth defined steel mechanically by a co-efficient representing the sum of its strength and ductility. With the object of having universally adopted names which should indicate the nature and the distinction between iron and steel, an International Committee

Fig. 534.—More Recent Annealing Furnace.

recommended the following: (1) That all malleable compounds of iron, with its ordinary ingredients, which are aggregated from pasty masses, or from piles, or from any form of iron not in a fluid state, and which will not sensibly harden and temper, and which generally resemble what is called wrought iron, shall be called weld iron (German, *Schweisseisen;* French, *fer soude*). (2) That such compounds when they will from any cause harden and temper, and which resemble what is now called "puddled steel," shall be called weld steel (German, *Schweiss stahl;* French, *acier soude*). (3) That all compounds of iron, with its ordinary ingredients, which have been cast from a fluid state into malleable masses, and which will not sensibly harden by being quenched in water while at a red heat, shall be called ingot iron (German, *flusseisen;* French, *fer fondu*). (4) That all such compounds, when they shall from any cause so harden, shall be called ingot steel (German, *fluss stahl;* French, *acier fondu*). Steel which will harden by heating to any temperature and using any quenching liquid is termed weld steel. That which will harden by being heated to redness and quenched in water is termed steel.

"Temper": the Content of Carbon.

Next to quality of steel—by which is meant the percentage of phosphorus, sulphur, silicon, manganese, etc.—the most important thing is "temper," or percentage of carbon. This "temper" must not be confounded with "tempering," with which it has nothing to do. For many purposes, indeed, temper is of more importance than quality. Nothing is more common than for steel to be rejected as bad in quality because it has been used for a purpose for which the temper was unsuitable.

Special Tool Steels.

There are also many peculiar kinds of steel. A special steel for taps, called mild-centred, is made by converting an ingot of very mild cast steel, so that the additional carbon only penetrates a short distance. These bars are afterwards hammered or rolled down to the size required, and have the advantage of possessing a hard surface without losing the toughness of the mild centre. Another special steel, somewhat analogous, is produced by melting a hard steel on to a slab of iron or very mild steel heated hot enough to weld with the molten steel so that a bar may be produced, one side of which is iron and the other side steel, the quantity of each being regulated as may be required. A steel specially adapted for tools used in turning chilled rolls and some other purposes is made by adding a percentage of wolfram, or, as the metal is more generally called, tungsten, sometimes with and sometimes without carbon, frequently to such an extent that it can be used without hardening in water. Special steel of this kind is the finest grained that can be produced, but it is so brittle that it is generally almost useless.

Care in Selecting Tool Steel.

The quality of a tool depends mainly upon the kind of steel used, the method of forming it, and the way in which it is hardened and tempered. A careless workman picks up any scrap of steel when a tap, a punch, or a cutter is to be made, regardless of the fact that the tool so made most likely will break, lose its edge, or in other ways prove untrustworthy. Tools should be made of cast steel, and the special quality known as "tool steel," which can be obtained in flat strips and in square or round rods, will always pay for the trouble of getting. The nearer the rough-drawn or rolled steel is to the dimensions of the finished tool, the better and stronger the tool will be. For a set of turning tools for a slide-rest, or for fitting a cutter bar, a rod of tool steel that, with the slightest cleaning up, will fit the tool holder should be obtained. Similarly, to make $\frac{1}{4}$-in. taps, it will be better to use $\frac{1}{4}$-in. round rod than to reduce $\frac{3}{8}$-in. rod by turning. For this purpose, tool steel is always made "full" in size, so that $\frac{1}{4}$-in. rough will trim up $\frac{1}{4}$-in. finished.

Hardening Steel.

Having suitably formed the tool of the proper material, remembering that tool steel must not be bent cold or worked at a low red heat, the next process is to harden and temper it. Hardening is effected by heating to redness and plunging into a cooling medium. There are two points here upon which much might be written: first, the exact heat required, and, secondly, the proper liquid in which to cool the steel. With regard to heating, it should be remembered that the lowest temperature at which the steel will harden gives the strongest and toughest tool; and the highest temperature to which the steel

use of the solution is to cool the steel rapidly; therefore the water should be cold, and the tool immersed in it quickly. The commonest hardening bath is clear cold water, though many workers use salt and water or brine. For obtaining extreme hardness, some profess to have found satisfactory a solution of 1 lb. of citric acid crystals in 1 gal. of water. For very thin articles a bath of oil is necessary. For hardening springs sperm oil is very satisfactory, and for hardening cut-

Fig. 535.—Fletcher-Russell Gas Muffle.

can be raised without burning it gives the hardest and most brittle tool.

LIQUIDS FOR HARDENING.

For hardening liquids, tallow, mercury, molten lead, brine, resin, beeswax, lard, and other agents have been recommended as having advantages over water or oil. These things, however, had better be left alone, and water only used, except in the case of very small watch tools, for which oil sometimes gives better results. The

ting tools raw linseed oil. Many successful hardeners use water that has been boiled. Further, many workers hold that small, odd-shaped pieces are not so liable to crack or to harden unevenly when the water in which they are dipped for cooling is slightly warmed.

CORRECT HARDENING HEAT.

The heat at which steel will harden depends upon its quality. The higher the quality of the steel, the lower the temperature at which it can be hardened. Tool steel will harden if heated to a dull red and plunged into cold water. Bessemer or other mild steel will only harden if heated to a white heat, and is then too brittle to take a cutting edge. As already stated, the lower the heat at which the tool is hardened, the tougher it will be; therefore, the heat required should be determined according to the use to which the tool is to be put. A drill, a tap, or a reamer is subjected to great twisting strain, and is required as tough as possible; thus a tap made of tool steel should be heated to a dull red only. A centre punch is required to be hard, and is subjected to no very severe bending or twisting strains; so it should be heated to a bright cherry-red. A lathe-cutter should be something between the two; being required hard and tough at the same time, it must be heated to a full red.

Source of Heat for Hardening.

For heating, a fire or a blowpipe flame should be used. A series of blowpipes is described in a later chapter, "Soldering, Brazing, and Riveting." In cases where large quantities of tools, etc., have to be hardened regularly, it may pay to get a gas-heated oven or furnace of the Fletcher-Russell make. Fig. 535 shows a suitable muffle which is made in at least seven stock sizes, the clear inside working space ranging from 2 in. × 2 in. × 4 in. to 20 in. × 6 in. × 30 in. The reverberatory furnace shown by Fig. 536 is made in one stock size—12 in. square, inside floor—but other sizes can be obtained. A Bunsen burner without blast is of little use.

Testing Hardened Steel.

When the tool is taken out of the water, the surface should have a whitish mottled appearance, caused by the partial scaling off of the black surface; if it is not mottled, the steel probably has not hardened. The usual test is with a file. If the steel is hard, the file will slip over the edge with a rattling sound: if soft, it will cut. Should it be still soft, it shows that the

steel was not heated sufficiently, and it must be tried again. Some workers prefer the hammer test; if no bruises can be made with a well-hardened hammer on the surface of a piece of hardened steel, it is safe to suppose that the latter is hardened to a proper depth. They assert that a piece of steel may be case-hardened merely, not sufficiently deep to be of service, although no file will mark it; while there may be cases in which the hardening is admirable but for a film on the surface, which may be easily filed. In some experiments a piece of hardened steel, having this unhardened envelope of decarbonised steel, was first measured very carefully; then all that could be filed away was removed and a second measurement made. In some cases the hardened piece was reduced in size as much as ·015 in., and in others as little as ·0005 in. Still, the piece of steel itself, with its skin off, was thoroughly hardened.

Hardening Annealed Steel.

The greater difficulty experienced in hardening steel that has been repeatedly annealed arises from the fact, probably, that the annealing is at too great a heat, or that the heat is too long continued. Either of these causes will decarbonise the surface of steel, leading to the opinion that such steel does not harden because it is capable of being filed.

Preventing Distortion when Hardening.

Steel expands considerably when heated to redness, and is contracted suddenly when dipped in the cold water. If heated or cooled unequally, it does not contract evenly, and the result is a warped or bent tool. To prevent this, place the steel in a uniformly warm part of the fire; or, if using a blowpipe, heat it carefully to the same colour from end to end. If it is long and thin, like a tap, drop it vertically into the water; for if dropped sideways, one side may cool more quickly than the other and warp it. A cutting tool that is hardened at one end only does not need these precautions.

Causes of Distortion.

The manner of dipping the steel has been shown to cause distortion ; steel also warps out of shape in the fire from uneven heating ; and it springs in heating, owing to an unequal and unnatural tension of some parts produced in forging. Difference in section at different points in a piece will cause more sudden cooling and contracting in some parts than in others. Some steels at the hardening heat will, in dipping, change in volume more than others. With different steels all the and unyielding, while the interior of the piece is still hot and largely expanded, there must be a change in shape or volume (or both) produced by hardening.

Hardened Steel too Brittle for Use.

Hardened steel, with very few exceptions, is too brittle for tools. Files and turning gravers for hand use are left quite hard ; but punches, taps, reamers, chisels, drills, lathe-cutters, etc., require tempering to take off the extreme brittleness.

Fig. 536.—Fletcher-Russell Gas Reverberatory Furnace.

different heats may be required for hardening, ranging from a yellow heat by daylight to a dark red by twilight. Finally, after considering all the tendencies towards distortion and change in volume, resulting from a variation in treatment, quality, or condition, there still remains this fact, that heating expands the steel, whilst sudden cooling, for hardening, greatly contracts it. As the outer portions cool first sufficiently to become hard

Unfortunately, this process also takes away from the hardness as well, and so a compromise has to be arrived at, giving to the tool as much hardness as possible and leaving it just tough enough to bear its work.

Tempering Steel.

The heating of hardened steel softens it and reduces its brittleness ; this is called tempering. If a piece of hard steel is brightened with emery cloth and heated over a fire, or upon an iron plate placed over a fire, it will first become of a pale straw colour; then this will turn darker and approach a red; next the red turns to purple and blue, and the blue becomes paler until it disappears and the metal becomes a dull red. At this point all hardness is gone. Thus the metal goes in succession through all the stages of hardness, and the process can be arrested at any point by taking it off the fire or the hot plate. Until the first pale straw colour appears, the steel has not lost any of its hardness—at any rate, not appreciably. At a pale straw, the extreme brittleness has gone, but it is still too hard to file, and will keep an edge well. This is the temper for a lathe-cutter, a mill-cutter, a chisel, a punch, or a drill. A rather deeper yellow, bordering on red, is the temper for a slender tap or a reamer. Deep blue is the temper for a spring, as at this point the steel has its maximum elasticity and strength, though it would not keep a cutting edge. Beyond this temper the steel becomes too soft for any tools whatever. Before tempering a tool, brighten it with emery cloth so that the colour can be well seen. This should be done in daylight, as by artificial light the colours are deceptive, and a full straw colour may be obtained before any change is observed.

- - -

TEMPERING STEEL IN MOLTEN LEAD.

Molten lead is a good heating agent for tempering steel articles of unequal thickness, as these can be heated, it is claimed, more uniformly by this method than by placing in an open fire or by supporting on an iron plate over a fire. Lead melts uniformly at a temperature of 612° F., and by alloying the lead with tin in varying proportions, as explained in the table herewith, an extensive range of temperatures may be obtained. In using such baths, cover the surface with powdered charcoal to prevent the oxidation of the molten metal. It will be noticed that the colours given do not quite conform in their corresponding tempers to those mentioned

previously. Steels differ so greatly in nature and the treatment they should receive that it is not strange that different workers should obtain different results by what, on the face of them, are identical methods.

Colour.	Articles to be Tempered.	Composition of Lath.		Temperature in degrees F.
		Lead.	Tin.	
Yellowish tint	Lancets	7	4	420°
	Other surgical instruments...	7·5	4	430°
	Razors, etc.	8	4	442°
Pale yellow ...	Penknives. and some implements of surgery	8½	4	450°
Straw yellow ...	Large penknives, scalpels, etc.	10	4	470°
Brown yellow	Scissors, shears, garden hoes, cold chisels, etc ...	14	4	490°
	Axes, firmer chisels, plane irons, pocketknives, etc. ...	19	4	509°
Light purple ...	Table-knives, large shears, etc.	30	4	530°
Dark purple ...	Swords, watch-springs, etc.	48	4	550°
Clear blue ...	Large springs, daggers, augers, fine saws, etc.	50	2	558°
Pale blue... ...	Pit saws, hand saws, and some springs ...	Boiling linseed oil		60.°
Greenish blue	Articles which require to be somewhat softer	Molten lead		612°

SPECIAL METHODS OF TEMPERING.

A convenient way to temper a large article is to lay upon it a small piece of brightened sheet steel, using the colour of this as an index to the temper of the article. Some tools require to be tempered to an even colour throughout, and these may be laid upon an iron plate over a bright fire and rolled over as they heat to ensure evenness; or they may be held by one end in a clean Bunsen flame or a spirit-lamp flame, and moved about, changing ends and "leading the colour" along until even. Another method is to heat an iron plate to redness in the forge, carry it to the light, and place the tool upon it until the colour appears. Do not try to temper a tool by holding it in an ordinary gas flame, as the smoke and deposit will prevent the colour being seen. A tap or a reamer may be thus tempered evenly from end to end, and then the temper of the handle end may be further let down to a blue, allowing the blue to run up to the commencement of the cutting edges, where it will fade off through red into the straw colour; this gives the handle end greater spring and toughness.

HEATING WITHOUT SCALING.

If it is necessary that a piece should be finished and polished before hardening, a method of heating will have to be used that does not scale the surface. The usual way to do this is to enclose the steel in a box or tube packed with animal black, and plugged up in such a way that when all is red-hot together the contents can be shot out into water.

TEMPERING SLENDER AND IRREGULARLY-SHAPED WORK.

When a long, slender, and irregularly-shaped piece of steel has to be tempered evenly throughout, a good plan is to obtain a heap of sand, place it on an iron plate over the fire, and stir it up until all is nearly red hot. Then, after brightening, immerse the steel in the sand, and move it about quickly. Withdraw it for a moment to see how the colour is getting on, and re-insert until done. In tempering, it is well to remember that to obtain a good colour, certain and easy to see, the tools must be brightened and polished up as smooth as possible, and should be entirely free from finger-marks and grease.

BLUEING STEEL.

Blueing is an operation quite distinct from tempering. It is done to give a good appearance and to preserve from rust. To obtain a good blue, the surface must first be polished, a burnished surface being no good. The colour may be obtained by the methods described above. For strawing or blueing very slender articles, a heap of powdered charcoal is placed in an iron pan and lit by a blowpipe flame or a gas-jet. When alight, it is gently blown and stirred up until it is all evenly glowing or smouldering. The articles are then immersed and treated as in the sand method. This is also a good way of tempering many small tools, drills, etc.

INFLUENCE OF FORGING ON DISTORTION.

It has been stated that steel will spring in heating, owing to an unequal tension of some parts produced in forging. This is an awkward defect. If it has received a greater number of blows on one side than the other, or the blows have been given with greater force on one side than the other, although the steel may be straight when dipped into the cooling bath, it is certain to become distorted during cooling in proportion to the inequality in number and force of blows given during the forging process. In connection with this a metal turner's method of preventing distortion may be noted. Assume that a forged taper steel bar is to be turned in a lathe and afterwards tempered without risk of distortion in the latter process. Having centred it, take a rough cut from end to end in the lathe; then make it a dull red heat throughout, and plunge it into dry slaked lime or charcoal dust, and allow it to cool. When it is cold take another cut down it in the lathe, and repeat the heating and cooling process. Unless the bar has had extra rough usage during forging, this will usually be sufficient. It may then be finished in the lathe and tempered, when it should be free from distortion; in case of doubt it may be heated and cooled even a third time before finishing. If the article is flat instead of round, and fitted by grinding or filing, the same process must be observed if great accuracy is essential. The cause of distortion is that the outer surface has become of a closer texture in some places than in others, and unless that surface is removed it is certain, when cooled during the tempering process, to contract more there than elsewhere.

HARDENING AND TEMPERING MISCELLANEOUS WORK.

The foregoing instructions, if followed intelligently, will suffice for the hardening and tempering of most steel tools and implements; but, of course, it is usual to modify the method adopted according to the exigencies of each case. To make this clear the following information is given.

TEMPERING AXES.

The eye of a manufactured axe is not made of steel, but of iron, which cannot possibly be tempered, steel being used only for the cutting edge and for the poll. But supposing that the axe is made entirely of steel, as in a home-forged job, even then the eye is never tempered, just as in the case of the shank of a cold chisel, the shank of a drill, or the tang of a file. It is the cutting edge only, and the steel for half an inch to an inch back from the edge, that is let down and quenched, and so tempered. The steel, having been hardened, is heated to a low red or to cherry red. So long as the steel is not overheated, either will do. Remember that a cherry red on a dull day will be a low red on a bright day, and that almost every separate brand of steel requires to be worked at a different temperature, and also requires different manipulation, hammering, etc.

TEMPERING PUNCHES AND DRILLS.

A punch or a drill is required to be hard only at its extreme edge, and may be tempered by heating it from the handle or shaft end to a blue, and allowing the colour to run up to the point, dropping it into water to arrest the heat the moment the point reaches a very pale straw; this gives a hard point and a tough and elastic shaft.

HARDENING CLOCK PALLETS.

Harden each end of the pallet separately; leave the middle soft, and, if necessary, bend it. There will then be no necessity for tempering.

TEMPERING TURNING TOOLS.

A lathe cutter or a chisel is hardened only at the point, and may be tempered by holding the shaft in tongs or pliers and heating the end until the colour appears. A mill-cutter should be laid upon an iron plate and very gradually warmed over a flame or fire. After tempering, the tools should be given their sharp cutting edges by means of the oilstone. Tools for metal turning are tempered usually to some shade of yellow, but the quality of the steel will greatly affect the ultimate result. Many tools are hardened and tempered at a single operation. The steel is heated and the point only plunged into water; the tool is then withdrawn, and the heat in the shank draws the temper. This method is only practicable after long experience, and it is not to be commended for general adoption.

SPECIAL METHOD OF HARDENING AND TEMPERING SMALL PIERCING PUNCHES AND DIES.

According to R. B. Hodgson, punches from $\frac{1}{16}$ in. to $\frac{1}{2}$ in. in diameter, and dies from $\frac{1}{4}$ in. to 1 in. in diameter, are hardened best by heating in a wrought-iron pipe about 1 ft. long and 2 in. in diameter, with one end closed. This may be done by welding a wrought-iron plug in one end of the pipe. The small punches or dies are placed in the pipe, which is thrust into the breeze fire. This method gives much better results than can be obtained when the flame of a fire is allowed to come in contact with the tools. When the punches or beds, as the case may be, are heated sufficiently, the pipe is removed from the fire, and the tools are tipped into a bucket of clean water containing a handful of common salt. Tools hardened in this manner will be found to be quite clean and ready for tempering. This may be done by placing the tools upon a wrought-iron plate, 1 ft. square by $\frac{1}{2}$ in. thick, heated over a gas stove; the small round punches would be rolled over the hot plate until the required colour appears, whilst small dies are best placed endways upon the plate. In the case of a small die the heat travels up from the bottom, so that when the cutting end or face is a straw colour the back will probably be blue; this will be an advantage, as it is only the cutting end that is required very hard, whilst the effect of tempering the back of the die will help to preserve it.

HARDENING MILLING-MACHINE CUTTERS.

One method of hardening milling-machine cutters requires the use of a large, high fire, in which the cutter is buried. Only enough blast to bring the work to the required heat must be used, and the heat should be uniform throughout the cutter. If the piece has not been annealed, remove it red hot from the fire, and allow it to cool off slowly until the red has entirely disappeared, when it can be placed again in the fire, slowly brought to the required heat, plunged in the bath of tepid water or brine, and worked round well until it stops "singing." At this point it should be removed and instantly plunged into the oil bath, and left there until it is cool, when the strain should be removed by holding it over the fire until it is warm enough to "snap" when touched by the moistened finger. It can then be laid aside, and the temper drawn at leisure.

HARDENING AND TEMPERING MILL BILLS OR PICKS.

Very few persons treat these implements successfully. As mill bills are subject to rapid jarring shocks on their cutting edges whilst being used on the mill-stones, they must be carefully forged at a low red heat with light blows to toughen the steel and give it a good grain. Overheating and heavy blows will ruin mill bills. In tempering their edges, regard must be paid to the fact that a hard, keen edge, supported by tough metal behind, is an essential. If made too hard in the supporting part—from $\frac{1}{4}$ in. to $\frac{1}{2}$ in. back from the edge—the brittle steel will snap here whilst dressing the stone. A method that has been recommended is as follows: Heat slowly to a dull red and cool the point for $\frac{3}{4}$ in. in clean rain-water; then let them down to blue, and quench outright. If slightly hard, let down a little more. Heating in molten lead and quenching in water at 62° Fahr. is also recommended for tempering.

FORGING MILL BILLS OR PICKS.

As suggested already, the method of forging exercises a great influence on the success of the tempering. There is no great secret attached to the forging of mill bills for cutting French-Burr mill-stones. The great object is to observe carefully that they are not overheated, and that the points are well hammered together before they are drawn down. Heavy blows should not be struck, and a dark blood-red heat should be the extreme limit. Many smiths allow the tool to get almost white hot, and consequently the "temper" (content of carbon) is burnt out of the steel. The form of mill picks may differ. If for cutting out the furrows, they should be square and pointed at an obtuse angle; but if for the skirt of the stone, they should be shaped at the point like a cold chipping chisel for cast iron. This is the practice for flour mills; for phosphate and mineral mills the square mill pick holds good for all parts of the stone. They should be worked on the anvil by being well and lightly hammered at the point so as to ensure toughness.

HARDENING AND TEMPERING TAPS.

Heat the tap in a fire or gas flame to a dull red, and plunge into cold water. This will make it too hard. To temper it, brighten one side on a bit of sandstone, and heat it in the middle through a gas flame. Watch the brightened side, and when a yellow tinge has travelled slowly to the point, cool in oil. It requires some experience to do it always with success. Cheap taps are, as a rule, not let down; they are made of inferior steel, and do not become very hard in the first instance; in a large box of such screwing tackle that was tested, every tap and die proved soft enough to file. Good steel can be hardened at one operation by plunging the tool at red heat into a vessel of water having about three inches of oil floating on the top; the shank may require tempering if it, too, has been made hard; but that is not always the case. The great point seems to be to get a good heat to the right degree for the steel, and equal all over the part to be made hard. Where several tools are to be hardened it is a good plan to heat them in molten lead, as that ensures the same heat in every part; if, however, any of the lead adheres between the teeth of the tool, that spot so covered will prove soft. Watchmakers, engravers, etc., are very fond of hardening and tempering their small tools at one operation by plunging them red hot into sealing wax repeatedly till cold. The reason seems to be that the sealing wax and oil cause the heat to be abstracted more slowly than if water were used, and therefore make the tool rather less hard and less brittle, so that it does not require the usual "letting down."

HARDENING PLATES OF LAWN MOWER.

The method of hardening lawn-mower cutting plates depends on whether they are of English or foreign manufacture. The majority of English plates are made somewhat similar to a skate blade—that is, with a layer of iron between two layers of steel. To harden such plates, get them to a blood-red heat, grip them in the centre with a pair of close tongs, and plunge them in the water edgeways, moving them about till cold, but taking care to move them so that they go through the water edgeways. In this case the layer of iron helps to keep the blade to its original shape. The majority of foreign plates are made of a different kind of metal, and to harden them without warping they must be cramped between two pieces of iron, made hot, and cooled out as in the former case. It is no drawback to a plate if it is slightly warped, for with a little packing when being screwed on it can be brought perfectly true. Some persons do not trouble to re-harden the plates, as they can be obtained very cheap ready hardened.

HARDENING WATCH PINION DRILLS.

Ready-made drills generally are too soft to cut steel watch pinions; they have to be re-hardened by heating the blades only in a flame and rapidly withdrawing them with a sudden jerk. This is called "flirting" them, and the sudden cooling in the air effects the hardening. Sharpen them before using, and lubricate with turps. Occasionally a pinion is found too hard to be drilled even by this method; it then has to be lowered to a blue temper.

HARDENING AND TEMPERING CYCLE CONES.

If the bicycle bearings are of cast steel, they are hardened by heating to a cherry-red and instantly plunging into cold water or oil. They are then brightened with emery cloth, and, to temper them, are carefully heated until they assume a medium straw colour. If left too light a colour they will probably chip. When made of Bessemer or mild steel they must be case-hardened. Most cycle cones are made from a special steel for which water or oil is the best cooling agent. With this class of steel, heat gradually to cherry-red and plunge into a bath of water having about 2 in. of oil on its surface. If the steel will not harden thus, use water only.

HARDENING FACE OF STEEL HAMMER.

There is some difference of opinion as to how this should be done, and three methods, each vouched for by practical workers, are given below. (1) To harden a cast-steel hammer so that the centre of the face will be as hard as the edges, heat the face to a cherry-red, and see that it is the same heat in the centre as at the edges. Then, when cooling the face in the water, keep moving it about until it is quite cold; brighten it up with a rub stone and let down to a deep straw colour. (2) For heavy hammers the point to be observed is to keep the centre of the hammer face as hard as the edges. Fig. 537 shows the method of hardening a two-faced sledge hammer. Water from the main or cistern is led to the faces as illustrated. If a heavy hammer red hot is kept still in cold water for some seconds, when taken out it will be found to be black at the edges of the face, but the centre of the face will be almost as red as when plunged into the water. It is therefore obvious that some method like that illustrated is necessary to cool the faces equally. (3) To harden the faces of a cast-steel hammer that is several pounds in weight, make the hammer red hot all over, then grasp it with a pair of tongs through the eye; slack one face of the hammer for about 1½ in. up until it is quite cold in water, then slack the other face for the same distance up.

thus leaving the centre of the hammer hot, to bring the faces to the right temperature for tempering. When the faces are cooled out, rub them bright with sandstone, and when they are of a deep straw colour cool them alternately in the water to arrest the softening process, and so continue until there is not sufficient heat in the centre of the hammer to make any difference to the faces. By this means the faces will be hardened and the eye of the hammer left soft. If there is not sufficient heat in the centre of the hammer to bring the faces to the desired colour, a piece of hot iron should be placed through the eye.

HARDENING AND TEMPERING CARRIAGE SPRINGS.

Take the back plate of the spring, heat it to redness, and cool it by plunging in water. When it is cold, re-heat the plate to temper it. To ascertain its right temper, rub a piece of wood along it till the wood flares. When letting the plates down, do not allow them to remain stationary in the fire, but move them backwards and forwards, working from the centre of the plate to one end ; treat the other end similarly. When it is tempered, and whilst it is hot, hold the plate over a saddle tool or in the jaw of the vice and give it a sufficient number of blows to bring it to the right compass. Having finished the back plate, let it lie on the fitting plate edgeways ; make the next plate red hot, place it in its position on the back plate, and whilst an assistant is gripping the two plates together at one end smartly pinch them together with the spring tongs ; then cool out in water, re-heat, and temper as before. Repeat the process with each succeeding plate until the spring is finished. When fitting the plates together, there should be a space between each pair of plates at the centre of the spring whilst the ends are touching.

HARDENING AND TEMPERING CYCLE SADDLE SPRINGS.

In hardening and tempering the spring of a cycle saddle, a good plan is to test a piece bent roughly to the shape of the spring before treating the finished article. The following will probably give satisfaction. Heat the spring gradually to a dull red on a clear fire or with a blow-pipe, plunge in soft water with the chill off, dry and heat in a Bunsen burner or ordinary gas flame, and "flare off" ; that is, when the spring is sufficiently hot, rub some flat or thick oil all over it, and let it burn off. The spring should be just hot enough to set light to the fat or oil. This may be done two or three times according to the thickness of the steel.

HARDENING ENGRAVED STEEL DIE PLATES.

The first and principal point for consideration when hardening engraved steel plates is the quality of the steel. These plates are made of die steel ; they are de-carbonised (or converted into iron) as far as the proposed depth of cut only, leaving the interior of the plate comparatively hard. They are then faced on both sides, the face on which the engraving is to be

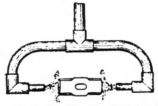

Fig. 537.—Hardening Two-faced Sledge Hammer.

done being highly polished. For hardening the die-plate, an iron box with close-fitting lid or its equivalent, a supply of ivory dust or turnings, and a forge fire supplied with bellows, must be obtained. The plate, after having been engraved, must be put into a warm place to " sweat " out the moisture, any dampness being carefully wiped away. Then place the ivory dust in the box, and blow up the fire until the dust is converted into charcoal, frequently stirring up the contents of the box to get rid of even a trace of dampness. Now put the plates in the box with a layer of charcoal, and ram down tightly, the object being to exclude all air from the plates. They may be placed in tiers, with layers of charcoal between them. The box must, of course, be of a size to accommodate the work to be hardened. The charged box is put into the middle of the fire, the coke heaped around it, and the fire blown up until the box shows an equal red heat all over the outside, when presumably the plates inside the box are at the same cherry-red. Some experience is required to determine this, and care should be exercised, for if the box is overheated the die-plates are quite spoiled, and if underheated the ivory dust is wasted, and the whole process has to be repeated. But assuming that so far all is successful, take off the lid, and plunge the box and its contents into a large cistern of cold water.

Tempering Engraved Steel Die-plates.

Then comes the tempering of the die-plate. While the box is cooling in the cistern of water, put an iron bar in the fire to heat. By the time this is heated the box in the water will be ready to handle. Take out the plate, and wipe it dry. To temper it, take the bar of iron out of the fire, place the plate fairly upon it, and watch in a good light for any change of colour. The colour of its temper will depend very much upon the nature of the work for which the die-plate is to be used. Thus, if pressure on the plate only is to be used, it may be a degree or two harder; but if the plate has to suffer the force of a blow, as from a drop-stamp, it must be softer, and preferably backed with iron.

New Process of Hardening Steel.

By this method, the metal is coated with a mixture of whiting and varnish, heated to a cherry-red, and then dipped for a few seconds in acidulated water. The steel is then dipped in rape oil for a slightly longer time, and is finally laid in a cooling bath of rock oil or a mixture of water and whiting. By dipping the steel first in the water the heat is drawn away from the outer layer, which thus becomes hard. Dipping it in the rape oil retards the cooling of the interior of the metal, and obviates the risk of cracks appearing. More or less fantastic methods of hardening and

tempering are being introduced constantly, but generally they are not to be recommended.

Case-hardening.

Generally, only iron is case-hardened, but steel, if it is mild or made by the Bessemer process, is also suitable. The "Mechanic's Workshop Handybook" says that any articles of good iron can be case-hardened to equal steel, but the hardening penetrates only skin-deep. There is no occasion to temper the iron at all after case-hardening, and any parts of the objects can be rendered hard, leaving the other parts in their normal condition. An ordinary fire will furnish all the heat required. In the first place, provide sufficient of the hardening compound, made as follows: Take equal parts of prussiate of potash, sal-ammoniac, and common salt: pulverise, and thoroughly mix.

The Process of Case-hardening Small Work.

The process is as follows: First make the iron hot, and spread the compound over the part to be made hard: again put the iron in the fire and fuse the powder, allowing it to run all over the parts to be operated on. Up to this stage the metal should be heated to only a moderate extent—say, just bordering on red heat. The compound may be applied several times: the more put on, to a certain extent, the deeper will be the hardening. This is a detail which practice alone will enable one to determine. Having thoroughly melted a quantity of the powder, and allowed it to soak in at a dull-red heat, raise the temperature to that required for hardening steel (a full blood-red), and quench the article in cold water. The surface which has been operated on by the hardening powder will now be as hard as hardened steel, whilst that part which has not had any powder applied to it will be as soft as before. This is the most simple process, and will be found very easy to perform. Prussiate of potash by itself will effect the hardening, but the compound is better. Practice will enable one to judge the exact quantity of powder required to produce the desired effect.

CASE-HARDENING LARGE WORK.

Case-hardening, when carried out on a more extensive scale, is effected by enclosing the articles to be case-hardened in wrought-iron boxes, together with leather shreds, ground raw bones, hydrocarbonated bone black and sal soda, and subjecting them to red heat for a period of from twelve hours, extending in 2 ft. 6 in. in diameter. For small work, tubes of wrought iron or old pulley bosses are used. The bottom of the box is covered with a thick layer of the hardening material, which may consist of bone dust, leather clippings, or hoofs, mixed with salt or charcoal powder. Care must be taken to give the forgings good support among the material, so that they shall not become distorted by their own weight while at a red heat. When the box is filled with alternate layers of metal and

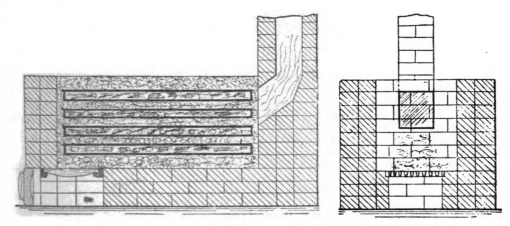

Figs. 538 and 539.—Longitudinal and Cross Sections of Small Case-hardening Furnace.

some cases to several days. At the end of that time — the carburisation having been effected—the articles are hardened by making red hot and suddenly cooling. Exposing the iron to a red heat for some hours, surrounded by leather cuttings or the parings of hoofs and horns, the whole being placed in a metal box of some kind, will likewise make the surface of the iron susceptible of being hardened on being cooled in cold water when at a red heat. The ordinary methods of case-hardening are quite inadequate when large wrought-iron forgings of irregular shape require to be treated. These forgings are box-hardened in the following manner. For the heaviest work, cast-iron boxes of circular form with cast-iron covers are used. They are of sizes suitable for the work in hand, ranging between 1 ft. and of material, the cover is put on, and luted with fireclay to make it nearly air-tight. It is essential that air be excluded. Then it is placed in a fire or, preferably, in a reverberatory furnace, for from ten to thirty-six hours. The time during which the box is exposed to the heat of the furnace mainly regulates the depth to which the metal will be hardened. The chemical activity of the hardening agents, however, influences the result. The addition of powdered yellow prussiate of potash is often an improvement. The forgings are turned out into cold water, and are thus hardened to a depth which ranges from $\frac{1}{16}$ in. to nearly $\frac{1}{8}$ in. ; but in the same forgings the depth of the hardening will not be quite uniform. For light articles, of course, a mere film of surface hardening is enough ; for heavy work the steely casing should penetrate to nearly $\frac{1}{8}$ in. Since hardening distorts the work, the minimum amount of penetration that is consistent with the purpose for which the forgings are required should be imparted to them—$\frac{1}{16}$ in., or a bare $\frac{1}{16}$ in., may be taken as a good average. The distorted outlines have to be corrected with an emery wheel or with emery paper.

FURNACE FOR CASE-HARDENING.

Fig. 538 is a longitudinal section and Fig. 539 a cross section of a small case-hardening furnace which it would be worth time and money to build if large quantities of work had to be treated daily. The furnace has an interior area of 5 ft. by 1 ft. 9 in., and is 2 ft. deep, but these dimensions may be varied to suit requirements. The furnace is built of brickwork, which must be good, sound, and well bedded. The outer case is lined with best quality firebricks fixed with fireclay. A flue is constructed at one end and carried into a chimney-stack at least 15 ft. or 16 ft. high, so as to secure a good draught. At the front end opposite the flue is the arrangement for the fire-hole. The bars are fixed in angle rails, as shown in Fig. 538, and open to the front, this opening being covered by a cast-iron door sliding in and out. Two openings of one brick (9 in. by 3 in.) should be left on each side of the brickwork to admit air; these openings should be closed by iron slides as described above and shown in Figs. 538 and 539. The straight-hatched portions indicate ordinary brickwork, and the curled-hatched portions the firebricks.

OPERATING CASE-HARDENING FURNACE.

The fire is started in the ordinary way, and when well alight is spread over the bottom and covered with new coke to the level of the first tray. The trays must be of good sheet iron. The hardening material, which, as already said, consists of leather, horn, bones, etc., forms in the tray a bottom layer on which the articles are placed. They are covered with the material and the tray is placed in the furnace and covered with a layer of coke. This operation is completed until the furnace is filled up and covered with coke. When the fire is clear and white hot, the iron slides may be placed in the side holes and the whole furnace left to burn for from four to four and a half hours. The articles are then taken out and immersed, if of steel, in oil. If the articles are of iron, they must be immersed in water. They will then be ready, if the hardening is successful, for polishing and finishing.

THE DEMENGE PROCESS OF CASE-HARDENING STEEL.

This has been adopted in many of the principal steel works in France, and though quite outside the scope of the amateur, except perhaps experimentally, may be briefly noted here. It consists in lining one of the vertical sides of the mould with carburising substances. One of the faces of the casting is thus carburised, the action being prevented from penetrating too deeply by casting the vertical side opposite to the carburising side.

DRILLING AND BORING.

HISTORY OF THE COMMON DRILL.

FORMERLY drills were made on occasion from square bar steel, which the blacksmith had twisted and flattened at one end to form a drill. The object of the

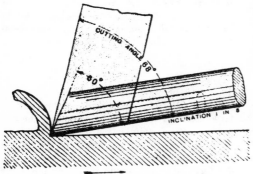

Fig. 540.—Diagram showing Angles of Cutting Edges for Tools.

twisted stem was to screw the cuttings out of the hole, and to some extent this succeeded, but not perfectly. The twisted square section revolving in the round hole had a tendency to crush or grind up the cuttings; and if they were once reduced to powder, it was difficult (especially in drilling vertically) for the drill to lift the powdered metal out of the hole. In most cases the lips of these drills were of such form that the cutting angle, or face of each lip, which ought to have been about 60° (Fig. 540), was 90°, or even still more obtuse; this being an angle which would scrape only, but could hardly be expected to cut sweetly or rapidly. Again, says Mr. Ford Smith, there were attempts to make the cutting angles of the two lips of much the same number of degrees as

that given by the twist itself in a good twist drill. This was done by forging or filing a semi-circular or curved groove on the lower face F of each lip (Figs. 541 and 542). For a short time lips thus formed cut fairly well, but a very small amount of re-grinding soon put them out of shape and made them of such obtuse cutting angles that good results could no longer be obtained from them, and they had to be filed into form again and re-hardened.

A DRILL'S CUTTING LIPS.

To arrive at the best results in drilling, each of the cutting lips should make the same angle with a central line taken

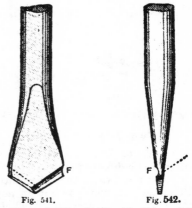

Fig. 541.   Fig. 542.

Figs. 541 and 542.—Drill with Grooved Lip.

through the body of the drill; in other words, the angles A and B (Figs. 543 and 544) should each have exactly the same number of degrees, say 60°. Fig. 545 is a view of the cutting end. The clearance angles also should be identical, and the leading point c should form the exact centre point of the drill. From practice it is found that if these proportions are not correct, the drill cannot pierce the metal it is drilling at more than about half the proper speed, and the hole produced will also be larger than the drill itself, as will be explained later.

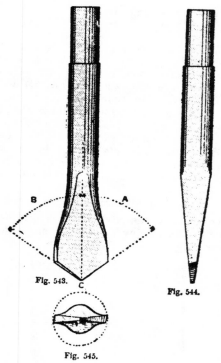

Fig. 543.

Fig. 544.

Fig. 545.

Figs. 543 to 545.—Drill with Lips at 60°.

Two Forms of Drill.

Two forms of drill are commonly used, one cutting in one direction only, as those used on a lathe do always, the other cutting in both directions, which is not an improvement even when used in most favourable circumstances. The cutting edge should be made so that the clearance or angle of relief is 3°, and to give more clearance is only to make a mistake. Tools used on the hardest materials, as well as those used on the softest, are all equally cutting to the best advantage with

only 3° of clearance. Those drills that are supposed to cut both ways, but which really do very little better than scrape away the material, are made so that the two bevels, which produce the cutting edge, enclose about 27°. A straight-shank fluted drill is shown by Fig. 546.

Steel for Making Drills.

The steel used in making drills should always be the very best obtainable for the purpose. The small quantity required for making a drill makes any question of relative cost per pound practically of no moment. "Silver steel"—the rods of bright steel wire sold in foot lengths—commonly serves as stock for making drills, but it is not the best. A good hand-hammered steel of square section is probably not to be surpassed. Hammering improves metal to a wonderful extent, and the hammer can be used satisfactorily only on square steel.

Heating Tool Steel.

Good steel is easily spoiled by incautious treatment in heating. Annealing the raw material first brings it under the influence of heat. A dull red—that is to say, a colour that does not show itself in bright daylight—should never be exceeded. The cooling must be watched very carefully, so that it is equal and gradual. If this is not attended to the steel may show faults subsequently. Properly annealed at a low red heat, the steel may be worked with a file quite easily. Overheated steel is always difficult to work, and when made up into tools it is a failure.

Forging Drills.

Large drills have to be forged under heat, but drills so large that they cannot be shaped cold under the hammer are seldom called for. A light hammer and a large number of gentle taps will spoil the steel at this stage; one or more cracks will be started by this treatment. The correct method of flattening a drill is to use a comparatively heavy hammer, and to strike one smart blow. This spreads the steel and does not crack it.

HARDENING DRILLS.

Hardening comes next, and is perhaps the most critical point in making nearly all tools. Success depends upon heating the steel to the least possible temperature to ensure its hardening when suddenly cooled in oil. What this temperature is must be found by experiment; it will depend chiefly upon the amount of carbon in the steel, and the higher the percentage the lower will be the temperature at which it will harden. Tempering may be dispensed with if the drill has been properly treated every time it has been through the fire. (See also the last chapter, p. 160.)

GRINDING DRILLS.

Extreme accuracy of shape is desirable, but it cannot be produced by hand-grinding; neither can a common drill, having a rough black stem more or less eccentric, be ground accurately, even by aid of a grinding machine with mechanism for holding it. To grind any drill accurately, it must be concentric and perfectly true throughout with the shank, as that part has to be held by the drill-grinding machine.

THE TWIST DRILL.

Nearly fifty years ago both the late Sir Joseph Whitworth and the late Mr. Greenwood, of Leeds, made some twist drills; but it is to be presumed that a large amount of success was not achieved with them, and for some reason the system was not persevered with. After that period the Manhattan Firearms Company, in America, produced some beautifully-finished twist drills, which, though of a superior description, were not durable; the two lips were too keen in their cutting angles, and were apt to drag themselves into the metal they were cutting, finally to dig in and to jam fast, and to twist themselves into fragments. Mr. Morse then took the matter up, and by diminishing by about 50 per cent. the keenness of the cutting lips of twist drills, made a great success of them. He used the grinding line A (Fig. 547), and an increasing twist, as denoted by the dotted lines. Fig. 548 shows the cutting end. In such a drill

of the standard length, and before it is worn shorter by grinding, the twist is so rapid towards the lips that the angle they present, or what may well be referred to as the angle of the cutting surface, is very nearly the same as had been established for cutters that act on metals, as in Fig. 540.

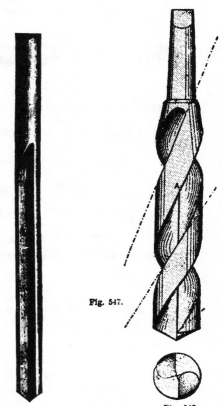

Fig. 547.

Fig. 548.

Fig. 546.—Straight-shank Fluted Drill.

Figs. 547 and 548.—Twist Drill with Grinding Line.

ANGLE OF TWIST IN DRILLS.

If, however, the angle of twist is made to increase towards the lips, it will, of course, decrease towards the shank, as in Fig. 547. The shorter the drill is worn the more obtuse the cutting angle becomes, and the less freedom will it have, supposing, of course, that the angle, when the drill was new, was the most efficient.

Suppose this decrease of twist were carried still further by lengthening the drill, a cutting angle of 90° would eventually be arrived at. The old common style of drill usually has a cutting edge which is so obtuse as not to cut the metal sweetly; on the contrary, there is a tearing action, which puts so much torsional strain on the drill that fracture is almost certain. For this reason the shape shown by Fig. 547 is

contraction in cooling as to be very soft and porous. In such cases it is perfectly impossible to prevent a common drill from running into the soft side. This sort of imperfect hole is most trying to the fitter or erector, and if it has to be tapped to receive a screwed bolt or stud, is most destructive to steel taps. The taps are very liable to be broken, and an immense loss of time may also take place in at-

Fig. 549.—Bit Stock Drill.

condemned. It is therefore obviously advantageous to adopt from the first the best cutting angle for all twist drills, and to preserve this angle through the whole length of the twisted part, so that, however short the drill may be worn, it always presents the same angle, and that the most efficient which can be obtained. (See Figs. 549 to 551, which show three shapes,

tempting to tap the hole square with the planed face. A twist drill, on the other hand, from its construction, is bound to penetrate truly, and produce holes which are as perfect as it is possible to make them.

Lips of Twist Drills.

An important step in twist drill manu-

Fig. 550.—Morse Twist Drill with Straight Shank.

the bit stock drill, Morse modern straight-shank drill, and the Morse modern taper-shank drill.)

When a Drill "Runs."

A common drill may "run," as it is usually termed, and produce a hole which is anything but straight. This means that the point of the drill will run away from

facture has been to fix a standard shape and angle of clearance for both lips. This angle might be tampered with if the regrinding were done by hand, and too much or too little clearance given. If too little clearance (Fig. 552), or in some cases none at all, is given to the drill, the cutting lips then cannot reach the metal; consequently they cannot cut. The self-acting feed of

Fig. 551.—Morse Twist Drill with Taper Shank.

the denser parts of the metal it is cutting, and penetrate into the opposite side, which is soft and spongy. This is especially the case in castings, where, for instance, a boss may be quite sound on the one side, while on the other a mass of metal may be full of blow-holes, or so drawn away by

the drilling machine keeps crowding on the feed until either the machine, or more probably the drill, gives way. Again, if too much clearance is given (Fig. 553), the keen edges of the lips dig into the metal, and embed themselves there, and of course break off.

EFFECT OF FEED ON SHAPE OF DRILL.

To give an idea of the extreme accuracy which must be imparted to a twist drill, it must be borne in mind that even a good

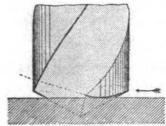

Fig. 552.—Twist Drill with Insufficient Clearance.

feed is only $\frac{1}{100}$ in. to each revolution ; and as two lips are employed to remove this thickness of metal, each lip has only half that quantity to cut, or $\frac{1}{200}$ in. This $\frac{1}{200}$ in. is as much as can be taken in practice by each lip in drills of ordinary sizes. Thus, if one lip of a drill stands before the other to the extent of $\frac{1}{100}$ in. only, the prominent lip, or portion of a lip, will have to remove the whole thickness of the metal from the hole at each turn. The lip of a drill will not stand such treatment ; the prominent lip would either break or become blunted too rapidly. To get over this difficulty, the feed would have to be reduced by one-half, or to $\frac{1}{200}$ in. per turn, which would mean about half the number of holes drilled in a given time.

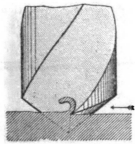

Fig. 553.—Twist Drill with Too Much Clearance.

GRINDING TWIST DRILLS.

The grinding line A (Fig. 547) was introduced in the United States to assist the

operator in keeping both lips of the drill identical. To arrive at this, however, is more than can be accomplished by hand-grinding, as not fewer than three points have to be carefully watched, namely: (1) That both lips are exactly the same length. (2) That both have the same clearance angles. (3) That both make the same angle with the centre line on the body of the drill. If these are not attended to, the drill lips may, for instance, be both ground so as to converge exactly to the grinding lines at the point or centre of the drill, and may still be of such different lengths and angles as to produce very bad

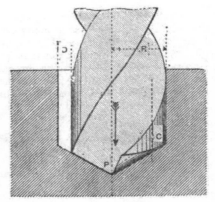

Fig. 554.—Twist Drill with Unequal Lips.

results in drilling. Fig. 554 is drawn in exaggerated form to show the ill effect of grinding one lip of a drill longer than the other. The grinding line is found to be more or less a source of weakness. It is therefore advisable to dispense with it, if possible ; and where a good twist-drill grinding machine is used, the grinding line is seldom or never looked at, and in that case is useless. If it is still desirable to have grinding lines (as when hand-grinding has to be relied upon), they should be made as faint as possible, and not cut deeply into the thin central part of the drill, so as to weaken it.

TEST FOR ACCURATELY-GROUND TWIST DRILL.

When a twist drill is ground accurately,

holes made with it in a well adjusted drilling machine will be so nearly the size of the drill itself that in many cases the drill will not afterwards drop vertically through the drilled hole by its own gravity. This is the most severe test that can be made of the accuracy of re-grinding, and of

each hole, in a hammered scrap-iron bar, till the drill pierced through it varied from 1 minute 20 seconds to 1½ minutes. The holes drilled were perfectly straight. The speed at which the drill was cutting was nearly 20 ft. per minute at its periphery, and the feed was 100 revolutions per in.

Figs. 555 and 556.—Drill Stocks.

the uniformity of all parts of the twist drill.

ADVANTAGES OF TWIST DRILLS.

The whole of the drilling in many establishments is now done entirely by twist drills. Since their introduction it is found that the self-acting feed can be increased about 90 per cent.; and in some engineering works the feeds in some machines have

of depth drilled. The drill was lubricated with soap and water, and went clean through the 2¾ in. without being withdrawn; and after it had drilled each hole it felt quite cool to the hand, its temperature being about 75° F. It is found that 120 to 130 such holes can be drilled before it is necessary to re-sharpen the twist drill. This ought to be done immediately the

Figs. 557 and 558.—Drill Stocks.

been increased by fully 200 per cent., and consequently three holes are now being drilled in the same time as one was originally drilled with the old style of drill and with old machines. It may be interesting to give a few results out of numerous tests and experiments made with the

drill exhibits the slightest sign of distress. If carefully examined, after this number of holes has been drilled, the prominent cutting parts of the lips, which have removed the metal, will be found very slightly blunted or rounded, to the extent of about $\frac{1}{100}$ of an inch; and on this length being

Figs. 559 and 560.—Drill Stocks.

twist drills. Many thousands of holes ½ in. in diameter and 2¾ in. deep have been drilled, by improved ½-in. twist drills, at so high a rate of feed that the spindle of the drilling machine could be seen visibly descending and driving the drill before it. The time occupied from the starting of

carefully ground by the machine off the end of the twist drill, the lips are brought up to perfectly sharp cutting edges again. When the drills run without any eccentricity, there is no pressure, and next to no friction, on the sides of the flutes, the whole of the pressure and work being taken on the ends of the drills. Consequently, they are not found to wear smaller in diameter at the lip end, and they retain their sizes, with careful usage, in a wonderful manner.

Small Drills.

Drills for small work, such as watch and

diameter. The manner of using them is made quite clear by Fig. 561. The Archimedian drill (Fig. 562) may be used instead of the drill-bow stock and drill.

Whirl Drill.

Fig. 563 shows a particularly useful drill that can be bought ready made, or made by the worker himself. The steel spindle

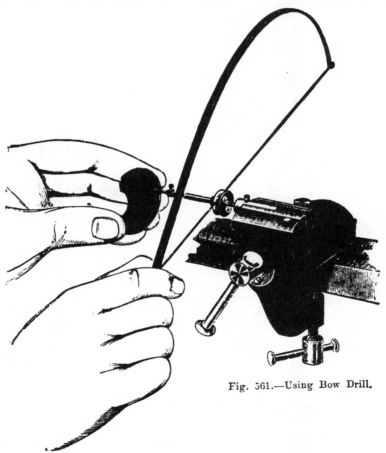

Fig. 561.—Using Bow Drill.

clock work, etc., are used chiefly with one of the forms of drill-stocks shown by Figs. 555 to 560. These are worked with a bow. Any number of drills can be used successively with the one drill-stock. These stocks vary from 3 in. to 4 in. long, and are bored to take drills $\frac{1}{10}$ in. to $\frac{3}{10}$ in. in

is 14 in. long and of $\frac{1}{4}$ in. diameter. Its bottom end may either be the actual drilling point or hold a little chuck in which interchangeable drill points can be held. The top end should be flattened slightly and have a hole drilled through it to take the thong or tape. The finger-piece is of hard wood—box or ebony—and is about 8 in. long by $\frac{7}{8}$ in. thick. Its edges should be rounded off; in Fig. 563 the piece is shown turned. Bore a hole through its centre large enough for the spindle to pass through very freely. A flywheel of wood or metal—the heavier the better—should

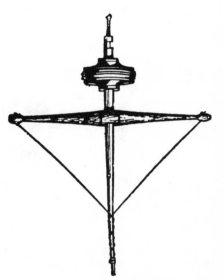

Fig. 563.—Whirl Drill.

Fig. 562.—Archimedian Drill.

be hammered over the bottom end for nearly 2 in. to make a tight fit. Tie a piece of tape or a leather bootlace to one end of the finger-piece, or pass it through a hole which may be drilled in the end, and knot it underneath. Then thread the hole in the top of the spindle and fasten the thong to the other end of the finger-piece, thus completing the drill.

USING THE WHIRL DRILL.

Having attached a drill point and ensured a good starting place, grasp the
finger-piece with the thumb and first finger

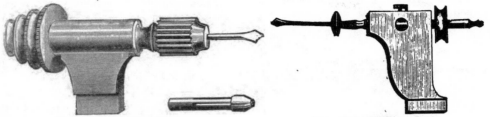

Fig. 564.—Drilling Lathe.

on the left side of the spindle, and by the
second and third fingers on the other side.
Revolve the flywheel until the thong is
twisted round the spindle as far as it will
go. Then, by giving a slight downward
pressure to the finger-piece, the thong will
untwist and turn the drill at a fair speed,
and on the hand being raised at the right
moment, will recoil in the opposite direction. Naturally, practice alone will enable the worker to obtain rapid and
continuous motion.

SIMPLE DRILLING ARRANGEMENT.

An appliance for small work is shown by
Fig. 564. The body of the appliance
passes through a steel tube, which is fitted

shank is hollow, and is cut longitudinally
into three jaws which hold the drill. The
milled cap shown tightens the jaws on the
drill, and internally is threaded so as to

Fig. 565.—Drilling Lathe.

screw on to the end of the shank from
which the drill projects. Each drill is
furnished with two sets of chuck jaws
for holding larger and smaller drill points ;
the drill points themselves ranging from
a very small size to $\frac{1}{4}$ in., which is about
the largest that could be used with good
effect. Similar arrangements in common
use are shown by Fig. 565 to 568.

DRILLING FRAME.

The drilling frame illustrated by Fig. 569
may be readily made, it being formed of
wrought-iron gas or steam pipes and fittings. The lower portion on which it rests
is a flange which has four bolt-holes in it,

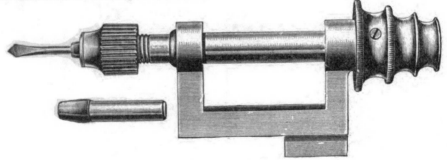

Fig. 566.—Drilling Lathe.

with a foot rebated for $\frac{3}{8}$ in. from the bottom. By this means it is grasped and
held in a vice. On one end of the shaft
passing through the tube is a three-speed
pulley, by which the drill is actuated by
means of a bow. The other end of the

the top part being screwed. Into this is
screwed a piece of pipe of the required
length, and upon it is screwed an elbow,
with another short length of pipe screwed
into it, and on the outer end of this a
T-piece is screwed. The T-piece should

have a tap run right through it, as sometimes a portion in the centre is left unscrewed. The screw for feeding the drill is a short length of pipe screwed right down the outside. The inside should be reamed out with a parallel reamer, and

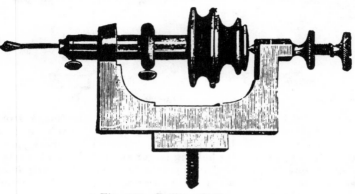

Fig. 567.—Drilling Lathe.

the drill spindle turned and fitted to slide in it. On the top of the screw a double handle for turning it is fixed. The lower end of the drill spindle is made larger, and bored for the drills to fit in, and a handle keyed on the top; the drills are prevented from turning by a small set-screw at one side. The flange is bolted down to a bench, and in order to give more width for drilling a parallel block of wood is fixed under the drill. The size of the pipes should be suitable for the size of hole to be drilled, but pipes $1\frac{1}{4}$ in. or $1\frac{1}{2}$ in. outside diameter will be a fair average size. The T-piece may be a forging, if it cannot be screwed right through.

MAKING DRILLING MACHINE.

Fig. 570 is an elevation of a drilling machine complete. The two wrought-iron uprights A should be $1\frac{1}{4}$ in. wide, like the

rest of the framework. Bend them first, care being taken to get the feet at right angles, and then cut them to length. Mark off the holes, two $\frac{7}{16}$ in. in diameter for $\frac{3}{8}$-in. bolts for the cross-bars. In one upright an extra hole must be drilled $\frac{3}{8}$ in. in diameter to take the hand-wheel shaft. This should be about midway between the $\frac{7}{16}$-in. holes, though the exact position depends on the diameter of the bevel wheels. Drill two $\frac{7}{16}$-in. holes in each foot for the holding-down bolts. The cross-bars B and C have $\frac{3}{8}$-in. holes through the centre to take the spindle F. The key-way in the latter can be cut by a $\frac{7}{16}$-in. cross-cut chisel, and afterwards cleaned out by a small square file. Next obtain a pair of bevel wheels of the same pitch, one wheel, if possible, having twice as many teeth as the other. The wheels should be drilled $\frac{3}{8}$ in., the key-way in the larger wheel on the vertical spindle being parallel, that in the

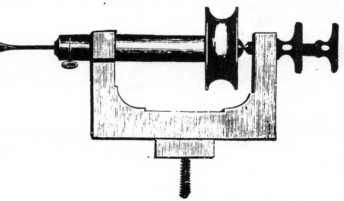

Fig. 568.—Drilling Lathe.

wheel on hand-wheel shaft E being slightly taper depthways. One end of the horizontal shaft must have a $\frac{7}{16}$-in. key-way, and the wheel should be knocked on and then keyed up by a small key, preferably with a head. At the other end, the hand-wheel, from 8 in. to 10 in. in diameter, is attached either by a screw or by a square on the shaft. The wheel on the spindle F must work easily when a small parallel key is placed in the slot. The frame being bolted up, make the upright stay D so that it will just go between the two cross-bars; drill a $\frac{7}{8}$-in. hole at each end, and put the stay in position. Now, with the spindle in position, with the wheel on as in Fig. 570, and with the other wheel in gear but off the shaft E, the $\frac{3}{8}$-in. holes in D and A can be marked off, and also the holes in the cross-bars B and C. For the feed gear, a piece of brass or wrought iron may be cut to shape (Fig. 571), and two $\frac{7}{8}$-in. holes and one $\frac{3}{8}$-in. hole should be drilled through

prove the appearance. Fig. 572 illustrates a form of drilling machine that is in common use among watch- and clock-makers. It is suitable for most small work.

USING LATHE FOR FINE DRILLING.

A worn lathe can be made to do fine drilling, no matter how badly it may be

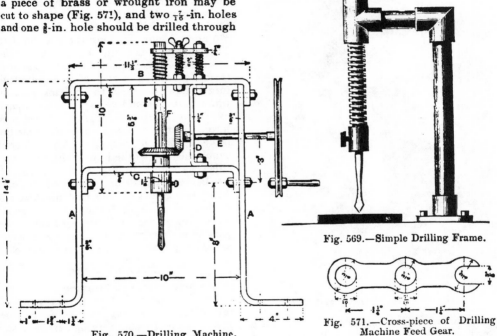

Fig. 569.—Simple Drilling Frame.

Fig. 570.—Drilling Machine.

Fig. 571.—Cross-piece of Drilling Machine Feed Gear.

it, the $\frac{3}{8}$-in. hole being cut out afterwards. Round the spindle is coiled some brass wire, coils also being wound round the two studs which are fastened to the top cross-bar by $\frac{3}{8}$-in. nuts. The two studs are screwed throughout the lengths. The feed is put on by a wing-nut on the centre stud, the springs bringing the spindle back when the wing-nut is released. A coat of black enamel over the fixtures will greatly im-

out of truth, or how loose the spindle may be in its bearings. So long as there is a turning motion at one end of the lathe bed, and the capability of pressing the work forward at the other end with either a hand-wheel or lever, all kinds of drilling, thick and thin, heavy and light, can be done, but there must be a handy chuck for holding drills and a variety of tools for drilling. If one of them is placed in the chuck and the jaws shut down solid upon it, and the lathe set in motion, it will be noticed that the drill does not run true, but a few blows with the hammer will soon fix this, so that at least the point will remain in one position except when the belt fastening passes the driving pulley when it vibrates for a moment. The body of the drill has a peculiar motion, and it is here an old lathe has the advantage—the drill is ready to enter wherever it finds a cavity to settle in, and the looseness of

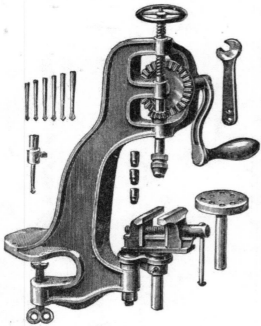

Fig. 572.—Drilling Machine and Accessories.

the spindle allows for an adjustment on a new centre ; besides, the drill is ground a little uneven, and is cutting a hole larger than the body of the drill, which is all the better, as this part is so badly out of truth. As the depth increases, a wobbling motion comes in at the sides of the drill, striking against the walls of the hole ; this does not matter—the drill can reach the full depth without breaking off in its socket or slipping in the jaws of the chuck. Whenever the edge is worn away, the

wrench may be hooked on to the feed lever, while with the pipe-wrench the chuck can be set up so solid that there is no need to use the three flat places (if they exist, and they are not recommended) ground on the shank of the drill.

MAKING D-BIT FOR LATHE USE.

The bits about to be described, D, rose, and enlarging, are adapted for the lathe, and it may here be pointed out that the work of the driller is greatly simplified in many cases when he has the use of a lathe. The D-bit is adapted for boring long or deep true holes of diameter too small to permit of their being bored by means of an ordinary bar or tool (see p. 185). Fig. 573 is an elevation, Fig. 574 a plan, and Fig. 575 an end view showing the cutting part. The general shape in section is semicircular, hence the name of D-bit. The tool is made from a piece of square steel, forged down to form the cutting part, at which end it should be, when in the rough, as shown in Fig. 576 ; a small projection is formed in the middle to accommodate the lathe centre in turning. The material should also be worked up at the corners or sides, as shown in Fig. 576, to permit of its being gauged or callipered to size. After being forged, the tool must be annealed to allow of turning and filing it. Annealing is accomplished by heating the part to a cherry-red and burying it in fine ashes or quicklime until it becomes cool.

TURNING D-BIT.

The process of metal turning will be fully dealt with later, but in the meantime a few hints on turning the D-bit may be given. A half-round forging can be turned only with the assistance of a slide rest. If, therefore, hand tools only are available, it will be necessary to forge the cutting end circular in section, and form the flat part afterwards, and this in the case of a moderately large tool would be tedious. In turning the tool, it is carried between the lathe centres in the usual way, and driven through a carrier fixed on the square end. The turned part must be made nicely parallel and finished smooth.

The turned half-round part should never be less than two or three diameters in length.

CUTTING END OF D-BIT.

The end of the D-bit may be roughly faced to assist and guide the act of filing up the front rake or clearance, as shown in Fig. 573. Not more than 3° or 4° of rake or clearance should be given, as the tool works steadier the nearer the front cutting edge is kept square to the axis of the tool. This edge also must be slightly slanted or bevelled, as shown in Fig. 574, and whatever radius is given to the cutting corner must be slightly increased as the corner is followed round so as to make sure of its clearing itself. The top side

plunge it in bodily for a moment, withdrawing it and immediately brightening the end with a piece of sandstone so that the colours may be noted as the heat from the partially cooled shank spreads back to the end which was gradually immersed at first. The temper should vary according to the material to be operated upon. For brass and cast iron a dark brownish straw colour will stand well, but for iron or steel it should be somewhat softer—say purple, or dark blue tinged with red. Always remember that the lighter the straw colour the harder the tool, and as the blue tints become paler so the temper softens, making the tool more unfit for hard work. In any case, immediately the colour is reached the tool must be plunged into the water

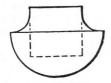

Fig. 575.—End View of D-bit.

Fig. 573.—Elevation of D-bit.

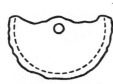

Fig. 576.—Forging D-bit.

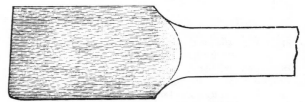

Fig. 574.—Plan of D-bit.

must be filed down to the centre line, and this will be facilitated by drawing a line across the end for guidance. For cast iron and brass, the upper surface forming the cutting edge may be left straight, as shown by the full line in Fig. 573, but for wrought iron the cutting angle should be reduced by filing the relief shown by the dotted curve.

TEMPERING D-BIT.

The D-bit, having been finished to size carefully, must be tempered. First heat the cutting end to a bright or cherry-red heat, and immerse it gradually in lukewarm water to a depth of about 1 in., and then

and finally cooled. If possible, use the tool while it is slightly warm, and if the material is hard it will stand better if used without grinding.

USING D-BITS IN THE LATHE.

D-bits are used in the lathe as follows: Suppose the work to be operated upon is set and fixed in the chuck; bore a short distance at the mouth of the hole to the exact size of the bit. A distance of about ½ in. will be sufficient, or just enough to cover the bevelling on the cutting edge and ensure a fair bearing for the bit to start upon. The tool should be held in the slide rest by the square shank, but it must be very carefully set to agree with the lathe centres. It may be used, supported, and fed or advanced by the poppet head, the centre in the end of the shank being utilised for this purpose, and the tool kept from revolving under the pressure of the cut by means of a spanner held in the hand, or a carrier fixed on the shank and supported by the **T** or slide rest. (These instructions will be better understood when a later chapter on lathework has been read.) In all cases the tool must be kept well lubricated from start to finish, and this in small deep holes will be readily accomplished by the aid of a syringe or squirt. The bit should never be allowed to become rusty or be otherwise neglected.

ROSE-BIT.

Figs. 577 and 578 give profile and end view respectively of a rose-bit: this is

cleared to the shank by a shoulder, must be made quite parallel; indeed, it is safer to make the cutting end a shade larger than the other end of the gauge part to prevent jamming. The conical part, out of which the teeth are formed, must also be turned, and the centres should be left in both ends of the tool for future convenience. The grooves shown running along the gauge part are for lubrication, and may be cut in the lathe by fixing a round-nosed tool on its side in the slide rest, and traversing the rest or saddle by hand. The teeth are formed by filing, and care must be taken that they are all truly cut to the cone. The square taper part must be well fitted to the drill spindle, as nothing injures a socket more than a badly-fitting shank. For such turned tools as this, the square should be preferably replaced by a turned taper secured by a key

Figs. 577 and 578.—Two Views of Rose-bit.

specially adapted for accurate work or where interchangeability is necessary. It is not suited for heavy cutting, being employed chiefly for finishing holes which have been rough drilled or bored within a fraction of the finished size. It is, however, better adapted for vertical work, such as in a drilling machine, than for the lathe, on account of the difficulty of lubricating it sufficiently while it is in a horizontal position.

MAKING ROSE-BIT.

For making this tool, obtain a piece of steel big enough to allow for turning. The end is tapered and squared, as in Fig. 577, to fit the spindle of the drill, if it should be required to use it in both machine and lathe. The material must be annealed according to the directions already given for the **D**-bit, and then turned up to the required size, just easy to the callipers or gauge. The gauge part, which is

or setscrew; it is a matter of difficulty to get a square driven bit to run true.

TEMPERING AND USING ROSE-BIT.

The tempering of the rose-bit may be carried out exactly as described for the **D**-bit. All scale must be cleared off, and the bit polished after tempering to secure a smooth hole. It must not be used at a high speed, especially for deep holes where expansion from heat might prove troublesome. When this tool is used in a lathe, it is held by means of the poppet head.

ENLARGING DRILL AND SUBSTITUTE FOR **D**-BIT.

A cheap and handy drill is shown in elevation, end view, and plan by Figs. 579 to 581. It is more suitable for enlarging holes which have already been bored, but it is a serviceable substitute for the more expensive **D**-bit. It consists of a piece of flat steel of suitable length and thickness, the latter being proportionate to the diameter of the hole, say about one-fourth of that diameter. The cutting end is formed like a common drill, but it need

slide rest, the boring of a small or moderate-sized cylinder presents but little difficulty. If the lathe has a proper four-jawed expanding chuck of suitable size,

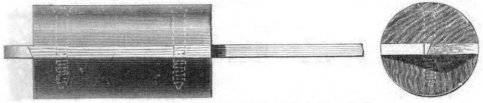

Figs. 579 and 580.—Two Views of Enlarging Drill.

not be brought to a point; the dotted line shown would give enough cutting edge. The tool is turned on the edges for a portion of its length, and pieces of hard wood are fastened on each side, these also being turned to the size of hole required. The pieces of wood keep the drill central and steady while cutting, and are secured by screws passing through the bit as shown. Where a heavy cut has to be carried, it may be necessary to cut clearance for the chips through the wood: but this is very detrimental to the latter, and it is better not to cut any clearance, but withdraw the tool and clean out the hole at intervals. This tool is held up to its work in the manner described for the D-bit. Tempering is also similar, except that it should not be made quite so hard on account of its being thin, thereby rendering the corners liable to give under a heavy or sudden cut. The centres should be left in the ends of this tool to facilitate the

the operation becomes quite simple and easy. First, the cylinder is set truly and gripped firmly by the back or bottom flange, as it is important that the front or top flange which receives the cover carrying piston-rod stuffing-box be faced before shifting the cylinder after boring. This also applies to all cases of facing and boring where truth is indispensable. If possible, there should be sufficient clearance between the back flange and face of chuck or dogs to allow of a hook tool being used for partly facing this one also before removal; this will be found very handy in resetting the cylinder when turned end for end to finish the flange. In boring, an ordinary hook tool may be employed, or a cutter bar, but either of these must be very strong and stiff, to prevent chattering. In the case of a small cylinder, and in the absence of a four-jawed expanding chuck, a suitable bell chuck may be advantageously used.

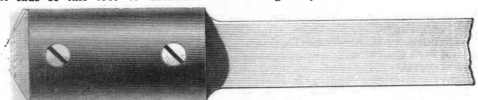

Fig. 581.—Plan of Enlarging Drill.

mounting of fresh wood packing from time to time.

BORING SMALL STEAM ENGINE CYLINDERS.

To those possessing a lathe with traversing, or self-acting, motion, or even a good

USING COMMON LATHE FOR BORING.

If the boring has to be carried out in a common lathe provided only with T-rests, the work is more difficult. D-bits, broaches or rose bits, or drills have to be specially prepared in accordance with the bore required, and this has to be done beforehand, while the lathe is unoccupied by the cylinder. The cylinder is first chucked and set, and with the hand tools it is bored out to size of drill for a short

there should be a squared end by which it may be driven. A saw kerf must be run up the centre, and a thin steel cutter fixed by screws, as shown. The thickness of this tool may be regulated by the saw cut,

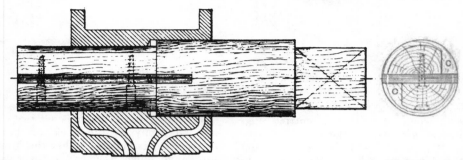

Fig. 582.—Section of Cylinder with Boring Tool.

Fig. 583.—Clearance for Borings, etc.

distance—say ⅛ in. or ¼ in.—so that the bit, broach, or drill may be started truly. The feed or advance of the tool is given by the poppet-head, and to facilitate this the centres upon which the tool has been prepared must be left in. The tool may be kept from turning, under the strain of cutting, by means of a carrier or spanner, and the edges must be kept well lubricated with oil during the whole process. The rate of feed must also be regular, as far as possible. One flange may, of course, be faced without shifting, but the other must be finished on a mandrel.

SERVICEABLE BORING TOOL.

In the absence of a first-class lathe, replete with expanding chucks, etc., and provided with special bits and drills, many cases of need must be met by the adoption of makeshifts, and for this particular job may be employed a tool similar to that illustrated in Figs. 582 and 583, where it is shown in the act of boring a cylinder. This simple but serviceable tool consists of a piece of hard wood—beech or oak—turned up truly in two sizes as shown, the larger equal to the finished bore, and the smaller to fit the rough bore as cast. Each of these sizes must accommodate the length of the cylinder. The

large diameter should be rather longer; and the shape is given in Fig. 584. It must fit nicely into the boring stem, so that when screwed up tight the cylindrical stem will not be distorted. Clearance, as shown at c c (Fig. 583), must be cut along the small part to allow the borings to fall away and prevent jamming in working.

OPERATING THE BORING TOOL.

The tool as above described may be

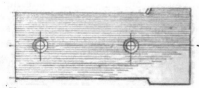

Fig. 584.—Outline of Boring Tool.

worked by means of a wooden cross handle fitted on the square, the motive power being supplied by hand, and the feed given by means of a suitable weight laid on the handle. The cylinder under operation may be held in a vice, or otherwise secured, provided always that there is room for the tool to clear through the bottom. After boring, the flanges must be faced on a mandrel. It is advisable to smooth the rough bore, as it comes from the foundry, with a round file, so as to give the tool as much fair play as possible. A fairly good bore can be obtained at one cut, but to ensure accuracy two operations with different-sized tools are desirable.

PACKING-UP LATHE HEADS.

When the cylinder is too large for the lathe, the heads can be packed up to meet the case. This is commonly done in prac-

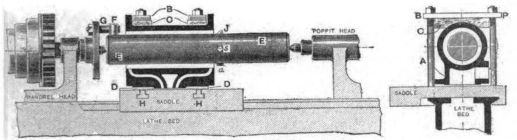

Figs. 585 and 586.—Boring Cylinder in Lathe.

BORING LARGE CYLINDERS IN THE SELF-ACTING LATHE.

The method about to be described renders necessary the use of a suitable self-acting lathe with a good saddle, and its advantage is that a practically parallel bore is obtained with a minimum amount of trouble. The wear of the tool in cutting prevents a perfectly true result being obtained when a large surface has to be cut. All that can be done is to make sure that the tools used for the finishing cut are hardened and tempered properly. Figs. 585 and 586 show a front elevation and part section and also cross section of the principal part of the lathe; the cylinder, shown in solid black, is mounted and fixed on the saddle ready for boring. For the sake of clearness, the cylinder is shown in section, and to economise space part of the lathe heads and bed is omitted. The cylinder is held down to the saddle by bolts A and straps or plates B. These straps bear on pieces of hard wood C, fitted to cylinder body. Liners, or packing pieces of hard wood, for keeping the cylinder at its proper height, are interposed at D. E is the boring bar, F its carrier, and G the driver. The T-shaped slots at H are for the holding-down bolts A. Figs. 587 and 588 give details of one method by which the cutting tool J is adjusted and secured in the bar E.

tice, and more especially in small shops where the choice of machines is limited, and their range small. Some engineers keep cast-iron packing pieces for this purpose, but hard wood answers quite as well,

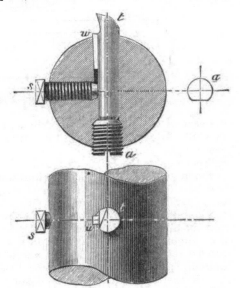

Figs. 587 and 588.—Section and Part Elevation of Boring Bar.

at a pinch. But take care that, when the lathe feed is taken from the leading screw, the heads are not packed up beyond the reach of the change wheels; where the feed is obtained from an independent back shaft, it is only necessary to lengthen the small driving belt as required. Figs. 585 and 586 represent the boring of a 3½-in. cylinder in a 4½-in. lathe, with a boring bar 2¼ in. in diameter.

Mounting and Setting Cylinder for Boring.

In mounting and setting the cylinder, proceed as follows: The slide rest having been removed, and the face of the saddle cleared, the cylinder is packed up to agree roughly with the lathe centres, and lightly bolted down by means of the straps B and bolts A. The bar is put in place between

be thrown upon the self-acting gear, and a breakdown may result. It will be found perhaps that the strain on the holding-down bolts A has sprung the saddle slightly and caused it to bind on the V's of the lathe bed. This must be watched, and the wedge piece in front of the bed adjusted accordingly.

Boring Cylinder and Facing Flanges.

Everything having been set and adjusted, a fair start on the boring may be made. Two cuts, a roughing and a finishing one, will be sufficient, and the feed should not be heavy, but at the rate of about forty cuts per inch. The roughing cut should be arranged so as to leave about

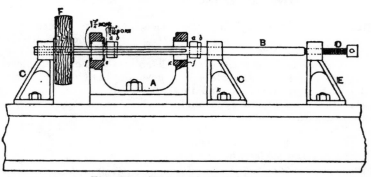

Fig. 589.—Boring Fast Headstock.

the centres, and the bore of the cylinder set carefully to it by means of a tool stuck in temporarily for the purpose, or by the callipers or V-foot scribing block. If it is found to be too low, small alterations can be effected by the insertion of strips of paper under the liners D; if it is too high these liners must be planed down, making allowance, however, for the compression of the liners. Lateral adjustment can be readily given by blows, where required, with a wooden mallet or block. Check the bore with the outside of the cylinder, to ensure the thickness being equal all round. The adjustment of the saddle upon the lathe bed is important. If this is neglected, and it happens to be slack, the bore will be oval, owing to side play. If it is too tight, undue strain will

$\frac{1}{32}$ in. for the finishing one. Unless a circle has been struck on the flange for guidance, the tool will have to be set to depth by trial, measurements being taken constantly with a pair of inside callipers set to size. The roughing tool may be of the shape given in Figs. 587 and 588, but the finishing one must be flatter on the cutting edge, not so pointed, but thicker and bluffer, so as to guard as far as possible against chattering. A slight alteration in the shape of the cutting edge of a tool makes a great difference in its performance. By substituting knife or side cutting tools for the boring ones, the flanges may be faced in position, but this must be carried out carefully to avoid shifting the job. The whole surface of the flange should not be attacked at once; but a narrow ring should first be faced, commencing from the bore, then another larger one brought to the first, and so on, shifting or lengthening the tool by degrees until the whole surface has been roughed down. During these operations the feed must be given by hand, and the rough facing should be done after the rough and before the finished boring cuts. If possible, the flange of the steam chest, or rather slide face, on which the cylinder is shown resting, should be secured by wedges or stops to prevent its twisting or shifting laterally upon the saddle. Before running the finishing cut through the bore, the holding-down bolts must be slacked off slightly but equally, so as to avoid, as far as pos-

bar should always be carefully drilled to fit those of the lathe, and the one running on the dead centre of the poppet must have a small hole drilled up a short way so as to clear the centre point, as well as to hold a small supply of lubricant; the centre must not be allowed to run dry or squeaky. This is important, as the finishing cut must be kept going without the least stoppage from start to finish. Fig. 587 is a section of the bar, showing the method of fixing and adjusting the tool. Referring to Figs. 587 and 588, t is the tool adjusted vertically by the screw a and secured by the set-screw s, the latter being further assisted by the wedge w, which fits into a suitable slot and tightens against a flat upon the

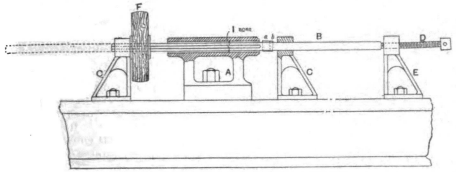

Fig. 590.—Boring Loose Headstock or Poppet.

sible, flattening or springing the cylinder; but in the case illustrated, this, owing to the proximity of the bolts to the flanges, and the small size of cylinder, would hardly be of appreciable extent. It becomes important, however, in the case of large cylinders. By means of a suitable mandrel, the flanges may be faced upon it by a distinct operation, and with considerably less trouble than with the boring bar.

THE BORING BAR.

The boring bar should always be as stiff as possible, and therefore as large in diameter as reasonable clearance for chips will allow. For large work, a block or head is keyed on the bar, and the cutters are carried in that instead of in the bar as illustrated. The centres in the ends of the

front of the tool, as shown clearly in the section to the right hand of Fig. 587. The set-screw s should be of steel, and slightly hardened at the point. The hole through the bar for the tool is first drilled, and then tapped at the bottom to receive the adjusting screw a. The hole for the set-screw is then drilled and the slot for the wedge cut, after which the set-screw hole must be tapped. Both of these screws must fit well into the holes so as to guard against their working loose. The wedge w must also be of steel, but it should not be hardened. The tool should fit well into its bed, not too tight, but just a nice driving fit. It can be backed out easily when necessary by removing the adjusting screw a, and using a drift. If, on account of limited clearance, there is no room for the head of the set-screw s, this may be taken out, and the tool held only by the wedge w, but this wedge must not be omitted, as on it depends the rigidity of the tool.

Fig. 591.—Template for Bracket.

BORING LATHE HEADSTOCKS.

In making a lathe from bought castings, the fast and loose headstocks (here distinguished as headstock and poppet) have to be drilled or bored, and such work is typical of many operations in everyday workshop practice. The castings of the heads are supplied either solid or cored to suit purchasers. Where a stout drilling machine is available, it is better to drill through the solid metal, at least in the fast head. It will be more convenient to most workers to have the poppet cored, though the fast head should be solid. The diameter of the cored hole in the former will be $\frac{3}{4}$ in., so as to admit a $\frac{5}{8}$-in. boring bar, and it should not be smaller. Two traverses will be necessary in boring, a roughing and a finishing cut.

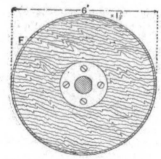

Fig. 592.—Pulley for Boring Bar.

RIG-UP FOR BORING HEADSTOCKS.

The rig-up for boring the headstocks is shown in Figs. 589 and 590, the first representing the boring of the headstock, and the second that of the loose poppet. The same accessories are used in each case. A larger boring bar might be used for the headstock, but it is not necessary to go to the trouble of making two bars. The headstock A (Fig. 589) being bolted firmly, the boring bar B with its cutters is revolved in bearings C, and fed forward by the screw D in the bracket E. The bar is of iron, though preferably of steel for stiffness; it is fitted with flat cutters a held in slots with wedges b. The bores are alike at each end, namely $1\frac{3}{4}$ in., tapering to $1\frac{1}{16}$ in. Round cutters could be fitted to the bar, entering into holes simply drilled across the bar, and would be more easily fitted. This would remove too much metal from a $\frac{5}{8}$-in. bar, which would be further reduced by the insertion of a pinching screw, the head of which, too, would have to be countersunk to pass through the hole cored in the poppet. Using flat cutters as shown, with the leading corners

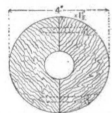

Fig. 593.—Driving Pulley.

rounded, the same tools serve for roughing and finishing, being driven out a little farther from the centre of the bar after the first traverse.

CASTINGS TO CARRY BORING BAR.

The castings C, in which the bar is carried, are pierced right through to receive the bolts, by which they are held to the bed with a shallow tongue between the shears. The brackets in front view should be broad enough to bear well across the shears. They may follow the outline of the template shown in Fig. 591. The metal in the bracket webs may be from $\frac{5}{16}$ in. to $\frac{3}{8}$ in. thick, and that in the foot $\frac{1}{2}$ in. or $\frac{9}{16}$ in. The holes to receive the bar must be bored accurately in respect of height of centres, of position midway between the bed shears, and of bore. To ensure this, fit the bases first to the bed shears; then use a template (Fig. 591) of sheet metal, filed at the bottom edge to the same shape as the feet of the heads, and having a hole bored in it of the same diameter as the bar. This should be stood against the faces of the castings c on the lathe bed and the holes carefully scribed off; then the castings can be drilled, and afterwards bored to the scribed circles, either in a lathe or under a drilling machine. The reversal of the template against the heads will indicate whether the centres are over the centre of the shears.

TAPERING BORED HOLES.

The holes in the headstock are roughed out parallel, but must be finished slightly taper, to permit of the driving in of the bushes. The finish to taper must be done

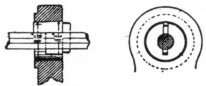

Figs. 594 and 595.—Finishing Cutter for Boring Headstock.

with tapered cutters (see Figs. 594 and 595). They are shouldered to bear against the bar to maintain the diameter, and to steady them. The taper in the holes is in opposite directions, being from the outer ends of the headstock in both cases. After the front hole, therefore, has been finished, with the headstock and brackets in the positions shown in Fig. 589, the hinder hole may be tapered by turning the bar end for end, and moving the bracket E to the left hand.

FACING-UP AROUND BORED HOLES.

The faces e and f (Fig. 589) around the bored holes must be trued up at right angles with the holes; just a light skim over is sufficient. This is effected (without moving the headstock from its place) by means of a facing cutter, shown enlarged in Figs. 596 and 597. It is fitted in the slots made for the boring cutters. It may

be made to cut on one face, or on both, as shown, the latter being the better arrangement. The very slight feed forward necessary is imparted in boring. These cutters are made from tool steel, fitted while soft. They need not be tempered, but may be merely hardened, as they have no real edges, and operate by scraping only (see the chapter on surfacing metals).

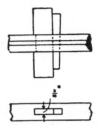

Figs. 596 and 597.—Facing Cutter.

DRIVING BORING BAR.

The boring bar B (Fig. 589) is driven through the pulley F, shown enlarged in Fig. 592. As the boring bar is fed forward by hand by the screw D (Fig. 589), the bar is pushed through the pulley, which is prevented from rotating round the bar by means of a flat filed on the bar as a sub-

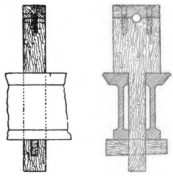

Figs. 598 and 599.—Wooden Bracket to Carry Boring Bar.

stitute for a key groove, and two metal plates are screwed to the pulley faces, having flats to match the flat on the bar. The bar always remains in contact with the bracket c. The alternative to this device would be to have a pulley as wide as the traverse of the bar, plus the width of the belt. The pulley F is driven from a wooden pulley made in halves (see Fig. 593) on the crank axle. The slowest speed on the driving wheel would be too great. Use a belt ¾ in. wide, and, as this is only for a temporary purpose, a pair of common luggage straps, fastened together, will do very well. The buckles permit of ready tightening up, and are not objectionable at the slow speed that is required for boring.

BORING POPPET.

The boring of the poppet A is illustrated in Fig. 590. The bore is so long that it will be necessary to shift the back bracket E forward when half the length has been bored, or else to make the screw D long enough to cover the entire traverse. The cutter and bar B will have a tendency to follow the cored hole, if the latter is a little out of truth. It may be advisable to take three light traverses to avoid wobbling. The cutter must be kept sharp and backed off, so that it shall only cut on the slightly rounding leading edge. It will assist the boring if the chips are blown out with bellows. The ends are afterwards skimmed over lightly with a facing cutter.

WOODEN BRACKETS TO CARRY BORING BAR.

If casting the brackets is troublesome, it is possible to substitute brackets of hard wood as shown in Figs. 598 and 599. This is a parallel chunk of beech, birch, or oak, fitted on the bed, and having a tongue coming down between the cheeks for fastening with a mortise and wedge. The hole for the boring bar may be bored in the solid with a centre-bit, and finished with a gouge. But it will be better to fit a wooden cap, as shown, and to cut the bearing out with a gouge in the block and its cap.

TAPS, SCREW-PLATES, AND DIES.

Screws Used in Metalwork.

ONE of the most useful contrivances ever invented is the screw in its various forms. It is employed for an infinite variety of purposes, and is used almost universally to hold together two or more pieces of material. It is used as a means of imparting motion in gearing, as in the worm and wheel, the worm being simply a fixed screw turning on its axis. A great many presses

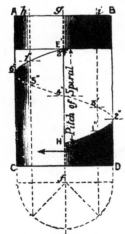

Fig. 600.—Formation of Spiral of Screw around Cylinder.

are worked by, and owe their great power of compressing substances into small bulk to, the screw, which in these machines is subject to great strains. The breech mechanism of large and small guns, where the strain is enormous, is kept in position by a screw. Then, again, the fine adjustments of the microscope and the almost inappreciable movements of the measuring blocks in the machine for measuring to the 1,000,000th part of an inch, owe their

minuteness and accuracy to the screw, which in these two cases is subject to a very slight strain. From the above instances—just a few of those that could be given—it will be seen between what extremes the screw can be usefully employed.

What is a Screw?

The following information applies solely to the screws used in metalwork. The forms of the threads vary according to the purpose and the material for which they are employed. Most persons have some idea of a screw or spiral, but many do not

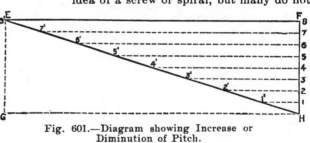

Fig. 601.—Diagram showing Increase or Diminution of Pitch.

know how to define it and how it is obtained. A screw has been defined as a short-pitch spiral, and a spiral as a long-pitch screw, but such definitions convey hardly any information. Below is shown how to obtain a spiral line practically. Take a cylinder, A B C D (Fig. 600), of any convenient size; also a rectangular piece of paper, E F G H (Fig. 601), whose length must be exactly equal to the circumference of the cylinder—that is, the points E G must coincide with F H respectively when the sheet is wrapped round the cylinder. Cut off the dotted portion E G H of this rectangle, and wrap the remaining triangle E F H again round the cylinder, making F H parallel with the

axis. The points E and F should now
coincide at E' F' (Fig. 600), and the line H' E'
in its winding round the cylinder forms a
spiral or screw, the pitch of which is equal
to the depth F H (Fig. 601). The pitch
of a screw or spiral is the distance the

Fig. 602.—Whitworth Screw Thread.

thread or line rises during one complete
revolution. For instance, if the cylinder
is turned once round in the direction of
the arrow, meanwhile following the line
from H' (Fig. 600), it will have gradually
risen from H' to F' or E', which distance,
as before stated, is the pitch of the spiral.
By increasing or diminishing the line F H
(Fig. 601) the pitch is increased or
diminished. By this means the pitch
can be varied from the smallest fraction of an inch, as in some watch
screws, to almost anything, if the body
round which it winds is large enough.
A good illustration of a long-pitch small
diameter screw is the egg-whisk, where,
by sliding a wooden nut backwards and
forwards along a twisted wire shaft, the
whisk rotates first in one direction and
then in the other ; spiral staircases, on the
other hand, are an example of big-pitch
screws. When once the method of obtaining a spiral line is grasped, the student
will follow with comparative ease the
drawing of screw or spiral threads, as
these merely consist of combinations of
spiral lines.

Shapes of Screw Threads.

Unwin defines a screw as " a cylindrical
bar on which has been formed a helical

projection or thread." These projections,
in order to suit the different purposes for
which they are intended, are of various
shapes, of which the four principal forms
are shown in section in Figs. 602 to 605 :
but the first two only—the V and the
square thread—are in very general use.
Fig. 602 shows the ordinary V or triangular thread, as used in all English workshops. The use of a common thread was
early recognised by our manufacturers to
be of the very greatest importance, and as
Sir Joseph Whitworth was the first to pro-
pose a uniform system of threads, it is
known by his name. The Americans use a
somewhat similar thread, called "the
Sellers," after the introducer.

The Whitworth Thread.

Fig. 602 shows the method of designing
the Whitworth thread. Draw two parallel
lines 0·96 of the pitch apart. Set off the
pitch on one of these lines, and at the
points so obtained form angles of 55°, as
shown. Again, draw two parallel lines
centrally between the first two, and 0·64 of
the pitch apart. The threads where these
lines cut them, or one-sixth the depth of
the thread, are rounded off at the top and
bottom. The rounding off facilitates the
cutting of the screw, and renders it less

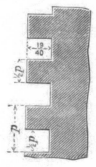

Fig. 603.—Square Screw Thread.

liable to damage. When these screws are
used in conjunction with nuts, they are
termed bolts, and one of their drawbacks
is that, as will be seen from the slope of the
threads, when tightened in their nuts,
the greater the force used to screw them

up the greater is the tendency to burst the nut. For wrought-iron pipes a finer pitched screw thread is used, and is known as the gas thread; it does not cut away quite so much of the metal of the tube. The only threads in general use are Whitworth's; they form the basis of the threads on the ordinary. coach bolts, and all users of screws and bolts should possess a set of Whitworth taps, embracing all those sizes commonly in use. The angle of these threads is 55°, and $\frac{1}{6}$ of the depth is rounded off the tops and $\frac{1}{6}$ off the bottoms of the teeth. The following table gives sizes and number of threads per inch up to one inch:—

Diameter in in.—$\frac{1}{16}$, $\frac{1}{4}$, $\frac{5}{16}$, $\frac{3}{8}$, $\frac{7}{16}$, $\frac{1}{2}$, $\frac{5}{8}$, $\frac{3}{4}$, $\frac{7}{8}$, 1.
Threads per in.—24, 20, 18, 16, 14, 12, 11, 10, 9, 8.

THE SQUARE THREAD.

Fig. 603 shows a square thread, and there are, compared with the V thread, so few used that there is no real standard, each screw being designed for its special work; but the pitch is generally twice that of the triangular thread for the same diameter of bolt. In designing it, the pitch is first settled, and this divided by two gives one-half for the thread and one-

Fig. 604.—Whitworth Screw Thread, Rounded.

half for the space, the depth of which is slightly less than one-half—or, say, $\frac{19}{40}$—of the pitch. These square-threaded screws are generally used to transmit motion. The surface of the thread being normal to the axis of the screw, there is with this form of thread no oblique pressure on the nut, as in the preceding (Fig. 602).

THE ROUNDED-SQUARE THREAD.

Fig. 604 shows a thread which is the same as that shown by Fig. 603, only the threads are completely rounded. This form is employed where the screw is subject to very rough usage.

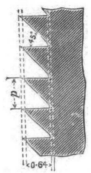

Fig. 605.—Buttress Screw Thread.

THE BUTTRESS THREAD.

Fig. 605 shows a form of screw thread called "buttress," which is used when the screw has to resist a pressure acting in one direction. Here the surface bearing the strain, as in the square thread, is normal to the axis, so that there is likewise no tendency to burst the nut. The method of designing is almost the same as in Fig. 602. Draw a vertical line, and set off on it the pitch. From each of these divisions draw horizontal and angular lines at 45°. Through the point where the horizontal line from one of the divisions cuts the angular line from the one above it, draw a line parallel to the first one, which will pass through all the inner angles of the teeth. Then, as in Fig. 602, draw two parallel lines, 0·64 of the pitch apart, centrally between the first two lines. These will cut off a portion of the angles, which should be rounded as shown.

DRAWING VARIETIES OF SCREWS.

It would, of course, take up too much time in most machine drawings to delineate screws accurately, and it would be almost impossible in some cases, owing to the smallness of the screw. On this account Figs. 606 to 612—which show the conventional methods of drawing threads on bolts—are drawn of varying size. Figs. 606 and 607 represent the ordinary V-threaded screw, the first (Fig. 606) being delineated by thin, parallel, equi-distant, slightly inclined lines. Fig. 607, being larger, shows a short thick line between

Fig. 606.   Fig. 607.   Fig. 608.

Figs. 606 to 608.—Conventional Screw Threads.

two fine ones; this gives a much neater appearance to it than if the lines had been the same. Fig. 608 represents a small square-threaded screw, shown by two fine parallel lines for the thread, while the space between the two threads is slightly increased. Figs. 609 and 610 show respectively a single and a double square-threaded screw, somewhat larger than Fig. 608, therefore shown in more detail. The threads in these are represented merely by straight lines, but more in detail than in Fig. 607. Figs. 611 and 612 show respectively double and single V-threaded screws, which are likewise formed by straight lines.

Right- and Left-handed Screws.

There are, of course, right- and left-handed screws; those here illustrated are all right-handed. A right-handed screw, when turned in the direction of the arrow (Fig. 610), will enter a nut placed at the end of it; a left-handed one would, however, work out. Another simple way of distinguishing between them is: hold the screw vertically, when, if the threads in passing round the body rise from the left-hand side to the right, then the screw is right-handed; if they rise from the right to the left, then the screw is left-handed. The spiral line (Fig. 601) is also right-handed.

A good illustration of a right- and left-handed screw working together is the adjustable coupling between railway carriages.

Requirements in a Screw.

The chief points required in a screw are —power, to draw the parts it unites into due contact; strength, to withstand any strains to which it may be subjected, and durability, to stand the wear and tear of fixing and unfixing. The characteristics of any particular thread are divided into pitch, depth, and shape. The power of a screw depends on the pitch, the strength and durability depending on the depth and shape of the thread. In selecting a thread for any particular purpose, no definite rule can be given for determining pitch, depth, or shape. It may be obvious that some are too coarse or too fine, and an equally apparent discrepancy of depth may be noticed, but the intermediate rate which has to be selected must be left to the judgment. Fine and deep threads are evidently unsuited for use in cast iron, as the material would crumble away; whereas coarse screws are not suited for fine adjustment or shallow holes. Whitworth's standard rates are now almost universally adopted for the ordinary bolts and nuts used in general engineering work, and

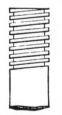
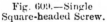

Fig. 609.—Single Square-headed Screw.   Fig. 610.—Double Square-headed Screw.

should be employed in all cases where their characteristics do not· virtually debar them. Particulars of sizes are given on the previous page.

Fine and Coarse Threads.

In most of the smaller sizes of threads, such as are used for model work and such like, the pitch is comparatively much finer, it being governed principally by reference to the depth ; the extra depth of a coarser thread would weaken the centre part of the thread to an inordinate extent ; while a finer rate, necessarily shallower, would be liable to wear out by friction. The shape of a thread is governed by the nature of the material into which the screw is to be put ; if it be homogeneous and tenacious, a sharp, angular thread will remain intact, but for material of a friable nature the tops and bottoms of the thread require to be rounded off, as are those manufactured by Whitworth, which are suitable for use in all kinds of metal work, cast-iron, wrought-iron, etc. There are three ways of describing the number of threads on a screw ; thus the $\frac{3}{4}$ in. referred to in table on p. 191 might be described as having 10 threads to the inch, being $\frac{1}{10}$ pitch, or as ·1 pitch ; these two latter measurements represent the distance apart of the threads. It will be well to bear in mind these three definitions when speaking of screws.

SCREW CUTTING.

Screws are cut in various ways : (1) In the screw-cutting lathe with a single point tool ground to the proper shape held in the slide rest, which is moved at the required rate by the leading screw and suitable wheels. (2) By a comb-screw tool used in a similar manner. (3) By a hand comb-screw tool used on the T-rest, to guide which requires considerable manual dexterity. (4) By dies held in a die-stock, and (5) by screw-plates.

SCREW CUTTING IN THE LATHE.

When one has the opportunity of using a screw-cutting lathe, the first method should be adopted in commencing to make anything like a perfect screw. With a single-point tool held in the slide rest a perfectly true thread (that is, not drunken) is cut, and the proper shape can be given afterwards, with dies, or, better still, a screw-plate, which leaves the work properly finished to the exact size. It is impossible to form a good screw of more than $\frac{1}{8}$ in. diameter, with a screw-plate only ; but,

having cut out the greater part of the metal and made a perfectly true thread, a properly-made screw-plate is the most convenient and effective tool to use for the purpose of perfecting the shape and diameter of any screw. By the second method the use of screw-plates or dies is entirely dispensed with, the screw being finished completely in the lathe ; this is the plan to adopt when cutting an out-of-the-way thread, and one has no plate or dies with which to regulate the size and shape. In the slide rest the comb-screw tool requires using carefully, or the whole thread may be torn off. Care must be taken that the process of screw-cutting is not continued too long, or the screw will be reduced to too small a diameter.

| Fig. 611.—Double V-threaded Screw. | Fig. 612.—Single V-threaded Screw. |

CUTTING SCREWS WITH HAND COMB-SCREW.

The third method is only available to those able to handle the hand-screw tools tolerably well ; the only means of learning to do so is practice. It may here be mentioned that all the inside screw tools sold at the tool shops were formerly made with the teeth slanting the wrong way, through being cut on a right-handed hub. They should be cut on a left-handed hub, which would give the teeth the requisite amount of lead in the right direction. Anyhow, made as stated, they were quite wrong for right-hand inside threads, being really cut as they should be for striking a left-handed thread ; few people seemed to be aware of this fact.

CUTTING SCREWS WITH DIE-STOCKS.

The fourth method noted above is the most usual and gives general satisfaction. By means of the die-stock, screws can be cut near enough for ordinary purposes, without the use of the screw-cutting lathe, varying in diameter from, say, $\frac{1}{8}$ in. to 1 in., which is about the largest size an amateur is likely to require. For this range of sizes three die-stocks would be necessary,

SOME FORMS OF TAPS.

Illustrations of two forms for taps are given by Figs. 613 and 614. Fig. 613 is Whitworth's pattern, and allows the shank of the tap to pass through the hole it has screwed. Fig. 614 has a shoulder at B, the

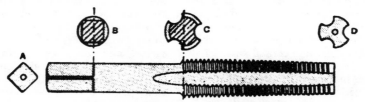

Fig. 613.—Whitworth Pattern Screw Tap.

thus: small size to take from $\frac{1}{8}$ in. to $\frac{1}{4}$ in., the sizes varying by 32nds of an inch, thus: $\frac{1}{8}$, $\frac{5}{32}$, $\frac{3}{16}$, $\frac{7}{32}$. Medium size to take from $\frac{1}{4}$ to $\frac{1}{2}$, varying by 16ths, would give $\frac{1}{4}$, $\frac{5}{16}$, $\frac{3}{8}$, $\frac{7}{16}$; the large size, to take dies from $\frac{1}{2}$ in. to 1 in., varying by 8ths, thus: $\frac{1}{2}$, $\frac{5}{8}$, $\frac{3}{4}$, 1 in. This would furnish a very good stock of screwing tackle for general use. Special threads, such as square, left-handed, double-threaded, etc., could be added at any time by making new dies.

SCREW MAKING WITH SCREW-PLATES.

The fifth method of making screws—by screw-plates—though in general use, is by no means to be recommended; the screws, being formed by a combination of squeezing, pressing, tearing, and burnishing, and sometimes a little cutting, are not to be put forward as models. Screw-plates do very well for small screws, from the smallest made, up to, say, $\frac{1}{8}$ in.; but for any

full size of the tap; this shape allows of a much larger square for the wrench: the collar B is left the exact diameter of the tap, and serves as a calliper gauge when drilling a clearing hole. This form of tap must be screwed all the way back after tapping a hole, whereas the other (Fig. 613) will pass right through. This saving of time, though important in the manufacturer's workshop, is perhaps too insignificant to be taken into consideration by amateurs. Sections of each form are given at A, B, C, and D.

MAKING SCREW TAPS.

With regard to making the various forms of screwing tackle, taps claim some attention first. Suppose the sizes previously mentioned have to be made (from $\frac{1}{8}$ in. to 1 in. inclusive), the very best quality of cast steel should be used; select that showing, where fractured, a uniform

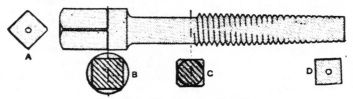

Fig. 614.—Shouldered Screw Tap.

larger size, the metal should be cut out before applying the screw-plate, which is, as before said, very convenient for regulating the diameter.

texture of not too fine a grain, and having a bluish hue; cut off lengths of the various sizes required and carefully anneal them; should the steel, in annealing, be made too hot, it will be rendered useless, and if not made hot enough it cannot be worked satisfactorily; heat it to a good blood-red, and let it cool gradually. Now file up the ends of each piece; see that they are the correct length, and centre punch them (a useful tool, known as a bell centre punch, is shown by Fig. 615). Put each between the lathe centres to make sure of their being punched truly central. They must now have the ends drilled to a suitable size and depth, say $\frac{1}{16}$ in. for the small, $\frac{3}{32}$ in. for the medium, and $\frac{1}{8}$ in. for the large sizes, drilling to about the same depth as the diameter of the drill; then chamfer, just to take off the sharp edge; now put each piece successively between the centres of the lathe, and with a graver true up the ends. These minute directions are not trivial, and must be carried out if good tools are required; otherwise the resulting taps will be comparatively useless. Having proceeded so far, there will be, reckoning three taps, a plug, taper, and entry for each thread, thirty-six pieces of steel of twelve different sizes; and before going any further, carefully look through each set of three and see that there are no defects, either in the material or workmanship. See that they are all of equal length and diameter, which must be just a trifle more than the nominal diameter of the finished tap.

ROUGHING-OUT SCREW TAPS.

The shape having been chosen, the next job is to rough out the pieces of steel as nearly to it as possible, measure off the length of thread, C D, Fig. 613, for each size, and mark it, also the length of square head of top, from A to B; when they are all marked, take a cut over the shank part, B to C, working as near the gauge lines as is quite safe, for up to the present only the taps are being roughed out. Before cutting the thread and finishing they must be again carefully annealed; the repetition of this process is necessary in consequence of the large bulk of metal removed since annealing the first time, and if not repeated the taps would go out of shape in the process of hardening; besides, it is

requisite to have the steel as soft as possible, in order to cut the threads satisfactorily.

SHAPING THE TAPS.

On putting the blank taps into the lathe after this second annealing, probably some may be found to have warped; they must be turned up true. That part of the tap which is to take the thread must now be got to size very carefully, a superfine cut

Fig. 615.—Bell Centre Punch.

file being used to eliminate all traces of the slide-rest tool. Now they are ready for being screwed; this can be done by either of the methods previously described. The best is to remove the bulk of the metal with a single-point slide-rest tool in a screw-cutting lathe, and finish with a properly-made comb-screw tool, also in the slide rest. When the thread is finished, it is easy to turn down the plain part of the shank to the size of the bottom of the thread by observing the scratches left on it by the tool when cutting the full depth of the thread; the best way to finish is with a superfine cut file. In a properly-made tap, as shown in Fig. 614, the shank c forms a calliper gauge for a drill to bore a hole the tapping size, and the collar B serves the same purpose for a clearing drill. A fine line should be turned at B to show the length of the wrench square, which is made to fit the tap wrench (Fig. 627, p. 200), of which there should be three sizes, each having two different sized square holes, so that taps from $\frac{1}{4}$ in. to $\frac{7}{8}$ in. fit the same wrench, the $\frac{1}{4}$ in. and $\frac{7}{8}$ in. in the smaller hole, and the $\frac{3}{8}$ in. to $\frac{7}{16}$ in. in the larger. On this principle all the taps belonging to one die-stock fit one wrench, and each lot forms a complete set so far as its range of sizes permits.

VARIETIES OF TAPS.

There are several shapes of taps in use, other than the now generally used machine made fluted, the small sizes being difficult

Fig. 616.—Tap with Four Flats.   Fig. 617.—Triangular Tap.

to flute, or rather they are usually made by people who do not possess proper appliances for cutting the grooves. Taps of square, triangular, half-round, and other section are variously estimated as best, but there is no doubt that a properly grooved tap is superior to any others, not only with regard to actual cutting, but for strength and durability; but as home-made taps are not likely to be so made it is advisable to describe the other means of obtaining a similar, though perhaps not quite such a satisfactory, result by means of the file.

FLUTING TAPS.

When one has overhead gear, a live spindle, and the necessary appliances for driving a circular cutter, taps should be fluted, using a circular-edged cutter, which cuts a semi-circular groove, the diameter of which equals one-half of the diameter of the tap; on cutting three

grooves with a cutter this size three-sixths of the circumference of the tap will be removed, and three-sixths will be left for the thread. In order to carry out the correct proportion it would require a different circular cutter for each diameter of tap. These cutters being rather expensive, it is seldom that any great quantity is to be found in an amateur's workshop, hence the comparative rarity of fluted home-made taps.

SQUARING TAPS.

If it is not possible to get them thus fluted, they must be filed square, or, better, with the overhead gear and the "live spindle" in the slide rest; holding the tap in position with the division peg, take a trip along at four equi-distant places. When making several taps, the following method of squaring them up will be found much superior to filing, but in this case also the overhead gear is required, though the multiplicity of cutters is dispensed with: Mount a "face-cutter"—that is, a circular cutter, having teeth on its front side or face—on the spindle, and put the tap to be squared between the lathe centres. The carrier being fixed to the throw of the point-chuck and the division peg in action, set the top slide of the rest to a slight angle to give the requisite taper to the tap, set the spindle in motion, and take a cut along the tap. Shift the division plate, and continue the operations till the four flats of the tap are finished; by this means absolute uniformity is ensured, which is not usually the case with filed taps turned out by amateurs. One face-cutter will serve for taps of any diameter, and, of course, by so adjusting the divisions three flats can be made instead of four. The plug tap is squared parallel, and the slide rest set to the requisite angle for the taper and entry tap, which should be cut down at the point, so that the measurement across the angles is just the same size as the bottom of the thread, which is left just full at the end c (see Fig. 614.) The sizes of the taps and the number of threads per inch should be legibly stamped on the shank of each one, and when they have all been smoothed up, they will be ready for hardening.

SHAPES OF TAPS.

Originally, the ordinary method was to file four flats or faces on the thread, giving four cutting faces at right angles to each other. A tap of this kind is shown in

Fig. 618.—Half-round Tap.

Fig 619.—Fluted Tap.

section by Fig. 616, but this form is bad, especially for tapping cast iron. The cutting angle is 135°, which is far too great—in fact, it hardly cuts at all, but acts more by squeezing the meal into the form of a thread. Cast iron crumbles to pieces under the pressure, and the resulting thread is very poor. Great exertion is required to turn a tap of this kind, and it is necessary to make the shanks extra strong, or they would be twisted off with the strain. Owing to the power used in turning them round, they are more readily broken than taps of a weaker section which cut more freely.

TRIANGULAR-SECTION TAP.

A great improvement on the above form is the triangular (Fig. 617). In this case the cutting angle is 120°. It does not, however, leave much room for the cuttings, and if much thread is left between the faces it is hard to turn. Backing off the thread a little improves this, but the clearance thus given must not be excessive, or the points will dig in and be broken off.

HALF-ROUND TAPS.

The half-round section (Fig. 618) has also been advocated. This gives a cutting angle of 90°. It cuts freely, and is said to stand well in spite of its apparently weak section. It, however, has only one

cutting face, and is easily displaced from its true direction when in use, which is a great drawback. In the fluted form having a radical face (as in Fig. 619), the cutting angle is also 90°. It has three cutting faces, which lessens the exertion required to turn it, and it is readily kept true in the hole being tapped. If the grooves are made as in Fig. 620, the angle approximates to 90°, and the taps are easier and safer to harden. Of course, if the cutting faces are made as in Fig. 621, the angle is keener, but when the motion of the tap is reversed so as to withdraw it, the extreme points of the thread are apt to break off. However, the larger sizes of taps are often made this way, because the thread being stronger they stand better. A great deal depends, too, upon the hardening—it is so easy to burn the points of the threads, and then they break at once. Up-to date taps have not three flutes, but four, as has been shown, and the engineer's usual set includes the plug tap (Fig. 622), the taper (Fig. 623), and the bottoming tap (Fig. 624). Smiths' taps are not quite the same, as is apparent from Fig. 625.

TAPS WITH DETACHABLE CUTTERS.

Patents have been taken out for taps having loose, or, rather, detachable, cutters, but these are only available for the large sizes, while the fitting is quite beyond the skill of the beginner.

MACHINE TAPS FOR NUTS.

Machine taps for tapping nuts are generally made as shown in Fig. 626, so that

Figs. 620 and 621.—Fluted Taps.

one tap only is required to cut a full thread. From the point a portion of the tap is turned down to the bottom of the thread. This part is inserted in the hole in the nut, and acts as a guide to the tap while the thread is being formed. As the tap revolves the nut is moved forward until the tap starts cutting. At the top end of the tap a portion about equal in length to the diameter is left parallel. The rest of the body is tapered. The tap is passed right through the nut, and the thread is thus cut at one operation.

water so as to make a miniature whirlpool in the bucket ; take the tap and plunge it perpendicularly, threaded end first, quickly but steadily, into the centre of the whirlpool. It this is done carefully there need be no fear of the taps going out of shape, if they have been thoroughly annealed twice, as directed. For the smallest sizes, to guard against the points of the

Fig. 622.—Plug Tap.

Taps for Die Cutting, etc.

When the tap is intended to be used for cutting dies, etc., more flutes usually are put in. The flutes are cut with an angular cutter, just leaving a round at the bottom of the grooves. About nine flutes may be put in a 1 in. tap of this class. These taps are commonly called " hobs," and are used to cut worm wheels, chasers, and similar work. The hob for cutting a worm wheel is made the counterpart of the worm, being

threads being burned and spoilt, rub a quantity of common soap on them before putting them into the fire. When all have been treated thus, with a file try each one at both ends to make sure that they are hard all through, and then get ready for tempering.

Tempering Taps.

Make both ends and one of the flats of the wrench-square quite bright with

Fig. 623.—Taper Tap.

the same diameter and same shape of thread. The flutes are then cut parallel to the axis of the worm. The threads are backed off a little, and the hob is hardened as described below.

Hardening Taps.

For hardening taps get plenty of water, say a bucketful ; add a handful of salt, and heat the taps one at a time in a clear fire

emery paper, and also, if square, one of the flats, or, if grooved, one of the grooves along the thread. See that these bright parts are quite clean. The actual tempering had best be done over the gas, using a Bunsen burner ; in order to heat the tap evenly, get a piece of stout metal to interpose between it and the flame—a piece of angle iron will answer the purpose very well—and make the taps hot enough to

Fig. 624.—Bottoming Tap.

till they are well red hot, but do not allow any of them to get hot enough to blister. When they are at the right heat, stir the

change the colour on the bright parts to a deep yellow—not purple—then dip in cold water.

SIZE ALLOWANCE IN HARDENING TAPS.

Hand taps usually are finished off the veriest trifle small, because hardening a tap makes it slightly thicker and shorter, so that it cuts a thread bigger in diameter

OTHER POINTS IN HARDENING TAPS.

Hardening is by far the most difficult part of making a tap, owing to the uncertain behaviour of the steel. Before the steel is used at all it should be thoroughly

Fig. 625.—Smith's Taper Tap.

than the tap was originally cut. Machine taps, however, are generally full, so that screws and bolts will be an easy fit. No definite rule can be given for allowances to be made for alteration in size during hardening, because different brands of steel vary so much. Experimenting with a piece of the steel will show the allowance. Then if the same brand of steel is always used, and the same process gone through in hardening, the same results reasonably may be expected, and the probable variation either in thickness or length allowed for. For all ordinary purposes, however, it does not matter, because the hole is tapped first, and then the screw is made to fit. Always use good steel for tap making, and, in fact, for cutting tools generally. It does not pay to use poor stuff, for it will not take a good even temper, and, consequently, will not wear well. There is no universal rule for

annealed, as has been said. If possible, a tap or reamer should be turned down to shape out of the solid, and not forged. If forging is necessary, the steel should be thoroughly annealed after it, then roughed out in the lathe and annealed again. Hammering sets up strains in the metal, which suddenly show themselves in hardening by causing the metal to bend. Roughing out even, if a heavy cut be taken, may cause strains, which may affect the satisfactory hardening. If the greatest accuracy of size and pitch is required in a tap, it is best to anneal after roughing out in all cases, whether the tap has been forged or not. As to the proper heat for hardening, much depends on the particular brand of steel being used. The temperature at which steel hardens is variable, and the best heat is undoubtedly the lowest which will give the requisite degree of hardness. Experiment only will decide what this

Fig. 626.—Machine Tap for Nuts.

the length of taps, but a proportion of about four times the diameter for the length of the threaded portion, and about four times for the length of the shank, answers well. It is better to adopt some system in making all kinds of small tools, so as to have the series uniform. The wrench square on the end of the shank is usually made one diameter long.

heat is. It is very little trouble to try a piece, and it may save a deal of time and money. Of course, there is only the eye by which to judge the heat, and some practice is required to tell properly when two pieces of steel have been heated to about the same degree. What smiths call cherry red is about right as a rule. Heat evenly, and take great care not to burn the thin cutting edges before the thicker part is properly hot. The fire should be clean and bright, but not too hot. Never try to harden a large tap in a small quantity of water. If a tap is plunged in in an inclined position, it is nearly sure to be crooked.

ANOTHER METHOD OF TEMPERING TAPS.

A good way of tempering taps, reamers, and tools of that class, when there are not many to do, is to get some pieces of tubing large enough to let the tap pass easily in. Get one of these pieces to a nice red heat, and then commence to heat the next, so that as one is being used and cools, the next is getting ready. The tap should be oiled all over, and held with a pair of tongs in the hot tube, which may rest on a **V** block on the anvil. The tap must not be held still, but moved about and turned round, keeping it meanwhile in the middle

DRILLING HOLES FOR TAPPING.

In drilling holes for tapping, a little clearance is generally allowed in excess of the diameter of the screw at the bottom of the thread. This allowance should really be greater for wrought iron than for cast iron, but in practice the same drill is used for both.

KEEPING TAPS SQUARE WITH WORK.

Although the hole intended to be tapped may be drilled quite square and true, the tap, owing to unskilful usage, may not follow the hole, and so may get out of its true direction. To avoid this, steady, even pressure should be exerted on both ends of the tap wrench while the tap is being started. Care should also be taken to watch the progress of the tap, and if the eye is not a sufficient guide, a small square should be applied in two different directions as a test.

Fig. 627.—Tap Wrench.

of the hot tube so as to temper it evenly. The colour must be carefully watched on the brightened parts until it reaches a deep yellow, a "nut brown," or "dark straw" colour, when the tap must be rapidly cooled in water. These colours are rather difficult to describe, but by examining a new tap or milling cutter are readily seen and should be carefully noted. If the tap is only intended to be used on brass, it may be left harder. Tempering should be done slowly. The slower the steel is heated the more uniform the temper is likely to be, and consequently the less likely are the cutting edges to be too soft, and the rest of the metal too hard. If the temper is let down too low, the only thing is to re-harden and try again. A few failures must be expected at first, as some experience is required to tell when the temper is just right. After tempering, the shank is polished in the lathe, leaving the colour on the flutes and about a quarter of an inch up the shank.

TAPPING HOLE THAT DOES NOT GO RIGHT THROUGH.

To tap a hole that is not a thoroughfare —that is, one that does not go right through—requires the most care, because if the tap breaks off short it is very difficult to get it out, and, besides, it often happens that the taper tap cannot be used at all. A $\frac{1}{4}$-in. Whitworth plug tap can be started in a cast-iron hole if the mouth of the hole is slightly enlarged, if the tap is slightly ground at the point, and if the hole is slightly too large to leave a full thread. The grinding of the point of the tap makes it a little conical at the end, but, at the same time, it must be done, so as to back off the cutting edges.

LUBRICANTS FOR TAPPING.

In tapping wrought iron and steel, oil should, of course, be used as a lubricant, but it is not necessary for cast iron and brass.

WHY TAPS BREAK.

Taps, especially small ones, should not be lent to a beginner, who is sure to break some, and it is not until this occurs that he can form any idea of how much force a tap will bear. Indeed, it might not be a bad plan to give a few old taps to him,

TAP WRENCH.

The wrench shown by Fig. 627 is made from a piece of iron ⅝ in. by ⅜ in., and 7 in. long, centred at each end, and the handles turned up to the pattern shown; the square holes are first drilled round and then punched out square; the whole thing,

Fig. 628.—Tap Wrench.

and let him screw on with the tap wrench until they are screwed in two. After this he will be able to work with more confidence, neither wasting time by too much caution nor working too fast, and, by using too much strength, breaking off a tap in the hole. When a tap breaks in ordinary use, generally there has been not only the necessary twisting force, but also an unconsciously exerted sideways bending force—either the wrench was forced harder with one hand than with the other, or the tap was not straight up, and was forced to one side to correct the fault. Taps sometimes break when being turned backwards to take them out. The cutting edges of taps and dies, in working, roll up in front of them like little shavings, which gather in the groove of the tool, and must be broken off when it is turned backwards; this breaking off requires even more force than the continuation of the cutting.

after being finished off, is case-hardened all over. Another tap wrench, or lever, is shown by Fig. 628. The square hole in the middle is to receive the squared part of the tap shank. In Fig. 628 the corners of the hole are relieved, so that the corners of the shank will not get the pull. The handles are round, and generally tapered. Their length varies according to the diameter of the tap. The wrenches are generally stamped with the diameter of the tap they fit.

TAP SOCKETS.

The socket shown in Fig. 629 is used when some obstruction prevents the wrench being turned round. The square hole A at the one end fits the shank of the tap. The end B is made square to fit the wrench, which by the length of socket may be lifted clear of the obstruction. In fact, the socket is just like an ordinary "box-key," and answers a similar purpose.

Fig. 629.—Tap Socket.

This is why a tap always should be turned halfway round and then back.

REMOVING TAP BROKEN IN HOLE.

When a tap is broken off in a hole, and cannot be got out with pliers or by rapping with a punch, it must be heated by a blow-pipe flame till it is soft enough to drill a hole up the middle of it, which hole can then be enlarged till the bit of the tap can be broken and the pieces got out.

TAPS FOR WATCH SCREW THREADS.

Taps from watch screw threads may be made from needles, but probably they would not last long. A tap should be made from the best steel; therefore get a length of tool steel wire of the correct size. From this cut off a suitable length, say 1½ in. Soften it by heating to a dull red and allowing it to cool slowly. Hold it in a pin-vice, and, resting it on a piece of boxwood, file it to a gentle taper until the end just enters the hole in a screw-plate; the wire may then be screwed into the latter, plenty of oil being used. When it goes hard, turn it back half a turn, then forward three-quarters of a turn, back

parallel cutting edges (see Fig. 631); but if it be parallel from end to end, there is nothing to steady it when entering and prevent its cutting the mouth of the hole tool large; hence some consider it better

Fig. 630.—Bit-stock Taper Reamer.

half a turn again, and so on, advancing slowly until a full thread is cut for a sufficient distance. Then file three flats upon it for the whole length of the thread, tapering the flats to the end, where they should meet in a knife edge and show only half a full thread. Harden the tap by heating to a red colour and plunging in cold water. Temper it by brightening one flat and heating it over a flame until it is of a pale straw colour. Then carefully smooth all three flats on an oilstone so as to leave good cutting edges. Finally, file some nicks in the soft end to indicate the number of the hole in the screw-plate to which it belongs.

REAMERS.

A reamer is used to smooth the hole left by the drill, and it is mentioned here because the making and use of a reamer have much in common with those of a screw tap. Some modern reamers are shown by Figs. 630 and 631, a shell reamer by Fig. 632, a rose shell reamer by Fig. 633, and an adjustable reamer by Fig. 634. Small

to taper the end slightly for a distance equal to at least the diameter.

NUMBER OF CUTTING EDGES IN REAMER.

With regard to the number of cutting edges, English reamers used to have three, but in the United States they had not less than four, and as many more as convenient. (There are ten in some of those shown by Figs. 630 to 634.) A single cutting edge only has been suggested, on the score that it is an easy matter to grind one cutting edge straight and true, and to re-sharpen it, and the cutter could be made adjustable for cutting diameter; a single tooth, it has been argued, can be fed more easily to its cut, since it will be easier to rotate and pass to its cut. In the case of taper holes, a single tooth can be fed much more easily, and will be free from the sudden release from the cut which reamers having much taper and a number of teeth are subject to, while a finer degree of cut can be taken. However, American reamers of improved design, such as those shown by Figs. 630 to 634, have many cutting edges.

Fig. 631.—Parallel Reamer.

holes must be finished by reamer, and up to about 2 in. in diameter all holes are best so finished. Points are: the shape of the reamer, how many cutting edges and what clearance they should have, and what the shape of each tooth, individually, should be. To ream parallel, the tool must have

CLEARANCE IN REAMERS.

The clearance is the next consideration. The more the clearance the more a given amount of wear will decrease the size of the reamer, but the easier the reamer can be moved or fed to its cut. The less the clearance (so long as there is some) the easier the reamer may be kept to standard size, but the more the power required to operate it and feed or push it to its cut. The more the number of cutting edges the

Fig. 632.—Shell Reamer.

more the clearance must be, which has been advanced as another point in favour of a single-tooth reamer. On taper reamers ample clearance is a decided advantage.

ANGLE OF REAMER'S CUTTING EDGE.

The angle of the groove or flute face at and near the cutting edges is the next consideration. Whitworth made the grooves half circles or nearly so; some make the faces radial, others an acute angle to the tooth. A radial face is thought to be too keen for good work—the tool loses its edge too soon; an obtuse angle is better, the degree increasing as the number of teeth decreases, but never varying more than about 40° from a radial line except for brass, for which a variation of 20° is permissible in a reamer having a single cutting edge. In discussing both clearance and radial face angle, it has been assumed that such clearance and angle are to be equal at all parts of the reamer's length, and to facilitate the sharpening

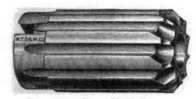

Fig. 633.—Rose Shell Reamer.

process this is very desirable; but if both the clearance and the angle be enough to enable the reamer to cut freely at the tapered entering end, the clearance may

gradually be reduced towards the other end, while the angle may be made more obtuse, which gives the very best form of reamer for standard work.

MAKING REAMERS.

The method of making reamers is similar to that followed in making taps. In turning they are usually left full, and ground to template size after hardening and tempering. If the reamer hardens somewhat crooked, it will often grind up to size. In fluting, cut a little deep, to allow for this grinding. More flutes are usually put in reamers than in taps; an inch reamer may have nine, or even more, depending on the nature of the work to be done with it. A good pitch for the flutes is from $\frac{1}{4}$ in. to $\frac{3}{16}$ in. It is desirable to put in an odd

Fig. 634.—Adjustable Shell Reamer.

number of flutes. If there are longitudinal grooves in the hole to be reamed, the flutes should be cut spirally. Reamers are somewhat given to chattering. An irregular spacing of the flutes in a great measure obviates this, and should always, if possible, be adopted. The difference of spacing must only be slight, about ·0025 being quite sufficient. The cutting faces generally are radial; the bearing surfaces should be narrow. In backing off, much clearance should not be given, but a little is necessary to allow the cuttings to get clear of the bearing surfaces, or they will make the hole rough. If too much clearance is given the chances of chattering are increased, which will cause the cutting edges to wear very fast, or perhaps break, and the reamer will be spoiled. For brass or gun-metal more clearance may be given than for iron. Reamers are hardened and tempered just as taps are treated, the temper being left quite hard. Thoroughly annealing the steel before using, and leaving no sharp corners in the flutes, will do much to prevent reamers bending or breaking.

Fig. 635.—Special Reamer or Broach

SPECIAL REAMER OR BROACH.

Fig. 635 shows a reamer or broach screwed at the end. In making such a tool for a ⅞-in. hole, the screwed end would be screwed with a good ⅞-in. Whitworth thread, say up to ⅞ in. from the end; here it would have a shallow recess ¼ in. wide, and then it would be squared down with a side tool in the lathe 1⅛ in. from the end. This forms the cutting edge. The reaming part of the tool, while it is being forged, must be kept large enough to be filed up in cants, whose section must be in the form of a pentagon or five-sided figure. This can be carried up for, say, 2 in., and to be large enough to ream out a hole which will nicely fit the unscrewed part of a tap bolt or turned stud. When the pedestal of a

ally to be parallel with the engine crank shaft, it is set in that position and marked with a scriber through the holes in the foot of the pedestal. The holes in the bed-plate then are drilled with a ratchet brace and a drill that will just go into an ordinary ⅞-in. nut; these holes are then tapped. Now, if the studs were put into their places and the pedestal tried on, it would probably be a very little out. The tool shown by Fig. 635 is to rectify this error. Place the pedestal on the bed-plate in the position in which the holes were marked. Take two old tap bolts, with their unscrewed parts filed down level with the threads so as not to touch the pedestal; these, if placed in the opposite diagonal corner holes and screwed down tight, will hold the pedestal steady. Screw the tool down the

Fig. 636.—Taper Die-stock.

other two holes until the face at B comes through the pedestal and touches the bed-plate. Having done this, take off the pedestal, and put in the two studs where the tool was; then do the same with the other two holes, put the studs in, and the job is finished. There will be no cause to fear the possibility of the pedestal shifting by the wear and tear of its working;

Fig. 637.—Stock and Die.

governor has to be fixed on to the bed-plate of an ordinary horizontal engine, the foot of the pedestal is drilled to suit the ⅞-in. studs by which it is to be bolted to the bed-plate, and as its pulley has gener-

it will be too good a fit, and exactly parallel with the crank shaft as required. To make such a tool or set of such tools screwed to suit the different sizes of studs would save time, money, and trouble in the majority of cases. The tool must be made of tool steel, kept soft until it is filed up, and then hardened and tempered.

MAKING DIE-STOCK.

Two forms of double-handed die-stocks for general use are shown by Figs. 636 and 637; whilst Fig. 638 illustrates a small single-handed die-stock suitable for threads of about $\frac{1}{4}$ in. diameter. In Fig. 638 a number of sectional views are shown, A being a section of the screw, B and C cross sections of the slotted part which holds the dies, D a section of the small part of the handle, E a longitudinal section of the slotted part shown in cross section by B and C, whilst F is a cross section of the end of the handle. For making such a die-stock, it is best to get a forging made nearly to shape, the iron being all the better for well hammering; but it is easy to use a piece of plain bar iron and cut the stock out of the solid metal, which is not such a long job as may be supposed. Get a piece of good sound iron $1\frac{1}{2}$ in. by $\frac{1}{2}$ in., and about 6 in. long; centre the two ends accurately, and drill in the centres with a small drill, run the iron in the lathe, and take a cut along it, from end to end, with a tool in the slide rest, so as to take the black skin entirely off from both edges; also, at the same time, true up the two ends. Put the work in the vice and file up the two sides flat and truly equidistant from the axial line. Having done so, mark out the shape of the die-stock on one side of the iron; replace the piece in the lathe and turn up that part forming the handle. So far the tool is shaped only roughly, and the handle need be turned merely as a parallel cylinder fully $\frac{1}{2}$ in. in diameter. Reverse the ends between the centres and turn the corners off at that end where the clamping screw is tapped through. It is advisable to drill the hole for this screw full tapping size, and to chamfer out the end deeply, so that when the tap is put through the hole the work will not be thrown out of truth between the lathe centres. In fact, this hole may be drilled, chamfered, and tapped as soon as ever the truth of the centres is assured;

it may be bored to a depth reaching the bottom of the slot in which the dies are to be fitted.

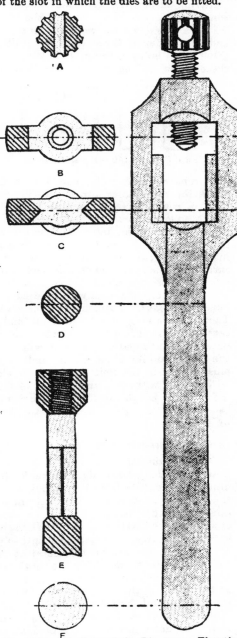

Fig. 638.—Single-handed Die-stock, Elevation and Sections.

SHAPING DIE SLOT.

Having turned the die-stock roughly to shape, mark out the slot for the dies. Scribe a line down the centre of one side, and centre-punch three dots for drilling three ⅜-in. holes through the iron, bore these holes, taking care to see that the

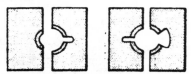

Figs. 639 and C40.—Die-stock Dies.

drill runs through truly, and with a file break the three holes into one. This will make a rough rectangular hole 1⅛ in. by ⅜ in. ; square it out carefully and accurately to ₇⁄₁₆ in. wide, and verify the central position of the slot continually by measuring form the two outer edges. Too much care cannot be bestowed on this part of the work, as absolute truth in the plain hole will assist very materially in guiding the bevelled angles on which the dies slide. These bevels may now be filed up, and a carefully scribed line drawn along from the outer edge and parallel with it will form the best guide for filing to. A small ridge must be left in the centre part of the slot to form a "witness," and when the four facets of the bevels are filed, the sides may be reduced in thickness from ⅛ in. to ₅⁄₁₆ in., or even ¼ in.

SHAPING THE STOCK.

Before beginning to reduce the thickness draw lines along both edges as guides for filing to ; then file from both sides, using the file at right angles to the length of the die-stock, and cutting out a groove equal in width to the length of the slot. When this is reduced nearly to gauge, file lengthways of the stock, to take off the metal at the corners, carefully preserving the continuity of the ½-in. cylindrical portion as turned at both ends of the die-stock, one forming the handle, and the other the boss for the screw to be tapped through. This part should be done carefully, as a blemish detracts very much from the look of the fin-

ished stock, though probably it would not impair its working qualities. Considerable caution must be exercised in using the file over these circular protuberances, or an ugly scratch will be made.

TURNING DIE-STOCK HANDLE TAPERING.

The handle may now be turned tapering, to give it a lighter and neater appearance. Set the slide rest over, so that it will make the cylinder about ⅜ in. in diameter at that part where the side pieces swell out, and ¼ in. at the extreme end ; turn it down to these dimensions, and file up the continuance of the cone to exactly the same angle. This will require even more care than the previously described cylindrical parts The terminations of these cylindrical parts of the slot should be bevelled off, as shown in the sectional sketch to the left of Fig. 638. It now remains but to file away the bevel in the slot at that end next the screw, so as to allow the dies to be put in, and the die-stock is finished. File away a distance of ⅜ in., that being the depth of the dies ; round off the end of the handle with a graver, and "touch-up" the boss for the screw. Smooth up the entire stock with emery paper, using oil with it, and when all the file marks are taken out of the flats, and the turning-tool marks out of the handle part, etc., the stock is ready for hardening.

CASE-HARDENING DIE-STOCK.

Being made of iron, the die-stock can be case-hardened only. The following process fulfils the requirements of such a job, but one of the processes described on pp. 164 to 166 can be used instead if desired. A compound is made of prussiate of potash, sal-ammoniac, and common salt, of each equal parts by weight (no great accuracy of proportions is necessary) ; pulverise each separately and mix thoroughly. This mixture should be put by in a stoppered jar, and may be labelled "Case-hardening Compound." Make the die-stock red hot, as hot as possible without scaling ; draw it out of the fire, lay it on a sheet of metal, and cover with the compound, rubbing on as much as possible.

When the iron gets cool, and there is a good coating of the compound adhering, replace it in a clear fire and re-heat gradually. When it is of a good cherry-red it may be quenched in clean water, and will become flint hard on the exterior surfaces only, where the compound has acted ; it is, however, better to give the article a second coating of hardening powder, reheating, as before, previous to quenching. A third coating is desirable when a particularly good job is required. The more compound got into the iron the deeper will be the hardening, but care must be exercised in heating, so as not to burn the metal and scale it. After the stock is thoroughly hardened it may be made bright again by the aid of emery paper. There is very little chance of the iron being distorted in the hardening process, but it is advisable to anneal it before hardening by putting it in the fire after the bulk of the material has been removed in roughing out. The work must be heated to a full blood heat and allowed to cool gradually.

DIE-STOCK CLAMPING SCREW.

The screw belonging to the die-stock is, as will be seen by Fig. 638, and by the section A, rather peculiarly shaped ; it is made of good cast steel ½ in. in diameter, and the head has, in addition to the one or two "tommy holes," a series of small grooves cut longitudinally on its diameter. These are to afford a grip for the fingers, so that the screw may be rolled in and out half a dozen turns between the finger and thumb when changing a pair of dies ; there should be about a dozen grooves, say ₁/₁₆ in. wide. A plain round screw head is difficult to turn, unless the fitting is very loose, but with the serrated head considerable force can be applied with the fingers only. Sometimes a flat thumb-screw is used, but the wings often come in the way when the die-stock is in use. This clamping screw should be hardened and tempered, and made to the exact length—whether too long or too short it looks equally bad.

DIE-STOCK DIES.

In the illustrations of screwing dies about to be referred to, the left-hand die of each pair is in every case the "top die," the die to the right being the "bottom"— that is to say, the die which is first placed in the stock ; the "top" one receives the pressure of the point of the clamping screw. Very often no attention is paid to the identity of "top" or "bottom," the dies being put in indiscriminately, but the strongest die should always be put in to take the pressure of the screw, and the weaker one is made correspondingly strong by being supported all along its end by the flat bottom of the die-stock slot. The dies are all drawn to size for cutting a ½-in. Whitworth thread. Fig. 639 shows a design very frequently seen. The dies are made on a tap ½ in. in diameter, and the bottom die contains half the circumference of the thread, the top one being filed off considerably, so much that the tops of the teeth of the thread in the die lie very nearly on a plane with the side flats. This top die only guides. The bottom die has a small groove cut across it to facilitate the escape of chips, but all the cutting is done by the two upper face corners of this die. There should be plenty of space left for the escape of chips between the two dies when they are closed to the proper size for finishing the screw. Dies of this form are very bad "guides" for the rate of thread, and do not cut quickly, but are very handy for finishing.

ANOTHER PAIR OF DIES.

Fig. 640 shows a pair of dies very similar,

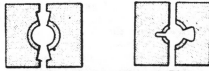

Figs. 641 and 642.—Die-stock Dies.

but the top one is made like the bottom die in Fig. 639, and the bottom die has a large slot with under-cut sides. These dies have nothing to recommend them particularly, though they are often to be seen. They are good for cutting threads of smaller diameter than that of the tap on which they were made. All the cutting is performed by the four corners of the dies, each one doing an equal share, but, there being no space for the escape of chips, these dies soon get clogged up, and have to be cleaned out. None of the cut-

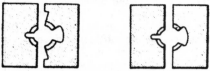

Figs. 643 and 644.—Die-stock Dies.

ting can be done by the edges of the bottom slot until the work, being screwed, becomes smaller in diameter than was the tap with which the dies were cut, and consequently the "shavings" do not find an outlet by this channel till then.

IMPROVED DIES.

Fig. 641 shows a pair of dies shaped properly for cutting, having a slight angle of relief at all the corners; these dies act first-rate, and there is plenty of room for the chips to escape through the slots on each side of the centre. There are no slots across the bottom of these dies, as is the case with all the others figured (excepting the top die of Fig. 639), consequently they will not cut any smaller than the full diameter of the tap on which they were made, and cannot be forced to do so. It is advisable to screw the pair of dies quite close together when tapping them, and thus make it easy to tell when the proper size is being cut by the dies in after use; it saves the trouble of gauging, etc. The best plan for making the clearance holes is to mark and drill the two holes, about ⅛ in. in diameter, as near to the outer corners as the V groove will allow. The dies must be held quite firm and close together whilst doing this, and the holes must be made just through the dividing line; the dies are then taken out of the stock, and the angle of relief filed up with a smooth, flat file.

THE BEGINNER'S BEST FORM OF DIES.

Fig. 642 shows the best form of dies that can be made without extra taps, and imple-ments that are usually dispensed with by beginners. It embraces all the good points of the preceding types, the top die being the same as in Fig. 640, whilst the bottom die combines in itself Figs. 640 and 641. These dies will cut threads of smaller diameter than the original tap, the top die being, of course, reduced in depth as described for the top die of Fig. 639. The cutting angle cannot be got out so easily by drilling, but may be easily managed by the aid of a small round file to cut a semi-circular groove, and the plane may then be filed up as before. A round file must always be used in making the corners of the slots of the bottom dies, as shown in Figs. 640 and 642 to 644, as the corners must be rounded out, not filed in to a sharp angle, as this would tend to cause a fracture across the steel when hardening. These sharply-cut corners are the principal causes for steel splitting, for if a piece of steel is cut in to a very sharp angle, not necessarily very acute, it may be broken with much greater ease than could be a piece more deeply cut, but having the angle blunt in its extreme corner. In breaking a bar of steel by first "nicking" round with a file, the sharpness of the nicking, rather than its depth, will prove effectual. Whenever a piece of steel is nicked in sharp, and then hardened, it is almost sure to crack at that part; therefore always be

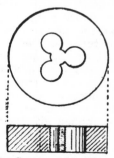

Fig. 645.—Screw-cutting Plate or Die.

careful to use a file with a rounded edge when working out the corners of pieces which have to be hardened. Triangular files can be bought, having round corners, and are very useful for filing out the v grooves in the sides of the dies, when these latter cannot be milled up in lengths, and cut off all ready for the die-stock.

DIES THAT LEAD WELL.

The dies shown by Fig. 643 differ from the previous dies, inasmuch as the two dies are cut with taps of different diameters, the top die being the same as that in Fig. 642, whilst the bottom die is cut on a master-tap double the depth of the thread larger than the nominal diameter. This plan secures a better lead for the dies, as the bottom die in this pair embraces the blank bolt to about half its circumference, and touches all that amount at starting, which is, of course, the time when good guiding is most wanted. The difficulty in making this sort of die is, that it necessitates making an extra blank top die for use in cutting out the bottom die figured with the large master-tap, and it is not very often that one feels disposed to spend the extra time and trouble entailed. Fig. 644 shows a precisely similar pair of dies, both being cut on a large master-tap. These dies lead excellently, but they are most "greedy" with the cutting just at the wrong time—that is, when the work is very nearly completed. They are the best dies for roughing down a thread, but another pair cut on a small tap would be better for finishing with. All Whitworth's dies are cut on large master-taps, and so are the majority of dies manufactured for sale ; but where only one pair of dies for each size is required, the cost of master-taps forms a large item in the cost of a set of screwing tackle. Taking all things into consideration, the dies shown by Fig. 642 are the best for general purposes, and represent the form of dies recommended for all sizes, though the "Whitworth," or "Patent Direct Action," are far superior when the size of the work entails an appreciable amount of muscular power to actuate the die-stock.

MAKING SMALL SCREW-CUTTING PLATES OR DIES.

To obtain good cutting angles in small plates or dies which are to cut brass screws $\frac{1}{8}$ in. in diameter by No. 7 B.A.S. pitch, drill either three or four equi-distant holes as shown in Fig. 645 around the threaded hole. By this means a cutting angle instead of a scraping action is obtained. Freedom of cutting and ease of withdrawal are assured by leaving the cutting faces of the threads very narrow ; in a die to screw $\frac{1}{8}$ in. diameter they should be rather less than $\frac{1}{16}$ in. wide. The dies will be turned and a hole drilled in the lathe quite true and concentric, and tapped, and then the clearing holes that are to form the cutting angles must be drilled. The threads may then be filled with soap and the dies heated to the lowest heat at which they can be hardened, which will be a low shade of red, depending on the quality of the steel. The hardening may be done either in water to which salt has been added or in oil. The

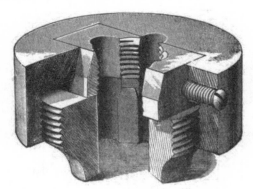

Fig. 646.—Adjustable Die and Collet.

colour at which to temper is a medium straw, previous to which the sides must be polished with emery cloth. Reduce to the colour for tempering on a heated bar or plate, turning the dies over until the straw colour is reached. This must be done gradually, otherwise the threads will become hotter than the body. Properly, the dies should be split on one side and a screw put in an encircling holder, much as in the types already described, so that the size of the threaded hole can be adjusted very finely, as well as when the dies get slightly larger by wear.

"LITTLE GIANT" ADJUSTABLE DIE.

"The "Little Giant" adjustable die, shown in general and sectional view by Fig. 646, has a collet or holder made of two pieces, one the cap and the other the guide, and the cap has a bevelled slot into which the bevel of the die is fitted. The

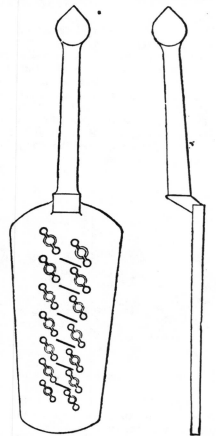

Figs. 647 and 648.—Two Views of Lancashire Screw-plate.

guide is screwed into the cap so as to throw the die into the bevel of the collet, and into line. To adjust the die, merely turn the screws at the end in or out as may be necessary.

USING DIE-STOCK.

In cutting a screw with dies, if the blank be of wrought iron, it may be turned slightly smaller than the finished size of the thread, because the action of the dies will not only cut the groove, but also squeeze up the thread. It is also very liable to bend the screw if much force is used. The squeezing up and bending action does not take place with steel, which is therefore turned the correct size before screwing. It is a very easy thing to produce a drunken thread with the dies in the die-stock. To avoid this, open the dies and close them on the middle of the screw, wriggling the stock about a little, so that it may place itself squarely on the screw-blank, and screwing up the dies the while till they seem to clasp it fairly. When this is the case, give the screw of the dies about half a turn more, to sink them into the blank, put some oil, and begin cutting. Watch the effect, as it is possible to cut a double thread. If a drunken thread is produced, it can only be corrected in a screw-cutting lathe.

SCREW-PLATES.

Screw-plates are used generally for making only the smallest sizes of screws, say from ⅛ in. downwards, although they are occasionally made as large as ½ in. One form is shown in Figs. 647 and 648. It consists of a flat plate of the best tool steel, varying in thickness from one-half to the full diameter of the screw it is intended to cut. Of course, if several sizes of holes are made in the one plate, the thickness is made suitable for the largest. This plate is drilled with a series of holes corresponding to the sizes of screws it is intended to cut.

OVERCOMING THE SCREW-PLATE'S DEFECT.

The chief fault of screw-plates is that the position of the cutting edges with regard to each other is fixed—that is, they cannot be separated and brought closer together. To attempt to cut a full thread by forcing a plate having a standard size tapped hole in it over the rod to be screwed would require the exertion of so much power, apart from the difficulty of getting the plate to start, that the rod or pin would be twisted and perhaps broken. To avoid this, several holes of different diameters are used for the one size of screw. The first one, which is, of course, the largest, merely marks out the thread. The others are used in succession, each one cutting the thread a

Figs. 649 and 650.—Screw-plate Perforations.

little deeper, until the last makes a full thread of the standard diameter. For the very smallest sizes of screws, two or three holes are used; for the larger ones, four, or even six. The holes belonging to one set are connected by lines, as shown in Fig. 647. These lines can be cut with a chisel before the plate is hardened.

MAKING SCREW-PLATES.

The holes, their number and diameters having been determined, are drilled in the plate, which has been previously planed or ground flat on both sides. The holes must not be drilled the same diameter as the screw they are intended to cut, but what is known as the "tapping size" of that screw—that is, the diameter at the bottom of the thread. The holes are generally opened out slightly on the starting side, to make the thread easier to start; though the last hole of a set, which is used more as a gauge than a cutter, is not. The holes are then tapped with a master-tap, which has more grooves cut in it than an ordinary tap, giving more cutting faces; but, with this exception, it is made and hardened in the ordinary way. The next step is to provide these holes with cutting edges. There are several methods of producing the cutting edges. The most usual is shown in Fig. 647 and in Fig. 649. Two small holes are drilled, one on each side of the tapped hole, and close to it. The three holes are then connected by cutting through the metal between them, either with a saw or a small file. This gives four cutting faces, two of which operate when the plate is moved round in one direction, and two when the motion is reversed.

This method gives good cutting edges and plenty of room for the cuttings. It is not liable to break in hardening, and is easily cleaned. Sometimes the small holes are omitted, but just the notches cut, as in Fig. 650; while occasionally the holes for the same size of screw are joined together by the notches, as in Fig. 651. This latter method has this advantage: If the piece of wire being screwed happens to break off short in the plate, a small saw can be readily used to cut the bit in two. This loosens it, and it is easily removed. The cutting edges having been made, each set of holes is stamped with the size of screw they cut. The plate is then ready for hardening, in which care must be taken to heat the plate evenly, and avoid burning the cutting edges. Temper slowly to dark straw colour. Two ordinary forms of screw-plates are shown by Figs. 652 and 653.

LARGE SCREW-PLATES.

Screw-plates have occasionally been made for $\frac{1}{2}$-in., and even larger, screws. In these sizes three or four notches are sometimes cut, giving six or eight cutting faces respectively; only half of them, however, acting at the same time, the particular faces cutting depending on the direction in which the plate is turned. For all sizes above $\frac{1}{8}$ in., however, the stock and dies are preferable. It is comparatively seldom that screws smaller than $\frac{1}{8}$ in. in diameter are wanted. However, when such tiny screws are often used it is far better to have a tiny die-stock adequate to their manufacture, and taps from $\frac{1}{16}$ in. to $\frac{1}{8}$ in., increasing by 64ths, will do all that can be required.

Figs. 651.—Screw-plate Perforations.

USING SCREW-PLATES.

The iron intended to be screwed by a plate is held generally in a vice. The point is filed a little taper to enable the plate to start more easily. After the point has been filed up, the plate is applied, the largest hole of the set being used first. Downward pressure, varying with the size of the screw, is necessary to start the thread, after which the plate moves for-

the holes should be cleaned, or the next time it is used the cuttings may bind, and break the thread in the plate.

OTHER POINTS IN USING SCREW-PLATES.

If the rod on which it is desired to cut

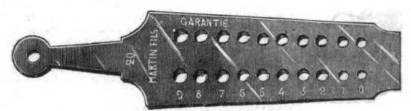

Fig. 652.—Swiss Screw-plate.

ward by its own action. Starting the thread is the great difficulty. The plate must be kept square with the rod being screwed, or what is known as a "drunken" thread will be made. A little practice will give the necessary skill. Of course, the more the point of the iron rod is tapered, the easier it is to start the thread; but, then, all this tapered portion may have to be cut off after the thread is finished, which is a waste both of material and time. Besides, in short screws there is not length enough to allow much taper-ing, so that it is better to learn to do with

a thread is steel, and is hard, or has hard places in it, it should be softened before the plate is applied, or the cutting edges will soon be worn out or broken. If the rod or pin to be threaded has been turned in the lathe, it is generally left in the chuck, the point slightly tapered and the plate pressed against it, the lathe mean-while being pulled round by hand. The screw-plate is sometimes used as a gauge to size when a number of screws have to be cut the same diameter. The screws are cut as nearly alike as possible, and the plate is then passed over them to ensure

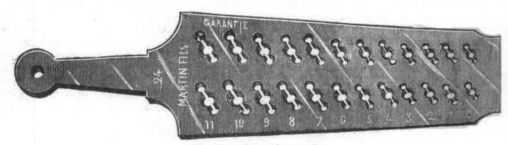

Fig. 653.—Swiss Screw-plate.

as little as possible. Oil is used as a lubricant with iron and steel, but is not required for brass. The plate is worked both forward and backward, and should not be forced on too fast, or the rod will be twisted. A thread can be cut with a plate right up to a shoulder or to the head of a screw. After the plate has been used,

their similarity. In small machines—such as sewing-machines and automatic tools, which are turned out in duplicate in large numbers—all the separate parts are made interchangeable, so that repairs are more easily executed. All the screws have therefore to be alike, and a screw-plate is a ready means of insuring this.

SOLDERING, BRAZING, AND RIVETING.

AFTER soldering, soft and hard, has been defined, soldering bits and stoves will be described. Solders of many kinds will then have their preparation explained, and

copper, brass, or silver, the term brazing is employed. Soldering may be defined as the uniting of two metals by means of a more fusible metal or metallic alloy; this easily fused metal, or solder as it is termed,

Fig. 654.—Ordinary Copper Bit.

recipes will be given for various kinds of fluxes. The process of soft soldering, the preparation of the bit, and the methods of soldering a variety of materials and articles, will then be entered into fully. The vexed question as to how aluminium may best be soldered will then be discussed. Blowpipe soft soldering will lead up to hard soldering, or brazing as it is termed, and various shapes of blowpipes, blowlamps, and other heating appliances will be described and illustrated. Then will come a description of the actual process of brazing, and several typical jobs, including cycle

is applied usually in sheet metalwork, with a heated copper-bit, miscalled a soldering iron, various fluxes being used to facilitate the melting and flowing of the solder, and to assist it to adhere to the two surfaces to be joined. Particulars of the requisite tools and materials will be given before describing the practical processes.

SOLDERING BITS.

Soldering bits differ in size and shape, according to the work to be done. Fig. 654 is the ordinary pointed soldering bit

Fig. 655.—Bent Copper Bit.

brazing, will be explained. The chapter will conclude with information on riveting.

SOLDERING DEFINED.

Soft soldering is a process of uniting metals by means of fusible alloys of lead and tin. When the uniting metal is hard, such as

used for general work; Fig. 655 shows a lighter tool, having a bent point. The bottoming bit (Fig. 656) is used for soldering round the bottoms of cylindrical vessels. A hatchet bit is illustrated by Fig. 657. A self-heating bit, employing spirit as fuel, is shown by Fig. 658. Ordinary soldering bits should have from 4 in. to 5 in. of copper, in addition to the copper riveted in the shank, as in constant usage the length diminishes by filing and

wire instead of a ferrule. A hole, the size of the round shank, should be bored about three-fourths of the way up the handle; the remainder should be bored smaller for the pointed end of the shank to come through about $\frac{1}{4}$ in., this end being either

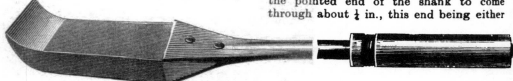

Fig. 656.—Bottoming Copper Bit.

drawing out. A copper bit that has a long shank fatigues the arm quickly, not only by the weight of the tool, but by the cramped position into which it throws the arm. For general use, the soldering bit should be about 16 in. long from point to

bent and clenched into the end of the handle or riveted with a small burr or washer.

Stoves for Heating Soldering Bits.

For heating the copper bit, some form of

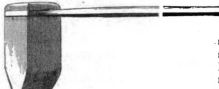

Fig. 657.—Hatchet Copper Bit.

extremity of handle. The handle should be made so as to afford a firm grasp and balance the copper.

stove will be required. There are several stoves made specially for the purpose. Fig. 659 shows a common form of fire-pot mostly used for outdoor work; a simple stove on the lines of this may be made easily by a metal plate worker. Fig. 660 shows a good form for indoor use. The top part lifts off; the front and outside grating are in one piece, and are easy of removal when it is required to clear out the ashes that drop through from the top

Fig. 658.—Self-heating Copper Bit.

Handles of Copper Bits.

Good handles are turned from sound beech, and are about 5 in. long and $1\frac{1}{8}$ in. to $1\frac{1}{4}$ in. in diameter. A groove is turned in the handle to receive a binding of copper

part; the whole stands in a cast-iron tray. Coke is the best fuel for the indoor stove, and charcoal for the outdoor one. Coal is not suitable, as it smokes the soldering bits.

GAS STOVES.

Gas stoves may be used for heating the bits, and have the advantages of requir-

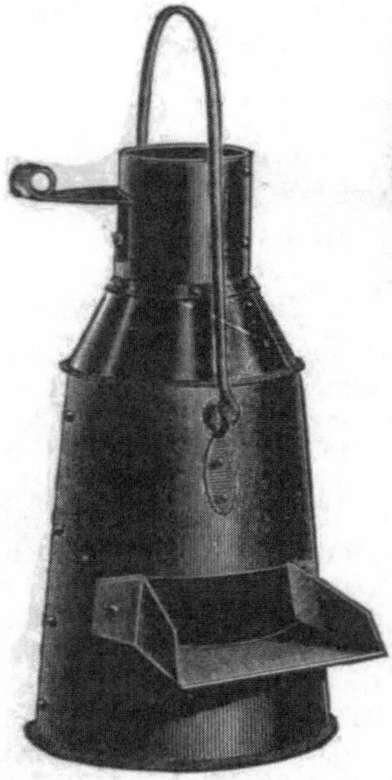

OIL STOVES.

Paraffin stoves for heating soldering bits have been used. The wickless stove is

Fig. 661.—Gas Stove for Soldering Bits.

Fig. 659.—Fire-pot for Soldering Bits.

ing but little attention, and of being soon made ready for use; with a gas stove, also, the solder is not so liable to be burnt

employed, and the arrangement partakes largely of the nature of a Roots blower. But a wick stove can also be used; and a Rippingille's paraffin stove, with a burner not less than 4 in., and a sliding tank, can be easily adapted for the purpose. Get a sheet iron chimney, as shown in Fig. 663, 3 in. high, to fit over the dome of the burner; rivet four legs on it, or make some arrangement so that it will not topple over. A medium-sized iron will get hot in about five minutes, and the stove is handy for many other purposes.

SELF-HEATING SOLDERING BIT.

Soldering bits heated by benzoline or spirit, such as that illustrated by Fig. 658, may be made with a small barrel-shaped reservoir, and this, in addition to holding the spirit, is used as the soldering bit

Fig. 660.—Stove for Soldering Bits.

Fig. 662.—Gas Stove for Soldering Bits.

off the bit through inattention. Suitable gas stoves are illustrated by Figs. 661 and 662.

handle. One end of the reservoir is fitted with a filling cap at the side, and from the opposite end protrudes the tube carrying the burner. To the tube end of the reservoir an iron clip is attached, and this secures an iron bar which stands out over the burner head. At the end of this bar the copper bit is attached and held either vertically or horizontally in the flame.

SELF-CONTAINED GAS-HEATED BIT.

A gas-heated soldering bit is shown by Fig. 664. It has a hollow handle to which is attached the copper, the flame passing through the hollow handle and striking the back and sides of the copper. At the other end of the handle, as shown in Fig. 664, is attached a flexible rubber tube which allows the bit to be used in any position desired. When worn, the copper only is renewed, and the iron handle does not have to be replaced. As the flame

Fig. 663.—Oil Stove Chimney.

strikes the back and sides of the copper, excessive wear on the point is prevented.

REST FOR SOLDERING BIT.

A rest for a hot soldering bit consists of what is known as a "chair" for the support of hot-water pipes. These chairs are made in different sizes, to suit pipes of various diameters, but one suited for a 3-in. pipe makes the best stand for a copper bit. It is easily obtained, and costs only a few pence. Fig. 665 shows the bit and rest.

REQUISITES FOR SOFT SOLDERING.

Soldering requisites, besides solder and flux, are a jar to hold the flux, a file or two, a scraping knife, and one or two brushes to apply the flux. A good brush for the purpose can be made from a bit of compo

tube; a knot of hair out of a broom is slipped into the tube, which is flattened with a blow of the hammer. A suitable "brush" also may be made by hammering the end of a piece of cane until the fibres separate (see Fig. 666).

Fig. 664.—Gas-heated Soldering Bit.

TINMEN'S SOLDER.

Tinmen's solder is composed of lead and tin in varying proportions. A solder suitable for rough jobbing and outdoor work is made by melting in a ladle 3 lb. of lead and adding 4 lb. of tin; mix well with a smaller ladle, remove the dross which will float on the top, and run into strips, in moulds made by bending some pieces of sheet-iron anglewise, and turning the ends so as to prevent the solder running out. These moulds are about 14 in. long, and are shown by Fig. 667. For a finer solder for general purposes, new work, etc., the proportion is 4 of tin to 2 of lead. For blowpipe solder add 1 part of bismuth to the alloy last mentioned, and run out in fine strips by means of a small iron ladle having a fine hole (see Fig. 668), or having a proper spout as in Fig. 669. Ladle out some molten solder, and, holding it over a flat iron plate, which the bottom of the ladle should just touch, cant the solder to flow through the hole, at the same time drawing the ladle from

Fig. 665.—Soldering Bit and Rest.

left to right; a fine stream of solder cools as it touches the cold iron plate, making sticks fit for use with the blowpipe. Another blowpipe solder is $1\frac{1}{2}$ of tin to 1 of lead made in the same way.

PREPARATION OF SOLDER.

In preparing solders, whether hard or soft, take great care to avoid two faults—a want of uniformity in the melted mass, and a change in the proportions of the constituents by the loss of volatile or oxidable

Fig. 666.—Cane Brush for Soldering Flux.

ingredients. Thus, where copper, silver, and similar metals are to be mixed with tin, zinc, etc., it is necessary to melt the least fusible metal first. When copper and zinc are heated together, a large portion of the zinc passes off in fumes. Soft solders should be melted under tallow, and hard solders under powdered charcoal, to prevent waste by oxidation.

SOLDERS OF VARYING FUSIBILITY.

Advantage may be taken of the varying degrees of fusibility of solders to make several joints in the same piece of work. Thus, if the first joint has been made with fine tinner's solder, there would be no danger of melting it by making a joint near it with bismuth solder, composed of lead, 4; tin, 4; and bismuth, 1; and the melting point of both is far enough removed from

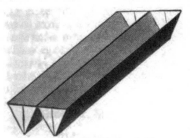

Fig. 667.—Mould for Solder.

SOFT SOLDERS.

that of a solder composed of lead, 2; tin, 1; and bismuth, 2, to be in no danger of fusion during the use of the latter.

The following table gives particulars of various soft solders, together with their

approximate melting points on the Fahrenheit scale :—

Lead.	Tin.	Bismuth.	Melting Point, Fahrenheit.
4	4	1	320°
3	3	1	310°
2	2	1	290°
1	1	1	255°
2	1	2	235°
1	1	2	205°
1	1	2	201°
8	4	15*	150°

* and 3 parts of cadmium.

Fig. 668.—Solder Ladle with Hole.

Bismuth, a constituent of most soft solders, is a metal having a grey-white colour with a red tinge; it burns with a blue flame when heated considerably, and is easily dissolved by slightly diluted nitric acid. When making any of the alloys mentioned in the table melt the metal with the highest melting point first, and then add the remaining metals; well stir the whole, and cast into ingots or strips for solder.

HARD SILVER SOLDER.

A silver solder for soldering copper, though here mentioned out of its place, is composed of 5 parts of copper, 3 parts of zinc, and 2 parts of silver. Melt the copper first, then add the silver, and lastly the zinc; directly the zinc is immersed, rapidly stir the alloy so as to render its

Fig. 669.—Self-skimming Solder Ladle.

composition equal throughout, and then cast it in a small ingot mould. The ingot is then rolled down to form a small sheet

equal to about No. 18 s.w.g. gauge in
thickness, and from this narrow strips are
cut as required. Ordinary solder may be
converted into fine solder by melting and
then adding the silver in the proportion
given above.

Solder Containing Phosphorus.

Special solders sometimes contain phosphorus, but there is a difficulty in adding
this to the other ingredients. A typical
recipe is:—Zinc 50·03 parts, tin 47·99 parts,
aluminium 1·76 parts, and phosphorus ·22
parts. When adding the phosphorus, first
mix it with the tin as follows. Take a
length of 1-in. gas barrel which has a
screwed cap at one end, the opposite end
being closed with a tin plug. Remove the
screwed cap, and, having carefully dried
between blotting paper the proper proportion of phosphorus, insert the latter in the
tube and replace the cap. Now put the
plugged end of the tube into the molten
tin ; this will melt the plug of tin and so
allow the phosphorus to come in contact
with the molten metal. The ingot of phos-
phor-tin formed could be alloyed with the
zinc and aluminium, or preferably with the
zinc alone, leaving out the aluminium altogether. When this solder is used on
aluminium, the phosphorus acts as a flux ;
consequently no other is necessary.

Making Solder Wire.

The following simple method of making
solder wire is very handy for small work.
Take a sheet of stiff writing or drawing
paper and roll it in a conical form, rather
broad in comparison with its length ; make
a ring of stiff wire to hold it in, attaching
a suitable handle to the ring. The point
of the cone should first of all be cut off to
leave an orifice of the size required. It
should then be filled with molten solder,
and held above a pail of cold water, and the
stream of solder flowing from the cone will
solidify as it runs and form the wire. If
held a little higher, so that the stream of
solder breaks into drops before striking
the water, it will form handy elongated
"tears" of metal ; when it is held still
higher, each drop forms a thin concave cup

or shell, and each of these forms will be
found to have its own peculiar uses in blowpipe work.

Adding Sulphur to Solder.

Brimstone (roll sulphur) can be added to
heated solder, in order to cleanse it from
impurities — principally zinc, which in
solder is fatal to a clean homogeneous
joint. As the sulphur melts, it unites with
the obnoxious metal to form a zinc sulphide, which rises to the surface in the
form of dross, which can be removed easily.
Care must be taken that the suffocating
fumes or gas (sulphurous acid) formed by
the heating of the sulphur are carried
away up the flue and not into the workshop.

Flux or Soldering Fluid.

Attention is now directed to the various
fluxes used in soldering. The object of a
flux is to remove impurities from the surfaces to be joined so that the solder can
come into contact with them. Soldering
fluid is the name often given to the fluxes—
raw and killed spirits—used in soft soldering. Raw spirit is muriatic, also called
hydrochloric, acid made by the action of
sulphuric acid on salt, and called by some
"spirits of salts." As it gives off visible
vapours, it is also called "smoking salts."
These fumes rust tools and take the colour
from clothes. Hydrochloric acid is used
by galvanisers and tinners as a bath in
which to prepare the black iron to receive
its coating of whiter metals after the oxide
is removed. Raw spirit has a strong destructive action on zinc, and is used as a
flux to solder that metal, and also to solder
galvanised ware. In soldering zinc the
raw spirit becomes "killed" by dissolving
the zinc, and forms zinc chloride.

Killed Spirit.

The flux in use for soldering nearly all
kinds of sheet metals is the zinc chloride
just referred to. To prepare it, take a
large earthenware pan and place a quantity
of clean scrap zinc on the bottom. Half
fill the pan with muriatic acid, when a
dense and irritating vapour rises, and continues until it ceases to bubble. When the

zinc and chlorine are united, the acid is said to be "killed." Put a slate over the pan and leave it for, say, twelve hours; then pour off the clear liquor and bottle for use. Killed spirit may be made on a small scale with a pennyworth of muriatic acid and a few zinc slating nails or a piece of sheet zinc. If the acid is not entirely killed by using an excess of zinc, the free acid present may cause a black stain to show on the work when soldering. To prevent this, some add a piece of washing soda or sal-ammoniac to neutralise the free acid. It is always best, after using strong spirits, to wipe with a wet rag, and to clean off afterwards. To solder brass, copper, or bright iron, or to tin metals before soldering them, use the killed spirit neat; but where there is but little oxide to overcome, as in tinware, use half water and half spirit; a pointed stick of wood or a small brush (see p. 216) will hold enough flux for soft-soldering jobs Sal-ammoniac, being the chloride of ammonia, forms a useful flux for tinning copper goods.

Resin Flux

To solder without acids, fluxes containing carbon are used; these, when heated, flow like a varnish over the surface to be soldered, thus protecting it from the oxygen in the air at a time when metal is most likely to oxidise. These, generally, are used after an article is tinned by an acid process. Resin is the principal flux of this kind, and, when powdered, is very useful for soldering bright tinware; it can be scraped off again, leaving a bright line of solder without fear of rust in its track. Resin is hardly a true flux, as it cannot remove oxide from the metals being dealt with, but, when used, it should be placed on rustless material, which it will preserve in that state. In using ordinary tinner's solder for uniting surfaces that are already tinned—such as tinned plate and tinned copper—resin is flux, but when surfaces of iron, brass, or copper that have not been tinned, are to be joined by soft solder, the soldering fluid is by far the most convenient. Resin possesses this important advantage over soldering fluid, that it does not induce subsequent corrosion of the article to which it is applied. When acid fluxes have been applied to anything that is liable to rust, it is necessary to see that they are thoroughly washed off with clean warm water, and the articles carefully and thoroughly dried. Powdered resin and oil are used for best tinware and for soft metal; the mixture can be wiped off while hot, thus avoiding the risk of scratching the work by scraping it.

Other Fluxes.

To solder pewter, Britannia metal, and other alloys having a low melting point, use Gallipoli oil or Venice turpentine, the latter being crude turpentine before the spirit has been distilled off. To solder heavy lead, Russian tallow is generally used, with or without resin, while for leadlight soldering a palm-oil candle is recommended as a suitable flux. A flux said to be perfectly non-corrosive, causing no injury to the health of the workman, was introduced a few years ago; it was made by dissolving 1 part lactic acid and 1 part glycerine in 8 parts of water. A solution or phosphoric acid in alcohol makes, it is said, an excellent soldering fluid, which has some advantages over chloride of zinc.

Soldering without Separate Flux.

Solder that can be used without a special flux contains the flux—resin—enclosed inside it. This can be melted with a match or red-hot poker. It is used sometimes by electric workers for small joints in wires, and is called solderine. It consists of a very soft solder tube filled with resin.

Tinning Copper Bits.

Before any soldering can be done, the soldering bit must be tinned. To do this, heat it in the fire to blood-red, grip it in a vice, and quickly file the four faces quite bright; dip the end of the copper bit in killed spirit, rub it on a piece of sal-ammoniac, hold some solder to the point of the bit, and melt a little on the lump of sal-ammoniac, rubbing and turning the bit at the same time. If it is hot enough, the solder will flow, and coat the face of the copper. Dip it again in the spirit, and the operation is complete. Another method of tinning the bit is to heat it as though for soldering. File the parts to be tinned, put some resin on a soft brick, and rub the filed parts on the brick and in the resin, which the heat of the copper will melt. While rubbing the copper on the brick, press a piece of "fine" or "half-and-half" solder on the part to be tinned, and a film of solder will adhere to the copper. The surplus resin will adhere to the brick, and the surplus solder will remain on the resin. To re-tin the bit when necessary, again file the copper and rub it on the brick. The resin and solder already there will suffice for several operations, and more resin and solder can be added as required. Tinning the copper in this way is best effected when it is barely hot enough to melt the solder. If it is too hot, the filed parts tarnish or oxidise before the resin and solder can be applied. Another method is to rub the filed copper on a block of sal-ammoniac and apply the solder, or it can be dipped in killed spirits of salts and applied to the solder. In these two ways the copper is eaten away very quickly by the sal-ammoniac or the spirits. Resin is the best flux to use.

Another Method of Tinning Copper Bits.

Another method of tinning a bit is with a piece of clean tin-plate, about 4 in. square, nailed to a piece of wood; heat and file the bit as before, dip the end in killed spirits, and then put a pinch of resin on the tin-plate, melt a little solder on it, and rub the bit briskly on the tin-plate. The solder will quickly flow on the clean part of the soldering bit. Do not make the copper bit red hot after tinning, or the whole process of filing and tinning will have to be repeated. It should be the aim of the workman to use the copper bits so that they remain tinned, re-tinning involving loss of time and the filing quickly wearing away the copper. Care should be taken that bits are not burnt in the stove

The Process of Soft Soldering.

In soft soldering it is essential to get absolutely clean surfaces. This cleanliness must be both mechanical and chemical. Mechanically clean the surfaces by scraping or filing, and before the air has time to oxidise the clean surfaces, chemically clean them by applying the acid flux. In dealing with corroded zinc, scrape the corroded surface with a sharp steel scraper, and then render the surface of the zinc clean and bright by rubbing it with emery cloth. The clean surface can then be soft-soldered with a copper bit, using tinman's solder of good quality, with hydrochloric acid as a flux. Sound soldering cannot be produced unless the metals, where they are to be joined together, are brought to the same heat. Consequently, in soldering a thin piece of metal to a thick one, it is necessary to hold the soldering-iron a long time on the thick piece before attempting to solder the thin piece to it. The obvious reason for heating the thick piece first is that, by reason of its bulk, it "holds" the heat longer than the thin piece. Therefore, when the thick piece is judged to be approaching the soldering heat, the thin piece is laid upon it, and fixed by applying the flux, melting the necessary amount of solder upon it by means of the soldering bit; hold the hot soldering bit firmly to the thin piece of metal, keeping this down also by a piece of wood. The usual sign that the soldered joint is sound is that air or gas bubbles arise from the melted solder, and when they discontinue, take off the bit, or rather slide it off on to a part still uncompleted. One essential point to be remembered is that the soldering bit must have a good face—that is, the part of the copper bit used for the purpose of rubbing on the metal and solder to make them of the necessary heat for joining must be well "tinned," or in other words, well covered with solder. This "face" is essential to good soldering, and if, in heating the bit (which should be done in a clear fire, or in a gas or oil stove), it is suffered to become red hot, the tinned face is oxidised by the heat and is destroyed, partly by sublimating and partly by amalgamating with the copper of the soldering bit. The face can then only be renewed by the old burnt face being filed off (it is extremely hard) and then adopting one of the methods of tinning already described.

"Flushing" Soldered Joints.

Expert solderers have a method of levelling and cleaning their soldered joints by "flushing"—an operation which is extremely simple in the performance when seen in practice, but exceedingly difficult to explain on paper. It may be, perhaps, best compared to the glazier levelling the putty with his knife.

Soldering Leak in Tin-plate Vessel.

The repair of a vessel—say a teapot—having a pinhole in the bottom, is a typical elementary soldering job. With a knife, scrape round the pinhole a clean place about the size of a sixpence. An old razor driven into a file handle makes a very good scraper, though, for some purposes, a more pointed one is required. Have the bit, solder, spirits, and spirit brush handy, heat the bit, hold the teapot in the left hand, dip the spirit brush in the spirits, press it against the sides of the jar to remove superfluous liquid, and apply a little to the place to be soldered. Dip the bit in the spirits to clean it, apply it to the piece of solder, and try to pick up a little with it. This may be found a little difficult at first, the solder running all over the bench instead of clinging to the bit. Hold the bit in a nearly horizontal position, and do not dig its point into the solder, but lay the side of the bit on it lightly near the point and draw inwards ; then, keeping the bit as near horizontal as possible, bring the point of it with the solder it has picked up to the hole in the bottom of the teapot. As it touches the leaky place lift the handle of the soldering bit, and the solder will flow off at the point ; move the bit just round the hole and the solder will run all over it. Remove the bit, and allow the metal to cool (blowing on it will accelerate the cooling), and the job is done. Try with water, to make sure that the repair is sound.

Soldering Leaky Kettle Spout.

When the spout of a tin teakettle leaks it should be treated as follows :—Dry it thoroughly, scrape it well a little farther each way than the leak extends, hold it in the most suitable position for soldering, which will be on its side, apply the flux, and solder it, laying the angle of the bit in the angle formed by the spout and body.

Soldering on Coffee Pot Knob.

To put a knob on a coffee pot lid, merely place it in the hole (after scraping, of course), and with the bit melt a little solder and the end of the knob together inside, the knob resting on the bench whilst soldering, and taking care not to keep the bit on it too long or the knob will melt away. A little stud is usually punched out and soldered inside over the place where the knob is fixed, to give a neat appearance.

Soldering Patches on Saucepan.

To patch a tin saucepan that leaks proceeds as follows :—Cut a piece of tin about 1 in. square, snip off the corners, lay it on the bottom of the saucepan just over the hole, mark round it with a brad-awl, scrape clean all the inside of the square thus marked, and ¼-in. all round the outside of it ; or, in other words, if the patch is 1 in. square, scrape a clean place 1½ in. square. If the saucepan is in good condition and scrapes bright and clean, it will be ready to solder ; but when the bottom is eaten by fire and rust it is difficult to get clean, and after scraping and cleaning it as well as can be done, it must be "tinned" to make sure of the solder flowing under the patch. To do this, simply rub the hot soldering bit over it with a little solder, applying spirits to make it tin easily. Then lay on the patch and solder it, drawing the bit first round the edges and then all over it, holding the patch down with something to prevent its shifting ; the solder flows underneath, and the patch is what is called "sweated" on. This description of putting on a patch applies also to teakettles, coffee pots, fish kettles, etc., with the exception that a large patch need not be sweated all over, but simply soldered all round the edge, letting the bit rest mainly on the patch, so as to draw the metal underneath.

SOLDERING COPPER OR BRASS.

For these metals the bit has to be hotter than for tin-plate. See that the part to be soldered is well cleaned as before. Use killed spirits as a flux. If there is any fancy brass or copper work, file it off quite smooth when finished. Then get a small piece of sulphate of copper and crush it very fine, and mix it with cold water to the consistency of paint. Then rub a little of this on the part of the work which is soldered, and wipe it off again with a dry rag. This makes the solder look like copper, and if it is done well it is hardly detectable on copper work.

SOLDERING SPOUT ON COPPER KETTLE.

To re-solder a spout on a copper kettle, first thoroughly clean, with a piece of emery cloth, the copper where the spout is to be inserted, and also clean the spout around its large end. Then tin the copper inside the kettle where the spout is to be soldered, and also the spout, using killed spirits as a flux. Pass the small end of the spout through the hole from the inside of the kettle, and press it up so that the small flange on the large end of the spout butts against the side of the kettle; then solder round the spout on the inside of the kettle, and leave a thin body of solder floated smoothly round where the join occurs, the same flux being used as for the tinning.

PATCHING COPPER KETTLE.

A copper kettle that is cracked along the edge of the bottom for about 1¼ in. can be repaired by putting a patch on the bottom and turning it up on the side. Scrape clean a place on the bottom 1 in. wide, and extending a little each side of the crack, scrape up the side ½ in. in the same manner, and then tin the places. Now take a piece of thin copper 1½ in. by 1 in., clean it both sides, and tin one side with the soldering bit. Fix such a job as this in the vice by the handle, as it leaves the hands at liberty, and place the piece of copper, tinned side down, on the bottom of the kettle, leaving about ⅜ in. overlapping to turn up on the side. Solder the piece on to the bottom, cut the overlapping piece to the same sweep as the sides, and with a light hammer rap it close up to the side, and then solder round that part; with a file or scraper clean off any superfluous metal that may have run the patch, and over if necessary, colour the solder with a little solution of sulphate of copper.

SOLDERING SOFT METAL TEAPOTS.

Soldering soft alloys will require some skill and practice to master, and it would be advisable for learners to practise on an old metal teapot before attempting more ambitious work. The metal must be scraped bright, because a mere clean appearance is no criterion that it is clean enough to solder properly; a light soldering bit should be used, as the bit cannot be rested on the work as can be done in soldering tinware. The weight of the bit being all on the wrist, it would be impossible to hold a heavy bit steady. The bit must be just the right degree of heat: if it is too hot, it may make a hole in the metal; if not hot enough it makes the work botchy. If a soft metal teapot leaks all around the spout, and has been recently used, the first thing to do is to dry it out thoroughly, or on applying the bit the solder will splutter about and bubble up instead of flowing nicely. Having dried it well, the next thing will be to scrape it, for which purpose the large blade of a penknife answers very well. Take some very fine running solder, and have ready some resin and oil for a flux; apply it round the spout with a little brush, heat the bit, apply a little solder to the teapot spout, and gently draw it round with the bit, scarcely letting it touch the metal of the teapot. The heat of the solder will melt the metal as it is drawn round; if the bit is fairly hot, do not try to make the solder run round very far, but keep picking more up, and reheat when it hangs. Supposing that the spout has been rounded successfully, the next thing is to clean off the job so that the repairs can hardly be detected. With a small half-round file about 6 in. long, work round the job lightly, taking care not to file the sides and spout of the teapot more than can be helped, and when it has been got fairly level and smooth go round it with emery cloth or a round-ended scraper. After this use a steel or agate burnisher.

REPAIRING PEWTER ARTICLES.

Vessels of pewter, Britannia metal, etc., are repaired by soldering. Pewterers' solder is composed of 2 parts of bismuth, 1 part of lead, and 1 part of tin. When making the alloy, melt the lead first, then add the tin and bismuth; sprinkle a little resin on the surface of the molten alloy to prevent oxidation, well stir it, and then pour the metal into an iron mould. When using the alloy, first well clean the article where it is to be soldered by scraping with a sharp knife, then rub a little tallow over the cleansed part. Melt a small knob of solder from the stick; place the knob on the part to be soldered, and, with a fine jet from a blowpipe, blow gently upon the solder until it flows over the part to be repaired and adheres to the pewter; smooth the edges of the patch of solder with a smooth file, and finish off with a burnisher.

SOLDERING METAL RIMS.

A rather difficult kind of work is the repairing of the rims and covers of china tea pots and hot-water jugs. These are extremely thin, and made of a very soft and fusible metal. The parts to be soldered must be scraped carefully; no streaks of uncleaned metal should be seen, very little flux should be used, and a very steady hand is required. Good solder which melts at a low temperature should be used. If these directions are attended to, there is little fear of failure or spoiling the work.

SOLDERING ZINC.

Zinc is a difficult metal to solder smoothly and well, even when new and clean. The strong or raw spirit of salt must be used, and the worker must try to keep the spirit to the part to be soldered, and not let it run all over the work; and after the seam joint, or whatever it may be, is soldered, the spirits should be wiped off when the work is cold, as the spirits have a very corrosive action. Soldering zinc is

unpleasant, because of the smell caused by applying the acid to the zinc. It will be found that the solder does not flow very well on zinc; this is because some of the zinc mingles with the solder and deteriorates it, and it will cling to and get hard on the soldering bit like a lot of dross. The bit must be cleaned at intervals if much zinc is being soldered; a light touch up with a file all round, and a rub on the sal-ammoniac with a little solder puts it all right again. Old zinc has to be got fairly clean with a scraper or file before attempting to solder, and the spirits should be used on it freely. Zinc is not a soft metal like pewter, but it will melt under the iron if thin and if the bit is very hot; practise will soon show the right degree of heat to work with. Smooth the soldering where

Fig. 670.—Shavehook.

rough with file and scraper. The plumber's shavehook (Fig. 670) is a good tool for cleaning off zinc after soldering, especially on long seams.

SOLDERING GALVANISED IRON.

The treatment of this is very similar to that of zinc, strong spirits being used as the flux. An article in zinc or galvanised iron that is to contain any liquid should be very carefully tried before being passed as finished, as though it may appear all right, actually it may be leaky. Galvanised iron is really steel coated with zinc.

REPAIRING CORRUGATED IRON.

The best way of repairing sheets of galvanised corrugated iron in which there are several large holes is to cover the holes with sound pieces of the same material. The pieces should be fitted to the holes, and then soldered in position. The iron must be well cleaned by scraping, or by other suitable means, where the soldering is to be done, and raw spirit (hydrochloric acid) used as a flux; the iron can then be soldered with ordinary tin-man's solder and a copper bit in the usual way.

REPAIRING SPOUT OF IRON TEAKETTLE.

Most wrought-iron kettles are tinned outside as well as inside before being japanned. If on scraping off the black this proves to be the case, and it has not been in use very long, the scraping knife will be all that is required to clean it for soldering; but if old and rusty round the spout, say, through having been put by when the leak was discovered, instead of being sent to be repaired at once, then the file will be required; a 10 in. half-round is about the most useful for these jobbing operations. Carefully clean all round the spout till there are no dirty places or rust spots. Then the successive steps are—apply the spirits, heat the copper bit to a good heat, apply the solder by holding a strip of it in one hand and melting a little on with the bit held in the other. Holding the kettle draw the bit round the spout and the solder will follow. Let the bit rest on the work, as it must be made hot before the solder will unite to it properly; a strong job cannot be made unless the parts to be united are as hot as the metal that is to unite them. For example, supposing a kettle is coated with fur to the depth of $\frac{1}{2}$ in. all over the inside, the solder cannot be made to flow freely, on account of the wet fur taking the heat from the bit, and although it might be possible to stick a little solder round the spout, a good sound job cannot be produced; but if the fur is cleaned out, there is no difficulty.

SOLDERING STEEL.

Steel is not a suitable metal for soft soldering, but still the following method answers fairly well. The work must be scraped perfectly clean, and the parts brushed over with killed spirits. Heat the work to drive off the moisture, taking care not to smoke it, or in any way to make the surfaces dirty; while the work is hot, try

to make the solder adhere; if it will not, add a little Venice turpentine, and use the copper bit. The proper application of heat, and cleanliness, are the essentials. When the surfaces are tinned, tie the parts together; using a little Venice turpentine or solution as flux, apply the hot bit, and add solder if necessary. Generally, merely brushing the surfaces with killed spirit, and drying them before tying together, answers if the hands and tools are clean and free from grease

SOLDERING BRONZE FIGURE.

To mend a broken bronze figure by solder-ing or "sweating," place a small quantity of solder on the face of the broken parts, hold the two together and heat from (pre-ferably) a Bunsen burner. When the solder runs, press the two parts together and hold them firm till set. The mark may then be covered over with bronze solution (ready mixed) to which a little black is added till the right shade is obtained. A very fine camel-hair pencil should be used for applying the solution. If the article has been well soldered there should be only the slightest mark showing the joint.

SOLDERING BRONZED ZINC ORNAMENTS.

The bronze or lacquer must be scraped off a little each side of the joint, and clean raw spirits and a fairly hot copper bit used, so as to melt the zinc slightly and cause it to unite firmly to the solder. Use a very little solder at a time, trim off with a file and scraper, and then brush over with powdered sulphate of copper (bluestone), and killed spirit.

BRONZING SOLDERED JOINTS.

To bronze the soldered joints on a brass fitting by a superficial coating, it is neces-sary to varnish the soldered parts with a thin coating of gold size, and, while the latter is still wet, dust over the moist parts with bronze powder until the whole of the soft solder is hidden. When the gold size has set hard, the bronze powder will be found to have firmly adhered to it, and the surface may then be rendered smooth by very lightly rubbing with a burnishing tool.

SOLDERING CATCH ON GUN-BARREL.

In soldering a catch on a gun-barrel it will be necessary to tin both the barrel and the catch, and then to bind the latter to the barrel with strong wire, though special cramps for the purpose are available; also bind the barrels for some distance from each side of the catch, making the ribs secure with wedges. To melt the solder, use heaters; these are generally made of copper with iron handles; or iron rods can be used, the ends being made red hot and inserted in the barrels. Cut some small slips of thin solder and place them on each side of the catch, using powdered resin as the flux, though some workers hold that the best flux for the purpose is sal-ammoniac. Baker's preparation can also be used as a soldering fluid. As soon as the solder melts, remove the heaters and cool the barrels.

DIFFICULTY IN SOLDERING ALUMINIUM.

Almost as soon as aluminium was prepared on a large scale, it was discovered that it could not be soldered with the ordinary alloy of tin and lead; but about 1885 M. Christofle, a goldsmith of Paris, found that either pure tin or pure zinc would unite aluminium, and could therefore be used as a solder for it. In practice, however, the tin soon falls apart, while zinc, by itself, is brittle, and discolours badly. The failure of tin is due to the fact that it forms an alloy with the aluminium, and this alloy is decomposed by the oxygen of the air. This decomposition is most noticeable in aluminium-tin alloys containing only 10 per cent. or less of tin. M. Hulot, therefore, tried copper-plating the ends to be joined, and then soldering the coppered surface in the ordinary way. The Tissier brothers were the first to employ alloys of aluminium and zinc, which, although found to be too brittle, have since been much used. In soldering aluminium, if the soldered part be immersed in water, the solder invariably separates from the

aluminium in periods varying from two to seven weeks. According to M. Moissan, aluminium containing the slightest trace of sodium forms a strong galvanic couple with most metals when immersed in water, and this may possibly account for this separation.

ZINC-ALUMINIUM SOLDERING ALLOYS.

The following are some of the zinc-aluminium mixtures which have been employed, one of the chief difficulties experienced in their use being the high melting point:—

1.

| Aluminium | ... | ... | 8 parts. |
| Zinc | ... | ... | 92 parts. |

2.

| Aluminium | ... | ... | 12 parts. |
| Zinc | ... | ... | 88 parts. |

3.

| Aluminium | ... | ... | 15 parts. |
| Zinc | ... | ... | 85 parts. |

4.

| Aluminium | ... | ... | 20 parts. |
| Zinc | ... | ... | 80 parts. |

In preparing these alloys, it is usual to melt the aluminium, and then add the zinc to it gradually. When this has melted some fat is added, and the whole mixture is stirred with an iron rod and poured into moulds.

SPECIAL FLUX.

The special flux for use with zinc-aluminium solders consists of a mixture of 3 parts copaiba balsam, 1 of Venetian turpentine, and a few drops of lemon juice. It is usual to dip the soldering bit into this mixture.

MOUREY'S ALUMINIUM SOLDER.

M. Mourey, another Parisian goldsmith, improved on these alloys by the addition of copper, and later by the addition of silver. His blowpipe solder is made up as follows:—

Tin ...	...	...	...	6 parts.
Zinc ...	...	...	...	3 parts.
Aluminium	...	...	...	2 parts.
Copper	...	...	...	1 part.
Silver	...	...	...	1 part.

ALUMINIUM SOLDERS CONTAINING PHOSPHORUS.

These solders probably owe their adhesiveness to the great affinity that phosphorus has for oxygen. If melted alloy containing phosphorus is placed upon aluminium there is naturally a tendency to absorb oxygen from the impure film on the surface of the aluminium, as well as from the air surrounding it. The presence

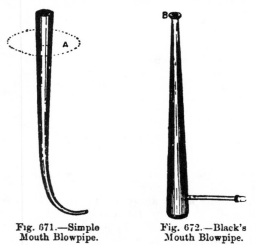

Fig. 671.—Simple
Mouth Blowpipe.

Fig. 672.—Black's
Mouth Blowpipe.

of zinc in phosphor solders, together with the absorption of air during the soldering process, probably accounts for the comparatively brittle nature and the dull colour of this class of solders. A solder should be free from brittle metals, as their presence in soft solders generally decreases the malleability of the alloy. The method of preparing solders in which phosphorus is an ingredient is explained on p. 218.

EASILY FUSIBLE ALUMINIUM SOLDERS.

Alloys of an easily fusible nature for soldering aluminium are recommended by the following authorities:—Frishmuth, of Philadelphia, says: Tin 95 parts, and bismuth 5 parts; or tin 97 parts, and bismuth 3 parts; while J. Richards recommends aluminium 2·5 parts, zinc 25·25 parts, phosphorus ·25 parts, and tin 72 parts. Other alloys for this purpose are: aluminium 1 part, tin 9 parts; or cadmium 5 parts, zinc 2 parts, and tin 3 parts. Also phosphor tin (in variable proportion); or tin 20 parts, and zinc 1 part; or tin 99 parts, and copper 1 part; or tin 90 parts, copper 9 parts, and bismuth 1 part.

HOW TO SOLDER ALUMINIUM.

Any of the solders given in the previous paragraph can be fused readily with a copper bit, which, to ensure success, should be of a wedge-shape bent round to, roughly, a quarter circle. Its edge is then at right angles to the aluminium, and, by lightly moving the bit backwards and forwards while in contact with the aluminium and flowing solder, the scraping action removes the impure film. The coated surface can then be soldered with an ordinary copper bit. When soldering, everything should be perfectly clean; the work, too, must be done quickly, as, if the surface is not coated at the first attempt, the aluminium is injuriously affected, and a good joint cannot be made unless the affected part is removed by scraping or filing, and the operation of soldering repeated. The difficulty of getting solder to adhere to aluminium is caused by a metallic film (probably an oxide) which forms on the surface of the heated metal, and which prevents ordinary soft solders alloying to form a strong joint. A flux might be used to render the surface of the aluminium pure during the soldering operation, or the film might be removed by mechanical means (by scraping, as just recommended), or a solder devised that would dissolve or combine with the film on the surface of the metal while both solder and aluminium were heated. The composition of a reliable flux for soft soldering aluminium has not yet been made public.

THE FILM OF OXIDE.

That it is an impure film that prevents

aluminium being soldered satisfactorily may be proved in the following way: Immerse a piece of sheet aluminium along the edge in molten tin and allow it to remain until it reaches the same temperature as the tin. On withdrawing

Fig. 673.   Fig. 674.

Fig. 673.—Trumpet Mouthpiece for Blowpipe.
Fig. 674.—Bone Mouthpiece for Blowpipe.

the metal, it will be seen that the tin has not adhered to any part of the edge. Again immerse, and then lightly scrape the immersed edge with a wedge-shaped tinned copper bit; the scraping will at once remove the impure film from the surface of the immersed metal, and union between the new surface and the molten tin will at once be effected.

"Sweating" Metal Together.

In many cases, tin, copper, brass, etc., may be united neatly without using the soldering bit, by filing or turning the joints so that they fit closely, moistening them with soldering fluid (see p. 218), placing a piece of smooth tinfoil between them, tying them together with binding wire, and heating the whole in a lamp or fire till the tinfoil melts. Pieces of brass may be joined in this way so that the joints are quite invisible.

Soldering without Bits.

A small blowpipe is sometimes substituted for the soldering bit, but is unsuitable for soldering tin cans, kitchen utensils, etc. Soldering with a bit—say, round the spout of a kettle—is done with a kind of rubbing or scraping movement. This brings the metals into more intimate contact, and causes them to adhere. Of course, a blowpipe flame cannot be used in such a way, nor can the solder be guided with the blowpipe so well as with the copper bit; but such jobs as soldering a pin hole, or anything like that, may be

managed with a blowpipe, though even then the bit is rarely dispensed with altogether. Lamps of special construction are also used (see p. 236).

Mouth Blowpipes.

The simplest blowpipe, for use with a bundle of rushes, a tallow candle, or a spirit lamp, is shown by Fig. 671. This is the common mouth blowpipe, and to prevent it slipping round in the mouth, an oval flange A is soldered on. In Black's blowpipe (Fig. 672), the saliva from the mouth is caught in the end of the tube instead of being projected into the flame or upon the work; after use, this blowpipe should be put in the flame for a moment or so to remove the lodged moisture. Mouth blowpipes vary in length from 9 in. to 12 in., the mouthpiece diameter is from $\frac{1}{4}$-in. to $\frac{1}{2}$-in., and the bore of the jet is $\frac{1}{16}$ in. or less. A platinum or glass nozzle gives the best flame.

Blowpipe Mouthpieces.

These vary in form and in material, and ferrules of gold, silver, horn, and ivory are used at times. For ordinary work, the mouthpiece may be lacquered and then plated with gold or silver. A trumpet mouthpiece (Fig. 673) may be made by turning a piece of wood to shape, and boring it centrally, to fit over the end of the blowpipe. A square mouthpiece held between the teeth prevents the blowpipe slipping, and enables the operator to have the use of both his hands. The bone mouthpiece (Fig. 674) fits the shape of blowpipe shown by Fig. 672, providing the blowpipe rim B is removed.

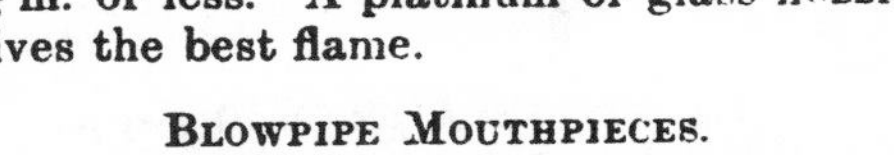

Fig. 675.—Fletcher's Mouth Blowpipe.

Fletcher's Mouth Blowpipe.

Fletcher's blowpipe (Fig. 675) is constructed especially for soldering. If the air jet that strikes the flame is hot, the power of the flame will be increased, and this result is obtained by coiling the small end of the tube so that a portion of it remains in the flame, whilst the jet still retains the correct position for impinging on the work.

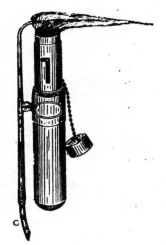

Fig. 676.—Spirit Lamp and Blowpipe.

SPIRIT LAMPS COMBINED WITH BLOWPIPES.

Ordinary mouth blowpipes rarely being longer than 9 in. or 10 in., the operator's eyes are brought uncomfortably near the work, and to overcome this difficulty, blowpipe lamps may be used. Fig. 676 shows a portable brass spirit lamp, to whose case a blowpipe is attached in such a way that it can be detached readily. The fuel is methylated spirit, or, preferably, benzoline, and the wick must be very absorbent. The flame, having been started, is regulated by the sliding tube. Of course, one hand must hold the spirit lamp, and this is a disadvantage. By lengthening the rubber air-supply tube c, the operator can keep his head as far from the flame as he pleases. A pointed flame is produced, but Cole's spirit torch and blowpipe (Fig. 677) gives a thick flame, about 6 in. long. Fig. 678 resembles Fig. 676, but the blowpipe is connected to a band E, so that its nozzle can be thrust into, or drawn from, the flame. From the same band a chain suspends a cap which covers the top of the lamp when this is not in use. The band

should be fairly tight, just capable of sliding along and keeping any position given to it. The case is a brass tube of from ⅜ in. to 1 in. bore; on its bottom end is soldered a disc, and under this is fixed a wooden handle.

GAS BLOWPIPES.

Of the many kinds of blowpipes in use, probably the gas blowpipe is the best. With a sufficient gas supply the power of the appliance depends entirely on the air pressure, and this, speaking generally, should not be less than ¾ lb. per sq. in. The production of such a pressure will cause the average adult little inconvenience, for, by straining, it is quite possible to blow through a tube with a force of something like 1½ lb. per sq. in., the average pressure being about half this. The lungs, however, are really very defective blowing instruments, a second-rate pair of bellows being far superior. When the air supply is produced mechanically, it is generally by double bellows similar to those of a smith's forge, or bellows fitted with an elastic reservoir. Worked properly the former

Fig. 677—Cole's Spirit Torch and Blowpipe.

would give a fairly constant supply of air, but the use of the rubber disc undoubtedly results in the pressure being far more even. Blowers are illustrated on pp. 232 to 236.

SIMPLEST FORM OF GAS BLOWPIPE.

In its simplest form the gas blowpipe resembles Fig. 679, the rubber tube connecting it with the gas supply being fixed

Fig. 678.—Another Spirit Lamp and Blowpipe.

on the pipe at the point of connection with the cock. To construct the appliance, one end of a piece of brass gas pipe of the required length, with, say, a ⅜-in. bore, is bent as shown, whilst at the back of the curve thus made a hole is drilled to admit a tube, A, 𝟷⁄₁₆-in. in diameter. This should have one end (see dotted lines) bent to correspond with the angle previously formed in the larger tube, whilst its other extremity should be bent upwards. Make

larger one, the ends being almost flush and quite concentric. Solder the parallel portions of the tubes together, and then fix a gas-regulating cock to the larger one, as in Fig. 679. The end is then connected to an ordinary bracket or burner by an indiarubber tube, G, and a short piece of tubing is fitted with a bone or other mouthpiece and attached to the projecting end of the air tube. This instrument will do any soldering, and will be equal to melting gold, silver, and brass, or brazing odd jobs in iron or steel. Of course, when used for the last-named purpose it would be in conjunction with asbestos cubes or other supports.

RELATIVE VOLUMES OF GAS AND AIR FOR BLOWPIPES.

The late Mr. Thos. Fletcher has said that, speaking roughly, but still sufficiently near to make a correct rule by which to work, a blowpipe requires one volume of gas to eight of air. If the gas is supplied at a pressure equal to 1 in. of water, and the air at eight times that pressure, then, to get the best effect, the area of the gas and air pipes should be equal. If the air supply is equal to 16 in. of water pressure, the gas pipe must be double the area of the air, and so on in proportion. Some makers assert that a better working flame is produced by using ten volumes of air to one volume of gas, but, of course, if the blow-

Fig. 679.—Simple Gas Blowpipe.

these pipes red hot where they are to be bent, and, if they are afterwards plunged in cold water, the material will to some extent be softened, and its tendency to split will be obviated. The smaller tube is passed through the hole in the bend of the

pipe is fitted with taps, the supplies can be adjusted easily. It will be found, however, that any practical departure from Fletcher's rule will result in a loss of power. As Kirk says, a blowpipe with a ⅛-in. air jet, if worked with an air pressure of 10 oz. per sq. in., that is 15 in. of water, will braze up to about ½ lb. total weight, or in other words will securely unite two pieces of brass each weighing ¼ lb. With the same pressure a ½-in. bore air jet will braze a total weight of about 2 lb., and so on in proportion. It will be understood that the

a T coupling and diminishing socket, an elbow, and one or two pieces of pipe. The air tube A (represented for the most part by dotted lines) passes through the dimin-

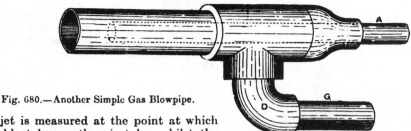

Fig. 680.—Another Simple Gas Blowpipe.

air jet is measured at the point at which the blast leaves the air tube, whilst the area of the gas supply is that of the annular space between the two tubes. When the air tube is thus carried inside the gas tube (see Fig. 680), the tool appears to be much larger than it really is, and this accounts for the fact that a ½-in. size blowpipe with the air tube fixed outside the gas supply

ishing socket until it almost reaches the nozzle of the blowpipe, with which it is concentric. By using the elbow D, the two supply pipes are brought parallel to each other, so that the indiarubber connecting tubes can be more easily held in the hand like reins, as by simply squeezing them the flame can be readily regulated. Sometimes, in cases of emergency, a plug drilled to admit the air tube is used in place of the socket. The plug is thrust into the end of the T socket, but in all cases it must be air-tight. This blowpipe can be used efficiently only in conjunction with a foot blower.

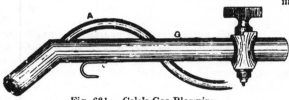

Fig. 681.—Cole's Gas Blowpipe.

is just as effective as one of the ¾-in. size, which carries the air tube inside the stem. All indiarubber tubing must be perfectly smooth inside, for if it is wired or in any way rough, the resultant friction will cause a loss of pressure. It should also be of as large a bore as is convenient.

COLE'S GAS BLOWPIPE.

With regard to up-to-date appliances, Fig. 681 shows a blowpipe manufactured by M. Cole, of West Didsbury, for use with a foot blower; it can be obtained in various sizes, but for cycle brazing, etc., a ½-in. gas tube and a $\frac{1}{16}$ in. air tube will answer.

Fig. 682.—Lever Gas Blowpipe.

LARGE AND EFFICIENT BLOWPIPE.

Fig. 680 shows a method of constructing a large and efficient blowpipe in a few minutes, the only materials required being

LEVER GAS BLOWPIPES.

The gas blowpipe (Fig. 682) has a lever under the control of the finger, by which the supply of gas can be regulated most easily. This appliance will be found extremely useful by all general braziers and cycle builders. In some blowpipes the gas is continually burning when the tool is not

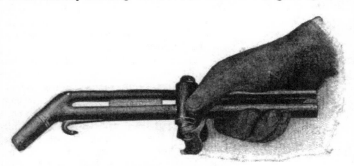

Fig. 683.—Fletcher's Gas Blowpipe.

being used, but with this appliance, when the gas is turned on at the main cock the pilot light at P is the only light burning until the tool is actually required. By pressing the lever when brazing, any desired quantity of gas can be obtained, but the moment the blowpipe is released the lever automatically allows only the pilot light to burn until the gas is entirely shut off at the main cock. In Figs. 679 to 682, A is the air supply and G the gaspipe.

OTHER AUTOMATIC GAS BLOWPIPES.

The automatic blowpipe, C pattern, made by Fletcher, Russell & Co., is in three sizes, and is suitable for all ordinary workshop practice. Both gas and air are controlled by a movement of the finger, so that, after using it for a few minutes, one acquires the knack of regulating the flame to a nicety. Fig. 683 shows a blowpipe fitted with Fletcher's improved control tap arrangement. This tap is useful, as the gas and air supplies are independent of each other. Fig. 684 shows an injector blowpipe, a device of great power.

THE USE OF THE BLOWPIPE.

The blowpipe can be used in the repair of brooches, jewellery, etc., in hardening and

tempering tools, besides rendering possible anhydrous analysis and the brazing of such articles as bandsaws and bicycles; in fact, a good gas blowpipe, combined with an efficient blower, obviates in many cases the necessity for a smith's hearth or furnace.

USING THE MOUTH BLOWPIPE.

Only comparatively small jobs can be done with the mouth blowpipe. A total weight of about ½ lb. could with ease be soft soldered, but when brazing, even on the best of supports, it would be found a difficult matter to unite anything exceeding 1½ oz. total weight. It is somewhat difficult to describe the art of blowing with the mouth, but the object of the operator is to send a continuous stream of air through the blowpipe and at the same time to breathe through the nostrils. It is thus possible to keep up a perfectly regular blast for quite fifteen minutes, though to one not accustomed to the blowpipe it would seem absolutely necessary to make an occasional break during such a period. In this case the lungs are the bellows, the mouth is a chamber, and the cheeks are elastic discs. If the mouth is filled with air until the cheeks are distended, it will be found possible to close the passage connecting the mouth and the throat, so that

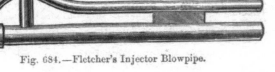

Fig. 684.—Fletcher's Injector Blowpipe.

the mouth will still be full, though the operator breathes through the nostrils. The blowpipe now being inserted between the lips, the muscles of the cheeks force a current of air through it, whilst the partial vacuum thus created in the mouth is filled by the air taken in from the lungs, which are, of course, supplied by the operator breathing through the nose. More air has now been forced by the elasticity of the cheeks through the blowpipe, so that immediately the lungs are replenished the passage between the mouth and throat is

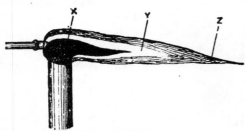

Fig. 685.—Blowpipe Flame.

opened, and the mouth again refilled; thus a continuous blast is maintained. One warning is necessary—do not inhale air through the blowpipe.

SOFT SOLDERING WITH THE MOUTH BLOWPIPE.

The article to be repaired is placed in such a position that the worker can blow on the spot to be soldered. Then the blowpipe is inserted in the mouth, a stick of solder is held in the right hand, and a candle or torch in the left hand. If a clear and regular blast is maintained, and the operator continually keeps touching the joint with the solder (at the same time using plenty of flux), the parts will soon be strongly united. A smoky flame will foul the joint, and as parts which are smoked will not unite with the alloy, they must necessarily be cleansed and blown over again. Smoky flames are produced when an ample supply of gas at a good pressure is employed in conjunction with a weak air blast. Under these conditions the flame is inefficient, wasteful, and of low temperature, whilst when the gas supply is poor and the air pressure powerful the flame is literally starved, the result being oxidation or the production of "scale." To obtain an effective and economical blowpipe flame, the air and gas supplies must be carefully regulated, so that the carbon shall be burnt with a smokeless flame.

THE BLOWPIPE FLAME.

This consists essentially of three parts. x (Fig. 685) is a non-luminous cone consisting of a mixture of atmospheric air and unburnt combustible gases (each with a low temperature); Y is a luminous cone composed of burning gases (carbon and carbonic acid being in excess); and z is a cone, the oxygen in which renders it less luminous and free from combustible materials, its temperature being exceptionally high, especially where, the cone comes in contact with the point of the cone Y. Because of its properties, z is technically termed the oxidising or outer flame, whilst Y is known as the inner or reducing flame, because when it is applied to some easily reducible substance—say oxide of lead—oxygen mingles with the unburnt carbon and produces carbonic oxide, the lead being thus separated or reduced.

THE HEAT OF THE BLOWPIPE FLAME.

The blowpipe flame is one of intense heat, even that produced by blowing a common candle being capable of melting metallic fragments when they are supported on a bed of charcoal. The pointed

Fig. 686.—Fletcher's Foot Blower with Reservoir above.

flame gives the greatest heat, and this can be produced simply by increasing or decreasing the space between the flame and the article to be soldered or the metal to be melted. If it is necessary that the heat should be felt over a greater surface of the metal, a lamp or torch similar to Fig. 677

or Fig. 678, pp. 228 and 229, may be employed, as, the wick and the aperture of the blowpipe being both larger than usual, a bush or sheet flame is produced.

Fig. 687.—Fletcher's Foot Blower with Reservoir Below.

SUPPORTS FOR BLOWPIPE WORK.

Of supports for the work under the blowpipe, one of the most common is a block of pumice-stone. Another common support is coke, which, being a fuel, augments the heat of the flame to a considerable extent. The heat is also confined and retained by using on the hearth asbestos nuts, firebrick, and charcoal; but the latter is chiefly employed for small work. Charcoal made from hard woods, however, will not do, for it generally throws up sparks, crackles, and splits immediately the bed begins to glow, thus displacing the object to be brazed. The most efficient kind is said to be charcoal from the willow, and this may be purchased in compressed blocks. A reliable support is a mixture of powdered charcoal and fireclay made into a doughlike compound by adding rice paste. This should be pressed into a mould (an ordinary firebrick hollowed out will do) and allowed to dry. For large blowpipes the charcoal must be in small quantity only, and it is said that for great heats it is better to use a solution of silicate of soda instead of rice-flour paste. In this case the charcoal may be in larger proportion, as it is not so liable to burn away,

owing to the protection afforded by the silicate of soda.

APPLYING THE BLOWPIPE FLAME.

When soldering with the blowpipe the beginner often makes the solder flow by impinging the flame directly on it. As a result it fuses, assumes a globular shape, and harmlessly rolls round on the flux to the cooler metal beneath; the body of the work should have been heated previously to a point hotter than the melting point of the solder. Mistakes sometimes result from uncleanliness, or from the use of poor fluxes and unsuitable solders. In most cases, however, failure is due to the heat not being applied at the right time nor in the right place. If the body over which the solder is to flow is heated by the blowpipe flame until it is a shade hotter than the melting point of the alloy, the metal will create a kind of attraction which will draw the solder into the joint. Thus the parts are united in a neat and lasting manner.

FOOT BELLOWS FOR GAS BLOWPIPES.

The larger blowpipes, illustrated by Figs. 679 to 683, can be used only in conjunction with a bellows, and three foot bellows,

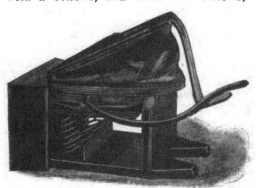

Fig. 688.—Fletcher's Foot Blower with Spring Reservoir.

made by Fletcher, Russell & Co., are here shown. Fig. 686 is particularly suitable for the purpose, though that illustrated by Fig. 687 has the advantage of a reversed reservoir, risk of injury to the rubber disc being thus avoided. In Fig. 688 is seen a spring reservoir in place of the indiarubber ball. In the cheaper form of blower illus-

Fig. 689—Griffin's Foot Blower.

trated by Fig. 689, the reservoir is hung up on the wall out of the way.

HOME-MADE FOOT BELLOWS.

The home-made foot bellows for blowpipe work, about to be described, is built on the principle of a single-action forge bellows. The dimensions given are convenient, but may be modified as required. A smaller

is a front elevation of the bellows with footboard removed, and Fig. 691 is a side elevation. Probably the only materials that will have to be purchased are the leather, spring, and two hooks, since any old box will supply the necessary wood. A piece of thin bellows leather will be required, and out of this cut two strips measuring 7 in. by 24 in. Turn or saw three discs 7½ in. in diameter, from ½-in. board, though the middle one might with advantage be a little thicker; they should be as free from knots as possible. The disc intended to form the top of the bellows should now be laid aside while the lower two are fitted with valves. To do this, make a hole through the centre of each with a 1 in. centre-bit; then get two small slips of thin wood about 1½ in. by 2 in., and cut off from the remaining leather two pieces of the same width as the wood slips, but ½ in. longer; glue the slips of wood firmly on to the smooth side of the leather, leaving the ½ in. projecting. Next tack them through the projecting part on to the disc, keeping the wood uppermost and central with the holes (see Fig. 692).

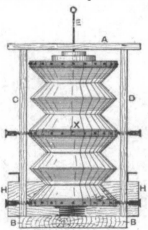

Fig. 690.

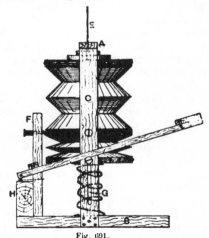

Fig. 691.

Figs. 690 and 691.—Home-made Foot Bellows.

size is not recommended, but perhaps an additional inch in the diameter of the disc would make the apparatus more useful, as it might then be used in conjuction with a small forge for very light work. Fig. 690

MAKING THE BELLOWS.

The two strips of leather are now sewn, or cemented with indiarubber solution, so as to form two cylinders through which the discs can just be passed, though one end of one cylinder should be very slightly larger than this to allow it to overlap the other cylinder on the middle disc. Place this disc in position and secure with a few

Fig. 692.—Bellows Valve.

tacks. Now from rather stout galvanised iron wire bend two rings of the same size as the discs. These may be soldered together, but will suffice if overlapped and well bound with brass binding wire. They should be sewn in four or five places inside and half-way down each cylinder, care being taken not to allow any leak through the stitching; the other two discs should then be secured at the ends so that the valve of the middle and bottom discs both work upwards. Now closely stud the leather round all three discs with large-headed tacks. After which a $\frac{3}{8}$-in. hole must be driven sloping upwards from the middle of the centre disc (see x, Figs. 690 and 692) till it breaks through into the top compartment; this is the wind hole. This completes the bellows, and it only remains to fix them up in a suitable frame for foot action, the construction of which is shown in the illustrations.

BASE-BOARD, ETC., OF FOOT BELLOWS.

The base-board B (Figs. 690 and 691) is $\frac{3}{4}$ in. or 1 in. in thickness, 9 in. in width, and about 13 in. in length. About 6 in. from one end two pieces of batten, 16 in. long (c, D), are then let in, and $\frac{1}{4}$-in. grooves, as shown at G (Fig. 691), are cut

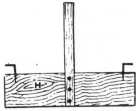

Fig. 693.—Bellows Foot-board Support,

through them. A block of wood H (Figs. 691 and 693), 11 in. long and 3 in. across, is then obtained, and two square hooks are

screwed in 1 in. distant from each end, while a square piece of wood F (Fig. 691), 9 in. high, is screwed to the middle of it to form a third support to the bellows. This block is then screwed on to the further end of the base so that 1 in. projects each side. The bellows are then fixed in place by three screws into the centre disc: two through the middle joints of the uprights c, D, and one at the same height through the pillar F. Two screws are now also put through the grooves into the lower disc, leaving about $\frac{1}{2}$ in. projecting, while washers are placed between the bellows and uprights so that the motion may be as smooth and even as possible.

COMPLETING THE FOOT BELLOWS.

A foot-board frame (Fig. 694) must now be made; the length of this is immaterial,

Fig. 694.—Bellows Foot-board.

but it must be wide enough just to pass easily over the uprights c, D. Holes are drilled for the reception of the hooks, and the frame is then finally put together in place, resting on the top of the projecting screws of the lower disc. A piece of wood A (Fig. 690) similar to the uprights is then nailed across their tops; and through a central hole in this a thin rod E is passed and fixed firmly into the centre of one upper disc to act as a guide; it is threaded through discs of lead whose weight will depend upon the wind pressure required. A good strong sofa spring, as powerful as possible, is then put under the bottom disc, and the machine is ready for use. All leakage in the bellows should be reduced to a minimum, and after the bellows are made the leather should be coaxed into neat folds and squeezed in a vice for some hours. Pan lids filled with lead make very good weights, and the wind can be conveyed in ¼-in. compo. pipe bent round the centre disc and up one of the uprights.

IMPROVISED BELLOWS.

Suitable bellows may be improvised for blowpipe work. Those shown in Figs. 695 and 696 will produce a steady current of air at a regular pressure. An intermittent current is of no use. Get two pairs of household bellows—the larger the better—and fix them up as shown in Figs. 695 and 696, in which A is the feeder, B the reservoir, and C the pedal. Construct a frame as shown, of 3 in. by 3 in. pine, and cut off the pipe of the lower bellows at D; then cut a hole in the top board to correspond with the hole in the upper bellows. Weights can be put on the top to give the pressure,

packed with asbestos, which serves as a filter and stops any impurity in the benzoline from getting to the burner. The flame can be lowered to a glimmer when not actually in use, thus saving the trouble of relighting. When the lamp is to be used, the regulator R (Fig. 698) is screwed up tight; and care must be taken to ascertain, from time to time, that the burner or nipple C is open and perfectly clean. If this becomes obstructed, it can be cleaned by unscrewing the tube T, and passing a fine steel wire through the hole, taking care not to enlarge the hole. The lamp should be completely filled with benzoline every time it is to be used. A little methylated spirits is poured into the basin A, and set alight. When the apparatus has become slightly warm, the regulator R is opened gradually. To extinguish the flame, the regulator must be screwed up tight. If

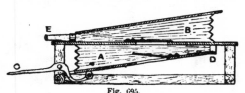

Fig. 695.   Fig. 696.

Figs. 695 and 696—Elevations of Simple Bellows.

and a small one will be needed to cause the bottom part to fall rapidly. Fix the two ends of a strong rubber spring to E and C, and thus complete the bellows.

BENZOLINE SOLDERING LAMP.

A soldering lamp sometimes is used in the place of a blowpipe, and it should combine perfect security with compactness and portability. The lamp illustrated by Fig. 697 appears to fulfil these conditions thoroughly. In this lamp only benzoline of good quality should be used, never ordinary paraffin. When the lamp is in use and the body of it is very hot, the inside pressure does not exceed three-fifths of an atmosphere, whether the regulator R (Fig. 698) is open or almost closed. Thus the danger of explosion, which is such a drawback to lamps which use ordinary paraffin, is avoided. The upper parts of the lamp are subjected to great heat and therefore are

any escape is observed round the screw of the regulator, R, the square P should be screwed up with the key supplied by the makers, so as to tighten the asbestos packing. The lamp above described is only one of a great number of such appliances, but it is fairly typical of them all.

METHYLATED SPIRIT SOLDERING LAMP.

A spirit lamp for soldering is best made from a brass tube 8 in. long by 1 in. in diameter, closed at the bottom. The tube is loosely filled with cotton wick, and spirit is then poured on and thoroughly saturates the wick, which should be pushed down so that it is flush with the open end of the tube. A soldering lamp ought not to be near a reservoir of spirit, which would easily take fire. If a reservoir is desired, make it of tinplate, closed, and with a tube 2 ft. or 3 ft. long to connect it to the lamp; it should also be supplied with a tap, and fixed at the proper height to give the required flow of spirit.

Hard Soldering or Brazing.

Brazing, as already explained, is another name for hard soldering, and differs from soft soldering in the fact that the uniting metal or solder has a higher melting point and so cannot be applied with an ordinary bit ; instead, a forge or a powerful blow-pipe must be employed to make the hard solder—known as spelter—flow into the

and use of the blowpipe having already been given, the brazing lamp and brazing hearth may here be described, leaving the forge fire for description in the next chapter.

Lamps for Brazing.

Brazing with spelter—that is, with hard solder—necessitates a far greater heat than soft soldering, and when a lamp is used to supply the heat, it must be much more powerful than the soldering lamp illustrated on this page. The fuel may be

Fig. 698.

Fig. 697.

Figs. 697 and 698.—General View and Section of Benzoline Soldering Lamp.

joint. Brazing is used where greater strength is required than can be given by soft solder, or when an article has to stand a degree of heat when in use that would cause soft solder to run.

Source of Heat for Brazing.

There are three or four usual sources of heat for brazing—the blowpipe, the coke or coal forge fire, the benzoline or paraffin brazing lamp, and the gas brazing hearth, this being an arrangement of blowpipes. Sufficient information on the construction

benzoline or paraffin, according to the special construction of the lamp, and there is generally a hand pump fitted for putting the fuel under air pressure. One of the most powerful lamps made is the Invicta No. 1, which gives a flame 18 in. long, and in parts $1\frac{1}{2}$ in. in diameter, this being capable of melting a $\frac{1}{2}$-in. copper rod in three minutes. Four pints of benzoline will give a full flame for an hour and a quarter, but by regulating the flame, such a quantity will last for several hours. A still more powerful lamp gives a 21 in. flame, 2 in.

thick, and holds 6 pints of benzoline. A typical paraffin brazing lamp can give a 12 in. flame ; under a pressure of 4 atmospheres 1 quart of paraffin lasts about 40 minutes, and under 2 atmospheres about 80 minutes. A larger size holds 3 quarts, gives a 20 in. flame, and its fuel lasts for 60 to 100 minutes, according to whether the pressure is four or two atmospheres.

GAS BRAZING HEARTH.

A brazing hearth with two gas blowpipes and a power blower is illustrated by Fig.

would not be very difficult to improvise such an arrangement, if Fig. 699 be taken as a guide.

FAN BLOWER FOR BRAZING HEARTH.

Figs. 700 to 702, which are to a scale of 1 in. to the foot, show a small fan blower that will provide a blast of 4 oz. pressure per sq. in., and will deliver about 60 cub. ft. at 2,200 revolutions per minute ; such a blower is suitable for use with any kind of brazing hearth, gas or coal. Fig. 700 illustrates the fan complete, Fig. 701

Fig. 699.—Gas Brazing Hearth.

699. This is made by Fletcher, Russell and Co., Ltd., and the stock size has a table 2 ft. by 2 ft., supporting a ribbed clay or coke bed, and two or three bricks which serve as supports for the work. For most purposes one blowpipe is sufficient. There are flexible connections from the air and gas supplies to the blowpipes, and altogether the appliance is extremely convenient for general work. It

shows the fan with one half of the outer case removed, whilst Fig. 702 is a section. As the whole of the fan is of cast-iron, patterns will have to be made ; but this is not very difficult, as core boxes are not required, and any ironfounder will cast the parts. The case A is in two parts, fastened together with bolts at the lugs B. The joint must be well fitted and made airtight with thick oiled brown paper. The disc c and the vanes are cast in one piece, the vanes having a curve as shown in Fig. 701. At the air inlet the vanes must be cut away as in Fig. 702, to enable them to cut into the incoming air. The vanes must run as close to the casing as possible, only sufficient clearance being left to allow them to revolve without touching. The disc is secured to the shaft either with a key or a set-screw as preferred. The bearings D are secured to the outer casing by the studs E; the shaft T is of mild steel; G is the pulley for a round belt, and H is a collar. Both collar and pulley are fastened to the shaft by set-screws, and set so that the disc may revolve without touching the case.

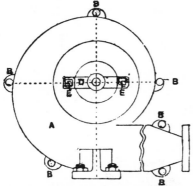

Fig. 700.—Fan Blower for Brazing Hearth.

From 3,000 to 3,300 ft.-lb. per minute, or $\frac{1}{10}$ horse-power, will be required to run the fan.

HARD SOLDER OR SPELTER.

A suitable solder for use in brazing small steel articles may be made in the proportions of silver 18, brass wire 2, copper 1. Melt in a crucible; when cold, hammer into a thin sheet, or granulate while molten by pouring into water. For small articles, a solder that will flow at a lower temperature than brass wire should be used. For brass, copper, or iron, equal parts of silver and brass are used at times, though for iron, copper, or very infusible brass, nothing is better than silver coin rolled out thin, which may be done by any silversmith or dentist. This makes decidedly

the toughest of all joints, and as a little silver goes a long way it is not very expensive. A commoner spelter for small

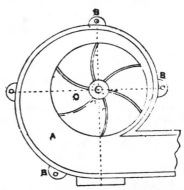

Fig. 701.—Fan Blower with Outer Case Removed.

articles of brass consists of 5 parts copper, 3 parts zinc, and 2 parts silver.

PREPARATION OF HARD SOLDER.

To obtain hard solders of uniform composition they may be granulated by pouring them into water through a wet broom. Sometimes they are cast in solid masses and reduced to powder by filing. Silver solders for jewellers are generally rolled into thin plates. Hard solders are usually reduced to powder, either by granulation

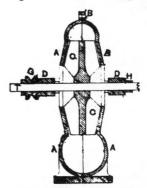

Fig. 702.—Section of Fan Blower.

or filing, and then spread along the joints after being mixed with borax that has been fused and powdered. It is not necessary that the grains of solder should be placed between the pieces to be joined, as with the aid of the borax they will "sweat" into the joint as soon as fusion takes place. The same is true of soft solder applied with a soldering fluid. One of the essentials of success, however, is that the surfaces be clean, bright, and free from rust.

Borax Flux for Brazing.

Borax is a true flux, and, as it is an acid, will combine with most of the oxides, which are usually bases. It protects the metal from the further action of the air, so that the solder is enabled to come into actual contact with the surfaces which are to be joined. Water dissolves only a very small quantity of borax properly. It is, therefore, not possible to mix borax powder and water in anything but a paste in which the borax sinks. This sinking of the borax does no harm, as, when hot, it melts and spreads all over the work evenly. For small articles requiring only a very little borax, such as spectacle frames, keys, etc., lump borax rubbed down with water on a slate produces a paste of much finer and more even consistency than does borax powder. Generally borax—lump borax, not powdered—is rubbed up on a clean slate with water until it is as thick as cream, or until it stands in ridges. The slate is 4 in. to 6 in. in diameter, and to roughen it, is sometimes scored across with the edge of a file. Make sufficient at one time for the job in hand, and apply it to the work with a borax pencil (camel-hair brush), taking particular care to cover completely all the surfaces over which the solder is to run.

Applying Spelter to Work.

For priming small work with the hard solder or spelter, small pieces of it (pallions) may be picked up with the borax pencil and laid along the seam. Gentle heating drives off the moisture, and, if the pallions have not shifted, the full heat can then be applied to make the solder run. Another way, after boraxing the soldering seam, is to have a narrow strip of solder, held in long pliers or anything suitable, and apply it to the work when this is

hot enough to make it run ; and as it runs the hard solder is moved along the seam. Should there be too little borax on the work, pass a piece of lump borax along it, and it will hold if the metal is hot enough. This is instead of leaving the work to get cool previous to applying more with the pencil. Borax is sometimes burnt and powdered before mixing with water, but that is only for cases where it is of great importance that the solder does not move.

Preparing Seams for Brazing.

If seams in brass are not required to stand much working after brazing, they may be joined edge to edge. When seams are formed in this way, little nicks, about $\frac{1}{2}$ in. apart, should be filed out along the edges, so that the solder flowing through the nicks during the soldering operation will render the joint sound. If the seam is to be worked after soldering, a small lap is necessary to ensure adequate strength. To form seams of this type, first thin the edge of the metal along the ends that are to form the seams, about $\frac{1}{8}$ in. in from the edge, so that when the two edges are lapped over each other their combined thickness at the seams will be the same as the single thickness of the metal at other parts. Cut a small cramp at the top and bottom of the seam, and fit the opposite edge in these cramps.

Brazing Brass.

After preparing the seams by either of the above methods, fasten binding wire round the articles so as to hold the pieces securely in position. Now powder some borax for use as a flux, and soak it in enough water to form a thick paste ; place a little of this along the parts to be soldered, and gently heat the article by some suitable means, such as foot bellows and blowpipe, so that it will expand equally, and not disarrange the seam ; increase the temperature until the metal is a dull red, and then take a strip of the hard solder (made of copper, zinc, and silver, as previously explained), dip the end in the borax, and, holding the opposite end with the pliers, rub the solder along the seam until a little melts off. Keep the solder in a molten state, and with a piece of wire flattened at one end gently rub the solder along the seam until every part is joined.

BRAZING IRON.

Small articles of iron may be joined in a similar way with equal parts of copper and zinc, but if the iron is to be hammered much after brazing, 2 parts of copper and 1 part of zinc would be more suitable. For use mix together equal parts of the borax paste and grains of spelter, and along the seams place sufficient of the mixture to unite them when melted. Some dry borax should also be kept ready at hand, so that a little may be taken and thrown on the spelter at any point where the material does not appear to be flowing freely.

BRAZING BRASS TUBES.

To braze together two brass tubes, reduce the diameter of the end that is to be brazed till the lengthening piece will just go on. Place on a layer of best soft brazing spelter, which should be in grains about the size of small pin heads mixed with some very fine dry powdered borax, and just damped sufficiently to hold to the tube; or brazing brass, obtainable in strips about ½ in. wide, could be held in a pair of small tongs and moved round the joint till all was well flushed. On applying heat to the tube and to the brazing brass, the latter will melt and will run well between the two tubes till quite flushed. When cold, the surplus brass must be dressed off and the tubes left level. This method can be adopted when the tubes of model locomotives have to be lengthened, the ends being left soft enough for drifting into the tube plate.

BRAZING TUBES TOGETHER WITH LINER.

An invisible and a much stronger joint could be made by inserting a liner, say 1 in. long and of a size to telescope in the tubes to be joined, placing ½ in. on each side of the joint; bring the ends of the tube flush and braze. If a tube to telescope is not to hand, use a piece of that to be joined, cutting sufficient out and closing up. Use

a soft solder, say 1 of copper and 1 of zinc, so that the solder runs before the tube.

BRAZING TUBES WITH OPEN FIRES.

For this, the heat must be applied evenly, so as to make the solder melt uniformly. This is done by moving the work about in the fire. When it is properly heated and the solder has flushed, the work should be removed from the fire, and after the solder has set it may be cooled in cold water without injury. Tubes to be brazed are generally secured by binding wire twisted together around the tube with the pliers. All tubes that are heated upon an open fire are brazed from within; otherwise the heat would have to be transmitted across the tube with greater risk of melting the lower part of the tube, the air in the tube being a bad conductor of heat. The two ends of the tube should be open so that the melting of the solder can be watched. In brazing long tubes the work rests upon the flat plate of the hearth, and portions equal to the length of the fire are united in succession.

BRAZING ELBOW PIPE.

Elbow pipes that are to be brazed are prepared as follows:—When cutting the patterns, leave, on one pipe only, a small clip at the side of the throat seam and one at the top of the back of the elbow. Then, upon the end of a mandrel, bevel the edge of the pipe with the clips, so that the second pipe, when forming the elbow, will overlap a little; also bend the clips so that they fit closely inside the overlapping pipe. Now, through the two clips, rivet the pipes together, so that they will not spring apart while brazing. With a flattened wire charge the elbow inside with spelter and borax along one side, and place it on the fire. Allow the elbow gradually to get to a very dull red all over, so that the sides will not open through unequal expansion. Then heat quickly to the melting-point of the spelter, and tilt the elbow so that the spelter will flow towards the throat. Charge the opposite side with spelter and flux, and repeat the first operation; put the top of the elbow in position on the fire, charge this part of the seam, and flow the spelter each way to meet that already applied.

Fig. 703.—Key Stem Dovetailed for Brazing.

BRAZING KEY STEMS.

In brazing together the broken parts of a key stem, first it is necessary to file the fractured ends quite true; this may entail the shortening of the key by $\frac{1}{4}$ in. or $\frac{1}{2}$ in., and as another $\frac{1}{4}$ in. will be lost in making the joint, it may be advisable to use another key bow having a longer piece of stem than the one which was broken off. With a warding file cut a dovetail on each of the ends to be joined, as shown by Fig. 703. A small half-round file will assist in making the edges true and square. The pieces must interlock perfectly and when

Fig. 704.—Brazing Guard for Use on Forge Fire.

this is the case, very lightly hammer the joint, around which then bind seven or eight turns of brass wire to act as spelter. Wet the joint, sprinkle powdered borax on it, and, holding the key in a pair of tongs, place it in a clear part of a forge fire made

with charcoal, small coke, or coal cinders, and commence to blow steadily the forge bellows or blower. Failing a forge fire, use a blowpipe, placing the key on a

Fig. 705.—Spatula for Applying Spelter.

piece of charcoal or pumice-stone whilst the heat is being applied. If the forge fire is used it is as well to support the key on a guard of thick iron plate (see Fig. 704) having a hole in its centre over which the joint to be brazed is placed. In this way the necessary local heating is obtained, and much labour in cleaning the key afterwards is avoided. On being heated, the borax swells and boils up, and should be pressed down with a spatula, previously dipped in cold water to prevent the hot borax adhering to it; a suitable spatula is made by flattening one end of a 1-ft. length of $\frac{1}{4}$-in. round iron rod, having at its other end

Fig. 706.—Key with Brazed and Dovetailed Stem.

an eye by which it may be hung when not in use (Fig. 705). With this spatula, also, powdered spelter may be added to the joint if required. When the brass wire begins to run, assist the flow by adding powdered borax, and when all the brass

(Fig. 708) to a bright heat (technically known as a spurtling heat), and slip it over the saw so that the splice comes between the jaws. When the brass wire melts and

Fig. 707.—Dovetailed Rod secured for Brazing.

has run into the joint, rub off superfluous molten metal from underneath and allow the joint to cool gradually. When cold, file up and clean the stem of the key until only a thin bright line of brass can be seen. Fig. 706 shows the finished key.

BRAZING METAL RODS.

The above description of brazing a key stem applies equally well to ordinary metal rods; when these are rather long, however, it is as well to bind the dovetailed pieces to a cranked wire, as shown in Fig. 707.

BRAZING BAND-SAWS.

By one method of brazing band-saws it is necessary to provide an iron heater, as in Fig. 708, the two arms of the fork being

Fig. 708.—Heater used in Brazing Band-saws.

at least 1½ in. long by ¾ in. wide, and welded and attached to a handle of ½-in. or ⅜-in. round iron, about 2 ft. long. A cramp (Fig. 709) is also required; it is made out of ¾-in. by ½-in. iron, and is thickened at the ends to take ⅜-in. set bolts (see Fig. 710). File each end of the saw for the length of two teeth, and fix the ends in the cramp as shown in Fig. 709, taking care that the saw is quite straight. Twist one loop of iron binding wire round the splice to hold it in place; then bind about a foot or more, according to the width of saw, of soft brass brazing wire round the splice. Moisten the whole with a saturated solution of borax, heat the iron

runs into the splice, remove the iron, let the saw cool to a dull red, and then quench in oil, afterwards filing up the spelter.

ANOTHER METHOD OF BRAZING BAND-SAWS.

Perhaps the most simple and reliable method is to use bright-hot tongs and black-hot tongs. File the ends of the saw

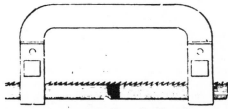

Fig. 709.—Band-saw in Cramps.

taper for the length of two or three teeth, so that when lapped one over the other they will be the thickness of the blade. Damp the ends, then place a little powdered borax and brass spelter between the ends that are being lapped. Heat a pair of heavy tongs in the fire until they become bright-hot, then close them on the joint until the spelter runs, as it will, if

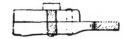

Fig. 710.—End View of Part of Band-saw Cramp.

the tongs are properly hot, in less than a minute. While slipping this pair of tongs off, another pair, made black-hot, must be quickly slipped on by one who has been holding them in readiness, and closed tightly on the joint. Remove these, hammer the joint tight, and clean up with a single-cut flat file. To set the teeth, lay the blade of the saw on a small steel anvil, the edge of which is bevelled. The teeth must overhang the bevel, and every alternate tooth is struck with a small hammer. When this is done, turn the saw, and treat the remaining teeth in like manner. To correct any irregularity in the set, the teeth should be side-jointed. This is done by placing a topping file longitudinally against the sides of the teeth, and lightly passing it over all the teeth on each side.

BAND-SAW REST.

A rest for the saw can be made from a piece of flat iron, as shown in Fig. 711,

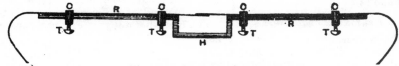

Fig. 711.—Band-saw Clamped to Rest.

where R is the rest. The part H may be held in a vice, or secured to the end of a bench with a clamp. The saw is placed on the rest, and held in position by means of four small clamps C. These clamps are tightened on the saw and rest by turning the little thumbscrews T. Fig. 712 is an enlarged view of one of the clamps. These may be made from $\frac{1}{16}$-in. or $\frac{1}{4}$-in. flat iron ; the rest is made from $\frac{1}{16}$-in. or $\frac{3}{8}$-in. flat iron, and must be perfectly straight. When brazing, keep the back edge of the saw fair with the edge of the rest.

CYCLE BRAZING.

Cycle brazing is fairly simple, but proficiency is obtained only by care and practice. Spoiling the tubes by overheating is the most general fault of a beginner ; in fact, skilled workmen will sometimes do it, especially on very thin tube. When borax is applied freely to the tubes near the joint there is less likelihood of overheating or burning of the tubes. The borax is a protection from the direct blast of the blowpipe, and its mere application or that of any flux must lower the temperature of a hot tube.

The tools and materials required are a gas blowpipe, a sheet-iron hearth or pan on which to place the work, and filled with small coke or breeze or with shaped firebricks, with which the work is backed up to retain the heat (see Fig. 699, p. 238, which shows a suitable brazing hearth). There will also be needed a small double-action foot bellows, or, if power is available, a small fan is preferable, as it gives a steady, continuous blast, whereas a bellows gives a more or less intermittent blast. An adjustable stand to hold the blowpipe in any desired position is very useful, as it leaves both hands free to manipulate the work. A much cheaper set of tools employs a brazing lamp instead of the blowpipe and bellows or fan. These lamps are fairly effective if they are of sufficient power.

SPELTER AND FLUX FOR CYCLE BRAZING.

A wire brush with which to clean off the superfluous borax, brass, and scale is very desirable, saving much labour in filing up as well as prolonging the life of the files. Some No. 3 spelter or brazing wire and borax, or one of the special brazing fluxes, also is necessary. Some braziers prefer spelter and others wire. If spelter is used, a steel or iron rod feeder will be required $\frac{3}{16}$ in. or $\frac{1}{4}$ in. in diameter and about 18 in. long ; it must be flattened at one end as in Fig. 705, p. 242, for applying the spelter. If brass wire is used, something is required to hold the short ends to prevent waste, such as a length of small bore steel tubing to fit the wire, sawn up for 2 in. Borax should not be purchased already crushed, but should be in lumps, when it may be crushed as fine as possible. It is an advantage to have the borax calcined, as it then settles down to its work at once when applied to the hot job, instead of rising and falling off, as uncalcined borax has a tendency to do.

STARTING CYCLE BRAZING.

One essential item to bear in mind is cleanliness. All parts to be brazed must be quite clean, and free from scale, rust, or grease. As a guide to the procedure, the brazing of all the joints on an ordinary cycle frame may be described. Before starting, have ready, close to the brazing hearth, two iron or tin boxes, one to hold the crushed borax, and the other the spelter mixed with about one-third of its bulk of borax.

BRAZING FORK ENDS AND LUGS OF CYCLE.

The back fork ends and lugs for the top of the backstays will be the first joints. When these have been fitted and cleaned and the insides of the tubes have been filed clean and free from scale, knock them in tightly and load the inside with about half a thimbleful of the mixed brass and borax. Shake this down to the joint, place it on the hearth, with the joint to be brazed, 1 in. or so lower than the other end, so that when melted the charge will flow towards the joint. Direct the flame to the thickest part of the lug first, so that the tube may be the last part to heat. As soon as the parts begin to turn red, with the feeder apply borax rather freely to the joint, so that it has a wet appearance before the heat is sufficient for the charge inside to run. Work the feeder, dipped in borax, round the joint until the brass appears on the outside, then quickly apply a little of the brass and borax mixture to the joint, and as soon as this has melted remove from the hearth, twist the tube round to distribute the charge, and rub quickly with the wire brush. When the lug is heated to nearly the melting point of the brass, direct the flame upon the tube by the joint; and if the tube gets too hot before the brass flows, apply borax freely to the part which is overheating. The tubes should not be allowed to attain a white heat, or they will be burnt and ruined.

BRAZING BACK FORK BRIDGE.

The back fork bridge should have the two short tubes cleaned and fitted to the lugs similarly, and brazed up before the back forks are fitted. Load the two tubes as before. Heat one joint at a time, below, so that the heavy part of the lugs gets the heat first, and proceed in the same manner as with the fork ends. It is best to place the work on the hearth so that the tubes have only a gentle slope upwards, otherwise it will be difficult to "draw" the inside charge through the joint from the outside. The brass will appear first at the underneath part of the joint.

Fig. 712.—Band-saw Clamp.

PROTECTING NON-BRAZED PARTS.

It will save time afterwards in fitting fork blades if the inside of the crown, where the blades are fitted, is painted over with blacklead mixed to a paste with oil. This should be used on all parts to be kept free from brass, borax, or scale, such as the thread of the bottom bracket, the screw holes, etc. It may also be painted over the tubes to within $\frac{1}{4}$ in. of the joints, as it will facilitate cleaning up the work and help to protect the tubes from overheating. Keep the blacklead mixture out of the joints themselves, or a sound braze will be impossible.

BRAZING FRONT FORK CROWN AND STEERING TUBE LINER.

The front fork crown should be cleaned out, and the end of the steering tube cleaned and fitted in tight. If a liner (see p. 241) is to be fitted to the steering tube, do this now, as the liner can be brazed at the same time as the fork crown. Shake some brass and borax down inside the tube so as to rest on the edge of the liner, and place the work on the hearth almost flat, with a slight inclination downwards of the end to which the crown is attached. To braze the crown joint and liner, direct the flame to the centre of the crown, which should be well backed up with coke or breeze to retain the heat. Apply borax freely round the joint until the proper heat is nearly reached, when the flame should be shifted up the tube to just about where the charge for the liner is lodging. As soon as this starts to melt (which can be seen by looking down the tube), shift the flame slowly downwards towards the crown, at the same time applying brass and borax to the joint and round the edge of the liner and tube. Brazing this end of the liner is faciliated if it projects about ⅛ in. from the end of the tube, thus affording a better lodgment for the brass than would be possible if the two tubes were quite flush. The liner can be filed off flush after the work is cool. When the brass has run well into the joint, turn the work over on the hearth and apply more to this side of the joint, and as soon as melted, remove from the hearth and brush clean.

FITTING TOGETHER CYCLE TUBES, ETC.

The head lugs can now be fitted to the head tube, and the top and bottom tubes to the head lugs. Fit also the down tube to the bottom bracket, the seat lug to the down tube, and the top and bottom tubes to the seat lug and to the front lug of the bottom bracket. Of course, all these lugs will require to be well cleaned on the inside; and the ends of tubes that enter them must also be thoroughly cleaned.

PINNING OR PEGGING CYCLE JOINTS.

Up to now the joints, if a tight fit, do not require pinning or pegging. However, it facilitates matters if the various lugs to be pegged are drilled before the tubes are fitted, as it is easy then to drill or punch the tubes through the holes in the lugs. When fitting these joints, be sure there is no spring on the various lugs; otherwise, when the heat is applied a distorted and unsightly frame will result. To avoid this, fit the head tube to the top and bottom head lugs, the top and bottom tubes to the head, and then the down tube to the bracket and seat lug. Now if the open

ends of the top and bottom tubes are at the same distance apart as the holes into which they are to fit in the seat lug and bracket, all should be right, but if they are not the proper distance apart, one or both tubes must be pulled at the head lugs until they are. Pulling the tube over may cause an opening in the end of the lug from which the tube is pulled. This should be caulked up with a caulking chisel, for the sake of appearance and to prevent the tube springing back again. The front part when correct may be joined up to the seat lug and bracket. Knock the tubes well home, see that the head tube is quite parallel with the down tube, peg up the joints, and knock a hole through the top of the down tube into the front of the seat lug by which to feed the top head lug and seat lug joints. If the frame is held against the work-bench, resting on the front of the top head lug, the hole can easily be punched through with a small chisel. Make it of ample size, say ⅜ in., as any superfluous borax or brass left from brazing can be shaken out easily afterwards, thus avoiding an annoying rattle in the frame. Paint with blacklead the thread of the bracket, the lubricator hole, the holes into which the bridge tubes fit, inside the ball-head lugs, and the ends of the tube. This part will now be ready for brazing.

BRAZING HEAD LUG JOINTS.

The top head lug joint should first be treated. Load through the ⅜-in. hole in the down tube with about half a thimbleful of brass and borax, shake it down to the joint, and place it on the hearth so that both head tube and top tube slope downwards towards the lug. Direct the flame to the thickest part of the lug and begin to apply borax to both the tubes round the joint, as, this lug being rather light and two joints being brazed at this heat, the lug and tubes will be hot before the borax has well entered the joints. As soon as the lug is hot enough, stop applying borax and feed with brass and borax mixture to both joints. After the first charge is applied, put some more borax on and then a second charge of the mixture, and as soon as this begins to flow, get ready to remove the frame from the hearth, as unless the joints are very good fits, the brass will run away and leave gaps if the heat is left on too long. As soon as the frame is removed, brush it well, and if any brass has got into the ball races, scrape it out with an old file. A file is also handy to clean off any scale or brass which has become too hard set to be brushed off. The next joint will be the bottom head lug, which will be a repetition of the top. Load the joint through the bottom bracket.

Brazing Cycle Bracket Joints.

The two bracket joints can now be brazed. As the two tubes entering this have open ends, they cannot be loaded, and all the work has to be done from the outside. Place the frame on the hearth with the bracket towards the operator and the tubes sloping upwards away. Back up well with the coke or breeze and direct the flame on to the bracket just below the lugs. Both joints are made at one heat, but the flame should be manipulated so that one joint gets hot just sufficiently in advance of the other to give time to make it properly, and as the bottom tube is generally of a stouter gauge than the down tube, it will be well to get this hot first. Borax should be fed on to both joints as soon as the heat has started, then feed with the brass and borax the same as for the head lugs. The brass may be assisted to run through the joints by applying a little borax to the ends of the tubes inside the bracket; avoid filling the thread with brass. These two joints are the most difficult to braze properly, and require quickness to get the various operations through before the tubes are too hot. If it is seen that, towards the close of the operations, either of the tubes is getting too hot, knock away the backing of coke or breeze from that part and apply borax freely. With smartness, two applications of brass, etc., should be sufficient, one as soon as the lug is hot enough, then borax and finally more brass; remove quickly from the hearth before the final charge has had time to flow and soak out of the joints. Brush this part well and scrape out the

threaded ends lightly with a file before the brass has time to set hard. The two holes in the rear of the bracket may also have the scale scraped out whilst the work is hot.

Brazing Seat Lug Joint of Cycle.

The seat lug joint may now be tackled if the top head joint is cool. If it is still hot, the charge intended for the seat lug may stay in the top end when put in. Load as before through the hole in the down tube, and then, with the finger over the hole, shake the charge back to the seat lug and proceed as with the head lug joints. Here again brazing will be facilitated by fitting the down tube so that it projects slightly beyond the lug.

Fitting and Brazing Fork Blades, etc.

When the joints enumerated are all brazed, the part can be set aside to cool whilst the front fork blades and rear forks are fitted and brazed to the crown and bridge respectively. A description of one will apply to the other. Fit the blades so that they are at the correct distance apart at their ends, without undue spring. Fit a stretcher such as a wheel spindle between the ends to keep them apart. Load each blade with a small quantity of brass and borax, peg up, and with the fork on its side, braze one blade at a time, turning the fork over on the hearth to braze the second blade. While the forks are prepared and brazed, the handle-bar, seat pillars, etc., may also be prepared and brazed. This does not require special description.

Brazing Bridge to Bottom Bracket.

The remaining parts are the back forks with bridge to the bottom bracket and the top stays to the back fork ends, if these are of the brazed variety. The two short tubes to fit in the bracket should be cleaned on the ends, the holes in the bracket being bright, and the two tubes loaded with a small charge knocked into place, pegged up, and brazed. Place the frame on the hearth with the down tube perpendicular and the fork ends supported so that their weight is not on the lugs of the bracket. Back up the two joints, direct the flame on to the nearer joint, braze this, knock away the backing of coke, and proceed with the other joint. Do not get the forks out of position whilst the joints are hot. If the backstays are bolted on or detachable, they can be fixed before the bracket joints are brazed; this will keep everything firmly in place. Otherwise a stay or stays should be temporarily fixed from the seat lug to the fork ends.

BRAZING TUBES WITH CLOSED ENDS.

All tubes to be brazed which have both ends closed up such as the front and back forks, the two short tubes from the bridge to the bracket, and the top backstays, should have a small air hole drilled on the in or under side as much out of sight as possible. Where both ends of a tube have been closed by firm, well-fitted lugs, the tube, if not provided with such air holes, has been known to burst, causing a serious accident. Even if this does not occur, a bad joint will be the result, as the charge will blow out of the joint till the hot gases have escaped.

OTHER POINTS IN CYCLE BRAZING.

After a joint has been brazed, do not on any account assist or hasten the process of cooling by water or otherwise, but rather retard where possible by plunging the work into ashes or powdered lime immediately after removal from the hearth. Finally, support the parts so that weight on a joint being brazed is avoided. All pegs should be touched up with borax and brass whilst the joints are being brazed, or they may shift or come out all together after the machine is finished and enamelled.

SILVER SOLDERING.

Silver solder is used a great deal in the arts, and owing to the sparing or careful way in which it is used most work requires but little finishing after soldering, so that the silver solder, although expensive, is in reality the cheapest solder in the long run. The silver solder is rolled into thin sheets and cut into narrow strips with the shears. The joints or edges to be united are coated with pulverised borax which has been heated or boiled to drive off the water of crystallisation. The small strips of solder are then placed with forceps upon the edges or joints to be united, and the work is heated upon the brazing hearth. The process of silver soldering upon the larger scale is essentially the same as brazing, already so fully treated. For hard soldering small work, such as drawing instruments, jewellery, buttons, etc., the blowpipe is almost exclusively used, and the solder used is of the finest or best quality, such as gold or silver solder, which is always drawn into thin sheets or very fine wire, and it is sometimes pulverised or granulated by filing; but more heat is required to fuse a minute particle than for a large piece of metal. In soldering jewellery the worker usually applies the borax or other flux in solution with a very small camel's-hair brush. The solder is rolled into very thin sheets, and then cut into minute particles of any desired shape or size, and then so delicately applied to the work that after soldering there is no excess to be removed by filing or scraping. The borax or other flux is removed by rubbing the work with a rag that has been moistened with water or diluted acids.

COMMON METHOD OF APPLYING BORAX IN FINE SOLDERING.

The flux used with both gold and silver solders is borax. Working jewellers generally rub a lump of borax on a piece of slate with a few drops of water (just as water-colours are ground) to a cream-like consistency. The solder is scraped clean, to remove all trace of oxide, cut into little pieces, and mixed with the borax. The actual process of soldering is modified to suit the article in hand. Usually the edges to be soldered are cleaned, wetted with the borax fluid, and placed closely in contact. If possible, the article is bound tightly together with binding-wire. This is fine wire of soft iron, made specially for such purposes. A piece of pumice-stone or charcoal is used to rest the work on whilst it is being heated.

GOLD SOLDERS FOR BLOWPIPE WORK.

Gold solders used on gold articles are made from gold of the quality of the article —say, 18 or 16 carats—to which is added $\frac{1}{12}$th of silver and $\frac{1}{24}$th of copper, or a larger proportion of silver and copper for ware of inferior fineness. The quality of the solder is always a trifle inferior to the metal on which it is used, so that the solder may melt at a lower heat than the article. The melting point of 18 carat gold is 1995° Fahrenheit, of 15 carat 1992°, and 9 carat 1979°, while easy silver solder melts at about 1802° Fahr. This shows that, although 9 or 15 carat gold could be used to solder 18 carat, it is not possible to use 18 carat to solder 15 carat. The same principle applies to silver and brass ; and the quality of the solder has to be known before any attempt should be made to carry out the actual soldering of an article. Another important point is that thin gold articles, like brooches, will not bear so hard a solder as the same quality of gold will do when made up solid, as in the case of a bangle ring. Solder for 18 carat and 15 carat is made thus : take 1 dwt. of the gold, and add 2 gr. fine silver and 1 gr. fine copper ; melt well together, and roll out thin. For 12 carat, the addition of 3 gr. fine silver and 1 of fine copper to the dwt. is advisable ; while for 9 carat the most useful solder is made from 1 part fine gold, 1 part fine copper, and 2 parts fine silver.

SILVER SOLDERS FOR BLOWPIPE WORK.

For a hard silver solder, melt together 5 dwt. fine silver, 1 dwt. 8 gr. fine copper and 8 gr. of spelter ; or else to the 5 dwt. of silver, add 1 dwt. 16 gr. of good quality brass wire or pins. These and some other alloys contain a volatile metal—zinc—and must have the silver melted before the brass is put in, plenty of borax being used all through the operation. For easy silver solder, take 10 dwt. fine silver and 5 dwt. brass wire or pins, and melt and flat to size 6 or 7. For solder easier still, use

spelter in place of brass wire. All these solders can be bought at any refiners' or jewellers' material shops.

HARD SOLDERING GOLD AND SILVER.

Once the proper solder is obtained, it is easy to do the actual soldering of plain, solid articles. All that is required is that the parts to be joined are held firmly in contact, iron wire being the usual means. These contact surfaces are scraped perfectly clean, all burr being removed as well. Next charge the joint with a paste made from lump borax rubbed up with water on a piece of clean slate. This paste is used of a milk-like consistency for gold solders, but for silver soldering it should be quite as thick as cream. Next charge the seam with pallions of solder, and heat with a blowpipe flame until the solder runs. The heat should be applied gently at first, for the borax will swell up, and is likely to shift the pallions of solder away from the places they are meant to remain in until flushed. Pallions of solder is a name given to the square or oblong pieces produced by cutting a piece of flat solder into a row of parallel teeth, then by cutting again direct across these ; their size is larger or smaller, according to the work in hand. It is better that three or four small pieces be put on the work than one large one. The solder should be scraped before being cut, for on the perfect cleanliness of all materials the success of the soldering depends, that is, if the heat is correctly applied. Solder will always run towards the point of greatest heat. Soldering is a very simple process in itself, but it becomes a most difficult art to practise with success in the numerous and varied conditions of jewellery making and repairing.

OCCASIONAL METHOD OF BRAZING LIGHT IRONWORK.

In brazing light ironwork, such as locks, hinges, etc., an occasional method is to cover the work with a thin coating of loam to prevent the iron from being scaled off by the heat. Sheet iron may be brazed at a cherry-red heat, with iron filings as the spelter, and pulverised borax as the flux. The spelter and flux are laid between the pieces to be united, and the whole is bound together with binding wire and heated to a red heat. Then it is taken from the fire, laid upon the anvil, and united by a stroke upon the set hammer. Steel or heavy iron may be united in the same way at a very low heat. · For brazing iron, steel and other light coloured metals, and also brass work that requires to be very neatly done, the silver solder is generally used because of its superior fusibility and because it combines so well with most metals, without gnawing or eating away the sharp edges of the joints.

"Cold" Soldering.

Soldering without heat, commonly called cold soldering, is a common process employed when articles cannot be exposed to heat, or when the work cannot be got at with either a copper bit or a blowpipe flame. The process of cold soldering can be extended even to joining two faces of dirty cast-iron together. Although the first preparation is tedious, a large quantity of the materials can be made at once, and the actual soldering process is simple and quick. The flux consists of 1 part of metallic sodium to 50 or 60 parts of mercury; this must be kept in a stoppered bottle, closed from the air. It has the property of amalgamating (equivalent to tinning by heat) any metallic surface, cast iron included. Metallic sodium alloys with mercury by being shaken up in a bottle with it. If this is too much trouble, the sodium amalgam can be bought ready made from any chemist or dealer in reagents. For the solder, make a weak solution of sulphate of copper (about 10 oz. to 1 qt. of water). Precipitate the copper by rods of zinc; wash the precipitate two or three times with hot water; drain off the water, and add for every 3 oz. of precipitate 6 oz. or 7 oz. of mercury; add also a little sulphuric acid, to assist the combination of the two metals. The finely divided copper combines with the mercury, and they form a paste, which sets intensely hard in a few hours; and, whilst soft, this paste should be made into small pellets, which harden, and have the property of softening by heat and again hardening in a few hours. When the pellets are wanted for use, heat one or more until the mercury oozes out from the surface in small beads, shake or wipe these off, and rub the pellet into a soft paste in a small pestle and mortar, or by any other convenient means, until it is as smooth and soft as painters' white lead. This, when put on the surface amalgamated by the sodium and mercury, adheres firmly and sets perfectly hard in about three hours. The joint can be parted, if necessary, either by a hammer and cold chisel or by a heat about sufficient to melt plumbers' solder.

Riveting Light Work.

Riveting light work—that is, metal from $\frac{1}{16}$ in. to $\frac{3}{16}$ in. thick—will now be treated, and then the riveting of heavy work will be described. The two kinds of rivets used are cup-head or snap (Fig. 713) and the flat or counter-sunk head (Fig. 714); the former is the better for the thinnest material, because the hole has not to be countersunk, and countersinking greatly weakens the thin metal. A flat head, unless countersunk, stands above the iron, keen and jagged, and looks unsightly, whilst the cup-head rivet, on the other hand, forms a neat finish to the work. The rivets may be of iron or copper, and either $\frac{1}{16}$ in. or $\frac{1}{8}$ in. in diameter.

Punching Rivet Holes in Thin Metal.

To punch a hole in a strip of thin iron, lay the strip upon the end grain of a hard wood block, and then drive through it, with a single blow from a hammer, a suitable steel punch, flat at its round cutting end, but made keen (square) at the edges by touching the end on the grindstone. The punched-out disc will become embedded in the wood, and the hardness of the wood will prevent much burr from forming. What there is must be filed off or hammered down, and the hole reamed out with any suitable tool, the special tool for the purpose being shown by Fig. 715.

OTHER WAYS OF MAKING RIVET HOLES.

A better way of punching out the holes is shown in Fig. 716, in which A is a piece of bar steel, having a narrow slit a cut at one end with a hack-saw. The steel punch B passes through a hole drilled right through the bar from one face to the other. The iron strip C is passed through the slit and adjusted, and a single blow of the punch upon it forms the hole, and the disc passes through the bottom part of the hole in the bar A. This method gives a

in the two strips which it has to connect, the tail projects for a distance equal to about $1\frac{1}{4}$ times its own diameter. That is to say, a $\frac{1}{8}$-in. rivet should project about $\frac{3}{16}$ in.—a very full $\frac{1}{8}$ in.—whilst a $\frac{1}{8}$-in. rivet should project about $\frac{4}{32}$ in. A firm bedding is obtained, such as that afforded by a small anvil or iron block, or even by the top of a vice-jaw; and upon this the head of the rivet is laid, while the tail is hammered over, using the narrow cross-pene of the hammer; a good quality $\frac{1}{2}$ lb.

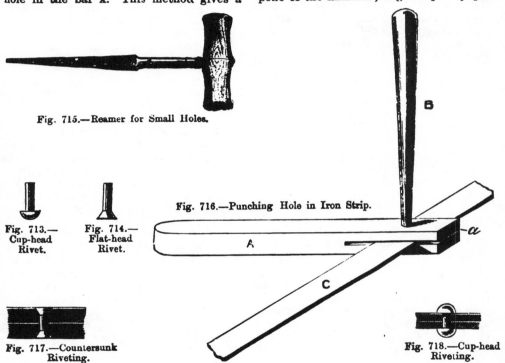

Fig. 715.—Reamer for Small Holes.

Fig. 713.—Cup-head Rivet.

Fig. 714.—Flat-head Rivet.

Fig. 716.—Punching Hole in Iron Strip.

Fig. 717.—Countersunk Riveting.

Fig. 718.—Cup-head Riveting.

keen, clear hole, and there is no need to use file or reamer for finishing it, because the strip C has a better bedding upon the iron than it would have upon a wood block. For drilling holes in thicker iron an archimedean drill-stock (Fig. 562, p. 174) is used.

THE ACTUAL RIVETING.

Light, thin work is riveted in the following way: Obtain a rivet of such a length that, when it is inserted through the holes

hammer is suitable. The object of the blows being to spread the metal, they are delivered sideways as well as vertically, so as to drive the metal outwards. Do this gradually and equally, otherwise the spreading-out will be one-sided, or the metal will split or merely bend over.

LIGHT CUP-HEAD RIVETS.

For countersunk rivets, the battered end is simply filed off flush with the surface of the metal, as in Fig. 717 ; but cup-head rivets are finished off as in Fig. 718, by means of the "snap" (Fig. 719). This is a piece of steel rod cupped out at one end

Fig. 719.—Snap for Rounding Rivet Heads.

to a semicircular form, and hardened. The hollowed end fits over the tail end of the rivet, and two or three blows upon the head of the tool finish off the rivet in a neat manner. All the above information in riveting applies to very light metal, such as bent iron work.

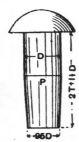

Fig. 720.—Ordinary Rivet.

RIVETS FOR LARGE WORK.

Rivets are used chiefly to connect together parts of plate iron and braced girders, boilers, wrought-iron tanks, etc., and are preferred to bolts because they give a better grip, and being nearly always fixed whilst hot, they contract in cooling and bring the plates together with great force. They are usually made of

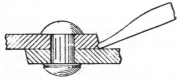

Fig. 721.—Caulking Riveted Joint.

wrought-iron or steel, and consist of a shank and a head (see Fig. 720). The lower half of the shank is tapered slightly, and the main diameter, D, is also made a

trifle less than the diameter of the hole it is to occupy, the difference being usually $\frac{1}{16}$ in. for rivets of $\frac{3}{4}$ in. diameter and upwards. When the shank P (Fig. 720) is in

Fig. 722.—Fullering Tool.

the hole, the part protruding is hammered either by hand or machine into the form of head required till it butts tightly against the plates.

CAULKING RIVETED JOINTS.

To tighten a riveted joint, caulking is resorted to ; this consists in knocking down the edges of the plate and heads of the rivet with a caulking tool as in Fig. 721. The tool is driven by hand hammering or by pneumatic devices, the plates then being punished less, but severe caulking on the rivet head diminishes the grip. A preferable method of making a tight joint is by means of a fullering tool (Fig. 722) with a thickness at the point equal to that of the plate. Bevelling the edges of the plates as shown to about 1 in 8 with the vertical, or about 80° with the horizontal, facilitates the operation. Caulking may be done on one side only, that is, on the same side as the riveting, but in the best work both sides are tightened. The finished caulking should appear like a parallel groove, about $\frac{1}{32}$ in. deep and say $\frac{1}{8}$ in. wide in a $\frac{3}{8}$-in. plate.

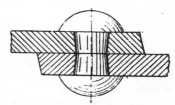

Fig. 723.—Riveted Joint with Coned Holes.

MAKING HOLES FOR RIVETS.

The holes for the rivets usually are drilled, as punching, although cheaper, injures the plates. (Punching holes in hot metal is fully explained in the next chapter.) This injury may be partly removed by annealing. A punched hole is

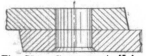

Fig. 724.—Countersunk Hole.

usually taper, but this may be an advantage, as the narrow ends can be put adjacent to one another (see Fig. 723), and the rivet shank then tends to keep the plates together irrespective of the heads. The burr which is formed during the process of drilling is removed, and if the edges are then countersunk slightly as shown in Fig. 724, the shearing strength of the rivets

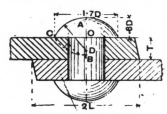

Fig. 725.—Cup-head Riveting.

is increased. The holes must not be countersunk more than ⅛th of the diameter or the rivets will have a tendency to separate the plates. In best work, especially for boilers, the holes are drilled when the plates are bent or flanged.

CUP-HEAD RIVETS.

The most common form of rivet head is

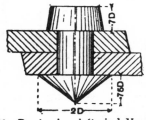

Fig. 726.—Pan-head and Conical Head Rivets.

shown in Fig. 725, and is known as the cup, spherical, button, or snap head. In draw-

ing this head, with o as centre, describe a circle with a diameter equal to that of the rivet, cutting the centre line produced at A and B. With the latter as centre and

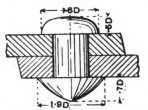

Fig. 727.—Modified Cup-head Rivets.

with the same radius, cut the top of the plate at c. Make A D equal to o c, and with D as centre describe the arc of the top of the rivet head through c. By Unwin's rule the diameter of the rivet equals 1·2 times the square root of the thickness of the plate, the hole being 4 per cent. to 20 per cent. of the diameter larger than the cold rivet.

OTHER SHAPES OF RIVET HEADS.

The upper part of Fig. 726 shows a pan-head rivet, whilst the lower part is a

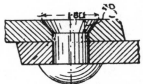

Fig. 728.—Countersunk Riveting.

conical head. Cheese heads have sides at right angles to the plate. Modified forms of cup or snap heads are shown at Fig. 727, the upper one of ellipsoidal and the lower of conoidal form. All proportions in Figs. 720 and 725 to 729 are given in terms of the diameter of the rivet. Countersunk heads are shown at Figs. 728

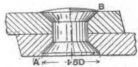

Fig. 729.—Modified Countersunk Riveting.

and 729. As the sharp edge of the counter-sunk head is liable to spring away from the

plate, a shoulder as at A (Fig. 729) is some·
times used, or, if convenient, a curve as
at B.

POWER FOR RIVETING.

Riveting is largely done by hydraulic
means, as hydraulic power can be con-

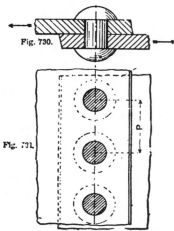

Fig. 730.

Fig. 731.

Figs. 730 and 731.—Single-riveted Lap Joint.

veyed to great distances, and stored till
wanted, without much loss. Also, the
steady and known pressure on the rivet
head, coupled with an increase due to
absorption of the momentum of the
accumulator weight at the moment of
closing, is just the action required, and
causes the rivet to fill the hole more completely than is likely to be the case with
steam or hand riveting where the action
is percussive. Riveting is also done by
pneumatic means, and makers assert that
by this means seventy $\frac{7}{8}$-in. rivets can be
driven per hour per riveter. Caulking
the rivet heads and plates is not considered
necessary if hydraulic riveting is done properly.

SINGLE-RIVETED LAP JOINT.

The simplest method of riveting together
two plates is by the single-riveted lap joint
shown in Figs. 730 and 731. The overlap
should not be less than three times the
diameter of the rivet. The following table
gives approximate proportions in inches
for various rivets in single-riveted lap
joints with different thickness of plates,
Rt and Rs being the efficiency per cent. of

the resistance to tearing and shearing
respectively :—

Thickness of Plate.	Diameter of Rivet.	Pitch.	Rt.	Rs.
$\frac{1}{16}$	$\frac{7}{8}$	$1\frac{3}{8}$	61 0	61·0
	$1\frac{1}{8}$	$1\frac{3}{8} + \frac{1}{16}$	58·4	71·8
$\frac{1}{8}$	$\frac{3}{4}$	$1\frac{11}{16}$	60 9	61 2
	$\frac{15}{16}$	$1\frac{3}{4}$	58·1	71·4
$\frac{3}{16}$	$\frac{11}{16}$	2	59·8	59·2
	$\frac{7}{8}$	$2\frac{1}{4}$	58·3	70·1
$\frac{1}{4}$	$1\frac{1}{16}$	$2\frac{1}{4}$	57·5	58·3
	1	$2\frac{1}{4}$	58·0	70·3
$\frac{5}{16}$	1	$2\frac{3}{8}$	58·3	58·7
	$1\frac{1}{4}$	$2\frac{3}{8} + \frac{1}{16}$	56·6	68·1
$\frac{3}{8}$	$1\frac{1}{2}$	$2\frac{1}{8}$	56 9	57·4
	$1\frac{1}{8}$	$2\frac{5}{8}$	55·2	66·7

The first and alternate lines in this
and subsequent tables represent the proportions for iron plates and rivets and
punched holes, whilst the other lines
give those for steel plates and rivets
with drilled holes. A single-riveted lap
joint may give way by the shearing
of the rivets, as shown by arrows
in Fig. 730, by the tearing of the plates
between the rivets, or between the edge
and the rivet holes, or by the crushing of
the rivet or plate. Experience has proved
that to avoid the cross breaking of the
plate between edge and hole, in all

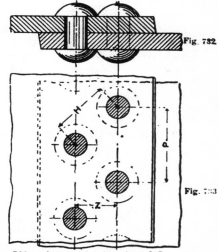

Fig. 732.

Fig. 733.

Figs. 732 and 733.—Double-riveted Lap Joint.

riveted joints the distance from the edge
of the plate to the centre of the rivet should be
equal to $1\frac{1}{2}$ times the diameter of the rivet.

Double-riveted Lap Joints.

Double-riveted lap joints are illustrated by Figs. 732 to 734. The rivets may be

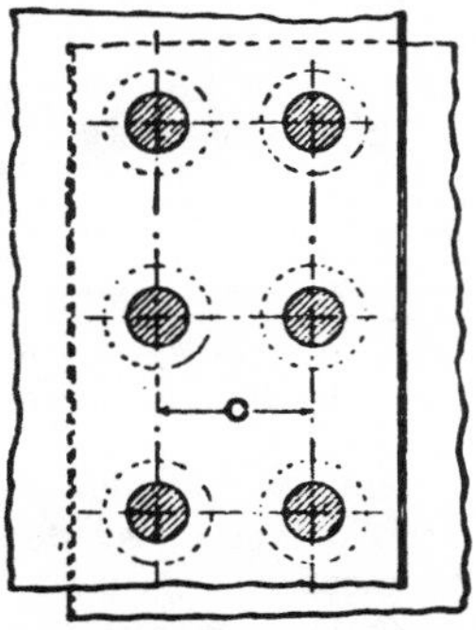

Fig. 734.—Double-riveted Lap Joint.

arranged either in a zigzag (Fig. 733) or in a chain (Fig. 734). The following table shows in inches the usual proportions of rivets in double riveting, *C* being the distance between the rows in chain riveting, and *Z* the same distance in zigzag riveting, the lap in each case not being less than one and a half times *D*.

Thickness of Plate.	*Diameter of Rivet.*	*Pitch.*	*Z.*	*C.*	*Rt.*	*Rs.*
⅜	11/16	2¼	1 5/16	1⅞	73·7	75·5
	¾	2⅜	1⅜	2	71·4	71·8
7/16	¾	2¾	1⅜	2	72·7	73·4
	13/16	2⅝	1 7/16	2⅛	70·5	69·0
½	13/16	2¾	1½	2⅛	71·7	72·1
	⅞	2 13/16	1½	2¼	68·9	68·4
9/16	⅞	3	1 9/16	2¼	70·8	71·3
	15/16	2⅞	1 9/16	2⅜	67 4	68·8
⅝	15/16	3⅛	1⅝	2⅜	70 0	70·7
	1	3	1⅝	2½	66·7	67·0
11/16	1	3¼	1¾	2½	69·2	70·3
	1 1/16	3⅛	1 11/16	2⅝	66·0	66·0
¾	1 1/16	3 7/16	1 13/16	2⅝	69.1	68·8
	1⅛	3¼	1¾	2¾	65·4	65·2
13/16	1⅛	3 5/16	1⅞	2¾	68·4	68 7
	1 3/16	3½	1 13/16	2⅞	64·8	64·6
⅞	1 3/16	3¾	2	2⅞	68·3	67·5
	1¼	3½	1 15/16	3	64·3	64·1
15/16	1¼	3⅞	2 1/16	3	67·7	67·6
	1 5/16	3⅝	2	3⅛	63·8	63·7
1	1 5/16	4	2 3/16	3⅛	67 2	67·6
	1⅜	3¾	2⅛	3¼	63.3	63·4

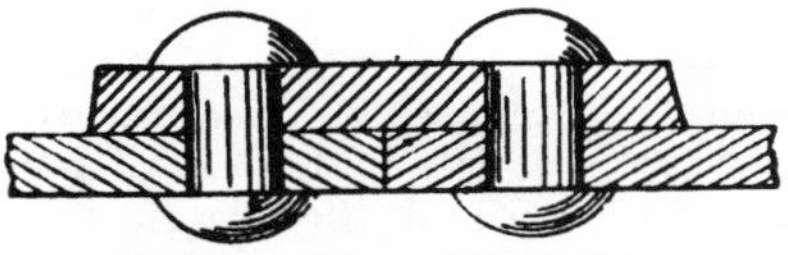

Fig. 735.—Riveted Butt Joint.

Professor Kennedy states, as the results of experiments, that H (Fig. 733) should be about 30 per cent. to 35 per cent. more than C (Fig. 734).

Cover or Butt Straps.

Another method of joining two plates is

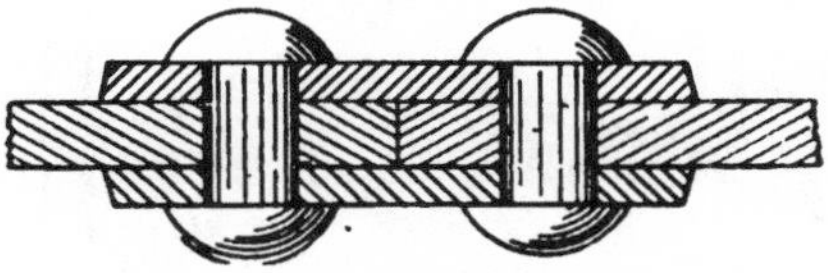

Fig. 736.—Riveted Butt Joint.

by one or more cover or butt straps (Figs. 735 and 736), the ends of the plates being butted against each other. This joint is treated as two single-riveted lap joints. Fig. 735 shows that the tension on the plates will tend to bend the cover strap; for this reason either the strap is made about 1⅛ times the thickness of the plate, or two cover straps are employed as in Fig. 736, when each is usually about ⅝ the thickness of the plates. The butt straps should be cut from the ends of plates and not from bars, so that the grains of the strap and plate may run in the same direction. The pitch in this case may be

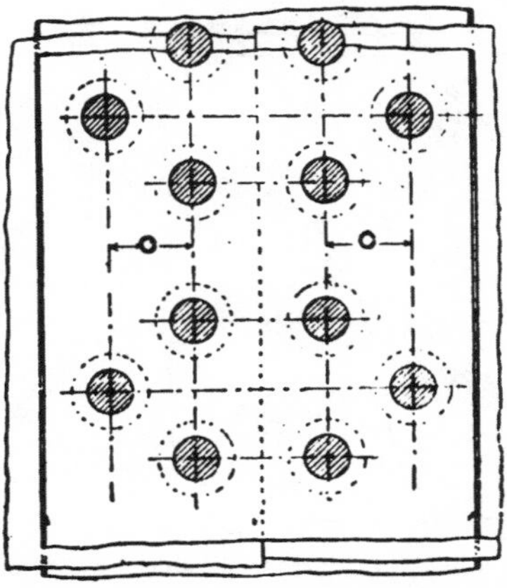

Fig. 737.—Double-riveted Butt Joint.

calculated as for a double-riveted lap joint. The diameter of the rivets in this kind of butt joint is usually ¼ in. more than the thickness of the plates when using iron plates and rivets, and 5/16 in. when using steel plates and rivets. Figs. 737 and 738 show a butt joint with double riveting, and in Fig. 737, on a section of the pitch, there are six areas of rivets sheared, namely,

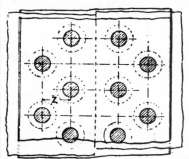

Fig. 738.—Double-riveted Butt Joint.

four whole rivets and two halves on each side of the plate. Similarly in Fig. 738 there are only four areas sheared. Generally, iron rivets in the last form may be $\frac{3}{16}$ in., and steel rivets $\frac{1}{8}$ in., larger than the thickness of the plate. Fig. 737 is the stronger form, and the diameter of rivets need only be $\frac{1}{8}$ in. and $\frac{3}{16}$ in. more than the thickness of plate for iron and steel rivets and plates respectively. In Fig. 738 the

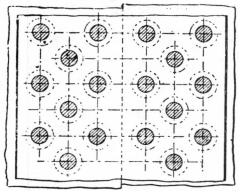

Fig. 739.—Treble-riveted Butt Joint.

thickness of the straps should be $\frac{5}{8}$ the thickness of the plates.

TREBLE-RIVETED BUTT JOINT.

Fig. 739 is a plan of a treble-riveted butt joint with two cover straps. The pitch in the outer side rows may be twice that of the four inner rows, or the pitch of the two outer and two inner rows may be double the pitch of the other rows. Each case, however, should be worked out on its merits, remembering that in a butt joint shearing must be considered on one side of the butt line only. In compres-

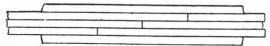

Fig. 740.—Step Joint of Heavy Girder.

sion, the butting of the plates against one another is not sufficient by itself. Even a very slight inaccuracy in workmanship will cause the thrust to be transmitted through the rivets and cover straps. In best bridge work, all joints are designed so that they can take the entire thrust. Unwin states that the only way in which compression joints may safely differ from tension joints is that the rivets may be more closely spaced across the plate. In the booms of large heavy girders the layers may be joined with one pair of plates, the joints being arranged in step fashion, as in Fig. 740, and the cover plates may be cut off as shown by Fig. 741. With wrought-iron, the tensile resistance of the plates is practically equal to the shearing resistance of the rivets, and the Board of Trade allows a shearing resistance of 17 tons per square inch for iron rivets, although good rivet iron should possess a shearing resistance of about 25 tons per square inch.

TREBLE-RIVETED LAP JOINT.

Fig. 742 shows a plan of a treble-riveted lap joint, the rivets being in zigzag fashion. They might be arranged in chain fashion, or with the outer rows at twice the pitch of the middle row, either chain or zigzag fashion. The strength of the joint is increased if either of the last two methods is adopted. When the pitch of the outer rows is double that of the inner rows, the edges of the plates are sometimes scalloped out to permit of the joint being more securely caulked. The following table gives dimensions for treble-riveted lap joints. As a rule, the diameter may be taken as equal to 1·27 times the thickness of plate for iron plates and rivets, and 1·59 times the thickness of the

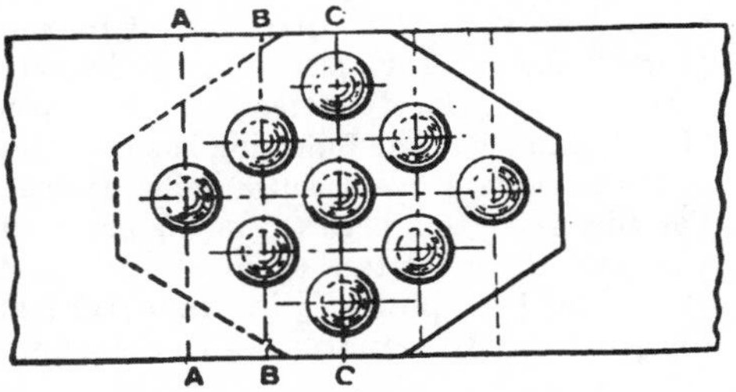

Fig. 741.—Lap Joint for Tie Bar.

plate for steel plates and rivets. Quadruple lap joints can also be worked out similar to the previous joints.

Thickness of Plate.	*Diameter of Rivet.*	*Pitch.*	*Z.*	*C.*	*Rt.*	*Rs.*
5/8	13/16	3 1/4	1 5/8	2 1/8	75·1	76·6
	7/8	3 1/8	1 5/8	2 1/4	72·0	73·9
11/16	7/8	3 1/2	1 3/4	2 1/4	75·0	75·0
	15/16	3 3/8	1 3/4	2 3/8	72·2	71·4
3/4	15/16	3 11/16	1 7/8	2 3/8	74·6	74·9
	1	3 5/8	1 13/16	2 1/2	71·4	71·8
13/16	1	3 7/8	1 15/16	2 1/2	74·2	74·8
	1 1/16	3 11/16	1 13/16	2 5/8	71·2	71·0
7/8	1 1/16	3 7/8	2 1/16	2 5/8	74·2	73·7
	1 1/8	3 3/4	2	2 3/4	71·0	70·4
15/16	1 1/8	4 5/16	2 3/16	2 3/4	73·9	73·8
	1 3/16	4	2 1/16	2 7/8	70·3	70·9
1	1 3/16	4 1/2	2 1/4	2 7/8	73·6	73·5
	1 1/4	4 3/16	2 3/16	3	70·1	70·3
1 1/16	1 1/4	4 11/16	2 3/8	3	73·3	73·9
	1 5/16	4 5/8	2 1/4	3 1/8	70·0	69·9
1 1/8	1 5/16	4 7/8	2 1/2	3 1/8	73·1	74·0
	1 3/8	4 1/2	2 3/8	3 1/4	69·4	70·4

Riveted Lap Joint for Tie Bar.

To design a lap joint for a tie bar of a type similar to that shown in Fig. 741, using steel plates 1 in. thick and steel rivets, the breadth of lap being 9 in., and nine rivets of equal diameter being used: D (Fig. 720) may be calculated to equal 1·16 in., or nearly 1¼ in. With the plate torn across A A, there is simple tearing, and the strength of this section compared with that of the solid plate equals 87 per cent., while the shearing efficiency of the plate at this section will equal 86·8 per cent. With regard to tearing at B B and shearing at A A, the joint has an efficiency of 83·7 per cent., and for tearing at C C and shearing at A A and B B, the efficiency is 90·1 per cent. The object of placing the rivets as shown is to ensure that the strength of the joint may be as nearly as possible equal to that of the original plate.

Size of Rivet Holes.

In some experiments made abroad to determine the influence of frictional resistances under varying conditions, the majority of specimens tested were butt joints with double cover straps, the straps and rivets being carefully planed. In some of the hand-riveted groups, the holes were reamed out carefully, and the rivets, when heated to a cherry red, fitted so well that considerable force was necessary to drive them home. The results showed that the effect of giving the rivet holes a slightly larger diameter than the rivet was to increase the load before any relative movement of the straps was observed. This elastic movement depends upon the frictional resistance of the plates and straps. The resistance to tension is made up of friction between the plates and the shearing strength of the rivets. When the load is removed, the strained rivets can only regain their original form by overcoming this plate friction, and the greater this friction the farther will the rivet be from its original state. With the rivets a loose fit in the holes, the frictional resistance in the hand-riveted specimens is

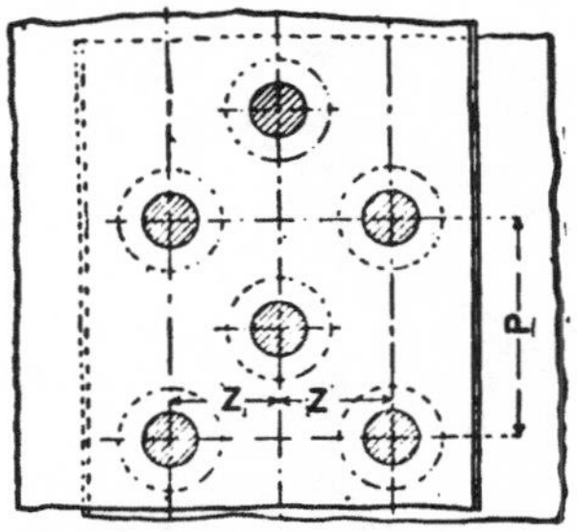

Fig. 742.—Treble-Riveted Lap Joint.

greater, although the permanent set is also greater. The increase in frictional resistance is not so marked in specimens having the straps thicker than the plate.

Hydraulically riveted specimens approach very nearly in their behaviour to the rivets which fit the holes exactly. The experiments also showed that repetitions of the load did not cause decrease in the frictional resistance.

DIAMETER OF RIVETS FOR BOILERS.

A list of diameters of rivets to be used with boiler plates of given thickness is here presented. The diameter of the rivet may equal 1·2 times the square root of the thickness of the riveted plate. On this basis the following list has been prepared :—$\frac{3}{16}$-in. plate, $\frac{1}{2}$ in. diameter ; $\frac{5}{8}$-in plate, $\frac{3}{4}$ in. diameter ; $\frac{7}{16}$-in. plate, $1\frac{3}{8}$ in. diameter ; $\frac{1}{2}$-in. plate, $\frac{7}{8}$-in. diameter ; $\frac{5}{8}$-in. plate, $1\frac{1}{8}$ in. diameter ; $\frac{3}{4}$-in. plate, $1\frac{1}{16}$ in. diameter ; $\frac{7}{8}$-in. plate, $1\frac{1}{8}$ in. diameter ; and 1-in. plate, $1\frac{1}{4}$ in. diameter. The following has been given as the practice of Lancashire boilermakers :— For $\frac{3}{8}$-in. and $\frac{7}{16}$-in. plates, $\frac{3}{4}$ in. diameter ; for $\frac{1}{2}$-in. and $\frac{9}{16}$-in. plates, $1\frac{3}{8}$ in. diameter ; for $\frac{1}{16}$-in. and $\frac{5}{8}$-in. plates, $\frac{7}{8}$ in. diameter ; and for $\frac{3}{4}$-in. and 1-in. plates, $1\frac{3}{8}$ in. diameter.

FORGING IRON AND STEEL.

IRON BARS FOR SMITHS' USE.

ROUND, square, and flat bars are the chief forms in which iron is used by the smith. The round bars are rolled in the following Wire Gauge sizes: 7, 6, 5, 4, 3, 2, 1; they range from $\frac{3}{16}$ in. up to $2\frac{5}{8}$ in., increasing by sixteenths of an inch; from $2\frac{5}{8}$ in. to 5 in., increasing by eighths; from 5 in. to $5\frac{1}{2}$ in. by quarters; from $5\frac{1}{2}$ in. to 6 in. by eighths; and from 6 in. to $6\frac{1}{2}$ in. by quarters; and each of these can be rolled to "full" and "bare," as well as to exact sizes. Thus sixty-eight different diameters of round bar, from $\frac{3}{16}$ in. to $6\frac{1}{2}$ in., are available. The squares increase from $\frac{3}{16}$ in. to $1\frac{5}{8}$ in. by sixteenths of an inch; from $1\frac{5}{8}$ in. to 4 in. by eighths; from 4 in. to $5\frac{1}{2}$ in. by quarters; and from $5\frac{1}{2}$ in. to $6\frac{1}{2}$ in. by half-inches; this gives fifty-one sizes in square sections from $\frac{3}{16}$ in. to $6\frac{1}{2}$ in. The flat bars range from $\frac{3}{8}$ in. to 12 in. wide, in almost all thicknesses from $\frac{1}{4}$ in. upwards, advancing by sixteenths and eighths of an inch. Flats in sixty-five different widths are obtainable.

QUALITIES OF SMITHS' BAR IRON.

Iron is made in four qualities—the "ordinary," or common, "best," "best, best," and "best, best, best." Good iron being rather costly, it is advisable to preserve odds and ends of iron and steel for small work and odd jobs. Inferior iron is not economical The quantities used in light work are so small that the saving is scarcely worth taking into account, while inferior iron is a constant source of anxiety. A good iron is silvery and clean-looking; and compared with it a bad iron is dull and dirty. A good iron is free or nearly free from flaws, but a bad iron always shows up flaws when brought to a red heat, these being due to the inter-mixed cinder and scale left by insufficient puddling. Some of the flaws may be removed by raising the iron to welding heat, and hammering it thoroughly all over.

IRON AFFECTED BY ITS CHEMICAL CONTENTS.

Carbon, manganese, phosphorus, and silicon change both the appearance and the physical qualities of iron. The foreign ingredients in wrought iron seldom exceed 1 per cent., being often only $\frac{1}{2}$ per cent., but they affect the metal to an extent that makes itself very evident at the anvil. The purest iron is the one most readily forged, whether hot or cold; iron that can be forged well while hot, but not when cold, is said to be "cold short," and this is due to very minute quantities of phosphorus, antimony, and silicon. Iron that develops cracks while being forged hot is said to be "hot short," and this may be due to a minute quantity of sulphur, whose amount may not perhaps exceed ·03 per cent., or it may be due to antimony.

FIBRE OF ROLLED WROUGHT IRON.

Wrought iron is in a fibrous condition, as is also, though in a lesser degree, mild steel. This allows of the metal being shaped under the hammer. A bar of iron placed across the anvil cutter (Fig. 743) and nicked around with a chisel, may be broken short off with little effort, and only a practised eye can detect by the appearance of the fracture, which is wholly crystalline, whether the iron is cast or wrought. The same bar bent without nicking until it breaks, or gradually torn asunder, shows the fracture wholly fibrous, the long, string-like fibres becoming drawn out as though the bar had been built up of innumerable fine strings of metal. Again, if a crank-shaft or a lever-arm breaks at a sharp re-entrant angle (that is, a more or less acute angle pointing inwards), the fracture will be crystalline. But if it

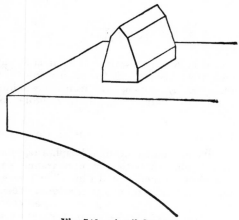

Fig. 743.—Anvil Cutter.

breaks at an angle whose faces are gradually merged into one another with a curve or hollow, the fracture will be fibrous.

DIRECTION OF FIBRE.

Owing to its fibrous character, wrought iron must be considerably stronger in the direction in which it is rolled A A (Fig. 744) than in the transverse direction B B. The difference is in the ratio of about 21 to 17 —that is, if it would require 21 tons per

slight extent in mild steel, where the rolling is only incidental to the shaping. The direction and relative strength of fibre have a most marked influence upon design

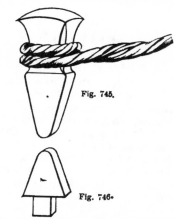

Fig. 745.

Fig. 746.

Figs. 745 and 746.—Top and Bottom Fullers.

in wrought iron. For example, the direction in which work is subjected to the greater stress, should always coincide with the longitudinal direction of the fibres. Again, curved work should not be cut from the solid, thus severing the continuity of the grain, if it is possible to bend it round and so preserve the fibres continuous. This shows why in many cases it is better to split or divide a bar, and bend or fork it, so preserving continuity of grain,

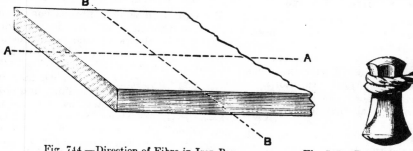

Fig. 744.—Direction of Fibre in Iron Bar.

Fig. 747.—Round-faced Fuller.

inch to break the bar through the line B B, across the fibres, 17 tons would suffice to break it along A A, along or with the fibres. This holds good only to a very

rather than to slot out or to weld on. It explains many practical points in the working of wrought iron as distinguished from the working of the homogeneous mild steels. It also shows the advantage of keeping the edges rounded and not sharp on fullering and similar tools (Figs. 745 to 748) by whose use the grain is not violently severed, but rather bent to shape.

LIABILITY OF IRON TO CRACK.

Inferior iron is much more liable to crack than is first-class metal. Lowmoor

Fig. 748.—Hollow Fullering Tool.

iron and the treble-best qualities of Staffordshire iron are comparatively close-grained and tough, but bad iron often shows cracks at the end (Fig. 749) when cut off with the hot set. (Hot and cold sets respectively are shown by Figs. 750 and 751.) These cracks are due to imperfect union, and to the presence of cinder in the iron. Sometimes hammering at welding heat improves such iron, but not if the composition of the iron is bad.

CONSOLIDATING IRON.

Work that is to be screwed or subjected

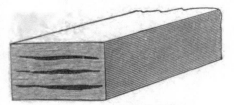

Fig. 749.— Cracks in Bad Iron.

to great stress or wear, should be forged as sound and close-grained as possible by hammering, or by consolidation, between top and bottom tools at welding heat.

Otherwise the fibres may become partly separated and the metal frayed, so that the open texture of the iron will collect grit and wear rapidly.

FORMATION OF THE FORGE FIRE.

Good forging can be done only when the forge fire is properly made. Four common

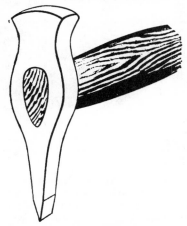

Fig. 750.—Hot Set.

types of forges, the last of them being gas-heated, are illustrated by Figs. 752 to 755. The "stock" is the term given to the mass of hard-caked coal on a smith's hearth, within which the heat is confined;

Fig. 751.—Cold Set.

to some extent it forms a reserve of fuel, but its chief purpose is to prevent radiation of the heat, and to concentrate it upon the work in the hearth. If it were not for the stock, much of the heat would go up the chimney, and the work would be

Fig. 752.—Portable Forge with Long-shaped Bellows.

oxidised more rapidly than it is, owing to its partial exposure to the air. The stock has two portions—one lying against the tuyere, described on p. 263, and hearth-back, the other placed opposite to the first in the direction of the coal and water bunks, the work lying midway between the two. The

Fig. 753.—Sturtevant Improved Hand Forge.

central portion of the fire is the only part replenished much in the course of a day's

work. The stock, though highly heated, does not burn away sensibly, because it is

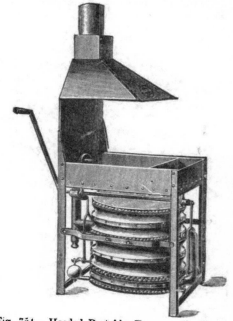

Fig. 754.—Hooded Portable Forge, with Circular Double-blast Bellows.

Fig. 755.—Fletcher, Russell & Co.'s Gas Forge.

protected from the direct action of the blast, and the upper portions are kept damp. Yet the inner faces, being in direct contact with fuel supplied from time to time to the central part of the fire, are at a glow with heat.

THE FORGE TUYERE.

The tuyere, already referred to, is a

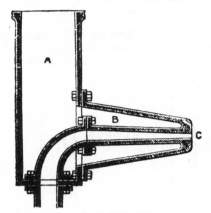

Fig. 756.—Tank Water Tuyere.

simple blast tube in small forges, with its nozzle thickened to prevent its destruction by the intense heat of the forge fire. One form is clearly shown in Fig. 752. In large forges, where the blast has to be powerful, the tuyere is water-jacketed as in Figs. 756 and 757. In the one system, a cast-iron tank A (Fig. 756) contains water, which circulates in the annular jacket B, around nozzle C. The tuyere

Fig. 757.—Pipe Water Tuyere.

shown by Fig. 757 is made of wrought and cast-iron, the water supply entering the annular jacket by means of pipe A, and leaving it by pipe B.

MAKING THE FORGE FIRE.

In making the forge fire, first the stock is built at the back and front of the hearth, and beaten hard with the slice (Fig. 758) or with the sledge; to prevent the tuyere hole from being choked, pass an iron rod into it for the time being. The fire is then lit in the central portions with a handful of shavings and a little coal, assisted by a gentle blast. For manipulating the fire, the poker (Fig. 759) and rake (Fig. 760) will be useful.

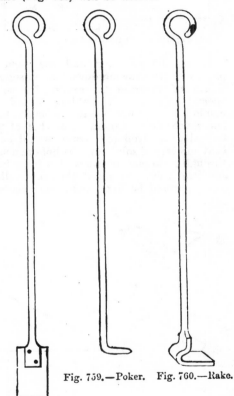

Fig. 758.—Slice.

Fig. 759.—Poker. Fig. 760.—Rake.

THE CHOICE OF METHOD : DRAWING-DOWN, UPSETTING, OR WELDING?

Forgings of unequal sectional area are formed by drawing-down, upsetting, or welding, or by a combination of all three. Generally, the choice of method is simply a question of ultimate results; in given circumstances one of the three may involve less work, or economise material, or may be the only one possible with material that happens to be in stock, or because there are odds and ends that it is desirable to use up; or, lastly, because it is the method most practicable with the tools and manual help available. However, it is often a question of relative dimensions. If there is a great difference

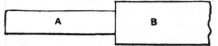

Fig. 761.—Square Bar Drawn Down.

between the enlarged and the reduced part, neither drawing-down nor upsetting would be resorted to, except for some special reason, but welding would be employed. An eye having a small hole and much metal around it, as that of the tie-rod of an iron roof truss, would have that end forged solid, and the hole punched through; whereas an eye with a large hole and relatively little metal, thus resembling a loop, would be bent round and welded

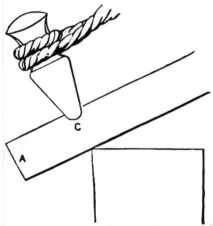

Fig. 762.—Nicking with Top Fuller.

An eye of medium thickness could be made in either fashion.

NICKING IRON BAR.

An iron bar is drawn-down in the following way. Suppose the portion marked A (Fig. 761) is to be drawn-down from a bar originally of the size of B. The bar is

laid across the edge of the anvil, and nicked at c (Fig. 762) with a top fuller. If both sides of the bar have to be drawn-

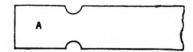

Fig 763.—Bar Nicked with Top and Bottom Fullers.

down, then a bottom fuller would be inserted in the anvil under the top fuller, and the bar would be nicked between them

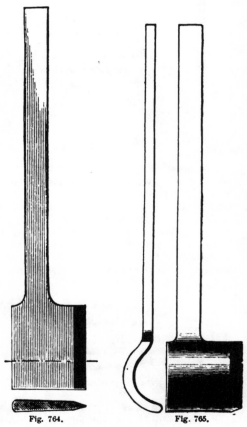

Fig. 764.
Fig. 765.

Fig. 764.—Smith's Straight Chisel.
Fig. 765.—Chisel for Cutting Circular Ends.

as in Fig. 763. Chisels (Figs. 764 and 765) or sets are not used in such nicking, for these divide the fibres of the metal, while the round-faced fullers simply alter their direction without breaking their continuity.

DRAWING-DOWN SQUARE BAR.

When the work has been nicked with the

Fig. 766.—Flatter.

fuller, the metal along A (Figs. 761 and 763) is drawn-down or thinned by a succession of blows from the hand hammer or sledge, with or without previous fullering. When fullering tools are used, the top fuller would be employed singly if only the one face required reduction; or in pairs, one above the other, if both faces had to be drawn-down. A succession of depressions is formed upon the surface of the spread-out work with ridges between, and

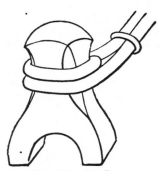

Fig. 767.—Top Swage.

these have to be obliterated with the hammer. Fullering and hammering lengthen the bar, and spread it sideways. If the bar is to be equal-sided, the spreading or widening has to be prevented by rapidly hammering the sides and faces alternately.

After every few blows on the faces, turn the bar quarter round during the brief interval between a couple of blows, and give several blows upon the edges as

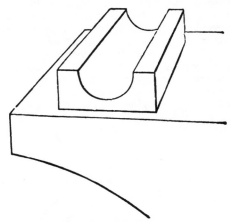

Fig. 768.—Bottom Swage.

a corrective to those on the faces. With practice, this rapid changing of the faces on the anvil is accomplished without damaging the rectangular form. Drawing-down is frequently done with the hammer alone, when the reduction in area is only slight. The process of thinning always begins at the end of the iron farthest from the smith, and proceeds towards him. One inch and a half or two inches is drawn-down at a time.

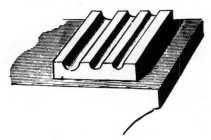

Fig. 769.—Bottom Swage.

SMOOTHING DRAWN-DOWN SURFACE.

The surface of the roughed down iron is not smooth, though a good smith can impart a fair finish to a flat surface with the hammer alone. The hammer should strike so as not to bruise the work much, and there is a knack in using it so that its edges do not mark the work, the blow being given by the central rounding portion of the face only. Striking fair with the middle of the hammer face, each mark serves to partly obliterate others, and leaves a fair surface, slightly wavy. With the assistance of a hammerman the surface can be smoothed more effectually by means of a flatter (Fig. 766), which the smith holds in the right hand and slides all over the surface of the work while the

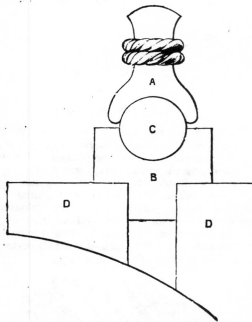

Fig. 770.—Round Bar between Top and Bottom Swage.

hammerman strikes it with the sledge. This leaves the work very smooth.

DRAWING-DOWN ROUND BAR.

The process of drawing-down a round bar is the same in principle, but slightly different in detail. The rod is nicked round with a fuller and drawn-down under the hammer, beginning as before at the end farthest from the hand. The metal is extended with the hammer only, and the

rod is rotated between each hammer blow. Finally, the work is smoothed by means of swages (Figs. 767 to 769), the work lying in a bottom swage of nearly semicircular form. Fig. 770 shows an anvil

Fig. 771.—Spring Swages.

swage fitted into the hole in the anvil while blows are struck upon the upper surface of the rod with a hand hammer, if the smith be single-handed, or, if he has an assistant, with the sledge upon a top swage; this swage must be a counterpart of the bottom. In Fig. 770, A indicates the top and B the bottom swage, with the bar of iron C between them; D shows the anvil

USE OF POWER HAMMER AND SPRING SWAGES.

By means of spring swages (Figs. 771 to 773), a worker, single-handed, can sometimes make use of the top as well as of the bottom tool. A steam hammer simplifies drawing-down and finishing. Top or bottom fullers need not then be used, for, the position of the shoulders having been marked on the bar, this is laid on the anvil of the steam hammer. Then the work is drawn-down under the tup or hammer, the bar being turned quickly during each period of ascent of the tup. If the bar is long and of large diameter, the drawing-down will still have to be done in short lengths, two or three inches from the end being drawn-down first. The bar being held perfectly flat, it leaves the hammer finished. For rod work, top and

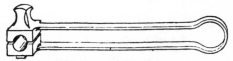

Fig. 772.—Spring Swages.

bottom spring swages are placed under the tup when finishing. The rod is drawn-down roughly between the anvil and the tup, being rotated rapidly between each blow, a rudely circular form being imparted. Before the reduction is quite complete, the spring swages, held by an attendant, are placed on the anvil, the work inserted between them, and rounded and finished by a few final blows of the tup on the top swage—the work still being revolved during each period of ascent of the tup.

Fig. 773.—Spring Swages.

BRUSHING OFF IRON OXIDE.

A scale of iron oxide forms rapidly on the hot metal during the above operations. The larger the forging, the greater is the quantity of oxide formed. It should be removed with a switch of brushwood as fast as it forms, otherwise the hammer blows drive it into the surface of the work, and form a rough scale, which is afterwards both unsightly and a hindrance to easy tooling in the vice or lathe. Where the forgings are large, a man stands by the steam hammer and brushes away the scale after every half-dozen or so blows. At the anvil the striker or the smith knocks the scale off immediately the iron is removed from the fire, and afterwards as often as may be required.

DETERMINING LENGTH OF IRON FOR DRAWING-DOWN.

In judging the length of iron to be allowed for drawing-down, the practised smith does not trouble to calculate in figures the exact length required. Yet when working expensive qualities of metal, it is well to remember a rule of simple proportion. The original section

of a bar bears the same proportion to a given reduced section that the length of the latter bears to that of the former. Thus if a bar originally 3 in. square has to be reduced to 1 in. square, 1 in. in length of the 3 in. bar will be taken for reduction to a 3 in. length of 1 in. square. If the reduced portion is tapered, or of unequal and varying dimensions, then the mean of the various sectional areas must be taken. Additional allowance must be made for ragged and perhaps burnt ends, and a trifle for inaccuracy in cutting off.

UPSETTING.

Upsetting, or jumping up, is an alternative of drawing-down, the metal being knocked or jumped up into a mass larger in area than the bar itself. It is a slower and more laborious process than drawing-down, and to upset a very moderate mass of metal will require several heats. Thus a big shoulder cannot be formed by upsetting; instead, a ring or collar, or a solid

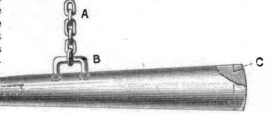

Fig. 774.—Monkey, or Swinging Pendulum Hammer.

mass of metal, has to be welded on. The hand hammer cannot be used in upsetting, except the iron be cut off sufficiently short to go endwise under it; instead, the monkey, or swinging pendulum hammer (Fig. 774) is used.

MONKEY OR SWINGING PENDULUM HAMMER.

A monkey, or swinging pendulum hammer (Fig. 774), is made of cast-iron, and is balanced carefully on its centre of gravity by the correct setting in position of the eye B, which is cast into the bar. The monkey is suspended by a chain A from a beam. The battering end is preserved from fracture, and from too rapid wear, by means of the wrought-iron band C, which is shrunk into a shouldered recess of dovetailed section. The wrought-iron handle or porter D is cast into the monkey, and has an eye, through which a small

Fig. 775.—Bar with Collar on End.

chain or rope E is passed; by means of this the monkey is pulled backwards after every blow. In use it is drawn back several feet from the perpendicular and then let go; being heavy it strikes the work with great force.

Upsetting a Collar.

A collar A (Fig. 775) can be upset upon the end of a rod whose original section is that of B. The end that is to form A is enclosed in the fire, but only the precise amount for upsetting is heated; the iron adjoining it is kept quite cool and black by heaping damp coal around it. The end is then brought to welding heat, and taken from the fire. Sometimes on removal from the fire, the extreme face of the heated

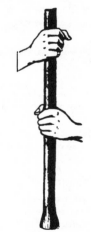

Fig. 776.—Jumping Bar Vertically.

end is dipped into the water trough to chill it, to bring it to a better condition for resisting the blows of the hammer; but this is not always done. For the actual upset-

ting, the bar may be held vertically in the two hands (see Fig. 776), and the white-hot end jumped down repeatedly upon the anvil face, or upon a cast-iron plate let

Fig. 777.—Upsetting Bar Horizontally.

into the ground alongside of the anvil stand. Another method of upsetting is to lay the bar horizontally, as in Fig. 777, upon the anvil face, holding it in one hand if light, and then hand-hammering the end to be upset. If heavy, the bar is held in both hands by an assistant, or slung in a chain from the forge crane, and upset with a sledge-hammer. When very heavy, it is laid upon the anvil face or upon a levelling block, and the swinging monkey is driven against it.

Other Points in Upsetting.

Three or four heats are often required to jump up a moderate mass of metal; therefore fairly exact dimensions are not at once obtainable by this method, as in the case of drawing-down. The jumped-up mass of metal, in spite of much care in localising the heat precisely where it is required, is very unequal, and quite without sharp shoulders. Considerably more metal has to be massed together than is actually required in the completed work in order to allow of symmetrical finish to

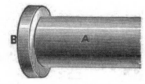

Fig. 778.—Bar with Shouldered End.

size. Upsetting tends to separate the fibres of the metal. It is therefore necessary to counteract this by hammering the jumped up portion at a welding heat. When the metal for the collar is massed in sufficient quantity, it is finished parallel in swages, and the square shoulder finished with a set hammer or flatter, or in swages, and the end with hammer and flatter. A collar can be formed upon any portion of a bar situated away from the ends by localising the heat in the position required, and then jumping up the metal at that particular place, until sufficient mass is obtained for finishing to size and shape. Any other sections can be heated, and the spreading out can be performed by upsetting in one direction more than in others.

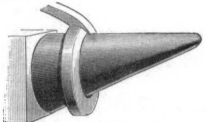

Fig. 779.—Forming Ring on Anvil Beak.

WELDING IRON AND STEEL.

Weldability is one of the most valuable properties of wrought-iron and mild steel. Welding is often the alternative of drawing-down or of upsetting. Correct heat and cleanliness are the chief requisites. At welding heat the metal is in a state of partial fusion on the surface, and is then extremely plastic; hammering will cause two surfaces so heated to adhere, and the joint possesses as much strength as the other parts of the metal. The welding heats for iron and steel and for different quali-

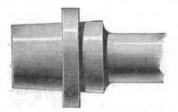

Fig. 780.—Ring Slipped over Mandrel.

ties of each material greatly differ. Any iron requires a much greater heat than steel, and the better the quality of the iron the higher the welding heat that it will stand without becoming burned. At a

welding heat iron gives off dazzling sparks; steel shows only an intense yellow, and gives off but few sparks. The ascertaining of the correct heat is a matter of experience entirely. To illustrate the process of welding more clearly, two plain examples, one a collared rod, and the other a plain straight rod, are given.

WELDING RING OR COLLAR.

Suppose that it is necessary to weld a collar B (Fig. 778) to a rod A. First cut

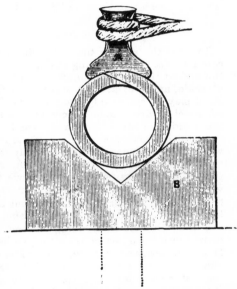

Fig. 781.—Finishing Ring on V-block with Hollow Tool.

off the rod A, and then prepare the ring B. For the last-named take a square bar, say $\frac{1}{8}$ in. larger than the finished section required, and, with a hot set, cut off one end diagonally or else fuller it down. Then bend the bar roughly into circular form over the anvil beak (Fig. 779), and cut it off to the proper length diagonally, to match the diagonal cut on the other end. The metal must have sufficient lap to allow for welding and for dressing off and finishing. If the ring is fairly true, it will be ready to go into the fire for welding; but if not, slip it over a mandrel (Fig. 780), and give the scarfed joint a neat appearance, either with the hammer alone or with a hollow tool (see Fig. 781). Then slip off the ring and flatten the faces (Fig. 782). This is precisely the plan that would be adopted in welding a separate ring.

Fig. 782.—Smoothing Faces of Ring with Flatter.

WELDING RING TO ROD.

To weld the ring to the rod to form the collar B (Fig. 778), the ring is slipped over the end of the rod, taking care to remove all scale, and they are put into a clear fire. Sand may be sprinkled over the work. but with a clear fire it is not necessary. When the welding heat is attained, which for wrought-iron is of a dazzling whiteness, when the iron seems ready to melt and particles appear ready to drop off, and sparks are thrown off rapidly, remove the work from the fire, place on a V-block B (Fig. 781), and hammer the scarf joint and the ring all round with a hand hammer, the rod with its ring being continually turned into fresh positions on the V-block. If a hammerman's services are available, the hollow tool A is used, and a few blows upon it consolidate and smooth the surfaces. Then the faces and shoulders are finished by means of a heading tool (Fig. 783) having a hole of a size suitable to take the rod, a few blows with hammer and flatter finishing off both the under shouldered face and the upper flat face.

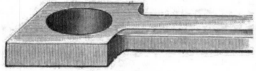

Fig. 783.—Heading Tool.

The circular part may have to be worked over again with the hollow tool to finish the surface, for the welding heat is soon past, and if the joint is not effected in the first few seconds it will be more or less imperfect.

PLAIN WELDED JOINT.

The scarfed joint is employed for a plain weld, and there must be plenty of metal to allow for hammering the joint together

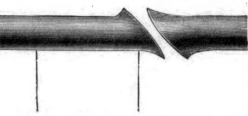

Fig. 784.—Ends Scarfed and Upset ready for Welding.

and for finishing it afterwards without reducing below correct sizes ; therefore, the ends of the bar have not only to be scarfed, but to be slightly upset. The meeting ends, which should be cut off square, are laid horizontally upon the anvil, and are then upset or beaten over while ready at a welding heat. This done, they are next laid over the edge of the anvil, and scarfed or beaten down diagonally with a fullering tool, the face of the scarf being made rounding rather than hollow. Both ends having been served precisely alike, they are heated to welding point. Lift the work vertically out from the fire, and do not drag it through the coal. Dirt will show as dark specks on the white-hot iron, and should be brushed off with a switch of brushwood. The smith and an assistant lay the scarfed ends together as shown in Fig. 784, and then two or three blows with the hand hammer cause the ends to unite. Turn the rod rapidly about on the anvil while consolidating the joint all round with hand hammers or sledges ; finish between top and bottom swages. It will be apparent that without the first enlargement or upsetting of the rods, the process of welding and swaging would have thinned the rod at the welded section below that of the other portions. How much to upset and how much to scarf experience alone will determine.

PUNCHING HOLES IN METAL.

Punching, drifting, and drilling are the three methods by which the smith commonly makes holes in metal. The first

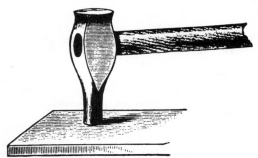

Fig. 785.—Round Punch on Thin Bar.

two are performed on red-hot iron and steel; the last, and sometimes the second also, on cold metal. Punches are circular, square, oval, oblong, and wedge-shaped, and their handles are of hazel or iron. A punch (Fig. 785) and bolster (Figs. 786 and 787) may be used together. Punches may make a clean, finished hole, or they may remove the bulk of the metal, leaving a small allowance for finishing with drift, reamer, or boring tool. The iron to be punched is brought to a suitable heat, full red or white, and laid across the anvil, and the punch is driven about half-way through by blows from a sledge or hand hammer. The punch is then withdrawn, and the iron is turned over and laid upon its opposite face. A dark, chilled spot indicates the position of the punch hole on the other side, and enables the smith to place the punch again correctly for piercing the

Fig. 786.—Circular Bolster.

metal so that the holes meet. To properly finish off the perforation, the work is laid upon a bolster or over the hole in the anvil, and the punch then passes freely through.

If the hole is deep, the hot iron closes and tightens around the punch, making it necessary to withdraw the punch at every three or four blows. Further, the heat of the iron makes the punch very hot, so that after every three or six blows it is necessary to cool the punch in water.

PUNCHING SLOTTED COTTER WAYS.

To punch slotted cotter ways in which

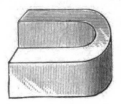

Fig. 787.—Horseshoe-shaped Bolster.

the section of the iron is not enlarged (see pp. 60 and 61, and Figs. 173 to 177) take a tapered oblong punch or, instead, a steel drift with rounding ends. Raise the iron to a welding heat, and properly support it according to its shape, upon a bolster or a bottom tool, and drive the punch half-way into it. Turn the iron over, cool the punch in water, and drive it in exactly

Fig. 788.—Punching Hole through Stout Pin.

opposite to the first position, until the openings meet at the centre of the bar.

The slightly tapered punch makes a rough and doubly tapered hole, into which a parallel drift or filling-piece is then driven, the hole taking the form of the filling-piece. The outside of the iron is smoothed and finished, and when the shape is completed the filling-piece is driven out.

PUNCHING HOLE THROUGH STOUT PIN.

The method of punching a hole through a stout pin is shown by Fig. 788, the pin

 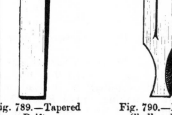

Fig. 789.—Tapered Drift.

Fig. 790.—Drift for Shallow Holes.

being 3¼ in. diameter; the hole measures 1½ in. by ½ in., and is made at one heat. The punch, A, has its body below the handle, about 6 in. long, and is tapered. The pin being brought to a white heat, the punch is driven almost through. Then the pin is turned over and the punch driven into the dark spot which has appeared, and the hole is thus completed. During the punching the tool has to be cooled several times in water. At the first stage of making the hole, the pin lies upon an ordinary bottom swage. At the second stage it lies upon a bolster, B, in form like a hollow swage, but pierced with a central hole, through which the drift can find a clear way. As the hole is thus roughly punched, the metal around it will be partly compressed, partly bulged; very little is actually driven out and removed by the punch. The bulging of the pin is corrected by hammering between top and bottom swages, and then the hole is finished by drifting, all being done during the one heat. During the punching, whenever the punch is withdrawn to be cooled, a little small

coal is strewn in the punched hole, to burn up the gas which would otherwise resist the passage of the punch.

DRIFTS.

Drifts are used for finishing holes when facilities for machining them are not available. They are smooth, and are driven through the punched holes in the red-hot metal to enlarge, shape, and smooth them. Drifts may be taper or parallel, and in section they may be circular, square, oblong, elliptical or polygonal to correspond with the shape of the finished hole. When two drifts are used, one tapers considerably as in Fig. 789, so as to enlarge the hole, and the other—the filler or filling-in piece—tapers but slightly, and imparts the finished dimensions. Smooth drifts are not often parallel, owing to the difficulty of removing them from a hole.

CUTTING DRIFTS.

The cutting drift has the action of a sharp punch, and in some cases that of a file. It does not press open a hole, but, by removing metal, smooths and finishes a hole already brought nearly to shape and

Fig. 791.—Drift for Deep Holes.

Fig. 792.—Drift for Smoothing Holes.

size. A drift for shallow holes is shown by Fig. 790, and one for deep holes by Fig. 791. These are filed to shape, hardened, and then tempered to a colour between brown and purple. The finer the work, the finer and closer together are the teeth of the drift, and they are bevelled to allow of clearance for the chips produced. Following the usual practice already set forth, the angle between the cutting edge and the face is greater for hard metals than for soft ones. The tool should be well lubricated with oil whilst in use, and the work should be properly bedded on a metal block, taking care to drive the drift straight. For a deep hole, withdraw the tool once or twice to clear the chips.

ALTERING SHAPES OF HOLES WITH DRIFTS.

Toothed drifts can be used for making round holes into square, hexagonal, etc. To square a round hole, insert a half-round plug and operate on the other half of the hole, little by little, interposing thin backings one after the other as the toothed drift enlarges the hole. When one half of the hole is shaped, fill it up and do the other half. This also applies to an

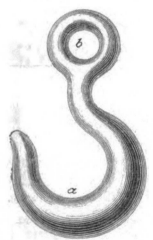

Fig. 793.—Crane Lifting-hook.

elliptical hole. The cutting drift of the special shape shown by Fig. 792 is used to smooth holes that do not pass right through the material. The tool illustrated is intended for a rectangular hole, to be smoothed in one direction, the face or edge A removing a thin shaving. For the second cut, a thin strip of metal is placed behind the drift; for the third cut, another strip, and so on

PRESERVING CONTINUITY OF FIBRE IN PUNCHING.

It has already been stated (see p. 61) that in punching holes it is necessary to take account of the direction of the fibre of the metal; otherwise the iron will become divided instead of spread out. Punching puts considerable tension on the fibres

Fig. 794.—Punched Tie Rod.

around the hole, with reduction of area. These points may be illustrated in a few examples. The crane lifting-hook (a, Fig. 793) is invariably bent round like the dip crank, and its sectional strength is preserved. If it were slotted the hook would break with much less strain. Instead of drilling a hole for the eye b of the crane hook, it is bent round and welded, or else punched. In drilling, the metal is severed; in punching, it is thrust aside and not divided. The punching, if done properly, preserves the continuity of the fibres, and, by bulging out the metal upon each side, preserves an equal section, and little or no upsetting is required. The punched rod (Fig. 794) is an illustration of the effect produced in the middle of a bar, and common in roof trusses. The eyes of hammers, and the cotter ways in bolts and rods, should always be punched. There is then no separation, but only a parting or spreading of the fibres. In

Fig. 795.—Welded Eye.

forging the eye of a winch handle (Fig. 795), instead of making a solid end and drilling and filing a square hole, the bar is bent round a mandrel, and then welded.

PRESERVING CONTINUITY OF FIBRE IN FORKED ENDS.

The continuity of fibre is preserved in large forked ends in the manner illustrated by Figs. 796 to 798. Small forked ends usually are shaped out of the solid but broad ends, like that illustrated, while

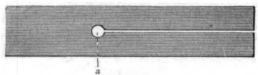

Fig. 796.—Bar Divided with Hot Set.

those of moderate width are formed by dividing the bar and then opening it out. Fig. 796 shows the bar from which the forked end has to be made. A hole is punched through at *a*; this does not sever the fibre, but merely thrusts it sideways. Then the bar is divided with a hot set from the hole *a* outward to the end. The set is driven first from one face half-way through the bar, and then from the other face to meet in the middle. The punched hole prevents all risk of the set splitting the fibres inwards beyond the hole. Then the bar is opened out, first with a wedge, afterwards with the hammer, as in Fig. 797, and finally finished as at Fig. 798. If the fork were cut from the solid, the fibre would be short; but being opened out and bent round, it runs continuously.

DIRECTION OF LAYERS OF IRON.

The direction in which the layers of iron occur has often to be considered in

Fig. 797.—Bar Opened Out.

forging. In a cotter key, for example, the layers of iron should be arranged not in the direction of rotation of the shaft as in Fig. 799, as the pressure would tend to shear the key off in the plane of the layers, but at right angles, as in Fig. 800; that is, the layers should radiate from the centre of the shaft, so that the pressure will tend to close them. Nuts should be punched at right angles to the direction of the fibre, otherwise the layers of iron are liable to become separated. The forked eccentric rod end (Fig. 801), forged solid and slotted out, should have the layers run not as engraved, but in the plane of the paper instead, otherwise the fibre is apt to open at *b*, and the forked end may fracture along *a*, *b*.

STEEL SUITED TO LIGHT FORGING.

Steel is homogeneous, that is, it is not laminated, and so is well adapted for light and delicate forgings, which have to be strong and rigid. Wrought-iron can be worked almost at fusing point, but each brand of steel works best at one particular temperature, though this never exceeds a

Fig. 798.—Finished Fork End.

full red heat. Steel polishes brighter than wrought-iron, and therefore is better suited to work where good finish is necessary.

DIFFICULTY IN FORGING STEEL.

Whether steel is more troublesome to work than iron depends on the nature of the forging, and on the quality of the steel. For small model work, steel is much the easier, it being more rigid and not having any grain to open out. In large and heavy forgings these qualities are not so important. Steel forgings of moderate size and thin sections have to be worked at a lower temperature, and thus require a larger number of heats, than iron; and for this reason the steel has to be worked rapidly, while it retains its heat. There is more initial difficulty in welding steel than iron, due to the differences in the

Fig. 799.—Key with Fibre Incorrect.

various brands, but once the best welding heat for any particular steel is known, there is no more trouble in welding it. Sand alone is used as a flux for welding iron; for steel, a mixture of sand and common salt is better.

Points in Working Steel.

There is a greater difference of opinion as to the working of steel than there is in regard to the working of iron. The practice of allowing steel to "soak" in the fire is approved by some and denounced by others. The temperature to which steel can be raised without burning it is also disputed. Steel varies in quality much more than wrought-iron, and its peculiar value is due to its chemical and molecular composition, slight differences in which cause great changes in the nature of the material; it must be remembered that slight and sudden alterations in temperature are alone sufficient to change entirely the molecular arrangement, and affect the working. A man accustomed to

Fig. 800.—Key with Fibre Correct.

work in one quality, chiefly of steel, and finding that a certain treatment answers satisfactorily, may discover that his method gives but poor results with another quality; consequently, he will condemn the steel, when, really, the fault lies in

wrong treatment. A bar of new steel should be experimented with so as to ascertain the best method of working it; otherwise failure to obtain the best results will frequently follow. The practice of upsetting steel is deprecated by some; but good steel will upset just as well as iron.

Burning Steel in Forging.

Steel should be made as hot as it will safely bear, but it must not be overheated, or it will be burnt. A higher temperature is allowable for large forgings than for light work generally. Burnt steel crumbles to pieces under the hammer like cast-iron, and shows a coarse granulated fracture. There are, however, degrees in burning; the steel may be burnt only slightly on the surface, so slightly that it does not fracture, or, if fractured, does not show this coarsely crystalline structure, yet its quality for cutting tools will be sensibly impaired. The temperature at

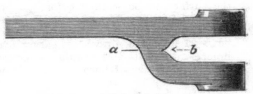

Fig. 801.—Forked End with Fibre Incorrect.

which overheating occurs varies, of course, with different qualities, but if scales form and fall off, the steel is, as a rule, overheated. It is almost impossible to restore burnt steel, though if it is burnt only slightly a good hammering may improve it. Steel, even more than iron, should be turned around in the fire to keep the heat uniform, and the blast should be slackened after the edges have become red-hot. Hammering should not be continued after the steel has lost its redness.

"Blue Heat" of Steel.

The "blue heat" temperature in steel corresponds with the "black heat" in iron. Experiments show that a steel plate heated, and allowed to cool, does not suffer any diminution of strength, but that while it is cooling, and while at a "blue heat," any hammering or bending injures the metal seriously. The "blue heat" corresponds with any temperature between about 470° and 600°. F.

WELDING CAST STEEL.

In welding cast steel, the flux may consist of borax $\frac{1}{4}$ lb., washing potash $\frac{1}{4}$ lb., and a small quantity of powdered white glass, all melted together, and, when cold,

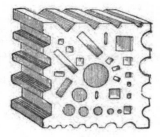

Fig. 802.—Swage Block.

pounded. This should be thrown over the surface to be worked before putting the steel into the fire, more being added afterwards as required. Cast steel should be kept from the air when heating over breeze—not coal—and should be heated carefully for fear of burning. The blows should at first be light.

SWAGE BLOCK AND STAND.

The smith's appliances are not complete without a swage block (Fig. 802). This is of iron pierced with numerous holes—round, square and oblong—and its sides are provided with circular and V-shaped grooves. In the position illustrated by Fig. 802 the grooves serve as bottom swages for circular, hexagonal and rectangular work. When lying upon the back face parallel with the one shown in Fig. 802, it serves as a bolster, upon which

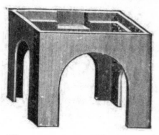

Fig. 803.—Swage Block Stand.

holes are punched and drifted, and as a heading tool upon which shouldered work is finished. The stand upon which the swage block is mounted may be as in Fig. 303; or, instead, it may consist of upper and lower cast-iron frames, with the upper one provided with strips, enclosing the block. The two frames are held together by four shouldered wrought-iron pillars, whose pins pass through holes cast in the frames, and are riveted over at top and bottom.

WORKING SHEET METAL.

INTRODUCTION.

THE metals with which the worker in sheet metal is chiefly concerned are copper, brass, and tinned steel, the last-named being known to him as tin-plate. Zinc,

Fig. 804.—Creasing Stake.

bronze, and pewter are employed occasionally, and aluminium is coming into popular use for small articles owing to its light weight and non-tendency to tarnishing. Sheet metal working consists of (a) preparing the pattern, (b) cutting and shaping the metal, (c) putting together the article

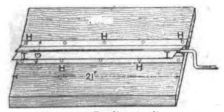

Fig. 805.—Beading Appliance.

by soldering, brazing, or riveting. The three processes last mentioned having already been described exhaustively, this chapter will be concerned with pattern development and metal shaping. A kind of metal plate work known as spinning will be treated separately in a later chapter.

BEADING TIN-PLATE.

Beading tin-plate by hand—that is, bending it into circular form of small section—is done as follows: Suppose the bead is to be made ½ in. diameter and

Fig. 806.—Section of Beading Appliance.

20 in. long. Cut the pieces of metal 1½ in. wide and the required length, and get a piece of ½-in. iron rod, 30 in. in working length, with a ring bent at one end with which to hold it. Now "edge" or bend slightly both sides of each piece of metal. Bend the pieces round the iron rod in a suitable groove in the creasing stake (Fig. 804); but if this cannot be obtained, make a ½-in. groove in a piece of tough, close-grained wood, in which to bend the metal. Work from one end to the other, turning it round at the same time.

Figs. 807 and 808.—Elevation and Section of Rod.

APPLIANCE FOR BEADING TIN-PLATE.

A simple appliance for beading tin-plate can be made as shown in Fig. 805, which illustrates the machine complete, with a piece of tin-plate in position ready for turning. Fig. 806 is a section of the machine without the rod. Fig. 807 shows the rod, and Fig. 808 the same in section. To make the appliance, get a piece of $\frac{7}{8}$-in. iron rod about 30 in. long, file 21 in. of it (from one end) until it is clean and bright, and tin it with a soldering-iron. To do this, brush over the filed part of the rod with killed spirits, and then coat it with solder, applying it by means of the soldering-bit. If

Fig. 809.—
Flat Hammer.

Fig. 810.—
Planishing Hammer.

the surface is rough, put the rod on the top of the fire until all the solder is melted, then wipe off all superfluous solder with a piece of tow or an old rag. When this has been done, bend the part not tinned to form a handle (see Fig. 807). Take a piece of xxxx, or about 24 gauge tin-plate, cut a strip 20½ in. by 1⅜ in., bend each edge slightly in the direction it has to be turned, and solder one edge securely to the rod. Bend the tin-plate round the rod so as to form a tube with a lap of about $\frac{3}{10}$ in.; leave the seam open slightly, as shown in Fig. 808, so that a piece of sheet metal can enter for about ⅛ in. Solder firmly at each end, and the rod is complete. To make the remainder of the appliance, get a piece

Fig. 811.—
Concave Hammer.

Fig. 812.—
Convex Hammer.

of tin-plate or brass about 20 gauge, 21 in. long, and 1½ in. wide; bend this into a tube with the seam open about ¼ in. or $\frac{1}{5}$ in. (A, Fig. 806), and see that the rod (Fig. 807) works freely inside. Solder this tube on a piece of strong tin-plate (B C,

Fig. 806), 21 in. by 3 in., having holes H punched in it, by which to screw it to a block of wood. Solder along each side a piece of tin-plate to support the tube, and the appliance is then finished.

USING BEADING APPLIANCE.

Fig. 805 shows the appliance with a piece of tin-plate T ready for turning. When

Fig. 813.—
Square-faced Hammer.

Fig. 814.—
Hollowing Hammer.

using this appliance keep the burr up with the left hand, press the tin evenly into the open seam of the rod (while it is in the machine), and turn gently two or three times with the right hand; then draw out the rod and bead. If it works too tightly, lubricate with some dry blacklead.

TINMEN'S HAMMERS AND MALLET.

Tinmen's hammers are of special shapes, and include the flat hammer (Fig. 809), the planishing hammer (Fig. 810), concave hammer (Fig. 811), convex hammer (Fig. 812), square-faced hammer (Fig. 813), hollowing hammer (Fig. 814), smoothing

Fig. 815.—
Smoothing Hammer.

Fig. 816.—
Pene Hammer.

Fig. 817.—
Riveting Hammer.

hammer (Fig. 815), pene hammer (Fig. 816), riveting hammer (Fig. 817), box hammer (Fig. 818), and the large block hammer (Fig. 819). The boxwood mallet (Fig. 820) should be from 2 in. to 3 in. in diameter.

WIRING TIN-PLATE.

When wiring tin-plate articles the method of folding the edges to receive the wires must vary with the shape of the body. In the simplest form of wiring a straight length of metal is wired outside along its length, the edge being afterwards turned to form a containing cylinder. The

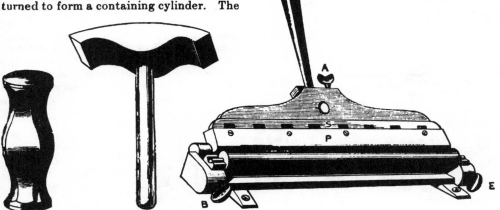

Fig. 819.
Block Hammer.

Fig. 818.
Box Hammer.

fold for the wire is made by bending the metal in a folding machine (Fig. 821), or, if a machine is not available, by working the metal with a smooth-faced mallet upon a hatchet stake (Fig. 822) until the section obtained resembles Fig. 823. Then the wire is held closely under the fold at the right-hand end of the body, and the metal is worked over it with the mallet upon any flat, smooth-faced tool.

Fig. 820.—Boxwood Mallet.

FOLDING MACHINE.

The folding machine already alluded to must be described briefly. It is held down to the bench by screws passing through four lugs (see Fig. 821). The sheet metal

to be folded is slipped over the front roller under the steel plate P, the amount of bending being regulated by the notched brass slide S, which is moved by the ad-

Fig. 821.—Folding Machine.

justment screw A. With the front roller high, a sharp fold is obtained, and when it is low, the fold is more rounded. Seven or more sizes of these machines are made, the smallest being 18 in. long and weighing 34 lb., and the largest being 48 in. long, and weighing a little more than 2 cwt. The most useful size for ordinary work is No. 1½, which is about 20 in. long.

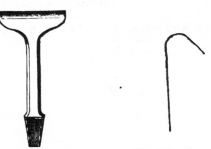

Fig. 822.—
Hatchet Stake.

Fig. 823.—Bent
Edge of Metal.

SMOOTHING WIRED FOLDS.

To smooth the fold the wired part is placed in a suitable crease on the creasing

iron (Fig. 804), and the fold is worked into
the crease with the mallet ; or the underneath edge of the closed fold is held upon
the edge of a square-faced tool, and the
fold closed and smoothed with the mallet

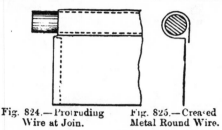

Fig. 824.—Protruding   Fig. 825.—Creased
Wire at Join.   Metal Round Wire.

while passing the work along the edge of
the tool. A third method is to pass the
wired part through rollers with a crease
that fits it closely. The most convenient
method is to pass the wired part through a
pair of "taking up" wheels in a wiring
machine. In bodies of this type, a short
length of wire protrudes at one end, and
there is a space of nearly equal length at
the opposite end, so that when the body is
turned the wire can be worked into the
space to avoid weakness at the seam. The
wire should be left out at the end that has
the fold underneath, as shown by Fig. 824,
otherwise there will be a difficulty in forming the seam.

WIRING STRAIGHT BODIES.

Straight bodies, handles, etc., are also
often wired in the crease, the article being
folded as described in the first case, and
the wire placed in position at the starting

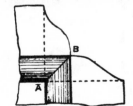

Fig. 826.—Wiring Sharp Corner.

end. The fold is then closed down flat,
and the metal counter-sunk in one operation by driving the fold over the wire with
the mallet into a crease in a crease iron,
until the section is as shown by Fig. 825.

WIRING SHARP CORNERS.

When wiring flat surfaces with sharp
angular corners—a kitchen blower, for example—the corner should be cut as shown
by Fig. 826. Then, after the edge has been
folded and worked over the wire, if it over-
laps at any part along the mitre line, a
piece of sheet metal should be cut so as
to fit in the angle at A. Next, with a fine
metal saw, cut down the edges until the
mitre line is clear (the piece of metal lying
in the angle preventing the saw scratching
the part of the article below the wire).
Then tuck the fold close under the wire,
and work round the wire lightly with the

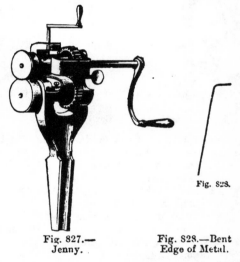

Fig. 828.

Fig. 827.—   Fig. 828.—Bent
Jenny.   Edge of Metal.

mallet until each corner has been closed to
form a neat mitre.

WIRING TAPERING CIRCULAR BODIES.

Tapering circular bodies with sides inclined at less than a right angle with the
ground are nearly always wired after being
turned, and after the seams have been
formed. The slant of the side being partly
in the direction in which the edge is first
thrown off, very little stretching is necessary to work the metal over for the fold.

THE OPERATION OF WIRING DESCRIBED.

In wiring small articles an edge of the
desired size is first thrown off in the jenny

(Fig. 827) to the form shown in Fig. 828. This is then worked over on the edging tool or half-moon stake (Fig. 829) to the shape shown in Fig. 823. The fold is then closed

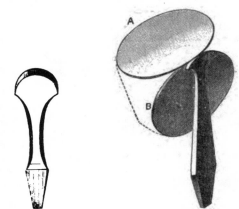

Fig. 829.—Half-moon Stake.

Fig. 830.—Edging Pan Bottom.

over the wire with a mallet upon a bevel stake, and smoothed in a wiring machine as previously described. (Fig. 830 shows a pan bottom in process of edging, A being the position at starting and B the position at finishing.) When the body is of large diameter, the edge can be brought right over to Fig. 831 by the taking-up wheels on the wiring machine. When the taper is in the reverse direction, and the body is to be wired round the smaller end, the folding is generally executed while the

WIRING TAPERING RIMS.

Tapering rims for the feet of articles are often wired from the inside. When this is done the rim is first turned, and seams are formed, and then a bead (Fig. 832) is

Fig. 831.— Bent Edge.

Fig. 832.— Beaded Edge.

Fig. 833.— Wired Edge.

formed in a swaging machine. Circles of wire of a diameter equal to the article are then turned, and one of these is placed inside the bead. The shoulder of the bead is placed on any smooth-edged tool, or on the edge of the bench, and the fold is closed over the wire with a round hammer. Fig. 833 shows the finished section.

WIRING CIRCLES IN THE FLAT.

When wiring circles or arcs of circles in the flat, an edge rather smaller than necessary to cover the wire completely is taken up, and the remaining amount is drawn over with the mallet during the wiring operation. If the full-size fold were taken up at first, and then closed down over the wire, the underneath side of the metal, which should be flat, would be slightly countersunk, the metal working up on the

Fig. 834.—Wired Circle.

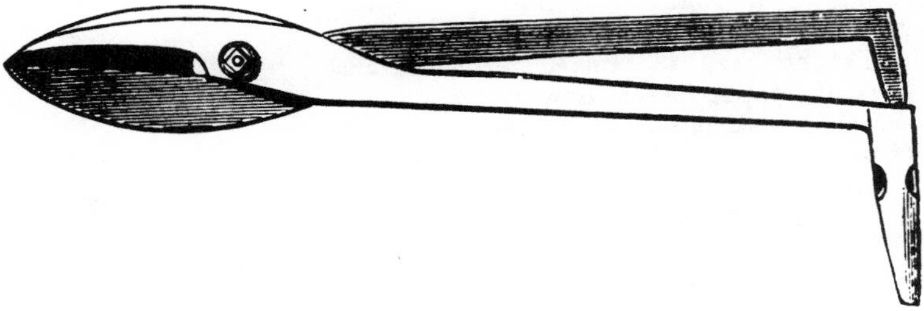

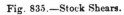

Fig. 835.—Stock Shears.

article is in the flat, the edge being worked over upon a hatchet stake (Fig. 822) with a smooth mallet, slightly round-faced. The wiring is then accomplished as in the first case.

underneath side along the part shown by the dotted line (Fig. 834).

CUTTING SHEET METAL.

Sheet metal is cut to shape with shears

which are made in two principal shapes, stock (Fig. 835) and hand or Scotch (Fig. 836). For small work, trimming large work,

beak irons (Fig. 847) are required for turning the various sizes of saucepan handles, water-pot spouts, etc., and Fig. 848 shows

Fig. 836.—Hand or Scotch Shears.

Fig. 837.—Straight and Bent Snips.

and for general use in repairs, straight snips, A, Fig. 837, and bent snips, B, Fig. 837, are indispensable. A small assortment of cold chisels should be obtained, and tinmen's pliers (Fig. 838) and round-nosed pliers (Fig. 839) will be found extremely useful.

TINMAN'S ANVILS AND STAKES.

It has already been shown that sheet metal is shaped principally by being bent over small anvils or stakes which fit into holes cut in the bench top. The tinman's

Fig. 838.—Tinman's Pliers.

anvil (Fig. 840) is steel-faced, highly polished, and practically flat; usually it is held in a special block, and on it plain surfaces are planished. The anvil stake (Fig. 841) is similar, but smaller, whilst the round stake (Fig. 842) is used in riveting patches on the bottoms of vessels. The convex stake (Fig. 843) is used in forming the tops of pepper boxes, flower boxes, etc.; seaming is done chiefly on the side stake (Fig. 844), which the pipe stake (Fig. 845)

Fig. 839.— Round-nosed Pliers.

resembles, being, however, shorter in the stem and longer in the arm. The funnel stake (Fig. 846) is used for shaping, grooving, and seaming funnels. Two or three

an extinguisher stake, a kind of small beak iron. The use of the saucepan belly stake (Fig. 849) is obvious. The tinman's horse (Fig. 850) has a hole at each end to receive the horse-heads (Figs. 851 to 853). The creasing stake, hatchet stake, and half-moon stake are illustrated by Figs. 804, 822, and 829 respectively.

FLATTENING SHEET METAL.

Before considering the principles underlying the flattening of sheet metal, a practical example may well be given. Assume that the plate to be set is "loose" at A B C D (see Fig. 854); to make it flat, the parts of the sheet opposite the buckled edge must be stretched with a setting hammer, used upon a large circular iron slab, known as a setter. The dotted lines upon the diagram indicate the places at which the blows are to be delivered, and a few addi-

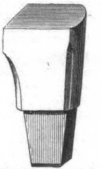

Fig. 840.—Tinman's Anvil.

Fig. 841.—Anvil Stake.

tional blows along the centre after the buckles are drawn out will stiffen the sheet. Buckling in sheets of metal is chiefly due to impurities in the metal, to a defect in the rollers, or to unequal annealing. One of the most embarrassing and perplexing of all jobs for a novice to undertake is that of flattening a piece of metal, and the great

Fig. 842.—Round Stake.

Fig. 843.—Convex Stake.

difficulty lies in the fact that in hammering the effect of the hammer blows cannot be seen. It appears that in proportion to the elastic power in a piece of metal so is its stubbornness in yielding to any flattening force. Elasticity is the property which enables metals to recover their original form after being subjected to pressure or tension—that is, after they have been contracted or expanded.

ELASTICITY OR SPRINGINESS IN METALS.

Steel is generally credited with a capa-

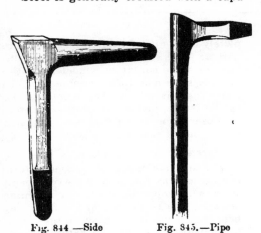

Fig. 844.—Side Stake.

Fig. 845.—Pipe Stake.

city for greater elasticity than any other metal, and is about the most intractable substance that the metal flattener has to

deal with. (Tin-plate is but tinned steel, it must be remembered.) Brass is only moderately springy; while lead seems to possess little or no elastic force. A good

Fig. 846.—Tunnel Stake.

Fig. 847.—Beak Iron.

Fig. 848.—Extinguisher Stake.

hand-saw may be bent so that its tip touches the handle, yet when released it will resume its original form. If a saw made of brass, exactly the size and shape of the steel one, were bent in the same way, its form would, undoubtedly, be affected permanently. If the saw were of lead, the degree of the permanent bend would be much greater than even in the case of the brass. It may be said that the elasticity of a metal plate has a certain relation to its thickness. The steel saw bent up double without apparent injury; the brass one did not. But, presuming the brass had been thin enough there would have been no difficulty in bending it to a very acute curve without altering its shape any more than the steel. Therefore, it may

Fig. 849.—Saucepan Belly Stake.

be taken for granted that, other things being equal, the thinner a piece of metal the greater its elasticity.

HAMMERING SHEET METAL.

With regard to the hammering of irregular metal with the object of making it regular, it is clear that the treatment will

Fig. 850.—Tinman's Horse.

vary greatly according to the character of the metal. As has been pointed out, the thickness of the metal must be considered. Very thin stuff must not be beaten with a heavy hammer, but always with a mallet, the blow from a mallet giving a much milder percussion. Occasionally there is a necessity for an intermediate kind of stroke—not so hard as the hammer, and yet a little harder than the mallet. To obtain this, a lead-faced hammer is used, or the face of the hammer is covered with a properly-grained piece of tough wood. (Instructions for making a lead-faced hammer will be given later.) A hammer of this kind has to be used also when the surface requiring to be beaten is to be preserved as much as possible from disfiguration. Not only does hammering extend metal, but it also tends to toughen it. In ordinary flattening, however, the extent of this toughening is inappreciable. If

Figs. 851 to 853.—Horse-heads.

the metal expands, there must be a corresponding diminution in thickness. It may be said that every blow of a hammer fulfils three offices: stretching, thinning, and hardening.

FORM OF SHEET METAL ALTERED BY HAMMERING.

With regard to the alteration of form caused by hammering, every plate has, of course, a particular shape which depends on the fact that its constituent atoms are taking a definite relative position. When, by hammering, the position of some of these is altered, it follows that the form of the plate is altered likewise. The impact on the top of the surface affects the under surface also; but the effect gets less as the metal gets thicker; to very thick metal these remarks do not apply. If a solid cubic foot of iron is struck the effect of the blow diminishes as it recedes from the upper surface, so that by the time it reaches the lowest particles the amount of the force is inappreciable. There is a difference between a blow struck in the centre of a plate and a blow which com-

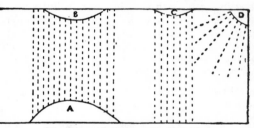

Fig. 854.—Flattening Buckled Plate.

pletely covers it. In the former case, the part actually struck is depressed or sunken; no mere concavity is made, and the plate is altered in form permanently. In many cases it will not be distinguishable by the senses, just as one could not detect by observation alone that an expansion of the metal had taken place.

DETERMINING ALTERATION IN FORM

Why the metal takes this particular form can be seen by the following experiment. With some ordinary rushes or burning material, smoke the top of the flattening bed, and leave it for a short time until the moisture that has appeared is dissipated. Now take a round-faced hammer, and gently place it vertically on the blackened surface. On removal, the black will be seen to have been disturbed, and to what extent should be noticed. Next, instead of lightly placing, give a swinging blow of moderate force, and mark the effect. The disturbance to the black will in this

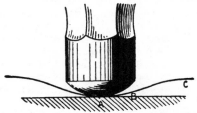

Fig. 855.—Sheet Metal Struck by Hammer.

instance be considerably greater than in the former case. The solution of this is found in the fact that both the face of the hammer and the face of the anvil are elastic bodies, and so are capable of being pressed slightly out of shape without permanent distortion. When the hammer struck the anvil its convex face was flattened, and the underlying body likewise suffered a depression of surface ; but both, being elastic, immediately regained their original form. Hence the larger disturbance of the black when the hard blow was dealt. What this has to do with hollowing a plate every time it is struck will now be shown. If a piece of metal is interposed between the two colliding surfaces it will certainly retard their elastic tendency. Under ordinary conditions, however, the force will not be annihilated.

What Happens when Sheet Metal is Hammered.

When a blow is struck, the metal is forced to a similar position to that depicted in Fig. 855. The deeper the metal is thrust—that is, the harder the blow— the higher will the adjoining metal raise

Fig. 856.—Sheet Metal after Striking with Hammer.

itself from the level surface. This uprising appears to be due to a lever-like operation to which the metal is subjected. If in Fig. 855 A is the power, B the fulcrum, and C the weight, a rude kind of lever action

will be understood. In Fig. 855, the position of metal and hammer when the force of the blow has attained a maximum is shown. A section only is represented, but the circularity of the action must not be disregarded. When the hammer is removed, the plate and anvil will be similar to Fig. 856, B acting as a pivot on which the heavier half of the plate bears the lighter portion above the level bed. Should the experiment be carried out on metal that will show effects very clearly, it will be seen that the bruise made on the metal may be divided into two parts—the actual point of contact and the surrounding annulus (Fig. 857). It will be seen that the bending of the metal is not dependent mainly on the form given to the part that is brought in direct contact with hammer and anvil. When the impact occurs, the surrounding metal is suddenly elevated, but in becoming so it has to contend with

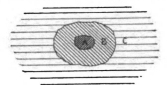

Fig. 857.—Bruise Caused by Hammer Blow.

those parts beyond the action of the force which are endeavouring to settle down by reason of their weight and inclination to retain their original form. This causes a very mild permanent bend of the metal all round the point of contact, which results in giving the metal a depressed form. The part struck by the hammer is not so distorted as would be imagined ; indeed, it can scarcely be perceived to be bent at all.

Procedure in Flattening Sheet Metal.

As to treatment in general, it will be found best to deal first with definite bruises or bends, which may be approximately neutralised by extensive rather than limited impact. This is also a good way of equalising surfaces or stresses. After this, most plates will be either centrally tight or loose, and a more special beating will follow, constantly testing by straight-edge, sight, and manual pressure. The standard, of course, is the perfectly plane surface, but this standard is never attained. A piece of metal is beaten flat, yet it is not difficult to see that its surface is far from being regular; and the very next operation to which it will be subjected is that of surfacing; a flattened plate would not bear the test of a surface-plate in ordinary cases without disclosing how far short it fell of an actual plane.

Bent Metal.

Metal not in a flat condition is termed bent. Bent forms of metal may be divided into two groups, which should be carefully noted, because the treatment of one is entirely different from that of the other. In the first group are all " bends," and in the second all " buckles." Sometimes it is difficult to say to which of the two classes metal belongs; but, speaking generally, parts of plates or of sheets of metal that are bent may be distinguished from buckled parts by not giving to slight pressure and by being of any shape or any size within reason, irrespective of the size or thickness of the sheet in which they are contained.

Buckled Metal.

A buckle cannot be of any size in a certain piece of metal. A little uprising may be punched in a piece of copper or iron— ⅛ in. thick, say—and the whole distortion not occupy more than ¼ in.; while to produce a buckle of this size in the same metal would be an impossibility. A "bend" may be made in almost anything; but to create a buckle there must be particular circumstances. Then, also, a buckle has a definite and somewhat uniform shape, while a bruise or bend may be any shape whatever. Commonly, buckles are known as "loose" metal; but bends are obviously not possessed of this quality, and may be said to be destitute of all spring. The main difficulty of all plate-flattening rests on the proper dispersion of what are generally known as buckles. Of course, it will be impossible to tell exactly where a buckle begins and where it ends, but certain experiments will assist in explaining the nature and disposition of a buckle.

First, take a piece of brass about 10 in. square, say, and of about 22 B.W.G. in thickness. This metal occupies a kind of mean between the extremes of malleability and elasticity; but brass varies in quality very often, some samples being tougher and more elastic than others. In all, however, general properties are the same. Assume the plate to be flat, and for beating it use a boxwood mallet of ordinary type—not a hammer. As the plate rests on the hammering surface it will be noticed that every part of it is in contact with the level anvil bed. If tapped lightly at any part, there is a deadness of sound indicating a total absence of vibration. Before beginning to hammer definitely, roughly mark on both sides of the plate, with pen and ink, a square about 2½ in. from the edge, putting also on the side about to be beaten a mark of some sort to distinguish it from the other side. In all malleting do not go beyond the margin of the squares. Now begin to beat regularly. After a short time, on examining the metal, it will be found to have sunk in the centre, as one would naturally suppose. And it is evident if the centre has sunk the edges have risen, for the metal will always rest on its lowest parts. This will be found, if tested, to be as stated. As the plate rests, if its edges be tapped lightly with the mallet, they will be found springy to the touch. Indeed, it will be easy to find what parts are in actual contact, and what are not, by delivering light blows of equal power on the surface of the brass, and noting difference in sound and feel. On turning over the plate there is a reversal of the conditions—those parts that were hitherto dead being springy, and vice versâ. On taking up the plate after the manner shown in Fig. 858, and slightly pressing the centre backwards and forwards, it is impossible to discover anything approaching a buckly nature. Probably the centre moves a little, but no sound is audible, as is usual with a piece of buckly metal turned from a concave to a convex position. But, even now, the germ of a buckle is present, as can be shown by plying the plate to and fro vigorously, when the customary sound will be heard ; a plane piece of metal could not, of course, give forth any sound. Now take the plate and beat smartly and regularly as before, on the same side, turning the metal as on a pivot to assist in giving every part its due impact.

EXAMPLE OF BUCKLED PLATE.

When hammering ceases, the plate can oe made to whirl on its hammered part. Turn it over, and each of the four edges will rest in close contact with the anvil, the central part being very much elevated. On being tested as before, the plate will emit the usual buckly sound when its centre is thrust in, but it has a capacity for bending strictly to one side only. This condition _of the centre represents a

loose. Thus so far can be seen how a plate, or part of a plate, may, by the same influence, be brought from a tight, unyielding plane to a fairly mobile state, and from thence onwards to a definite tight or bent condition.

ALTERING ONE-SIDED BUCKLE.

To continue the experiment: By continuing the beating as hitherto, nothing could be learned beyond what has already been reasoned out ; so turn over the plate, so that the reverse of the marked side is uppermost. Before beating, see that there is some indelible mark to distinguish one side from the other. Pressing the brass well to the bed, regularly distribute a few mallet blows, and the plate will now

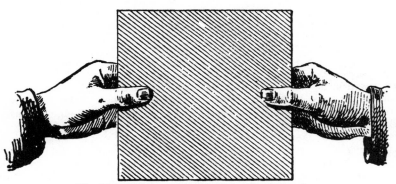

Fig. 858.—Taking up Metal Plate after Flattening.

buckle, but is only a one-sided example ; for, remember, the centre, if pressed inwards and then released, springs back instantly, like the bottom of a common oil-can. It does not stay where pushed or even return slowly, but rebounds with a vigour characteristic of all one-sided buckles.

ANOTHER EXAMPLE.

If the plate is further beaten as before, the quickness with which it would recover, and the force necessary to spring it, would gradually increase, until a point would be reached where any pressure inwards would alter the form of the metal. In such a case the centre would be better described as bruised rather than buckled, bent rather than sprung, and tight rather than

disclose a new feature. Previously the bulge was totally one-sided ; now it has a slight tendency to bend the opposite way. On pushing in the unmarked side it will be found that resistance to pressure has diminished, the brass springing back with feebler power. A little more careful malleting, and the centre is very slow indeed to recover ; finally, it does not recover at all. In examining the metal, it is always presumed to be held horizontally by the edges. The buckle has changed from one-sided to double-sided. It may also be regarded as loose, as it stands ; tight buckles of this character will be met with later on. With a little skill a loose buckle can be made to balance at equal distance from the neutral line of the plate—that is, the extended centre would incline no more one way than the other. As the plate remains at present, it will be concave on the non-inked side as long as it is kept horizontal, but not otherwise. This can be proved by gently raising the metal at one end. No sooner has it reached a certain inclination (╱) than the sunken centre

Fig. 859.—Loose-balanced Buckle.

springs upwards. Lower the plate, and the concavity returns. To understand this, bear in mind that the hammered part of the plate naturally has weight, besides being what is termed "loose"; and consequently the metal keeps depressed as long as it is maintained horizontal. The middle part really hangs suspended on the edges all round, though retaining a latent power to push itself upwards and remain in that case supported on the edges, provided only that its centre of gravity be shifted past a certain limit. This is done when the plate is tilted—the downward pressure is diminished, and, in consequence, proves insufficient to overcome the innate inclination of the metal to bulge upwards.

PLATE WITH TRUE BUCKLE.

A little more beating on the same side of the metal, and the plate will have to be held nearly vertical before the central part springs outwards. A few more blows from the mallet complete the true buckle. The plate has now to be held some slight distance beyond the vertical to make the centre thrust outwards; while if it is moved back to the other side it will be

noticed that, when at a similar angle, the centre will rebound and sink that way. So, no matter to which side the plate is inclined, when at a particular angle from the perpendicular, the centre will push itself downwards (see Fig. 859). This condition of the extended part of the plate is termed a loose-balanced buckle, for very obvious reasons. If, after getting the plate into this condition, the hammering is still continued on the unmarked side, the balance will be made unequal again, the plate bulging most in the direction of the beating; and to get the centre to shelve downwards with the marked side uppermost, the plate will have to be held at a very acute angle (╱), while the slightest distance past the vertical would be sufficient to throw the bulge the opposite way. And so one might work on till the plate would go hollow only when held horizontally, and further to a point where the influence of weight would be insufficient to sink the centre. The plate must be forced now in order to get it to go either way; if it were hammered much longer, the buckle would develop from a tight condition into what may be termed a definite bend.

BALANCING A BUCKLE.

Sometimes it is difficult to balance a buckle, and through irregular hammering the metal will get decidedly out of order. This cannot be wondered at. When it is considered that a blow of no great force from the mallet is sufficient to disturb the equilibrium of a balanced buckle, it will be seen that delicate treatment is needful when approaching the poising condition. Should one overhammer the uninked side, the plate must then be reversed, and a few light blows delivered on the opposite side, which will probably set it right. Occasionally, however, beating on both sides, as judgment dictates, will be necessary to get the buckle balanced. Shallow buckles can be balanced very accurately. A few degrees past the vertical is enough to drive to right or left, as the case may be. The centre here would be quite shaky. It would appear that the shallower depressions may be balanced with greater fineness than those of deeper form. Probably nicety of adjustment decreases as the hollow deepens, because very low buckles do not admit of balancing at all, being too large in substance or surface to pass the neutral line of the plate without assistance ; and that is one reason why they have been designated tight. Thoroughly loose buckles will always be known by their capacity for sinking inwardly, when held in a favourable position, by sheer force of natural weight.

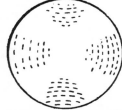

Fig. 860.—Metal Plate with Tight Centre.

Flat, Bent, and Buckled Metal Compared.

Bent metal resembles flat metal in being free or unrestrained. In a plane piece of metal, as well as in bent, every part rests composedly in its own particular section. Not so, however, with buckled metal, which is termed restrained metal. If it were possible to cut out the centre of the buckly plate mentioned previously—the centre including the whole range of the buckle—it would be so extended by this liberty that it would not be possible, without considerable difficulty, to fit it back again. The task would be like trying to fit a circular piece of metal into a hole smaller than itself, which could be done only by puckering up or restraining the circular piece.

Stretching or Extending Metal Plate.

It has been seen that hammering cannot contract, but can only extend, and that it always has a tendency to cause a depression on the surface which is struck. If, then, a flat plate had to be merely stretched or extended, not altered in form, the most rational way of working would be to beat regularly each side of the plate alternately. When one surface has re-

ceived a certain number of blows, a certain depression will have been made. Then, if the plate be turned, and a like treatment given to the opposite side, the metal will be righted again. This is the best way of working when a part of a piece of metal is to be expanded without bending it in any way.

Metal Disc with Tight Centre.

When a round piece of flat brass, 3 in. in diameter, and about the thickness of blotting-paper, is beaten at the edge all round, regularly, with a small mallet, a twist is generated, as shown in Fig. 860. The rim has been enlarged so that it is not in harmony with the centre. When placed on the anvil, two parts of the edge press decidedly down, and two stick decidedly up. The centre is tight and the margin loose. Working on principles stated, if the centre be sufficiently extended the plate will settle flat. This is proved by extending the centre of the plate, not by beating, but by means of heat. Hold the metal by the edges centrally over a candle flame for a short time, then lay it on the anvil, and the brass shows itself approximately flat. As the metal gets cool, it will be seen that those parts of the rim that originally curved upwards come up gradually until the plate is in its former condition again. Of course, the heat gives the metal only a temporary, and what is required is a permanent, condition. With the mallet, therefore, steadily beat the centre of the plate until its original form is regained—noting, by the way, that when this stage is reached the plate is necessarily slightly larger than it was before, owing, of course, to the hammering treatment it has received.

Other Points in Flattening Metal.

To produce symmetry where distortion exists, all parts cannot be treated uniformly. This must be obvious. If the circular piece of the distorted brass disc had been heated to the same degree all over, the disc would not have changed its form. Every part would have been stretched—the rim as much as the centre. Similar results would have followed if the brass had been beaten all over alike. The next point to notice is that the beating on irregular parts should be proportionate to their irregularity. The brass disc was flattened because the central part was made equal with the margin. But if malleting the centre is continued for some time, the result is that a buckle is generated there, as shown in Fig. 861. In beating the metal each side is treated alternately. This shows that "tight" and "loose" are only relative terms, dependent on each other for their force and meaning. These examples will find a parallel in many samples of distorted metal met with in ordinary work.

PUNCHES FOR SHAPING SHEET METAL.

Before proceeding to describe the miscellaneous examples of sheet metal work

Fig. 861.—Metal Plate with Loose Centre.

with which this chapter will conclude, a few of the most commonly used punches may be illustrated. Fig. 862 is the metal plate worker's solid punch, and Fig. 863 the hollow punch, the former being made in about 10 sizes, from $\frac{1}{16}$ in. to $\frac{7}{16}$ in., and the dimensions of the latter ranging from $\frac{3}{8}$ in. to $2\frac{1}{2}$ in. The groove punch or groover, Fig. 864, is used in laying seams together. The rivet set, Fig. 865, is a kind of punch, and its purpose is to draw rivets through two or more thicknesses of metal; of the sixteen sizes made, three or four only will be found sufficient.

FORMING HALF-SPHERICAL STUDS.

Half-spherical studs, such as are soldered to the bottoms of water-cans and other vessels, can be made by punching sheet

metal with a large hollow punch on a lead block about $1\frac{1}{2}$ in. thick and from 8 in. to 12 in. square. The blank is then hollowed with a stud hammer in a hollow made in

Fig. 862.—Solid Punch.

Fig. 863.—Hollow Punch.

the lead block; or, instead, they are worked in a stud boss (Fig. 866) with a special punch (see Fig. 867).

OIL COOKING STOVE.

The smaller oil stoves are to be obtained so cheap that it is scarcely worth while to make them, but the larger ones are well worth making, and will be found extremely convenient, particularly during the hot weather. The first example of metal plate work to be given in this chapter, therefore, may well be a large oil cooking-stove, whose oven will admit anything not larger

Fig. 864.—Groove Punch or Groover.

Fig. 865.—Rivet Set.

than $9\frac{1}{2}$ in. by 11 in. by 8 in., whilst two or three saucepans can be boiled on the top at the same time as the oven is in use.

THE STOVE BURNER.

The only part of the stove likely to prove troublesome in the making is the burner, and considerable care is required to secure both safety and efficiency. One pattern only is described, as, once the principle of construction is grasped, the making of other patterns and sizes will not present much difficulty; but this principle must be thoroughly understood before any attempt is made to vary the pattern, otherwise there may be a dangerous failure. Fig. 868 illustrates the principle on which the particular form of oil burner used in cooking-stoves is constructed. A is the wick-tube; B is the section of the "strainer," a piece of perforated tin-plate; C is the chimney, almost invariably made of metal. with a small sight-hole in it, covered with mica : D is the section of a curved piece of sheet metal called the "dome," which has

F g. 866.—Stud Boss.

Fig. 867—Stud Punches.

a slit in the centre for the flame to pass through.

PRINCIPLE OF OIL STOVE BURNER.

Hot air always rises and cold air sinks, so, when the wick is lighted, the air in the stove chimney, getting heated, rises, and flows out of the top, no matter how tall the chimney may be or how tortuous the passage. It may be straight, or may have elbows in it, or may be bent at right angles and round corners; but if it dips downwards at any part the lamp will most likely burn badly. The hot air passing out of the chimney is immediately replaced by cold air entering through the perforations in the strainer. The latter moderates the rush of air, steadies the flame, and prevents it from being blown out by any stray puff of wind. It also forms with the dome, when the ends are blocked up, a sort of chamber, in which the air is partially warmed before it comes in contact with the flame. The dome compels all the air which is drawn up into the chimney to pass through the narrow slit ; and the air, thus forced into contact with the flame, is partly

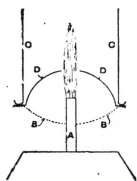

Fig. 868.-—Section of Oil Stove Burner.

burnt, thus increasing both the light and heat.

BURNER WICK-TUBES.

It will be best to make the burner first. The dimensions given are calculated to suit a $4\frac{1}{2}$-in. wick, and if any other size is wanted it is only necessary to increase or reduce the measurements correspondingly in the direction of the width of the wick ; but it is better to make a duplex burner than to make the wick wider than $4\frac{1}{2}$ in. Very wide wicks rarely work smoothly and evenly. Make the wick-tubes of stout tin-plate; each tube of two pieces. with the sides overlapping, as shown in section in Fig. 869. When one half is slid into the other a space large enough to admit the wick freely should remain. The length of the tube from top to bottom is $3\frac{1}{4}$ in., and. of course, the neatest end will be put uppermost in the burner. In one of the halves of the wick-tube holes $\frac{1}{2}$ in. by $\frac{1}{8}$ in. must be cut to admit the winding-wheels, and the tops of these holes must be $1\frac{1}{2}$ in.

Fig. 869.—Section of Wick Tube.

from the top edge of the wick-tube. One hole should be put $\frac{1}{2}$ in. from each side of the tube, and the intervening space should have not less than one hole to the inch.

BURNER WINDING-WHEELS.

On a piece of brass plate as thick as a new florin mark out $\frac{1}{2}$-in. circles for the winding-wheels. Drill all the centres to

Fig. 870.—Winding-wheel.

admit tinned hard-iron wire, $\frac{1}{8}$ in. thick, or a size larger. Cut out the circles, and file teeth in them, as shown in Fig. 870. A shows part of wheel marked out; B, wheel notched with three-sided file; C, finished teeth. Several discs can, of course, be threaded on a piece of wire to keep them together, and be filed out at once. The labour of making these wheels can be lightened somewhat by buying the discs ready stamped out, and, if expense is no object, the wheel-cutter will cut the teeth. Probably, also, brass pinion-wire can be had as large as required, or nearly so, and with suitable tools, slices may be cut off the end to produce the wheels. Great accuracy is not required in the spacing of the teeth, the only points to observe strictly being that the teeth are all the same length, and that there are no burrs or sharp corners to tear the cotton. Thread the wheels on a piece of the wire, and

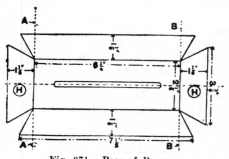

Fig. 871.—Base of Burner.

solder them on, one wheel opposite each of the holes in the wick-tube. Put the halves of the latter together, and solder them, first clearing off all burrs and sharp edges

likely to catch or fray the wick. If the top is not square with the sides, correct it by filing.

BURNER BASE.

Cut out a piece of tin-plate to the dimensions shown in Fig. 871. If any other size of burner is being made, the dimensions between the dotted lines A and B must be altered accordingly. Bend up the edges to form a tray, and solder the corners inside. The long shot must admit the wick-tube without shake or forcing; and the round holes must be large enough to admit the winding-wheels. Solder the wick-tube in position with $1\frac{1}{4}$ in. of the top projecting, taking care that it is upright. Make two $\frac{3}{4}$-in. tin-plate discs, each having a hole in the centre.

Fig. 872.—Bearing for Wick-winder.

BURNER WICK-WINDER.

Put the winder in position, thread one of the discs on each end of the wire, and solder them over H (Fig. 871), so as to cover up the holes. These small discs should be soldered only temporarily until it is seen that the winding-wheels work smoothly and without catching in the slots of the wick-tube. The wheels should project into the tube equally—far enough to dig about half-way into the wick; and the winder must be parallel with the top of the wick-tube. Considerable care is required over this part of the work in order to ensure a smooth and even winding action. The small tin-plate discs on the ends of the winder-handle facilitate adjustment. The winder is still too weak to wind up the wick without bending. It must therefore be strengthened by a piece of stout brass, bent and soldered as shown in Fig. 872, so as to form a bearing near the middle of the winder. A very wide wick may want two or more of these bearings. A thick disc of brass as large as a halfpenny, and having a milled edge, must be soldered on the end of the winder. In default of a milled edge, notching with a file will answer well. A cross-shaped piece of brass would afford a

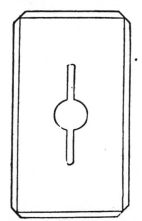

Fig. 873.—Bottom of Burner Base.

better grip to the fingers and do away with the difficulty of milling. Both burners must be made to wind to the right to raise the wick; the winder must not project more than about 2¼ in. from the burner.

Burner Bottom.

For the bottom of the burner, a piece of tin-plate with a slot shaped as shown in Fig. 873 must be cut out, of such a size that when ¼ in. of the edge is turned under all round, and the burner placed on top with the wick-tube threaded through the slot, ½ in. of the tin will be seen all round the burner. Solder it on, but do not solder

Burner Strainer.

The frame for supporting the strainer must be cut out like Fig. 874; and when it is bent into a rectangle across the dotted lines, and the join soldered, the feet A should come just on the edges of the base of

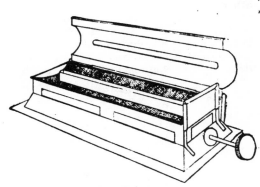

Fig. 875.—Oil Stove Lamp.

the burner as in Fig. 875. The frame is soldered on. The strainer is made of perforated tin-plate, which is sold in sheets 14 in. by 10 in. of various degrees of fineness. A good size is 14 holes to the inch linear. Coarser perforation, although cheaper, would not be so effective; finer can be had, but it would be too fragile, besides being more liable to get clogged up with dust when in use. Cut out a piece of the perforated tin as long as the supporting frame, and, having cut out a long slot to admit the wick-tube, bend the strainer to a semicircular form, and drop it into position. The edges which stand above the long sides of the frame must be bent down-

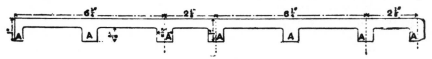

Fig. 874.—Strainer to support Burner Frame.

the wick-tube to it. The semicircles are cut out of the piece shown by Fig. 873 for ventilation, and to allow of the escape of the oil, that would otherwise accumulate in the burner and smell badly.

wards until horizontal; and, if more than ¼ in. projects, cut the surplus off. With spots of solder, tack the strainer to the supporting frame and wick-tube here and there, just to keep it in position.

BURNER DOME.

Make the dome out of sheet brass as thick as can be bent to the curve required. A slot is cut in the top $\frac{3}{8}$ in. wide, and $\frac{1}{4}$ in. longer than the width of the wick. The top of the dome is $\frac{5}{8}$ in. from the top of the wick-tube. The two long sides are bent out horizontally. One edge rests on the turned-out edge of the strainer, and the other is bent round a piece of the $\frac{1}{8}$-in. wire to form a hinge. The ends of the wire are turned down, and, after being bent to fit the sloping side of the base of the burner, are soldered to it. The entire length of the dome-edge need not be

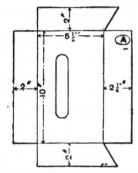

Fig. 876.—Oil Stove Reservoir Pattern.

used for the hinge; 1 in. at each end is sufficient. If all is correctly done, the dome will lie flat over the strainer with its ends flush; the slot in the dome will be exactly over the wick, and the dome itself will be capable of being thrown back to allow of trimming and cleaning (see Fig. 875). Both ends of the dome are still open. The back end will be closed by the chimney when the lamp is in position; and the front end must be closed by a piece of tin-plate bent round a stout wire frame soldered to the end of the burner-base as in Fig. 875.

OIL STOVE RESERVOIR.

To form the reservoir, cut out a piece of tin-plate like Fig. 876. Bend up the sides, and solder the corners inside very carefully, as they cannot be subsequently got at in case of leakage. Solder the brass fillers to pieces of tin tube, and, having cut

the tube to the proper angle, solder it on the reservoir over the hole A. The hole in the top of the reservoir should be $\frac{1}{2}$ in. larger all round than the wick-tube. Make

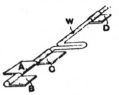

Fig. 877.—Lamp Catch.

the bottom of the reservoir to snap on, and solder it carefully. Solder on the handle, observing, as regards both handle and filler, that the reservoirs are rights and lefts. To fix the burner in position, little fastenings like Fig. 877 must be made, A being soldered on the projecting base of the burner, and B, C, and D on the top of the reservoir; and if all is properly done the burner will be held down truly and firmly. D prevents the wire W being entirely withdrawn, and perhaps lost.

OIL STOVE CHIMNEYS.

The remainder of the oil stove had best be made of the best sheet charcoal iron, which has a bluish, smooth surface. The common coke iron, with a grey, rough, blistered surface, is much cheaper, but will

Fig. 878.—Oil Stove Chimney.

crack if bent sharp, and the sheets are almost invariably buckled badly. Make the chimneys $10\frac{1}{2}$ in. high after the bottom edge is wired, and before the top edges are turned over, and of such dimensions in horizontal section that the chimneys would fit easily over the dome as regards its width, and tightly—or, better, would just not slip over—as regards its length from back to front. The bottom edges must be wired with the ⅛-in. wire along the back and sides, but not on the front; the latter

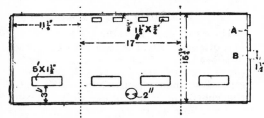

Fig. 879.—Stove Body Pattern.

has a semicircular piece cut out of it to agree with the curve of the dome. The join in the chimney ought to be a folded one, but riveting will do, provided the join is close. A 1¼-in. circle for a sight-hole is cut out of the front, with its centre 3½ in. from the bottom, and a similar hole is cut in an oblong piece of metal. The latter is temporarily riveted over the hole in the chimney so that the two holes coincide, and a piece of mica, not glass, is slipped in between. The whole arrangement is shown in Fig. 878; the top edges, however, must not be turned down until the chimney is finally fixed in the oven. The lamp can then be lighted; when, on standing the chimney in position on the dome, the flame should be steady and white.

OIL STOVE BODY.

The body of the oil stove can now be taken in hand. Cut out a piece of sheet iron to the dimensions given in Fig. 879. Bend it up at right angles along the dotted lines to form the sides and back. Turn outwards at right angles ¼ in. of top and bottom, and wire the two remaining edges, turning them outwards also. The edges are not included in the measurements. The five holes help to keep the lamps cool, and the others to create a draught up the chimneys when the stove is in use. The spaces A and B will provide places for hinges when the edge is wired. Cut out the bottom of the oven (Fig. 880) to fit. The front edge

is wired, and the holes for the chimneys should not be cut clean out, but ½ in. of the edges should be turned up as shown, both to get a closer join and to prevent juice or gravy from running out of the oven down to the burner. The nearest sides of the chimneys are 9½ in. apart. Slip the chimneys into the holes provided, and rivet the oven bottom in position at such a height from the bottom of the stove that the lamps can be slid in and out without catching against the wired edge. Rivet only temporarily at present.

OVEN TOP AND OTHER DETAILS.

Make the top of the oven just the same as the bottom, except that the chimney holes must be cut out clean, and the measurement from back to front must be somewhat greater, as 2⅜ in. of the front is turned up at right angles, and ⅜ in. of the turned-up part is turned out horizontally, forming a hollow top to the stove 2 in. deep. Temporarily fix it to the stove, and insert the chimneys. If all is right, the latter will be upright, and there will be a turned-out edge all round the top. The front edge, however, will not project beyond the wired edges of the sides more than $\frac{1}{16}$ in., if as much. Make and fix the bottom of the stove by lapping the edges over those of the sides and back, and wire the front edge. Small feet of some sort must be put on the back to compensate for the unevenness caused by the thickness of the wired edge in front. Insert the lamps

Fig. 880.—Bottom of Oil Stove Oven.

taking great care that the chimneys rest on the tops of the burners, and then mark round the chimney-tops which project beyond the top of the oven. Take out chimneys and oven-top and bottom all together, and, without separating them, turn the marked-off portion of the chimney tops outwards; which done, all can be replaced, the riveting finished, and the mica clamps (removed to permit of the insertion of the chimneys) permanently fixed. The top of the stove should then present the appearance of Fig. 881. During riveting the lamps must be frequently slipped in and out, for fear of their becoming bound or

Fig. 881.—Oil Stove with Top Removed.

too loose. There must be no holes, due to imperfect fitting, to admit soot or smoke to the oven.

Oil Stove Top.

The top of the stove is fitted on like the bottom, except that the edges are turned over all round. A circular hole, about 5 in. in diameter, is cut out over each chimney-top (the precise spot does not matter), and the top strengthened across the middle by a sort of girder, made as follows:—A strip of sheet iron, 2 in. by 9 in., has $\frac{3}{4}$ in. of one side bent at right angles, and the other side wired. The unwired edge is riveted to the underside of the stove top, thus enabling the stove to support a large saucepan without caving in. Before finally fixing on the top, provision must be made for saucepans smaller than the holes in the top of the stove, as the lamp will go out if anything is stood flat down on the chimney-top. The usual thing is a perforated cast-iron plate with feet that are bolted to the oven-top; but, if this is impracticable, a very good substitute would be two lengths of stout iron rod fixed about 1 in. above the chimney-tops.

Saucepan Hole Covers for Oil Stove.

Two covers of sheet iron must be made for the holes in the top of the stove, for use when nothing but baking is being done. Unless the heat is compelled to pass over the top of the oven, and out at the back, baking will be almost impossible. These covers ought to be sunk in the centre, so as to fit in the holes without sliding about, and they should have a ring, or something similar, by which to lift them.

Oil Stove Oven Door.

The oven door must fit well to prevent loss of heat. Bend up the edge of a piece of sheet iron so that, when dropped in the place intended for the door, it fits neatly all round; then wire the edges. Holes for the ventilator must be carefully marked, and cut out as in Fig. 882; and similar holes must be cut out of a circular piece of metal, which is riveted to the back of the door over the holes. Riveting must be done over a washer, so that the ventilator can be moved. A short length of wire is riveted into the movable part, and projects through a curved slot in the door, to serve as a handle for opening and closing the ventilator. Circular sight-holes must be cut in the door to correspond with those in the chimneys, but they should be put a little higher up, so as to enable the observer to look down to the flame. Make two frames of sheet brass, like those in

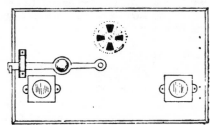

Fig. 882.—Oil Stove Oven Door.

Fig. 882. Bend up $\frac{1}{16}$ in. of all four edges, and then bend the lugs outward. Put a piece of glass in each, and rivet them on the door. Rivet over washers, to facilitate removal if the glass breaks.

Oven Door Latch.

The oven door latch is of cast brass. Make two wooden patterns like Figs. 883 and 884, $\frac{1}{4}$ in. thick, and with the bend in the ends, as shown, to avoid the wired edges of the door and side of the stove. The latch may be of the dimensions shown, but the size should vary with the size of the stove. A small latch looks rather paltry. A few incised lines improve the appearance. File up the latch casting,

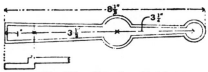

Fig. 883.—Oven-door Latch.

and, having fixed on a brass knob, rivet the latch to the door with a large copper rivet and washer, not so tight as to prevent its moving. Rivet over the latch a band of brass to keep it in place. About ¾ in. will be enough movement to allow.

Hingeing Oven Door.

The door can now be fixed on the stove. Make two hinges of tin-plate like B (Fig. 877, p. 294) on a piece of wire. Pull them open, and push them on to the places where the wire is left exposed on the side of the oven. Tin the edges of the door where they come against the hinges, and, having placed the door in exactly the position it is intended to occupy when closed, run some solder along the hinges, so as to fasten them to the edge of the door. A thin table-knife inserted under the hinges will keep them up against the door whilst they are being soldered.

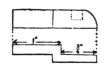

Fig. 884.—Oven-door Catch.

Completing Oil Cooking Stove.

The catch for the latch can now be riveted on the side of the stove, and a shelf (movable, of course) fitted in the oven. Fig. 885 shows the finished stove with the door removed. If the best charcoal iron has been used, no better finish

could be had than the natural surface of the iron. In any event, black varnish must not be used, as it smells for a long time when the stove is in use. Black-lead the stove in the ordinary way. There is no heat to spare in this stove. Every little aperture lets in cold air and lessens the power of the oven. If there are any holes due to bad fitting in corners and the like, they may be filled in with a little squeezed fireclay.

Stove Plate-warmer and Boiler.

Some makers fit plate-warmers and a boiler. The plate-warmer is simply a narrow sort of cupboard made of sheet metal, and riveted to the side of the stove; and the plates are slipped in on their edges, and the door closed. There is no real need for such an arrangement, as

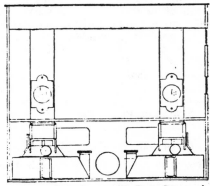

Fig. 885.—Oil Stove with Door Removed.

plates can be slipped between the chimneys and the sides of the oven. If desired, a plate-warmer might be fixed on the back of the stove; there is more room there. Although there may be little use for these plate-warmers, considered as such, they serve as "jackets" to the oven, and keep the heat in. Stoves are frequently made with shallow trays of sheet metal, about ½ in. deep, riveted to the sides, which answer the purpose better, perhaps. The boiler is a tin-plate tank having a lid and tap; and it is hung on the side of the stove, so as to get the heat from the side of the chimney.

POINTS IN USING OIL COOKING STOVE.

With regard to the lamps, never cut the wicks; simply rub off the charred portion with the finger, rubbing from the centre of

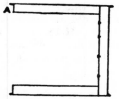

Fig. 886.—Section of Oil Stove Oven.

the wick to the sides so as to spread it out. This is most important, as rubbing from the sides to the middle tends to narrow the wick, and leaves spaces at the sides, down which the flame is apt to flicker dangerously. As a safeguard to some extent, the wick tubes may be prolonged nearly to the bottom of the reservoir. The wicks can be turned up very high without smoking, if it is done gradually. When in use the stove should not be placed so that the hot air is discharged into the room, as the hot, dry, vitiated air from the lamps is most injurious. It is well always to have a kettle or saucepan of water on the top, whether it is wanted or not, as the steam keeps the air moist.

OVEN FOR ORDINARY OIL STOVE.

The oven about to be described is a

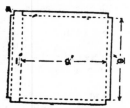

Fig. 888.—Top and Bottom Pattern for Oven.

separate attachment to an ordinary oil stove, and its construction may prove preferable to that of the more ambitious piece of work previously dealt with. In designing such an oven the chief consideration should be regular heating all round

the stove. This is best obtained by having an air jacket which distributes the heat. The ovens rest on top of the stove, the same as a kettle or saucepan, but clips can

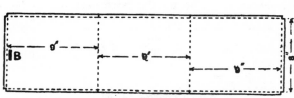

Fig. 887.—Body Pattern of Oil Stove Oven.

be riveted to the bottom of the oven to fit the top of the stove, to prevent it shifting when the oven door is opened and closed. The heat is regulated by the wicks, or, in the case of wickless stoves, by the size of the flame. The oven shown in section by Fig. 886 can be of tin-plate or plain iron. The heat soon discolours tin-plate, and it does not look well after having been used a few times. With iron, the smooth side should be outside, and will then need no blacking or other finish.

OVEN BODY.

The drawings of the oven (Fig. 886 to 889) are to scale. Fig. 886 is a section of the oven, Fig. 887 the body pattern, Fig. 888 the top and bottom pattern, and Fig. 889 the outer case pattern. The oven space is 9 in. by 9 in. by 8 in. First cut out the

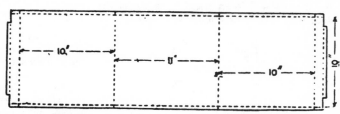

Fig. 889.—Outer Case Pattern of Oil Stove Oven.

body of the oven. A piece of No. 24 B.W.G. charcoal iron, 2 ft. 3½ in. by 8½ in., is marked out as shown in Fig. 887, the corners being notched square and nicked ¼ in. at the ends of the dotted lines. On the left-hand side, ¼ in. from the folding line, cut out a slot B, $\frac{1}{8}$ in. by $\frac{3}{4}$ in., for the door catch. Fold the $\frac{1}{4}$-in. edges up square and bend on the dotted lines the reverse way, that is, with the $\frac{1}{4}$-in. edges outside. Next cut out two pieces, $9\frac{1}{2}$ in. by $10\frac{3}{4}$ in.,

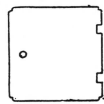

Fig. 890.—Oil Stove Oven Door Pattern.

for the top and bottom of the oven, and notch $\frac{1}{2}$ in. by $\frac{1}{4}$ in. at the back, and $\frac{1}{4}$ in. by $\frac{1}{4}$ in. at the front, as shown in Fig. 888. Fold up $\frac{1}{4}$ in. all round to fit the edges on Fig. 887, and knock them up. Then on the hatchet stake bend on the dotted line to make the faces A (Fig. 886), and flatten out the fold on the ends of the faces. Cut a piece of iron 10 in. long and $1\frac{3}{4}$ in. wide and fold the two long edges at right angles, but in opposite directions, thus ⌐⌐, the folds being $\frac{1}{4}$ in. and $\frac{1}{4}$ in. respectively; rivet this to the middle of the oven back, 1 in. projecting at the top and bottom. This prevents the oven dropping at the back, and holds out the back of the outside case when paning down the top and the bottom.

OVEN CASE.

The oven case may now be proceeded with. If made of iron the body can be

Fig. 891.—Door Catch Riveted to Handle.   Fig. 892.—Door Catch.

cut in one piece as shown by Fig. 889, but if of tin-plate, the back and sides are made separately. The $\frac{1}{4}$-in. edge at the top and the bottom is folded square outside, and a $\frac{1}{4}$-in. edge on the ends is folded square inside, and then bent square inwards on

the dotted lines. The oven can now be slipped in the case and a rivet put in at each corner. Next cut out the top and bottom 1 ft. by 11 in., notch for paning down, fold the front and the side edges right over, and slide on the bottom. Then edge up the back on the hatchet stake and pane down all round.

OVEN DOOR.

Next make the oven door (Fig. 890). Cut the material so that when wired it just clears the top and the bottom of the case and measures $\frac{3}{4}$ in. wider than the oven. Notch the corners, cut out two spaces for the hinges, and wire it all round. Cut two pieces 1 in. wide for the hinges, slip in the slots, sink them in the crease iron, and cut them off $\frac{1}{4}$ in. from the wire. Punch two holes in each for rivets, place the door in

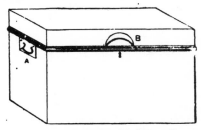

Fig. 893.—Deed Box with Lid or Cover.

position, and mark the places where the holes come. On a piece of iron about 2 in. wide and $\frac{1}{2}$ in. thick fixed in the vice, put a narrow piece of hard wood. Then place the oven in position and punch the holes. Remove the wood and rivet on the door. Slide on the top of the case, edge up the back, and pane down. Obtain a brass cupboard knob E (Fig. 891), cut a piece of stout iron or brass C (Fig. 892) $\frac{1}{2}$ in. by $1\frac{1}{4}$ in., punch a small square hole in it $\frac{1}{4}$ in. from the wide end, and bend as at D (Fig. 892). Then cut the screw of the knob to $\frac{3}{8}$ in. long and file it (except $\frac{1}{16}$ in. from the knob, which is left round) to fit the square hole in C. Punch the hole in the door for this round part, insert the knob, add a washer, and rivet the piece C (Fig. 892) to the shank of the knob. This completes the oil stove oven.

DEED BOXES.

The deed boxes shown by Figs. 893 and 894 can be made of No. 22 B.W.G. tin-plate or of "terne" iron, the latter being ordinary sheet iron coated with an alloy of lead and tin. The body of the box shown by Fig. 893 has five pieces—two sides, two ends, and bottom, the cover being in one piece. The sides are cut ¾ in. longer than the proposed length, and 1¼ in. deeper. The ends are cut ⅜ in. longer and 1¼ in. deeper. The extra 1¼ in. is for a feather-edge at the top, and allowance for the seam at the bottom. First cut two rods of ¹⁄₁₆-in. iron, each equal in length to the combined length and width of the box, and with these rods wire the ends of the box and sink them in the crease iron. Set the compasses to ⅜ in. and mark along each

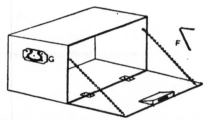

Fig. 894.—Deed Box with Fall-down Front.

wired edge; bend them in the angle bender or over a hatchet stake and beat them down on the crease iron, having the creased wired edge face downwards in an appropriate groove of the tool. Now withdraw the rods, and wire the sides in a similar manner, and leave the projecting ends to stand at equal distances from each end of the feathered edge. Set off the edges for the corner seams in the usual manner, and bend the projecting rods of iron at right angles to the front. Push the wired edges of the ends of the box over the projecting rods of one side, and beat up the seams on the end of the box. Then push the projecting rods of the other side into the wired edges of the ends of the box and beat up the other two seams; see that the bottom is true across, and set off a ¹⁄₁₆ in. edge. The bottom is cut ⅜ in. longer and wider than the outside measurements of

the bottom edges of the box, and is edged, paned on, and worked up in the usual manner.

DEED BOX LID OR COVER.

If the box, when completed, is to be as

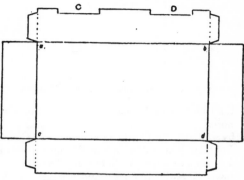

Fig. 895.—Deed Box Lid Pattern.

represented by Fig. 893, a cover will be necessary, a pattern for which is shown by Fig. 895, where $a\,b\,c\,d$ is a plan of the top of the box. The depth of the cover is allowed outside this rectangle, plus working edges, and it should be notched as indicated; c and D are to receive the hinges. Bend to shape, beat the lap edges over the corners as shown by dotted lines, solder the corners, and wire the edges with ¹⁄₁₆-in. rod iron. Two straps of iron bent over the exposed rod at c and D, when sunk in the crease iron, form the hinges.

FIXING DEED BOX LOCK.

Fig. 896 is a pattern for the support of

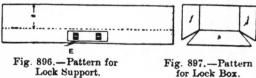

Fig. 896.—Pattern for Lock Support. Fig. 897.—Pattern for Lock Box.

the top of the lock, which is soldered inside the front of the lid. The depth e is equal to the difference between the depth of the lid and the depth of the feather-edge of the box, including the thickness of the lock-plate E. This plate is not fixed yet; for the present, bend the pattern along the dotted lines to a right angle and solder it in position. The lid can then be fixed by soldering the hinges to the box outside, after which the hinges are bent over the top of the box inside and soldered there also. Fig. 897 is a pattern for the lock box which is to carry the lock. Bend the three sides *f*, *h*, and *j* to shape, solder the corners from the inside, and solder the lock in.

the top of the box, and the pin of the lock in the centre of the hole. Get a strip of brown paper to cover the top of the lock, and push the lock-plate in position. This will prevent solder falling into the lock, and will also prevent the lock, plate, and lid being all soldered together. With a hot bit melt a small quantity of solder on the plate and inside the lid where it is to be fixed; alternately heat one and then the other, and instantly, when the solder is running on each, close the lid and turn the

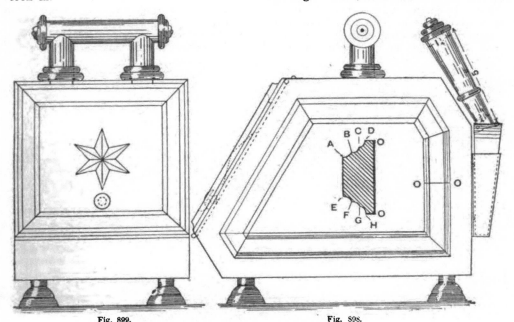

Fig. 899.　　　　　　Fig. 898.

Figs. 898 and 899.—Side and Front Elevations of Coal Vase.

KEYHOLE, ETC., OF DEED BOX.

To find the position of the keyhole, mark a line down the centre of the front of the box, hold the lock with its top flush with the top of the box inside, and press it to the box, when the pin of the lock will dent the front of the box and thus indicate where the keyhole is to be cut. Punch a hole a little larger than the barrel of the key, where the dent is, and a smaller hole beneath it, and finish with a small chisel and warding file. The lock can now be soldered in position, with its top level with

key. If the work has been properly done, the lock will work smoothly; when cold, unlock, and trim off the surplus solder inside. A pair of box handles should be riveted to the ends as at A (Fig. 893), and at B is soldered a small crescent-shaped clip with which to raise the lid. If required, two small pieces of chain are soldered inside to the lid and the box to prevent the lid falling back.

FALL-DOWN FRONT DEED BOX.

The box shown by Fig. 894 has a fall-down front, and is made somewhat differently. The sides and ends are cut the same length as for the other box, but only ¾ in. deeper, instead of 1¼ in. In this case, the edges are simply wired, and the box is put together in the usual way, except that the wired edge projects inside the box instead of outside. Bend four strips of iron to the shape of F (Fig. 894), two equal in length to the side of the box, and two to the end, and solder them inside the box close up to the wired edges, so that a projection will be formed on which the lid may rest. The lid is flat in this instance, and is wired with ⅟₁₆-in. rod-iron, and should then fit the wired opening of

COAL VASE.

The coal vase shown by the side elevation (Fig. 898) and front elevation (Fig. 899) may be made of sheet brass with copper moulding at the sides and door. The pattern for the body would be a rectangle 30 in. by 8¼ in., with allowances for wiring at the ends, and for an edge along each side. Wire the edge forming the opening at the bottom of the door along the outside, and the end for the top of the opening along the inside. Fold up the long edges at right angles to the outside of the body, and then bend the body to sharp angles at distances to suit the side pattern (Fig. 900). Cut out this pattern, making

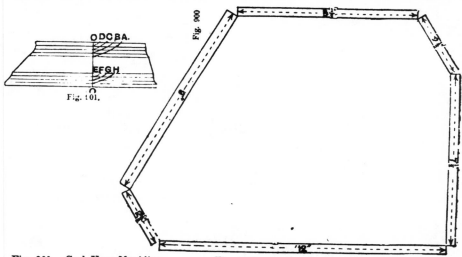

Fig. 901.—Coal Vase Moulding.   Fig. 900.—Pattern for Coal Vase Side.

the box not too tightly, and so rest on the projection fixed for it. It will be better for the box to carry the hinges in a box of this description, which should be riveted as well as soldered to the lid. The handles are fixed at G (Fig. 894), and two chains are soldered to the box and lid to carry the latter and any weight which may be temporarily placed on it. The lock (in this case a cupboard lock) is fixed and a slot is cut in the projecting ledge inside the box to allow the bolt of the lock to shoot home. Clean off with whiting and turps, polish with dry whiting, and give two coats of good black paint.

allowances for a double edge upon each side except the front, where the allowance should be for wire. Notch the side as shown, and fold the edges over along all the sides to fit the square edge on the body. Wire along the front inside.

MOULDING FOR COAL VASE.

Now prepare the moulding for the side, as in Fig. 898. Draw the outside lines of the moulding of width O O (Fig. 898). Join the inner and outer angles by straight lines, as shown on Fig. 898, to obtain the mitre lines. Next draw a section of the moulding, as shown. Divide the curved part of the section into any convenient number of equal parts, and through the division points draw lines parallel to the side of the moulding to cut the mitre lines; from the points along the back mitre, draw lines parallel to the side of the back

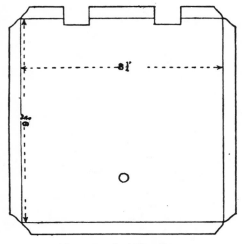

Fig. 902.—Coal Vase Door.

moulding, and continue these by the same method along the remaining mouldings. To work the pattern for the moulding, upon any straight line o o (Fig. 901) mark off divisions o D C B A, E F G H o corresponding to those on the section (Fig. 898). Draw parallel lines through these points, as shown, producing them at each side. Now transfer the lengths from the line o o (Fig. 898) to the top mitre line on the back piece of moulding, to the lines on the pattern with corresponding letters. Take the lengths from o o (Fig. 898) to the lower mitre line, and transfer these to the pattern, marking them off from the line o o (Fig. 901) on the lines with corresponding letters. Join the points on each end, as shown, to complete the pattern for the back section of the moulding. The patterns for the remaining pieces are worked in the same way.

COAL VASE DOOR.

When making the door, cut the pattern (Fig. 902), and wire it with stout wire, so

that it will not easily wear through when plated at the spaces left for the hinge. The moulding on the door is of the same section as that on the sides, and the patterns are worked in the same way. To work the mouldings to the section, fold the sides up at right angles along the lines A E (Fig. 901); then, with the dresser used upon a piece of round bar-iron, work the curved part of the moulding outwards. Now set the outside edge of the curved part down square upon the hatchet stake to finish the moulding. With a smooth file touch up the ends of each piece, and fit them accurately. Place the set of mouldings in position with the inside upwards, and solder them along the mitre lines on the inside. The moulding can be soldered to the sides in position, but, if the bit is very hot, the side of the vase will probably buckle, thus spoiling the appearance of the vase. Now pane down the sides upon the body, and knock them up upon a square head fitted in a tinman's horse, and an upright square-headed stake.

DOOR HINGE AND ORNAMENTAL STAR.

The hinge for the door is cut from stout copper (see Fig. 903). Pass its square end through the space left after wiring at the

Fig. 903.—Coal Vase Door Hinge.

Fig. 904.—Ornamental Star.

top of the door, bend it over the wire, and close it down neatly with the edge of a square-faced hammer. The copper star (Fig. 904) for the centre of the door is bent downwards upon the hatchet stake on each line drawn from the outer angles to the centre. Now alternately place each line from the inner angles to the centre on the hatchet stake, and bend the metal outwards; then with a small bright hammer work the lines up smooth and sharp. Fix the star and the cast knob on the door by soldering, or by any other convenient method.

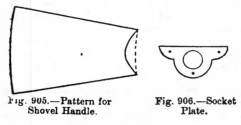

Fig. 905.—Pattern for Shovel Handle.

Fig. 906.—Socket Plate.

COAL VASE HANDLES.

The handles of the coal vase are of brass tube cut to mitre with the cross tube ; they may be brazed or soft-soldered together, and then brass wire rings soldered around the ends. The feet are small hollowed discs of brass soldered together, a flat disc being soldered in the bottom one ; or the latter may be wired round the edge. A small cylinder is soldered on the top of the smaller hollowed disc and the feet soldered in position.

socket plate (Fig. 906). The large end of the handle has wire rings, as shown in Fig. 898. To work the shovel pattern (Fig. 907) to shape, first bend the sides up square along the inner lines, then with a small stud hammer pitch up the curved back

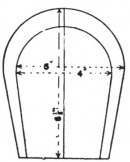

Fig. 907.—Shovel Pattern.

upon a hollowing block, working it up as nearly as possible square. Now with a wedge-shaped mallet work round the

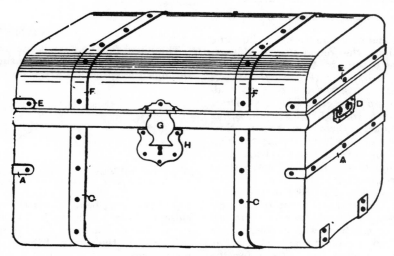

Fig. 908.—Travelling Trunk

SHOVEL FOR COAL VASE.

The pattern for the shovel handle (Fig. 905) is a frustum of a right cone with some material pared away at the bottom to tilt the handle when brazed in the hole in the

shoulder of the hollowed part upon the end of a mandrel, commencing at the shoulder and working round and outwards until the curved back is worked up straight ; then smooth it with a bright, flat-faced hammer.

Bend the socket plate to fit the shovel, and rivet it in position.

COMPLETING COAL VASE.

Make a box to carry the shovel as shown on Fig. 898, and rivet this to the back. Punch holes in the front of the top of the vase to correspond to those in the hinges, and fasten the latter by small nuts and bolts. The top handle could be attached in a similar way, or by soldering. All solder showing should be cleaned off by

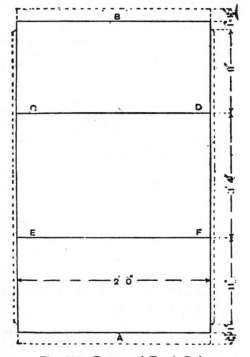

Fig. 909.—Pattern of Trunk Body.

filing or scraping, finishing with fine emery cloth, the vase then being scoured free from all scratches with a fine emery and oil paste, and afterwards polished with a clean, dry cloth. If a polishing machine is available, the spare solder and scratches could be removed by sanding, and the vase then polished with lime and a rag mop.

TRAVELLING TRUNK.

The travelling trunk illustrated by Fig. 908 can be made of No. 20 B.W.G. terne iron. It is 2 ft. long, 1 ft. 4 in. wide, and 1 ft. 5 in. deep, including the cover. First make the body of the trunk, which is in three pieces.

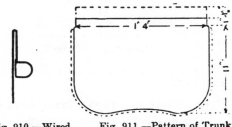

Fig. 910.—Wired   Fig. 911.—Pattern of Trunk
Feather Edge.   End.

In trunks of this description the sides and bottom are usually in one piece, which is set out as shown by Fig. 909. The small edges along the sides, each $\frac{3}{16}$ in., shown by dotted lines, are for the seams. The allowances A and B (Fig. 909) are for wiring and feather edges. Having cut out the pattern, roll it well, and wire with $\frac{1}{4}$-in. rod-iron along the dotted lines A and B, leaving at each end a length of wire equal to half the width of the trunk. Set compasses to 1 in., and mark along the edges just wired. Bend over a sharp stake and beat down to the body; then sink in a crease-iron so that in section it will be as shown by Fig. 910. The body can now be shaped by bending along the lines C D and E F over a round mandrel until the sides are at right angles to the bottom. The last can be arched over a large half-round stake. A $\frac{3}{16}$ in. edge, for the end seam, is now set off at each end outwardly and at right angles to the body. Up-end the body on a piece of iron, mark round, and allow $\frac{1}{16}$ in. extra all round, and sufficient for wiring and a feather edge at the top. Now cut out the pattern for the end (see Fig. 911); then mark and cut out for the other end. The wired feather edges are made as previously described, with the exception that in this case the wire is withdrawn. If the wire is rubbed over with blacklead previous to wiring, it can easily be withdrawn. The $\frac{3}{16}$-in. edges are next set off at right angles to the ends. The projecting wires of the sides are beaten down at right angles and pushed into the empty wired edges of the ends, after which

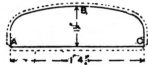

Fig. 912.—Pattern of Cover End.

the seam edges are paned together and beaten up.

TRUNK ANGLE PLATES AND OTHER DETAILS.

Four angle plates for protecting the bottom of the trunk should now be made of 1¼-in. by $\frac{3}{16}$-in. flat iron. Make them 3 in. each way, drill two holes in each half, and rivet them to the sides and bottom. Two lengths of 1¼-in. by ⅛-in. iron, cut 3 in. longer than the width of the trunk, are bent at each end, drilled, and riveted to the sides and ends of the trunk, as shown at A (Fig. 908). Two bands of the same iron, equal in length to the combined depth of each side and the width of the bottom of the trunk, are also fixed at c

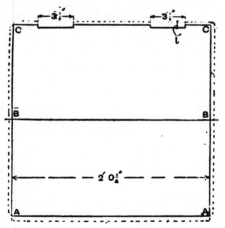

Fig. 913.—Pattern of Trunk Cover.

(Fig. 908). A pair of fall-down plate handles are riveted to the ends at D (Fig. 908). The bead on which the cover rests when closed can be made larger if desired,

by bending lengths of iron to the required size in a semicircular crease or to a square section, in which case the corners should be cut to the correct angle, and strongly soldered in position over the wired edge. This completes the body of the trunk, but the lid remains to be made.

TRUNK LID OR COVER.

The cover or lid of the trunk can be made in three pieces, the top and sides in one piece, and the two ends separate. First cut two ends as in Fig. 912, allowing ⅜ in. for seaming around A B C, and ⅝ in. for wiring along the straight edge at the bottom. Set off at right angles a $\frac{3}{16}$-in. edge around A B C, and set off in an opposite direction the ⅝-in. edge along the straight edge. Fig. 913 shows the pattern for the top and sides. Draw line B B, and

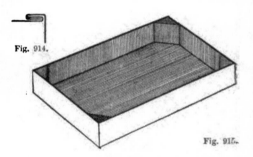

Fig. 914.—Section of Seam.
Fig. 915.—Rectangular Drainer.

at right angles each way set off lines A C equal in length to the distances A to B and B to C in Fig. 912. All other dimensions are shown in Fig. 913. Allow $\frac{3}{16}$ in. extra for seaming along each length A B C, and ⅝ in. extra for wiring along A A and C C, as shown in Fig. 913. The notches are for the hinges. Roll the pattern to take out creases, and bend it to shape. Set off the edges for the seams previously described for the body, and beat the ends on. Set off the wiring edges and wire with ¼-in. iron rod, which should be in two pieces, with the joints meeting in the centre of each end. Fig. 914 shows a section of all the seams.

COMPLETING TRAVELLING TRUNK.

For the hinges, cut two pieces of iron 6 in. long and equal in width to the bare length of the notches. Bend these over the wire which is exposed, and sink them in the crease iron. Two bands of iron of the same strength as those of the body are cut, bent, drilled, and fixed to the ends and sides of the cover as at E (Fig. 908). The bands F, of the same iron, are also fixed. The cover is then dropped on to the body, the hinges being inside. They are soldered to the body on the outside, and then bent over the top and strongly soldered on the inside. The lock G and H is locked and laid in position in the centre of the trunk. The part G is cranked, if necessary, to allow it to rest on the moulded bead. The holes should then be carefully marked and punched, after which the lock can be screwed in position. When this has been done the trunk can finally be cleaned and painted to the desired colour.

Fig. 916.—Tray of Drainer.

RECTANGULAR DRAINER.

Sheet metal drainers, for washed glasses, etc., are largely used by publicans, and to prevent waste in retailing oil or other liquids, by chemists, oilmen, and others. Fig. 915 represents an ordinary rectangular drainer, which can be made of zinc, tinplate, or tinned iron. It consists practically of a shallow tray, with four corner supports, and a perforated top. A pattern for the tray is shown by Fig. 916, where A B represents the length of the drainer, B C the width, and C D the depth. Allowance for wiring is not shown, but is additional. Notch the corners as indicated, and bend the sides and ends along the

dotted lines at right angles to the bottom, and beat over the corner laps a, b, c, and d. Now solder at the corners, and set off the wiring edge, and then wire it with say No. 8 or No. 10 B.W.G. wire.

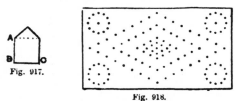

Fig. 917.

Fig. 918.

Fig. 917.—Corner Support of Drainer.
Fig. 918.—Perforated Drainer Top.

CORNER SUPPORTS OF DRAINER.

A pattern for the four corner supports is shown at Fig. 917, where A B represents the depth, and B C the width. The depth should be at least ¼ in. less than the finished depth of the tray, to allow the perforated top to fit in position. Bend the pattern along the dotted line in Fig. 917, until the top is at right angles to the front, and solder the pieces in the corners of the drainer.

PERFORATED TOP OF DRAINER.

For the perforated top (Fig. 918) the design is first set out, and then the holes are made with a ⅛-in. punch on a hard block of wood, and the burrs are flattened down with a mallet. This flattening process will probably cause the top to buckle, but judicious planishing on a bright anvil will soon straighten it. The top must now be strengthened by wiring, though a better finish and neater appearance are obtained by soldering on ⅜-in. beading, which is cut at the corners to form mitre joints. The top can then be freely put on or taken off.

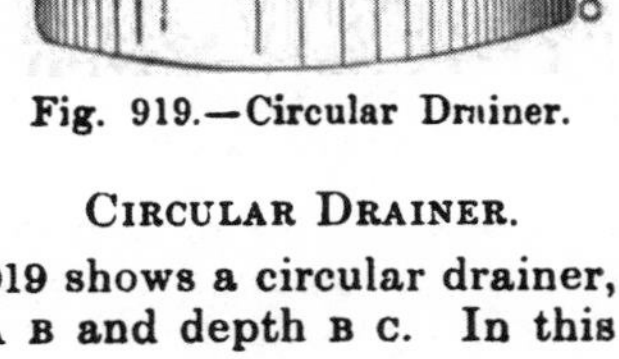

Fig. 919.—Circular Drainer.

CIRCULAR DRAINER.

Fig. 919 shows a circular drainer, of diameter A B and depth B C. In this case a strip of metal is cut equal in length to the circumference of the drainer plus lap for the seam, and equal in width to the depth plus the wiring edge. This is wired, bent

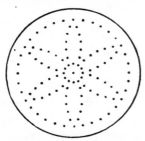

Fig. 920.—Perforated Drainer Top.

to shape, and soldered together. A disc of metal is cut so that when it is edged it tightly fits the rim to which it is soldered. Three supports will be required, and they must be bent to fit the circular body to which they are soldered at equal distances from one another. The perforated top (Fig. 920) is made like the other top, except that beading the edge is accomplished with one length of beading, necessitating only one joint. However, this is far more difficult than beading the rectangular top. Of

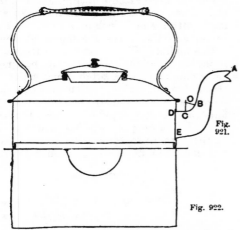

Figs. 921 and 922.—Elevation and Half-plan of Square Kettle.

course, other shapes of drainer could be suggested, but the two here described are those generally used, and other shapes are

modifications of them. Sheet zinc is recommended where water is to be used, but for oil, tin-plate will be suitable.

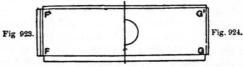

Figs. 923 and 924.—Half-patterns for Side and End of Kettle.

SQUARE COPPER KETTLE.

If the body of a square kettle (Figs. 921 and 922) is to measure, when finished, 9 in. by 9 in. by 3⅝ in. deep, the patterns for the sides and ends would be rectangles, those for the ends measuring 9½ in. by 4 in., and those for the sides 9¼ in. by 4 in., the last pieces being notched for box seams to be formed at the corners, as shown by Figs. 923 and 924, the former representing half the side pattern, and the latter half the end pattern. The allowance along the top edge is for folding up square, so that the top may be paned down upon it, and that along the bottom of the patterns is for folding an edge over into which the edge of the bottom is fitted, as shown at the bottom of Fig. 921. This method affords better protection from wear than when the bottom is joined to the body by a knocked-up seam. Each notch at F and F' (Fig. 923) equals two folds, and each at G and G' (Fig. 924), one fold only. A hole should also be punched in the front end of the kettle through which the spout is inserted.

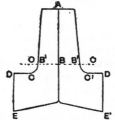

Fig. 925.—Spout Pattern.

PATTERN FOR KETTLE SPOUT.

When working the spout pattern from the elevation (Fig. 921), first draw two lines at right angles to each other at B

(Fig. 925). Make the top of the spout pattern along the line at A equal to the circumference of the spout at A (Fig. 921), and make A B (Fig. 925) equal to the length of the curve A B (Fig. 921). Through the

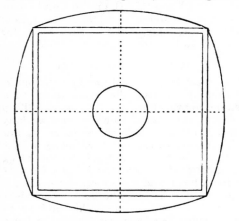

Fig. 926.—Pattern for Kettle Top.

point B (Fig. 925) draw a line at right angles to A B and make the distance from B' to B" equal to the circumference at B (Fig. 921). Set off the radius O B (Fig. 921) on the pattern as B' O, B" O (Fig. 925), and with the points O as centres and O B' as radius, draw arcs of circles. Then transfer the length B C (Fig. 921), and set it off on the pattern as B' C' B" C. From the points C and C', draw lines parallel to B' B" and make the distances C D and C' D', equal to C D (Fig. 921). From D and D' on the pattern, draw lines at right angles to C D, and make D E and D' E' equal to the semicircumference of the spout at D E (Fig. 921). From E to the centre of the pattern draw

Fig. 927.—Pattern for Kettle Lid or Cover.

lines, which are inclined at the same angle to D E as the base of the spout makes with D E (Fig. 921). A small notch cut where

these lines join the centre line completes the pattern.

TOP OF SQUARE KETTLE.

To draw the pattern for the top, first draw a square of 10½ in. side, and add to

Fig. 928.—Rim of Kettle Lid or Cover.

the square curved allowances for hollowing, as shown by Fig. 926. A circle 4 in. in diameter for the cover hole should also be marked at the centre of the pattern. Fig. 927 is the cover pattern with an allowance for hollowing and edging, and Fig. 928 is a frustum of a cone, which forms the cover rim, the dotted line being the allowance for an edge.

KETTLE HANDLE.

The uprights for the handle may be cut from stout copper band to the shape shown by Fig. 929. The holes for the rivets should then be drilled, and also one at the top of each upright through which the spindle could pass to the opposite end. The spindle carries the wood handle, the uprights also being bent to the shape shown by Fig. 921. Instead of making the uprights as described, a wood pattern could be prepared for use in making suitable copper castings.

Fig. 929.—Kettle Handle Upright.

MAKING KETTLE SPOUT.

To make the kettle spout, begin by thinning the long edges by hammering; then with a stretching hammer, used upon a beak iron, set inwards the two arcs of circles forming the throat. Bend the spout round upon the same tool along the edges B' and B'' (Fig. 925) and work it round until the edges overlap. Fasten some binding wire round the spout at the top and bottom, arrange some spelter and borax along the inside of the seam, and braze it. Then work over the underneath part of the spout until the edges E and E' overlap, and then braze this seam. Now load the spout with lead, and bend the neck at the top of the spout by working it round with a round-faced hammer used on the hollowing block. While the spout is still loaded, file off all spare spelter that may have run through the seams, and with a small hammer work the metal smooth and the spout to the shape required when finished. The lead is then run out from the spout, a small edge or seating is thrown off round the larger end, and the top is cut with the snips to shape. Then the spout is tinned inside.

Fig. 930.　　　　Fig. 931　　　　Fig. 932.

Fig. 930 to 932.—Sections of Seams.

MAKING SQUARE KETTLE BODY.

When making the body, fold up the edges along the top and ends square on the front and back, as shown at G (Fig. 930), and fold the edge along the bottom right over upon the opposite side or inside the body, as shown at the base of Fig. 921. Fold the long edge at the top of the sides square and outwards, and the bottom edge as for the front. The ends of these pieces are folded over inwards, as shown at F (Fig. 930). To form the seam at the corners, hook the end of a side over the end of the front, as shown by Fig. 931, place the two pieces upon any flat-faced tool which has a straight edge, and then work until a seam is formed, as shown by Fig. 932, this method of working being repeated for the remaining three corners. The bottom pattern is a square $9\frac{1}{2}$ in. by

$9\frac{1}{2}$ in., notched to allow for a $\frac{1}{4}$-in. edge being turned up square on each side ; when this is done, the bottom is placed edge downwards inside the body, and pushed down into the bottom folds until it rests as shown by Fig. 921. The bottom and seams are then soldered down and along each side of the body, which is then ready for the top to be paned on to it.

HOLLOWING KETTLE TOP.

The kettle top is hollowed on the block. Begin by working the metal down to a slight depth along the four sides. Then take a bullet-faced hammer, and working in a rather deep hole on the block, hollow the corners to the depth required for the top. Then place a side of the top along an edge of the block, and using a hammer with a large and rather flat face, hit the side until it is brought flat and straight along the edge. Repeat this process on each side. Again using the bullet-faced hammer, work along the sharp shoulder formed by forcing the sides down, and work the metal upon a flat place on the block until the curve of the shoulder of the top is of equal height all the way round it ; then carefully work over the whole of the hollowed surface until it is rendered smooth.

COMPLETING KETTLE TOP.

If the four edges of the top do not rest quite flat, they should be pared true with the snips before the next operation. This consists in marking the top along each edge with an edging machine, the distance from the edge of the top to where it is marked being made equal to the amount required for the flange, usually about $\frac{1}{2}$ in. Now with the top upon a flat surface, place the edge of a hatchet stake on the crease formed by the edging machine along one side, and drive down the tool with the mallet until the flange is set down square ; repeat this operation on the remaining sides. An edge is next taken up on each side, so that the top will fit closely over the edge at the top of the body, and this edge may be paned down upon the body edge to form a seam, as shown where the top joins the body in Fig. 921.

KETTLE LID OR COVER.

The small cover is made by first turning the frustum of a cone forming the cover rim (Fig. 928), and fitting it rather tightly to the wired rim fixed in the top, and soldering the overlapping edges. A small flange is then thrown off round the top edge. The circle (Fig. 927) is then hollowed slightly all over, an edge on it being taken up so that it will fit over the flange of the rim; this edge is closed down to form the seam shown in section by Fig. 921. The knob is soldered in position from the inside of the cover, and a little hollowed circular stud is then placed over the patch of solder, and very neatly soldered to the inside of the cover; or the knob may be fixed with a small nut screwed on the threaded shank of the knob, in which case a stud would not be necessary.

COMPLETING SQUARE KETTLE.

Before the top is paned down as described in the paragraph before the last, the hole for the cover is cut out with circular snips, and a narrow wired rim of the same diameter as the hole is inserted in it, its lower edge being worked over to the inside of the top with a round-faced hammer. Rivet the handle upon the top, solder over the rivet heads on the inside, and place the top in position on the body, and then pane it down smoothly. It is then soldered along each edge to render it sound. Now from the inside of the kettle push the spout through the hole punched for it; the larger end butts close against the front of the kettle; then solder it strongly there from the inside.

REMOVING DENT FROM KETTLE.

To remove a dent from a kettle insert the head of a small round-faced hammer through the cover hole, and knock the dent outwards; then hold the face of the hammer up against the bruised part, and go over the outside lightly with a flat-faced bright hammer until the metal is quite smooth.

SAUCEPAN AND KETTLE LIDS.

Saucepan and kettle lids are made either by stamping, or by hollowing on a wooden block by hand. By the first method, metal of the proper size and shape is placed upon a die, and a casting of similar shape to the die is raised between parallel guide bars and allowed to fall upon the disc. This operation is repeated until the metal assumes the desired shape. The second method is to place the metal on a block, and then to hollow it out by means of a series of blows from a round-faced hammer; the blows are delivered regularly round the edge first, and then in a series of concentric circles as far in towards the centre as may be desired. The cover is smoothed by lighter blows delivered over the surface with regularity, or by a series of radial strokes upon a planishing wheel.

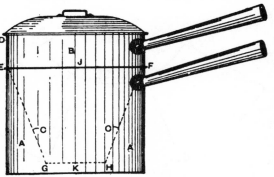

Fig. 933. —Porringer or Milk Saucepan.

When hollowing ovals, as with a kettle lid, the sides of the oval do not require so much hammering as the ends; both round and oval discs are usually worked in " tacks " of four or six discs, according to the strength of the metal used.

PORRINGER OR MILK SAUCEPAN.

A porringer or milk saucepan in which milk, porridge, etc., may be cooked without fear of burning is illustrated by Fig. 933. It consists of an outer and an inner saucepan, the outer saucepan containing water, which, when boiling, cooks the food in the inner saucepan. It can be made of tin-plate, and the pattern for the outer saucepan A is a rectangular-shaped piece of metal, with a length equal to its circumference, the width being equal to the proposed depth, plus working edges. This is edged for a grooved seam, wired, turned to shape, and grooved, and the bottom is beaten up and a handle is riveted on as illustrated.

THE INNER SAUCEPAN.

The body of the inner saucepan is made of two pieces, the top B and bottom part C (shown dotted). The part B has also a rectangular-shaped pattern, having a length equal to that of the outer saucepan, but with a width equal to D E, plus the working edges. This is made without a bottom, but it has a creased edge set off the bottom edge to take the top of C, the

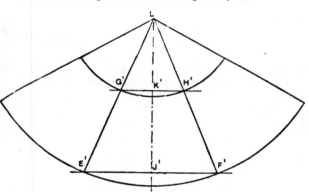

Fig. 934.—Bottom Part of Inner Saucepan.

other portion of the body. The method of setting out a pattern for C is shown in Fig. 934, where E′ F′ is equal to E F (Fig. 933), whilst G′ H′ (Fig. 934) and J′ K′ are respectively equal to G H (Fig. 933) and J K. Lines are now drawn from E′ (Fig. 934) to G′ and from F′ to H′ and produced until they meet at at L. With L as centre and radii respectively equal to L E′ and L G′ (Fig. 934), draw arcs of circles. The larger arc is made equal to 3½ times E F (Fig. 933), and then joined to the centre L with straight lines, thus giving the pattern required. Working edges are additional, and must be allowed.

COMPLETING THE PORRINGER.

This pattern, after being edged for a

grooved seam, is bent to shape and grooved together. An edge is set off the wider end so as to fit tight in the creased edge on B (Fig. 933), after which it is paned together, and a bottom beaten on the smaller end of C. The handle is riveted in position as before, and the soldering is now done on

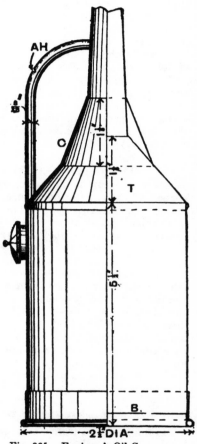

Fig. 935.—Engineer's Oil Can.

the inside, resin being the flux. A lid is now required, and the rim is made to fit each saucepan separately, so that for convenience one lid answers for either saucepan. A hollowed disc is next creased and paned on the rim, after which a small handle is riveted in the centre, as shown in Fig. 933.

ENGINEER'S OIL-CAN.

An engineer's oil-can, in its general form, is represented by Fig. 935. The body is about 5½ in. deep and 2½ in. in diameter. The spout may be from 6 in. to 10 in. long, tapering from ½ in. to ⅟₁₆ in. The body should be cut about 8 in. by 5½ in., the hole for the brass feeder screw being punched in the flat. When rounding the bodies— which may be done on a smooth piece of

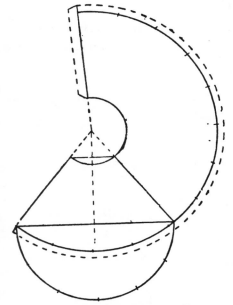

Fig. 936.—Pattern for Oil-can Top.

2-in. shafting—turn them first one way, then the other. This is termed "breaking them in," the object being to avoid the appearance of unsightly ridges on the body when it is soldered. Sometimes the body is wired round the bottom; when this is the case, it is done before turning. The usual way is to wire a separate piece, about ¾ in. deep, and solder it round the bottom, with the wire inside, after the bottom is put in the body. Fig. 935, however, shows the piece wired on the outside. In laying the seam, allow ¼ in. for lap; solder neatly on the outside. The top is set as shown by Fig. 936, a semi-

circle been drawn on the base. Divide this into six parts; when the outer curve of the pattern is drawn, set off twelve of these divisions round it, and it will fit the body. Cut it out, round it up, solder the seam, and edge it all round for soldering to the body. Mark out the boss in the same manner as the top. Let the spout be 1½ in. at one end and about ⅜ in. at the other. To round it up, slightly bend both edges to give it a start, and finish on a steel spindle. Cut out the bottom, edge it on the shaft, and solder it in. Cut out the piece for placing round the bottom, and wire it with 12-gauge wire. Round it up and solder it on. Solder the screw in and the top on. Drive a hole in near the edge for the brass tube. Solder on the spout, then slip the boss into position and solder it. Make a hole in the spout for the brass tube, which can also be soldered. Bend the end of the spout on the head of a mallet by gently taping with a suitable hammer. The letter references in Fig. 935 are explained thus: A H, air-hole; B, bottom; C, boss; and T, top.

SHOWER BATH.

A shower bath, which can be fixed in the corner of a room, is here described. It can be made portable if desired, and can easily be converted to a fixture. The cistern should stand on a wooden platform fixed to the walls of the room. If practicable, this should be fixed at least 7 ft. high from the floor. A cubic foot of water (about 6 gal.) is sufficient for a good shower. The proportions of the cistern can therefore be suited to requirements as long as it contains a cubic foot of water. Make the cistern of galvanised iron with the edges at the top wired, have a grooved or riveted seam, and knock the bottom up. If a square or triangular cistern is preferred, make the body in one piece with the seam at one of the corners. A 2-in. valve, with a nipple end, should then be soldered in the bottom of the cistern with the nipple end projecting downwards. The top of the valve inside the cistern is afterwards connected to a lever and chain so as to work on the same principle as a w.c. cistern.

SHOWER BATH ROSE.

Fig. 937 shows the rose which makes the shower, and which is screwed on the nipple end of the valve; it should be of zinc or

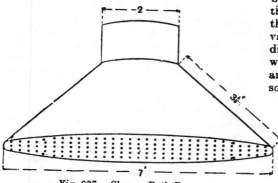

Fig. 937.—Shower Bath Rose.

tinned brass. Cut the pattern (Fig. 938), and bend it to shape over a funnel stake. It is then soldered and beaten smooth and round over the same stake. Cut a disc $\frac{1}{2}$ in. larger in diameter than the largest diameter of the body just described, and hollow it slightly on a hollowing block. Crease the edge so as to fit the body

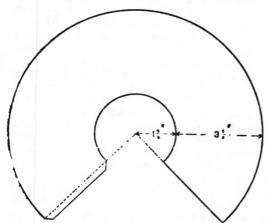

Fig. 938.—Pattern of Shower Rose.

tightly, but, before fixing, mark with the compasses a series of concentric circles, starting about $\frac{3}{4}$ in. from the edge. A number of holes should now be made with

a very fine punch round these circles at intervals of about $\frac{1}{4}$ in. This can be done over a smooth block of lead or a block of hard wood. Knock down the burrs on the block used for hollowing, fix the disc to the body, and solder round. A collar threaded inside to fit the nipple end of the valve must now be fixed to the smaller diameter of the rose. This can be done by working a small edge on the rose piece and pushing on the collar, afterwards soldering it in position.

COMPLETING SHOWER BATH.

A $2\frac{1}{2}$-in. hole is bored through the cistern platform to allow the nipple of the valve

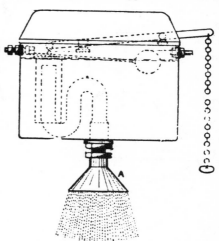

Fig. 939.—Cistern and Shower Rose.

to project. Fix the cistern in position, and screw the rose on from underneath (see Fig. 939). The person stands in a shallow pan, large enough to catch the full shower. If the shower bath complete is to be a fixture, a waste pipe should be connected to the standing pan and conveyed outside. A ball tap should be connected to the top of the cistern, to which a $\frac{1}{2}$-in. water service is conveyed, and an overflow pipe may be connected to the cistern and conveyed to the waste pipe of the standing pan. A good coat of red-lead paint will preserve the cistern; the rose and the pan should be treated with bath enamel. A piece of oilcloth fixed to each side of the corner will prevent the walls being splashed. The cistern should be cased in, and, if desired, a curved rod of iron can be fixed to the casing and two sliding curtains suspended.

ends round to the same curve as the hollowed part. A few blows from a flat-faced hammer, delivered upon the centre or flat part of the blower, may be necessary to set it so that it will be free from twist.

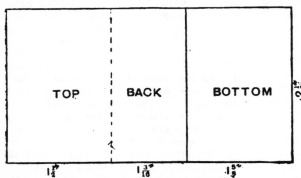

Fig. 940.—Pattern of Tobacco Box Body.

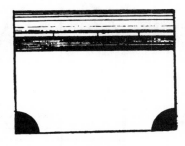

Fig. 942.—Plan of Tobacco Box.

BLOWERS FOR REGISTER STOVES.

In making a blower for a register stove, begin hollowing it by working round the circular part with a series of regular blows from a bullet-faced hammer, and holding the edge of the metal over a shallow hole in the hollowing block. This will curve the metal to a slight depth round the edge. Then bend the metal about 1 in. from the edge along the straight part, so that it makes a sharp angle; this will keep the bottom stiff while the remainder of the hollowing is done. Now begin on the circular part again, and work round from the edge in towards the centre, in a series of concentric circles, working it, if necessary, in a deeper hole

BRASS TOBACCO BOX.

A piece of No. 22 B.W.G. brass, about 6 in. long and 4 in. wide, will be sufficient to make an oblong tobacco box. It will require tinning inside, as tobacco would soon turn the surface of the brass green. Mark the body out as shown in Fig. 940. Cut down the dotted line: the smaller piece is for the lid; the other piece is for the bottom and back, and should be bent by means of a ¾-in. round iron rod to the shape shown in Fig. 941. Next cut two pieces for the ends (Fig. 941) and one for the front (Fig. 942, which is a top view). Bend the front piece as shown in Fig. 941, and then solder these three pieces together inside. Now make the hinge as shown in Fig. 943. This should be of thin sheet

Fig. 941.—End Elevation of Tobacco Box.

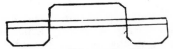

Fig. 943.—Tobacco Box Hinge.

than that used for commencing in. When the blower is hollowed to the depth necessary, go over the hollowed part again with a series of light regular blows until it is rendered smooth. Knock out smooth the break along the bottom, and then bend the

copper, or wire not thicker than 16 B.W.G. The centre piece of the hinge should be soldered to the lid—inside, of course; the end pieces are soldered on the inside to the back. Solder the front and ends to the bottom and back, and the box is ready for the lid. Before fixing this on, put a narrow rim on the underside to fit nicely in the box, and if a slight projection be raised on the outer surface of the front part of the rim, this in turn to fit into a corresponding depression or hollow in the

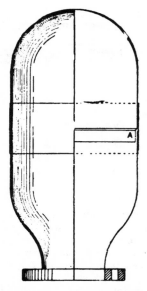

Fig. 944.— Half Elevation and Section of Air Vessel.

front piece of box, a neat fastening will be obtained. Next solder the lid in position. Open the lid, and then solder firmly on the inside. Now solder pieces of thin sheet copper at the corners, and one or two narrow strips along the joint. When finished, clean off all superfluous solder on the outside.

COPPER AIR VESSEL.

When making an air vessel in copper, some difficulty is usually experienced in working in the neck at the base of the vessel, for when the lower part is made from a frustum of a cone with a brazed seam, the brazing is apt to break during the working of the metal. In the method here described the vessel is made in nearly equal halves, both the top and bottom parts being worked from the flat and then joined by the seam at the middle A (Fig.

944) as shown. The top or dome half is simplest to work. To make it, begin by cutting out a circle equal in diameter to the distance round the centre section of the dome, and hollow this roughly to the shape of a half ball. Next insert a round head D (Fig. 945) on the end of the mandrel C, this being fastened by a clamp to the bench, and with a wedge-shaped mallet work on the head a hollow seating B. Continue driving the copper down on the head, hitting the metal on the shoulder as indicated, and working round and outwards towards the edge until the curved side is straight as indicated by the dotted line E. Thoroughly anneal the bowl, and form a second narrow hollow seating as before, and work this down to the edge, repeating the annealing and the working until the dome is shaped as in Fig. 944. Dip the dome in nitric acid and then rinse in water, and, if the metal is free from black spots, planish the cylindrical part with a bright flat-faced hammer on a mandrel. The hollowed part of the dome at the top is also planished on a round head fitted in the top of a perpendicular stake.

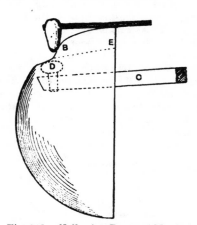

Fig. 945.—Hollowing Dome on Mandrel.

LOWER PART OF AIR VESSEL.

To make the lower part of the air vessel, begin by hollowing a circle of suitable size to as great a depth as possible in the hollowing block; then work the metal in at the middle from a circle equal to that required for fitting the inside of the flange shown at the base of Fig. 944, and gradually tuck the metal in by the method described for working the dome until the

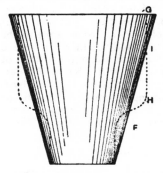

Fig. 946.—Lower Part of Air Vessel.

copper is worked to the cone shape shown by Fig. 946. Then work the copper in at the base up to the point F to form the straight sides for the neck of the vessel, and, commencing at I, work in the top part of the cone until the diameter at G is equal to the diameter of the vessel at the middle.

OTHER PARTS OF AIR VESSEL.

The remaining part of the vessel from I to H is stretched out to the shape shown by the dotted lines, a flat-faced hammer being used on the round head. The whole of the surface, after cleaning, is then hammered over with a bright planishing hammer. The seam at the middle of the vessel may be formed by a knocked-up and grooved seam, by a brazed seam, or by a soft-soldered seam A (Fig. 944), lapping the lower half over the upper half, and fitting tightly over this joint a copper band and sweating the whole together. The flange is then fitted tightly round the neck, and strongly soldered in position. To polish the vessel by hand, form a thin paste of fine emery and oil, scour all marks out of the copper with this, clean off all the oil, dust over with crocus, and finish off bright with a smooth cloth.

TABLE TEA URN IN SHEET METAL.

A tea urn which can be made either of copper or of tin-plate is shown by Fig. 947. If copper is used, it should first be cleaned, tinned, and planished on a bright anvil. It will be seen, on reference to Fig. 947, that the body of the urn is a frustum, or portion of a right cone. To find the volume of a complete right cone, multiply the area of the base by the height (both in inches), and divide by 3; this gives the answer in cubic inches. As only a part or frustum of a right cone is dealt with, the volume of that part of the cone which is cut off must be subtracted. The answer should be divided by 277 to give the result in gallons, since there are approximately 277 cub. in. in 1 gal. of water.

SETTING OUT URN PATTERN.

Having decided on the diameter of the top and bottom, and the height, set out the pattern as in Fig. 948. First draw an indefinite line A B, and at a distance

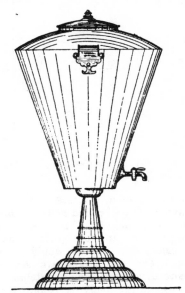

Fig. 947.—Table Tea Urn.

equal to the depth of the urn body, draw another (K L) parallel to A B. From the line A B draw the perpendicular line C D. Along the line A B, set off C H and C I, each equal to half the larger diameter of the urn. Similarly along the line K L set off E F and E G, each equal to half the smaller diameter. Join H F and I G, and continue the lines until they intersect as at D. Then with D as centre and D C as radius, describe an arc of a circle ; and with D as centre and D E as radius, describe another. From C set off along the arc C A' and C B', each equal to three and one-seventh times the distance from C to H. Join A' to D and B' to D, which will give A' B' F' D', the pattern required. The working edges must be arranged to suit the kind of seam intended. If a grooved seam is required, a small edge on each side must be allowed as shown by

TOP OF URN.

The top is a circle of metal cut $1\frac{1}{2}$ in. larger in diameter than the largest diameter of the body. It is then hollowed with a hollowing hammer on a suitable block of wood until its diameter is reduced to $\frac{3}{4}$ in. larger than that of the urn. This allows of a $\frac{3}{16}$-in. edge being taken up to fit the urn body tightly. Before fixing it, a hole equal in diameter to that of the required cover rim should be cut in the centre. The rim can be made by cutting a strip of metal equal in length to the circumference of the hole, plus, say, 1 in. for lap, and equal in width to the required

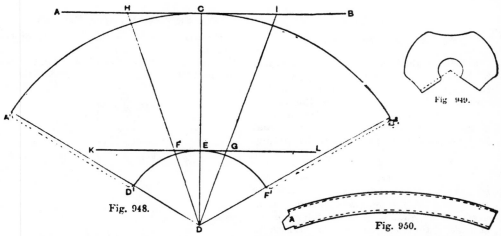

Fig. 949.

Fig. 948.

Fig. 950.

Fig. 948.—Setting Out Tea Urn.   Fig. 949.—Boss for Urn Tap.   Fig. 950.—Pattern for Rim of Urn Lid.

the dotted lines in Fig. 948. The pattern should now be well rolled so as to " break " the metal thoroughly. This, when properly done, obviates any unsightly creases in the finished article which would otherwise appear. The two edges are set off in opposite directions over a hatchet stake, after which the body is rounded either through rollers or over a half-round stake, so that the two edges clip each other. The seam is next grooved either inside or outside. It is better to groove it inside when practicable, as it allows the top and bottom to fit better, in addition to improving the general appearance of the urn.

depth of the rim, plus sufficient for a wiring edge. This is then wired, turned, and swaged, after which it can be fixed in position by beating the swage edge over to the inside of the top, and then soldering. The top can now be fixed on the body and soldered.

HANDLES, TAP, ETC., OF URN.

A pair of suitable handles, either of cast iron or cast brass, are then riveted to the body as shown in Fig. 947 ; a body of solder should be floated over the rivet heads inside. The hole for the tap should be punched about 2 in. from the bottom of the urn body and in the centre of the two handles. A suitable tap is a $\frac{3}{8}$-in. or $\frac{1}{2}$-in. brass or plated bib tap with an ebony lever handle. The tap can now be soldered in, and should be strengthened by soldering a boss from the tap to the urn, as shown in Fig. 947. A pattern of the boss is given by Fig. 949. The bottom is a disc hollowed and edged in the same manner, but not quite so deeply, as the top, and then fitted and soldered.

STAND OR FOOT OF URN.

The stand or foot of the urn consists of a taper tube and a series of hollowed concentric circles, increasing in diameter by steps as shown. The pattern of the tube can be set out as explained for the body. After rounding and seaming it, which should be done over a beak-iron or a taper mandrel, a number of swagings are raised upon it, including one which should be about $\frac{1}{16}$ in. distant from the edge of the largest diameter. This should be let in the smallest hollowed circle, so that the swaging rests upon it. The $\frac{1}{16}$-in. edge should be beaten over and afterwards soldered from underneath. Each succeeding circle must have its inner diameter edged so that the next smaller circle fits tightly over it. After fitting, the pieces should be strongly soldered together from the inside. The stand can then be fixed centrally through a deep hollow boss to the bottom of the urn body with a good sweating of solder. A disc of metal for the base of the stand should be $\frac{3}{8}$ in. larger in diameter, and edged so that it fits tightly inside the lowest hollowed circle. Before this is soldered in position the stand can be loaded with sand if preferred, which will reduce the possibility of its being accidentally upset.

STRAINER OF URN.

If a permanently fixed strainer is required it can be made by cutting an oval-shaped piece of metal, which must afterwards be perforated, hollowed, and soldered over the tap hole inside the urn. A movable strainer can be made by bending a piece of stout perforated metal about 5 in. to a semicircular section. On a sharp-edged stake set off a $\frac{1}{4}$-in. edge outwards on each side, and solder on each end perforated pieces of metal cut to shape and edged to fit the semicircle and the rounded body of the urn, against which they must fit closely. Two small angle bars, of the same length as the strainer, are next soldered inside the urn, so that the strainer slides down freely. A small ring should be soldered on the top of the strainer, so that it can be conveniently handled.

LID OF URN.

To make the lid of the urn, cut a strip of metal to the pattern shown by Fig. 950, bend it round so that it fits the wired cover-rim freely, and mark it. Notch the lap A

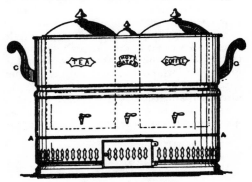

Fig. 951.—Gas-heated Urn.

(Fig. 950) and set off an edge to the bottom dotted line over a hatchet stake. Beat down this edge on the rim, turn this round so that the edge is inside, and solder. An edge represented by the top dotted line is then set off outwardly at right angles to the rim, over which a hollowed disc of metal is paned on. A knob of china or ornamental metal is afterwards screwed or soldered through the centre. If the urn is made of copper, scrape off all superfluous solder, clean with powdered emery and oil, and finish with crocus powder; if it is made of tin-plate, clean it with whiting and turps, afterwards polishing with dry whiting.

GAS-HEATED URN FOR TEA AND COFFEE.

The urn for tea, coffee, and hot water illustrated by Fig. 951 can be made of sheet copper, brass, or tin-plate. The tea and coffee containers are, roughly speaking, two cylinders suspended in the urn proper from the top, while the hot water

Fig. 952.—Dovetail Seam for Brazing.

occupies the remaining space underneath the bottoms and around the sides. The contents of the urn are kept hot by a Bunsen burner, which can be made by bending a piece of $\frac{3}{4}$-in. gas pipe (iron) to the shape of the bottom of the urn, but smaller in length and width by about 2 in. This is perforated at intervals of, say, $\frac{1}{2}$ in. ; one end is plugged up, and an air supply hole is cut underneath, diametrically opposite to the perforations, at about 2 in. from the other end, which is fitted with a gas jet capable of receiving a $\frac{1}{2}$-in. supply. An alternative method is to connect together with a $\frac{1}{2}$-in. brass tee two cast-iron Bunsen gas-rings, but the former method is the cheaper.

MAKING THE URN.

If possible, the urn should be made of No. 24 B.W.G. sheet copper, mounted with brass fittings ; these are brass taps, handles, knobs, and name-plates. To ascertain the length of the body, cut and

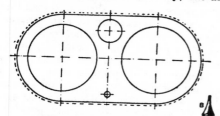

Fig. 953.—Plan of Urn Top.

bend a strip of metal to the plan, and add $\frac{1}{2}$ in. for the lap of the seam. The body of the urn is then cut equal in length to that just obtained, and equal in width to the proposed depth of urn, plus working edges

One end is notched for a dovetail seam for brazing (see Fig. 952), the hole for the door at the bottom and the perforations are cut, and the pattern is then bent to shape and brazed. In preparing the seam for brazing, lap the dovetail notches alternately under and over the straightedge and flatten down. Use borax as a flux and a rather soft spelter. Trim the seam, thoroughly clean the inside of the body, and tin it with block tin, using sal-ammoniac as a flux ; after which it should be planished, if convenient, to give a better finish.

OGEE MOULDINGS ON URN

Two ogee mouldings are now worked on, one at about the centre and one close to the bottom ; these greatly enhance the appearance of the urn, besides considerably strengthening it. A half-round swage is sunk in at A (Fig. 951) between the two other swages where the bottom of the urn

Fig. 954.—Pattern for Urn Door

is to be fixed, thus forming a ledge on which the bottom rests.

TOP OF URN.

Edges are now set off the top and bottom of the body in an outward direction, and are wired with tinned wire. For the top, the pattern is first cut to the dotted lines in Fig. 953, which shows the working edges. The pattern is tinned and planished before the holes are cut ; then it is edged to fit and soldered on. The small hole shown in Fig. 953 is for a steam jet B, which is soldered in position before the top is fixed. This is always open, and allows the steam from the boiling water to escape instead of boiling over at the lid.

URN HANDLES, NAMEPLATES, AND BIB TAPS.

A pair of handles, each consisting of two brass supports with screwed ends and nuts to fit, and an ebony cross-bar, are fixed in position, as at C (Fig. 951), and strongly floated over with solder inside. The nameplates are now cut out of strong sheet brass, and are polished and stamped or

holes are now punched inside each marking, and the plates may then be secured by

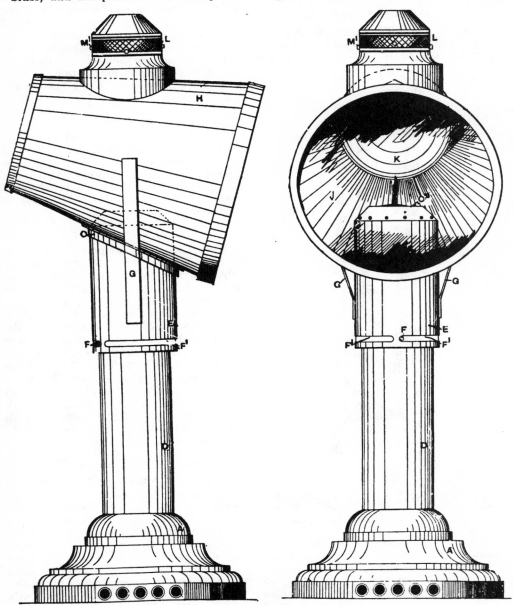

Figs. 955 and 956.—Elevations of Candle Lamp.

engraved. Hold these in position and lightly mark around them. Two or more

soldering from inside. Three holes are cut to suit three $\frac{1}{2}$-in. brass bib taps, which have ebony handles and tinned ends, but these are not fixed yet.

TEA AND COFFEE CONTAINERS.

The tea and coffee containers are now made; patterns are not needed for these, as they are simple rectangles, equal in length

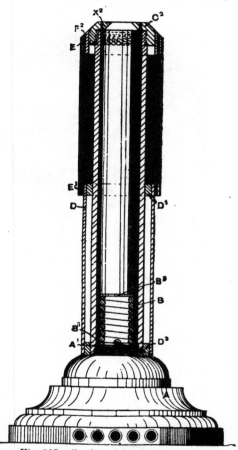

Fig. 957.—Section of Candle Lamp Pillar.

to the circumferences of the holes, plus allowances for seams, and equal in width to the proposed depth, plus allowances for wiring edges and bottom seams. Both sides of each should be tinned, after which they are wired, turned to shape, and seamed together. The bottoms should be tinned on both sides, and beaten up, and then strongly soldered inside, the solder-

ing being continued along the seams also. Drop these through the holes in the top until they rest on their wired edges, and mark the tap holes. These are now cut out, and the containers are finally fixed by soldering the wired edges to the top of the urn. The taps are then soldered both outside and also inside the containers. The central tap, of course, is only fixed to the outer case.

URN DOOR.

The door is cut and notched as at Fig. 954, p. 320, and then planished, swaged to match the urn, and wired. Two hinges are attached at D, and these are riveted to the urn. The knob for the door is fitted with a tongue of metal by which the door is fastened.

COMPLETING THE URN.

The bottom of the urn proper, which also should be tinned and planished, is cut and edged to fit tightly inside; it should be pushed up from the bottom of the urn so as to lodge on the swage projecting inside the urn as at A Fig. 951, p. 319, and there strongly soldered. In making the lids, allow a rather large working edge for each hollowed disc, as the lids should be very deep to give the best effect. A brass knob is screwed and then soldered in the centre of each cover; and finally the urn is cleaned, first with emery cloth, next with emery powder and oil, and afterwards with crocus powder and oil, finishing by polishing with dry crocus powder.

RE-BOTTOMING TIN-PLATE PAN.

When a pan requires re-bottoming, and assuming that it has an iron rim round the old bottom, first knock off the latter with a hammer and chisel, cut the edge of the body level at the bottom, and throw off an edge of about $\frac{3}{16}$ in. To do this, turn the pan upside down. Hold the body against the edge of an iron weight. Keep the bottom of the pan slightly above the weight, and knock back the edge gently with a mallet. If the pan is rusty or dirty, clean it by scraping inside and out, paying particular attention to the seams; do this before edging, and clean it about 1 in. from the bottom upward. When it is edged, cut a piece of tin about **xxx**, $\frac{3}{8}$ in. larger in diameter than the distance over both edges. Now edge the piece cut out in the same way as the bottom of an ordinary canister. Slip it over the edge on the bottom of the body, and hammer it over to

Fig. 958.—Candle Lamp Ventilating Ring.

it. Get it well down, place the pan on the end of a round bar or pipe of suitable thickness, and gradually fold the two edges over with a mallet until they lie evenly against the body of the pan. Of course, the bar must go inside the pan. The same result can be obtained by holding a round, flat weight inside. This operation is known as knocking-up, or double-seaming. Solder it inside. To make the hoop, round the iron to the size, put it on, mark it off, and rivet it. Now rivet three or four pieces of tin to the hoop in such a manner as to be able to bend them over the top edge of the iron, and so enable the hoop to be soldered to the body.

Candle Lamp or Safety Candlestick.

The candlestick shown by Figs. 955 and [956, p. 321, is intended for use as a library candlestick or a dark-room lamp for photographers. The hood or shroud prevents the flame coming in contact with any inflammable articles. The whole interior of the lamp must be plated in order to get the maximum reflection, and, if thought necessary, the exterior may also be plated. The metal employed in its construction is soft brass, either wrought or cast.

Pillar of Candle Lamp.

Fig 957 is a sectional elevation of the pillar and base. The pillar should be made first. It consists of an inner tube B of $\frac{7}{8}$-in. bore and about 7 in. long, screwed with a convenient gas thread at each end—an inside thread at the base to receive the screwed bush A′, and an outside thread at the top to screw into the bevelled cap C². Around the base a small ring D² should be brazed on, and a similar one, D′, about halfway up the tube B. The outer tube D, which is of $1\frac{3}{8}$-in. bore and $3\frac{3}{4}$ in. long, passes over these two rings, and must be fitted in position and brazed. When the tubes B and D are together, their bottom ends should be faced off flat in order to get a good seating on the loose base A. The outer tube D is shown in elevation in Figs. 955 and 956.

Hood or Shroud of Candle Lamp.

The shroud consists of the coned tube H, one end of which is closed by the reflector K, and the other by the glass door J (see Fig. 956). Attached to the shroud is the tube E. The whole fits over the pillar as at present made, and is fixed by passing the slot F¹ over the pin F, as seen in Figs. 955 and 956, and also in Fig. 958. This forms a bayonet joint, and permits the shroud to be detached. It will be seen that by using this bayonet joint the shroud can be adjusted annularly, so that the light rays may be thrown over any given area without moving the whole candlestick. G G are

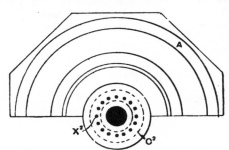

Fig. 959.—Half-plan of Candle Lamp Base.

stays. The tube E should be of about $1\frac{3}{8}$-in. bore, and brazed into each end are the rings E¹ and E², Fig. 957; E¹ is shown in Fig. 958, and E² is similar except that the pins for the bayonet joint are omitted. The holes shown in these rings are for ventilation, and must on no account be left out; they are best drilled with a clear $\frac{3}{32}$-in. drill. Fig. 959 is a half-plan of the base.

Making the Hood or Shroud.

The shroud pattern is shown by Fig. 960, and the side, end, and front elevations of the shroud after brazing by Figs. 961 to 963.

whilst Fig. 966 shows a section of the outer and inner cones, and Fig. 967 is an elevation of the inner cone. The arrangement consists of a cone M, the walls of which are perforated and the base coned out so that

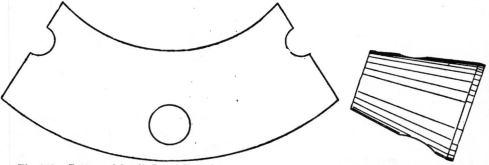

Fig. 960.—Pattern of Candle Lamp Shroud.

Fig. 962.—Side Elevation of Shroud.

The shroud is made of 16 B.W.G. soft brass cut out to the pattern shown by Fig. 960, and when finished should measure in diameter 2⅝ in., tapering to 4⅜ in., and 4½ in. long. Before bending, pieces should be cut out as indicated in Fig. 960, the centre hole being for the tube E to pass through, whilst the semicircular notches will form the ventilating shaft for L, Fig. 955, when the metal is bent and the ends are joined. The brass should be cleaned and polished before bending, and joined by brazing.

the smoke may be freely drawn up, fitted inside the casting L, and secured in position by the two pins M¹, Figs. 955 and 956.

Reflector, etc., of Candle Lamp.

Figs. 968 and 969 show the reflector, which is situated at the narrow end of the shroud, to which it should be brazed. Fig. 968 is a sectional elevation, and shows the construction of the rim—the part that is brazed to the inner portion of the shroud.

Fig. 961.—End Elevation of Shroud.

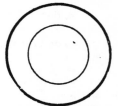

Fig. 963.—Front Elevation of Shroud.

The flange O (Fig. 955) should be brazed to the tube E and to the shroud H; the ventilating cap L is also fixed by brazing. Figs. 961 to 963 show the shroud before the flanges are brazed on.

Candle Lamp Ventilator.

The ventilating arrangement is shown in elevation and plan by Figs. 964 and 965,

This reflector, with the other parts, should be prepared for electro-plating before finally putting together. Figs 970 to 972 show details of the loose cap which, when in position on the shroud, contains the glass. It will be seen from Fig. 970 that the rim of this flange is fitted with a bayonet joint, by means of which the glass and the cap are kept in position on the shroud, which is thus rendered light-tight.

THE CANDLE SPRING.

A spiral spring on which the candle rests is shown in section B^1 (Fig. 957, p. 322): one end is attached to the brass bush A^1, seen in elevation, Fig. 973, p. 326. The

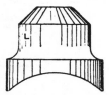

Fig. 964.—Top Ventilating Cone.

method of securing the spring is shown in Fig. 974, one end being bent to form an eye, a hole drilled and tapped in the centre of the bush A^1, and a small set-screw passed through the end of the spring into

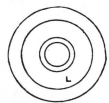

Fig. 965.—Plan of Top Ventilating Cone.

the bush. A small disc of thin brass B^3 (Fig. 957, p. 322) is attached to the free end of the spring, and upon this the candle rests. When the cap C^2 is screwed to the pillar, the spiral spring presses the

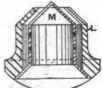

Fig. 966.—Section of Ventilating Cone.

candle upwards against the cone on the under side of the cap. The bush A^1 is fastened to the pillar tube B by means of the pin D^2 (Fig. 957).

CANDLE LAMP BASE.

The base should be about $4\frac{1}{2}$ in. in diameter and sufficiently heavy to stand firm.

A hole 1 in. in diameter should be drilled in the centre and tapped with a gas thread to suit the brass bush A^1. This bush has a number of air holes drilled through it, and provision for ventilation must also be made in the base. If the base is octagonal,

Fig. 967.—Inner Ventilating Cone.

as shown by Fig. 959, the best way is to drill holes through four diametrical sides.

VENTILATION OF CANDLE LAMP.

Carriage candles must be employed in this lamp, as all other kinds are too soft and generally unfit for use. To insert the candle, take off the shroud by releasing the bayonet joint and unscrew the cone cap C^2. The course of ventilation is as follows:—Air passes through the ventilation holes X (Fig. 973) in the base, and up through holes in the bush A^1; then into the tube B, and out through the holes X^2 in the cone cap, and surrounds the flame in the shroud. Ventilation is also secured by air being drawn up through the holes in E^1 (Fig. 957) at the base of tube E, up through the black

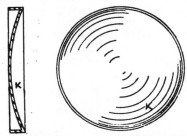

Figs. 968 and 969.—Section and Elevation of Reflector.

space forming the bore of this tube, and out through the holes in E^2 into the shroud. The interior of the cones L, M should be oxidised and blackened before use, whilst all other parts may be electro-plated at a small cost, which will add to the value and appearance of the candlestick.

Naphtha or Oil-gas Lamp.

Lighting by means of oil-gas lamps is cheap and effective, especially where

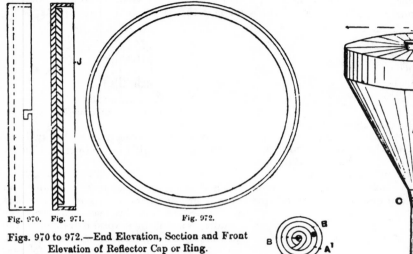

Figs. 970 to 972.—End Elevation, Section and Front Elevation of Reflector Cap or Ring.

a large torch is required for open-air purposes. The lamps described below can be made of any sheet metal, but brass is most durable, and gives a better general appearance; if brass is used, it should first be tinned on the side that is to be the inside of the lamp. Fig. 975 shows the most popular shape of lamp. The pattern for the part marked A is represented by Fig. 976, and is practically a semicircle whose radius is equal to the largest diameter of A. The radius of

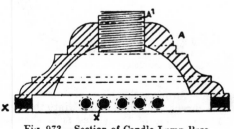

Fig. 973.—Section of Candle Lamp Base.

the smaller arc is equal to the proposed diameter of the pipe—usually $\frac{3}{8}$ in. brass

tube—which is connected to the reservoir. Allowances, shown by the dotted lines, must be made for working edges.

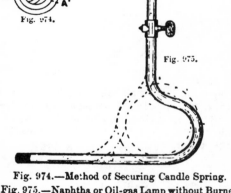

Fig. 974.—Method of Securing Candle Spring.
Fig. 975.—Naphtha or Oil-gas Lamp without Burner.

Making the Naphtha Lamp.

The pattern is turned to shape over a funnel stake and soldered. A creased edge is set off the larger diameter into which B fits tightly. The pattern for B is a rectangle, two sides of which are equal to the proposed depth (here 2 in.), with long sides equal to the proposed circumference plus lap for the seam; this can be rolled to shape and soldered. Before attaching it to A, a disc is cut $\frac{1}{8}$ in. larger in diameter and hollowed on a wooden block with a hollowing hammer, so that when edged it will fit tightly over B. A hole is punched in the centre for a feeder screw about 1 in. in diameter, through which a small hole has been drilled to allow the vapour of the heated oil to escape and so ensure safety. This is then soldered in position, and the hollowed disc can also be soldered to B. Cut next a piece of ⅜-in. brass tube 6 in. long, screw one end so that it will run on a ⅜-in. brass plug tap, and solder it in A from the inside. The boss c (Fig. 975) can be cut, turned to shape, and soldered; its

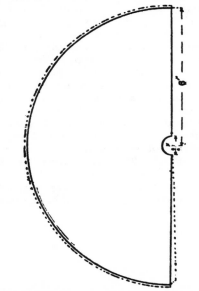

Fig. 976.—Pattern of Conical Body of Naphtha Lamp.

pattern will more or less resemble Fig. 938, p. 314; but the dimensions will not be the same. A hook D is made of No. B.W.G. wire, plated with sheet brass to form a hinge, and soldered in position.

Naphtha Lamp Pipe and Burner.

The shape of the pipe below the tap is a matter of choice (an alternate design is shown in dotted lines), but the pipe should be screwed at each end. A suitable burner is a Wells oil-gas lamp burner, a diagram of which is shown in Fig. 977; this is serviceable and effective, superseding the old-

fashioned type entirely. The flame issues from the vapour nipple A and strikes a spreader, which performs the double service of diverting the flame through the grating and vaporising the oil as it passes

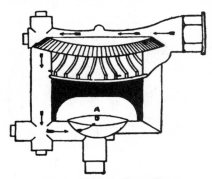

Fig. 977.—Naphtha Lamp Burner.

along the top and down to the burner proper in the direction indicated by the arrows. Before fixing the burner a piece of brass gauze is rolled about half a dozen times round a piece of wire (bent to Fig. 978 for the purpose of withdrawal) and pushed in the burner pipe, after which the pipes, burner, and tap can be screwed together, the joint being made with white lead. To clean the nipple, a pricker (Fig. 979) of No. 16 B.W.G. steel wire is required, the point A being sharpened.

Another Shape of Naphtha Lamp.

Fig. 980 shows another shape of lamp body. The pattern for the rim is a rectangle, two sides of which are equal to the proposed width, and two equal to the circumference plus ½ in. for seam. A hole is punched in the centre of this for the feeder

Fig. 978.—Wire for Burner Gauze.　　Fig. 979.—Burner Nipple Cleaner.

screw, after which the rim is turned to shape and soldered together. The feeder screw is next soldered in position from the inside. A hole for the burner supply pipe is punched diametrically opposite the feeder screw, and the shorter piece of pipe, previously screwed at one end for the tap, is soldered in from the inside. Two hollowed discs are made and soldered to

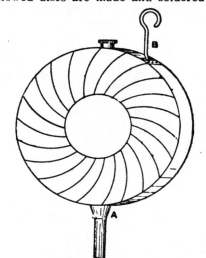

Fig. 980.—Another Shape of Naphtha or Oil-gas Lamp.

the rim, as described in the case of B (Fig. 975). The boss A (Fig. 980) and the hook B are next made and fixed, and the pipe, tap, and burner are screwed together. Of course, other shapes of lamp bodies could be suggested, but experience has proved these two to be most useful and convenient for all ordinary purposes.

MANAGEMENT OF OIL-GAS LAMP.

A few hints on the management of oil-gas lamps are here offered. Good results can be obtained either from petroleum or naphtha, the light always depending largely on the quality of the oil. The burner should always be kept clean, especially the vapour nipple, which should never be allowed to clog. Always have a small hole through the feeder screw which will allow the vapour of the oil in the lamp body to escape freely, in the event of its becoming slightly heated through conduction, as is sometimes the case; this is a safeguard against accidents. To light the lamp, turn on the tap gently until the oil spots or drips out of the vapour nipple A (Fig. 977), and hold it in and over a large flame, so that the burner will get sufficiently hot to vaporise the oil passing through it. Then the vapour or oil-gas will ignite at the nipple and remain so lighted, after which the size of the flame can be gradually regulated by the tap.

REPOUSSÉ WORK.

What Repoussé Is.

Repoussé work is the art of raising designs on sheet metal. The metal may be flat or otherwise, and of various shapes, and the ornamental design or pattern is formed from the back by means of hammers and punches. In the past, repoussé work occupied an important place in artistic craftsmanship, and some of the most famous artists of the olden times manipulated metal in this way. With the rapid advance in machinery, the stamp and press have done a large amount of the work formerly hammered up by hand. Where a large number of articles of the same pattern are required, a die is prepared, and they can be produced at a tenth of the cost of hand repoussé work. Thus they are alike in every detail, whilst the hand-worked articles are never exactly alike, and therefore must always have a higher value for the artist. For many years, it is said, the exceedingly simple processes were kept quite a trade secret, and jealously guarded against all intruders, though how this could be so is difficult to understand. Even the tools could not be purchased, it is said, it being customary for the "embossers," as they used to be called, to make them for themselves as occasion arose.

Application of Repoussé.

Amongst the articles to which repoussé work may be applied are tea trays, dishes, plates, salvers, card trays, ornamental shields, plaques, vases, tankards, cups, candle sconces and rings, alms dishes, spectacle cases, match boxes, flat candlesticks, menu holders, fancy racks, tea-pot stands, glove and handkerchief boxes, crumb trays and scoops, photo frames, barometer frames, finger plates, lock plates, ash trays, brush and mirror backs, letter racks, panels for firegrates, fender curbs, coal boxes, cabinets, bellows, and book mounts.

Metal for Repoussé Work.

The metal may be brass, bronze, copper, pewter, silver, or gold, but ordinarily copper is used. Cold rolled sheet copper should be used for best work; for commoner work use ordinary sheet copper. A suitable copper is known as soft soft-bright; on this only a very small pressure is put in its final finishing on the rolls. When metal is hardly soft enough, it may be improved by annealing. A metal which is particularly suitable is an aluminium-bronze alloy known as sun metal, both its colour and degree of softness being adapted to the work. Metal should always be properly prepared by being carefully planished and freed (by facing) from the scarcely visible flaws that would afterwards mar the work. The metal should also be selected for its softness, for some rolled metal, especially brass, which varies greatly in composition, is exceedingly hard and reliable to crack.

Repoussé Hammer

The hammer should consist of a light steel-faced head, as shown in Fig. 981, upon a slender but strong handle. It should be held by the right hand, the thick oval end being in the palm and the fore-finger resting upon the upper part of the handle. Another shape of repoussé hammer is illustrated by Fig. 982.

Repoussé Mallet.

Some repoussé workers think that a box-wood mallet (Fig. 983) is preferable to the usual hammers for those not accustomed to the use of hammers, as it gives for the same weight more striking surface, and is not so liable to slip and hit the fingers. Of course, for more advanced workers a

Fig. 981.—Repoussé Hammer.

properly shaped hammer with lancewood handle is much better. A mallet is very useful for flattening distorted metal, which should be beaten on a flat stake. Fig. 984 is an oddly shaped mallet in common use for repoussé work.

TRACING PUNCHES.

The tracing punch, two shapes of which are shown in Figs. 985 and 986, is gently and rapidly tapped as it is steadily drawn

Fig. 982.—Another Repoussé Hammer Head.

along the outline of the work, as shown in Fig. 987. It must be a well-made tool, properly pointed, without being too sharp, or the worker will be under a great disadvantage. The tool should be held by the thumb and first finger of the left hand, the second finger keeping it in position, and the third resting upon the work, and used as a guide (see Fig. 987). Much depends upon the setting of the tools ; if

Fig. 983.—Repoussé Mallet.

too sharp, they will cut through the brass and spoil it, and will not travel well ; and if blunt, they will cause irregular bruises. The tool should travel from left to right, and continuously, not with jerks or broken

lines. Two straight and two curved tracers are required, one large and one small of each kind.

Fig. 984.—Another Repoussé Mallet.

OUTLINING OR CUTTING-IN PUNCHES.

With these punches the design on the plain metal is outlined by a process of indenting. A suitable tool is shown by Fig. 988.

Figs. 985 and 986.—Tracers or Tracing Punches.

RAISING PUNCHES.

The raising punches are used after the metal has been turned face downwards upon the cement, and are hammered into

Fig. 987.—Using Tracing Punch.

the back of the design to produce on the face the required height and form. The ends of such punches may be as in Figs. 989 and 990, and the designs produced as in Fig. 991.

MATTING PUNCHES.

The matting punch (Fig. 992) usually has a square or oblong face, cut with diagonal or other lines, and is used to form the

Fig. 988.—Outlining or Cutting-in Punch.

matted ground that contrasts so effectively with the fluted or other polished portions of the work. It is made in a variety of patterns, coarse and fine (see Fig. 993) to suit the style of the work required.

OTHER USEFUL PUNCHES.

Variously shaped fluting punches (Fig. 994) will come in handy, and Fig. 995 shows

Fig. 989.—Raising Punches.

one of the many beading punches required; these are most useful tools, beads being largely used in chased work. Other designs of punches that may probably be wanted are shown by Figs. 996 to 998. Fig. 999 shows a tool that is often useful; it is known as a riffler, its ends being single-cut files.

SNARLING IRON.

The snarling iron for raising patterns on hollow vessels is a long rod (Fig. 1000,

Fig. 990.—Raising Punches.

p. 333) whose upper end reaches into the tankard, vase, cup, bowl, or other object. The rod slides through an iron ball B, and is fixed firmly in the vice, the ball resting on the top as shown. The work is now pressed against the upper end of the rod,

which is struck with the hammer at A, the effect being that the vibrations of the rod raise the prominent parts to the desired height. This work needs considerable skill. It gets its peculiar name of snarl-

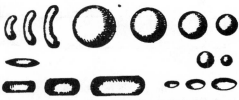

Fig. 991.—Designs Produced by Raising Punches.

ing from the snarl or noise occasioned by the process. If more convenient, a bent iron mounted in the vice jaws can be used, as in Fig. 1001, without the iron ball previously described. The end of the rod is variously shaped, as suggested in Fig. 1002.

PLIERS AND OTHER APPLIANCES.

The pliers illustrated by Figs. 1003 and 1004 are for the purpose of holding metal

Fig. 992.—Matting Punch.

over a gas flame or fire, to remove any adhering cement. For working small articles in thin metal, a circular leather bag sewn together round the edge and filled with sand (Fig. 1005, p. 334) is used. A tin-plate bowl filled with pitch composition (Fig. 1006) is useful in chasing up. It should stand in a ring of thick leather, riveted as shown by Fig. 1007, or in an iron ring bound with rope, as shown by Fig. 1008. With these stands the bowl may be tilted to any angle, and this gives ease in working. A pot or kettle for the pitch

Fig. 993.—Designs produced by Matting Punch.

must be obtained (Fig. 1009 shows a suitable shape) and a ladle (Fig. 1010) will be useful.

SUPPORTING THE METAL.

The metal has to be cemented down to a wooden block before work can be begun. A solution of pure pitch and plaster-of-Paris in equal parts is used for this

Fig. 994.—Fluting Punch.

purpose. The pitch is melted in the pot (Fig. 1009), and the plaster is added by degrees and stirred with the iron ladle (Fig. 1010) until it has all disappeared. The plaster should be thoroughly dry, dampness causing the pitch to boil over.

Fig. 995.—Beading Punch.

A little Russian tallow added also softens the cement and prevents the metal springing from it ; beeswax has a similar effect. The cement is poured, while fluid, over a wood block 1½ in. or 2 in. thick, and about 12 in. or 14 in. by 9 in. or 10 in., the coating being a little larger than the metal

Fig. 996.—Rounded Punch.

will cover (see Figs. 1011 and 1012). Then metal and cement should be left to cool. Some workers use a lead support instead of pitch. To do this obtain some lead as used in tea chests ; or get some soft rolled sheet lead, obtainable from a plumber, and melt it down. The pitch block is easier and

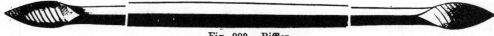

Fig. 999.—Riffler.

better to work on than lead, and produces better results, and if carefully used it need not give any more trouble than the lead.

TRANSFERRING DESIGN TO METAL.

Some ability in drawing is necessary to sketch the ornament to be raised on the metal. Having drawn the design, make a tracing on fairly thick paper, and with a sharp needle prick the tracing to produce a series of small holes. The pattern is then ready for use. When the metal and cement are cool, the first operation is to outline the pattern on the metal. Grease the metal, and place the traced pattern on it. Then with a pounce bag (Fig. 1013)

Fig. 997.—Rounded Punch.

dab the pattern all over, and the design will be marked with the white powder which passes through the holes in the tracing. These white lines should then be marked with the tracer. Take the tracer punch, and for making small and short lines, hold it as in Fig. 1014, p. 334, grasping the punch with the thumb and first finger of the left hand, using the second finger to keep it in position, whilst the third finger, pressing on the metal, guides the punch. If long curved lines are necessary, a better result is gained, it will prob-

Fig. 998.—Double-groove Punch

ably be found, by keeping all the fingers off the metal.

REPOUSSÉ WORK ON WOODEN SUPPORT.

As a medium to the introduction to the higher branches of repoussé work, there is a general opinion in favour of the less ambitious embossing of brass on a wooden basis, generally termed repoussé on wood.

This forms a pleasing and useful hobby for those who carefully follow it up, even if the more difficult and costly repoussé on pitch, or repoussé proper, is never attempted. Many pleasing and artistic articles may be repousséd or embossed by the simple methods followed in working on wood, and a delicacy of finish be thus attained; but, of course, the elaborate results of the higher branch are unattainable. When working on a wooden basis without pitch, the procedure

and with the straight tracer mark in the straight lines, doing the curved lines with the curved tools, giving light blows with the hammer or mallet. Having gone carefully over the design, mark out the outer or boundary line B (Fig. 1015), in this case four straight lines making the square. This being done, and having got the lines fairly straight, proceed to mat or dot in the background work; this should be done very carefully, the dots being distributed evenly, and not running into one another.

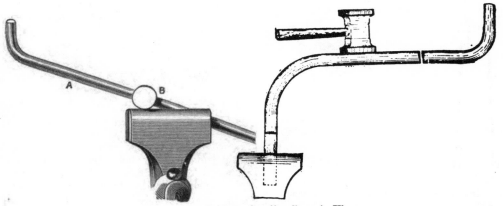

Figs. 1000 and 1001.—Snarling Irons in Vices.

may be as follows:—Lay the design on the brass, say a simple leaf, as Fig. 1015; place a piece of carbon paper between the brass and the design; trace over the design carefully with an agate or ivory point, or hard blacklead pencil, and so duplicate the drawing on the brass. Turn back the

The design or leaf will now appear to be slightly raised.

REPOUSSÉ ROSETTE.

Fig. 1016 shows a rosette in the first stage as marked with a tracing punch, according to the method first given. Fig. 1017 shows the rosette, with the central

Fig. 1002.—Shapes of Snarling Iron.

Fig. 1003.—Small Pliers.

design, and go over the lines with a sharp steel point, such as a sharpened knitting-needle. Place the wood upon a rug or mat upon the table, and begin to mark in these lines at the top left-hand corner,

and surrounding circles hammered up with the circular punches; and Fig. 1018 shows the rosette finished, with the central circle punched to represent the small seeds, and with the hollows round the centre marked to show the divisions.

RUNNING LEAF PATTERN.

The running leaf pattern (Fig. 1019) requires more care in working. The leaf

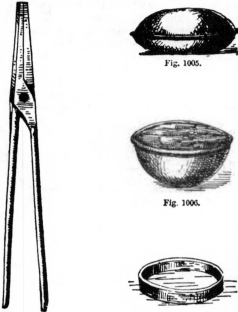

Fig. 1005.

Fig. 1006.

Fig. 1004.   Fig. 1007.

Fig. 1004.—Large Pliers. Fig. 1005.—Leather Sand Bag. Fig. 1006.—Pitch Bowl. Fig. 1007.—Stand for Pitch Bowl.

and balls, having been lined on the metal with the tracer, should be raised. This is done by turning the metal over, and using

Fig. 1008.

Fig. 1009.

Fig. 1008.—Another Stand for Pitch Bowl.
Fig. 1009.—Pitch Kettle.

the raising tool as instructed; in working the leaf, punches of different shape will

be required. The stems of the plant and the central vein of the leaves are then hammered up with a narrow punch, and

Fig. 1010.—Pitch Ladle.

the smaller veins traced in, care being taken to make the curves clear. Having once more turned the metal face upwards, any redundant form is corrected with the modelling tools, which are very similar to the raising tools. Before removing the metal from this position, texture, surface,

Fig. 1011.—Traced Plate Fixed to Wood Block.

and matting tools (various forms of which have been shown already) may be used to improve the design, and to mat in the background. The beaten-up plate when removed from the block will be somewhat out of shape, but can be flattened by beating on a flat stake with a mallet. Matt

Fig. 1012.—Section of Plate and Block.

punches are used by being worked over the surface and struck sharply with the small hammer, as has been explained.

CHASING.

Flat chasing is the decoration of flat surfaces without the use of the snarling

Fig. 1013.   Fig. 1014.

Fig. 1013.—Pounce Bag.
Fig. 1014.—Holding Tracing Punch.

iron. In relief chasing on hollow vessels, the snarling iron is first used as instructed on p. 331, the work is filled with a mixture of pitch and a little resin and plaster-of-Paris, poured in hot. and when quite cold is ready to be operated on from the front.

Fig. 1015.—Simple Leaf Pattern

The jug, dish cover, or whatever article is being operated upon, may now be said to be solid; the pitch having hardened gives

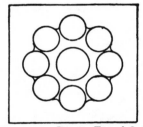

Fig. 1016.—Rosette Traced Out.

a stability to the article, and yet yields sufficiently to enable the metal to take the impressions of the various punches with

foot of the operator, who can now work with ease, the cord holding the article firmly while the sandbag prevents undue vibration.

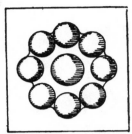

Fig. 1017.—Rosette Hammered Up with Circular Punches.

POINTS IN REPOUSSÉ WORK.

In repoussé work generally, and in modelling the human figure in particular,

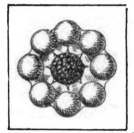

Fig. 1018.—Rosette Finished.

it is well to avoid very high relief, as this often gives vulgarity to the work. Bas-relief—low relief—is more suited to re-

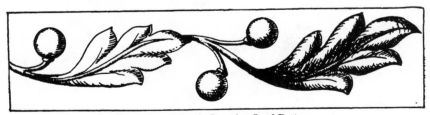

Fig. 1019.—Simple Running Leaf Pattern.

which the work is finished from the face. The solid block is now laid on a sandbag (see Fig. 1005), and a cord passed round it, through a hole in the bench, and under the

poussé work than is the high relief seen in ivory carvings. The face or mask requires very careful attention, as well as the hands and feet; these are important points.

Fig. 1020.—Repoussé Design for Book
Cover or Box Top.

Fig. 1021.—Repoussé Design for Side of Bellows.

Fig. 1022.—Repoussé Design for Handkerchief Box.

Drapery should be cut sharp and clear, falling in graceful folds, showing the con-

Fig. 1023.—Book Cover or Thermometer Frame Repoussé Design.

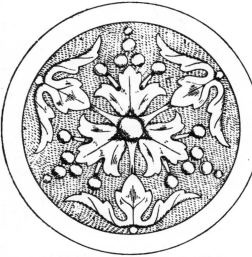

Fig. 1024.—Tray or Box Cover Repoussé Design.

tour of the body beneath. There should be no indefinite outlines; the figures

should be crisp and sharp, in contradistinction to painting, where this effect is avoided.

ELEMENTARY GEOMETRICAL DESIGNS.

To obviate the difficulty of starting on designs which are too advanced for beginners, many suitable elementary patterns are given on pp. 336 to 339. Their particular application as ornaments for domestic purposes is indicated in each case. A careful, painstaking working out of these patterns, even a repetition of the same exercise to acquire practical skill, will not

Fig. 1025.—Door-plate or Glove-box Cover Repoussé Design.

only yield satisfactory results, but prove a valuable basis for more advanced work. Simple geometrical designs are illustrated by Figs. 1020 to 1022, the first being adapted for a book cover or box top, the second for the side of a bellows, and the third for the top of a handkerchief box.

FREEHAND DESIGNS.

The designs Figs. 1023 to 1027 are equally suitable for repoussé on wood and repoussé on pitch. The tools required for them are few in number, and elementary

Fig. 1026.   Fig. 1027.

Fig. 1026.—Door-plate or Glove-box Cover Repoussé Design.   Fig. 1037.—Finger-plate Repoussé Design.

in style. The design Fig. 1023 is suitable for a book cover, for a box top, or even as the setting of a metal-faced thermometer. Fig. 1024 is applicable to a tray or any circular object. Figs. 1025 and 1026 are intended as door-plates, but they will do equally well as covers for glove-boxes. A door-plate design of a different character is shown by Fig. 1027, and a conventional design for almost any application is illustrated by Fig. 1028.

THE NECESSITY FOR SIMPLE EFFECTS.

In the early stages of repoussé work the learner should not attempt patterns abounding in small details, the great

thing being deftly to work the sheet brass or copper into those bosses or bulbous-shaped excrescences which allow much play of light. It may here be remarked that a wooden tool-handle makes a very useful tool for this purpose. As a preparatory exercise, the design for a finger-plate in Fig. 1029 deals largely with this effect. When the varying levels have been raised the details may be proceeded with. The design in Fig. 1030 offers plenty of scope for patient work, and, if neatly finished, is very graceful in design. The overlappings of the various members of the pattern are easily accomplished, and are very effective. A much larger article may be de-

Fig. 1028.—Conventional Repoussé Design.

corated by the pattern in Fig. 1031, a design for a salver or circular tray. The masses are very bold, and there is an absence of detail which makes the task of executing it not too prolonged or exacting. But if Fig. 1032 is attacked, there is plenty of opportunity for the repoussé worker to give evidence of progress in the art, the pattern being somewhat crowded with a variety of detail.

wide. These plates should be filed up square so as to fit together at the corners (see Fig. 1034). An angle piece of brass makes a corner, and may be either ornamental, as shown at Fig. 1033, or plain. The front and end plates are fixed to these corners by rivets or bolts and nuts. The

Fig. 1029.—Finger-plate in Repoussé Work.

REPOUSSÉ FENDER CURB.

A fender curb fitted complete and ornamented with repoussé work is shown by Fig. 1033. The brass or copper plates forming the curb may be of No. 16 or No. 17 B.W.G. metal, cut out to the size required and hammered perfectly flat and then ground smooth. The curb should fit round the outer edge of the tiled hearth. Assuming it to be 4 ft. long by 1 ft. deep,

repoussé work is executed in the ordinary way, the design being carefully drawn out and perforated, and set on the plate which is traced on the front, and hammered up from the back on a pitch block.

COMPLETING FENDER CURB.

The top may be of half-round brass tubing sawn up the centre of the flat side, and driven on to the plates as shown in

Fig. 1030.—Finger-plate in Repoussé Work.

it may be 5 in. or 6 in. high, the latter dimensions giving more room for the repoussé ornament. Then the brass front plate should be cut 4 ft. long by 6 in. wide, and the two end plates 1 ft. long by 6 in.

Fig. 1035. The corners should be mitred carefully and brazed so as to make a firm joint. The knobs are screwed on, as shown in Fig. 1036. Two or three small plates should be prepared, and fastened to

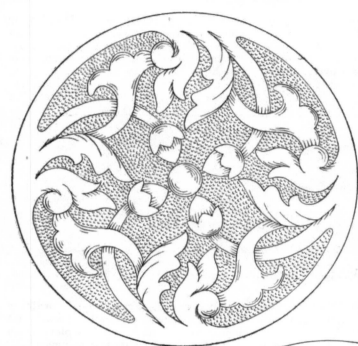

Fig. 1031.—Circular Tray
Repoussé Design.

tubing and the ogee moulding are smaller, as the front
may be only 3½ in. or 4 in.
high ; it fits under the bars
of the firegrate, and it should
be bent to the same shape.

OTHER FENDER FRONTS.

Figs. 1039 and 1040 give
other designs for the repoussé
work of the front plates and
angle corners, and Figs. 1041
and 1042 give corresponding
designs for the ash-pan
fronts. When the work has
been fitted together it must
be taken apart and polished
and lacquered, then carefully
put together and finished.
If the work is required to be
of antique copper it need not
be ground or polished, but
if passed slowly over a bright
coke fire a variation of colour
will be given to the copper

and has a pleasing effect.
the front and end plates
with small bolts and nuts.
The bottom moulding may
be a drawn ogee, mitred at
the corners and brazed and
fixed to the plates with
round or flat-headed screws.
Or, if preferred, a flat
tubing ⅞ in. by ⅜ in. may
be used, as shown at Fig.
1037, which illustrates a
tube corner pillar affixed
with screws instead of the
angle corner plates. This is
a little more difficult to fit
but has a very good effect.

REPOUSSÉ ASH-PAN FRONT.

The ash-pan front to
Fig. 1033 is shown by Fig.
1038. It is executed in the
same way, but the top

Fig. 1032.—Circular Tray Repoussé Design.

It should then be lacquered and the colour thus made permanent.

REPOUSSÉ PICTURE FRAME.

An ornamental design for a frame in re-

finely matted in. The framework immediately surrounding the picture should be left plain, the border should be of straight lines, and the regular geometrical pattern contained between them should be very

Fig. 1033.—Fender Curb in Repoussé Work.

poussé work is shown by Fig. 1043. A suitable material would be copper, brass, pewter, or even silver, No. 19 or No. 21 B.W.G., though the last - named would be costly. A sheet of metal about 16 in.

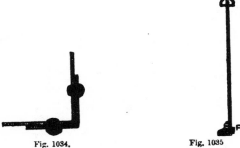

Fig. 1034.

Fig. 1035

Fig. 1034.—Horizontal Section of Fender Curb Corner.

Fig. 1035.—Section of Fender Curb Front.

square will be required, though the exact size of the shield illustrated is 14 in. long by 15 in. wide. The aperture is shown 6½ in. long by 5 in. high; though this would depend upon the size of the picture to be

Fig. 1036.—Vertical Section of Fender Curb Corner.

exhibited. The pattern is a very showy one, and need not be repousséd too highly, while the ground should be somewhat

neatly executed. When finished, the edges surrounding the aperture should be turned back, say ⅛ in. or so, in order to show a smooth inner edge, and the outer edge of the whole frame could be similarly treated. When finished off, the shield-like design could be attached to a thin hardwood board of a corresponding size by means of brass tacks or screws,

Fig. 1037.—Horizontal Section through Corner Pillar.

and two glass plates might be affixed for hanging purposes. In order to secure the picture in its proper place, some thin strips of metal could be tacked on three sides at the back.

Fig. 1038.—Ash-pan Front in Repoussé Work.

REPOUSSÉ LOCK PLATES.

The door of a house is not always a very picturesque object in itself, especially when it has weak panel mouldings, the handle and escutcheon for the keyhole seeming to stand quite alone and looking poor. Now if these are surmounted by a

Fig. 1039.—Fender Front in Repoussé Work.

Fig. 1040.—Fender Front in Repoussé Work.

Figs. 1041 and 1042.—Ash-pan Fronts in Repoussé Work.

Fig. 1043.—Picture Frame Repoussé Design.

plate in polished or bronzed metal they can be made decorative in themselves, and will give the panelled door a richer effect. Repoussé work is especially suited for this

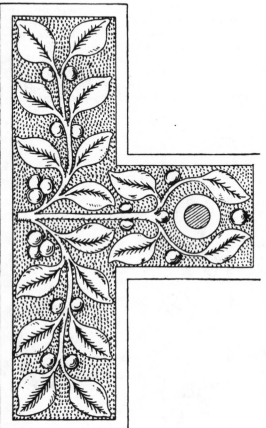

Fig. 1044.—Lock Plate Repoussé Design.

purpose. The lock plate should be not quite as wide as the stile of the door, so that there is a clear line of the door stile on each side of the plate when in position. Figs. 1044 and 1045 show conventional designs for lock plates. Fig. 1046 gives an illustration of a simple form of plate, designed so that its length may be increased if required by an addition to the stem. The plate may be from 18 in. to 24 in. in length and 7 in. to 9 in. wide at the central portion.

PREPARING LOCK PLATE.

Make a paper template of the door, carefully marking the position of the knob, spindle, and keyhole. Then on this paper template draw the design in outline, and very carefully trace this on thick tracing-paper; then perforate the ornamental lines of the outline, and the pattern is ready. A piece of brass, copper, or German silver of No. 18 or No. 19 B.W.G. is

Fig. 1045.—Lock Plate Repoussé Design.

required, and if it is to be polished have it hammered flat and ground; it is then ready for working on. In cutting out the metal, allow about ½ in. in the width and the same at each end for the rising of the metal. Oil the face of the plate, and on it place the pattern, dusting it over with a whiting bag ; this transfers the pattern to the plate. Put this, back downwards, on the pitch block, and trace round

Fig. 1046.—Lock Plate with Simple Repoussé Ornament.

the ornament with a steel tracer. Reverse the plate on the pitch block, and with the various punches raise the pattern, giving to the leaves and stems a variation of shape. If when taken off the block it seems that some chasing from the front would improve the effect, again put the plate, back downwards, on the pitch block, and do the necessary work on the front. Then carefully cut the plate to pattern, drill the holes for the fixing screws and the handle spindle, and the plate will then be ready for polishing and lacquering.

LOCK PLATE DESIGNS.

Fig. 1046 shows a plate with raised work at the ends only. Fig. 1047 illustrates a similar plate, but with the ornament of a finer character and needing more care to keep the scroll lines of good shape. In this design a half-round bead is carried round the edge, which is broken in places by a small ball ; this enables the edge lines to be kept straight more easily than they otherwise would be. If preferred, the ground flat portion of the plate may be

Fig. 1047.—Lock Plate with Repoussé Scrolls.

matted with a punch, a simple dot punch producing excellent effects. Fig. 1048 gives an illustration of a lock plate of a similar square shape, but with the raised pattern of a more ornamental character.

Fig. 1049 shows a still more ornamental lock plate, in which the leaf work is made the feature of the design, and the raising of the various portions of the leaves should be varied with all possible skill, so as to give them the best effects. The square form is altered by the projecting leaves. The matt ground is the small circle pattern, which, when well done, suits the

polished, but may simply be passed over a fire when raised, and then lacquered with silver lacquer or zapon. The plates as described may be used with the handles at

Fig. 1048.—Lock Plate with Repoussé Stem, Flower, etc.

Fig. 1049.—Lock Plate with Rich Repoussé Ornament.

other ornament, but any matt may be used, according to taste. The circular balls suggest the positions of the fixing screws.

FINISHING LOCK PLATES.

If these lock plates are preferred in antique copper, they need not be ground and

present on the doors, which are, of course, fixtures. Brass or copper handles may be purchased from any ironmonger who stocks cabinet brass foundry, and such handles are indicated on Figs. 1048 and 1049. Also the plates may be bronzed any suitable colour.

REPOUSSÉ DISH.

A design, two-thirds full size, for a dish or any similar article, is shown by Fig. 1050. Having already made a careful drawing of the design with which the plate the outline, taking care not to cut the metal in this fashion, **V**, but rather make an indentation thus: **U**. This, if rightly executed, will have marked the design firmly upon the reverse of the plate. Putting it upon its face, it can be beaten from

Fig. 1050.—Dish, Tray or Plaque in Repoussé Work.

is to be ornamented, the next thing to do is to pounce the sketch, and by this means transfer its lines to the sheet of copper. At this stage go over the lines carefully with Brunswick black or ink, to lessen the danger of lost lines. Then begin to chase the back until it is raised sufficiently on the front to enable the modelling to be done with freedom. Having got the ornament into an advanced state, proceed to hammer the plate into its concave shape before finishing. This is done by laying it

Figs. 1051 and 1052.—Repoussé Finger Plates.

Fig. 1053.—Pipe-rack in Repoussé Work.

upon a bed of pitch, and beating steadily with the hammer. The degree of delicacy of both labour and design can only be determined by the metal used. The design shown by Fig. 1050 would be suitable for copper, while for work of a more dainty finish silver would be preferable.

REPOUSSÉ FINGER PLATES.

The finger plates shown less than full size by Figs. 1051 and 1052 will be found good examples, and will give every opportunity for good modelling and dexterous handling of the tools. If the work is done well the results will be very pleasing.

REPOUSSÉ PIPE-RACK.

The pipe-rack in repoussé work illustrated by Figs. 1053 and 1054

Fig. 1054.—Repoussé Pipe-rack showing Positions of Holders.

is suitable for working either in one large sheet with the groundwork matted in, or, better still, in open repoussé work with the groundwork removed by means of the metal fretsaw, and the edges filed smooth all round. In the view of the pipe-rack given in Fig. 1053, it is supposed to be unpierced and mounted on a backboard of some hard wood. The interstices between the pattern are matted, but the inner space, on which the pipe-holders are fixed, can be left smooth and bright. The extreme dimensions of the design, which can be worked in copper or brass of about No. 21 S.W.G., are 17 in. by 15 in. The pattern

as in Fig. 1054, the backboard should be only 11 in. by 9 in. These backs can be fixed by drilling holes in the repoussèd sheet at suitable places, and then either hammering in some round-headed brass rivets or driving in some brass screws.

METAL CANOPIES FOR GRATES.

The appearance of any ordinary black grate may be very considerably improved by fitting on it a canopy in brass or copper; this will also give a little colouring to the general effect. These canopies are easily made and fitted. Figs. 1055 and

Fig. 1055.—Grate Canopy in Repoussé Work.

Fig. 1056.—End View of Grate Canopy.

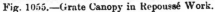

margin is 3 in. all round, thus leaving an inner oblong 11 in. by 9 in. The brass rings for holding the pipe-stems are 1 in. in diameter and $\frac{1}{8}$ in. deep, and are soldered on 2$\frac{1}{2}$ in. from the top. The semicircular holders for the pipe-bowls are 1$\frac{3}{4}$ in. in diameter and about 3 in. wide, and are fixed 2 in. from the bottom. The centre of the holder should fall exactly below the hole pierced for the ring. Fig. 1054 shows the positions of these. It will be best first to cut one of these holders in cardboard, and, when the exact size is obtained, to use it as a template for cutting the pieces of metal. If two small projecting pieces are left on the inner sides, holes to receive them might be made and additional strength secured. The design worked as in Fig. 1053 will take a backboard of corresponding size : but if worked

1056 show a canopy which may be fitted to any grate having a fixed curtain. To make this canopy get a piece of No. 18 or No. 19 B.W.G. sheet copper or brass of the size required, say 17 in. or 18 in. wide by 12$\frac{1}{2}$ in. high. Hammer this flat on a bright anvil, and have it ground, when it will be ready for raising in the centre. This is done as follows: Cut a piece of wood (as close-grained as possible) to the size of the portion of canopy to be raised. Hammer the metal plate into this hollow to get the shape required, and then hammer level on a half-circular stake iron. The plate may then be fitted to the grate by screwing to the cast-iron front with round-head brass or copper screws.

REPOUSSÉ WORK ON CANOPY.

The canopy is now ready to receive the repoussé ornamentation. Trace out the in the lathe covered with emery powder, the canopy being held against it, and finish off with a calico dolly and crocus powder in the same way. Then when lacquered,

Fig. 1057.—Grate Canopy with Repoussé Pattern on Raised Centre.

pattern on thick tracing-paper and perforate the lines. Oil the plate, place the tracing-paper on it, and dust through with whiting, which transfers the pattern. Put the plate on the pitch-block, and trace the marked pattern with a steel tracer. Then reverse the plate on the pitch-block, and hammer up the pattern with suitable punches. When taken off the pitch-block the plate will require straightening; this is done with a wood mallet on a flat wood-block. If the pattern has strained the

either with colourless lacquer or zapon, it is ready to be fixed in place on the grate. Fig. 1055 shows the ornament on the flat portion of the canopy only. A half-round beading should be fixed on the bottom of the canopy to stiffen the bottom edge.

FURTHER DESIGNS FOR CANOPIES.

Figs. 1057 and 1058 illustrate other designs on the same shape canopy. Fig. 1057 may also be made with the raised

Figs. 1058 and 1059.—Elevation and Section of Canopy with Repoussé Pattern on Raised Centre.

metal, flatten the plate by hammering the edge on a bright anvil.

FINISHING CANOPY.

The canopy is ready now to be finished. For polishing bright use a buff wheel fixed

portion of the canopy plain, and having the ornament only on the flat portion. Fig. 1059 shows the section of canopies illustrated by Figs. 1057 and 1058. The canopy illustrated by Fig. 1060 is constructed in a different way. In this form the sheet metal is cut with a chisel straight down each side, 2 in. from each end and to 2 in. from the top, and the central portion is bent outwards from the top as shown in Fig. 1061. It will be found easier to work

another design for the central portion of the canopy. A half-round or ogee pattern beading should be fitted to the bottom edge of the canopies as before described.

Fig. 1060.—Fitted Repoussé Work Canopy.

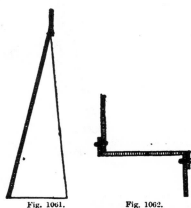

Fig. 1061.   Fig. 1062.

Figs. 1061 and 1062.—Vertical and Part Horizontal Sections of Fitted Canopy.

if the flattening of the plate, the grinding, and the repoussé work in raising the pattern, are done before the plate is cut. Cut two triangular pieces of metal, turning the edges down about ¾ in. or ⅞ in. from each edge, as shown in Fig. 1062, the wide portion of the plate being the distance by which the front of the canopy is to stand forward. The front edge of one side of the triangular pieces may be cut with a

CANOPY WITH CIRCULAR ENDS.

The canopy (Fig. 1064) is of a form that requires a little more fitting work. It is circular at the ends (see Fig. 1065), and is fitted to a flat plate cut out in the centre to the size of the raised portion. The central portion, after being hollowed out to size, has the outer edges, the top, and

Fig. 1063.—Repoussé Centre for Fitted Canopy.

Fig. 1064.—Repoussé Canopy with Circular Ends.

chisel to the pattern shown in Fig. 1060 or Fig. 1063, and then the triangular plates fixed to the canopy plate with bolts and nuts or riveted. Fig. 1063 illustrates

the two sides turned over flat, about 2 in., to fit on the plate, to which it is attached by bolts and nuts or rivets. This flat edge should be cut out to pattern as shown in

Fig. 1064, and the shape of the central raised portion is shown at Fig. 1066.

Fig. 1065.—End View of Canopy with Circular Ends.

Fig. 1066.—Horizontal Section of Canopy with Circular Ends.

service. As far as possible, useful articles should also be ornamental, and the crumb tray shown in Fig. 1067, when not in use, if placed on the sideboard or chiffonier, helps in the artistic decoration of the room. The tray is made of sheet metal, brass, copper, or German silver, and useful dimensions will be 9 in. wide at the mouth and about 9 in. deep altogether. The sheet metal should be No. 17 s.w.g.

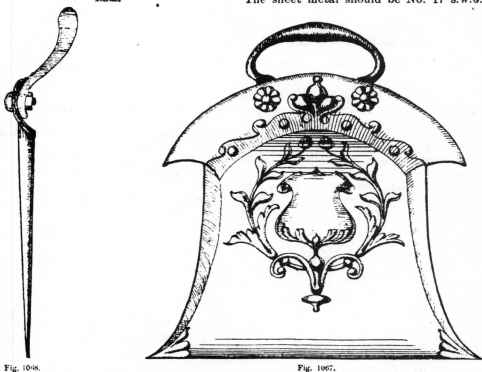

Fig. 1068.

Fig. 1067.

Figs. 1067 and 1068.—Elevation and Section of Crumb Tray in Repoussé Work.

CRUMB TRAY.

The crumb tray and brush are amongst the indispensable articles of household

if brass or copper, or No. 18 if German silver. The plate, being cut out to the size required, should be hammered flat and ground, then the two sides and top edge may be turned over, as shown in the section, Fig. 1068. A strip of similar metal should be cut to the circular shape, and then hollowed, so that the tray portion may fit over it. These two portions may be fastened together with round-head bolts and nuts, or may be riveted together,

Fig. 1069.—Crumb Tray Handle.

but this work cannot be done until the lower or tray portion has been repousséd to the desired pattern.

REPOUSSÉ WORK ON CRUMB TRAY.

The top edge of the tray, with its centre ornament, will need to be placed on the pitch block to have the bossed portions hammered up, as shown in Fig. 1067. The centre ornament should be carefully drawn on tracing paper, then pricked round, and, being laid on the metal in the required position, should be dusted down in the usual way. The lines thus dotted on the metal should be lined round with a tracing chisel. The plate is then ready to be reversed and placed on the pitch block, the raised portions being hammered up

date of the year in which the article was made.

CRUMB TRAY HANDLE, ETC.

Fig. 1069 gives an enlarged view of the handle, which may be cast either in brass or copper, and fixed to the tray, as

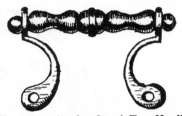

Fig. 1071.—Another Crumb Tray Handle.

shown in Fig. 1068, with ornamental bolts and nuts (Fig. 1070). Fig. 1071 shows an alternative handle, having a turned wood centre instead of being all metal. A very good effect may be produced by executing the different parts of the tray in two coloured metals, say brass for the tray, and copper for the handle and upper circular portion, but this must be left to the taste of the craftsman. The tray may be made in one piece of metal, but in that case there will be more difficulty in hammering up to shape.

CRUMB BRUSH.

The brush itself, with ordinary wood back, may be purchased of any brushmaker or ironmonger, and Fig. 1072 shows its ornamental metal back. A piece of

Fig. 1070.
Fig. 1070.—Bolt and Nut for Crumb Tray Handle.

Fig. 1072.
Fig. 1072.—Repoussé Plate for Crumb Brush.

with the various shaped punches to produce the desired effect. In the ornament illustrated by Fig. 1067, a shield is shown in the centre, and the centre of this shield may bear the artist's monogram, or the

sheet metal of No. 18 s.w.g. or No. 19 s.w.g. should be cut to the outer shape of the wood back of the brush, and hollowed, so as to fit firmly down on it, holes being drilled to fasten it with round-head screws. The pattern must then be re-poussèd as described before for the tray. The repoussé finished, the metal may be polished and lacquered, the parts of the tray fitted together, and the brush back attached to its wooden support.

lar form, $\frac{1}{2}$ in. wider than the size required, to allow for the working of the repoussé pattern on it. A section should be cut out of the ring, which, when the plate is bent to shape, will give the correct bevel. The repoussé work should be

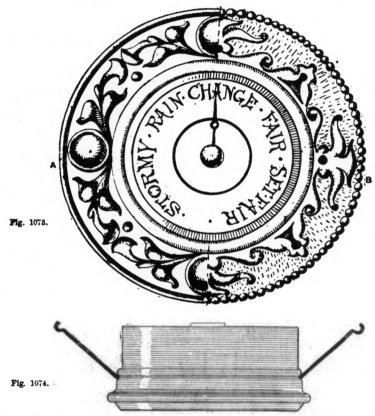

Fig. 1073.

Fig. 1074.

Fig. 1073.—Barometer fitted with Circular Repoussé Frame, showing Alternate Designs.
Fig. 1074.—Section of Barometer Frame.

CIRCULAR BAROMETER FRAME.

The circular aneroid barometer (Fig. 1073) is provided with a ring to hang it to the wall, and a frame may be made and fitted to it in several ways, two of which are here described. At A is shown a circular piece of brass, copper, or nickel silver, which may be fixed round the barometer as illustrated in section in Fig. 1074. The brass plate should be of circular chased when the circular plate is flat. In B (Fig. 1073) the groundwork of the ornamentation is shown matted; this is executed by punching all over the ground with a circular, three-cornered, or square punch. This gives a different shade of colouring to the metal, and may be equally well adopted for pattern A. When the design is impressed, the plate should be bent round the barometer between the mouldings, as shown in Fig. 1074, and the two

Fig. 1075.

Fig. 1076.

Fig. 1075.—Barometer on Repoussé Back Plate. Fig. 1076. Plan of Barometer and Back Plate.

edges brought together with two brass bolts and nuts ; this will secure it in place, and will be very effective. It may be left

Fig. 1077.—Coal Cauldron with Repoussé Ornament.

the dead colour of the metal or polished bright, but in either case should be lacquered with ordinary lacquer after being heated, or coated with zapon, applied cold.

may be fixed to the wall in the ordinary way, the aneroid barometer being fitted to the centre of the back plate and fixed to it with screws. The back plate may be of sheet brass, copper, or German silver, of No. 17 or No. 18 s.w.g., leaving a full $\frac{1}{2}$ in. of metal outside the ornament. The plate should be hammered flat and ground, and

Fig. 1079.—Upper Ornament of Coal Cauldron.

is then ready for the repoussé pattern to be placed on it. The barometer should then be fitted, the plate being polished and lacquered as before described ; the barometer may then be screwed in place ready for use. The plates may be bronzed by

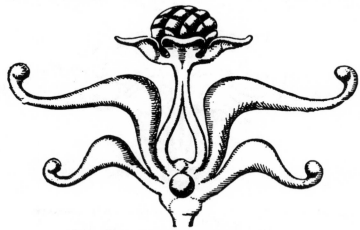

Fig. 1078.—Lower Ornament of Coal Cauldron.

ANOTHER BAROMETER FRAME.

Figs. 1075 and 1076 show another plan, the aneroid barometer being fixed in the centre of an ornamental metal plate which

electro-deposition to appear like antique bronze, verdigris bronze, or relief bronze, but, of course, this is entirely a matter of individual taste.

COAL CAULDRON ORNAMENTED WITH REPOUSSÉ.

The antique-looking metal coal box illustrated by Fig. 1077 can be made by utilising one of the cast-iron cauldrons that may be purchased from any ironmonger at a small cost. It should be selected with an opening about 12 in. in diameter, and as smooth in the casting as possible. This should be rubbed all over with stone or brick, and then with emery it may be forged from square iron, drawn out at each end to form the hooks, and hammered in the centre to a flat curve, as shown at Fig. 1081, which is a section on A B (Fig. 1080). In the centre a ring is formed by twisting the thin portion. If the handle is made in copper, prepare a wood pattern of half the handle; then make two castings, and braze them together in the centre at the ring junction. The hooks at the ends of the handle will clip the arms on the cauldron.

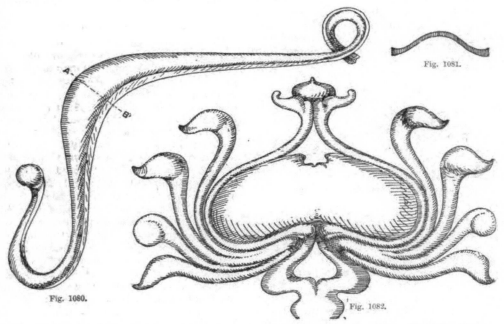

Fig. 1080.—Coal Cauldron Handle. Fig. 1081.—Section of Handle. Fig. 1082.—Alternative Design for Lower Ornament of Coal Cauldron.

cloth, so that it may have a smooth surface, and is then ready for the ornaments (Figs. 1078 and 1079). It has cast on it two arms, one on each side, connecting the rim to the body, and these will be utilised in fitting the handle.

HANDLE OF COAL CAULDRON.

The handle of the coal cauldron may be of wrought-iron or copper, and is shown enlarged at Fig. 1080. If in wrought-iron

ORNAMENTING COAL CAULDRON.

The outer surface of the cauldron should next be ornamented with bright metal. Copper or brass sheet of, say, No. 17 or No. 18 B.W.G. should be cut out to the pattern, the bottom being left long enough to cover the face side of the short legs of the cauldron. These plates should be made bright, and then repousséd as shown in Fig. 1078. Three of these ornaments will be required, one over each leg of the cauldron and between them, but on the rim of the cauldron, should be affixed the designs shown on Fig. 1079. These should be turned over so as to clip the top, and thus decorate the rim inside and outside. To fix these plates to the cauldron, small holes should be drilled in the metal for screws and nuts. The plates require a little setting to fit the sides of the vessel, and this may be executed with a wooden mallet or a piece of soft wood. When fitted, the ornaments should be removed, polished, and lacquered, and the cauldron enamelled dead black, with egg-shell finish. On this the brass or copper ornaments will look well. Or, if preferred, the cauldron may be painted any colour with enamel paint. Fig. 1082 gives an alternative pattern for the repoussé work of the lower portion of the cauldron.

Fig. 1083.—Ornamental Coal Tongs.

ORNAMENTAL TONGS.

Fig. 1083 illustrates a suitable pair of tongs to go with the cauldron. These may be executed in wrought-iron, the rose covering the joint being of polished copper. The handle is shown of hook form, so that it may fix itself over the upper rim of the cauldron and be secure. These tongs should be made of best iron, $\frac{5}{16}$ in. round, flattened out to a boss to form the joint, a turned circular pin passing through the centre, and being secured by the two copper roses. The lower ends are forked to hold the coal, and the tongs should have two coats of dead black.

BELLOWS ORNAMENTED WITH REPOUSSÉ.

The ordinary bellows, with the necessary leather-work flap, etc., may conveniently be purchased from ironmongers, for it is easier to do this and then to ornament it than to go to the trouble of making the bellows throughout. Fig. 1084 shows a metal plate which may be repoussed and then fixed on the front of the wooden bellows, thus producing a highly decorative effect. To work this plate, hammer a piece of No. 18 or No. 19 B.W.G. brass or copper flat on a smooth-faced anvil, and grind it to get a fine surface. Ordinary bellows as purchased are not of the shape illustrated, but of a more regular scroll or semi-heart shape. The covering plate may be similar, the pattern being arranged so that the fixing screws get a firm hold.

REPOUSSÉ ON BELLOWS PLATE.

Having prepared the plate, rub it over with an oily rag. Trace the pattern, as shown in Fig. 1084, on stout tracing paper, perforate the pattern on the edges very carefully, place the paper on the plate, and dust it over in the usual way to transfer the pattern to the plate; mark round with the tracer, when the plate will be ready for raising. Turn it over with the face side downwards, and fix it on the pitch block; then, with the various shaped punches, hammer up the pattern, this process requiring great skill with the pattern shown. The plate should next be released from the pitch block and fixed again on it with the face downwards; then chase it with a chaser, and chisel on any points needful to enhance the general effect. It will then be ready for polishing and lacquering, after which it may be screwed on the front of the bellows. The centre flat portion is shown with a ground or matt. This is executed by using a dot or other shaped punch when the plate is on the block, care being exercised to do this work regularly. This matted ground enhances the artistic value of the ornamental portions of the pattern.

COMPLETING BELLOWS ORNAMENTATION.

The fixing of the plate to the wood is shown at Fig. 1085; it is secured by round-headed screws, and the appearance is better if these are cross-nicked, as shown in Fig. 1084. Fig. 1086 shows a nozzle for the bellows. As purchased these are often very poor in shape, but it is easy to cut

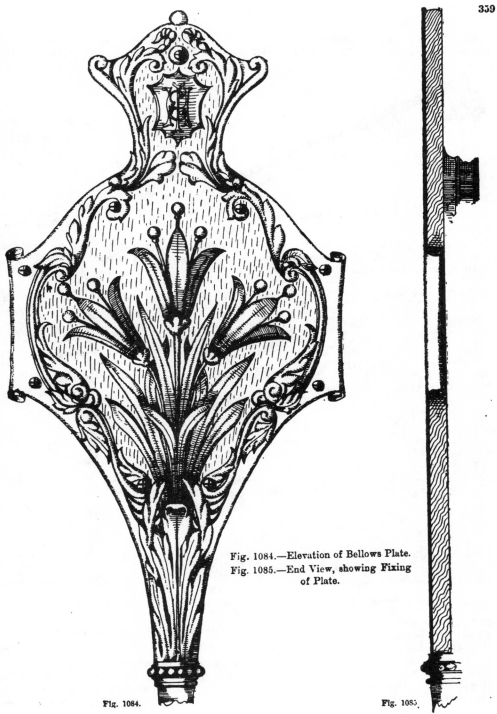

Fig. 1084.—Elevation of Bellows Plate.
Fig. 1085.—End View, showing Fixing
of Plate.

Fig. 1084.

Fig. 1085.

them off, tap the upper portion, and screw a new nozzle on it. Any brassfounder

Fig. 1088 is illustrated an alternative shield containing a date instead of a

Fig. 1086.—Metal Nozzle of Bellows.

Fig. 1087.—Ring for Bellows Handle.

Fig. 1088.—Alternative Shield Design.

would make one from a turned wood pattern. Fig. 1087 shows the handle ring with a plate made also of brass and re-

monogram. This work may be bronzed, or left as ordinary copper or brass.

Figs. 1089 and 1090.—Elevation and Section of Ornamental Flat-front Grate.

pousséd in the way described for the plate ; this is fixed to the back handle of the bellows for hanging on a nail or hook. In the centre of the handle portion of Fig. 1084 is shown a shield with monogram ; this should be altered to the monogram of the craftsman executing the work. At

WROUGHT-IRON AND REPOUSSÉ METAL GRATES.

In many of the houses built at the present time the cheapest fittings are used for grates. These can generally be very easily removed, and may be replaced at a comparatively small cost by grates made as here described. Thus Fig. 1089 illustrates a flat-fronted grate, to fit an ordinary marble or slate mantelpiece 36 in. or 38 in.

may be of ⅜-in. or ¼-in. round iron, and should be fixed to the framework with round-headed bolts and nuts as illustrated in section by Fig. 1090.

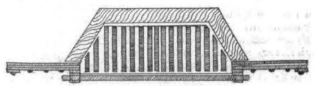

Fig. 1091.—Plan of Flat-front Grate.

wide by 38 in. high. The construction is as follows:—The fire portion should be 18 in. wide, and the sides and top 10 in. wide. The framework is of flat bar-iron 1¼ in. wide by ¼ in. thick, the ends of the bars being bent over each other and fastened together with round-headed rivets or round-headed screws and nuts. The ornamental standards of the fire-bars should

REPOUSSÉ PLATES.

The framework made as above requires filling in and ornamenting. This is done with three brass or copper plates bearing a repoussé pattern. These plates, being cut to the size after ornamentation, are fixed with rivets or with screws and nuts. Behind the front, sheet-iron plates of

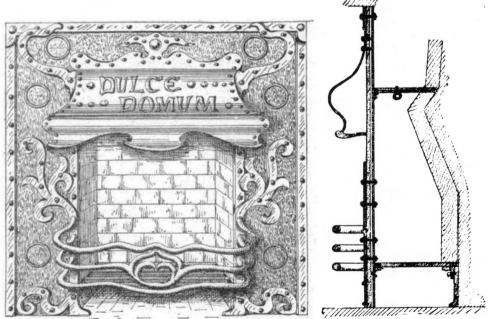

Figs. 1092 and 1093.—Elevation and Section of Grate with Ornamental Canopy.

be made of flat iron 1¼ in. by ¹⁄₁₆ in. thick, and turned over so that the horizontal bars pass through and are fixed firmly to the standards by screws. The horizontal bars

the same size as the repoussé plates, and fixed to the framework with them, are used to preserve the brass or copper plates. The repoussé design may, of course, be according to taste ; that illustrated in Fig. 1089 is Old English in character, and presents a style that is much in favour.

IRON GRATING.

The grating is next to be fitted at the bottom of the bars and built into the grate-back ; also the fall door at the top, to open and close, is required for the smoke. These are shown to be built into the brickwork at the back, being fixed to the front by angle-iron. The grating shown in Fig. 1089 is made of square iron bars built into the brickwork, but a cast-iron grating, as shown in plan in Fig. 1091, may be adopted, and may be obtained from any iron-foundry. This grate is built-in with ordinary fire-bricks, which, for effect, may be placed herring-bone fashion at the back. The connection to the flue is made in the ordinary way. The ash-pan is constructed of a sheet-iron pan $3\frac{1}{2}$ in. deep of the shape of the brick opening, and with an ornamental brass or copper plate fixed to the front with a wrought-iron handle. This arrangement hides the ashes.

ANOTHER DESIGN FOR ORNAMENTAL GRATE.

Fig. 1092 illustrates a similar grate, the frame being made of iron of the same size, while the brass or copper plates are similarly fitted. The fire-bars are of differing shape, and are riveted to the back standard. The canopy improves the appearance, and is useful in carrying off the smoke. It is made of sheet brass or copper, bent to shape, riveted at the corners, and fixed to the frame with bolts and nuts. The repoussé ornament and letters must be worked on the plate before the riveting is done. Fig. 1093 gives a section of this front, with an alternative method of fixing by using a solid fire-brick back, which may be obtained through any ironmonger. The grating in this case is fitted to the front with angle pieces, but stands on an iron foot at the back, which also carries the fire-brick back. The flue door at the top is here shown to slide, instead of lifting up and down. The connection to the flue is made in the ordinary way.

FINISHING ORNAMENTAL GRATES.

The framework of both the grates described above should be coated with a dull Berlin black. The brass or copper plates may be polished or left smooth-rubbed if an antique appearance is required. They may, of course, be lacquered, and instructions on lacquering will be given in a later chapter (see pp. 375 to 384).

ORIENTAL DECORATIVE BRASSWORK.

INTRODUCTION.

A FEW years ago repoussé, or beaten-up metalwork, was extremely popular; but "Oriental" or, as it is sometimes called, "Cairene" brasswork, is less difficult than

Fig. 1094.—Cutting Tool for Oriental Brasswork.

repoussé work, and involves less labour. A repoussé worker, or anyone who has previously done any sort of metalwork, would quickly pick up the way to do it, and the appliances he might already possess would again be of service to him.

AN ADVANTAGE OF ORIENTAL WORK.

Beaten work, too, is imitated so much now by stamped goods that it is difficult for the inexperienced eye to detect the difference between articles done by hand and those produced by machinery. To the expert, however, the marks of the tool in genuine work form a criterion, and if these are absent the article can probably be classed as worthless from an artistic point of view. Decorative incised brasswork cannot be copied in the way mentioned, which surely must be reckoned as a great advantage in these days of spurious imitations.

KINDS OF DECORATIVE BRASSWORK.

Different classes of metalwork are done in foreign countries, each nation having its own characteristics. For instance, Benares work differs from that done in Cairo, and the brass trays of Algiers are different from those of Tangiers, although the way in which they are worked is similar. The two kinds of metalwork to be met with in Eastern cities are: incised or engraved work in which the outline of the pattern is cut out and removed, and a rough sort of repoussé lightly beaten up from the back; in the latter case, however, the pattern or design is first chased on the metal.

TOOLS FOR DECORATIVE BRASSWORK.

Figs. 1094 and 1095 illustrate the two kinds of tool which are necessary for doing both, and it will be seen that they are entirely dissimilar, one having a sharp point for cutting into the metal, whilst the other has a rather blunt edge so as merely to indent the metal while following the outline. The way to do both styles of this work will be described later.

INCISED OR ENGRAVED WORK.

In incised or engraved work, the design is simply engraved on the metal and the pattern is finished when the groundwork or matting has been introduced, no beating up from the back being required. It may be thought, however, that plain chasing, without being repoussé, would answer the purpose just as well and present as

Fig. 1095.—Chasing Tool.

good an effect; but this is not so, as the outline, being only depressed and not cut out or incised, is not sufficiently defined, and the general appearance of the decoration is not so brilliant. If the two classes of work are compared, the difference will be at once noticed, and the fact of the pattern being engraved constitutes the chief beauty of this kind of decorative brasswork.

The Real Oriental Brasswork.

The metal work done by the Orientals is mostly of a rough and uneven character —that is to say, the lines do not always meet as they ought to, and frequently run into one another. The workers appear to

Fig. 1096.—Ornamented Finger Bowl.

study not accuracy of outline, but rather a good general effect, and therefore their work will not bear looking into closely. The patterns are executed to a great extent from the eye after a few circles have been marked, and are done very quickly. The designs are probably handed down from generation to generation, and do not on the whole present much variety. Here, however, this crude and rough style is not advocated ; it is apt to lead the beginner into a slovenly method of working, from which there will be difficulty in escaping. Not much beating up is necessary, as the relief is low, and the modelling of the leaves, etc., not closely attended to ; therefore, the work is comparatively easy for the beginner. Some of the Arabic trays worked in this fashion are very effective. Grotesque animals, heads, and figures of men often form a part of the design. These are surrounded, or the spaces between filled in, with Arabic characters, generally representing passages taken from the Koran. These characters, which are barely suggested in Fig. 1096 above, impart an originality to the work which is not found in other patterns.

Skill in Drawing Useful.

Some ability in drawing is a useful acquisition for doing metal work, as it helps considerably in the setting-out of designs and also in transferring them to the brass or copper. The chasing or engraving, as the case may be, is accomplished more satisfactorily by anyone able to draw curves or scrolls correctly, as he will not tolerate a design that is at all faulty in outline. In designing, an aptitude for drawing is of course essential, but if a good design is obtained and transferred to the metal, it may be sufficiently well carried out even if the worker knows but little of drawing.

Adaptability of Oriental Work.

Finger bowls for dessert are very pretty when ornamented in this way, and can be easily cleaned. Fig. 1096 is an illustration of one from Cairo. It is ornamented with grotesque heads and birds, the design being very roughly carried out, yet the effect is good. The outline is simply chased with the tool shown in Fig. 1095, and is not beaten up from the inside. Many of the large brass trays seen in England have been purchased at Cairo or in other Eastern cities from the Arabs or sellers of brasswork. All sorts of pretty ornaments may be made of brasswork. Good door-plates, which are expensive to buy if hand-chased, can easily be done by anyone who understands metalwork, and even if only roughly executed, have a better appearance than the ordinary wooden or china ones.

Beginning the Work.

Start with small pieces of work, not with something out of all proportion to the operator's skill. Large pieces can be done when the method of handling the tools has been mastered. Incised work is more suitable for small articles than repoussé, as

Fig. 1097.—Arabic Hammer.

bold patterns, or those adapted to large ornaments, are not so well represented as when beaten up from the back. Fine matting or groundwork takes a long time if carefully done, but is well worth the trouble and looks better than the more or less crude and imperfect work seen on Arabesque trays.

Tools for Incised Work.

The necessary tool for doing the incised work seen on brass trays is illustrated by Fig. 1094. The point of this tool has three facets, and it must be remembered that if

Fig. 1098.—Holding Hammer and Graver.

the point is too sharp it is liable to snap off, in which case it can be resharpened on an ordinary oilstone. Usually a slight rubbing on each facet in turn will render it serviceable again, and care must be taken not to overdo the process. Unlike engraving proper, which is performed generally by pressing the tool against the metal with the palm of the hand, in this brasswork a small hammer, illustrated by Fig. 1097, is used. This is the kind of hammer used by the Arabs, and is $6\frac{3}{4}$ in. long; the head being $\frac{3}{8}$ in. square at the front; it must be capable of being easily wielded, as a number of sharp blows have to be delivered to drive the point along the outline.

How to Use the Tools.

When these tools are obtained, the first thing is to learn how to hold them properly, as quite a different manner has to be adopted from that required for repoussé work. The graver must be grasped by the left hand with the thumb and first three

Fig. 1099.—Matting Hammer.

fingers, and the hammer is held in the right hand by the thumb and first and second fingers, as in Fig. 1098, from which there will be no difficulty in acquiring the correct method. In ordinary chasing,

such as is usual in repoussé work, the chasing tool is grasped in the same way with the left hand, but instead of being slanted is held nearly vertical to the metal. The hammer also is held with the ball of the handle resting in the palm and the forefinger stretched along the top to steady it. To make the tool follow the outline, blows with the hammer are given, the tool being at the same time somewhat inclined. A much simpler method, suitable to the rough beaten work, is to give one blow to the tool and then move it on to the end of the line and give another blow; this, however, does not produce a continuous outline.

Matting or Grounding.

Having outlined the pattern, the next thing is to put in the groundwork or matting, for which a hammer as used for repoussé work (Fig. 1099) is necessary; it is $9\frac{1}{4}$ in. long. For beating up from the

Fig. 1100.—Matting Tools.

back, or for introducing matting in which heavier blows are required, a still heavier hammer of the same pattern is recommended. These hammers can readily be purchased at any art metalworker's shop. The tool for matting or grounding, used by workers abroad, is often a piece of steel broken in two, the fracture presenting a rough surface, which is used for producing a fine matt. In this way it is possible for the beginner himself to make a most effective tool. But a variety of matting tools, ranging in price from 1s. to 2s. each, can be obtained. Three or four of the finer kinds are sufficient to begin with. Fig. 1100 shows the tools, and Fig. 1101 the patterns made by them, the numbers exactly corresponding, and they are given in the order in which they are most generally employed, beginning at No. 1, whose use will now be described.

THE "FREEZING" TOOL.

No. 1, Figs. 1100 and 1101, shows a "freezing" tool, which produces a fine groundwork in which the separate marks made by the tool cannot be distinguished. It is generally employed for filling-in small spaces in borders or between leaves, etc. ; and it has also a good effect for giving relief in the centre of a design. There are two ways of holding this tool : either placing the point directly on the metal and giving a tap with the hammer, running the impressions one into the other ; or else raising it a little, resting the little finger

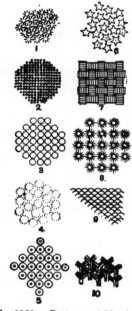

Fig. 1101.—Patterns of Matting.

on the brass. In the latter method a succession of light blows should be given, the tool springing up each time. By these methods two different patterns can be produced by the same tool. No. 2, Figs. 1100 and 1101, is a similar kind of tool, the point of which makes little dots on the metal. When this matting is done uniformly, and the impressions are made to join, a good dead effect is produced.

THE RING TOOL.

The ring tool No. 3, Figs. 1100 and 1101, is made in several sizes, and is used according to the space to be covered. The same applies to the ornamental ring tool No. 4. In working with these tools it is advisable that each circular impression should be clear and separate as far as possible. If the whole of the circle cannot be shown owing to want of space, a portion only should be inserted, the tool being held sideways, care being taken not to mark the design in so doing. No. 5 is a ring tool with a dot in the centre.

OTHER MATTING TOOLS.

No. 6, Figs. 1100 and 1101, is a star which produces a fine pattern, and looks effective when a change of matting is desired. No. 7 is a tool by which is obtained a groundwork extremely effective if adapted to suitable designs. The point is composed of several straight lines. If an impression with it is made and the tool then held at right angles to the first position, a kind of basket-work pattern is obtained. No. 8 is a tool which can be used for covering large spaces. No. 9 is a half-round tracer by which a matting which resembles engine-turning can be obtained.

STRAIGHT TRACER.

No. 10, Figs. 1100 and 1101, shows a straight tracer which produces the cleverest pattern of the whole—the matting frequently observed on Arabic trays. It will perhaps be found by the beginner the most difficult of all tools to use, as it requires practice before it is introduced neatly and well. The tool must be held upright and three blows given with the hammer, and then held at an obtuse angle, and three more lines, close together, imprinted on each side, so that the ends of the lines touch each other. It will thus be seen that a fork is made composed of nine lines. The pattern must be continued until all the spaces are filled in. In small corners it is not necessary to be very exact ; short lines crossing one another, representing the matting, will suffice.

POINTS IN USING MATTING TOOLS.

The matting tools illustrated by Figs. 1100 and 1101 should be sufficient for most workers, but there are others that can be purchased with which excellent patterns

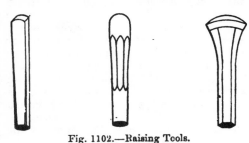

Fig. 1102.—Raising Tools.

can be made. Matting should be varied as much as possible. For example, the centre may be matted with a freezing tool and the outside portion with a more open pattern, such as a large ring tool or the Arab matting. In borders, a fine grounding looks best, as the spaces to be covered are usually small. By studying other kinds of metal work all sorts of hints on other varieties of matting may be gathered, as previously stated. Fig. 1101 illustrates a small piece of the pattern made by each of the matting tools, the numbers agreeing with Fig. 1100. By copying these the learner should experience little difficulty in reproducing any of the styles in his own work. In doing all kinds of matting the tool must be held upright with the third finger resting on the metal to steady it and prevent slipping. This is likely to occur especially when a single large impression has to be made, such as a ring. To avoid it press rather hard with the tool

Fig. 1103.—Tool Receptacle.

at the same time as the blow is delivered with the hammer.

RAISING TOOLS.

Having completed the matting, nothing further is required in the case of incised work. If repoussé work is required to be done, raising tools, illustrated by Fig. 1102, will be suitable for rough beaten work. One or two sizes of each would be sufficient for the beginner.

TOOL RECEPTACLE.

A holder for the tools when these are not in use is shown by Fig. 1103; the tools should be kept with their working ends uppermost. The holder is $2\frac{1}{2}$ in. high, 2 in. in diameter, and is made of tinplate. The top part should be cut out first, the edge being turned over and beaten down. It is then formed round a stake, and the two ends soldered together. The bottom is a circular piece of tinplate soldered to the cylinder.

PITCH BLOCK.

Metalwork is commonly considered a dirty process. The use of pitch has probably

Fig 1104.—Pitch Pan.

given rise to this notion, and if improperly handled of course this material makes a mess; but there is no necessity for this if the directions to be given are adhered to. In some rough metalwork the aid of pitch or cement upon which to lay the plate is not necessary, but in incised work it must be employed, or else the tool will not cut. In fact, with all methods of decorating metal it is far better to place it on a pitch block; the work is then firmly fixed, and is more easily managed. A pan, Fig. 1104, is required into which to pour the pitch. This, when filled, is a solid block, and the metal to be chased is placed upon this, rough side downwards, so that it adheres in every part. It is thus firmly fixed, and this is better than placing on a flat sheet of lead or wood, a device often resorted to by those who have a dislike to the use of pitch. Working on lead or wood is really more difficult, and is liable to cause the piece of metal to "buckle" or become uneven, and it is difficult to make it level again. The use of pitch is recommended for all kinds of metalwork. The oblong pan with two handles is the most useful shape. An ordinary baker's tin, or dripping pan, is suitable, or one could be made of galvanised iron. It should be about 14 in. long by 9 in. wide, and about 1½ in. deep.

THE PITCH OR CEMENT.

The cement for filling the tin consists half of common pitch and half of plaster-of-Paris; about 4 lb. of pitch and the same of plaster will be sufficient to fill a tin of the size given. Another recipe for the cement is: Brickdust 8 oz., resin 1 oz., and linseed oil one tablespoonful. This composition is softer and more tenacious than the first, and is more suitable should the work require to be beaten up from the back. In this case the metal has to be very tightly held down on the pitch block, or else it will spring off, dragging some of the pitch with it. For mixing the cement an iron kettle about 8 in. in diameter (see Fig. 1009, p. 334) will be required. A handle can be made with a piece of wire as shown. An iron ladle is handy for taking out the pitch when melted, or it may be poured straight into the pan. A stick or an iron spoon will do for mixing. The pitch should

Fig. 1105.—Spatula.

be first dissolved with gentle heat, and the plaster then added by degrees. If too much plaster is suddenly thrown into the cement it is apt to boil over. The kettle may be warmed on a slow fire or a small stove. When the pitch and plaster have been thoroughly incorporated, pour the liquid into the pan, and allow it to cool until it is of the consistency of putty before the metal is placed on it.

OTHER ACCESSORIES.

To deaden the noise caused by working on metal, cover the table on which the pitch block is placed with a piece of thick green baize or an old cloth. A cushion (see Fig. 1008, p. 334) can also be used to work on. This is an iron ring, with strips of list wrapped around it until it is about 2 in. thick. Another method is to use a canvas bag filled with sand. To remove the metal at any time, the surface of the pitch must be heated, a gas blowpipe or painter's blow-lamp being used for the purpose. Other useful appliances are a pair of hand shears, hand shears with curved blades, a wooden mallet, a spatula (Fig. 1105), and a hand drill for boring holes in the metal. The spatula is used for pushing the pitch over the edges of the metal after it has been placed on the block, but an old knife blade will serve the purpose almost as well. When either of these tools is used it should be dipped in cold water to prevent the pitch sticking to it. A pair of sleeves made of alpaca or linen, with a piece of elastic run in at each end, and reaching to a little above the elbow, are useful while the artist is working.

THE CHOICE OF METAL.

Brass or copper is the best metal for the beginner. Incised work is more suitable to the former, as it is not so tough, and is easier to cut than copper. Copper being soft, and therefore readily raised from the back, is better for beaten-up work. Bold designs rather than intricate patterns should be selected for working in copper. Brass is the most suitable material for incised or repoussé work. The best for Oriental brasswork is "best brazing brass," though the commoner or yellow brass is generally used in the manufacture of trays, and can be worked perfectly well. The former, however, being largely composed of copper, attains a more brilliant colour when it is polished.

POLISHING SHEET BRASS.

The sheet of metal as purchased is unpolished, and it must be polished on one side before the design is chased. A polishing lathe is the best apparatus for this purpose, emery wheels being employed, and the final gloss being obtained by means of Sheffield lime. A very fair surface can be produced by using rottenstone and lard oil. A drop or two of the oil is first poured on the plate, then a little rottenstone dusted on, and rubbed with a pad of cloth, or cloth wrapped round a piece of wood. A piece of thick brown paper should be placed underneath the brass to prevent the grease from soiling the bench or table, and the hands should be protected with a pair of chamois leather gloves. The rottenstone makes very fine scratches, which are scarcely perceptible to the eye. The surface of the metal, however, often has scales upon it, and the polishing lathe is the only means by which these can be removed.

Cutting and Levelling Sheet Brass.

The sheet of brass being properly polished, a portion large enough to take the design, and allowing about $\frac{1}{2}$ in. all round for resting on the pitch, must be cut off with the hand shears. A square is useful for measuring, the size being marked on the metal with a scriber. If the plate gets out of shape and unlevel, it should be made level on the felt board, using the mallet. This must be done carefully or the plate will " buckle," and the best way to remedy this is to give blows with the mallet round the edges of the raised part. This will expand the metal and allow it to settle into its place again. A common hammer should never be used for levelling, as it will cause dents in the surface, and these, when once made, cannot be entirely removed. It is not necessary to get the metal absolutely level, as any slight unevenness will disappear after the design has been worked and the article finished. The brass should not be beaten too severely whilst flattening, as it is apt to get hard. A few light blows will accomplish the purpose more readily than much hard work. Wood or lead answer very well as a levelling block, and will not damage the metal.

Transferring Designs.

When the brass is prepared, the design must first be drawn on strong tracing paper and then transferred by means of carbon paper to the surface. A little spirits of turpentine smeared over the surface of the metal with a piece of rag or cotton waste, and then allowed to dry, will make the lines stand out clearer. Having placed the paper design on the metal, put a weight on the edge to keep it from slipping. Then insert the carbon paper between the two and go over the outline carefully with a tracer, which may consist of a steel point, or even a bone point, ground moderately fine. The transferring is sometimes done by rubbing red chalk over the back of the design, but the carbonised paper gives the clearest outline. Do not press too hard with the tracer, otherwise the paper will be cut. After the outline has been carefully traced, it can be scratched in, if desired, with the tracer, should any portions become blurred while working. The outside line, which has to be cut out afterwards with the hand shears, and is apt to get rubbed out in cleaning off the pitch, should be scratched in.

Fixing Metal to Pitch Block.

The next process is to fix the metal on the pitch block, for which purpose a blowpipe or blow-lamp is required. Having warmed the surface of the pitch until it is almost liquid, place the metal on it, and with the spatula push the cement over the edges, taking care that it does not go too far and spoil the pattern. The spatula should be dipped in cold water to prevent the pitch adhering to it. When the plate is firmly fixed on the block, put a weight on it until it cools. In the case of a tray that is slightly raised in the centre, a good plan is to pour some liquid cement on to the back until it is almost level with the edge so that no spaces occur underneath where the air can get in. If this happens the metal will sink in when it is being matted, and it must be taken off and fastened afresh to the block.

Chaser's Pitch Bowl.

The above method is the best for flat work, but if bowls and other hollow articles are to be chased, a chaser's bowl (Fig. 1006, p. 334) will be required. This is of cast-iron, and resembles a cannon ball hollowed out. When this bowl is filled with pitch the article to be worked is fastened down upon it in the same manner as on a pitch block. The cushion (see Fig. 1008, p. 334) is useful for receiving the bowl, which is thus easily managed and can be turned in any direction. If two hollow articles, such as finger bowls, are to be worked, it is a good plan to fill them both with pitch, and, having warmed the surfaces, to stick them together. They can then be placed between two pieces of wood in a vice whilst being chased, the chaser's bowl in this case being dispensed with.

Engraving the Design.

When the cement has cooled sufficiently, the engraving of the design can be done. The method of holding the hammer and

Fig. 1106.—Easy Curves for Chasing.

tool is shown in Fig. 1098, p. 365. Pressing the tool rather hard upon the metal, give a tap or two with the hammer until the point sinks a little into the brass. Continue until the outline is engraved, making a clean cut and removing the chips of metal as the tool proceeds. A few easy curves, as illustrated by Fig. 1106, should be attempted when beginning, any waste piece of metal being used for the purpose. Circles or straight lines are probably the most difficult to cut well and neatly, and time and perseverance are necessary before the beginner will accomplish these satisfactorily. In going round a sharp curve the tool must be raised slightly; it must be driven much more slowly, and the blows of the hammer must be lighter and more quickly delivered. The work should be moved as required. A bold cut, rather

than merely scratching the surface of the metal, should be aimed at, as it gives a better effect to the design and makes it stand out. This also applies to ordinary chasing.

Completing the Design.

When the outline has been completed, the groundwork or matting should be introduced, as previously described. When the matting has been carefully inserted, the work is finished. Should any small portions of matting be omitted, after the work has been removed from the pitch block, they may be filled in by placing the metal on a sheet of lead; but if the space is a large one, the plate should be fixed again to the cement. To beat up the design from the back, the brass must be laid again on the pitch block face downwards, the raising tools (Fig. 1102) being used. A hammer, heavier than the one used for chasing the outline, will be required. As the amount of relief produced cannot be seen, the work may have to be taken up several times before the result of the modelling is satisfactory.

Removing Metal from Pitch Block.

After the design has been worked, the metal must be removed from the pitch block; this must be done whilst the pitch is hard, and no attempt at heating it should be made. Take a hammer and chisel, or a screwdriver, and chip off the cement, holding the plate round the edges, and taking care not to let the tool slip and damage the surface of the design. Having removed as much as possible of the pitch, put the end of the chisel underneath the metal and lever it off the block. It is a good plan, after this operation, to warm the surface of the pitch in order that it may be ready for the next piece of work.

Removing Pitch from Sheet Brass.

The pitch adhering to the plate must now be removed; the best appliance for doing this is a blowlamp. A small paraffin stove will do; or the metal may be warmed over a fire. When its underside begins to smoke, take a piece of cotton waste and rub off the pitch. A few drops of paraffin poured on will facilitate the process considerably. It is well to hold small pieces of work with a pair of long pliers to avoid burning the fingers.

CUTTING DOWN BRASS PLATE.

When the pitch has been removed from both sides of the metal, put it aside to

Fig. 1107.—Candlestick and Sconce.

cool. Then take the hand-shears and cut as near as possible to the outline, which should have been previously scratched in with a scriber; the bent snips are useful for cutting curves, or, if necessary, these parts can be filed. As the metal will have become uneven in working, it must be rendered level again by means of the wooden mallet, which will probably not be a difficult operation, as warming the metal while cleaning off the pitch will have annealed or softened it, thus making it more malleable.

BRASSWORK CANDLESTICKS.

Brasswork candlesticks can be constructed of various shapes; the one illustrated by Fig. 1107 is easy to make. Fig. 1108 shows the tray of the candlestick in the flat. The first thing to do, when the metal is engraved, is to shape the curves of the edge with a half-round file, and finish with a finer file. The sconce is a piece of brass tubing which is skimmed in the lathe; it is 1¾ in. long, and may have a flange at the top or be simply straight. Sconces with flanges look best, but are rather difficult to make owing to the metal being liable to split. Mark off about ¼ in. all round for the pins whilst the sconce is in the lathe, and, with a metal saw, remove two opposite pieces as shown at A (Fig. 1107). The pins, which must be exactly opposite each other, should be filed round. Before engraving the metal, a small circle, the size of the bottom of the sconce, should be

drawn in the centre and scratched on the metal. With a hand-drill bore two holes slightly larger than the size of the pins; fit the latter in, and turn them down with a hammer level with the back. This should be done when the whole has been finished and lacquered as described in the next chapter.

CANDLESTICK HANDLE.

To make the handle, cut out a strip of brass 6¼ in. by 1¼ in., and at about 1 in. from one end snip off ¼ in. from each side to form the part for fixing to the tray of the candlestick, and round off the corners with a file. Turn the edges over as far as the cut portion with the aid of a jenny, or beat with the wooden mallet over a hatchet stake. Bore three holes in the handle as shown in Fig. 1107, and also in the tray for the brass pins. Turn up the scallops of the tray all round over the semicircular side of the stake with the mallet, taking care not to dent the metal. Then rivet on the handle. Insert a brass pin in one of the holes in the handle and in the tray, and with a pair of nippers cut off the end almost flush with the bottom. Turn the candlestick over, resting the head of the pin on a suitable iron stake, and, by means of hammering, rivet over

Fig. 1108.—Candlestick Tray before Bending.

the other end until the pin holds the handle and tray together. Proceed in this manner with the other two pins, when the handle should be firmly fixed. A piece of dark cloth glued to the bottom gives a better finish to the candlestick. A brass candle-ring can be added to prevent the grease falling on the decorative work.

ORNAMENTAL BRASSWORK TRAYS.

Trays in various shapes afford great scope to the worker in metal, but those with plain edges are difficult to make, as they are spun in the lathe by a method to

Fig. 1109.—Square Tray.

be explained in detail in a later chapter. In taking a tray off the pitch block, a tap or two round the edge with the wooden mallet should be given, and it can then be easily removed. If the matting has caused the centre to become depressed, a few taps with the mallet at the back will put it right again. A tray can be easily made to the design shown in Fig. 1108, but on a larger scale. In this case it must be worked first in the flat and turned up afterwards. A piece of brass, 1 ft. square, allowing about ¾ in. for the edge, can be made into a serviceable waiter, suitable for visiting cards, etc. Square trays (as in Fig. 1109) are easier to construct than round ones. The sides can be turned up over a wooden block or the iron stake. A wooden block is best, as there is then no danger of damaging the metal.

ORNAMENTAL PEN-TRAYS.

For pen-trays a longer and narrower block must be used to turn up the sides. The triangular pieces at the corners must be cut out with the shears, and when the sides are turned up the edges will meet

Fig. 1110.—Pen-tray before Bending.

and can be soldered together. They can also be formed in the same way as the square tray. There are other shapes for trays, but those illustrated are the simplest. Fig. 1110 shows the pen-tray in

the flat, with the corners marked for cutting.

SCONCE BACK.

Fig. 1111 shows a simple form of sconce back which is easily made. When the

Fig. 1111.—Old Style of Sconce Back.

design has been worked in the flat, the bottom portion is placed in a wooden vice, or between two pieces of wood, and the back turned over square with the mallet. The sconce is then fixed on in exactly the same way as for the candlestick (Fig. 1107). This article can be made larger, so as to accommodate two

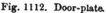

Fig. 1112. Door-plate.   Fig. 1113. Lock-plate.

or more sconces. A hole by which to hang it up should be bored at the top.

DOOR AND LOCK PLATES.

Door-plates made in brass or copper are effective, and Fig. 1112 is a design which answers for incised work or chasing. A hole at each corner should be bored to take ball-headed brass screws for fixing. A rigid style of pattern should be chosen for door-plates, and the bolder it is the better. Fig. 1113 is a design for a lock-plate.

PHOTOGRAPH FRAMES.

Fig. 1114 shows a photograph frame of some dark wood such as walnut or mahogany, with brass screwed on. The

Fig. 1115.—Ornamented Bellows.

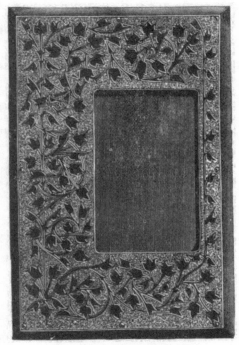

Fig. 1114.—Photograph Frame.

metal must be worked first and then cut to the proper size. The centre portion, which should be a little smaller than the photograph, can be easiest removed with the bent snips after a large hole is bored in one corner; the brass should be finished off with a file. The back, which should be about ½ in. larger all round than the metal, should be planed smooth and the edges chamfered. The opening to take the photograph must be cut out

Fig. 1116.—Frame for Barometer or Clock.

with a fine fret-saw, so as to keep the wood intact. Plane this down on one side and fasten it to the back of the frame with small brass hinges, also fixing on a catch to prevent it opening. A support can be fastened to the back with a hinge, and after the brass has been screwed to the face the photograph frame will be complete.

ORNAMENTED BELLOWS.

Fig. 1115 is a design in incised work for ornamenting bellows. The piece of decorative brasswork should be fastened on with brass screws.

CLOCK OR BAROMETER FRAME.

Fig. 1116 is a design for a round frame to hold a small clock or barometer. The brass is first cut out and worked and then fixed to the wooden frame; walnut is the best wood for this. The brass, which should exactly fit the circular part intended for it, can be fixed with small screws, or it can be glued to the wood. A brass ring by which to hang the frame should be screwed to the back.

DESIGNS FOR INCISED BRASSWORK.

The designs best adapted for incised work are those consisting of simple outlines. A flower, for instance, if accurately represented on paper, would not be suitable for this work; it must be conventionalised before it can be rendered satisfactorily on the metal. Geometrical patterns, in which easy curves predominate, are the best for the beginner. In improvising original designs, the quarter or half should be drawn first and then repeated. For instance, if it is intended to draw a pattern for a waiter, a circle the required size should be first struck out with the compasses and then divided into as many equal parts as desired. When the design has been accurately drawn in one of these divisions it is repeated all round. If there is a border to go round the edge, the same plan must be adopted, or else, when the pattern is completed, the ends will not meet. The designer should bear in mind that repetition, if not carried too far, plays an important part in nearly all decorative art; besides, effective designs could not be invented easily without its means.

FINISHING, LACQUERING, AND COLOURING BRASS.

FINISHING BRASS.

THERE are several methods of finishing brasswork, some articles being dipped and lacquered, others dipped and burnished, others polished and lacquered, and others polished, burnished, and lacquered.

DIPPING BRASS.

For dipping brass, one or two small lead-lined boshes or tanks, to hold the acid solutions, will be needed, also a swilling bosh for clear running water, with a few large earthenware jowles, an earthenware basket, a number of pairs of brass or copper tongs, and copper hooks. A basket composed of cane, and made by the blind at Edgbaston, Birmingham, has been used in the place of the earthenware one with every success, and bids fair to supersede it. On no account must iron be used in any of the following operations, as it discolours the articles. These are first put into a bath of pickle composed, generally, of the liquids containing solutions of salts of copper and zinc from previous operations, as this acts very slowly and does not cause pitting or roughening of the surface of the metal. If the old solutions are not used, a pickle must be made of from 5 to 20 parts of sulphuric acid to 100 parts of water. Into this bath the articles are placed till all discoloration from sulphur, annealing, or oxidation is removed, when they must be strung on wires ready for the dipping proper.

NITRIC ACID DIPPING BATH.

Nitric acid, which forms the dipping bath, exerts a more powerful action on zinc than on copper, hence the surface of dipped brass assumes a warmer shade, owing to the removal of more zinc than copper. The colour may be varied to a limited extent by using acid baths of different strengths. Nitric acid, too, with varying quantities of nitrous acid, produces different shades. Care must be exercised in the selection of the nitric acid or aquafortis for this part of the operation, as there are several qualities of acid in the trade. One, perfectly white, does not usually contain nitrous acid, and does not cleanse well, especially if freshly mixed; another, which is darker in colour, acts too energetically and pits the article dipped in it; a third, of a pale straw-yellow colour, is preferable. A little sawdust put into the strong nitric acid will generate nitrous fumes by the decomposition of the organic substance of the sawdust previous to diluting with water. If the article is not bright enough at the first dip, it must be dipped again and again, but between these operations it must be thoroughly well washed or swilled.

DEAD DIPPING.

Dead dipping, or the process by which the articles are dipped to produce a dead yellow surface, is performed as follows:— When the article has been thoroughly well pickled and swilled it is placed in stronger acid, producing a frothy appearance. This is rinsed off by washing in water, and the article is dipped for a few seconds in strong nitric acid, and washed in a bath containing dissolved argol or cream of tartar, and finally dried in sawdust. The articles in this process must be treated individually, not strung together. After the dipping the article is ready for burnishing or for lacquering.

BURNISHERS.

Burnishing is generally the most expeditious of all methods of polishing brass, and gives the greatest lustre to the finished article, and by its means the marks left in polishing are removed. The forms of the tools used are various, the shapes being made to suit the projections or hollows of the work under operation. (This is fully explained on pp. 145 to 148.) The burnishers are of steel and of stone. The latter is usually bloodstone, and the best is of a steel grey colour and contains the largest proportion of iron. The tools must be exceedingly well polished; this is usually done with emery, finishing with a leather bob covered with crocus.

BURNISHING BRASS.

The articles being burnished are covered with water, soap-suds, or vinegar water. The operation of burnishing is very simple, but the result will depend entirely on the skill of the workman. The burnishing tool must be pressed hard against the part of the article being burnished, and must be moved with a gliding motion backwards and forwards over the work. The article is then dipped in soapsuds and dried in hot sawdust, and it is then ready for lacquering.

POLISHING HIGH-CLASS BRASSWORK.

Before beginning to polish brass all marks of the file must be removed. Having used a superfine Lancashire file to smooth both the edges and the surfaces, take a piece of moderately fine emery paper and wrap it tightly, once only, round the file. By having many folds round the file the work becomes rounded at the edges, and this shows poor workmanship. Some use emery sticks, made of pieces of planed wood, about ⅜ in. thick and ¼ in. wide, quite flat on the surfaces. They are covered with thin glue, and the emery is powdered on to them, and then allowed to dry hard. Common work is rubbed over with emery cloth, but this will not do for good work. The paper folded once round the file is used in a similar manner to the file, and when the file-marks disappear, and the paper is worn, a little oil should be used to make it cut smoother. The edges and surfaces being prepared to this extent, the edges must be finished. To effect this, take a piece of flat, soft wood, and apply to its surface a little fine oilstone-powder; be sure that it is quite clean, as it is very annoying to make, just as the work is finished, a scratch that may be perhaps so deep that it will require filing out. When the work has been finished with the stick and oilstone-powder, use a clean buff with rottenstone and oil; then dry rottenstone alone should be used, this ensuring a bright, clean and flat surface. The surfaces will next have all the marks of the file taken out, and be got quite smooth with emery paper. The best and most ornamental way of finishing such work is by curling; water-of-Ayr stone, used with water, will remove such scratches as may be left from the emery paper; also it produces a cloudy surface previous to using the charcoal. File one end flat, and having a basin of water handy, dip the charcoal into it, and by curling it round and round in all directions a surface is formed having a most distinctive appearance. It is necessary now to rub with a piece of slate-pencil brought to a point. This is also used with water, and is moved in small circles, which are not regularly formed, but are interlaced in all directions.

LACQUERING BRASS.

Lacquering consists simply in applying varnish to the surface of the metal to prevent loss of colour, and should be done within an hour of dipping and burnishing. The articles necessary are as follows:—A lacquering stove, a few flat dishes, basins, and brushes; the implements must be kept on the stove, which is described later. The two forms of lacquering are known as the cold and warm methods. By the cold method the lacquer is evenly and thinly applied with a flat camel-hair brush over the surface of the article, which is then placed on the stove to dry. The warmth of the stove is allowed on the article for a minute or two only to set the lacquer, which finishes the work. In most brasswork, however, the articles must be heated on the stove to the temperature of boiling water; this drives off all moisture, and evaporates the spirit contained in the lacquer. If the work is too hot it oxidises, while if it is too cold the appearance is streaky, and the lacquer does not set thoroughly. This difficulty can be overcome only by experience, so different are the natures of the materials, the qualities of the lacquers, and the effects. The lacquer is brushed on with a flat camel-hair brush, as before, but great care is necessary or the work will have a most patchy and unsatisfactory appearance when it is finished.

Brush for Lacquering.

Some beginners almost invariably get their lacquering streaky. This is often due to the separation of the brush—that is, the hair of the brush, instead of keeping in one broad surface, spreads out in two or three portions pointed at the ends; thus the parts of the metal between the points do not get their share of lacquer. When this is the case better brushes, with a good body of hair, should be bought. In the opinion of a practical worker, the best kind of brush for applying lacquer to metal is one made in the following manner:—Take a piece of wood, a little broader than the work to be lacquered, and shape it to resemble the handle of a whitewash brush. With a thick saw cut a slit into it edgewise, and take a narrow strip of clean flannel, as long as the wood is broad, and fold it the longest way; then take a piece of white nankeen cloth, and fold it round the outside of the flannel; put them both in the saw kerf in the wood, with their folded edge outward, and fasten the cloth to the wood by means of screws passing through the side. Before fastening tight, put a piece of straight wire, about a quarter of an inch thick, through the bow of the folded cloth, and pull the cloth tight against the wire so as to make it smooth and straight. After the cloth is fastened tight in the wood, the wire is withdrawn, and the brush is specially fitted to be used for lacquer. The woollen cloth holds the lacquer, while the nankeen cloth prevents it flowing too freely, and also presents a smooth surface to the metal that is to be lacquered, while it prevents any particles coming off the woollen cloth on to the lacquered surface. A brush of this kind must not be dipped into a bowl of lacquer, but the lacquer must be put on to it with a common brush. Such a brush coats large flat surfaces very evenly. First coating the work with alcohol, or very thin lacquer, causes the lacquer, when applied, to flow more easily and regularly. The brush must be laid on the work very lightly and with a slight curved motion at the beginning of the stroke, so that it will miss the sharp edge of the work by which a portion of the lacquer would be pressed out and flow irregularly over the edge. The brush must then be drawn straight, and with equal pressure along the surface of the metal, and be lifted off the instant it reaches the other edge. In moderately broad surfaces a brush the full breadth of the work should be used; but in very wide surfaces, and where there are a number of large holes in the work, a brush is difficult to use.

Lacquering Stove.

The stove usually consists of an iron pan of any convenient diameter by about 4 in. deep, and supported on legs into which the exhaust steam from an engine may be conveyed and used. In other forms the heat is obtained from a Bunsen burner, which is more convenient, as the heat may be regulated as required. A lacquering stove heated by gas might be made as follows:—Obtain a sheet of iron ⅛ in. thick, 3 ft. long, by 2 ft. 6 in. wide. Mark 6 in. margin all round, and bend down, except one side (see Fig. 1117), this side being only 1 in. deep. Cut another piece 2 ft. 2 in. by 2 ft., and bend it up in 1 in. round three sides. Place this piece on the bottom of the other piece, and rivet the two together; this will make a shallow iron box 2 ft. square and 6 in. deep. The burner may be either round or rectangular. Fletcher's triple concentric burner is a good one; it allows of three separate heats. It must be lighted and placed inside the box, care being first taken that the airway to the burner is outside, so that the air to the burner may not be impeded. A stove of this description will give excellent results. The stove can stand on a pair of light iron legs; or a table might be made of light angle-iron framing.

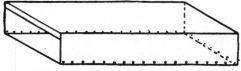

Fig. 1117.—Simple Lacquering Stove.

LACQUERS FOR BRASS

At the present time most of the work is done by means of ready-made lacquers, several firms preparing special lacquers, as zapon. The following will be useful for those who prefer to make their own:— Pale lacquer: 1 gal. methylated spirits of wine, 5 oz. of shellac, 4 oz. of gum sandarach, and 1 oz. of gum elemi. Mix well and expose to a gentle heat for a day or two, strain off, add ½ gal. of spirit to the sediment, and treat as before described. To make a green lacquer, add 6 oz. of turmeric and 1 oz. of gum gamboge to the pale lacquer when mixing. For pale gold lacquer: to 1 gal. methylated spirits of wine, add 10 oz. seedlac bruised and ½ oz. red sanders. Strain when thoroughly dissolved. Highly coloured lacquer may be prepared as follows:—To 2 qt. spirits of wine, add 2½ oz. shellac, 2 oz. gum sandarach, and ½ oz. gum elemi. Mix and keep gently warmed for two or three days; strain, colour with dragon's blood, and thin with 1 qt. spirits of wine. All lacquers must be kept in tins or stoneware, free from light and heat; light causes the lacquer to become darker in colour, and heat causes evaporation of spirit and consequent change. Preparations are given in the table on p. 379, and will be found reliable.

ADDITIONAL RECIPES FOR LACQUER.

The various fancy colourings may be given with any of the aniline colouring matters dissolved in alcohol. For making the undermentioned lacquers only the simplest appliances are requisite, such as tins, bottles, large funnels, filters, filtering paper, stirrers, measures, and other common workshop appliances. Gold lacquer: Seed lac, 4 parts; gamboge, 1 to 2 parts, according to depth of required colour; dragon's-blood, ½ part; dried spirit, 36 parts. Deep gold lacquer: Seed lac, 10 parts; turmeric, 4 parts; gamboge, 4 parts; dragon's-blood, 1 part; dried spirit, 80 parts. Pale brass lacquer: Bleached lac, 12 parts; aloes, 2 parts; gamboge, 1 part; dried spirit, 100 parts. Bronze lacquer: Lac, 4 parts; sandarach, 2 parts; gum acaroides, 2 parts; gamboge, 2 parts; dried spirit, 80 parts.

USING ZAPON.

Zapon is a substitute for lacquer, and is now used by most of the leading brass manufacturers. It is made and sold by the Fredk. Crane Chemical Co., 22 and 23, Newhall Hill, Birmingham, and has been found far superior to ordinary lacquer, both in ease of application and in appearance. For figured work especially it is suitable, as it preserves the finest hair-lines in all their metallic sharpness. It does not set so quickly as lacquer, which is an advantage on plain, highly polished surfaces. In using zapon, it is very important that as much care is taken in cleaning the work, and freeing it from all grease, polishing material, etc., as in ordinary lacquering; and the zaponning should be done in a room which is free from dust and draughts, and the articles should not be handled, as the fingers leave marks, as well as grease, on the polished surface. Articles to be zaponned do not require to be heated, another advantage, but the liquid is best applied in a warm room, and the articles should not be quite chilly. There are two ways of applying zapon—namely, by dipping and brushing. It is recommended to dip all articles that will admit of so doing, but in cases where, from consideration of expense or from the size of the article, this is not practicable, brushing may be resorted to. In this case, "dip" zapon should be flowed on with a full brush, care being taken not to brush over any part a second time until it has become dry, when any part that is at all defective may be easily touched up. Wave marks or brush marks will entirely disappear when dry. Should the articles

It can be mixed as desired, an old cup being used for the purpose. A small quantity of the gold colour will suffice.

PREPARING REPOUSSÉ WORK FOR COATING WITH BRASSOLINE.

Before lacquering with the brassoline, the work must be prepared. As the polish of the metal may be tarnished in cleaning off the pitch, the article should be given a final rub up with rottenstone and oil as previously described. Now pour a few drops of paraffin on the surface and wipe over with a clean rag to remove any

LACQUERS.

Description.	Spirits of Wine.	Shellac.	Gamboge.	Cape Aloes.	Turmeric.	Saffron.	Dragon's Blood.	Annatto.	Gum Sandarach.	Gum Juniper.	Amber.	Extract Red Sanders.	Alcohol.	Turpentine Varnish.	Gum Elemi.
	Gal.	Oz.	Oz.	Oz.	Oz.	Oz.	Oz.	Oz.	Oz.	Oz.	Oz.	Dr.	Pt.	Oz.	Oz.
Simple ...	1	8	—	—	—	—	—	—	—	—	—	—	—	—	—
Fine Pale ...	1	8	—	1½	—	—	—	—	—	—	—	—	—	—	—
„ „ ...	1	8	¼	—	¼	1	—	—	—	—	—	—	—	—	—
„ „ ...	2	8	8	—	¼	2	¼	¼	4	—	—	—	—	—	—
Pale Gold ...	2	16	—	—	16	—	¼	4	4	—	—	—	—	—	—
Pale Yellow ...	1	16	1	2	—	—	—	4	4	—	—	—	—	—	—
„ „ ...	1½	4	4	—	—	1	x½	¼	—	—	2	¼	—	—	—
„ „ ...	1	3	2	—	1½	—	—	—	—	—	—	—	1	—	—
Pale ...	1	8	¼	—	1	¼	—	¼	—	—	—	—	—	—	—
„ „ ...		3	—	—	1	—	¼	—	—	—	—	—	—	—	—
Gold ...	1	24	—	—	8	1	—	1	—	—	—	—	—	—	—
„ „ ...	1	6	¼	—	8	—	—	—	1½	—	—	—	—	12	—
„ „ ...	1	8	¼	1	10	—	—	—	2¼	—	—	—	—	—	2
„ „ ...	—	1	¾	—	—	2 gr.	1	—	2	—	—	—	1	—	¼
Deep ...	½	2½	—	—	—	—	x½	—	2	—	—	—	—	—	—
Red ...	1	8	—	—	—	—	4	16	—	—	—	—	—	—	—
Green ...	1	8	¼	—	2	—	—	—	—	—	—	—	—	—	¼

peel after coating, the cause is grease. "Brush Zapon," it may be said, should be applied as thin as possible; therefore, wipe the brush free from all excess, precisely as in ordinary lacquering. Brush once only over the surface to be covered, and allow it to dry; a second coat may then be given if desirable, but it is not necessary if the work is covered. Zapon can be had colourless, and in various colours, gold, green, etc.

USING BRASSOLINE.

Brassoline is prepared by the makers of zapon, and is just as useful in its way. For finishing brass repoussé, or other decorative work, it is well to obtain two bottles, one containing an almost colourless fluid, and the other a deep gold shade.

dirt or grease. Then dip the article in aquafortis, or dipping acid; and the stronger this is the better, especially for old brass. Large pieces of work are held with a pair of brass tongs, and the smaller ones with a piece of copper wire twisted round. When the acid has changed the colour of the brass and the article is quite bright and clean, remove it and plunge it into clean cold water; it is advisable to have two or three vessels of water in which to rinse. It should then be dried in hot sawdust or with a cloth.

ANOTHER METHOD OF PREPARING REPOUSSÉ FOR LACQUERING.

Another method of dipping is, when the work is wiped clean with paraffin, to take a slice of lemon and well rub the surface. Then let cold water run over it until every particle of lemon is washed off. This is a mild kind of dipping, and in cases where the brass is only slightly discoloured will have almost the same effect as the former plan. When dried the article is ready for lacquering. This need be done on the bright side only, the back of any article that is mounted on wood, such as a door-plate, not requiring it.

Applying Brassoline.

Brassoline can be applied to repoussé work whilst this is cold, but it is better if slightly warm. Having mixed the lacquer to the desired shade, take a little on a brush, and, the article having been warmed, apply the lacquer with even strokes, and do not go over the same ground twice. Care must be taken that the surface is entirely covered with lacquer; if not, the spaces omitted will soon tarnish. When this operation has been completed, heat the article again and put it aside to harden. If properly applied the coating will preserve the article bright for years. If the lacquer shows rainbow tints, too little has been laid on; a moderate quantity on the brush produces the best results. Do not finger the metal more than necessary, either before or after the process of lacquering, or the brilliancy of the polish will be injured. A brass plate to be mounted on wood should be lacquered before it is fixed, otherwise some of the lacquer might damage the frame. Candlesticks are better lacquered before the sconce has been fitted on; this also applies to sconce backs. If the brush becomes clogged with lacquer it must be washed with soft soap, which will soon render it soft. Methylated spirit is useful for removing lacquer and for cleaning out the cup.

Varnishes for Polished Brass.

The following varnishes, to be applied like lacquer, are useful for polished brass-work:—Add 1 part white lac to 5 parts alcohol; or, add 1 part white lac and 1 part gum mastic to 7 parts alcohol; or,

add 8 parts white lac, 2 parts gum sandarach, and 1 part Venetian turpentine to 50 parts alcohol; or, add 6 parts gum mastic, 2 parts gum elemi, and 12 parts Venetian turpentine to 64 parts alcohol.

Renovating Lacquered Brass and Copper.

Assume that an article composed of lacquered brass and copper requires to be cleaned, burnished, and relacquered. First remove the lacquer by brushing with an ordinary scrubbing-brush and strong boiling soda water. Then wash off with hot water, and polish with flour emery powder, crocus, and oil. Finish with dry crocus or very fine whiting. A calico dolly may be fixed to a lathe and the polishing done more easily. Then lacquer as previously described. Another method of cleaning off old lacquer is to place the metal articles in a boiling solution of carbonate of soda or potash, 1 lb. to a gallon of water. To remove the old lacquer, swill in clean water. Then dip in commercial aquafortis quickly several times till of a golden colour, swill each time in clean water, and add a pinch of cream of tartar to the last swilling. Dry out in hot sawdust. Burnish the bright parts with a steel burnisher, using a little oxgall to lubricate. Dry out in sawdust as before. Heat on a hot plate, and lacquer in the usual way. If the whole of the old lacquer is not removed, the result will be a most unworkmanlike patchiness and a generally unsatisfactory appearance.

Decay of Lacquered Surfaces.

Lacquered surfaces keep best in a genial, dry atmosphere, for where moisture from any source is present oxidation is accelerated. Where the air enveloping lacquered metal is repeatedly varying in temperature, there is often a deposit of moisture apart from any dampness of the place, and a distinct deposit of damp acts like a weak acid. When the rain beats on bare, polished brass, oxidation is greatly accelerated, and it is much the same with that which is lacquered. Where there is variableness of temperature the weight or size of the lacquered metal is a great factor in attracting moisture. Cold surfaces condense the water in warm air. A hollow lacquered brass would yield to any variation of temperature. But a solid ball would not accommodate itself so quickly to the temperature of the air surrounding it; consequently it would be covered with vapour. When moisture condenses, at first there is a steam-like settlement, which chills the transparency and lustre of a polished surface. The steaminess is caused by innumerable points of dampness pitching on the surface. These points are separate, and are regular in distribution and shape. It does not much matter whether the lacquered metal rests horizontally or vertically when the moisture is in this first stage, because it remains quite stationary in all positions, and the injurious effects of such deposit are always uniform. But when the moisture increases, these points swell in size and merge into one another. The moisture takes a somewhat globular form, but the uniformity of the deposit is disturbed, the lower parts of the globular points being thicker than the upper when the deposit rests on an inclined or a vertical surface. Moreover, on good clean surfaces of lacquer the moisture will be almost as inadhesive as it would be to grease. In the next stage the moisture takes to running in little streams, and adhesiveness is overcome by bulk. Naturally, in the line of running there is a marked transparency. The moisture is now most irregular, but it may safely be said that this stage is not reached by most lacquered surfaces.

How Lacquer Decays.

Now, in all evaporation there is shrinkage of the separate particles, but no separation. Deposits remain the same in number and shrink in bulk; also, they always tend to have a circular margin in the return, especially just before disappearing. The evaporating moisture clings round any little excrescence in the surface of the lacquer, and, in fact, scratches or dust determine to some extent both the situation and the shape that the globules take in evaporating. Speaking generally, the higher any articles are situated, the greater will be the decay of the lacquer on them. The large mural memorial brasses in churches may often be seen with the oxidation on them gradually deepening from their base upwards. So, also, lacquered brass gas-piping will show verdigris at the top, while the lower parts will be unaffected. The hot air ascends, and, besides being charged with vapour, has a tendency, by its temperature alone, to weaken the resisting power of the lacquer. What may be termed natural decay of lacquers takes place very slowly. After a lapse of fifteen or twenty years, or even longer, some lacquering will look almost as bright as when put on.

Bronzing Metals.

The term bronzing embraces the processes by which colours are imparted to metals through the agency of chemical means. At the present day the Japanese and French lead in bronzing, particularly the Japanese, who use several solutions of chemicals, by means of which articles on being boiled assume a very beautiful patina or coating. The following is the plan most generally adopted:—To 1 gal. of water add 438 gr. verdigris and 292 gr. sulphate of copper, and boil the articles in this. Brass takes a variety of tints from brown to black, according to the bath used. If the brass be dipped for a few seconds in a warmed neutral solution of crystallised acetate of copper it will show an orange tint. If the brass be dipped in a solution of copper, the result will be a greyish green; while if it be dipped in a solution of chloride of antimony for a single instant it will assume a beautiful violet. In all these cases the temperature of the article at the time of immersion will influence the result. If the article be boiled in a solution of copper sulphate, it will assume a moire appearance.

Brown to Deep Red Bronze.

A bronze that imparts to brass a shade from brown to a deep red is composed as follows:—Dissolve 2 oz. nitrate of iron and 2 oz. hyposulphite of soda in 1 pt. of water. When this is dissolved, immerse the articles in the bronze till they are of the required tint.

PALE GREEN TO OLIVE GREEN BRONZE.

A bronze that imparts to brass a shade from pale green to a deep olive green is made as follows:—To 2 parts of water add 1 part of perchloride of iron. Well mix and immerse the brass in it. To give a dark green tint, immerse the articles in a saturated solution of nitric acid and copper, and afterwards heat them.

USEFUL BRONZE FOR SMALL ARTICLES.

A very useful bronze for small brass articles is made from 1 part oxide of iron,

3¼ dwt. of hydrochloric acid must be well mixed in 1¾ pt. of water. After the article has been dipped it must be well washed and dried in hot sawdust. For warmer shades the following must be used at 60° C.: Add together 6½ dwt. each of chromic acid, hydrochloric acid, and potassium permanganate, 1 oz. 12 dwt. of sulphate iron, and 1¾ pt. of water. For browns of all shades the article must be dipped in a solution of nitrate of iron; the strength of the solution will determine the colour. For violet shades use a solution of chloride of antimony. For olive green, black the surface with a solution of iron and arsenic dissolved in hydrochloric acid, and polish with a blacklead brush, and coat with a lacquer made as follows:

	Water.	Nitrate Iron.	Perchloride Iron.	Hyposulphite Soda.	Nitric Acid.	Nitrate Copper.	Oxalic Acid.	Potassium Cyanide.	Muriate Arsenic.	Permuriate Iron Solution.	Copper Chloride.	Nitrate Tin.
	Pint.	Oz.	Oz.	Oz.	Oz.	Oz	Oz.	Oz.	Oz.	Pt.	Oz.	Oz.
Brown to black	8	2½	—	—	—	—	—	—	—	—	—	—
„ „	8	—	2½	—	—	—	—	—	—	—	—	—
Brown to red	8	8¼	—	8	—	—	—	—	—	—	—	—
„ „	8	—	—	8	¼	—	—	—	—	—	—	—
Brownish red	8	—	—	—	—	8	8	—	—	—	—	—
Dark brown	8	—	—	—	2	—	—	8	—	—	—	—
Blue	8	—	—	10	—	—	—	—	—	—	—	—
Steel grey	8	—	—	—	—	—	—	—	8	—	—	—
Black	8	—	—	—	—	—	—	—	80	8	—	—
Brown to red	8	16	—	16	—	—	—	—	—	—	—	—
Olive green	1	—	10	—	—	—	—	—	—	—	—	—
Dark green	1	—	—	—	1	4	—	—	—	—	—	—
Black	8	—	—	—	—	—	—	—	—	—	2	2

1 part white arsenic, and 12 parts hydrochloric acid. The article must be well cleaned to remove dirt or grease, and the bronze should be applied with a brush. The process may be retarded at any point by well oiling the article, after which it may be varnished or clear lacquered.

BRONZING BY SIMPLE DIPPING.

The following liquids will be found useful for the bronzing of brass articles by simple dipping, and will give any tint to dark red. Immerse for half a minute in a cold solution prepared as follows: 6½ dwt. potassium permanganate, 1 oz. 12 dwt. of ferrous sulphate of iron, and

to 1 part shellac varnish add 4 parts turmeric and 1 of gamboge. For green grey, immerse the article in a solution of copper chloride. For steel grey, immerse the article in a dilute boiling solution of chloride of arsenic. For blue, immerse in a strong solution of hyposulphite of soda. For black, immerse in a solution of copper chloride and nitrate of tin. The heat to which the articles are subjected will materially alter the colour. It is, therefore, imperative to try the articles in solutions first, and so to arrive at a definite shade, meanwhile noting carefully all essential particulars. The above tabulated list shows the bronzing mixtures at a glance.

BRONZING BRASS BRACKETS.

Fancy brass brackets, such as gas brackets, are usually only dipped in a nitric acid bath and burnished. If the dipping does not give the desired brightness, the brackets are dipped again and again, and thoroughly washed and dried between each dipping. If the finish is not then suitable, the brackets may be dead dipped; this gives a dead yellow surface, which, after the prominent parts are burnished, presents a very artistic appearance. To dead dip, after well pickling the articles, place in stronger nitric acid till a frothy appearance results; then wash in water and dip for a few seconds in the strongest nitric acid. Wash in a bath containing a little dissolved argol or cream of tartar, and dry in warm sawdust; then burnish the articles and lacquer in clear lacquer. A different but equally pleasing appearance may be given to the brackets by bronzing. A bath that imparts to brass a shade from brown to a deep red can be made by dissolving 2 oz. of nitrate of iron and 2 oz. of hyposulphite of soda in 1 pt. of water. Immerse the articles in this till they are of the required tint. For a shade from a pale green to a deep olive green, add 1 part of perchloride of iron to 2 parts of water. For a dark green tint take 1 pt. of water, 1 oz. of nitric acid, and 4 oz. of nitrate of copper. When dry the articles may be burnished in the usual way in part, or plain lacquered with a clear lacquer, or they may be plain varnished, according to taste.

BRONZING COPPER.

A bronzing solution for copper may be made by dissolving in vinegar 2 parts of verdigris to 1 part of sal-ammoniac, the solution being heated and diluted with water until there is no precipitate visible. The copper should be thoroughly cleaned and rendered free from grease, and the article then is immersed in the above boiling solution, and afterwards rinsed in cold water and dried in sawdust. This process is repeated until the surface is of equal colour and the desired tint is obtained. Copper can be bronzed by simple immersion in a solution of 5 drachms of nitrate of iron to 1 pt. of water. A bronze similar to that on tea urns is produced by thoroughly cleaning the surface and coating it with a paste of crocus powder and water applied with a brush, and then holding it in a sheet of iron over a clear fire for about a minute. A deep bronze can be obtained by substituting finely powdered plumbago for the crocus. Rubbing the metal with a solution of potassium sulphide (liver of sulphur) produces, when the coat of solution is dry, the appearance of antique bronze.

COLOURING COPPER REPOUSSÉ WORK.

The colour seen on repoussé work is due to chemical action, and the most common solution used is composed as follows: 220 gr. of verdigris, 550 gr. of copper sulphate, 5 fl. dr. of acetic acid or vinegar, and 1 gal. of water. Boiling the articles in this will give them a brownish-red colour. Various shades of brown to nearly black may be obtained by painting the desired parts with ammonium sulphide, or even by allowing sulphur fumes to play on them. Various shades from purple to red may also be given to copper by gently heating it and allowing to cool in the air. The better plan, however, would be to obtain some colouring matter such as the zapon colouring mediums, which may be had in a great variety of tints; paint it on the articles cold, and then allow them to dry.

COPPERING LACQUERED BRASS.

In treating lacquered brass so that it will have the appearance of copper, the brass must first be boiled in a strong solution of soda to remove all lacquer and dirt, and then polished and washed clean. Now get a roll of fine iron wire; in the centre of this place the articles and dip in a vessel containing dipping acid, which may consist of diluted sulphuric acid. Then wash off and lacquer as previously described.

OXIDISING BRASS.

Brass is oxidised rapidly by heating with an electric current, or by making it the anode in a solution through which an electric current is passing, but the effect thus obtained would not be such as is generally desired. Brass can be blackened by coating with a thin layer of platinum deposited from a weak solution of platinum chloride without the aid of a battery or other generator of electricity. This does not oxidise the brass, but will produce the desired effects. The following methods are in general employment.

BLACKENING BRASS.

There are several methods of obtaining the black surface seen on optical and scientific instruments; a description of two will suffice. The article must first be dipped till bright in the nitric acid bath, then well rinsed in clean water till all acid is removed, and finally placed in the following mixture till it has assumed the black tint needed:—Take 1 part of sulphate of iron, and add 1 part of white arsenic and 12 parts of hydrochloric acid. It must next be well rinsed in clean water, dried in sawdust, and polished with black-lead, after which it may be lacquered pale green. The other method, and one generally adopted, is first to polish the article with tripoli, and afterwards to wash it well with a mixture made as follows:—To 1 part of nitrate of tin add 2 parts of chloride of gold dissolved in a little water and acid. Place the article in the wash for a few minutes, remove it, and wipe with a clean linen rag. Excess of acid increases the intensity of the black.

ANOTHER METHOD OF BLACKENING BRASS.

The following method of blackening brass will also be found very good, and is the same as that adopted in oxidising silver articles. Give the article a light silver-plating by deposition, in a similar manner to ordinary cheap electroplated goods. Then prepare a solution made as follows. Dissolve in a little acetic acid 2 dwt. of sulphate of copper, 1 dwt. of nitrate of potash, and 2 dwt. of muriate of ammonia. After warming the articles, apply the solution with a camel-hair pencil or immerse in the bath, then expose them to the fumes of sulphur in a closed box. This may readily be done by placing in a tin biscuit-box a red hot iron bowl, such as the bowl off a small lead ladle, in which are a few pieces of sulphur. Hang the articles on a rod across the tin, and close the lid. It will be necessary to do this where there is a fairly good draught to carry off the sulphur fumes which are very obnoxious.

COLOURING BRASS BLUE-BLACK.

To obtain on brass a blue-black shade, dissolve copper turnings in nitric acid until the latter is saturated—that is, until it will dissolve no more; immerse the brass articles in this bath, dry clean, and heat moderately over a clear coke fire, free from smoke. Repeat this process till the correct shade is obtained. Another bath that gives a blue-black shade consists of 1 pt. of water, 5 dr. of iron perchloride, and 1 oz. of hyposulphite of soda. Dip the articles in the solution till they are of the desired tint, then swill and dry in clean sawdust.

LATHES AND LATHEWORK.

THE SCOPE OF THIS CHAPTER.

IT being of the greatest importance that the tyro in metal turning should understand the construction of a lathe and the functions of its every part this chapter a lathe will be explained, this being followed by a description of the simplest accessories with which the plainest metal turning can be accomplished. The centering of ordinary rod metal, the shapes of simple hand-turning tools, and some elementary processes will also be described,

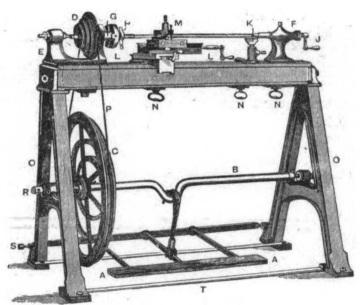

Fig. 1118.—Simple Metal-turning Lathe.

on lathework may appropriately begin with a description of a simple metal-turning lathe, and attention will be directed to Fig. 1118, which has been specially prepared to show the parts of such a tool. A typical workshop lathe with back-gear and one of the popular American bench lathes will next be described and illustrated, the method of testing and adjusting and then will come particulars and detail illustrations of many types of chucks—the appliances by means of which work is held for turning. The principal lathe attachments, the slide-rest, the division plate, and overhead motion will be shown and their functions made perfectly plain. The shape, use and principles governing the construction of slide-rest tools will be fully

gone into. The reader will then be in a
position to understand the construction
and advantages of lathes fitted up for
screw-cutting and other advanced turnery,

HAND LATHES.

With lathes that are driven by hand-
power, the ordinary drill-bow is the

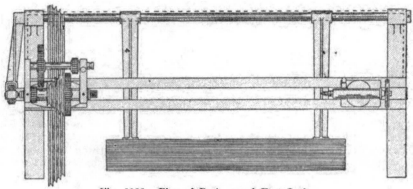

Fig. 1120.—Plan of Back-geared Foot Lathe.

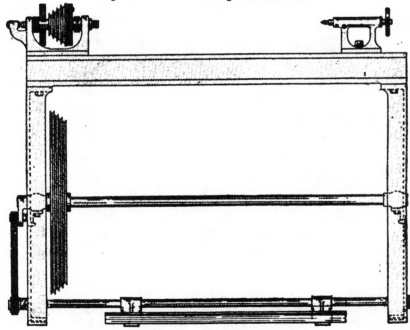

Fig. 1119.—Elevation of Back-geared Foot Lathe.

and illustrations of such lathes will be
given, to be followed by particulars of milling and slotting attachments and the
methods of making and using fly cutters.

means most frequently employed. Several
illustrations of such lathes are given on
pp. 173 to 176. Some hand lathes are
driven by means of a wheel, turned by the

left hand whilst the turner manipulates the turning tool with his right hand. The turns used by watchmakers is an example of the former; the throw used by clockmakers is an example of the latter. Hand lathes, which, a generation ago, were considered necessary for producing the most delicate turnery, have been superseded to a great extent by power-lathes. A lathe driven by other than hand-power, but having no slide-rest, is sometimes called a hand lathe, but the term is wrongly applied in such a case. Probably the fact of the machine being adapted for the use of hand tools only has suggested the erroneous term.

FOOT LATHES.

With lathes that are driven by foot-power, this is generally applied to a treadle, which is attached to a crank axle carrying a fly-wheel. Foot lathes have a wider field of use than either of the classes mentioned in the previous paragraph; as, obviously, it is a great advantage for the operator to have both his hands free.

POWER LATHES.

Power lathes are driven by steam, gas, electricity, wind, water or other power. These lathes involve the use of costly driving machinery, and for this reason their use is impossible to a great many workers.

SIMPLE METAL-TURNING LATHE DESCRIBED.

Fig. 1118 shows a very simple form of lathe, though the addition of a slide-rest removes it from extreme simplicity. A heavy iron stand fitted with treadle motion supports the actual lathe bed. A is the treadle working crank axle B, on which is the grooved flywheel C. From this a twisted gut band P drives the grooved pulley D, the whole of the casting E containing this pulley being known as the mandrel headstock or fast headstock. There are always two headstocks or "heads," the other being shown at F and going by the name of the loose headstock,

back headstock, poppet, or tailstock. The headstocks provide the two centres which support the revolving work, these centres being on the end of the mandrel, G, in the fast headstock, and on that of the barrel in the loose headstock. H is a carrier for transmitting the motion of the mandrel to the work. The barrel in the loose headstock can be drawn in or out by means of the handle J. For plain turning of the simplest kind, the tool is held by hand and is supported on a rest somewhere about midway between the centres. This rest, K, is shown near the end of the lathe bed L, where it is placed out of the way while the slide-rest M is in use. The slide-rest is a moving tool-holder, and its functions will be described in detail later. It will be noticed that whereas the

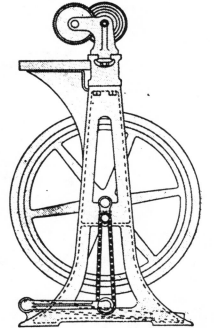

Fig. 1121.—End View of Back-geared Foot Lathe.

mandrel headstock E is bolted permanently to the lathe bed, the loose headstock, hand tool rest and slide-rest are movable and can be secured by the screw handles N just where they are required for use. The bed L and the side standards o must be substantial, for, should the metal bend appreciably—and the many clamping screws and bolts cause a tendency to do so—the centres are thrown out of alignment, and inaccurate work is

WELL-DESIGNED FOOT LATHE WITH BACK-GEAR.

Metal turners' lathes are known in great variety, but the ordinary workman, ama-

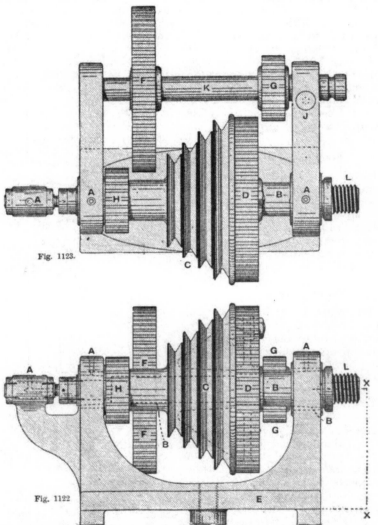

Fig. 1123.

Fig. 1122.

Figs. 1122 and 1123.—Plan and Elevation of Back-geared Mandrel Headstock.

produced. R and S are the cone pointed screws between which the crank axle and the treadle shaft respectively revolve, and T is a tie-rod to strengthen the stand.

teur or professional, uses a machine constructed on the lines of Figs. 1119 to 1121. Fig. 1119 is a front elevation, Fig. 1120 a plan, and Fig. 1121 an end elevation.

To make clear the shape of the parts of the lathe, reduced working drawings of the 4½-in. headstocks are also given. It may be said that many workers dispense

oil holes A, mandrel B, inside of pulley C, web of fixed gear-wheel D, and the tail-pin E, have their shapes and positions indicated by dotted lines. The end bear-

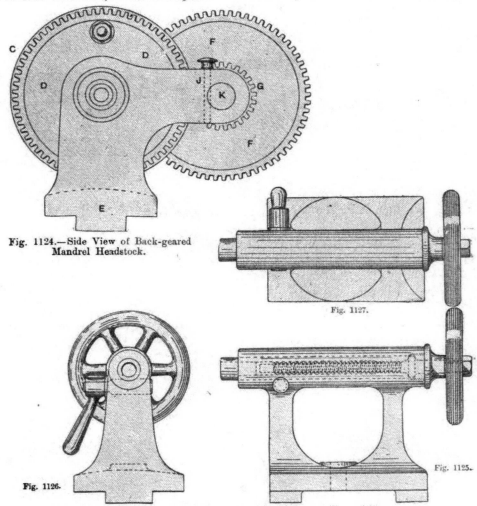

Fig. 1124.—Side View of Back-geared Mandrel Headstock.

Fig. 1127.

Fig. 1126.

Fig. 1125.

Figs. 1125 to 1127.—Front Elevation, Side Elevation, and Plan of Poppet.

with the back-gear (a species of change-speed-gear), in which case the mandrel headstock resembles Fig. 1132, which is shown on p. 391.

MANDREL HEADSTOCK AND BACK-GEAR.

Figs. 1122 and 1123 are elevation and plan of the mandrel headstock; the collars,

ing of the mandrel against the tail-pin is lubricated with oil supplied through a hole drilled in the centre of the tail-pin and continued vertically through the casting (see the dotted lines in Fig. 1122). Hard steel collars are fitted into the holes bored in the casting; over the rear end of the mandrel is slipped a hardened cone, and this is adjusted to a correct bearing by means of the pair of lock-nuts. Fig. 1124 shows the face of the mandrel headstock, and the fixed gear-wheel D, the back-gear F, and its pinion G. (The pinion carried by the mandrel is shown by H in

headstocks; thus a 4½-in. headstock measures 4½ in. from x to x, Fig. 1122.

BACK HEADSTOCK, TAILSTOCK, OR POPPET.

This is illustrated by Figs. 1125 to 1127,

Figs. 1128 and 1129.—Stepped Pulley Mandrel Headstock, and Poppet.

Fig. 1122.) The vertical pin J serves to keep the back-gear spindle K in position, to bring the wheels and pinions in or out of gear as may be required. The letters in Figs. 1122 and 1124 have the same references as those in Fig. 1123, which is a plan or top view of the back-geared mandrel headstock showing more clearly the general arrangement of the toothed wheels, etc. D H and F G are the gear-wheels and pinions carried by the mandrel and back-gear spindle respectively. The screwed

Figs. 1125 and 1126 being face and side views, and Fig. 1127 a plan. The dotted lines in Fig. 1125 show the interior construction of the sliding barrel. The screw used for moving the barrel in and out is left handed and square-threaded, the better to resist pressure, and it is just long enough to force out the cone centre when the barrel is drawn inwards. This is a great convenience that should always be arranged for in all lathes. It saves time and trouble in removing the cone centre.

Figs. 1130 and 1131.—Back-geared Stepped Pulley Mandrel Headstock, and Poppet.

nose of the mandrel is shown at L in Figs. 1122 and 1123; all the three figures are one-third full size. Height of centre from the bottom of the headstock is invariably the foundation of size in lathe

BACK-GEARING AND SYSTEM OF DRIVING.

The distinctive feature in the mandrel headstock illustrated by Figs. 1122 to 1124 —the back-gearing—consists in the arrangement of the wheels to give a slower speed to the mandrel, and correspondingly greater power over the work. Back-gear usually consists of a wheel and pinion on the mandrel, and a similar wheel and pinion on a shaft fitted in bearings parallel

the usual ratio in foot lathes, but the reduction is excessive, and causes a big difference between the fastest speed of the back-gearing and the slowest speed of the single-gear. The larger the diameter

Figs. 1132 and 1133.—Grooved Pulley Mandrel Headstock and Poppet.

with the mandrel. The driving pulley with **V**-grooves C, Fig. 1123, is fixed to the pinion H, but both revolve freely on the mandrel. The wheel D in front of the pulley is keyed to the mandrel. The other wheel F and pinion G are both keyed to the steel back-shaft K, which forms their axle and runs parallel with the mandrel. When in gear for slow speed, pinion H drives the wheel on the back-shaft, and pinion G in turn drives wheel D, keyed to the mandrel. The reduction in speed obtained by this gearing will

of the work, the slower must be the speed and the greater the power of the lathe mandrel; and it is for this reason that it is desirable to have a speed-changing device. By throwing the back-shaft K out of gear and attaching the wheel on the mandrel to the pulley behind it by means of a sliding bolt, the back-gear headstock is used as though it had a single speed pulley. The driving connection between the grooved flywheel, actuated by the pedal, and the grooved driving pulley is a twisted gut or other kind of endless

Figs. 1134 and 1135.—Back-geared Grooved Pulley Mandrel Headstock, and Poppet.

depend upon the relative proportions of the several wheels and pinions. Wheels having sixty, and pinions having twenty, teeth are frequently used, and these give a reduction of one-ninth. This is about

band. Gut bands are easily joined up by means of the hook and eye, as will be shown, this allowing of simple adjustment when it is necessary to take up any slackness after the belt has stretched.

SPEEDS FOR TURNING AND BORING METALS.

The speed at which work is turned varies with the kind of metal. It may be said that in turning cast iron the speed of the work past the tool may be 150 in. to 190 in. per minute; for wrought iron, 260 in. to 280 in. per minute; for yellow brass, 300 in. per minute; and for chilled rolls, 36 in. per minute. In boring, the following speeds are recommended:—For cast iron, 80 in. per minute, and for wrought iron, 140 in. per minute. For screw-cutting in steel a suitable speed is 90 in. per minute, but it depends on the nature of the material—Bessemer steel, for instance, being turned or screw-cut at a higher

Fig. 1136.—Hook and Eye for Gut Band.

speed than cast steel. To determine approximately the peripheral speed of a job in inches per minute, multiply its diameter in inches by 3⅛, or by 3·1416, and by the revolutions per minute.

TYPICAL HEADSTOCKS.

The detail, construction, and functions of the mandrel headstock and the tail headstock are explained elsewhere in these pages. For the present, it is instructive to compare the four sets of these appliances shown in general view by Figs. 1128 to 1135. In the single-geared headstock, Fig. 1128, the steel mandrel has reversed conical necks running in gunmetal bushes, whilst the driver chuck is clearly illustrated. The double-geared headstock, Fig. 1130, provides a change of speed, the back gear being brought

into engagement by an eccentric motion. Both the single- and the double-geared headstocks have flat pulleys for belt driving. Fig. 1132 shows a simple headstock with grooved pulleys for a gut band, the mandrel bearings being of hard steel in bushes of a similar material. The headstock shown by Fig. 1134 has grooved pulleys and back gearing, the latter being coupled up when necessary with the aid of an eccentric. Figs. 1129, 1131, 1133 and 1135 illustrate the four loose headstocks or poppets.

THE HAND TOOL-REST.

The hand or T-rest K, Fig. 1118, should have its socket bored at right angles to the sole, which should be planed with a dovetailed slot. According to "Lathe Work," if the lathe bed is double flat the sole of the hand-rest stands direct on it; of a V-bed, it should have a cast-iron foundation plate, shaped to fit the bed on the under side, and flat on the top. The screw which clamps the T should have a handle like that of a bench vice fitted to it, as it so often requires to be shifted to suit the work in hand. A similar permanent tommy is also desirable in the screw which clamps the back centre barrel, as it is so much more handy to be able to fix these parts without the trouble of finding the mislaid tommy on each occasion. The T itself for general use may be about 2 in. long on the top, and should be flat and level; in use it will be continually pitted, and must be filed up smooth again. For turning long cylinders by hand a much longer T is used, measuring as much as 5 in. or 6 in. For still longer rods, the rest may be a straight bar supported near its ends in two T socket-holders; by this plan a rest, reaching the entire length of the bed between the centres, can easily be fitted up. In turning work of short length the T-rest is sometimes found to be in the way, and a ⌐-rest is used instead; this is made of an angle piece, one leg fitted to the rest socket, and the other filed flat on its top surface. The ⌐-rest is often used with its point towards the work, thus giving a rest of about ¾ in. in length, very convenient for short work. **T** and **⌐**-rests are usually made of cast-iron, but wrought-iron is sometimes used, and this is the better material, especially for the latter shape.

DRIVING BANDS.

For bands, gut fastened with hook and eye is most serviceable. Round leather bands have the merit of being cheap, but a really tight band cannot be obtained with leather of small diameter, whereas gut bands are to be depended upon if the gut is of the best quality, and the hooks and eyes are of the right sort. Many a

the first one or two threads of the hook easily with his fingers. When the hook begins to tighten on the gut, insert a length of steel as a lever and screw the hook home; but do not overdo it, or the thread that has just been formed on the gut will be stripped, and very shortly the hook will come off. Treat the eye in the same way. Sometimes the gut projects somewhat inside the hook and eye so that they cannot be fastened. Should this be the case, with a red-hot wire sear away the surplus of gut. If the gut does not project inwardly, and the hook and eye will fasten properly, there may be no need for searing.

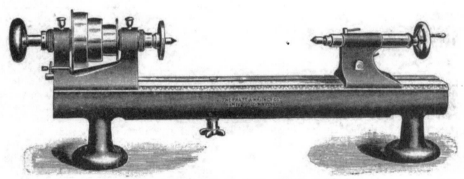

Fig. 1137.—American Bench Lathe for Small Work.

good length of gut band is spoilt for its special purpose by carelessness in measurement, and unskilful screwing home of the hook and eye (see Fig. 1136). The hooks and eyes should be screwed $\frac{1}{32}$ in. smaller in diameter than the gut band. Examine the threads inside, and if they are not sharply cut or are partially stripped, reject them. To fit the band, take the length of gut and pass it over the two wheels (noting well if the band has to be crossed), and draw it tightly enough to take out the coils; then cut it off, leaving an interval of about $\frac{7}{8}$ in. to be filled up with the length of the hook and eye. This is for a ¼-in. gut band, but the same proportions will suit the fitting of any size of band. With a sharp knife pare the ends of the gut so as to slightly point them towards the end, for about ½ in. (the slant must be very gentle) until the worker can coax on

BENCH LATHE FOR SMALL WORK.

A most useful bench lathe for small work, including watchwork, is illustrated by Fig. 1137. It is a Pratt and Whitney construction, and is a typical American machine of its kind. It will be noticed at a glance that it is arranged for driving by means of a flat leather belt, the three steps of the pulley being rounded only just enough to prevent the belt from running off. The bed is of such a length (32 in.) that an 18 in. rod can be turned between the centres, the swing of the lathe being 7 in.; that is, it is of "3½ in. centre." The swing over the slide-rest, when this accessory is fitted, is 4½ in. The bed is of such section (shown in detail by Fig. 1138) that the various attachments can be fitted on by clamping without cramping or distorting the metal. The **T**-rest is much as usual, and, together with the two head-stocks, is clamped to the bed by means of **T**-bolts with eccentric binders which permit of easy and rapid adjustment. The

Fig. 1138.—Section of Bench Lathe Bed.

tailstock is of precisely the shape adopted in larger lathes, but the headstock is of that type which the American small lathes have made popular. A separate paragraph will be devoted to it.

MANDREL HEADSTOCK OF SMALL AMERICAN LATHE.

The cone driving pulley has three steps, respectively 4½ in., 3½ in., and 2½ in. in diameter, to take a 1 in. belt, and the flange on the largest step is divided with two rows of holes, one of 60 and the other of 48. Its spindle, bushings, and back collet are of the special design shown in section by Fig. 1139, being made of tool steel, hardened and tempered. The bearings are double cones, having angles of 5° and 45°, and are protected by dust caps. The hollow spindle takes chucks ⅝ in. in diameter, the hole through the draw-in spindle being 0·44 in. in diameter,

to a standard size to receive index-plates or wheels to be used in connection with a screw-cutting device.

BACK-GEARED MANDREL HEADSTOCK FOR SCREW-CUTTING.

The mandrel headstock illustrated by Fig. 1140 is in many respects identical with that forming a part of the lathe shown by Fig. 1137. The principles of its construction are the same, but it has a few additions not shown in the other headstock. Dealing first with the details that serve to elucidate those of Fig. 1137, A B C show the three steps of the driving pulley which is mounted on the hollow mandrel or spindle J; this mandrel acts as a sleeve to the tubular rod, J'. When this rod is operated by the hand-wheel F, the split chuck G is drawn in or out—in to grip the work which is to be turned, or out to allow of taking the work off the lathe. The chuck is described in detail on p. 408. The connection between the tubular rod J' and chuck G is by means of internal screw threads on one and external threads on the other, as shown at H. With reference to the extra features not illustrated in Fig. 1137, but shown in Fig. 1140, it may be said that D and E represent respectively a spur wheel and a pinion on the mandrel and engaging with the back-gear in the ordinary way, as explained previously on p. 391. For purposes of screw-cutting, there is a fixed pinion K on the mandrel J, this driving the spur wheel L, on the same spindle

Fig. 1139.—Section through Mandrel of Bench Lathe.

whilst the largest wire that can pass through the spindle in a split wire chuck has a diameter of ⅜ in. (These details will be better understood when Fig. 1140, showing a similar spindle and other similar details, is described.) A point chuck, not a split chuck, is shown in Fig. 1137. The left hand end of the spindle is ground

with which is the pinion M. This transmits motion to a guide screw (not shown) which moves the slide-rest along the bed, so causing the tool to cut a spiral path on the work revolving between the lathe centres. P P show a lever, rod, and eccentric for attaching and detaching the headstock to and from the lathe bed.

It is now assumed that a lathe—new or second-hand—has been bought, and that it is to be tested before doing any important work upon it. In order to turn out accurate work, the lathe and its appliances must be adjusted accurately. Many lathes are placed on the market without regard to perfect coincidence in the various parts, and an immense amount of labour has to be spent on them to make up for original errors, which are always being increased and distributed throughout the machine, in the natural process of use. To turn out a really accurate machine, each part must be separately and independently accurate and true, and when put together all the parts must be perfectly coincident with each other. Otherwise the errors of the component parts, though possibly slight taken independently, will form in the aggregate serious defects.

LATHE BED IN WINDING.

It is not unusual to find a lathe bed, after being planed up, to be "in winding," that is, having an axial twist, or curved high or low in the middle. These defects are caused by the method of planing, sometimes through the casting having been bolted down on the bed of the planer too tightly, in such a manner as slightly to bend it. Another cause is that a casting is apt to go out of shape when the outer skin is removed, this

skin having been acted on by the sand of the mould when the metal was poured. If a long square bar of cast iron, tolerably

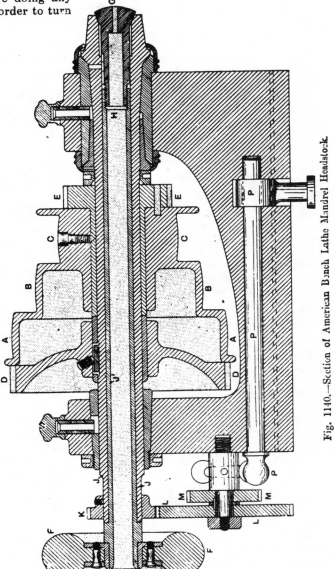

Fig. 1140.—Section of American Bench Lathe Mandrel Headstock.

straight, is fixed on the planing machine and carefully planed on one side only, when the bar is unchucked it will probably be found that the side which had been planed perfectly true is curved or in winding. Now, if the bar is again fixed, with the planed side downwards, the bolts and packing being arranged so that the bar is not distorted in the slightest degree, on planing the upper side all over and unchucking the bar the first side will be found to be considerably out of straight. The cause of this will be easily understood. That portion of the casting which has been quickly cooled by contact with the sand in the mould is in a different state of molecular tension from the inner part, and hence, when the skin is removed, and with it the tension, the entire bar will be affected.

PLANING LATHE BED.

To plane a lathe bed properly, the whole of the top part should be roughed over first, then the casting turned over, and the bottom done, and then the top finished; by this means, and with very careful chucking, there will be some prospect of getting a fairly true bed.

BOLTING LATHE BEDS TO STANDARDS.

Lathe beds which may be true in themselves are sometimes distorted when being bolted on to the standards; if these are not fairly fitted to the bed something will be twisted out of shape when the bolts are screwed up. The headstocks should be fitted on to the bed with great nicety, so that they will shift easily, but bed down firmly on a solid bearing without any shake whatever. The entire bearing surface of the headstocks should rest on the bed, and not merely the corners, as is sometimes the case; they should also bed down easily, not so as to require drawing with the holding down bolts before they take a fair bearing. The remarks made about headstocks are equally applicable to the bottom of tool-rest, slide-rest, boring collar, and other apparatus fitted to the bed. The importance of having a true lathe bed cannot be too highly estimated.

ALIGNMENT OF CENTRES.

The line of centres must be quite parallel with the bed, and perfectly straight. If the poppet headstock, barrel, and holes for bearings of the mandrel are not bored exactly opposite, and parallel with both the top and side bearings of the bed, the accuracy of the turning will be uncertain. Headstocks that are properly fitted to a bed, and bored with a proper boring bar, which slides in bearings, will be true; but as the boring of each hole is sometimes done quite independently of any other, the axial continuity of the mandrel and poppet barrel is not always to be relied on. The following are particulars of the tests to be applied for finding out the defects in a lathe.

TESTING LATHE BED.

In the first place the bed itself should be tested for straightness and squareness —to see that it is not " in winding "; the method of doing this is by the aid of a good straight-edge and winding strips. These are a pair of metal strips, similar to straight-edges, which are laid across the bed at right angles to its length and at various parts, any slight twist of the surface being shown in a much magnified form by the length of the strip. If the bearing of the headstocks, etc., is between the cheeks of the bed, the inner edges must be tested with the straight-edge in the usual manner. If the bed is a " V and flat " then the V must be carefully verified with the straight-edge, and if the outer edges of the " double flat bed " are bevelled, as in screw-cutting lathes, for the slide-rest carriage to travel on, these bevels must be gauged for straightness. In a screw-cutting lathe it is necessary to see that the interior space between the cheeks, which forms the bearing of the headstocks, is parallel with the outer bevels forming the guide for the slide-rest; this is best done by callipering. Lathe beds will sometimes be found to be out in this respect, the cause being that the outside and inside of the cheeks have been planed at different chuckings on the planer, though the dogs which hold the casting down should be so placed that they are out of the way of the planing tool for both inside and outside cuts.

Testing Alignment of Headstocks.

The headstocks must next be tested. Be sure that no dust or dirt has got in between the headstocks and the beds, put on the point chuck, see that it runs true, and try how it matches the point in the poppet head. This point should of course be quite true with the exterior of the poppet, or loose headstock, barrel, and be turned exactly to the same angle as the cone in the point chuck—60°. The angle made by a three-cornered file is the proper amount of taper for these points. It is a mistake to try to test the oppositeness of the points by contact, as sufficient accuracy cannot thus be ensured. The best plan is to use a narrow parallel straight-edge, and to adjust the distance apart of the cones, so that the opposite edges of the straight-edge rest against the opposite sides of the cones. The test is applied on both sides, and above and below. This will readily show whether the points are opposite.

Testing Parallelism of Mandrel Bearings.

The parallelism of the mandrel bearings with the top and inside of bed will be shown by fixing a rod of metal in a chuck as long as will allow the back centre to stand on the bed with the point clear. This rod of metal must be centred, and its centre should coincide with the point of the poppet or loose headstock. The rod must be strong enough to bear its own weight without drooping at the unsupported end, or the test will show false. Another method of testing the holes bored in the headstocks is by putting the mandrel head to the right of the poppet head and turning the barrel round to protrude from the back end of the poppet head; then, if the point of the poppet comes exactly opposite the tail pin of the mandrel, or centre of mandrel if it is in double bearings, the continuity of the

axial line of centres in a straight line parallel with the bed is proven. If there is a true face plate, or even a large cup chuck, fitted to the mandrel, the levelness of its bearings—that is to say, the parallelism of the mandrel axis with the top of the bed—can be tested by means of a true square placed on the bed and against the face of the chuck. If the surface of the face plate forms a right angle with the top of the bed, then the mandrel is the same height from the surface of the bed at both ends.

Testing Boring of Loose Headstock.

The truth of the boring through the poppet or loose headstock may be tested very accurately by means of gauges. Put

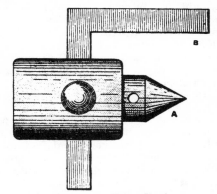

Fig. 1141.—Point Chuck.

the barrels, which must fit very nicely, into the hole, and gauge the distance from the bed, and the position with regard to side-bearing surfaces, etc. Then, without shifting the headstock, take out the barrel, and put it into the other end of the casting, and gauge again. Any error will soon be perceived, as the sense of touch is marvellously keen. This is proved by the ease with which minute differences, which cannot be detected by the eye, are easily shown by the feel in callipering.

Testing Fast Headstock.

The mandrel headstock can be tested in a similar way, but in that case a piece of hard wood must be turned to take the place of the mandrel fitting into both holes of the casting—the front collar hole and the back, or tail pin hole—a short parallel piece of exactly the same diameter being left projecting at both ends ; this short piece forms the part to gauge from,

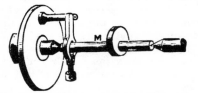

Fig. 1142.—Disc Point Chuck and Carrier.

in the same way as did the projecting poppet barrel in the other headstock. It may happen that the tail pin is not quite straight, or the centre in which the back end of the mandrel bears may be out of truth with regard to the cylindrical part which fits in the casting. In such cases this test will be of no avail ; though, of course, the tail pin would be the part which should be corrected.

Truing Lathe.

Defects in the truth of the fitting up of a lathe constitute a large item in the difference between good and bad tools. Having looked through the lathe generally, and found out where the faults are greatest, it will be easy to judge how to set about rectifying them. A little filing and scraping of the various parts, skilfully done, will often make a vast deal of difference to the working capabilities of a lathe, and will entail but little labour, which will be amply repaid by the greater accuracy of the work produced by the tool. To make a thoroughly good job of the correcting process, the bed itself will claim attention first. After that is made true, the back-centre is operated on next, and then the fast headstock, which can be wedged up to position if it has only one fair bearing surface.

The Point Chuck.

Certain accessories must be described before explaining the methods of simple turning. The point chuck is the most commonly used chuck, and it can only be employed when the rod of metal is supported at the other end by the back-centre, the work then being said to be turned " between centres." Other names for this chuck are driver chuck, running-centre, and take-about-chuck. The male centre, A, Fig. 1141, or point of the chuck, can be replaced by a female or hollow centre to take a piece of work that has pointed ends. Clockmakers, it may be said, turn all their arbors with pointed ends, whereas engineers use the male centre. The point chuck is employed whenever possible, and is quite suitable for use in the simple manipulations about to be described. The driving arm B passes through the chuck, and catches the tail of the carrier fixed to the metal rod, thus communicating the mandrel motion to the work. The point chuck as shown screws on to the mandrel nose, but large lathes may have the centres fitted into the mandrel itself, in which case a metal disc, usually of cast-iron, and having a short straight driving pin screwed into it to catch the tail of the carrier, is screwed on to the mandrel nose. Such an arrangement is

Fig. 1143.—Ordinary Carrier.

shown in Fig. 1142, in which M is a metal rod carrying an eccentric in course of turning. Other, and less simple, chucks will be left for description until after the use of hand tools in rough turning has been explained ; see p. 408 and forward.

THE CARRIER.

The driver B, Fig. 1141, imparts motion to the work by means of a carrier, Fig. 1143, through the eye of which the end of the rod metal is passed, the set screw, grip screw, or clamping screw then being screwed up until the carrier is secured firmly to the work. As the driver chuck revolves, it strikes against the tail of the carrier and imparts the rotating motion to the work.

DOUBLE-DRIVING CARRIER.

A carrier with two tails, so as to be double-driving, is illustrated by Fig. 1144, it being particularly suitable for use with a Clement's driver, this having two arms. Its advantages are that the two tails equalise the strain, and reduce the thrust upon the lathe centres. This carrier is claimed to allow of truer work being produced, to save time, and to be specially useful for work of large diameter. The position of the grip or clamping screw prevents its being sheared off through improper use as a driver, an accident to which the ordinary carrier is liable. The carrier is wholly of steel and is made in a range of sizes from No. 00, holding work ⅛ in. to ¾ in. in diameter, to No. 6, holding work 3⅝ in. to 4 in. in diameter.

CENTERING METAL ROD FOR TURNING.

The centering of the work is the first operation in the process of turning, and the accuracy of all subsequent operations is dependent upon its being done as perfectly as possible. In the first place, never turn a piece of work, no matter how trivial the job, without drilling and countersinking the centres, as the lathe centres then retain their truth for a much longer time than would otherwise be the case. To centre a piece of work properly, it should be, if a long shaft, straightened with a hammer or under a screw press, judging of the straightness merely with the eye; then, with a centre punch, make a dot only deep enough to hold on the centres. It can then either be revolved

and marked with chalk on the full side, and drawn over with the centre punch, or it may be " square-centred," which is done by having a centre ground square, leaving four cutting edges, as will be described. To centre a piece of metal rod so that it will run absolutely true is considered an impossible task by many; and for this reason ⅞-in. material is used for turning ¾-in. cylinders, though the labour involved in removing the superfluous ⅛ in. is quite wasted. Actually, it is not more difficult to centre a rod truly than to set it between the centres considerably out of centre, or at least the trouble of getting an absolutely true centre is comparatively inappreciable.

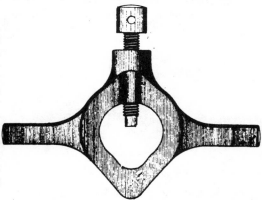

Fig. 1144.—Double-driving Carrier.

PREPARING METAL ROD FOR TURNING.

When about to turn a rod of metal between centres, select a suitable piece of material, just sufficiently large to clean up to sufficient size. See that it is straight, and cut off a suitable length; if the entire length is to be used for one particular object, cut off just enough and not more than is required. Of course, an error of, say, ¹⁄₁₆ in. in length, when it is on the right side, does not matter, but a superfluous ¼ in. will involve much unnecessary labour. The rod cut off, file up its ends perfectly flat, and at right angles to the length of the rod. If the end faces are at an inclination it is very difficult to judge correctly by the eye the exact centre. When the ends are squared the centres must be found and marked.

MARKING CENTRES ON METAL RODS.

There are several ways of doing this, but the one generally most easy is to mark the place with a fine centre punch. For rods of small diameter the centre may generally be judged by the eye with sufficient accuracy for practical purposes. The rod is fixed upright in the vice, and a light dot punched with a centre punch. The ends are then reversed, and the other end is dotted. The rod is then put between the centres of the lathe, and the amount of its eccentricity noted by turning the rod with the thumb and finger, at the same time applying a piece of chalk so that it touches the most prominent side of the rod. This is marked with the

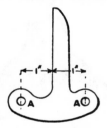

Fig. 1145.—Centre Square.

chalk, and the rod replaced in the vice. The punch is applied in such a direction as to drive the hole towards the most prominent or the marked side, and the work again tested. The indentation made by the centre punch should be as small as possible, so as just to hold on the lathe centre. By continued trial between the lathe centres and subsequent correction with the centre punch, the centre punch indentation can be brought to the exact centre of the rod. When this is accomplished, the centre is deepened by striking the centre punch in the direction of a straight line with the axis of the rod. By means of a blunt-pointed punch and a prick punch—one having a sharp, acute angled point—the indentation may be made sufficiently deep to hold the centres during the process of turning; but in all cases it is advisable to drill in the centres

with a small drill, which will prevent the centre point of the lathe " bottoming " in the hole, and thus causing the centre to run out.

MARKING CENTRE WITH CENTRE SQUARE.

The end of the rod metal can be centred by means of a centre square, of which there are many shapes, though all are alike in principle. It generally has three arms, the centre one bisecting the angle formed by the two others. The rod is placed in the fork or angle of the device, and a bisecting line drawn across the end by the aid of the guiding arm. Then the rod is shifted a quarter turn and the process repeated. It can now be seen that two lines drawn across the circular end, and both of them quite central, must bisect each other in the centre of the circle. One form of centre square is shown in Fig. 1145. It really has but one arm, and this is the straightedge bisecting the end of the rod, the purpose of the other arms being served just as well, if not better, by two steel pins, A, riveted to the blade, which is of steel, $\frac{1}{16}$ in. thick. The pins project the distance of one diameter, namely, $\frac{1}{4}$ in. The straightedge accurately bisects at a right angle the centre line of the two pins.

CENTERING WITH BELL CENTRE PUNCH.

Another method of centering rod metal is to use a bell centre punch (see Fig. 615, p. 195); this is a centre punch fitted to the apex of a hollow cone. The cone is put on the end of the rod, and the punch brought down to touch the work, a blow from a hammer causing the point to make an indentation, which, theoretically, would be in the centre of the rod. Practically, however, the dot may be quite eccentric; this will be so if the end of the rod is not exactly square with its length, or if the punch is not held perfectly straight with the rod.

DRILLING IN CENTRES OF METAL ROD.

This drilling in should be done in every case when the material is likely to be again put in the lathe on some future occasion. To attempt to turn any but the very commonest of work on plain punched centres is most unwise. To show the effect of drilling in the centres before straightening the work, a glance at Figs. 1146 and 1147 will suffice. Fig. 1146 shows

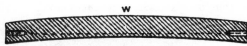

Fig. 1146.—Bowed Rod with Drill Centres

the section of a rod very much bowed, but correctly centred at each end. The centres are shown drilled in very much deeper than they should be in practice, the better to illustrate the effect. When the rod, Fig. 1146, is put between the centres on the lathe, the part w will, of course, run very much out of truth. If the rod is then straightened to run true in the middle, as shown by Fig. 1147, the direction of the drilled centres will not be in a line with the axis of rotation. The dotted lines in Fig. 1147 show the direction of the holes. If the work is rotated very long, and the centres become worn much deeper, the work will run out of truth. An inspection of the two illustrations (Figs. 1146 and 1147) will make the effect obvious.

STRAIGHTENING METAL ROD.

In many cases a rod can be straightened more accurately after it is mounted between the lathe centres; but it should always be done before drilling in the centres. Usually it is best to cut off a piece of rod the length required, to

STRAIGHTENING ROD WITH SCREW PRESS.

The best appliance for straightening a rod is a screw press made to traverse the bed of the lathe, and in the best make having wedge blocks with a **V** on the top that can be raised, and moved almost close together or far apart, as the crook in the shaft is short or long. While

Fig. 1147.—Effect of Straightening Drilled Bowed Rod.

the shaft is revolving, mark the full side with a piece of chalk every three or four inches for the full length; this shows where to begin, the long crooks being always straightened first. Sometimes it should be bent more than is necessary to bring the exact spot true, as, when the full side or crooks are on opposite sides, bending the other crooks will bring it back. It is always necessary to loosen the centres when using the press; the wedges should only be brought to bear and not forced under the shaft too tight. To become an expert at this part of the work requires practice, and skill can only be attained by closely watching the effect of each blow or strain of the hammer cr press upon each part of the shaft.

ANOTHER METHOD OF CENTERING METAL ROD.

A method of centering usually adopted where large quantities of rod pieces have to be centred, is to fix a stiff short drill

Fig. 1148.—Turner's Arm-rest.

straighten it tolerably true, and then to centre it truly at the ends. When it is mounted in the lathe, observe if the rod is crooked, and finally straighten it before drilling. After the centres have been drilled they must be countersunk, of which more will be said later.

in the back centre socket. This drill must be quite rigid, and it forms a centre. The work is then run between the centres, dots having been first punched at the ends, their central position being in this case of little importance. The rod has a carrier put on one end, and the lathe is set in motion. The drill will now cut if the back centre be brought forward, but this must be carefully avoided till the central position of the hole has been assured. This is got by placing a tool—the hook of the ordinary arm-rest (Fig.

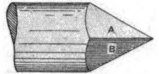

Fig. 1149.—Ordinary Square Centre.

1148) serves the purpose very well—against the revolving rod in such a position that the highest side will bear heavily against the tool, whilst the "scant" or least prominent side barely touches. The point of the drill on which the work revolves, being a cutting edge, will gradually cut in the direction of the high side of the work, the tool held against it forcing the rod in the opposite direction. To prevent the work from becoming loose between the centres, and at the same time to assist the cutting of the drill, the back centre must be kept up to its work by slightly advancing the barrel of the poppet.

THE SQUARE CENTRE.

The drill fitted into this poppet barrel is usually called a square centre, and is often made of the shape shown in Fig. 1149. The end is filed up to a four-sided point, and then hardened and tempered. The facets, A and B, may be

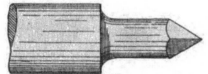

Fig. 1150.—Square Centre Reduced to Economise Grinding.

ground to make the point sharp again after being blunted by use. A considerable quantity of metal has to be ground away each time this square centre requires sharpening, and in order to reduce

this labour the form of the centre is frequently modified to that shown by Fig. 1150, in which the diameter is reduced very considerably. If the centre be kept short, as it should be, the reduction of diameter does not weaken it ap-

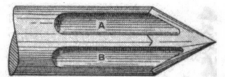

Fig. 1151.—Grooved or Fluted Square Centre.

preciably. In Fig. 1151 is shown another modification, the sides of the centre being fluted (see A and B) so as to reduce the grinding surfaces. By means of the square centre, the hollow made by the centre punch will quickly become drilled at the exact centre. Sometimes a bar of metal, having a < shaped notch in its end, is fixed in the slide-rest tool-holder, and brought to bear against the end of the rod. The rod rests between the forks of the < and its highest side, continually impinging against one or other side of the notch, the bar being advanced by the slide screw till the rod revolves in the notch with an equal bearing at all points. This is an excellent plan to be commended, especially when large numbers of rods of similar size require to be centred. One end being centred, the rod is reversed in the lathe, and the other end centred similarly. The rod must have centres drilled in with a small drill, and countersunk to ensure the work running true for any length of time.

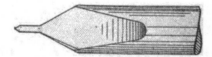

Fig. 1152.—Combined Drill and Countersink.

COUNTERSINKING THE CENTRES.

When preparing a number of rods, they are all centred first, then drilled, and finally countersunk. Some tools combine the drill and the countersink, the most simple form being shown by Fig.

1152. It is made simply by flattening a cylindrical piece of steel, as in making an ordinary drill. The end is then shaped to the form shown, the small part being the drill and the large part forming the countersink. Such a tool must be used

Fig. 1153.—Rose Countersink.

with care, as the drill point is very liable to breakage through being jammed in the hole by the shavings.

OTHER COUNTERSINKING TOOLS.

The rose countersink (Fig. 1153) has its end turned to the correct cone, and then about six teeth are filed on it; it cuts very well. The combination (Fig. 1154) is like Fig. 1153, but is drilled through the centre to take a small twist drill, this being fixed at any desired point by means of the screw shown at s. It may be removed for sharpening, or replaced if broken. The ordinary form of countersink is shown by Fig. 1155, and works very well in practice; it is easy to make, and it lasts a long time. The cone point is filed away to the diameter line, leaving A; a small facet, as at B, is sometimes made for clearance. This cone point should always be made exactly to match the lathe centres. The effect of grinding this countersink is shown by Fig. 1156, the point becoming a curved line. When this curve extends further than the

Fig. 1154.—Combined Rose Countersink and Twist Drill.

diameter of the small hole bored up the centre, the countersink ceases to act properly. A countersink made of good steel and well hardened will last with fair usage a very long while, so that an anticipation of the result shown in Fig. 1156 need not occasion much alarm.

MOUNTING METAL ROD BETWEEN LATHE CENTRES.

Having accurately centred the work, the next operation will be to mount it between

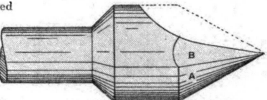

Fig. 1155.—Ordinary Countersink.

the lathe centres, and to see how the rod runs for truth all along. Most probably, the part about midway will be more or less out of truth, owing to the rod not being quite straight. The high point must be marked, and the rod removed and straightened by a blow of the hammer; this must be done on the anvil or some such tool. On no account should the work be struck with the hammer whilst between the lathe centres, as much damage may result to the lathe through the incautious use of a hammer. Besides, the

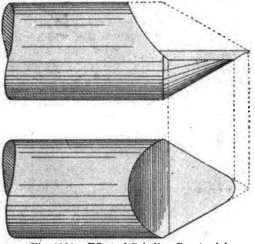

Fig. 1156.—Effect of Grinding Countersink.

work is too springy to straighten when held supported by the centres only. The chamfer countersunk in the end of the rod must fit the lathe centre. If the chamfer is either a greater or a less angle than the cone on which it has to work, the bearing will be only at one edge. The illustrations, Figs. 1157 and 1158, show this. In Fig. 1157 the hole is shown

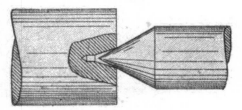

Fig. 1157.—Angle of Countersinking Too Acute.

countersunk at a much more acute angle than the centre point. Consequently, the bearing is only at the extreme end of the rod. Having but a small amount of bearing surface the centre would, of course, wear rapidly. In Fig. 1158 the hole is shown countersunk at a much wider angle, and the effect is practically the same, so far as a small bearing surface is concerned. The opinion that countersinks should be made to an angle that is not precisely like that of the lathe centres is absurd. The angle of 55° has some advocates, but 60° is as good, and is a more convenient standard. A gauge is easily made by cutting a notch with a triangular file, which will be 60°, and this angle for lathe centres should be universal. Fig. 1159 is intended to represent, in section, a rod of metal centred and countersunk properly, the countersink being shown on the right. The central hole is drilled in sufficiently

TRUING ENDS OF METAL ROD.

Supposing the rod to be properly centred, and quite straight, and the centres drilled in and chamfered, it will be advisable, before starting the turning proper, to make the ends of the rod quite true. Square off the extreme end faces of the work, and whilst this is being done the rod may be made the proper length, if any special predetermined length is required. The object in turning the ends true is to get an equal surface round the

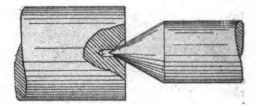

Fig. 1158.—Angle of Countersinking Too Obtuse.

chamfer in which the cone point takes its bearing, which would not be the case if the rod ran between centres with its ends out of flat. The centres of the work should always be kept lubricated, for if allowed to get dry they will cut and tear, and most likely entirely spoil all previous centrality of the chamfer.

HAND TOOLS FOR METAL TURNING.

Tools for metal turning are of two kinds, those held by hand and those held in the slide-rest; the former have their metal shanks so shaped that they can conveni-

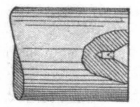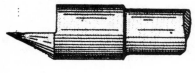

Fig. 1159.—Correctly Formed Hole and Countersink.

just to clear the point of the cone. The end is intended to be trued up on the lathe, for which purpose $\frac{1}{16}$ in. is allowed.

ently fit into wooden handles, whilst the latter have stems of heavy square-section steel, the same metal as that forming the cutting edge. The present fashion, how-ever, for both hand and slide-rest tools is to form the cutting points on separate small pieces of steel and hold these up to

volume. The graver, the most general tool for metal turning, is a bar of square steel, usually from $\frac{1}{4}$ in. to $\frac{7}{16}$ in. square, though smaller and larger tools are used.

Fig. 1160.—Hand Turning Tools for Metal.

their work by means of more or less heavy holders, examples of which will be given. Some hand tools in common employment for metal turning are illustrated by Fig. 1160, samples of six tools suitably handled being shown by Fig. 1161.

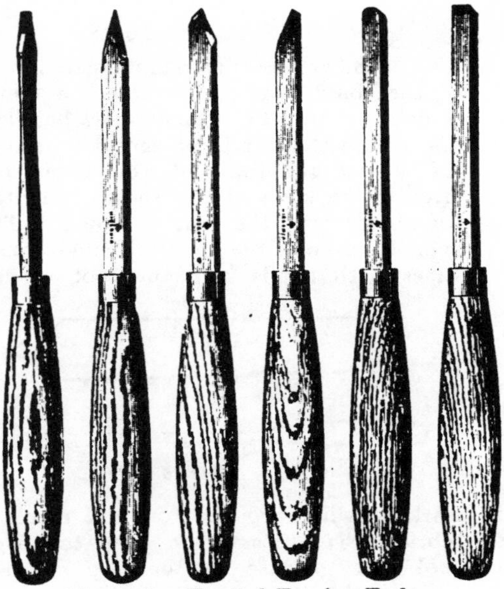

Fig. 1162 shows two views of a graver. All the flats are sometimes, but two are always, ground quite smooth. The end is ground off diagonally, the cutting edges being formed by the smooth sides meeting at the point (see Fig. 1062). The angle made by the diagonal diamond-shaped end with the shank varies to suit the

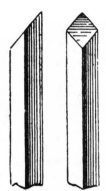

Fig. 1162.—Graver.

material that is to be cut, from 60° to 70° being about the usual limit.

Fig. 1161.—Handled Turning Tools.

GRAVER FOR METAL TURNING.

The particulars of metal turning tools given in this and some of the following paragraphs are reproduced with slight verbal alterations from "Lathe Work," a book written by the editor of this present

THE ROUTER.

A router (Fig. 1163) is a $\frac{3}{16}$ in. graver whose end has been ground to the shape shown. It is used for making the rough-ing cuts on brass. The tool is ground square to an angle of 90°, so that both the upper and lower edges will cut.

TRIANGULAR TOOL.

The triangular tool (Fig. 1164) is also much used. It is generally the same size as the graver, and is made from a worn-out triangular file by merely grinding the faces to remove all the teeth, leaving sharp edges at the angles, all of which are 60°. The end is ground off obliquely, leaving a point at one angle, but the side edges of the tool are generally used for cutting with; and in this respect it principally differs from the graver, which is used only at the end.

ROUND-NOSED TOOLS.

Round-nosed tools, which are made of strips of steel of various widths and thicknesses, having the ends ground off to a semi-circular shape, are used for

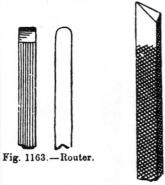

Fig. 1163.—Router.

Fig. 1164.—Triangular Tool.

hollowing out the exterior of metal work and turning curvilinear grooves. Tools of a similar construction with the ends shaped to various patterns are largely used for turning beadings and mouldings of various kinds. Numerous tools, ground to particular forms, are employed for special purposes, but of these little need be said here. Their form can be understood from Fig. 1160.

THE HEEL TOOL.

The heel tool, Fig. 1165, is another form of tool used in rough turning iron; the shaft of the tool is bent at nearly right angles, that it may be held either above

or below the shoulder of the workman as preferred. Some variation is made in the form of these heel tools, and they are sometimes made pointed instead of round upon the cutting edge.

Fig. 1165.—Heel Tool.

PARTING TOOL.

This is used for cutting off work in the lathe, and is shown handled in Fig. 1161, the parting tool being that shown on the extreme left. To give clearance to the blade of the tool, this is a little thinner at A, Fig. 1166, than it is at B, the difference being about $\frac{1}{32}$ in., ample for the purpose.

USING TURNING TOOLS.

With the graver and triangular tool most of the rough turning by hand on metal is done. These tools have short handles, and in use the left hand generally grasps the T-rest and the tool, the fingers encircling the stem of the socket, and the thumb clasping the tool to the T. The right hand holds the handle, though sometimes both hands hold the tool. When

Fig. 1166.—Parting Tool.

working with wrought-iron and steel, a lubricant is necessary to keep the edge cool and to ease the cutting. Water answers the purpose, but soapy water is better, and perhaps quite as good as oil, though much cheaper. Cast-iron, brass, and gun-metal are turned dry.

SIMPLE TURNING.

The way in which the graver is applied to the work is as follows:—The tool is laid with one angle on the T-rest, the point being towards the back centre, and the handle at an angle with the line of centres. The lathe being set in motion

the cylinder, cutting away all the ridges and reducing the surface to one level. This flat tool may be from about $\frac{1}{2}$ in. wide and $\frac{1}{8}$ in. thick up to double these

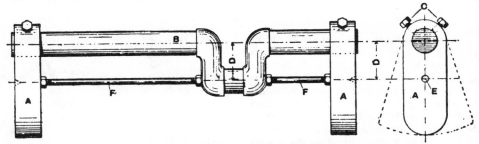

Figs. 1167 and 1168.—Turning Engine Counter-shaft Pins.

and the graver brought as near as possible to the work, it is firmly indented on the rest, and by bringing the handle towards the right the point is made to cut the work; the operation is assisted by turning the graver slightly over towards the left. This action makes a narrow groove on the work, and when the handle is so far to the right that the tool is disengaged, the graver is shifted along the rest to recommence the same process. The work turned by this means will consist of a series of grooves more or less irregular, but the concentric truth will be correct. In the same way the triangular tool may be used to produce a like result. To further finish the cylinder, after it has been made as straight as can be with the

dimensions; it can be used indifferently with either edge to cut. Cylinders with straight surfaces, whether parallel or coned, are generally finally finished by filing whilst in rapid revolution in the lathe, a fine file being used. To produce an extra smooth surface emery paper wrapped round the file is afterwards applied, and by this means a very high finish can be given to the work.

TURNING ENGINE CRANK-SHAFT.

One method of turning the crank-pin of a small crank-shaft is shown in Figs. 1167 and 1168, the former being a front elevation and the latter a side view. Iron slabs, A, are fastened, one at each turned end of the shaft B, by set screws C. The

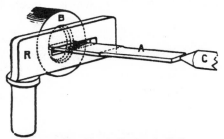

Fig. 1169.—Simple Boring in Lathe.

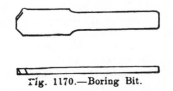

Fig. 1170.—Boring Bit.

slab is centred at E, so that D represents the throw of the crank. Sometimes the hole in the slab is larger than the turned end of the shaft; the hole is then packed so that the distance D between the centres can be adjusted. To stiffen the system, long bolts at F are introduced, being jammed tight by nuts at the ends. The slabs are often to the shapes shown by

graver, a flat tool is used. This somewhat resembles a carpenter's chisel ground off square at the end, so that no bevel exists. This tool is applied, end on, to the dotted lines in Fig. 1168. The centres of the slabs and of the crank-pin must be in line, the positions being set by the aid of vee-blocks, plumb-bob, and scribing block.

SIMPLE BORING IN THE LATHE.

One of the most simple methods of boring the end of a rod is shown diagrammatically by Fig. 1169. The slotted plate R replaces the ordinary T of the tool rest, and serves to steady and, indeed, to firmly hold the boring bit A (shown separately by Fig. 1170). The rod or cylinder to be bored is indicated at B, and is supported at one end by a chuck fixed on the mandrel nose, and it is the work, not the bit, that revolves during the operation of boring.

Fig. 1171.—Boring Collar. Fig. 1172.—Drill Pad.

The bit is kept up to its work by the back centre C of the loose headstock, which is advanced as the feed necessitates. The boring collar (Fig. 1171) is used in somewhat the same way as the slotted plate R, in Fig. 1169. It bolts down upon the lathe bed, and the disc can be rotated until one of the coned collars that suits the diameter of the work is brought into position, in which it is fixed by means of a peg. The coned hole forms an end bearing for the work. The drill pad (Fig. 1172) is another appliance of use for boring, etc., in the lathe. It is fitted in place of the cone centres and keeps the work up to the drill held in a drill chuck.

SPLIT CHUCK.

Of all the chucks for holding small rods, wire, etc., while being turned, the hardened steel split chuck, as used in the lathe headstock illustrated by Fig. 1140, is the best. It must be allowed that scroll chucks answer very well for holding material such as metal wire and metal rods from which screws, etc., are to be made, and they are useful for holding drills ; but they do not give absolute

Fig. 1173.—Split Steel Chuck.

truth, such as can be obtained with the split steel chuck. Fig. 1173 is a general view of such a chuck, one of a set of thirty-seven sizes having holes varying by only $\frac{1}{100}$ in. ; the smallest has a hole of $\frac{4}{100}$ in., and the largest of $\frac{4}{10}$ in. Much smaller chucks of the same kind are made for watchwork lathes, and they will hold a small needle perfectly true while being turned. The chucks fit into a perfectly true hole in the lathe mandrel 15 millimetres (a full $\frac{1}{2}$ in.) in diameter, the hole running right through the mandrel. A tube is put in through the tail end of the mandrel and has a wooden hand-wheel fitted to it by which to hold and turn it ; this tube also fits the bore of the mandrel, and at its front end where it meets the chuck it has an internal screw fitting on to the screw B at the end of the chuck (see Figs. 1173 and 1174). Fig. 1174 is a section of the chuck. The keyway at A receives a little pin, which projects into

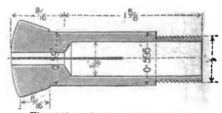

Fig. 1174.—Section of Split Chuck.

the bore of the mandrel and prevents the chuck from turning. As the tube is turned the screw threads draw in the chuck, and the conical part C, which fits a similar internal cone, causes the chuck to close upon anything (not smaller than $\frac{1}{100}$ in.) placed within it. The sides of the cone are at an inclination of 20° to the centre line. Fig. 1174 gives a section of the chuck

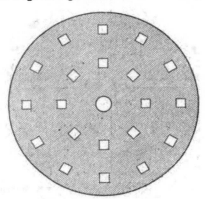

Fig. 1175.—Twenty-hole Face-plate.

with its principal dimensions, which are metrical, but are here given in fractions of an inch. The pitch of the screw is 1 millimetre, very nearly 25 to the inch. These chucks are interchangeable, and are made of the best tool steel, hardened at the front end and ground true. The smallest sizes are ground true in the hole by tiny steel pin-like laps, charged by rolling them in diamond dust; the chucks are

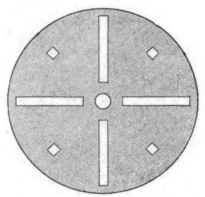

Fig. 1176.—Four-hole and Four-slot Face-plate.

placed in a slowly revolving mandrel whilst the little lap runs at a great speed in touch with one side of the hole till it is quite true.

FACE-PLATE CHUCKS.

These are next to the point chuck in simplicity. Illustrations of them are

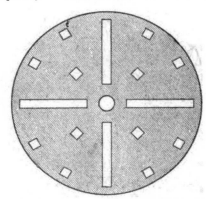

Fig. 1177.—Twelve-hole, Four-slot Face-plate.

given, all of them being one-sixth full size, Figs. 1175 to 1188 representing face-plates 12 in. in diameter, and showing the most usual methods of perforating face-plate chucks by boring and by slotting. The holes are often square, being made fully large to take the squared shoulders of $\frac{1}{2}$-in. bolts. Slots are of the same width, and round holes are made fully $\frac{1}{2}$ in. diameter to take $\frac{1}{2}$-in. hexagon-

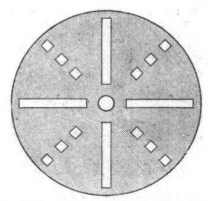

Fig. 1178.—Twelve-hole, Four-slot Face-plate.

headed bolts. To briefly describe the illustrations of the face-plate chucks, Fig. 1175 shows a face-plate with two circles of square holes, the larger having twelve, and the smaller eight holes. Fig. 1176 has four radial slots and four square holes interspaced, and shows the plan of per-

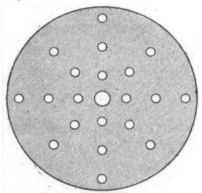

Fig. 1179.—Face-plate with Twenty Round Holes.

forations usually adopted in the commonest class of work. The four square holes are sometimes omitted. Fig. 1177 has four radial slots and a dozen square holes placed in two circles. It is like that shown by Fig. 1175 with slots replacing eight of the square holes. Fig. 1178 has four radial slots and a dozen square holes interspaced and on radial lines. In Fig. 1179 there are twenty round holes spaced to give a great variety of positions to the bolts when in use. The eight square holes in Fig. 1180 are interspaced in two

can be gathered from Figs. 1181 and 1182, which are drawn to the same scale as the other face-plates and illustrate the bottom

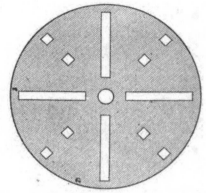

Fig. 1180.—Eight-hole and Four-slot Face-plate.

bracket of a cycle in position for boring out to receive cups.

FURTHER FACE-PLATE CHUCKS.

The chuck shown by Fig. 1183 has eight radial slots, four similar to those shown in several others of the illustrations, and interspaced are four other and shorter radial slots. Fig. 1184 has twelve radial slots, eight of which are equidistant near the circumference, and about one-fourth the diameter of the chuck in length; four others about the same length, but placed

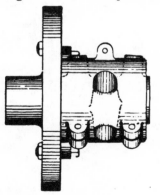

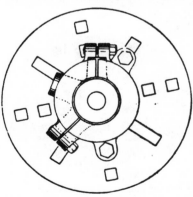

Figs. 1181 and 1182.—Side and Face Views of Cycle Bottom Bracket on Face plate.

circles and there are four slots; this is intermediary to Figs. 1176 and 1178. The manner of attaching work to such a chuck

near the centre, are interspaced, and corresponding to those are four ribs on the back, forming, with the last-named four slots, eight equidistant alternate slots and ribs spaced between the outer eight equidistant slots. This method of per-

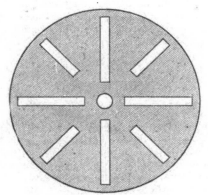

Fig. 1183.—Eight-slot Face-plate.

forating is often seen in American face-chucks. In Fig. 1185, there are four radial equidistant slots, and running parallel with two opposite radial slots are four others. These latter allow angle plates and other attachments to be bolted upon the face-chuck, and shifted to and from the centre as may be required to suit the work. Fig. 1186 has eight slots, four radial in the usual positions, and the other four arranged, half as shown in Fig. 1183, and half as shown in Fig. 1185.

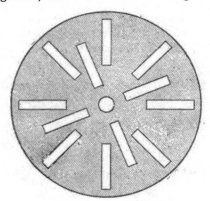

Fig. 1184.—Twelve-slot Face-plate.

This gives the facility mentioned in connection with the last named figure. Fig. 1187 has five slots and six square holes,

and is an arrangement perhaps more fanciful than useful. Fig. 1188 has six equidistant slots and thirty-six square

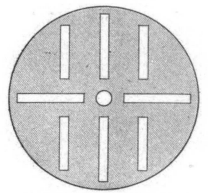

Fig. 1185.—Eight-slot Face-plate.

holes, giving a very wide choice for placing bolts.

PERFORATIONS OF FACE-PLATE CHUCKS.

To sum up the attributes of the several peculiarities of perforations as shown in the above face-chucks, the radial slots are convenient because they allow bolts to be moved any desired distance, however minute, within the limits of the slots. The square holes are convenient for holding dogs, which cannot slip as they may

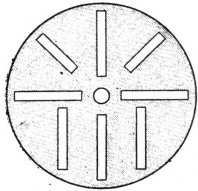

Fig. 1186.—Eight-slot Face-plate.

do when held in a slot. Parallel slots allow a bracket or a bar to be bolted to the chuck, and slid to or from the centre any distance within the limits of the slots. The slots and the square holes are often only cast in the chuck.

Fig: 1187.—Six-hole, Six-slot Face-plate.

FACE-PLATE ATTACHMENTS FOR HOLDING WORK.

Work is held in position for turning on a face-plate by means of brackets, screw dogs or clamping plates. A bracket is illustrated by Fig. 1189; this is bolted to the face-plate, and, in turn, the work is bolted to the bracket. Three or four screw dogs (Fig. 1190) can be bolted to a face-plate, the work being held by the pressure of the screws shown, whose cylin-

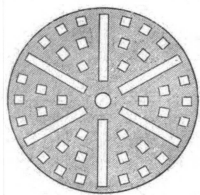

Fig. 1188.—Thirty-six-hole, Six-slot Face-plate.

drical heads are perforated for rotating them by means of a tommy. Fig. 1191 is a clamping plate which is held to the face-

plate by one of the bolts shown, the work being clipped between clamping plate and face-plate by means of the second bolt. The clamping plates illustrated by Figs. 1192 and 1193 have similar uses. Fig. 1194 shows a device helpful in clamping to

Fig. 1189.—Face-plate Angle Bracket.

the face-plate any flat piece of work having a boss or projection requiring to be turned or otherwise treated. The boss projects through the hole in the clamping plate, and bolts pass through the slots shown and hold the work to the face-plate.

SOME USEFUL CHUCKS.

A cup chuck is illustrated by Fig. 1195; this screws on to the mandrel nose, and the worked is forced into the cup. When furnished with screws for holding the work in the cup, such an appliance is known as a bell chuck (see Fig. 1196); there may be eight screws, as shown, or only four. The drill chuck (Fig. 1197) has a smaller opening than the cup chuck, and also is longer, but otherwise is very similar, the drill, however, being held in position by pinching with a steel set screw. The square hole chuck (Fig. 1198) has its tapered end driven into the mandrel opening, the work being held in the square hole.

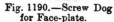

Fig. 1190.—Screw Dog   Fig. 1191.—Clamping
for Face-plate.   Plate.

FOUR-JAW CHUCK.

The four-jaw chuck or dog chuck (Fig. 1199) is most useful for holding all descriptions of work within its capacity. Many but unsuccessful attempts have been made to produce a cheaper substitute for this somewhat expensive appliance. In the four-jaw chuck shown by Figs. 1200 to 1203, each jaw has independent motion, this allowing the chuck to hold work irregular

Fig. 1192. Fig. 1193. Fig. 1194.

Figs. 1192 to 1194.—Clamping Plates.

in form as well as that which is symmetrical; also, it allows any work to be chucked eccentric—often a necessity, and generally a convenience. There are so-called self-centering chucks, having their jaws actuated all together by a scroll, or by a toothed wheel driving screws simultaneously by means of gearing, but they are inferior to chucks with independent jaws, notwithstanding statements to the contrary usually coming from interested sources. For many purposes self-centering chucks are quite useless, though for repetition work they save much time. The construction of the four-jaw chuck is shown by Figs. 1200 to 1203 (two-fifths full size), from which measurements can be taken. The chuck is 10 in. in diameter, being the full size that will swing on a 5-in. centre lathe, but the dimensions shown are sufficient for a chuck 12 in. in diameter if fitted to a moderately heavy

screw would be decidedly weak for a 10-in. chuck for use on a heavy back-geared lathe. Briefly, the chuck illustrated is one supplied by the Sir Joseph Whitworth Co., fitted to their 5-in. centre

Fig. 1197.—Plain Drill Fig. 1198.—Square Hole
Chuck. Chuck.

foot-lathe, and it would be hard to improve it either in design or dimensions, except with respect to the length of jaw relative to the traverse, as through some blunder in the chuck from which the drawings were made it is rendered incapable of gripping work of certain size, as will be explained subsequently. Fig. 1200 shows the cast-iron chuck both front and back, and is drawn so as to make all parts clear. The ragged line running slantways across the circle divides it, and those parts shown above this line belong to the front or face of the chuck, while the parts shown below belong to the back. The four radial slots, which form the slides for the four jaws, are each shown differently furnished, so as to get all the details of construction into one view.

One of the Chuck Jaws.

The upper slot A, Fig. 1200, shows the jaw and the traversing screw both in position as they would be in the complete chuck. The letter A appears on the front of the jaw, which has one step only. This

Fig. 1195.—Cup
Chuck.

Fig. 1196.—Eight-screw
Bell Chuck.

Fig. 1199.—Four-jaw Chuck.

6-in. lathe. The screws which set the jaws in motion are $\frac{1}{2}$ in. in diameter, strong enough for a 12-in. chuck, but a smaller

jaw is made of wrought-iron, and is case-hardened after being shaped. Two holes, drilled into the face of the jaw, and tapped with a $\frac{3}{16}$-in. diameter thread, receive the bolts by which extra biters can be fixed to the usual jaws for temporary or special uses. The section of this jaw,

CHUCK TRAVERSING SCREW.

The slot on the left is marked B; the jaw is entirely removed from this, and the traversing screw is drawn in elevation.

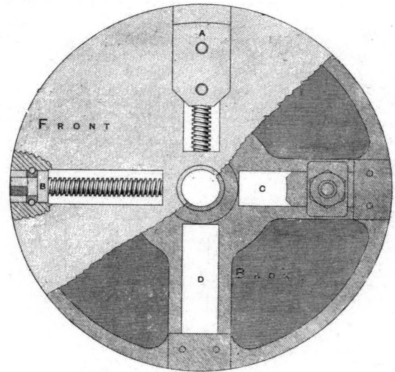

Fig. 1200.—Part Front and Back Views of Four-jaw Chuck.

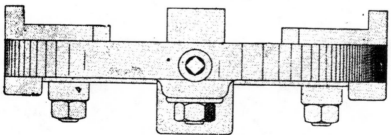

Fig. 1201.—Plan of Four-jaw Chuck.

shown on the right just by the letter E, Fig. 1202, shows the depth of these holes and other particulars. Fig. 1203 shows this jaw in section at right angles to the view given by Fig. 1202.

This screw is $\frac{1}{2}$ in. in diameter, and has a square thread eight to the inch cut left-handed. The ordinary $\frac{1}{2}$-in. Whitworth V thread having twelve to the inch, and right-handed, would be suitable, perhaps almost equally so. The use of screws cut indiscriminately left-handed or right-handed, for actuating sliding parts, is very mischievous, and serious mistakes often

screws in a direction contrary to that intended. Frequently when the screw used to actuate the poppet barrel is cut "wrong-handed," so far as affects the habit of the user, a breakage is caused, either by letting work, suspended on the cone centres, fly out when the intention is to tighten its bearings, or by unduly pressing a delicate tool when the intention is to relieve it entirely of strain. A small portion of the face of the chuck, Fig. 1200, is shown broken away at the end of the screw B, as indicated by the section lines, to lay bare the cast-iron on a plane with the centre of the screw. This shows the fitting of the steel ring which confines the end motion of the traversing screw, and which is held by two tapering steel pins. In the sectional view of the screw, Fig. 1202, this ring is shown removed. Each of the traversing screws has a short pivot on its inner end. These pivots freely enter shallow, flat-bottomed recesses made in the cast-iron, as shown in Fig. 1202. Each screw has a collar, like B, Fig. 1200, and these are fitted to the holes bored radially through the cast-iron at four equidistant places opposite the ends of the slots. The steel ring already mentioned is bored to pass freely over the squared end, and to bear against the collar of the traversing-screw, and the motion of this screw is thus confined by the ring which bears against the collars, and the flat pivot end of the screw which bears against the bottom of the shallow recess made at the inner end of the slot. The inner side of the collar is made to lie flush with the outer end of the slot, as shown at C, Fig. 1200; this is shown also near the top in Fig. 1202.

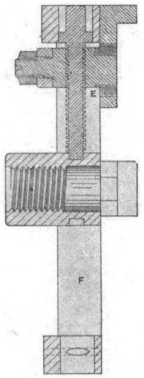

Fig. 1202.—Section of Four-jaw Chuck.

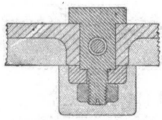

Fig. 1203.—Section of Chuck Jaw.

result from the practice. When one has become habituated to working a lathe with, say, left-handed traversing screws, and then shifts to a lathe with right-handed screws, he is very liable to operate the

RIGHT-HAND SLOT OF FOUR-JAW CHUCK.

The slot on the right is marked C, and is shown with the traversing screw removed, but the jaw remains in its place. The back of the jaw, as here shown, has a $\frac{1}{2}$-in. screwed stalk, and on this a hexagon nut. Under the nut is a washer in the form of a rectangular wrought-iron plate, two sections of which are shown in Figs. 1202 and 1203. This plate is conveniently made from a piece of bar iron of suitable section. It has a groove cut across it, and is carefully fitted over the squared part of the jaw, so that when the nut is screwed home the washer becomes firmly fixed against the jaw, and forms, as nearly as can be, a solid piece with it. Then the jaw should slide in the slot, fitting tightly against the face of the chuck on one side and against the back ribs on the other. In use, directly the jaw is brought up to its work and exerts pressure, it binds tightly against the chuck; but the fitting must be so close and held so strongly that the jaw does not tilt up in the least under the influence of this pressure.

LOWER SLOT OF FOUR-JAW CHUCK.

The lower slot is marked D, and is shown empty. This part is the back of the chuck, and the ribs which strengthen the casting alongside the slots may be seen and measured. The slot is $1\frac{3}{8}$ in. wide and 3 in. long; each rib is fully $1\frac{5}{8}$ in. wide. The entire chuck is $\frac{3}{8}$ in. thick, except where the slots are. It has its rim formed by a ring also $\frac{3}{8}$ in. thick and 1 in. wide. At these four places, where the large holes are made to admit the traversing screws, the rim is strengthened by additional thickness at the back, as shown in Figs. 1200 to 1202. The thread for the mandrel-nose is cut in a boss placed in the centre for that purpose. From this boss, ribs run to the outer rim on both sides of the slot (see Fig. 1203). The thickness of the chuck at that part where the word "back" appears is $\frac{3}{8}$ in., while the outer rim shown projects $\frac{5}{8}$ in. from it, and the ribs project $\frac{3}{4}$ in. These ribs, therefore, stand $\frac{1}{8}$ in. higher than the outer rim, and this is a convenience when planing them all level. In Fig. 1203 the sectional form of these ribs may be seen; the portion of this illustration that is section-lined most openly represents the cast-iron body of the chuck.

TRAVERSING SCREW RINGS, ETC.

The rings which confine the end motion of the traversing screws are fixed by tapering steel pins put in from the face of the chuck. The positions for these pins are shown near the outer ends of the slots B, C, and D in Fig. 1200, and near the bottom of Fig. 1202. These pin-holes are bored through on the diameter line of the hole in the casting, so that half the thickness of each pin trenches into the edge of the collar. These steel pins may be about $\frac{1}{16}$ in. in diameter, and about $1\frac{1}{2}$ in. long.

CENTRAL HOLE OF CHUCK, AND OTHER DETAILS.

Fig. 1202 shows the chuck cut through diametrically, and is a vertical section of Fig. 1200 taken through the centre. The thread by which the chuck is screwed on to the lathe nose is shown in the middle. That part of the thread which is beyond the nose is turned away, making a plain hole the full diameter of the thread. It is usually a good plan to have large holes in the centres of surface chucks, because when work is fixed upon them and holes are bored in the work, the boring may proceed without interruption right through the work and into the chuck, and even the mandrel nose itself may be bored into, unless there is an open space for the drill to come out in when through the work. In the upper part of Fig. 1202 a traversing screw is shown in section, it being similar to the one shown at B in Fig. 1200, except that the ring is removed. The jaw is shown in section just above E. In Fig. 1203 the jaw is shown in section also, but at right angles to the section in Fig. 1202. These two last named views of the jaw, together with that marked A in Fig. 1200, all show dimensions of a jaw.

THE FOUR JAWS OF THE CHUCK.

The four jaws for this chuck are all alike. Forgings of good wrought-iron should be used for making them, and they should be thoroughly case-hardened. See that the length of the jaw from the inner biting end to the biting surface of the step is not greater than the length the jaw can travel in its slot. If the acting surfaces of the jaw are farther apart than the total travel of the jaw in its slot, there will be certain sizes midway in the range of the chuck's capacity that cannot be held by the jaws, that will be too large to go between the points of the jaws when these are fully opened, and also which the stepped biters will not reach when the jaws are quite closed. This defect is present in the chuck illustrated; but it may be quite easily corrected by reducing slightly the length of the jaws.

FURTHER DETAILS OF THE CHUCK.

The lower half of Fig. 1202 shows the traversing screw removed, and the recess to receive the pivot end of the screw is shown. F indicates the side of the slot. The position of the tapering pin, which

square hole. Fig. 1203 shows a section of a jaw and of part of the cast-iron chuck taken at right angles to a slot. The substance of the casting can be gauged from this, and the dimensions of the jaw can also be taken. The circles near the middle of this illustration represent the traversing screw shown in section. This chuck grips objects by their edges, the jaws being adapted especially for this purpose. A substitute for the four-jaw chuck is made by attaching independent jaws, having usually only a small amount of traverse, to ordinary face-plate chucks.

DIE CHUCKS FOR WIRE AND RODS.

Die chucks are used for holding wire

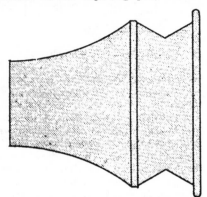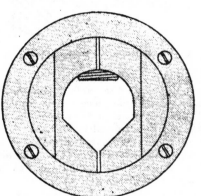

Figs. 1204 and 1205.—Elevations of Bennett's Die Chuck.

holds the ring, is dotted in near the bottom of this illustration, the full curved lines showing the entrenchment of the pin in the ring. Fig. 1201 shows the edge of the complete chuck, all the jaws being here drawn in their respective places. Near the middle of this illustration the squared end of a traversing screw is shown. The diameter of this end is $\frac{1}{2}$ in., and a $\frac{3}{8}$-in. square can be made on the $\frac{1}{2}$-in. round part, leaving the corners only slightly rounded. The hole around this screw is $\frac{7}{8}$ in. in diameter, and it allows a box spanner to be used for actuating the screw. This spanner has a round handle about 5 in. long, and tapering from about $\frac{3}{8}$ in. at the end to $\frac{1}{2}$ in. at that part where it joins the eye with the $\frac{3}{8}$ in.

and rod materials in the lathe. Chucks useful for this purpose are very numerous, and the principles on which they are constructed differ widely. Figs. 1204 to 1232 show a number of die chucks, many of the illustrations presenting end views of work gripped in the chuck by screw or die. Four distinct forms of die chuck are illustrated, but as there are shown different methods of construction in the details of each, as well as the different designs, the method of making a much greater variety is made clear, and several important modifications will be suggested.

BENNETT'S DIE CHUCK.

Figs. 1204 to 1206 represent in elevation and section a die chuck which pinches the work and fixes the die simultaneously by means of one screw. It is perhaps not suited for general jobbing, but with moderately careful treatment it is a very

for small diameters; the die is inverted for medium diameters, and removed, as in Fig. 1205, for large diameters. By this arrangement the motion of the sliding

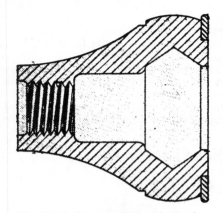

Fig. 1206.—Section of Bennett's Chuck.

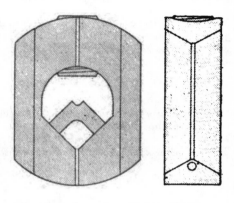

Figs. 1208 and 1209.—Sliding Die of Chuck.

enduring and useful appliance. Fig. 1204 is a side view of the body of Bennett's chuck, the sliding die-piece being removed. Fig. 1206 is a sectional view of the same, cut through horizontally, to show the shape of the V-grooves, in which the die-pieces slide. The thread for attaching it to the lathe-nose is shown, as also at the right-hand corners a section of the steel ring fixed to the front of the chuck by four small countersunk headed screws (see Fig. 1206). The steel ring strengthens the chuck by holding the two sides of the body against the side pressure of the die. The outer edge of this ring extends to the full diameter of the most projecting portion of the chuck, so as to form a guard against anything catching an eccentric projection, and the edge is well

die across the face of the chuck is not great, and the pinching screw, Fig. 1210, projects but very little, even when gripping work of the largest diameter the chuck will admit. Fig. 1208 shows a front view of the sliding die and Fig. 1209 an edge view. The sliding die is made somewhat shorter than the diameter of the body, so that it never projects beyond the body when shifted diameter-ways to suit any size of work. The steel ring, as shown in Figs. 1204 to 1206, is, therefore, large enough to circumscribe the greatest projection; but if the ring were to be used on any of the other die chucks that will be shown, it would require to be much larger in diameter to extend at least as far as the projecting screws. The face view of the chuck (Fig. 1205) indicates

Fig. 1207.—Chuck Die.

Fig. 1210.—Pinching Screw.

rounded to make it harmless if the hand of the user happens to come against it when rotating. The loose die, Fig. 1207, shown in position by Fig. 1208, is used

that the steel ring is fastened to the body of the chuck by four small screws. The ring fits over the two segmental side portions forming the face of the body, and is screwed against a flat shoulder behind. Thus the ring prevents the sides from spreading apart under the influence of the jamming action of the die. The central opening shows the full capacity of

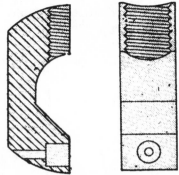

Figs. 1211 and 1212.—Half of Sliding Die.

the chuck for receiving work. The pinching screw has its inner end showing above this opening in Figs. 1205 and 1208. One half of the sliding die is shown in vertical section by Fig. 1211, a view of the inner side being given by Fig. 1212; the threaded portions show the matrix for the pinching screw (Fig. 1210). The plane against which the work is pinched is illustrated in Figs. 1211 and 1212, the circle in the latter showing the hole bored to contain a stiff spiral spring tightly coiled when in its place. This, pressing the two halves of the die outwards against the V-groove in the body, holds the die by friction in the V-grooves when the wedging force of the pinching screw is inoperative. A small hole, shown in Fig. 1209 near the angle of the V, is drilled through the die whilst the two halves are together. When these are afterwards separated, the

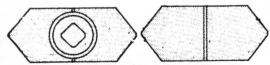

Figs. 1213 and 1214.—End Views of Sliding Die.

larger hole is bored some distance into the inner face of each half by means of a pin-drill, which bores the enlarged holes concentric with the original small one.

The stiff spiral spring is made of round steel wire, coiled to the correct diameter to fit the hole, and so long that the coils come near together, but do not touch when the two halves of the sliding die are together. This spring will hold the slider in the body by pressing the halves outwards against the V-grooves in the body. The two ends of this sliding die are shown in Figs. 1213 and 1214, the former showing also the end of the pinching screw. The central square is a hole through the screw, as shown in Fig. 1210, to receive a square-ended key.

CHUCK KEY.

This key (Figs. 1215 and 1216) is something after the style of the keys used for

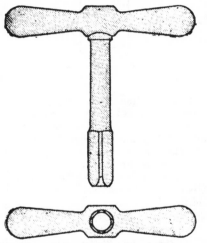

Figs. 1215 and 1216.—Chuck Key.

opening railway carriage doors. It must be powerful to tighten the chuck pinching screw efficiently. The end of the key and the square hole in the screw should be chamfered, so that the former is readily inserted in the latter. The squares themselves, both internal and external, should taper only to a very trifling amount.

THE DIES IN BENNETT'S CHUCK.

The circular ends of the die shown in Fig. 1208 should be curved to the same radius as the diameter of the body of the chuck, and not to the smaller circle given

by the length of the die itself. This will
give the die a larger bearing surface in the
V-grooves of the body, whilst it will not
in the least add to the working diameter,
as a moment's reflection will prove. For
the same reasons, the same method will
be applicable to the sliders of the other
chucks illustrated. The extra die has

tioned that all these die chucks are commonly set for the work to run true by
the aid of a hammer, which tool also
serves to straighten the wire projecting
from the chuck. The body of this chuck
is shaped as shown by Fig. 1217, and it
has a diametrical dovetailed groove across
the face in which the slider fits. The

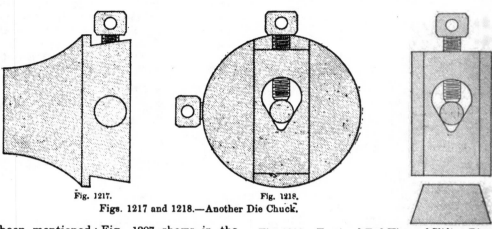

Fig. 1217.

Fig. 1218.

Figs. 1217 and 1218.—Another Die Chuck.

Fig. 1219.—Front and End Views of Sliding Die.

been mentioned; Fig. 1207 shows in the
order here mentioned, starting from the
left, an end view, side view, and the end
form of the finished die. The end view
indicates the way to get the die shaped
from a piece of square bar steel by drilling a hole through it in the position
shown by the dotted circle near the
middle of the square. The small groove
in the upper corner is intended for wire
from the smallest diameter, say $\frac{1}{8}$ in., that
the chuck will hold, to $\frac{3}{16}$ in. The larger
groove in the lower part is for rods
from $\frac{3}{16}$ in. to $\frac{6}{8}$ in. diameter. Rods larger
than this, up to about $1\frac{1}{8}$ in. diameter, are
held by the main die as in Fig. 1205.

ANOTHER DIE CHUCK.

Figs. 1217 to 1219 show a chuck which
has the end of its pinching screw impinging directly on the work; this causes the
work to be badly marked. The small
amount of surface contact does not afford
a good hold, and the work often becomes
loosened whilst it is being acted upon by
the turning tool. This chuck seems to
stand rough usage, and it may be men-

shape and size of this dovetailed groove
are shown by the end view of the slider
given below the front view, Fig. 1219. The
work (indicated by a circle) is pinched in
the angular part of the heart-shaped hole
by the pinching screw shown by Fig. 1220.
The slider is shown to be shorter than the

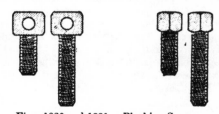

Figs. 1220 and 1221.—Pinching Screws.

diameter of the body of the chuck, and
its ends are square, but this is not necessarily so. A hole is tapped through the
side of the body at right angles to the
dovetailed groove, and a pinching screw,
the shorter of the two shown by Fig. 1220,
is inserted so as to fix the slider. The
point of this screw should not impinge

direct on the slanting side of the slider, which would not afford a good hold. The hole should be enlarged from the inside by means of a rose cutter in the manner described in the following paragraph.

ENLARGING HOLE WITH ROSE CUTTER.

The rose cutter is threaded to suit the end of an arbor which fits the hole. The arbor is put through the screw hole, and the rose cutter screwed on it inside the dovetailed groove. By fixing the arbor in an ordinary drill chuck and revolving it in the lathe, the hole in the die chuck can be recessed by drawing this latter away from the mandrel. This can often be managed by hand alone; but, if not,

filed off aslant to correspond with the slanting side of the slider. The whole face of this disc will then bear against the side of the slider when the end of the pinching screw impinges against its back, and forces it up to a solid bearing. In this way the slider will not be indented, nor the end of the pinching screw dubbed up so that it cannot be withdrawn. Indentation prevents the nice adjustment of the slider, and dubbing up happens if the point of the screw is not hardened. The two screws indicated in Fig. 1218 are shown separately by Fig. 1220; the shorter one is tapped through the body of the chuck to fix the slider, and the longer one is the pinching screw tapped through the

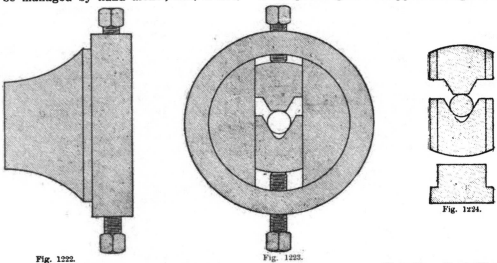

Fig. 1222.

Figs. 1222 and 1223.—Two-die Chuck.

Fig. 1223.

Fig. 1224.

Fig. 1224.—Chuck Dies.

the work may be attached to either the slide-rest or the back centre. The inner end of the screw-hole is usually enlarged before the hole is threaded; though with moderate care a threaded hole need not be damaged by the process. The enlargement should be to at least the full diameter of the screw; but somewhat larger is better, and it may be about ¼ in. deep.

OTHER DETAILS OF THE CHUCK.

A disc of metal is turned to fit the enlarged hole, and the outer side of it is

slider, and its end holds the work. Each screw has the last thread or so turned away at the ends, which should be made flat, and hardened. The capstan-headed screws may be replaced by those shown by Fig. 1221, which have square heads, in cases where a key is preferred to a tommy.

TWO-DIE CHUCK WITH INDEPENDENT DIES.

Figs. 1222 to 1224 show a two-die chuck in which the screws which press the independent dies together, and thus grip the work between them, are tapped through an iron ring. The body of this chuck differs but little from that of the one last described. The diametrical groove across its face is of the shape shown by the end view of one die at the bottom of Fig. 1224. This rectangular

Fig. 1225.—Pinching Screws.

shape is better than the dovetail form, as in this there is no way of adjusting the fit if the die becomes too slack. The motion of one die is much greater than that of the other when changing to a different size of work, and so the pinching screws are of different lengths, as shown in Fig. 1221. If a pair of dies is used, both of which are notched alike, they will of course each travel an equal distance when changing from one size to another. The ring in which the pinching screws are fitted is commonly of iron, and the two dies shown by Fig. 1223 have to accommodate themselves to the whole range of sizes that the chuck will hold. It is a good plan to fit the ring on the body sufficiently loose to allow it to be removed without much

done. If the long, projecting screws (Fig. 1221) are considered to be in the way (they are not so in practice) they may be replaced by those shown by Fig. 1210, or Fig. 1225. The latter form of screw saves at least double the length of the head in diameter, and may be made even much shorter.

IMPROVED DIE CHUCK.

The chuck illustrated by Figs. 1226 and 1227 has a jaw in place of the pinching screw, to grip the work. The long bearing surface afforded by the jaw causes the chuck to hold the work much more firmly than do the chucks previously shown, and the work is not bruised so much as it is by the end of the pinching screw. The body of the chuck has a diametrical dovetail groove across its face; the shape of this groove is shown in Fig. 1228, which is an end view of the slider which fits this groove. The headless pinching screws are drilled up and notched across to receive a screwdriver having a tit in the centre of its blade. This tit centres the blade at once, and makes the tool easier to handle, and at the same time strengthens the hold on the screw. The screwdriver is best

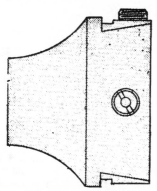

Fig. 1226.—Side View of Improved Die Chuck.

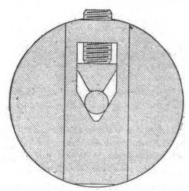

Fig. 1227.—Face View of Improved Die Chuck.

trouble, and then to fit three pair of dies, and make each pair specially suitable for a more limited range of sizes. Then dies suited to the diameter of the work to be held are placed in the groove, and the ring is pushed on its place. The operation of changing the dies is very quickly

with a handle of the shape shown in Fig 1215. The hole tapped through the side of the body of this chuck has its inner end recessed to take a disc between the screw's point and the slider. The shorter screw on the left of Fig. 1225 shows this screw. The segments shown in Fig. 1227 may be made apart from the body and fixed with screws, using steady-pins to accurately locate the segments as explained.

Fig. 1228.—End View of Slider.

SLIDER AND OTHER DETAILS OF IMPROVED CHUCK.

The slider is the most peculiar feature of this chuck, and is shown in front view by Fig. 1229, in vertical section by Fig. 1230, and in back view by Fig. 1231; the jaw apart is represented by Fig. 1232. The slider is fitted to the dovetail groove across the face of the body, and a heart-shaped opening is made in its centre. This hole is enlarged and made rectangular towards one end—the upper one in the illustrations. In a wide, shallow groove in the back of the slider (see Fig. 1228) the jaw (Fig. 1232) is fitted to slide freely but without shake. The jaw proper—the part that bites—projects into the rectangular hole just mentioned. The jaw is free to move parallel with the length of

groove in the slider. A pinching screw tapped through the end of the slider impinges upon the back of the biter, and so gives the means of gripping the work. The flat plate part of the jaw is bored through just below the biter, large enough to allow any work within the capacity of the chuck to pass through it. The larger circle in Fig. 1231 represents this hole; and it should be indicated, but is not, in Fig. 1230. The pinching screw to actuate this jaw is illustrated on the right of Fig. 1225.

MAKING DIE CHUCKS.

The construction of four kinds of die chuck having been minutely explained, some description may be given of how to make them. The body part of any one of the chucks which screws on to the lathe-nose would be the first piece to make. The main point for consideration is the length of the body part—that is, the amount of projection from the nose. When the lathe mandrel is tubular, and so allows space for rod work to project behind the chuck, the chuck should be made snug; but with a solid mandrel, it is often very convenient to have space

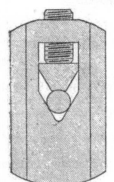

Fig. 1229.

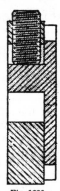

Fig. 1230.

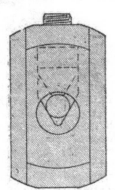

Fig. 1231.

Fig. 1232.

Figs. 1229, 1230 and 1231.—Front View, Vertical Section and Back View of Chuck Slider. Fig. 1232.—Chuck Jaw.

the slider, the motion being limited by the biter touching the top and bottom of the hole. When the slider is placed in the body of the chuck, the bottom of the dovetail groove bears against the flat back of the jaw, and confines it to the

within the body of the chuck to admit some surplus material projecting behind the die. Excessive overhang is, however, always detrimental in any chuck, and it is hard to assign any good reason for not making the lathe mandrel tubular. Most mandrels will bear boring out to a fairly useful diameter, even though the hole may not be made entirely through. The body of the chuck may be an iron, bronze, or brass casting, and a pattern for the foundry must be made. The casting, when obtained, is chucked in a four-jaw chuck such as is shown in Fig. 1199, p. 413, or in any other chuck that will hold it firmly. The end that is to be screwed on to the mandrel must be outwards. The hole for the thread is bored, and the end is turned true and flat. Cutting the thread by which to screw the chuck on the mandrel is a process that needs care and proper tools. The hole should be finished by means of a tap that precisely matches the thread on the nose. The

break away raggedly if it is filed off somewhat chamfering towards the cut, as is often done on metal when planing up a sharp corner. The tool will then cut to the last, up to the moment of dropping out of cut, instead of breaking off just the extreme end of the shaving, and with it the corner of the work. This refers to cast-iron; but there will be no trouble with any of the brassy metals, which have greater tenacity.

SHAPING SLIDER GROOVE OF CHUCK.

The form and position of the groove to receive the slider should be determined and marked on the body, and precautions taken to get it truly central. If the form of the groove is carefully turned in the

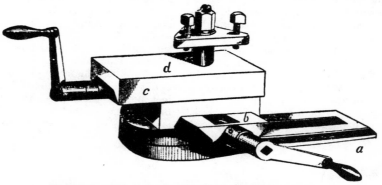

Fig. 1233.—Slide-rest with Main Slide Parallel with Work.

mouth of the hole should be enlarged by boring out to the full diameter of the thread, so as to clear away at least one complete turn of the thread, and thus allow the chuck to screw up to the shoulder. In connection with this point, there is no valid reason for grooving a thread on the mandrel-nose. When the casting screws on the mandrel right up to the shoulder and touches all round, it may be taken off. Being now screwed to fit the nose, it is put on its place and turned true and smooth all over. The diametrical groove may give some trouble in turning the face and a short distance back. Care must be taken not to get the edge of the groove damaged where the tool drops out of cut. The edge will not

middle of its length by the aid of a tool fixed in the slide-rest, it will give the best gauge lines for working to. For accuracy and speed in cutting out the groove to shape, use a cutter mounted on a geared drilling spindle. The body of the chuck is screwed tightly on the mandrel, which is held by the index peg, so that the groove lies horizontally. A suitable cutter is introduced, and the spindle packed up, to give the necessary width to the groove. A trip is then taken through the roughly-cast groove. The body is then rotated exactly half-way round, and again held by the index peg, and a second trip is taken through, which will make the two sides of the groove correspond. By packing up the spindle a small amount, and taking another cut along each side of the groove, it will be still further straightened and smoothed. When spindle and cutter are not available, other and less satisfactory means may be employed. Planing and slotting are both applicable, but the tools used in these processes will not give a symmetrical sectional form, unless great care is taken to set each tool very nicely to correspond with the fellow one which operated on the other side of the groove. The work of the planer or shaper will generally have to be finished by filing. The groove may be made true and smooth by filing only, and small sheet metal templates will be a great help. A cutting-punch or drift may be used, and will produce a good result; but this tool is too costly to buy or to make for use in making but one chuck. A drift for regulating the form of the groove after it has been filed nearly to size and shape can be filed up from a bar of steel, and the teeth made on it by filing notches, which need not be deep, and the drift need not be hardened if intended to be used only once. Good support is wanted for the body of the chuck to rest against, so as not to damage it when driving the drift; but it is better to force the drift through by screw pressure, if this can be managed.

Material for the Chuck Dies.

The dies for these chucks may be made of steel, iron, or brass. If steel is used, the question of hardening becomes an important one. These dies are awkwardly shaped pieces for hardening, and when left soft, steel dies are not so good as case-hardened iron ones. Good iron is a very suitable material, and when the jaws are made of it, they can be case-hardened, either all over, or only at those parts where actually required to withstand wear and tear. The simple process of case-hardening by means of yellow prussiate of potash is sufficient for this purpose. Brass dies are generally almost useless unless the working parts are properly lined with steel; and it is more troublesome to do this than to make the dies from the solid metal. Case-hardened iron may be selected as best when all things are considered.

Making Holes in Chuck Dies.

The hole in the die through which the work is to be passed, and there held by the pinching screw, is generally made somewhat heart-shaped. When two dies grip the work between them, the parts bite at three or four places round the work, and are generally tangential to it. When making the heart-shaped hole, the largest diameter the chuck will admit is the gauge for the large end of the heart. The apex is made by drilling a small hole, just beyond the circumference of the large one, the size of the smallest work the die is intended for. Straight lines tangent to these two circles give the angular sides, and by these angular sides the work is held. The angle formed here is usually about 60°. In some cases, however, it is 90°, as shown in Fig. 1205. The large hole and the small hole are both bored when the chuck is on the mandrel nose and the slider is in place; and the slider is shifted along its groove the necessary amount, after boring the large hole, to bring it into correct position for the small one to run truly central. A triangular file will soon remove the material between the two holes and make the sides straight.

Making Pinching Screws.

Chuck pinching screws must be made from good cast steel of suitable temper. The ends that will nip the work or the slider must be hardened and tempered, and the entire screw is all the better for being so treated. Sharp angles should be avoided at any part where the diameter changes—for example, where the head and the screwed part join; and with this precaution there is little fear of breakage in hardening the screws. Unless properly annealed during the process of manufacture, the screws are very likely to go out of shape, and perhaps the thread-rate will alter in hardening. The tempering of the screws is easily done by flaring off with oil. The screw threads should be cut to the Whitworth standard.

Advantages of the Slide-rest.

The first addition to the plain lathe—that is, a lathe with fast and loose headstocks and **T**-rest only—will be the slide-rest; then comes the division plate and index; and then the overhead motion. Slide-rests are probably the most useful, and, in fact, form an indispensable adjunct to any metal-turner's lathe. (Illustrations of lathes with slide-rests appear on pp. 385 and 442.) Although much work can be done on the lathe with the aid of

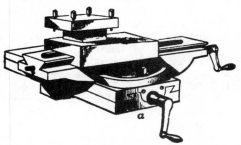

Fig. 1234.—Slide-rest with Main Slide at Right Angle to Work.

hand tools, yet a slide-rest greatly enlarges the capabilities of the machine. A properly made, securely fixed slide-rest will hold a tool perfectly rigid, no matter how irregularly the work may act upon it; for instance, with the use of a slide-rest the corners may be turned off a square rod of metal as easily as cylindrical work can be turned by the aid of hand-tools alone. Work which has a break in the continuity of its cylindrical surface is not easily turned with hand-tools. In turning a parallel cylinder with hand-tools considerable care and constant gauging are requisite to guard against cutting too deeply at some part. With a slide-rest there is no need of such care, for, when the tool has been properly adjusted, a perfectly true and parallel cylinder will be produced by simply feeding the tool up to the work. Slide-rests offer a mechanical hand to hold the turning tool, and the relative value of mechanical work and hand-work will be appreciated by mechanics.

SLIDE-TURNED AND HAND-TURNED WORK COMPARED.

The advantage of slide-turning over hand-turning will be best appreciated by comparing work done by the two methods. A plain cylindrical rod—say, for instance, a piece of shafting—may be mounted on a lathe fitted with a slide-rest, and turned up perfectly true, parallel, and smooth. Compare the truth, the parallelism, and the smoothness of the bar with those of a hand-turned bar, and note the superiority of the machine work. Slide-turning, too, takes less time than hand-turning.

TYPICAL COMPOUND SLIDE-REST.

A typical compound slide-rest for use with medium heavy work is illustrated by Fig. 1233. Its base-plate a is planed underneath to fit the lathe bed and to carry the various slides level and square; the holding-down bolt passes through the slot in the base-plate. A strong traversing slide or main slide, b, is operated by means of the screw and winch-handle illustrated; above this is the surface-slide or top slide c, whose upper surface, d, constitutes the tool-plate. The tool-holder shown above, d, is a triangular plate with a pinching screw in each of its three corners; through a central hole loosely passes a spindle, round which the plate may be swung to occupy any desirable position. The tool is held under the triangular plate by two of the screws, which can be tightened down upon it at any angle to suit the particular work in hand; the third screw acts simply as a support to the third corner of the triangular plate and assists in keeping it horizontal. The stud-bolt (the central screwed spindle with its nut) confines the upward travel of the triangular plate and allows of pressure being exerted on the tool. The points of the three screws must be harder than the threaded portions, otherwise the hard steel tools would flatten or upset the ends of the screws. A point to be noticed is that the base-plate a has its length at a right-angle with the work, the main slide b, consequently, being parallel with the work. Such a slide-rest as the one illustrated is not suited to heavy work, as its attachment to the lathe bed is not sufficiently strong, and there is too much of the appliance overhanging the front of the lathe.

A Stronger Compound Slide-rest.

The appliance represented by Fig. 1234 is of a more compact nature, and is superior in other ways, chief of them being that the bottom or main slide is at a right angle with the work and that the baseplate *a* has a projection fitting into the lathe-bed; these add strength to the appliance and make it of use for heavy work. The general arrangement otherwise is much as before, but the top slide has an angular motion, and a curved slot and screw for producing angular and tapered work. The tool-holder is here an immovable square plate, forming a part of the upper plate. It has four pinching screws, one at each corner, and there is thus a choice of four positions for the tool, which is held under any two of the screws. Though the choice of positions is more limited than is the slide-rest previously shown, this appliance is in general respects much stronger and more efficient.

Construction of Slide-rest.

Slide-rests are made in very many designs, ranging from very light ones for ornamental work in wood, ivory, etc., to very massive ones for shaping metals. The one about to be described is suited for the general heavy work that can be done on a foot lathe, and is made in the style common on foot lathes that have not a saddle fitted to the bed. It is modelled somewhat after the design of the orthodox ornamental slide-rest, and it has the lower slide parallel with the lathe bed. The slide-rests of heavy self-acting lathes have this lower slide at right angles with the bed. The following instructions on the making of a slide-rest, as well as the information in this chapter on the shapes and uses of face-plate, four-jaw, and die chucks, originally appeared in articles contributed by the editor of this work to the "English Mechanic." This description will make clear every detail of the appliance, and will serve also as a guide to any more or less skilled worker who may wish to make it.

Working Drawings of Slide-rest.

The drawings of the slide-rest about to be given are all three-quarter full size; but for building such a slide-rest to suit a lathe of given size, one has only to draw a scale to suit, and by measuring the drawings with this scale it will be seen what the actual dimensions should be. The drawings (Figs. 1235 to 1248) have been made from a slide-rest fitted to a 5-in. centre lathe; the main slide measures just 14 in. long in the original slide-rest, but to accommodate the drawings within these pages it has been necessary to show the slide broken in the conventional manner. Fig. 1235 is the top view or plan, Fig. 1236 the side view, and Fig. 1237 the front view; Fig. 1238 shows the main slide, and Fig. 1239 the top slide.

Adapting Drawings of Slide-rest.

The adaptation of the drawings to suit any required dimensions is very simple. The side view (Fig. 1236) shows the under side of the tool-gripping jaw exactly 5 in. above the base of the slide-rest; that is to say, the under side of the jaw is precisely on a level with the lathe-centres. To get the relative dimensions of all the parts to suit a slide-rest for a lathe of any size, carefully take the distance from the base to the under side of the jaw, as illustrated, and let this distance represent the height of centres of the lathe for which the slide-rest is to be made. Divide this distance into a number of parts representing the number of inches in the height of centres, and with these divisions construct a scale with which to measure all parts of the drawings, and thus obtain the dimensions of each detail. Awkward fractions will be avoided by somewhat modifying the dimensions. For example, the main slide, being 14 in. long in the slide-rest for 5 in. lathes, would be 11¼ in. for 4 in. and 16¼ in. for 6 in. lathes if strict proportion were to be observed; but in practice it will be better to make the former 11 in. long and the latter 17 in. long. The screws must always be made to a regulation thread size, selecting the one which most nearly agrees with the measurement shown by the scale. In short, when measuring the illustrations with the specially constructed scale, the

The traversing screws used in these slide-rests are almost without exception

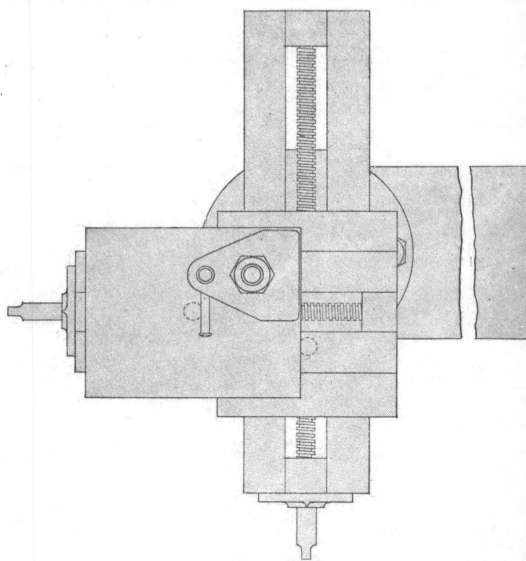

Fig. 1235.—Plan of Slide-rest.

dimensions should be read, and the parts constructed, to the nearest whole numbers, avoiding any attempt to call a dimension " bare " or " full."

square-threaded, and are commonly cut with ten threads per inch, which gives $\frac{1}{10}$ in. traversing motion for each complete rotation of the screw. This same rate of thread is often preserved even when the diameters of the screws differ considerably. Eigth threads per inch sometimes suits equally well, and is sometimes even more convenient in use, and the coarser thread is generally preferable in practice.

this the round base of the main slide rests. Fig. 1238 shows a section at this part. The bottom of the sole must be made flat and smooth, and the dovetail groove must be got out straight and smooth also; for this, milling does much

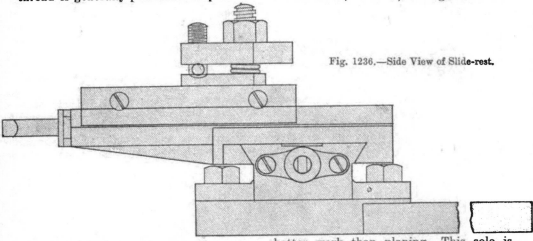

Fig. 1236.—Side View of Slide-rest.

SOLE OR BASE-PLATE OF SLIDE-REST.

The whole appliance is made of cast-iron, with the screws, etc., of steel, while the nuts of the traversing screws, the

better work than planing. This sole is conveniently worked by chucking it on a face-plate and facing the bottom; after this the dovetailed groove is made true with a circular cutter, operated by a

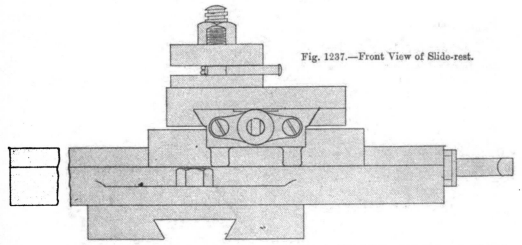

Fig. 1237.—Front View of Slide-rest.

jaw, and the plate, etc., are of wrought-iron. The sole or base-plate has a dovetail groove running from end to end in the bottom. One end is circular, and on

strong spindle driven from the overhead. The sole is turned over, and fixed to the face-plate with the middle of its circular end true. The top of this is turned level, and a small hole is made exactly in the centre, and the T-groove is turned as shown in the sectional view of the main slide (Fig. 1238). This T-groove is to take the heads of two bolts, one of which is shown in plan and elevation by Fig. 1240. These bolts have hexagon nuts, shown in the side and front views, and hold the main slide to the sole. When these nuts are slack, the main slide can be turned completely round on its base. The T-groove is preferable to curved slots in the round base-plate of the main slide; such slots would allow only a moderate angular motion, whilst the T-groove allows the main slide to be placed at any angle, and it is often very handy to place it parallel to a face-plate; that is, at

MAIN SLIDE OF SLIDE-REST.

The main slide is an iron casting standing on a circular base-plate, and the dovetailed sides for the saddle to fit upon are undercut to form an angle of about 55° with the top. For a 5-in. lathe this main slide is 14 in. long, and to make it straight, smooth, and parallel on all its bearing surfaces needs good workmanship. This could be done by filing, but it is a tedious process. The hole in each end (the two holes form bearings for the traversing screw) must be made quite true with the slide, otherwise the nuts will bind on the screw as the saddle is moved along the slide. One way of getting these holes true is to bore them before surfacing the slide, and to chuck

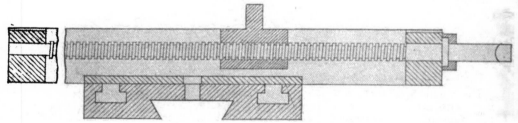

Fig. 1238.—Main Slide of Slide-rest.

right angles to the position illustrated. When placed this way, a tool can be traversed across the full diameter that the lathe will admit. Two round holes are made in the base-plate of the main slide to allow the bolts to pass through, and of course a hole must be made in the sole, into the T-slot, large enough to admit the heads of the bolts. These heads are somewhat flattened on opposite sides to admit them to the T-slot, and yet not allow them to rotate under the action of screwing up the nut. The small hole in the centre of the circular part of the sole is fitted with a steel pin which has a slight shoulder, as shown in Fig. 1238, and is driven tightly into the sole. The main slide has a hole bored in the centre of its base to fit this pin, and it should do so without shake, so that angular adjustment can be made easily and accurately.

the slide by the boring when surfacing the top and edges, the bottom being turned true and flat afterwards. A better way is to bore these holes after the saddle has been fitted, and then to make an iron template, about ⅛ in. thick, to wedge in the dovetails of the saddle. Bore a hole about the middle of the template to correspond with the required position of the bearings of the traversing screw. The saddle is placed slightly overhanging one end so as to hold the template, and the hole in the template is used to guide a rose-bit, which cuts the hole in the main slide for the traversing screw. By shifting the saddle to the other end of the main slide, and using the same template, placed inside out relatively to its former position, the rose-bit will cut a hole in that end, which will exactly correspond with the one first made. Though both holes may be more or less out of the middle of the main slide, this will not affect the working of the saddle, provided both holes are out the same distance and in the same direction, which they will be if the template has been properly used.

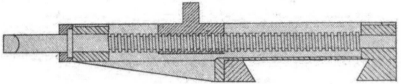

Fig. 1239.—Top Slide of Slide-rest.

SADDLE FOR MAIN SLIDE.

The saddle for the main slide is made in one piece with the top slide (Fig. 1239). It is always advisable to have as few parts as possible, so as to give the greatest rigidity to the structure. The top slide is 9 in. long for a 5-in. lathe. It is also of cast-iron, and a section of it is shown by Fig. 1239. The top slide must be got up true, on the top and on the dovetailed edges, in the same manner as the main slide. The saddle is fitted to the main slide so that the two sides are precisely at right angles one to the other. At the back end of the top slide a strip is cast solid with the saddle, and this strip is undercut to fit the dovetailed edge of the main slide. Opposite this strip a small fillet is cast, and the loose strip, shown by Fig. 1241, bears against this fillet, and is held on by four binding screws, one of which is shown above the

to fix this strip, the two middle ones screwing into the bearers of the slide. This strip should fit quite snug, and at the same time allow the saddle to move along the slide without shake, when the screws are finally tightened up. The holes in the strip, through which these binding screws pass, are elongated to allow some lateral motion to the strip during the process of fitting, and the ultimate bearing of the strip should be with its back edge against the fillet. The screws are intended only to draw the strip close against the saddle, and they must not impede the slight lateral motion necessary for adjustment.

TOP SLIDE TRAVERSING SCREW.

The traversing screw of the top slide is commonly made the same pitch and diameter as the screw of the main slide; but it is not unusual to make the former right-handed and the latter left-handed. This reversal of the threads is adopted to give the same effect, as to direction of traverse in the saddles, by the same apparent motion of the handles. The comfort of using either a right-handed or a left-handed traversing screw becomes simply a matter of habit, and when one

Fig. 1240.—T-groove Bolt.

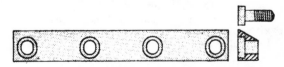

Fig. 1241.—Loose Strip for Slide-rest Saddle.

end section of the strip in Fig. 1241. This loose strip does not need any adjusting screws, as it is fitted to bear against the fillet with which it is intended to remain in close contact after it has once been fitted. Four binding screws must be used

has become accustomed to either one, the other seems awkward to manipulate. Persons who use various lathes intermittently frequently complain of the change in the direction of the traverse with the same direction of rotating the screws. This confusing alteration is caused by the indiscriminate use of right-handed or left-handed screws for traversing. This is especially apparent in the traversing screws in lathe poppet barrels, some being right and others left. The screw in the poppet and the one in the slide parallel to the line of centres should always be made to correspond in direction of thread.

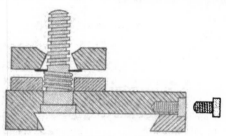

Fig. 1242.—Section of Top Saddle.

TOP SADDLE.

The top saddle is a rectangular plate of cast-iron, as shown in the top view of the complete slide-rest. It has a strip, cast solid with it, running the entire length of one side, as shown in the section of the top saddle (Fig. 1242). The other strip is separate (see Fig. 1243), and is attached by three binding-screws. It is held up to the dovetail bearing by two adjusting screws, which are tapped into the edge of the saddle, and recesses are countersunk to admit their heads. A section of the screwed-on strip is shown in Fig. 1243, which also illustrates one of the binding screws. At the right-hand of Fig. 1242 one of the adjusting screws is shown, and also the hole into which it screws. The top saddle is finished bright all over. It can be filed up flat and true, but milling or planing is preferable.

STUD-BOLT OF TOOL-HOLDER JAW.

Near one corner of the top saddle, on the side where the solid strip is cast, in the position shown by the plan, Fig. 1235, a hole is made to take the main stud-bolt for the tool-holder jaw. This steel stud-bolt is shown in Fig. 1242. It has a square thread, which is greatly superior to a V-thread for this purpose; this thread has

to withstand severe strains, and wears badly if not carefully treated. The stud-bolt is driven into the top saddle from the under side up to a shoulder, and it is necessary to cut away a small arc from the solid strip, at its most prominent part, under the hole to allow the head of the stud-bolt to pass. The diameter is kept large, especially near the head, for strength. If the shank of the bolt is turned away under the head it is likely to bend under usage. Before the slide-rest is put together finally, a small hole must be drilled, and a steel pin inserted to prevent the stud-bolt rotating in the saddle under the screwing action of the nut. This small hole is drilled half in the cast-iron and half in the steel bolt-head; it is not shown in the illustrations. If this is not done, the bolt will at some time work loose and rotate in the saddle with the hexagon nut jammed upon the thread. Should this occur it is generally necessary to destroy the stud-bolt in order to get it out of the saddle—the costly penalty for not taking proper precaution.

SLIDE-REST TOOL-HOLDER.

The tool-holder shown on the slide-rest in Figs. 1235 and 1236 is an adaptation of Willis's pattern. The jaw is of the form illustrated, is made of wrought-iron, and should be case-hardened. The plain, flat underside of the jaw is sometimes an unsatisfactory gripper, especially when it is called upon to grip anything that is lumpy, and not quite flat. It is well to put two

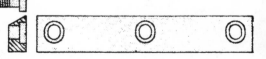

Fig. 1243.—Screwed-on Strip.

small steel pins, with round heads just like common snap rivets, in the corners of the jaw. The pin heads, hardened and tempered, form teeth which are always sure to bite on any tool put under the jaw, even though it may be slightly uneven, whilst the heads being smooth and rounded will not injure tool stems that are finished smooth, as that of a drilling spindle, for example. These suggested teeth are not shown in the drawings. Sometimes that part of the jaw which grips the tools is made with chequered teeth, like those on the jaws of a vice; but this does not give the certain grip at two far distant points,

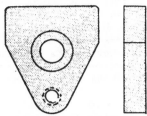

Fig. 1244.—Tool-holder Jaw.

such as the pin-heads afford. The jaw is shown apart in Fig. 1244, top and edge views being given. The hole which passes over the stud-bolt is first drilled through, and is then tapered from the under side, as shown in the section. The top side is then hollowed to a circular curve, and the under side of the hexagon nut, which fits the stud-bolt, is rounded to the corresponding radius. This arrangement allows the jaw to move freely by means of a species of ball-and-socket joint, to a certain extent from its normal horizontal position, and still to maintain sufficient bearing surface against the hexagon nut, and all round it. Thus the jaw accommodates itself to any small unevenness in the heights of its three corners. The corner of the jaw, shown at the bottom of Fig. 1244, is tapped to take the steel pinching screw, Fig. 1245. The case-hardened iron plate used under the jaw is illustrated by Fig. 1246. It is bored to go over the stud-bolt sufficiently large

Fig. 1245.—Tool-holder Pinching Screw.

to allow a steel spiral spring between them. This spring keeps the jaw lifted against the hexagon nut, so that space for the tool is always clear. A very thin washer, shown

76

in the section of top saddle, Fig. 1242, is placed between the under-side of the jaw and the spiral spring to prevent this latter from entering the tapered hole in the jaw, and so jamming in it. The part of the plate coming under the pinching screw's head is hollowed out to receive it.

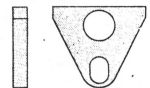

Fig. 1246.—Tool-holder Jaw Plate.

ATTACHMENTS OF THE TRAVERSING SCREWS.

The details connected with fitting up the slide-rest may now be noted. The traversing screws for the slides are both shown in the sectional views, Figs. 1238 and 1239. The end motion in each of these is confined by a collar near the handle end; this collar butts against the casting on its inner side, and its outer side is confined by a plate. These plates are alike for both screws, and one is illustrated by Fig. 1247, which gives front, section, and back views, and shows also one of the two binding screws for fixing the plate to the cast-iron slide. Wrought-iron plates wear the best, but brass and gun-metal have a better appearance. The two plates should be made together. First bore the large central hole in each, then turn the collars and mark the screw holes. Pin the two plates together by these holes and file up the edges, re-

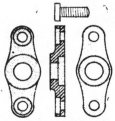

Fig. 1247.—Traversing Screw Plate.

versing the plates to see that both edges are uniform in contour, and correcting by filing where necessary. The central hole is countersunk with a flat pin drill to admit the collar of the traversing screw, and the holes for the binding screws are similarly countersunk, but from the opposite sides, to admit the heads of these binding screws. The cast-iron slides are tapped to receive the screwed portion of these screws, and their positions are shown on the front and side views of the complete slide-rest. There is a method of confining the end motion of the traversing screws by means of nuts on the small ends instead of with plates, but this will not be gone into here. The nuts for the traversing screws are both shown in section in Figs. 1238 and

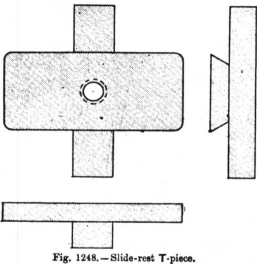

Fig. 1248.—Slide-rest T-piece.

1239. They should be made of wrought-iron for wear with steel screws, and each has a round stalk, by which it is attached to its saddle. A hole bored in the saddle receives this stalk, which is fitted to it and driven in tightly. In the plan of the complete rest, Fig. 1235, dotted circles show the positions of the holes in the saddles. One is in the middle of the top saddle, and the other is in the bearer of the top slide. Neither hole needs to be bored through, and the top surfaces of the castings will look better if these holes do not show. Tapping these nuts after they have been fitted is a process that demands care. The nut is first fitted by its stalk to the hole in the saddle, and then the projecting part

is shaped to pass freely in the groove of the slide. The position of the hole, through which the traversing screw is to pass, is marked by scribing with a needle through the holes in the castings. Each end of the nut is marked, and a hole is drilled through it, and then tapped. Unless great precaution is taken, the nut will not travel smoothly along the screw, and the saddle will be fast in some places and slack in others—always an annoyance. The method of securing the handles on these traversing screws is simple and convenient. The screw-shank is turned true, and has its end flattened by cutting away from opposite sides, as illustrated in Figs. 1238 and 1239. The handle has a pipe-stem drilled to a depth equal to the length of the screw-shank, less the flattened part, and a mortise hole is cut through the pipe-stem to admit this flattened part, which is in this way held.

PINCHING SCREW AND OTHER DETAILS OF TOOL-HOLDER.

The steel pinching screw for the tool-holder is shown by Fig. 1245. It has no enlarged head, but a hole is drilled diametrically through it near one end, and fitted with a steel "tommy" rod to act just like the handle of a vice. This is shown in the top and front views of the complete rest, Figs. 1235 and 1237. The lower end of the pinching screw is rounded off very freely, and it bears in the hollow of the under-plate mentioned before. The usual way of putting this screw in its place is with a head on the top, usually hexagonal, to be turned with a spanner; but by turning it upside down, as shown here, space is saved, and the parts are kept more snug. The method of gripping any tool with this Willis pattern tool-holder is apparently not generally understood, as frequently a large spanner is applied to the hexagon nut on the main stud-bolt to cause the jaw to bite. This is a wrong method. The tool having been laid in its place, the hexagon nut should be screwed up or down with the fingers to bring the jaw approximately level. The smaller pinching screw is then brought into use, and the final grip given by means of the "tommy" bar in this screw. Very little brute force is necessary if the principle of the tool-holder is understood.

SLIDE-REST T-PIECE.

The cast-iron T-piece, Fig. 1248, fits the groove in the sole of the slide-rest, and also the space between the cheeks of the lathe-bed. Its purpose is to hold

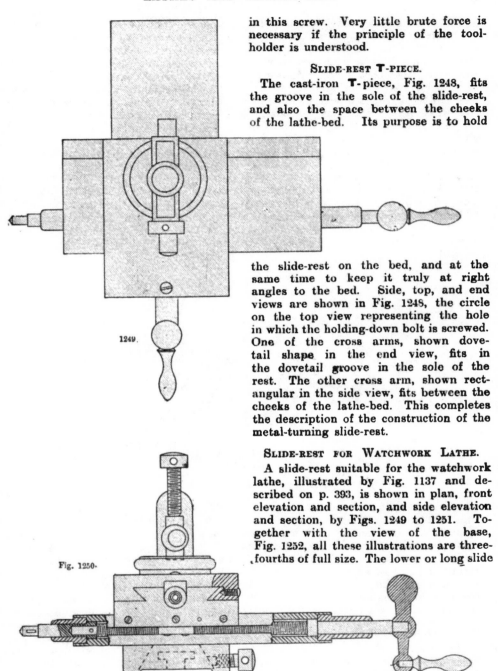

1249.

the slide-rest on the bed, and at the same time to keep it truly at right angles to the bed. Side, top, and end views are shown in Fig. 1248, the circle on the top view representing the hole in which the holding-down bolt is screwed. One of the cross arms, shown dovetail shape in the end view, fits in the dovetail groove in the sole of the rest. The other cross arm, shown rectangular in the side view, fits between the cheeks of the lathe-bed. This completes the description of the construction of the metal-turning slide-rest.

SLIDE-REST FOR WATCHWORK LATHE.

A slide-rest suitable for the watchwork lathe, illustrated by Fig. 1137 and described on p. 393, is shown in plan, front elevation and section, and side elevation and section, by Figs. 1249 to 1251. Together with the view of the base, Fig. 1252, all these illustrations are three-fourths of full size. The lower or long slide

Fig. 1250.

Figs. 1249 and 1250.—Plan and Front Elevation and Section of Watchwork Slide-rest.

is 3⅝ in. long, and the upper or short saddle is 2⅜ in. long. The base of the rest fits on a shoe ordinarily under the **T**-rest. The three screw-heads shown in the base, Fig. 1252, hold the lower, or main, slide to a tapering disc which is fitted in the base, is free to swivel round somewhat stiffly, and may be fixed by the pinching screw shown in the front view. The crank handle on the traversing screw of the long slide fits both ends, and may be used on either, according as the main slide is placed parallel to the lathe centres or at right angles to them. The saddle of the main slide is traversed by means of the nut fastened to the saddle by two screws. This nut passes along a groove cut in the edge of the lower slide.

upon a column, for instance, are so correctly spaced that no inequality can be detected by the finest measurement. The teeth of cogged wheels, too, are produced with wonderful accuracy. All this can be done by means of the division plate and index peg. Assume that a stand has been turned, and it is wished to mark the holes for the insertion of six legs or ornaments ; or, that a blank has been turned up for a six-sided nut, and it is desired to mark six equidistant lines upon it, to guide the worker in filing it up. This can be done rather awkwardly with the dividers; but it will take some time, and the work will not be very exact, and there will be the trouble of repeating the process for every nut or other object requiring division.

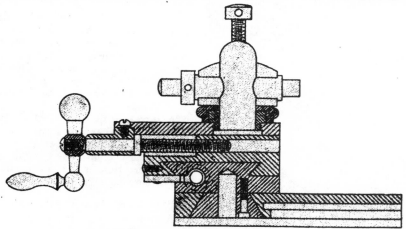

Fig. 1251.—Side Elevation and Section of Watchwork Slide-rest.

The top slide and its saddle are shown in Figs. 1249 to 1251. The tool-holder is a peculiar arrangement; a tube bored through eccentrically, and capable of being rotated in the pinching device, affords a certain amount of adjustment for height. An adjustable ring, shown in section in the side view, is also provided, to give further vertical adjustment.

DIVISION PLATE AND INDEX PEG.

The beautiful regularity with which ornaments and other enrichments are spaced round turned work is often a source of wonder to the uninitiated. The flutes

But suppose that instead of dividing the work, the pulley on the mandrel is divided into six, drilling a small hole at each division, and fixing a spring pointer with a little peg that would enter the holes and hold the mandrel fixed at any one of the positions; there will now be a division plate and an index peg, and it could be used to divide any number of circles into six divisions. This device allows of equally spaced divisions being placed around circular work, enables the worker to determine and draw angles upon flat surfaces, and, by means of the peg, to hold the work fixed whilst lines are scribed upon it, or whilst it is operated upon with drills or revolving cutters.

NUMBER OF HOLES IN DIVISION PLATE.

How many divisions the plate may conveniently have will be determined partly by the class of work intended to be executed, partly by the size of the pulley, and partly by the number of rows of holes for which there is room. For wheel cutting, rather larger holes will be required than for ornamental turning; but in every case the number chosen should have the greatest number of divisors; 360 is a most

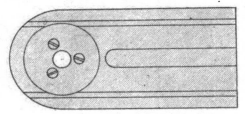

Fig. 1252.—Watchwork Slide-rest Base-plate.

useful number, having more divisors than any other not much greater than itself; and it was for this reason that it was chosen for the number of degrees in a circle. There is some difficulty in getting in so large a number, since the pulley of a 5-in. lathe will not be much more than 7 in. diameter. If the centres of the holes are put only $\frac{1}{16}$ in. apart, and $\frac{1}{4}$ of that distance is allowed for the hole, and $\frac{1}{4}$ for the space between the holes, there will be holes of $\frac{1}{24}$th, and spaces of $\frac{1}{48}$th; then the circumference required will be: $360 \times \frac{1}{16} = 22\frac{1}{2}$ in., corresponding to a diameter of a little more than 7 in. Such small holes are only fit for ornamental work. If the holes are $\frac{1}{16}$ in. in diameter, and the spaces are $\frac{1}{32}$ in., the holes must be $\frac{3}{32}$ in. from centre to centre, then $\frac{3}{32} \times 360 = 33\frac{3}{4}$ in. gives the required circumference, and corresponds to a diameter of about $10\frac{3}{4}$ in. of pulley, which would require the centres to be 7 in. high. Holes of $\frac{1}{24}$th are too small for any but the finest work; holes of $\frac{1}{16}$ in. will do for small wheel cutting; holes of $\frac{1}{12}$ in. are good for metal work, and will suffice for any-

thing likely to be required of a division plate. In ornamental and other work, it will be necessary to divide the circle into 2, 3, 4, 5, 6, 7, 8, 9, 10, etc., parts; and, therefore, the most useful numbers of holes will be those which can be divided by as many of these numbers as possible, without leaving any remainder. Such numbers can be found by multiplying together two or more of the above numbers. For instance, 8 multiplied by 9 gives 72 holes, a very good number, since it contains multiples of 2, 3, 4, 6, 8, 9, 12, 18, 24, 36. Here there are all the first-named numbers except 5, 7, and 10. Now 5 is already contained by 10; if then 7 is multiplied by 10, another number—70—is obtained, which will allow of the circle being divided by any number up to 10. This will suffice to show why these numbers are chosen, and also that the chief difficulty is with the prime numbers, 1, 3, 5, 7, 11, 13, 17, 19, etc., which are not divisible. The first four of these are included, the second four are seldom required, except in wheel cutting. It is possible—though probably not worth while—to include them by placing them all in one line, or circle, of holes. Beginning from one zero hole, divide the circle first into 11; then beginning from the same zero, divide the same circle into 13, then into 14 (for 7), then into 17, and lastly into 19. If this be done on a $6\frac{1}{2}$-in. circle, with $\frac{1}{16}$-in. holes, they will not clash. It would be well, however, to choose higher numbers than 70 and 72. Sometimes there are a good many rows of holes. In the Holtzapffel ornamental turning lathes the numbers chosen are 360, 192, 144, 120, 112, and 96; there is no necessity to have 96 as well as 192, except that it sometimes facilitates the counting; also, many of the chuck wheels are divided into ninety-six divisions, which makes it convenient to have a circle of that number on the pulley. If there be not room for the 360 row, substitute a row of 180 (half that number) between 192 and 140, as that would, at any rate, enable the workman to divide a circle into divisions of 2°. If there be room for only three rows of holes, 180, 144, 96, or 180, 96, 84, would do well.

DIVISION PLATE HOLES AND PEGS.

Cast-iron is the best material for a division plate, being most durable and the least likely to be indented or bruised ; gun metal is the most usual material, and it shows up the numbers and marks well ; be carefully rounded, lest it should scratch the plate In many division plates the holes are simply drilled straight in, about $\frac{1}{16}$ in. deep, whilst the peg is slightly coned, so that it only bears upon the mouth or edge of the hole. The holes

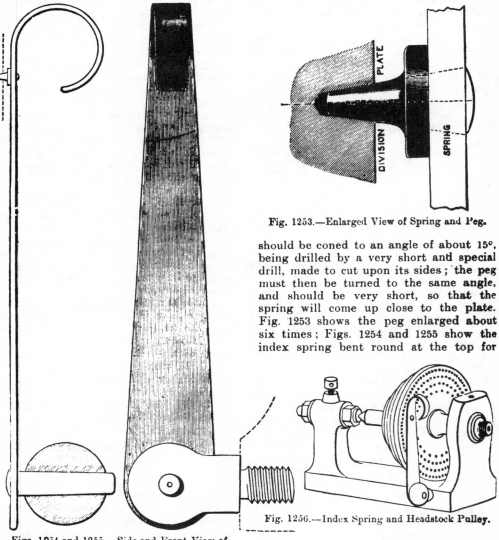

Fig. 1253.—Enlarged View of Spring and Peg.

should be coned to an angle of about 15°, being drilled by a very short and special drill, made to cut upon its sides ; the peg must then be turned to the same angle, and should be very short, so that the spring will come up close to the plate. Fig. 1253 shows the peg enlarged about six times ; Figs. 1254 and 1255 show the index spring bent round at the top for

Fig. 1256.—Index Spring and Headstock Pulley.

Figs. 1254 and 1255.—Side and Front View of Index Springs.

yellow brass is too soft for such a purpose. The point of the index peg should the finger, the peg being riveted in. At the bottom, as shown in the full-size illustration, Figs. 1254 and 1255, a slightly tapered pin is riveted into a second hole,

4½ in. from the first. This tapered pin fits into an iron or steel ball, which is screwed into the base of the headstock, seen dotted in Fig. 1255. The shoulder of will require to be about $\frac{1}{18}$-in. thick. The application of this arrangement, known as the plain index, to the pulley is illustrated in Fig. 1256.

Adjustable Index Peg.

Figs. 1257 to 1259 show a simple form of adjustable index peg; it can be made from the plain one just described, by heating the top of the spring, and bending it to the form of Fig. 1258. The peg of the plain one is driven out, and two holes are drilled above and two below the original hole into which the peg has been riveted; these five holes are then filed into one oblong slot, as shown by Fig. 1258. The new peg (Fig. 1259) is then turned and fitted with a nut, n, to bind it fast to the spring, and prevent the slightest possibility of shake. The screwed part has two flats, f, filed, one on each side, so that it would fit rather tightly into the long slot, and could be moved up and down by the screw. A thoroughfare hole is then drilled through the top of the spring, and then, with a drill of the tapping size, continued through the body of Fig. 1259; then the

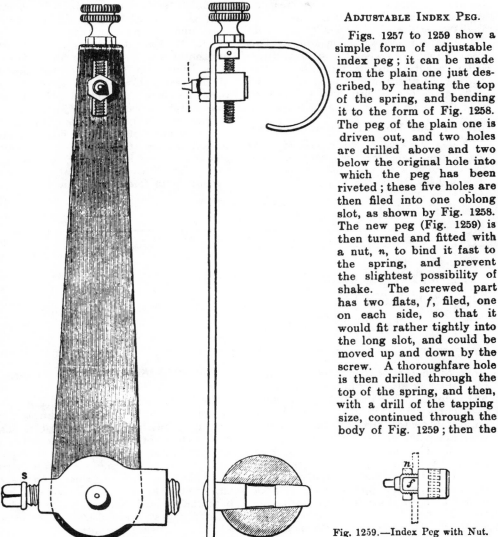

Fig. 1259.—Index Peg with Nut.

Figs. 1257 and 1258.—Back and Side Views of Adjustable Index Peg Spring.

this ball would have to be gradually turned away until, when screwed up hard, the spring comes upright when the peg is in position, as at Fig. 1254. The spring itself

adjustment steel screw in one piece, with its milled head, is fitted, and the small retaining collar fixed, and pinned through as close as possible, so as to avoid all looseness. The milled head has its collar divided into about ten divisions, and there is a mark on the spring to correspond. The advantage of the adjusting index peg over the plain one is considerable. It will often enable the workman to bring the flute, tooth, or pattern to correspond with a part already finished. The ball in Fig. 1258 is slightly different from that in Fig. 1255, the screw s being added. This screw being fixed when the work of dividing is begun, prevents the possibility of placing the peg in a hole in another row; it fixes the conical fitting in the ball, so that it can no longer act as a hinge.

SETTING OUT HOLES ON DIVISION PLATE.

The zero marks of the circles of holes are placed together, but they are not placed upon a radial line, because then it would be impossible to pass from one

which corresponds to the numbers of holes in one's own division plate. Eleven convenient numbers have been chosen and written down in the column headed "No. of Holes"; then comes, in a horizontal line, the row of "Divisors," containing every number up to 20, except the primes 11, 13, 17, 19. Looking along the first horizontal line underneath the divisors, it is found that a circle of 360 can be divided without remainder by twelve of these numbers. Numbers 240 and 180 come next in value, since they can be divided by eleven of the divisors; 120 by ten; 300 and 144 by nine, etc., 112 and 70 have the fewest, but they contain multiples of 7. It would, however, be necessary to continue the table to make the comparison quite fair.

TABLE OF DIVISIONS.

No. of Holes.	Divisors.														
	2	3	4	5	6	7	8	9	10	12	14	15	16	18	20
360	180	120	90	72	60	—	45	40	36	30	—	24	—	20	18
300	150	100	75	60	50	—	—	—	30	25	—	20	—	—	15
240	120	80	60	48	40	—	30	—	24	20	—	16	15	—	12
192	96	64	48	—	32	—	24	—	—	16	—	—	12	—	—
180	90	60	45	36	30	—	—	20	18	15	—	12	—	10	9
144	72	48	36	—	24	—	18	16	—	12	—	—	9	8	—
120	60	40	30	24	20	—	15	—	12	10	—	8	—	—	6
112	56	—	28	—	—	16	14	—	—	—	8	—	7	—	—
96	48	32	24	—	16	—	12	—	—	8	—	—	6	—	—
72	36	24	18	—	12	—	9	8	—	6	—	—	—	4	—
70	35	—	—	14	—	10	—	—	—	—	5	—	—	—	—

circle to another without slightly turning the pulley, or altering the adjusting screw of the index. The zero holes from which each circle counts should be placed upon an arc of a circle struck upon the pulley with the point of the division peg. The rows of holes should be at least ¼ in. apart, that there may be room for figures of a visible size, and for the dots, and other marks made opposite every 5th, 4th, or 3rd hole, to assist in the counting, by enabling one to put the index peg in every 5th, 4th, or 3rd hole without making a mistake. The table of divisors about to be given can be made by the worker for himself, and it can be continued from where it is left off here. It would be a good plan to copy it out on a card, varnish it, and keep it by the lathe for reference; or, at any rate, that part of it

HOW TO MAKE DIVISION PLATE AND INDEX PEG.

The divided head arrangement (Fig. 1260) differs somewhat from that illustrated by Fig. 1256, but, of course, resembles that in consisting of holes drilled equidistant on rings on the face of the headstock pulley. All lathes designed for general work should have at least two rings with twelve and twenty holes. At the side of the headstock a stiff spring (Fig. 1261) has a pointed peg at one end and at the other a slightly tapered pin, fitting into a projecting stud (Fig. 1262) screwed either to the side of the headstock or to the side of the lathe bed. When in use the pin is knocked tight in its fittings, and the other or dividing peg is pressed into one or other of the holes in the pulley, which must be machine drilled for accuracy.

STUD FOR INDEX PEG SPRING.

For the stud, Fig. 1262, secure a piece of $\frac{3}{4}$-in. stick brass in the scroll chuck, and turn it up clean for about $1\frac{1}{2}$ in. Then for a length of $\frac{1}{4}$ in. reduce the end to $\frac{1}{4}$ in. in diameter and cut a $\frac{1}{4}$-in. thread on it, and with the parting tool cut it off to

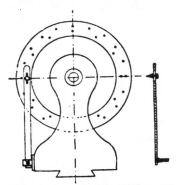

Figs. 1260 and 1261.—Division Plate and Spring.

1 in. long. Now drill and tap a $\frac{1}{4}$-in. central hole in a piece of brass 2 in. by 1 in. by $\frac{1}{4}$ in., remove the burrs with a file, slightly countersink the hole, and screw the stud in it. Then rivet it in, hammering the projecting end of the screw. Lastly, file the riveted end flush with the surface. Now approximately square the edges of the flat piece, and mark off the holes in it for the screws and in the stud for the tapered hole. The holes in the flat body of the projecting piece are drilled with a No. 2 B.A. clearing drill, and then countersunk. Two No. 2 B.A. steel taper head screws, $\frac{1}{4}$ in. long, will be used to screw the piece in position by the holes just drilled.

MAKING INDEX PEG SPRING, ETC.

The spring (Fig. 1261) is made of steel $\frac{1}{8}$ in. thick and $\frac{1}{2}$ in. wide. Its length is determined by the radius of the lathe and the position in which it is proposed to screw the stud. If the lathe is of 6 in. centre and the stud is to be at the side of the headstock, the length will be about $7\frac{1}{2}$ in. The tapered pin should be made of $\frac{1}{2}$-in. round steel, $1\frac{1}{2}$ in. long. File the ends flat and centre them. To do this,

secure the steel in the scroll chuck, and with a graver turn out the centre at each end. Then, with a No. 6 B.A. drill secured in the lathe, drill up each end for about $\frac{7}{16}$ in. Then in the centres turn one end taper to suit the hole in the stud. Let this be 1 in. long, joined to the thicker part by a hollow curve. Turn the other end for $\frac{1}{4}$ in. to $\frac{1}{4}$ in. diameter and thread it, taking care not to injure the previously tapered pin. Make the other peg from the same sized steel. Nip it in the scroll chuck, and turn down the end to $\frac{1}{4}$ in. diameter for a length of $\frac{1}{4}$ in. Drill and tap the end with a No. 4 B.A. thread for a depth of about $\frac{3}{8}$ in. Now leave about $\frac{1}{8}$ in. of the original size, reduce the steel for a length of $\frac{1}{2}$ in., tapering it down to a point. When it is quite cut off, chuck it in the scroll by the drilled end, finish the tapered point, put in a hollow curve, and file the peg to a point. Make a No. 4 B.A. cheese-headed steel screw $\frac{7}{16}$ in. long, the head being $\frac{7}{16}$ in. thick and $\frac{3}{8}$ in. in diameter. This secures the peg to the spring. Drill and tap a $\frac{1}{4}$-in. hole at one end of the spring and screw in the tapered pin. File the end round to the curve of the pin head and

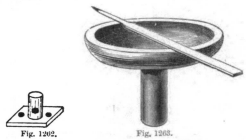

Fig. 1262.
Fig. 1263.
Fig. 1262.—Stud for Index Peg Spring.
Fig. 1263.—Scriber for use with Division Plate.

file down the end of the screw flush with the surface of the spring. At the other end put a slot $1\frac{1}{2}$ in. long and $\frac{1}{4}$ in. wide, for the drilled end of the peg. The near end of the slot will be $4\frac{1}{2}$ in. from the centre of the tapered pin. Mark the slot on the spring and drill a number of $\frac{1}{16}$-in. holes close together; remove the joining metal with a small round file, finishing the slot with a flat file, and

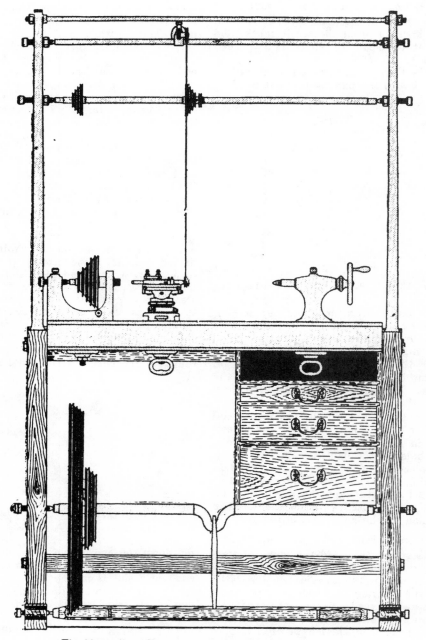

Fig. 1264.—Front Elevation of Lathe with Overhead Attachment.

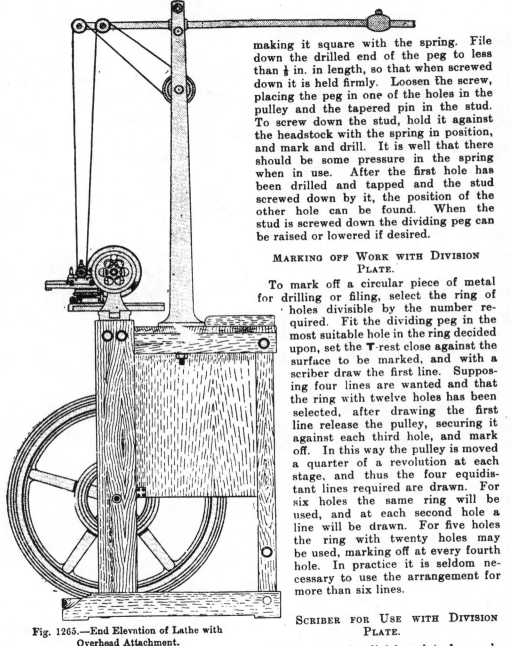

Fig. 1265.—End Elevation of Lathe with
Overhead Attachment.

making it square with the spring. File down the drilled end of the peg to less than ⅛ in. in length, so that when screwed down it is held firmly. Loosen the screw, placing the peg in one of the holes in the pulley and the tapered pin in the stud. To screw down the stud, hold it against the headstock with the spring in position, and mark and drill. It is well that there should be some pressure in the spring when in use. After the first hole has been drilled and tapped and the stud screwed down by it, the position of the other hole can be found. When the stud is screwed down the dividing peg can be raised or lowered if desired.

MARKING OFF WORK WITH DIVISION PLATE.

To mark off a circular piece of metal for drilling or filing, select the ring of holes divisible by the number required. Fit the dividing peg in the most suitable hole in the ring decided upon, set the T-rest close against the surface to be marked, and with a scriber draw the first line. Supposing four lines are wanted and that the ring with twelve holes has been selected, after drawing the first line release the pulley, securing it against each third hole, and mark off. In this way the pulley is moved a quarter of a revolution at each stage, and thus the four equidistant lines required are drawn. For six holes the same ring will be used, and at each second hole a line will be drawn. For five holes the ring with twenty holes may be used, marking off at every fourth hole. In practice it is seldom necessary to use the arrangement for more than six lines.

SCRIBER FOR USE WITH DIVISION PLATE.

When using the division plate for marking off work, something is required to guide the scribing point; the top of the T-rest is indeed sometimes used, but a more accurate method is to use the simple appliance shown by Fig. 1263. This is only

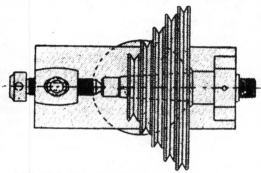

Fig. 1266.—Plan of Headstock.

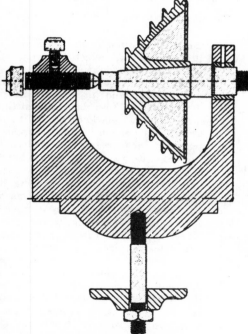

Fig. 1267.—Section of Headstock.

will provide a support and guide for the scriber seen lying upon it (Fig. 1263), which will keep its point in one plane. By this means lines can be struck upon uneven surfaces, if only the scriber point be first adjusted to the exact height of the centres of the lathe; the band is simply thrown off the pulley, and the lathe is at rest. But, when drilling is to be done, or some revolving cutters are to be driven, then the treadle and wheel are available for driving the drill or cutters by means of the overhead motion.

USE OF OVERHEAD ATTACHMENT.

The overhead attachment is another extremely useful addition to a lathe. When

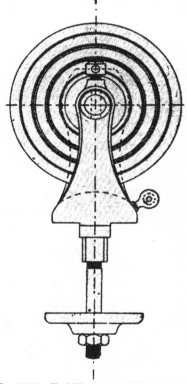

Fig. 1268.—End Elevation of Headstock.

a flat-rimmed saucer of hard wood, about 3 in. diameter, having a round stem or pin which fits the socket of the T-rest. If the hole in the T-rest socket be vertical, and at right angles to the lathe bed and the line of centres, then the saucer

the work is revolved in the plain lathe and cut by fixed tools held either in the hand or in the slide-rest, the result is naturally of circular section. But if, while the work is firmly held in a lathe as before, the tool itself, instead of the work, is rotated, and, while so rotating, is brought to act upon the work, an infinite variety of different forms can be produced. The

depend one upon another, and the overhead is of no use without the slide-rest and division plate (attachments already fully described). All revolving tools being held in the slide-rest, require to be set in motion from the treadle and fly-wheel, and this is the function of the overhead motion; it should be noted, however,

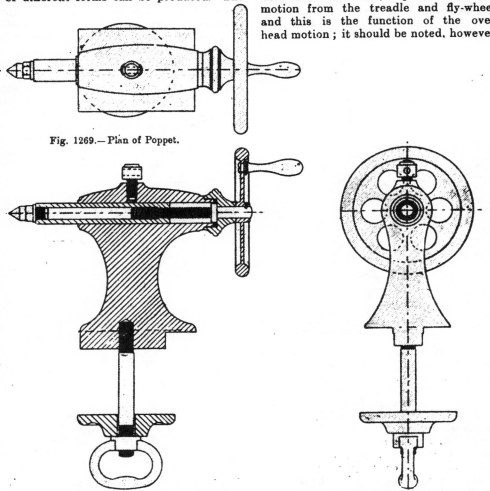

Fig. 1269.—Plan of Poppet.

Fig. 1270.—Section of Poppet.

Fig. 1271.—End Elevation of Poppet.

revolving tool may be a drill or a single-point "flying" or "fly" cutter, or it may be a serrated disc, something like a small circular saw, called a milling cutter; all these are carried by frames, in which they revolve, the frames being grasped and held and guided by the slide-rest. All these different appliances, it is obvious,

that the revolving tools do not remain in one fixed position like the mandrel, but they must be capable of being applied to their work in several positions, and of being moved whilst at work without throwing off the band or cord by which they are driven from the pulley over the lathe.

LATHE WITH OVERHEAD ATTACHMENT.

The systems of overhead gear available are many, it being customary to utilise peculiarities of workshop design in arranging a convenient rig-up; thus a wall, ceiling or beam may offer itself as a good place for securing the gear, and when the

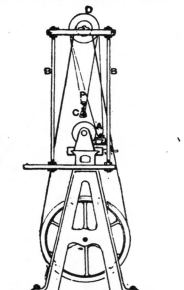

Fig. 1272.—Overhead Arrangement for Lathe.

position of the lathe is likely to be permanent such an advantage should not be neglected. When, however, the lathe has to be shifted about occasionally, it is usual to fit the overhead gear to standards projecting from the lathe itself, and this is probably the arrangement found most convenient by the average worker. The standards may form part of the lathe frame castings, or they may be cast separately and secured by bolts; this latter method is the one shown in Figs. 1264 and 1265, which are respectively front and end elevation of a lathe with overhead gearing. Such a lathe is particularly well adapted for plain and ornamental metal turning. Figs. 1266 to 1268 show the fast headstock in plan, section and end elevation, and Figs. 1269 to 1271 show similar views of the tailstock or poppet. There is no back-gearing, as there is in the lathe

shown by Fig. 1119, p. 386, but this is desirable. A brief description will show how the overhead motion is obtained. For ordinary turning, of course, the grooved flywheel on the treadle-driven cranked axle drives by means of a gut band the grooved pulley contained in the headstock. When a revolving cutter is to be used, the driving band is thrown off, and lengthened by the addition of a piece of gut so as to drive the grooved pulley fixed on the overhead spindle to the left (see Fig. 1264). On the same spindle is another grooved pulley, but capable of sliding and being fixed just where required. This second pulley drives the revolving tool, held in the slide-rest, by means of a second gut band, which, to obtain tension, is carried

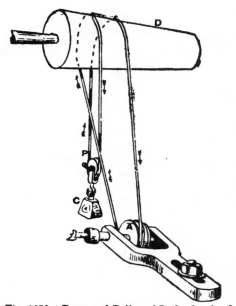

Fig. 1273.—Drum and Pulley of Lathe Overhead.

over two pulleys secured to the end of a level working on a pivoted rocking bar, as shown in the end view, Fig. 1265. This lever is weighted, the sliding weight being adjustable as regards its longitudinal position on the lever by means of a set-screw. The finer or more delicate the cutting tool in use, the less must be the tension, and the nearer to the rocking bar is the weight placed, and vice versâ. The lever with its pulleys can slide along the rocking bar in any direction, to allow of the tool being fed to the work either fixed between the

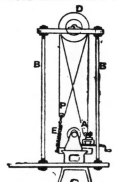

Fig. 1274.—Lathe Overhead with Spring instead of Weight.

lathe centres or held on the mandrel nose. The headstock grooved pulley is kept stationary by inserting a peg into one of its face holes.

ANOTHER OVERHEAD ARRANGEMENT.

A simple and efficient way of setting up an overhead shaft and fittings for driving revolving cutters is shown in Figs. 1272 and 1273, in which A represents the cutter. The uprights B may be of gas piping, the lower ends being fixed to the table and the upper ends having a cross-bar to carry the bearings of the overhead shaft, with drum D; a set of these supports is required at each end of the lathe. The chief part,

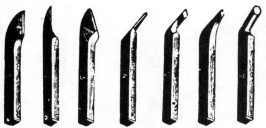

Fig. 1275.—Solid Slide-rest Tools.

however, is the tightening device. This consists of a pulley P and hook with weight C. Fig. 1273 shows the arrangement clearly. The gut band or belt should be

long enough to pass over the drum and pulleys. The weight keeps the band tight in whatever position the slide-rest may happen to be. Fig. 1274 shows a modification with a spring E instead of the weight. The lower end of this spring should be fixed to the carriage of the slide-rest. The arrangement with the weight is easier to construct.

Fig. 1276.—Solid Slide-rest Tools.

SHAPES OF SLIDE-REST TOOLS.

Slide-rest tools of the solid kind, now being superseded by small cutting edges held in "cutter bars," are of many and various shapes, twenty-one of the principal kinds being illustrated by Figs. 1275 to 1277. Five round-section tools having a cutting edge at each end are shown full size by Fig. 1278; these are made of Stubs steel, and are specially adapted for small engineering work, watchwork, etc.

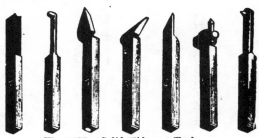

Fig. 1277.—Solid Slide-rest Tools.

ANGLE FOR APPLYING SLIDE-REST TOOLS.

With regard to the cutting action of such tools, Fig. 1279 shows a tool having its cutting edge applied to the work on the line of centres, which is the correct position. Fig. 1280 shows a tool incorrectly applied at a point above the centre.

Fig. 1281 shows a tool also incorrectly applied much below the centre. Perhaps the consideration of the effect of each successive cut on the relative position of

above the centre. As soon as the work becomes reduced to ¼ in. in diameter it will be impossible to make the tool act, because the whole of the work will be

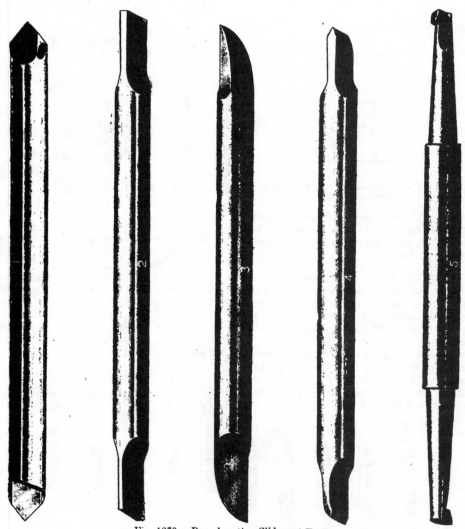

Fig. 1278.—Round-section Slide-rest Tools.

the tool to the work will suffice to show the error of applying tools either above or below the line of centre. Suppose that in Fig. 1280 the work is 1 in. in diameter, and that the point of the tool is ⅛ in.

below the cutting edge. In Fig. 1281 suppose the tool to be ⅛ in. below centre, and as soon as the work is reduced to ¼ in. in diameter the tool cannot be made to cut it further, because the work will be entirely above the cutting edge. As a matter of fact, the tool would in both cases cease to act effectively long before the work was so much reduced. It will

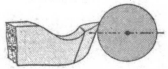

Fig. 1279.—Turning Tool Applied Correctly.

be seen that, during the passage of the tool from the periphery towards the centre, the angle that the cutting edge makes with the work is continuously changing. The face of the tool should form a tangent to the work on which it operates, and this position can only be maintained at all diameters when the cutting edge is applied on the line of centre.

Proper Angle for Turning Tool.

The position of the tool at Fig. 1279 is better explained by reference to Fig. 1282, of which illustration "Lathe Work" gives the following particulars:—The angles best suited for the particular material to be operated upon are most desirable in slide-rest tools. It will be understood that the cutting edge will penetrate best when it is thinnest; other considerations, however, prevent the adoption of this rule unreservedly; and for metal work tools are found to act best when the faces form the cutting edge at an angle of from 60° to 90°. The face of the tool coming next to the work requires to be ground at a slight angle, leaving the point prominent to prevent the whole face

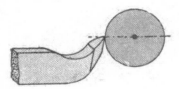

Fig. 1280.—Turning Tool Applied Above Centre.

touching the work, and so by the friction greatly increasing the labour of turning. When this requirement is satisfied the face should be as upright as possible, and

77

3° from the perpendicular suffices. This applies equally to tools with acute edges used on wood, though in using a knife edge the face of the tool itself usually

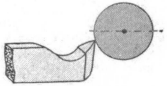

Fig. 1281.—Turning Tool Applied Below Centre.

rests against the work it has to cut, and there is no angle of clearance. Fig. 1282 shows tools correctly applied for cutting both wood and steel. The slide-rest tool, with a strong cutting edge suited to dividing hard metal, and the acute wood-turning chisel, each have the lower face angle placed in the same position with regard to the work. Therefore it is only the upper face, which wedges back and curls or breaks off the shaving, that is altered to agree with the different nature of the materials. The line of centres is shown at A B, and at precisely the height of this line should be the point of the tool fixed in the rest. Here it may be advisable to point out that tools must be packed up with parallel strips, otherwise the relative position of the angles is interfered with. The edge of the metal turning tool is formed by the meeting of the faces A X and D X; A X being parallel with A B, and D X 3° from the perpendicular makes the angle of the point to be 87°. This is the most obtuse angle usually employed, though for some purposes, where a scrap-

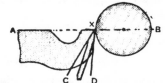

Fig. 1282.—Angles of Turning Tools for Wood and Metal.

ing action is required, the top face is bevelled off downwards to make the edge even more blunt. The edge of the soft wood chisel is formed by the meeting of the faces c x and d x, enclosing 25°, still keeping the lower face precisely in the same relative position. The tools might be applied at any part of the circle even vertically above it, so long as the same relative position is maintained; but the slide-rest as ordinarily constructed necessitates the application of the tool on a level with the centres.

APPARENT CONTRADICTION IN WOOD TURNING PRACTICE.

When hand tools are used, the position of the cutting edge relative to the work may be altered by simply rising or depressing the handle. The tool may be applied at any point on the periphery of the work, and so long as the cutting edge forms a tangent to the circle the principles before advocated will not be violated. When a wood-turner is seen to work with the turning tool applied somewhere about the top of the work, the theory of " applying the tool at the line of

Fig. 1283.—Turning Tool Shaped Incorrectly.

centres " appears to be practically contradicted, yet still holds good. The tool still forms a tangent to the circle, and as the work becomes smaller in diameter the position of the tool is altered by raising or lowering the handle within certain limits, and then the T-rest is shifted. It is only when the tool is moved perfectly horizontally, as it is with an ordinary slide-rest, that the point must be applied at precisely the same height as the lathe centres.

INCORRECTLY SHAPED SLIDE-REST TOOLS.

Slide-rest tools for metal turning have been made on the erroneous principle exemplified by Fig. 1283. In such a tool, the cutting edge is considerably above the top of the shank, and, as a rule, the tool cannot be used on an ordinary lathe. The top of the metal turning

slide-rest, for a foot-lathe of about 5 in. centre, is usually ½ in. below the line of centres; very rarely is the top of the rest ¾ in. below the centres, even in a 6 in. lathe. Yet slide-rest tools have

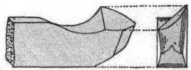

Fig. 1284.—Turning Tool with Slight Side Rake.

been made of ⅜ in. and ¾ in. steel, and have the cutting edges above the top of the shank even then. Consequently, the tools cannot be used unless tilted up from the back, so as to bring the point down to the work, and at the same time destroy all the relative angles of the cutting edges. Fig. 1284 shows a tool having the point depressed considerably, and with but slight side rake. In use this tool would be forced away from the work by the action of cutting. The slight side rake would tend to help the feed of the tool along the work. The cuttings made by this work would roll over slightly towards the right. If the tool had no side rake the cuttings would roll over on themselves, forming a ball, or breaking off as soon as they touched the work. The action of this tool is more scraping than cutting, for though the side rake gives a certain amount of cutting action, the front rake does not give any. Fig. 1285 shows a tool having no front rake, but a considerable rake on the leading side. With this tool the shaving would curl off in the direction

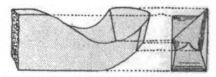

Fig. 1285.—Turning Tool without Front Rake.

shown by Fig. 1286. This tool would have no tendency to dig into the work from the front, but would be drawn into the cut by the side rake. If the slide-rest moved freely in a direction parallel to the cut, the tool would probably be broken by the self-feeding action just alluded to.

RAPID-CUTTING SLIDE-REST TOOL.

For the rapid removal of a large quantity of material, a tool having both front rake and side rake is best. A tool of this kind is shown by Fig. 1287, the form of shaving that it would make being illustrated by Fig. 1288. In this tool the action of cutting would tend to draw the tool into the work. When the tool was brought up to its work and began to cut, it would be drawn into the work as far as any back-lash, if the slide-rest tools would allow. Such a tool requires care in manipulation, as, if the slide-rest fitting is very slack, the tool will probably be broken as soon as it is brought into the cut. Frequently, however, the lathe would be stopped before any actual breakage occurred.

Fig. 1286.—Shaving made by Tool without Front Rake.

ADVANTAGES OF CUTTER-BARS.

Tool points held in cutter-bars are generally preferable to solid tools, because they allow saving in cost of steel, in cost of forging, in grinding, and in storage room, and they give better results. The cost of steel at, say, 6d. per lb. is very heavy in a large shop. A tool point weighs but a fraction of a solid tool. Solid tools require frequent re-forging, as they become ground away, and this includes re-tempering, and it is always a pity to have to draw the temper of a good cutting tool. Tool points once forged and tempered never require the periodical renewal of these operations, their section and their cutting angle remaining the same until by repeated re-grindings the tools are worn away. The solid tools are piled up in a mass on the end of the

lathe or in a drawer, while the tool points for a machine can all be contained in a little box. Better steel can be used for tool points than for solid tools because of the smaller quantities required, and better work can consequently be done.

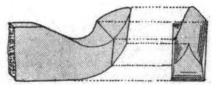

Fig. 1287.—Turning Tool with Front and Side Rake.

TOOL-HOLDERS OR CUTTER-BARS.

There are dozens of slide-rest tool-holders, used to a greater or less extent; but the principal and best kinds are not numerous. Several important functions are required in a reliable holder, and some of the best holders, simple though they seem, have involved a good deal of experiment and partial failure before achieving popularity. Workmen long accustomed to solid tools do not take kindly to holders at first, so that only the best holders can win their way into favour. A tool point will not answer for the heaviest cutting which is done in the machine. Even though it may not slip under the stress, it becomes too hot to keep its cutting edge, whereas the bulk of metal in the solid tool conducts away the heat as soon as it is generated. But, with this exception, the tool-holder is applicable to all classes of turning, to much boring, screw-cutting, planing, and shaping, and in a modified form for slotting.

Fig. 1288.—Shaving made by Tool with Front and Side Rake

HAYDON CUTTER-BAR.

The Haydon tool-holder, shown in Figs. 1289 to 1292, is one which has been very much used by metal turners in years past, but chiefly for light cutting. It is

unsuitable for heavy cutting, owing partly
to the form of the tool, parly to the overhang of the bar, and partly to the limited
range of the operations of the tool point.
The Haydon tool-holder, or cutter-bar,
is shown in elevation by Fig. 1289, in plan
by Fig. 1290, in longitudinal section by
Fig. 1291, taken on the centre plane P P
(Fig. 1290), and in part cross section by
Fig. 1292, taken on the plane indicated by
Q Q (Fig. 1290). Fig. 1292 shows the
end of the bar with the tool removed,
with a section through the strap. The

point. Since the front or clearance angle
remains, or may remain, constant for all
metals, that in a modern tool-holder is the
one which is usually fixed, and the top
face alone is ground. This holder was
designed on the basis of the exposition of
the operations of tools with double-cut-
ting edges, which was given by Professor
Willis, and which was only applicable to
tools of the graver or diamond-point type.
But this class of tools is scarcely used at
all now, except for light hand turning;
and Haydon's holder was designed speci-

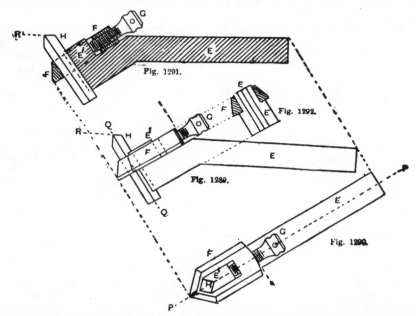

Figs. 1289 to 1292.—Elevation, Plan, Longitudinal Section, and Cross Section of Haydon Cutter-bar.

body, E, of the bar is made of iron or
steel. E' is a stop or boss around which
the strap or sling F fits, and which receives the pressure of the clamping screw
G to retain the tool H in position. One
end of the bar and one end of the strap
are V-grooved to pinch the tool H, which
is of square section. The bar is so constructed that the angle which the longitudinal axis of the tool makes with the
horizontal is 55°. This angle has certain
advantages, but it necessitates the grinding of both angles on each separate tool

ally for this type. Round tool points,
it may be remarked, might be used
in such a holder, but then the clearance angle would have to be ground;
and the bar could not then be used for
heavy cutting, the overhang being too
great for stability. But the overhang is
inherent in the design of the bar, because
of the angle at which the tools are placed.
But, further, the overhang of the tool
cannot be increased when needed for
reaching into narrow holes and recesses,
for parting off, and other operations which

are constantly requiring to be done. The tool cannot be thrust out farther from the bar without coming above the line of centres R, which is impracticable. Then,

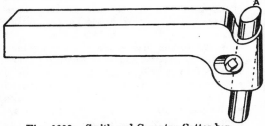

Fig. 1293.—Smith and Coventry Cutter-bar.

again, turning of much recessed portions and neat radiusing cannot be done with this holder. Right- and left-hand bars can be used for many classes of work, but even then there is a lack of mobility in the bar, and the previously-named objections still hold good.

SMITH AND COVENTRY CUTTER-BARS.

The two kinds of tool-holder manufactured by the Smith and Coventry firm are shown respectively by Figs. 1293 to 1297, and by Figs. 1298 and 1299. The first is the older form; the second is designed

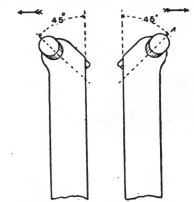

Fig. 1294 and 1295.—Right- and Left-hand Views of Smith and Coventry Cutter-bar.

to do a wider range of work than the first. The tool point A (Figs. 1293 and 1297) is of round steel, set and pinched in a drilled hole in the holder at such an angle

as makes a constant clearance angle of 1 in 8. The first holders were made straightforward only, but their use was limited; hence the adoption of the right- and left-hand arrangement shown in plan by Figs. 1294 and 1295, each holder traversing in the direction indicated by its arrow. By the adoption of this device the clearance angle is maintained practically constant for that portion of the tool which is operating. The top faces only of the tool points require to be ground, and one constant cutting angle of 50° has been adopted for all wrought metals, and one of 60° for all cast metals, a gauge (Fig. 1296) being used to ensure uniformity in this respect. The tool point is pinched with a set-screw, shown in

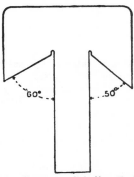

Fig. 1296.—Gauge for Grinding Tool Points.

section at B (Fig. 1297), and this is sufficient to withstand the stress of heavy cutting. The maximum depth of cut possible with any tool is equal to one-half its diameter. The tool steel in Fig. 1298 is also made to the dotted section shown above A for side cutting, being used then right- and left-handed. The same plan is often adopted in other tool bars of similar type.

SIDE TOP RAKE IN CUTTING TOOLS.

There is an important angle in tools cutting transversely—side top rake. If there is no side top rake, or sloping away of the cuting face from the direction in which the tool is travelling, then there is no true cutting, but scraping only. In the Haydon bar the top face of the tool has to be ground to impart this rake. In the tool-holders shown by Figs. 1294 and 1295 the bevel given to the cutting face of the point, conjointly with the angle at which its axis is set in the holder,

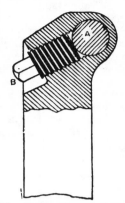

Fig. 1297.—Section through Cutter-bar.

imparts the requisite top rake. This is seen clearly in the general view, Fig. 1293. In the tool-holder designed by Professor Willis the tool point was of circular section, held at an angle of 55° with the horizontal, and the top and front faces were ground to give every variety of cutting angle, and also to furnish the necessary front or clearance angle.

THE BABBAGE CUTTER-BAR.

The Babbage tool-holder holds the tool point at an angle with the horizontal suitable for giving a constant front rake, and it is made to swivel to any angle in a horizontal plane, being gripped in notches between top and bottom grips or washers, and is tightened in any position with a central bolt.

SWIVELLING CUTTER-BAR.

The swivelling tool-holder shown by Figs. 1298 and 1299 was devised in order to accomplish all classes of work which could not be performed by holders previously made. The clearance angle of 1 in 8 corresponds with an angle of about 7°, which is excellent. One of from 3° to 4° is recommended by some writers; but in practice many tools have been found working well in which the clearance angle is nearer 10° or 12°. This, however, is too much for durability, and is not at all necessary, as anything between 4° and 7° is quite suitable. This clearance angle is adopted for nearly all modern tool-holders. The cutting angle is then either left constant for all metals or alloys, or it is left to the discretion of the workman to alter it by grinding the top face. The tool points A, Fig. 1298 and 1299, has a section resembling that of a truncated wedge, the sides and end of which make a clearance angle of 1 in 8. The clearance angle is therefore constant, and the cutting angle also is constant, being 68°. This can be altered by grinding the top face, but experience proves this is seldom necessary. The tool A is swivelled and securely clamped in this way:—It is supported in a recess in the swivelling rest B, which is prolonged to form the clamping-bolt C. The tool is pinched above in the collar D. Being therefore carried in the swivelling piece B, it is free to revolve, and the pinching of the bolt C tightens it between B and D in any required position. Being movable also lengthwise, it can be thrust into awkward corners and recesses. The cutting point can be ground vee'd, or round, or flat as required. The clearance angle always remains the same. For

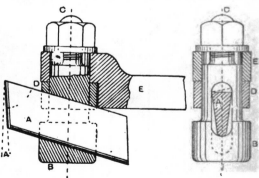

Figs. 1298 and 1299.—Swivelling Cutter-bar.

all ordinary work, one of these swivel holders is sufficient. All the tool points used with them are of the same **V** section, but some are ground for turning, some for parting, etc.

BASHALL'S CLIMAX CUTTER-BAR.

In Bashall's "Climax" tool-holder the tool is held in a circular body, and this is clamped between a bottom seat and top washer, the opposed faces of which are bevelled. A bolt tightens up the seat and washer round the circular body and tool. The holder is therefore capable of swivelling to angles with both horizontal and vertical planes. It turns straight forward, or right- or left-handed, the clearance angle remaining constant. Or by means of the circular body it can be set to give side rake, either right- or left-handed, or set to suitable angles for cutting screw-threads. Smith and Coventry's tool-holder is made also, if required, with a circular shank to swivel in a seat on the slide-rest, to suit any angle required for screw-cutting.

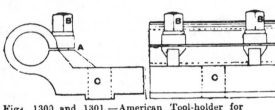

Figs. 1300 and 1301.—American Tool-holder for Slide-rest.

OTHER CUTTER-BARS.

In Tangye's tool-holder the end of the bar is formed into a cylindrical boss, and this is pierced with three slots to receive the tool point, which is roughly of a **V** shape. The tool is set at the angle to give front clearance. It is pinched with a set-screw through the top of the boss. In Allen's tool-holder the tool point is round, and set at an angle in the cutter, similarly to the round tool points in Smith and Coventry's bar; but it is gripped by means of a wedge behind it. To prevent slip, the back of the tool and the face of the wedge are similarly serrated. Sagar's tool-holder holds the tool by means of two wedges placed reverse ways, one of which is formed as a bolt and tightened against the other next the tool. In Newton's swivelling tool-holder the tool is held at an angle suitable for front clearance,

leaving the top face to be ground as required. The tool passes through a slot in the body of the boss very similar to that in Figs. 1298 and 1299. But the tightening of the nut pulls it up against the edge of

Fig. 1302.—Boring Bar.

the bar instead of against a collar. The face of that portion of the holder is chamfered off to correspond with the angular fixing of the tool.

CUTTER-BARS FOR INTERNAL WORK.

Nothing has yet been said about the forms of tool-holders which are used for internal work. For the most part these are plain bars, slotted to receive a tool point, which is pinched with a set-screw. The details are, however, subject to various modifications.

AMERICAN TOOL-HOLDER FOR SLIDE-REST.

An American tool-holder to be attached to the slide-rest for holding tool or cutter-bar has a tongue to fit into the **T**-slot in place of the regular tool post. It can be made to fit an English slide-rest by leaving off the tongue and planing it flat on the bottom. Figs. 1300 and 1301 show two views of the tool-holder. In beginning to make it, the base of the iron casting being planed, the hole should be bored with a boring bar between the centres of the lathe with which the holder is to be used. A $\frac{3}{4}$-in. hole is about right for a $4\frac{1}{2}$-in.

Fig. 1303.—Split Bush for Holding Boring Bar.

centre lathe. The slot A (Figs. 1300 and 1301) is cut with a hack-saw, and clamping screws are shown at B. The dotted lines at C indicate the bolt hole for fastening the holder to the slide-rest. Fig. 1302 shows a ¾-in. steel boring bar, which should have a total length of about 10 in. A ¼-in. tapped hole carries a grub screw, and a corner of the bar is filed off. The hole for the cutter should be drilled, the cutter being of $\frac{1}{16}$-in. square tool steel. Fig. 1303 shows a split brush to hold a ½-in. bar; it has a milled end to facilitate removal. Several such bushes should be made to accommodate a variety of bars, and also one or more with the holes eccentric to the centre of the bushing to hold small steel. By that means it is easy to place the cutting point of the tool at any height required.

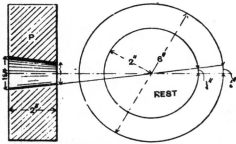

Fig. 1304.—Turning Tapered Hole in Piston

TURNING TAPERED HOLE IN ENGINE PISTON.

Taper-turning with the slide-rest is not too difficult for an ordinary worker. In making an engine it is usual to fit the piston-rod to the piston by means of a taper. In Fig. 1304, P is one form of piston in section. At the back or small end the hole is 1 in. in diameter; at the other it is 1½ in.; its length is 2 in. Mount the piston in a chuck, the small end of the hole being kept outwards. By this means the facing for the nut may be fixed up and the hole bored, both at one chucking; moreover, if the piston is to have a junk ring, all the turning may be done before slackening the jaws of the chuck. Face up for the nut, which must be measured from corner to corner if the facing is to be sunk, and bore the small end about $\frac{1}{16}$ in. less than 1 in. Bore right through, and then set the rest. If the latter has a turned portion below the

bottom slide, it will probably have degrees and minutes marked on the surface. Do not mind these; the degrees can only be used with tables compiled for the purpose. As the hole is ½ in. larger at the

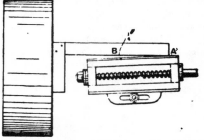

Fig. 1305.—Setting Slide-rest with Square.

back than at the front, the taper, taken upon the diameter, is ½ in. in 2 in. Calliper the turned portion of the rest. If the diameter is 6 in., and if the taper increases ½ in. in 2 in., then in 6 in. it will increase 1½ in. Fig. 1304 makes this clear. Divide 1½ in. by 4, which gives ⅜ in., and to this distance the zero mark must be moved in order to cut the hole ½ in. in 2 in. If the slide-rest does not turn on a centre pivot, but is simply bolted across the bed with no means of adjustment except by slacking the whole thing, the operation is even simpler. Hold a square on the faced part of the piston, and let the blade extend down the planed side of the slide, as shown by Fig. 1305. Make a chalk mark near the end of the slide, next the chuck, as B; 2 in. further along make another, A. Bring up the blade of the square, and move the rest until the distance between each chalk mark and

Fig. 1306.—Puddock.

blade of square equals one half of taper or ¼ in. in 2 in. When the blade touches at A, the distance from B to the blade should be ¼ in. If proper care is exercised, it will be possible to bore the hole right away without troubling about the back. Set a pair of callipers to 1 in., and keep trying them to the small end until it is exactly that diameter. It is most important to keep the tool at the height of the lathe centre, or the hole will not be to the correct taper.

portion; set the rest by either of the two methods which have been explained, and turn the tapered portion to about $\frac{1}{8}$ in. full of the required size. Put a narrow grooving or parting-tool in, and reduce the 1½-in. end of rod to the exact dimensions; run the slide-rest back 2 in., and

Fig. 1307.—Milnes Self-acting Slide and Screw-cutting Lathe.

TURNING TAPERED END OF PISTON ROD.

To fit the piston rod, the slide-rest may be used in the same position. As, however, it is not always convenient to start right away to fit the rod, the following will be the procedure if the rest has to be set over again:—Turn the part of the rod to be screwed right up to the tapered

cut another narrow groove exactly 1 in. in diameter. This will give the taper exact. Put a puddock in the rest; this is a piece of wood of the same section as lathe tools, as shown at Fig. 1306, and has a groove cut up about 1½ in. from one end, and a small piece of steel, similar to a small leg of a pair of callipers, pivoted in this groove. Run the slide-rest in until, by applying the movable arm of the puddock to the groove, it will touch. Lift the arm up so that it is clear of the work, run the rest to the other groove, and try whether the arm just touches the grooved portion. If it does, the taper may be particular end which touches. When attempting this task for the first time, it will be as well to mark off the taper at least $\frac{1}{16}$ in. further along the rod (away from the crosshead) until it has been tried in its place, when, if it is a fair fit, the

Fig. 1308.—Birch's Screw-cutting Foot Lathe.

finished, the rod screwed (allowing $\frac{1}{16}$ in. for draw), and tried in its place. If the grooves have been cut to the exact size of hole, and if the rest has been adjusted to correspond, it should show a bearing all along the taper. If it is harder at one end than at the other, take a trifle off the rod may be turned up to the right mark. It must be quite understood that the engine piston and rod have been selected merely because they are well known; the method of turning tapers with the slide-rest is exactly the same, no matter what the job may be.

MILNES' SELF-ACTING SLIDE AND SCREW-CUTTING LATHE.

The plain lathe and its attachments having been described, some particulars of more complicated machines for executing more advanced work may be given. Fig. 1307 shows the Milnes self-acting slide and screw-cutting lathe, a high-class tool having a removable gap piece. The gap in a 3 in. lathe accommodates work having a swing of 11 in., an increase of 5 in.; in a 4½ in. centre lathe, a piece 1 ft. 6 in. can be swung in the gap. The mandrel headstock has an eccentric motion for bringing the back gear into

BIRCH'S FOOT SCREW-CUTTING LATHE.

Birch's foot screw-cutting lathe of 3½ in., 4½ in., 5½ in., or 6 in. centre, is shown by Fig. 1308. It is strongly built, and is well adapted for both light and heavy work. The beds are deep, of strong section, and well ribbed. The fast headstocks have a four-speed cone pulley for gut, and the mandrels have parallel bearings running in bushes adjustable for wear. There are reversing wheels for right- and left-hand sliding and screw-cutting, and the back gears are thrown in and out of gear by an eccentric motion. The carriages have a cross-slide arranged

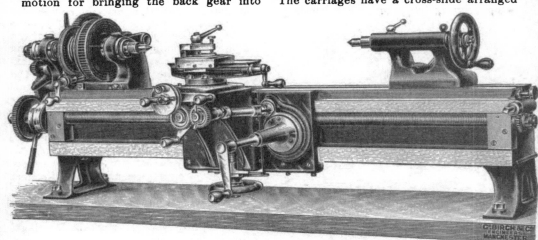

Fig. 1309.—Birch's Front Slide Bench Lathe.

action, and the steel mandrel revolves in hardened conical bearings running into steel bushes. There is a reversing motion, with machine cut steel cluster wheels, for cutting right- and left-hand screw threads. The self-acting sliding saddle is strong, and has a recessed cross-slide and ⊥-slots for boring purposes. There is hand surfacing motion, and a quick return by steel rack and pinion. The compound slide-rest is graduated for turning taper or conical work to any desired angle. The Willis universal tool-holder is fitted. The steel leading screw, gear-driven from the mandrel headstock, goes the full length of the bed. This lathe is a good tool for cutting long spirals.

to run clear out at the back without undoing any screws, so that boring tables can readily be fitted in their place. The cross-slides are made extra long to cover the screw. The carriages are fitted with a quick hand traverse motion. Brass index plates are fitted to the standards showing the change-wheels necessary for screw-cutting. The compound slide-rests have swivel base fully graduated for angle and taper turning. The lathes are provided with back stays on carriage, set of machine cut change-wheels, face-plate, driver-plate, two ordinary centres and one square centre, hardwood treadle board and back board, and set of hardened working screw-keys complete. An overhead for driving milling cutters, etc., held in the slide-rest, is provided by two substantial uprights of oval section, carried on the back of the standards ; two cone pulleys, and an arrangement of guide pulleys for tightening the driving band, are fitted.

BIRCH'S 4 IN. FRONT SLIDE BENCH LATHE.

The operations that can be performed on such a lathe as is shown by Fig. 1309 are endless. The range of work comprises, besides turning, such operations as boring, milling, gear cutting, slot drilling, slotting, etc., etc. For boring and

Fig. 1310.—Birch's Capstan Head.

milling the work is fastened upon the boring table, a boring bar or cutter arbor being placed between the centres ; the vertical slide of the carriage is extremely useful for adjusting the height of work, or setting the depth of cut. Milling can also be done by using the wheel-cutting and milling attachment fitted to the cross-slide of lathe, the work being carried between centres. By using this attachment either spur, bevel, worm, or spiral gears can be cut, and it can also be used for fluting taps and reamers, slot drilling, etc. By the use of the shaping attachment a gear can be key-wayed before it is taken out of the chuck, and internal gears, etc., can be cut by it. The bed is deep and well ribbed to secure rigidity,

the front having a slide on which the carriage travels, the carriage moving out of the way past the headstocks when not in use. The centre line of the headstocks is brought forward towards the front of the bed in order to suit the position of slide-rest. The fast headstock has hardened steel mandrel with parallel bearings running in bushes of special metal, adjustable for wear. The thrust arrangement is of special design, so arranged that the adjustment does not vary if the mandrel gets warm with running, the arrangement being enclosed and running in oil. The mandrel stands clear out at the back of the headstock so that a draw-in spindle for split chucks can be fitted. The cone pulley can be locked to the spindle by a concentric friction arrangement, keeping the mandrel in balance whether the back-gear is in or out. The loose head is locked to the bed by an eccentric arrangement, and the barrel is arranged to push out its own centres. The boss of the hand-wheel can be fitted with an adjustable micrometer, reading to $\frac{1}{1000}$ in., for use in drilling, etc. The carriage is long and has large wearing surfaces. The guide screw is placed inside the V's on front of the bed out of the way of chips and dirt. The guide screw nut is of gun-metal. The compound slide-rest is carried by a knee bracket with vertical slide on carriage, the vertical slide being found very convenient for use in milling, boring, etc., and also for adjusting the height of tool. The cross-slide is made so that it will run out at the back without undoing any nuts, etc., leaving the knee bracket clear for fitting attachments. The compound slide-rest has its base fully graduated for angle and taper turning. The swing frame has a special locking arrangement allowing it to be locked in any position. The screw of the vertical slide can be fitted with an adjustable micrometer reading to $\frac{1}{1000}$ in. if desired for use in milling. This lathe is also made with the following additions :—Self-acting surfacing motion to cross-slide, driven by worm on guidescrew ; universal joints and telescope shafts to compensate for the travel of the vertical slide ; and overhead motion.

TESTING SCREW-CUTTING LATHE.

With regard to a lathe having a saddle, slide-rest, and leading-screw, the tests should be applied as described for the plain lathe (they are of more importance now), and then the worker should proceed to test for parallel sliding. To do this,

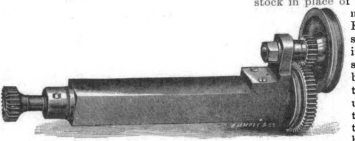

Fig. 1311.—Slot Drilling and Milling Frame.

take a bar about as long as the lathe will admit, and of, say, $\frac{3}{4}$ in. or 1 in. diameter; centre it and turn up a length of $\frac{1}{2}$ in. at one end. Take off the carrier, and, turning the bar round with the left hand, screw up the tool till it grazes the turned part; now take out the bar, rack the saddle to the other end, put in the bar end for end, and try whether the tool will graze the turned part as before. This is a very severe test, and if $\frac{1}{64}$ in. out of truth it would not be of great consequence on a 5 ft. bed, as the tool might wear that amount while turning up a long bar. To test for surfacing, take the largest face-plate and lay a straight-edge across it to see whether it is flat. Put it on the mandrel and place a tool in the slide-rest so as just to touch it, the cross slide of saddle being at one extremity of its course. Now move the saddle slide to the opposite extremity and try whether the tool still touches as before. It should do so if the lathe is worth having, but this point, like the last, is usually attended to, and there are means of adjustment. As stated above, these tests should be preceded by those described on pp. 396 to 398 for a plain lathe.

CAPSTAN HEAD.

The capstan head is a very useful addition to any lathe; it is used for making small screws, pins, etc., when the more expensive capstan lathe is not available. Fig. 1310 shows Birch's capstan head, this fitting in the barrel of the loose headstock in place of the centre. The attachment, as illustrated by Fig. 1310, is made in two sizes, the smaller one carrying six tools, and the larger seven. The revolving part being carried at an angle, the tools not in use stand up clear of the barrel, so that it is not necessary for the barrel to overhang the headstock. The holders for carrying screwing tools are arranged to slide in the head. Altogether, this is a most ingenious tool.

SLOT DRILLING AND MILLING FRAME.

The slot drilling and milling frame and the slot drills and milling and fly cutters shown by Figs. 1311 and 1312 respectively are for use in the ordinary slide-rest, for slot drilling, key-way cutting, and light milling. The appliance illustrated is made by Birch, and is single geared, for heavy work. The large gear is made with a grooved pulley attached for driving direct when using fly cutters for brass, wood, etc. With this apparatus is supplied a vertical rest attachment which fits on

Fig. 1312.—Slotting and Milling Tools, etc.

the top of the cross-slide, in place of compound slide-rest, and has a vertical swivel, which carries the compound slide-rest, and allows the top slide to be used in any position from vertical to horizontal, thus largely increasing the range of work. This, also, is a most desirable addition.

MILLING ATTACHMENT FOR LATHE.

The little machine described below is intended for those who possess a lathe with slide-rest, and desire a simple and inexpensive attachment capable of tooling flat surfaces, sawing or grooving metals, cutting teeth in small wheels, and work of a similar nature. As the spindle runs on dead centres, the labour of fitting adjustable cone or parallel bearings is dispensed means of the cross-sides. The upright bracket A carries the head B, having a projecting curved arm which supports the outer end of the spindle C. The head can be raised or lowered by turning the handle K, and fixed at any height by tightening the nut H at the back. The driving chain is long enough to permit the head to be raised to its highest position, and is kept

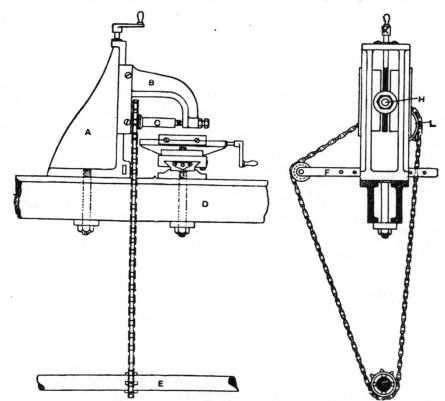

Figs. 1313 and 1314.—Elevations of Milling Attachment.

with. Figs. 1313 and 1314 show the machine in position and the method of driving. The motive power is transmitted from the flywheel shaft E of the lathe round the lathe-bed D by means of a chain to a toothed wheel L, running on a stud, which forms one of the centres supporting the spindle. The work is clamped to the top of the slide-rest and fed under the cutter by tight by an adjustable spring guide F. The cutter revolves towards the front of the lathe. The attachment as described is intended for a 4½-in. centre lathe, but it may readily be adapted to suit other sizes.

UPRIGHT BRACKET FOR MILLING ATTACHMENT.

The upright bracket shown in Figs. 1315 to 1317 is cast with two faces at right angles to each other, and provided with chipping strips. The lower face has also a snug in the centre to fit into the space in the lathe-bed. The other face has a slot ⅞ in. wide down the centre, which may

The long face of the casting may now be levelled by chipping and filing. Should, however, the face of the casting be fairly level, the simplest plan would be to grind the scale of the face and two edges on a

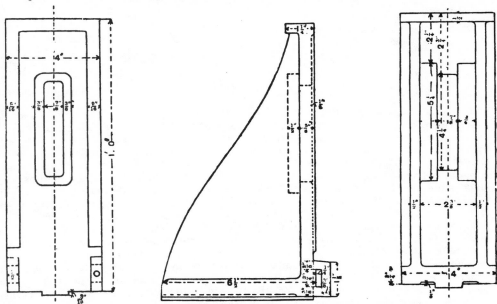

Figs. 1315 to 1317.—Elevations of Upright Bracket or Milling Attachment.

either be cast in or cut out afterwards, and on each side and at the back of this slot is a projecting strip, ½ in. thick and ¾ in. wide. There are also two lugs at the bottom in which slots will be filed to carry the chain guide. The casting is strengthened by two webs ½ in. thick, tapering outwards to $\frac{7}{16}$ in. The slot and strips should also be slightly tapered to enable the pattern to draw from the sand. In putting the pattern together care must be taken to have it square and the sides parallel. The head is shown in Figs. 1318 and 1319. The centre of the arm does not coincide with the centre of the sole, the object being to bring the spindle over the middle of the slide-rest cross-slide.

Surfacing Foot of Bracket.

Mark the centre of the foot of the upright, and drill and tap a ⅝-in. hole there.

grindstone. It may then be filed flat over the strips, so that when placed on a surface plate, or on the lathe-bed, it will rest

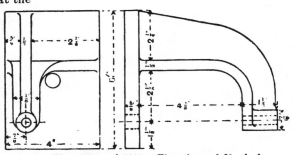

Figs 1318 and 1319.—Elevations of Head of Milling Attachment.

firmly without rocking. Now smear a few touches of red lead and oil on the surface plate, and rub the face of the casting upon it so as to indicate the points of contact. These spots must be lowered by filing until an even surface is obtained. The edges are to be filed parallel to each other and square to the face.

FITTING BRACKET TO LATHE-BED.

Draw a centre line and mark off the

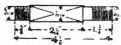

Fig. 1320.—Screwed Stud.

slot if this has not been cast in. To remove this metal, drill a row of $\frac{7}{8}$-in. holes close together, and form these into a slot by chipping and filing. The upright should now be clamped, foot upwards, with a bolt passing through the slot, and the snug made to fit in between the cheeks of the lathe bed. The easiest way to do this is to mark off $\frac{1}{8}$ in. at each end of the snug on both its sides, and chip away the metal between these marks $\frac{1}{16}$ in. deep. This will reduce the amount of metal to be removed by the file. The snug must be square to the surface of the upright. The parts of the foot that will rest on the lathe-bed must also be filed level and square to the other face and its edges. The strips at the back of the slot may require a little chipping to make them parallel to the face of the upright, and of equal thickness.

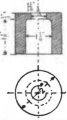

Fig. 1321.—Section and Plan of Washer.

HEAD OF MILLING ATTACHMENT.

Now take the casting of the head (Figs. 1318 and 1319) and mark the centre of the boss on the arm; clamp the sole against the face-plate of the lathe so that the centre of the boss runs true, and drill a $\frac{3}{8}$-in. tapping hole through the boss, con-

tinuing it through the sole also. Face up the boss and remove from the lathe. Make a mandrel to fit these holes, drive it through, and put in the lathe between centres, allowing the pin of the driving chuck to engage with the arm. Then face the sole of the casting quite flat. The man-

Fig. 1322.—Screw for Head of Milling Attachment.

drel should not project more than necessary beyond the casting or it will be liable to spring and cause chattering. Remove the mandrel, finish the surface by filing if necessary, and make the edges square and parallel, and the sole of exactly the same width as the face of the upright. The hole through the side is to be enlarged to $\frac{5}{8}$ in. tapping, and both holes then tapped to size. Now mark off a $\frac{5}{8}$-in. hole in the centre of the sole, and see that the hole will be opposite the slot in the upright when placed in position; then drill and tap $\frac{5}{8}$ in. diameter. The stud is made from $\frac{3}{4}$-in. square iron or soft steel, turned as shown in Fig. 1320. The shorter end should be screwed very tight in the hole in the centre of the head, with one of the flat sides of the stud uppermost. The washer (Fig. 1321) may be cast from a pattern or turned from a piece of round iron to the dimensions shown. The head may be clamped against the upright, keeping the edges of each together. Cut two

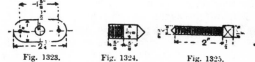

Fig. 1323.　　Fig. 1324.　　Fig. 1325.

Fig. 1323.—Plate for Screw. Figs. 1324 and 1325. —Centres for Milling Attachment Spindle.

pieces of strip iron 6 in. long, 1$\frac{1}{4}$ in. wide, and $\frac{3}{16}$ in. thick; file them flat, and screw them on each side of the head with $\frac{1}{4}$-in. countersunk screws so as to embrace the edges of the upright. The head should move up and down freely without side-shake when the nut is loosened.

The next thing to be done is to drill the
holes for the screw which works the head.
Scribe a line across the top flange of the
upright opposite the centre of the slot, and
mark off a ⅜-in. hole at a distance from the
face that will allow the screw to clear the
metal below the flange. Bolt the head in
its highest position, and drill a ⅜-in. tapping hole through the flange, continuing it
through the square stud below. Care
must be taken to have these holes in a
line parallel with the sides and face of the
upright. The hole in the flange is enlarged
to ⅜ in. diameter, and then recessed to a
depth of $\frac{7}{16}$ in. The hole in the stud
should be tapped. Make the screw as
shown in Fig. 1322, and file a square on
the end to take a small handle. The plate
to prevent the screw rising is shown in
Fig. 1323; it is made from $\frac{7}{16}$-in. wrought-
iron, and may be fastened down by two
$\frac{1}{16}$ in. screws. The two centres and the
cutter spindle should be made from mild
steel, and afterwards case-hardened. The
centre shown in Fig. 1324 must be screwed
very tight into the hole tapped in the
head. The outer centre (Fig. 1325) should

flywheel shaft of the lathe; and as it will
probably have a hole considerably larger
than the shaft, it should be staked on
with four keys, as shown in Fig. 1314. File
four flats on the shaft at equal distances

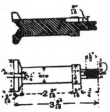

Figs. 1328 and 1329.—Section and Plan of Spindle.

round its circumference, and four key-
ways in the chain wheel; then key on with
four keys, adjusting these so that the
wheel runs true. The other chain wheel
has eleven teeth, and must be cast. To
make the pattern, get a piece of beech or
sycamore cut across the log, and turn it
to the shape and thickness shown in Figs.
1326 and 1327. The exact diameter depends on the pitch of the chain. To find
this, measure the distance from the end of
one link to the corresponding end of the
next. The pitch diameter of the wheel is
got by multiplying the pitch thus found
by the number of teeth required, and

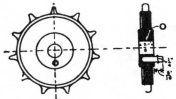

Figs. 1326 and 1327.—Chain Wheel or Sprocket.

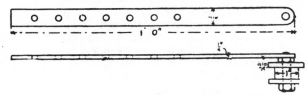

Figs. 1330 and 1331.—Chain Guide.

be screwed to fit the hole in the boss without shake. A lock-nut ¼ in. thick should
be face up for this at the same time. The
points of the centres must be turned to an
angle of 90°, and may be tested with the
square.

CHAIN AND CHAIN WHEEL.

For driving, obtain a light bicycle
chain, also a chain wheel with eight teeth,
as used for the back wheels of bicycles.
This chain wheel is to be keyed on to the

dividing by ϖ. To get the diameter over
the tops of the teeth, add to this the depth
of tooth. Thus, if the number of teeth
equals 11, the depth of teeth equals $\frac{3}{16}$ in.,
and the pitch equals $1\frac{1}{16}$ in., then the out-

side diameter will be $\dfrac{1\frac{1}{16} \times 11}{\varpi} \times \frac{5}{16} = 4\frac{11}{32}$ in.

When the pattern is turned, the pitch
circle is struck out and divided into eleven
parts, and the teeth are marked. The
intervening spaces are then cut away and

the teeth finished off to gear with the chain. The casting is chucked centrally in the lathe, and a hole bored to fit the stud (Fig. 1324), which forms one of the centres for the cutter spindle. A brass bush may be fitted in if desired, the hole

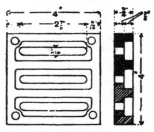

Figs. 1332 and 1333.—Plate for Holding Milling Work.

being bored larger for this purpose. Mount the chain wheel on a mandrel and turn up all over. Dress the teeth with a file, and taper them at the points so as to gear with the chain easily. Bore a small oil-hole in the boss.

CUTTER SPINDLE.

The spindle (Figs. 1328 and 1329) is a mild steel forging, and should be centred true ; deep centres are required, the square centre being ground to an angle of 90º. Face both ends to length, then recess the flanged end about $\frac{1}{4}$ in., and deepen the centre until, when the spindle is placed in position, the flange comes within $\frac{1}{32}$ in. of the boss of the chain wheel. Drill a $\frac{1}{16}$-in. hole about $\frac{1}{8}$ in. deep in each centre, and turn down to size. Thread the end, and face a nut and two or three washers to fit, the latter being $\frac{7}{8}$ in. diameter, $\frac{1}{8}$ in. thick, and $\frac{3}{8}$ in. bore. Drill the hole for a steel pin, shown in Fig. 1328, about $\frac{3}{8}$ in. deep and $\frac{1}{16}$ in. diameter, the pin projecting about $\frac{1}{16}$ in. to prevent the cutters turning on the spindle. Next file a notch in the flange, $\frac{1}{4}$ in. wide and $\frac{1}{16}$ in. deep, as shown. Place the spindle between centres, and mark on the boss of the chain wheel the position for a steel pin, shown in Figs. 1326 and 1327.

CHAIN GUIDE.

The chain guide (Figs. 1330 and 1331) is made from $\frac{1}{4}$-in. by $\frac{3}{4}$-in. iron, 1 ft. long. A $\frac{3}{16}$-in. hole is drilled and tapped at one end, and a screw and lock-nut are fitted to carry the boxwood roller shown in Fig. 1331. The two slots in the upright (Figs. 1315 and 1316) must be drilled and filed out to allow the bar to slide easily, and six or seven holes are drilled about 1 in. apart and tapped to fit a $\frac{1}{4}$-in. screw, which holds one end of a spiral spring, the other end being fastened to one of the projecting snugs (Fig. 1315). This spring should be about $1\frac{1}{2}$ in. long, to stretch to 3 in. or so.

MAKING MILLING CUTTERS.

The cutters are made from best tool steel, their size and shape depending on the special work to which they will be put ; one $1\frac{1}{4}$ in. or $1\frac{1}{2}$ in. diameter and $\frac{1}{4}$ in. thick, with rounded teeth, will be useful for grooving or roughing work. Another, the same size, with square teeth, can be used for cutting key-ways and finishing sur-

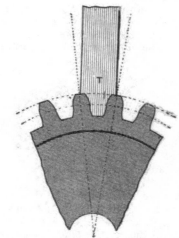

Fig. 1334.—Wheel Teeth and Fly Cutter Template.

faces. Also a circular saw of small diameter is invaluable for splitting bushes, nicking screws, cutting bars, etc. The blanks may be cut from strip steel such as is used for making scrapers, then bored $\frac{3}{8}$ in. diameter to fit the spindle, turned on a mandrel, and the thickness reduced towards the centre to give clearance to the teeth. These are formed round the edge with a half-round file, a smooth file being used to finish them. Next file a key-way in the cutter to fit over the pin in the

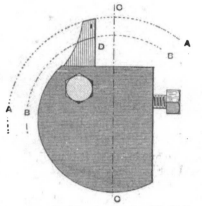

Fig. 1335.—Position for Shaping Fly Cutter.

spindle, and heat to a fairly bright red in a clear fire ; remove with a bent wire, and plunge edgeways into cold water. When quite cold, remove it and brighten the sides with emery cloth. To temper the cutter evenly requires care. Get a piece of ¾-in. round iron about 6 in. long, and turn ½ in. at one end down to ⅜ in. diameter, to pass through the hole in the cutter easily. Heat this bar red-hot, and hold it upright in a pair of tongs and place the cutter on the end. Now watch carefully the bright surface, and, as soon as a dark yellow colour is seen all over, drop the cutter at once into cold water. The teeth may be sharpened on a small emery disc running at a high speed in the lathe.

MEANS OF HOLDING MILLING WORK.

For holding work on the slide-rest, in conection with the appliance described, a small parallel vice will be found most useful ; or a plate may be made as shown in Figs. 1332 and 1333, milled flat on both sides and screwed to the top of the slide-rest with four countersunk screws. The work can be clamped down by T-headed bolts passed through the slots and turned

half round. If the top of the slide-rest is not level it should be milled. The illustrations are all drawn to scale.

MAKING FLY CUTTERS.

The fly cutter is particularly advantageous when a special shape or form requires to be reproduced exactly at long intervals. A good example of the application of the fly cutter is in producing some special or odd shape of gear-tooth, or one of very fine pitch. For making fly cutters a template T (Fig. 1334) must first be formed ; this must fit the space and one-half the thickness of each tooth, as shown by the dotted radial lines, and from this a tool may be made that can be held in a lathe-rest or milling machine vice, the tool being carefully filed or oilstoned to exact shape. This tool is to be used in making cutters only, and becomes, in fact, a standard. To produce the working cutters, which are pieces of steel let into an arbor, there are several methods. The blank piece of steel may be set in an arbor, as in Fig. 1335, and cut to shape by the standard or master-tool previously made, setting it true with the axis, as in Fig. 1338, p. 468, for use on the work. The cutter must be placed as in

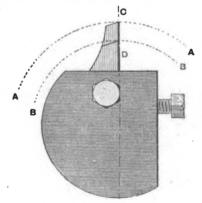

Fig. 1336.—Position for Using Fly Cutter.

Fig. 1336 to be used, the difference of position on the two arbors giving to the cutter the necessary clearance, as will be seen from the dotted arcs A A and B B. These correspond in the two figures, C representing the centre line of the arbors. It will be observed that the front face D may be ground to sharpen the tool without actually altering its shape. In moving the

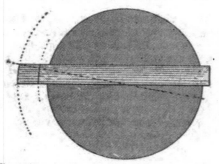

Fig. 1337.—Another Method of Producing Cutter.

cutter from one position to the other in the two arbors, it has, however, been caused to cut the work to a slightly different shape from that of the original template. But the change of position in the two arbors causes the cutter to produce a slightly different form from that of the master-tool.

ANOTHER METHOD OF MAKING FLY CUTTERS.

Another method of using the master-tool to make a cutter that may be ground

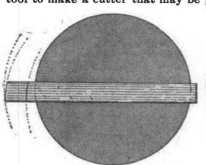

Fig. 1338.—Method of Using Fly Cutter.

without altering its shape, and yet have the necessary clearance, is shown in Figs. 1337 and 1338. The blank cutter is set in a slot at an angle to the axis of the arbor, as in Fig. 1337, and is cut to shape with the master-tool. The best method, therefore, is that shown in Figs. 1339 and 1340, the slot being central through both arbors,

and the front face D of the cutter being on a line with the arbor axis. The clearance is obtained in this case by producing the cutter in an arbor of smaller diameter

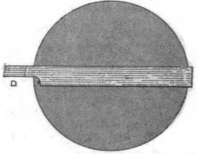

Fig. 1339.—Best Method of Making Fly Cutter.

than the one in which it is used, the dotted arcs in Fig. 1340 showing the clearance. In this method the shape cut by the tool will correspond exactly to that of the template from which the master-tool was shaped. It is obvious that this principle of tool construction is highly advantageous whenever some particular form or shape

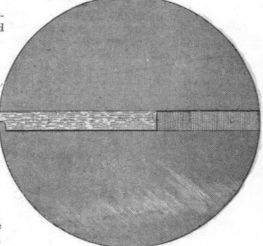

Fig. 1340.—Method of Using Fly Cutter.

is to be preserved. But in the case of curves or irregular forms to be frequently produced a circular cutter may be made from the cutter produced in Fig. 1341.

FLY CUTTERS FOR USE ON ONE SIDE ONLY.

By the above method the cutter would produce a circular cutter shaped on both sides ; but in many cases this would be unnecessary, and even in the case of gear teeth one side only of the cutter may be used, formed to the shape of one side of a tooth, and thus save much work. This is shown in Figs. 1342 and 1343 and in Figs. 1344 and 1345, in which the holders H are right and left, so that the cutter may be leaned in either direction to give the

CLOCK WHEEL FLY CUTTER.

Fly cutters for clock wheels have a single blade, and are generally shaped as shown

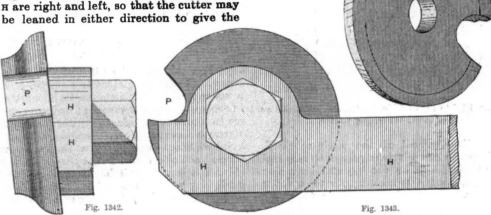

Fig. 1341.

Fig. 1342.

Fig. 1343.

Fig. 1341.—Circular Fly Cutter.   Figs. 1342 and 1343.—Holder for Right-hand Fly Cutter

necessary clearance and enable the cutter to be used as either a right- or left-hand tool. Instead of having only one opening.

in Figs. 1346 to 1348. The first of these figures is a side view showing rake and top clearance ; the second is a back view

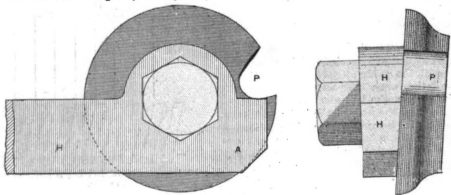

Figs. 1344 and 1345.—Holder for Left-hand Fly Cutter.

P, to give the cutting edge, the circular cutter may nave two, so that one edge may be used for roughing out the work and the other for finishing it.

showing side clearance, and the third is a front view. The centre hole is made to fit the spindle of the cutting frame, and the blade is made of as short a radius as possible. Such cutters must be run at from 150 to 200 revolutions per second. To make them, turn and file up from tool steel to the shape shown. Shape the blade to the space between two teeth of

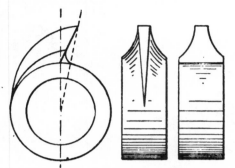

Figs. 1346 to 1348.—Side, Back, and Front Views of Clock Wheel Fly Cutter.

a well-cut wheel of the pitch required. When the blade is nearly down to size and well backed off on the top and both side curves, harden it and temper to a straw colour. Then finish the blade to shape and size by polishing down with a soft steel polisher charged with oilstone dust and oil mixed to a paste. Use the polisher like a file. This leaves it a clean grey. Polish the face and side curves with the same polisher cleaned and charged with red-stuff and oil (watchmakers' ordinary red-stuff). A polished cutter will cut a polished tooth.

SCREW-CUTTING IN THE LATHE.

This chapter will conclude with an enunciation of the principles of screw-cutting in the lathe, work that can be attempted by the turner who has mastered the construction and the uses of the lathe and its principal accessories. The descriptions of screw-cutting lathes previously given will make the following instructions easily comprehensible. The functions of the leading screw or guide screw referred to in those descriptions must first of all be explained. A (Fig. 1349) represents the leading screw of a self-acting lathe, and B is a rod upon which screw-threads have to be cut. The problem is how to utilise the leading screw A, whose pitch or rate of

thread is constant and invariable, for the cutting of threads of diversified pitches upon rod B, running between lathe centres. This evidently resolves itself into a question of the relative rates of revolution of A and B. The headstock mandrel, on which the work B is centred, drives the leading screw through the medium of change-wheels. The problem, therefore, is to impart the required ratio of speed from the mandrel, through the change-wheels, to the guide screw.

GRAPHIC EXAMPLES OF RELATIVE RATES OF REVOLUTION.

In Fig. 1349, a is the pitch of the guide screw. The pitch is equal to a thread and a space; or to the distance from centre to centre of adjacent threads; or from centre to centre of the spaces; or, in the most accurate language, the longitudinal or axial distance covered by a thread in a single revolution round its cylinder. Say this screw is of $\frac{1}{4}$ in. pitch. The screw is encircled by a clasp nut. If a screw of $\frac{1}{4}$ in. pitch, b, is to be cut, the clasp nut and tool

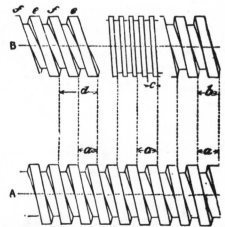

Fig. 1349.— Diagram of Leading Screw and Screw Threads.

point must be made to traverse the distance a and b in equal times. In other words, A and B must each make a complete revolution simultaneously. Clearly also, if a thread of $\frac{1}{8}$ in. pitch c (one-half the pitch of A) is to be cut, the rod B must make two revolutions while A is making one. But, on the other hand, if the thread *d* is to be of ½ in. pitch (twice the pitch of A) then B must make but one

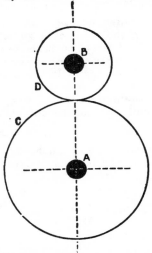

Fig. 1350.—Change Wheels for obtaining Smaller Pitch than that of Leading Screw.

revolution while A makes two. (Note that screw B is double threaded—that is, *e* is one thread and *f* another.) The problem, therefore, is resolved into one of relative rates of revolution and of relative traverse of the cutting tool in equal times. The rates of revolution of A and B are clearly in inverse proportion to the pitch of A and B. That is to say, to cut a screw upon B of half the pitch of A, then B must revolve at twice the speed of A; and to cut a screw on B of twice the pitch of A, then B must revolve at half the speed of A. Hence the change-wheels used must be selected so that the numbers of their teeth shall bear the same proportions as the screw rates of the guide screw and of the screw to be cut; and they must be so fixed that the inverse relations required shall be obtained. Therefore, if the rates are equal, wheels of equal size will be used on the spindles both of the guide screw and screw to be cut. But to cut on B a screw, *c*, of ⅛ in. pitch (half that of A) a wheel must be put on the mandrel B half the size of the

one put on A (as in Fig. 1350). Then wheel c (Fig. 1350) will make only half a revolution to a complete revolution of wheel D, and the leading screw A will be retarded in relation to the rod B by just one-half, and a thread of ¼ in. ÷ 2 = ⅛ in. will be cut on B. Conversely, to cut a thread of ½ in. pitch *d*, on B—that is, twice the pitch of A—a wheel D, will be put on B (Fig. 1351) having twice the number of teeth of the wheel c on A, and the rate of traverse of the leading screw A will be double that of the rod B. These are the fundamental facts, which, better than any set rules, should be well understood by the screw-cutter.

FRACTIONAL PITCHES.

Precisely the same inverse relation must exist when cutting fractional threads. A (Fig. 1352) shows the same leading screw of ¼ in. pitch. At *b* one and a half threads are cut in the length of the pitch of the guide screw. At *c* one thread is cut in the space occupied by one and a half threads of the guide screw. To cut *b* a wheel must go on the guide screw A, having one and

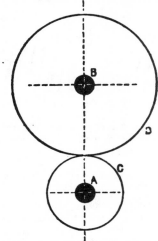

Fig. 1351.—Change Wheels for obtaining Larger Pitch than that of Leading Screw.

a half times the number of teeth of that on the lathe mandrel B. To cut *c* a wheel must go on the lathe mandrel B, having one and a half times the number of teeth of that on the guide screw.

DRIVER AND DRIVEN WHEELS.

The wheel that is put on the lathe mandrel B is called the driving or driver wheel; the one that goes on the guide screw A is termed the driven wheel. The same ratio must subsist between the number of teeth (or diameters) of the driving and the driven wheels as exists between the pitch of the guide screw and the pitch of the screw to be cut. The same thing is said in another way, convenient to remember, thus: When cutting threads of finer pitch than that on the leading screw, the smaller wheel must drive and the larger be driven. When cutting threads coarser than that on the lead-

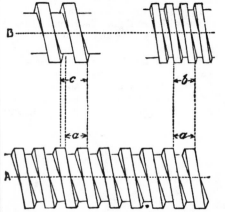

Fig. 1352.—Diagram Illustrating Fractional Pitches.

ing screw, the larger wheel will drive and the smaller be driven. These are axioms to be committed to memory.

SIMPLE AND COMPOUND TRAIN OF WHEELS.

When there is only one driving wheel and one driven wheel, the gearing is known as a simple train of wheels. But it often happens that several wheels are required to cut a screw of very fine or of very coarse pitch, and these compound trains, as they are called, sometimes occasion difficulty as to the correct relative locations of the several wheels. It is necessary to employ a compound train when the numbers representing the ratio of the leading screw and

of the screw to be cut extend beyond the limits of an ordinary series of change-wheels. Those limits are commonly wheels of twenty and one hundred and twenty teeth. The ratio of these is $\frac{120}{20} = 6$, and when limited by these wheels, the ratio between the pitch of a guide screw and that of a screw to be cut on any lathe cannot exceed six, unless a compound train is employed. In a lathe having a guide screw of $\frac{1}{4}$ in. pitch—four threads to the inch—twenty-four threads to the inch mark the limit that can be cut with a simple train, thus: $\frac{24}{4} = 6$. In a lathe having a guide screw of $\frac{1}{2}$ in. pitch—two threads to the inch—the limit is twelve threads per inch, thus: $\frac{12}{2} = 6$.

COMPOUND SCREW-CUTTING TRAIN DESCRIBED.

To give a clear idea of the use and arrangement of a compound train, suppose that a screw is to be cut having twenty-four threads to the inch in a lathe whose leading screw has two threads to the inch. The ratio is $\frac{24}{2} = 12$, and consequently, if a single train were used, the driver wheel —the wheel on the lathe mandrel—would have to make twelve revolutions to each revolution of the driven or guide screw wheel. Wheels of 240 and 20 teeth, therefore, would give this ratio: $\frac{240}{20} = 12$; the 20-toothed wheel would be the driver, and the 240-toothed wheel the driven. But as there is no wheel of 240 teeth in any ordinary set — it would probably be too large for gearing-up between the centres of the lathe mandrel and the guide screw— the supposed 20 and 240-toothed wheels must be broken up into factors, the ratio of whose products will have the ratio $\frac{240}{20}$, and in that case use is made of the intermediate stud on the rocking plate. The numbers are broken up, and the details of the process are as follows:

$$\frac{20}{240} = \frac{2 \times 10}{20 \times 12} = \text{adding cyphers,} \ \frac{20 \times 10}{200 \times 12}.$$

As there are neither 10, 200, nor 12-toothed wheels, try again ; add cyphers to 10 and 12, and then halve 200 and 100, thus :

$$\frac{20 \times \overset{50}{\cancel{100}}}{\underset{100}{\cancel{200}} \times 120,}$$

the same ratio between the numbers above and below the line being preserved, until the numbers obtained are not duplicates of one another, and are included in the ordinary sets of change wheels. So that if 20 and 50 drive, and 100 and 120 are driven, the ratio $\frac{100 \times 120}{20 \times 50} = 12$ will be obtained, this being the ratio required.

of the headstock gear as is necessary to ilustrate the arrangements of a compound train. A is the lathe mandrel ; B the guide screw ; C the cone pulley through which the lathe mandrel is driven ; D the large wheel of the back gear ; E the rocking or quadrant plate pivoted upon the end of the guide screw ; F the first driver wheel ; G the first driven wheel ; H the second driver wheel ; J the second driven wheel ; G and H are commonly termed the stud wheels ; K is the stud. As driver and driven wheels of various diameters are used for cutting screws of various pitches, the quadrant plate is provided with curved

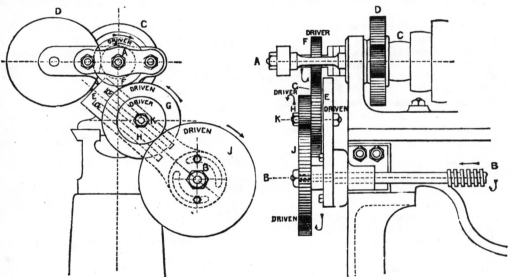

Figs 1353 and 1354 —Compound Train of Wheels for Screw-cutting.

ARRANGEMENT OF COMPOUND TRAIN OF WHEELS.

One of the drivers and one of the driven wheels must be placed upon the stud of the rocking plate. But the use of the term stud wheel, except when it happens to be an idle wheel employed for changing the direction of motion, as in cutting left-handed screws, causes confusion ; and here only the terms " driver ", and " driven " will be employed. The relative arrangements of these wheels are shown in Figs. 1353 and 1354, together with just as much

slots, to permit of free pivoting around the end of B. The wide range of radius permitted by these curved slots, and the choice of any position in the two straight slots in the rocking plate, permit of such range of position for the stud K that all the wheels from 20 to 120 teeth, and even higher, can be geared up for compound trains. It does not matter which of the drivers goes on the lathe mandrel and which on the stud. Neither does it matter which of the driven goes on the stud nor which on the guide screw, because this only amounts to a transposition of the numbers, and does not affect the products or the ratio. Thus, in the example last given, 20 might go on B, and 50 on A, and 100 might go on C, and 120 on B.

RECAPITULATION OF THE PRINCIPLES OF SCREW-CUTTING.

The fundamental principles are, in brief, as follows: The ratio existing between the guide screw and the screw to be cut must be the same as that between the driver and the driven wheel in a simple train; or, in a compound train, between the products of the drivers and of the driven wheels. This, therefore, is the fundamental principle of inverse ratios. Again, to cut pitches finer than that on the leading screw, the smaller wheel or wheels will drive, and the larger be driven. To cut pitches coarser than that on the leading screw, the larger wheel or wheels will drive and the smaller be driven. When arranging the wheels of a compound train, one driver is put on the lathe mandrel, one driven is put on the guide screw, and finally one driver and one driven are put on the stud.

FINDING SIZES OF CHANGE-WHEELS.

In practice, the common method employed is to write down in the form of a vulgar fraction the number of threads per inch of the screw to be cut, and deduce suitable wheels, having the same ratio as the numerator and denominator of the fraction. The numerator then corresponds with the driver, and the denominator with the driven wheels. This method covers all possible cases that can arise in practice, and is in harmony with the principles just laid down. Thus, with a leading screw of $\frac{1}{4}$ in. pitch—four threads per inch—it may be required to cut screws, say, having respectively $1\frac{1}{2}$ threads, 2 threads, $6\frac{1}{2}$ threads, and 20 threads per inch. The vulgar fractions expressing these relations will be as follows:— $\frac{4}{1\frac{1}{2}}, \frac{4}{2}, \frac{4}{6\frac{1}{2}}, \frac{4}{20}$. The relations being clearly set down, proceed to deduce suitable change wheels from the fractions. In two of these cases there are integers and fractions combined, but as these cannot be worked with, they must be turned into

either whole numbers or decimals, thus: Multiply the entire vulgar fraction by the denominator of the fractional number, or turn the latter into a decimal. In this way, $\frac{4}{1\frac{1}{2}}$ becomes $\frac{4}{1\frac{1}{2}} \times 2 = \frac{8}{3}$; or, as a decimal, $\frac{4}{1\cdot5}$; and $\frac{4}{6\frac{1}{2}}$ becomes $\frac{4}{6\frac{1}{2}} \times 2 = \frac{8}{13}$; or, as a decimal, $\frac{4}{6\cdot5}$. Wheels are deduced from ratio relations of this kind either by the addition of cyphers or by multiplying by 5 (because the numbers of teeth in the change wheels advance by fives), and then by halving, doubling, or otherwise breaking up or increasing the number first obtained, in equal proportions—that is, care must be taken to increase or diminish the numbers in numerator and denominator in exactly the same proportions. Now, taking each of these ratio numbers in succession, the following statement is obtained:—

$\frac{4}{1\frac{1}{2}} = \frac{8}{3} \times 5 = \frac{40}{15}$; or, adding a cypher, $\frac{8}{3} = \frac{80}{30}$; or, further increasing by one-half, $\frac{120}{45} = \frac{\text{driver.}}{\text{driven.}}$ Again, writing :— $\frac{4}{1\cdot5} \times 5 = \frac{20}{7\cdot5}$, and, multiplying by 2 $= \frac{40}{15\cdot0}$; or, adding a cypher to $\frac{4}{1\cdot5} = \frac{40}{15\cdot} = \frac{\text{driver.}}{\text{driven.}}$ Again, writing :— $\frac{4}{2} \times 5$ $= \frac{20}{10}$, and multiplying by 2 $= \frac{40}{20}$; or, adding a cypher to $\frac{4}{2} = \frac{40}{20} = \frac{\text{driver.}}{\text{driven.}}$ Again, $\frac{4}{6\frac{1}{2}} = \frac{8}{13}$ $\times 5 = \frac{40}{65}$, or writing :— $\frac{4}{6\cdot5}$, and adding a cypher $= \frac{40}{65} = \frac{\text{driver.}}{\text{driven.}}$ Again, $\frac{4}{20} \times 5 = \frac{20}{100}$; or, adding a cypher to $\frac{4}{20\cdot} = \frac{40}{200}$, and halving (because there is no 200-toothed wheel in a set), $= \frac{20}{100} = \frac{\text{driver.}}{\text{driven.}}$

A glance at the numbers shows that there are inverse relations between the guide screw and screw to be cut and the driver and driven wheels, and that in the case of the first two threads of $1\frac{1}{2}$ and 2 per inch respectively, which are coarser than the guide screw, the larger wheels drive and the smaller are driven, thus causing a more rapid revolution of the guide screw than of the lathe mandrel, and consequently a rapid longitudinal traverse of the cutting tool relatively to the work in the lathe. In the case of the second set of threads of 6¼ and 20 to the inch, which are finer than the guide screw, the smaller wheels drive and the larger are driven, the guide screw revolves slowly, and the tool is retarded in relation to the work. But putting the relation of the guide screw and screw to be cut in the fractional form shown, and remembering that the numerator gives driving and the denominator driven wheels, it is hardly possible to go wrong, because the fraction shows at once which wheels should go on the mandrel and which on the guide screw.

COARSE SCREWS.

Say that a screw of three-inch pitch, and then one having thirty-five threads to the inch, have to be cut. For cutting screws coarser than one-inch pitch, it is necessary to introduce into the vulgar fraction the number of threads contained in that distance of the leading screw, which corresponds with the pitch or length of the screw to be cut. Here, having a leading screw of four threads per inch, and a screw to be cut of three-inch pitch, write $4 \times 3 = \dfrac{12}{3}$; (twelve represents the number of threads in the length of 3 in. on the guide screw). Then, add a cypher $\dfrac{12}{3} = \dfrac{120}{30}$ driver driven, and these will cut a screw of three-inch pitch.

FINE SCREWS.

To cut a screw of thirty-five threads per inch, write $\dfrac{4}{35} = \dfrac{40}{350}$ driver driven. But there is no wheel of 350 teeth, so factors for other numbers must be obtained. Thus, for example:—$\dfrac{40}{350} = \dfrac{5 \times 2 \times 4}{5 \times 7 \times 10}$, adding cyphers and cancelling $\dfrac{\overset{25}{\cancel{50}} \times 20 \times 40.}{50 \times \underset{35}{\cancel{70}} \times 100.}$ One ^{50}

above the line is cancelled, because there are not two 50-toothed wheels in a set, and to preserve the ratio, cancel 70, and have 35 below the line ; and then 25, 20, and 40 will drive, and 50, 35, and 100 be the driven wheels.

PROVING RATIOS OF CHANGE-WHEELS.

Mistakes will sometimes occur in working out the sizes of the wheels, especially in compound trains ; and it is therefore always safer to prove the accuracy of the results before beginning to cut the thread. There are two methods by which the wheels can be proved, each involving a reference to first principles. One has already been laid down. The same ratio must exist between the driver and driven wheels as between the guide screw and screw to be cut. Thus, taking the first example of the coarse thread, the product of $\dfrac{120 \text{ driver}}{30 \text{ driven}}$ must equal $\dfrac{12 \text{ guide screw}}{3 \text{ screw to be cut}} = 4,$ which is correct. Again, in the second example of the fine thread and compound train, $\dfrac{4}{35} = \cdot114$, and $\dfrac{25 \times 20 \times 40}{50 \times 35 \times 100} = \cdot114$. The other rule is :—Divide the driven by the driver ; or, if a compound train, the product of the driven by that of the driver. Multiply the quotient by the number of threads of the guide screw ; the result will equal the number of threads in the screw required to be cut. This is essentially a proportion or ratio sum, thus :—As driver : driven : : guide screw : screw to be cut. In the first example, a screw of three-inch pitch ;

$$\dfrac{\text{driven } 30}{\text{driver } 120} = \cdot25$$

$$\underline{\quad 4 \text{ guide screw}}$$

$$1\cdot00$$

$$\underline{\quad 3 \text{ inches} = \text{length of pitch}}$$

$$3\cdot0 \text{ inches} = \text{screw to be cut.}$$

In the second example :—

$$\dfrac{\text{driven } 50 \times 35 \times 100}{\text{driver } 25 \times 20 \times 40} = \dfrac{8\cdot75}{4 \text{ guide screw}}$$

$$\underline{35\cdot00 \text{ screw to be cut.}}$$

These examples will be sufficient.

MILLIMETRE PITCHES.

Millimetre pitches must now be referred to. Only the few possess lathes that have a guide screw of a metrical pitch; yet it is desirable sometimes to cut threads of millimetre pitches on ordinary English lathes. It is very simple, and is done as follows: The metre is equal to a length of $39\frac{3}{8}$ in.—more exactly, 39·37079 in., and this contains 1,000 millimetres. A leading screw of the same length of $\frac{1}{4}$ in. pitch contains $157\frac{1}{2}$ threads = 157·5, so that the ratio between the two stands thus: $\frac{157\cdot5}{1,000}$, or reduced, $\frac{157\cdot5}{1,000} \div 2\cdot5 = \frac{63}{400}$. With a 63 prime wheel, therefore, millimetre pitches can be cut in a lathe with a leading screw of $\frac{1}{4}$-in. pitch. The error caused by using such a wheel will amount to only $\frac{1}{340}$ of an inch in the metre length. This number 63 must be multiplied by the number of millimetres in the pitch of the screw. If the screw is of 3 mm. pitch, multiply 63 by 3; if of 10 millimetres pitch, multiply 63 by 10, the number 400, of course, remaining constant. Evolving the figures in the usual way, write in the first case named $\frac{63 \times 3}{400} = \frac{63 \times 30}{4,000} = \frac{63 \times 30}{40 \times 100}$. Since 3 mm. is finer than the pitch of the leading screw, the wheels 63 and 30 will drive and 40 and 100 will be the driven. In the second case named $\frac{63 \times 10}{400} = \frac{63 \times 100}{4,000} = \frac{63 \times 100}{40 \times 100}$; as there are not two wheels of 100 teeth in a set, alter $\frac{63 \times 100}{40 \times 100}$ to $\frac{63 \times 50}{20 \times 100}$; and since 10 mm. is coarser than the pitch of the leading screw, the wheels 63 and 50 will drive, and 20 and 100 will be driven. The proof of the correctness of the wheels lies in the fundamental rule already twice given. In the first case, that of $\frac{63 \times 30}{40 \times 10}$, the ratio of the driven and driving wheels, is $\frac{63 \times 30}{40 \times 100} = \frac{1,890}{4,000} = \cdot4275$ and the ratio of the guide screw and screw to be cut is $\frac{63 \times 3}{400}$

$= \frac{189}{400} = \cdot4275$; therefore the wheels $\frac{63 \times 30}{40 \times 100}$ are proved to be correct. In the second case, $\frac{63 \times 50}{40 \times 100}$, the ratio of the driving and driven wheels, is $\frac{63 \times 50}{20 \times 100} = \frac{3,150}{2,000} = 1\cdot575$; and the ratio of the guide screw and screw to be cut is $\frac{63 \times 10}{400} = \frac{630}{400} = 1\cdot575$; therefore the wheels $\frac{63 \times 50}{20 \times 100}$ are proved to be correct.

POSITION OF SLIDE-REST SADDLE IN SCREW-CUTTING.

The position of the saddle for beginning a cut is sometimes a matter of indifference; but often there is only one position at which it must be set for beginning a cut, as, for instance, in the case of screw threads that are neither multiples nor aliquot parts of the pitch of the guide screw, and also in the cutting of multiple-threaded screws. It is usual in these cases to make a certain mark upon the change wheels, or to place some kind of stop on some portion of the lathe as a guide for the starting of the saddle for each cut. An aliquot number will divide a given number without leaving any remainder; thus, 7 is an aliquot part of 21, 4 of 16, and so on.

DOUBLE AND TRIPLE THREADS.

In the case of double and triple threads, the mandrel wheel is chalked in two and three equal divisions of teeth respectively, and these marks are brought successively opposite the corresponding space in the driven wheel for the starting of the cut for the second and third threads respectively. The rocking plate of the lathe is lowered for the purpose.

RANGE OF CHOICE IN CHANGE-WHEELS.

There is much range of choice possible in an ordinary set of change wheels. A set (shown at the front of Fig. 1307, p. 457) usually numbers 22 wheels. There are 21 wheels, ranging from 20 teeth to 120, advancing by 5 teeth; and there is one duplicate, usually a 40 or 60, for cutting threads of the same pitch as that of the leading screw. Having obtained the required ratio between the driving and driven wheels, it is very necessary sometimes to be able to substitute some wheels for others in compound trains; either because two wheels having equal numbers of teeth occur in both drivers and driven, or because some wheels which came out in calculation do not happen to be included in the set. It becomes necessary to halve, double, or take away, or add fractional parts, and the ultimate result is not affected in the least, provided that the total products of the drivers and the driven wheels are altered in exactly the same ratios. Thus, in the case of $\dfrac{50 \times 20}{100 \times 20}$, as there are not two wheels of 20 teeth in a set, the numbers both above and below the line must be altered. Halving is impossible, because the wheels do not run so low as 10 teeth. Doubling gives either two of 100 teeth or two of 40 teeth, which may not be in one set. But two of the wheels can be increased by one-half thus:—

$$\frac{50 \times 20}{100 \times 20} = \frac{75 \times 20}{100 \times 30},$$

75 and 30 being substituted for 50 and 20, which are cancelled. In this way many new combinations are often possible.

PRIME NUMBERS.

In certain cases none of the change-wheels in a set is available for cutting the thread required. This happens in the case of some of the prime numbers, as these cannot be broken up into factors, and in these cases it is necessary to have a change wheel having a prime number of teeth. This does not often occur in ordinary practice. The prime numbers above 23 up to 100 are 29, 31, 37, 41, 43, 47, 53, 59, 61, 67, 71, 73, 79, 83, 89, and 97.

SPINNING METALS ON THE LATHE.

Lathe for Metal-spinning.

Metal-spinning is the art of shaping by pressure a quickly rotating piece of sheet metal. Sheet metal, being ductile, can be shaped by bending, hammering, embossing, etc. The lathe has to run at a fairly high speed. A wood-turner's 6-in. centre lathe is suitable, but it must be strong to resist the pressure; consequently the mandrel should rest in two bearings. The screw on the nose must be coarse and the

however, be turned out on an ordinary lathe. For heavy work the lathe may be driven by a belt 1 in. or 1¼ in. wide, the pulley having three or four speeds. If driven by power, of course the usual overhead gear, with fast and loose pulleys, is applied. A lathe driven by gut or cord is suitable for small work only. Fig. 1355 shows a section of the headstock of a spinner's lathe. The tailstock is of the usual design, but it should be long and accurately centred.

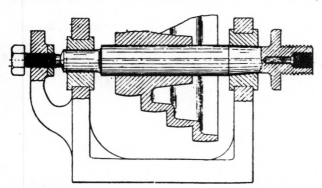

Fig. 1355.—Headstock of Spinner's Lathe.

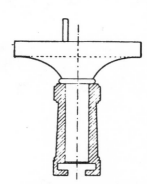

Fig. 1356.—Modified T-rest.

collar broad. It would be a great advantage if the mandrel were bored and screwed with an inside thread, in which could be fixed a short piece of hardened and tempered steel rod, itself tapped to take a ¼-in. steel rod; the reason for this will be made clear later. The bed should preferably be made of two well-seasoned beech planks 6 in. by 2½ in. and 2 in. apart; if possible, have two flat bands of iron 2 in. by ₁⁶₀ in. screwed on, with the inner edges quite parallel. Very good work can,

Professional Metal-spinner's Lathe.

The professional metal-spinner's lathe is a heavy one, generally of 8-in. centres, capable of being raised to 12 in. by means of four wooden blocks made for the purpose, and always at hand in case of need; the lathe of an oval spinner is never smaller than 12-in. centre. The mandrel of an ordinary spinning lathe is 2 in. in diameter, carrying a wide shoulder of 4 in., with a

2-in. nozzle of four threads to the inch. It is furnished with a back-centre of the ordinary type, very strongly made, and calculated to withstand a heavy strain

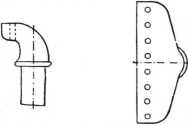

Figs. 1357 and 1358.—Parts of Modified **T**-rest.

when the slide is advanced some 8 in. or 9 in. towards the nose of the lathe head. The lathe has as well a hand rest some 8 in. in length at the top, and flat, being drilled with holes $\frac{1}{2}$ in. apart, in which round taper steel pegs are inserted in turn, to give the necessary leverage to the burnishing tool while the operation of spinning is in progress. The lathe is usually fixed up on a pair of $2\frac{1}{2}$-in. planks, set up

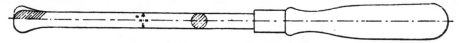

Fig. 1359.—Tool for Metal-spinning.

edgeways on firm supports let into the wall, and supported by legs, 4 in. by 3 in., at intervals; the surfaces of the planks have strips of sheet iron screwed on their faces the better to stand wear and tear. The lathe is driven from the overhead or underneath shafting, as may be most convenient. The spindle of the lathe is furnished with

Fig. 1360.   Fig. 1361.
Fig. 1360.—Tool for General Spinning. Fig. 1361.
—Tool for Staffs, Beads, etc.

fast and loose pulleys, and is driven by a 3-in. leather band from a 3-ft. pulley.

MODIFICATION OF HAND TOOL REST.

As the spinning tools bear on the work with considerable pressure, the tee-rest

should be modified (see Figs. 1356 to 1358). The tee-piece is furnished with a number of $\frac{1}{2}$-in. holes, $\frac{3}{4}$ in. from centre to centre and not more than $\frac{3}{8}$ in. from the front edge. A steel peg 3 in. long, and tapered

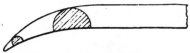

Fig. 1362.—Tool for Staffs, Beads, etc.

half of its length to fit easily in the holes without dropping through, can be shifted from hole to hole, and thus furnishes a movable fulcrum or resting-place for the tools.

TOOLS FOR SPINNING METAL.

The tools used besides the ordinary turning tools are of two kinds: spinning tools or burnishers, and finishing tools. The first are used for all metals that can be spun, the last only for such metals as will stand turning, namely, brass and German silver. Figs. 1359 to 1364 show a selection of burnishers for work of small and medium size, Fig. 1359 showing a complete tool. Tools Figs. 1359, 1362, and 1363 should be from 15 in. to 18 in. long; the other ones will be large enough if 12 in. to 15 in. long. Figs. 1362 and 1363 may be of $\frac{1}{2}$ in. square

Fig. 1363.—Tool for Staffs, Beads, etc.

section, or $\frac{5}{8}$ in. round. Figs. 1359 and 1360 are for general use; Figs. 1361 to 1363 are for staffs, beads, etc.; while Fig. 1364 is for smoothing larger surfaces. All tools should be highly finished on the working surface and then hardened, not tempered, so that the end is nearly glass-hard, while the other part is only moderately hard and not liable to snap. Figs. 1362 and 1363 should be tempered to sherry yellow about 3 in. along from the end. The other loose tools are few in number, consisting of wood-

Fig. 1364.—Tool for Smoothing Large Surfaces.

turners' gauges and chisels, inside chasing tool for mandrel nose (although blocks sawn to octagonal shape, and screwed to the regulation 2-in. nose, are now to be obtained at certain timber-yards), a smaller pair of screwing tools—say for ¾-in. holes for building up models when of unusual height—a small metal pot of oil, with which to lubricate the revolving

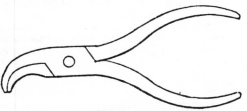

Fig. 1365.—Spinner's Beading Tool.

metal, a pair of 12-in. compasses, callipers, a straight-edge or a square, ordinary oil-can, Turkey oilstone set in a wood block long enough to lie across the lathe bed, and an assortment of seasoned beech blocks, squared and cornered ready for turning up.

Fig. 1366.—Tool for Finishing Spun Metal.

Keeping Spinning Tools in Order.

To preserve the tools, a piece of thick sole-leather 4 in. by 6 in. is nailed on a board with the rough side upwards; on this some putty powder is sprinkled. The tool is wiped free from oil and then the

end is rubbed to and fro to get a high polish; in a short time grooves will be worked in the leather, each fitting its particular tool. The brighter the tools, the easier and better is the work turned out.

Figs. 1367 to 1369.—Tools for Finishing Spun Metal.

Spinner's Beading Tool.

Besides the spinning tools proper, many spinners use a beading tool (Fig. 1365). This is shaped like a pair of pliers, but the

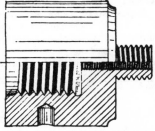

Fig. 1370.—Reducing Nozzle.

insides and edges of the jaws are rounded and smoothed so as not to tear the metal.

Tools for Finishing Spun Metal.

Brass and German silver can be finished after spinning by turning; the tools used,

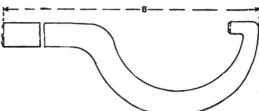

Fig. 1371.—Reducing Nozzle Chuck Key.

of ₅⁄₁₆-in. square steel, are illustrated by Figs. 1366 to 1369. The tools are ground down to a thickness of from $\frac{1}{32}$ in. to $\frac{1}{64}$ in. and the edge left square as at A (Fig. 1366). The length of each tool, excluding handle, is 9 in. to 10 in. Other tools, varying with the work, are used by the spinner, and for a power-driven lathe the tools are larger.

Fig. 1372.—Section of Spun Metal Tray.

MOULDS OR CHUCKS FOR SPINNING.

Either wood or metal can be used for the moulds or chucks. Spun articles are not often finished in one process; they generally require annealing, sometimes more than once, before being finished. The preliminary spinnings are almost always done on wooden chucks roughly turned to shape, but the last or finishing spinning is done on a chuck accurately turned to template. For the preliminary work any odd bits of wood, if large enough, will serve; and, generally speaking, when a chuck has done service for one article it may be altered for another. For cheapness, use common red beech, well seasoned; hornbeam, well seasoned, is best, however. For the finishing chucks, use hornbeam for larger work, and boxwood or metal for smaller. Lignum-vitæ is sometimes used. Boxwood generally meets all requirements of the spinner; when thoroughly well seasoned it will shrink very little, is easily turned, and is naturally smooth and hard. Metal chucks may be of cast iron, brass, or bronze. When using boxwood and metal for smaller articles, the screw on a headstock would necessitate rather large pieces of wood or large castings; therefore a reducing nozzle (Fig. 1370) is used. The central hole is tapped $\frac{7}{8}$ in.; in the cross-hole fits a key (Fig. 1371), by the aid of which the chuck can easily be fixed or removed.

ELEMENTARY METAL-SPINNING.

To assist the reader in forming an idea as to the means employed to press the metal to shape, a number of conventional illustrations (Figs. 1372 to 1377) are given. These are not drawn to scale, but show the appliances and suggest the processes employed in one method of spinning. Fig. 1372 is the section of a spun metal tray, the pattern, mould or chuck shown in Fig. 1373 being supposed to be turned to the same shape. The mould is secured to the headstock centre in a suitable manner. The metal is held as in Fig. 1374, by the wooden cone C, the surfaces of the metal B in contact with the mould A and holder C being treated with resin to afford a better grip. The holder is shown separately by Fig. 1375. The burnishing tool is brought to bear upon the metal as in Fig. 1376, the pin rest, Fig. 1377, assisting

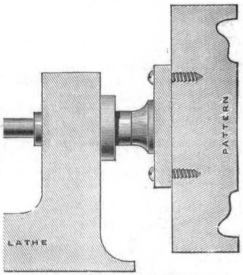

Fig. 1373.—Chuck Mounted for Spinning Tray.

in holding the tool in its correct position, and in obtaining the necessary leverage.

METALS USED IN SPINNING.

Of spinning metals, iron, zinc, and aluminium are generally stamped and are seldom used in any but rather shallow shapes; silver is hammered, so there remain copper and its alloys, brass, and German silver. Britannia metal is largely used for teapots, cruet-frames, candlesticks, etc., and is cast, stamped, and spun, but although worked with comparative ease, is hardly suitable to the novice, who may deal with copper and its alloys, brass, and German silver.

centre again, applies the various tools, in the first place very gently, until, by a succession of touches, the sheet is bent over the mould or chuck. The article is afterwards, in a similar manner, placed reversed against another chuck, and, by the application of a burnisher in each hand, the rounded, spherical, or other form is perfected. Some of these articles used to be partly formed in a die previous to the spinning; in others, in which the

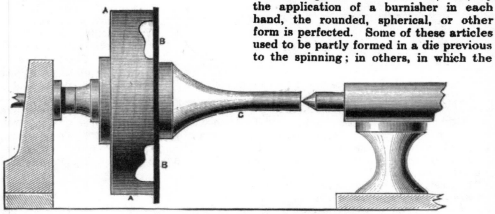

Fig. 1374.—Metal in Position for Spinning Tray.

THICKNESS OF METAL FOR SPINNING.

Britannia metal for spinning varies in gauge from six to eighteen; ten, twelve, and fourteen (sheet-metal gauge) are those most employed. The ingots into which the metal is cast are of standard weight and size, and are rolled from 10 in. to 18 in. wide in a continuous sheet, each width and thickness being suitable for a given piece of work.

SPINNING BRITANNIA METAL.

When spinning Britannia metal, a piece of the sheet metal is cut out with shears into a circular shape and of the size required. This is placed against the end of the revolving chuck; and against the metal outside is applied a circular piece of wood, having a small coned centre indented in it, in which the nose of the back-centre finds a place, and which is firmly screwed upon it. The metal being thus placed and rapidly spinning round, a piece of wood is then pressed against the edge with one hand, while the back-centre is slightly unscrewed, and the revolving sheet made roughly central. The spinner, having tightened up the back-

metal is made to overlap the mould, the mould is cut through with a parting tool to release the article, the lathe is stopped, and the fast ring of metal cut through, and put aside for re-melting.

Fig. 1375.—Turned Piece for Holding Metal whilst Spinning.

SPINNING HARDER METALS.

In the spinning of articles in the harder metals, such as copper, brass, German silver, and silver, the metal needs to be frequently annealed, and to be well lubricated while revolving in contact with the tools; otherwise the various processes are identical, though for hard metals greater care and skill are needed in the manipulation of tools; as, unlike the pliable Britannia metal, which grows soft under the friction of the burnishers, the harder metals grow still harder with a tendency to crack, which can be counteracted only by repeated annealings.

SPINNING COPPER.

Copper is one of the easiest metals to spin in the lathe, because it is pliable and can be annealed straight off when it becomes hard. The tool must bear on the metal with firmness, but it is best not to take very large feeds, but to mould the metal gradually. It is of great advantage to hold a piece of hard wood against the back of the blank, particularly in the

can be kept of the same thickness throughout. If the blank is fixed to the chuck by

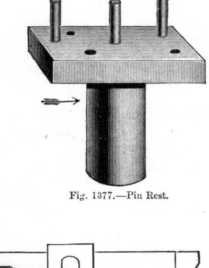

Fig. 1377.—Pin Rest.

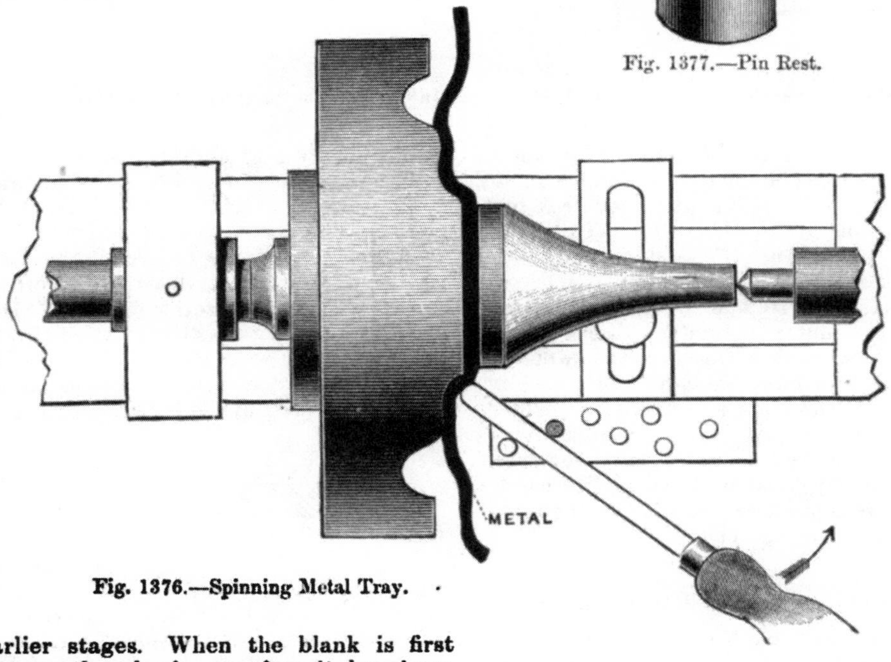

Fig. 1376.—Spinning Metal Tray.

earlier stages. When the blank is first put on the chuck, or after it has been annealed, it feels very soft and yielding, but after a short time it gets harder, and it is not wise to work it too hard. The tool should not be moved from centre to circumference only; that would tend to draw the metal away from the centre and make it thinner there and more liable to break. When the tool has travelled from the centre outwards, let it travel back again to the centre; in this way the metal

a screw through the centre, turn the chuck gradually during the spinning and anneal rather often.

CHUCK FOR SPINNING PEPPER CASTER MOUNT.

Practical metal-spinning may now be described. A very simple article to spin is the top of a pepper caster. It consists of the mount fixed on the bottle top, and the perforated lid, which fits tightly on the mount as in Fig. 1378. Make the mount first; here the outer sizes of the mouths of the bottles are guides. These vary, but if the difference in diameters is not more than $\frac{1}{16}$ in., one chuck will do; choose the largest bottle as a guide.

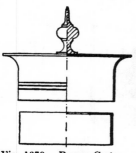

Fig. 1378.—Pepper Caster.

SPINNING PEPPER CASTER MOUNT.

Slip a suitable bung D on the rod c (Fig. 1379), then a metal blank E; put the rod in the hole in the chuck until the plug grips the rod, and put the lathe in motion. In Fig. 1379, and in all similar illustrations, either the upper or the left half represents

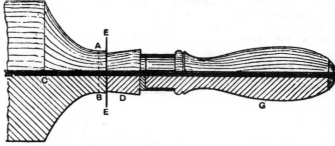

Fig. 1379.—Blank in Position for Spinning Pepper Caster Mount.

Rough down a bit of good hornbeam in the lathe to a diameter of about 3 in. by 4 in. long. Fix this to the headstock by a screw well up to the collar. Turn it to shape as in Fig. 1379, which shows the free end of the hornbeam chuck, the distance from A to B being a shade larger than the outer diameter of the caster mouth or neck. Through the wood is bored a hole large enough to take a $\frac{1}{16}$-in. steel rod, threaded at one end to grip in the loose steel plug which fits in the headstock. This rod should be well tempered and securely fixed in a strong file handle G by riveting; it serves to keep the metal blank fixed to the chuck.

BLANKS FOR PEPPER CASTER MOUNTS.

On a piece of No. 25 s.w.g. copper, brass, or German silver sheet, with compasses strike circles with a radius of $\frac{3}{4}$ in. for the mounts, and mark the centre by indenting it by a slight tap on the compasses. Cut out the blanks with shears. Punch the central hole by the tool shown in Fig. 1380, which is turned from $\frac{3}{8}$-in. steel rod 4 in. long, and hardened and tempered. By fixing the point of the punch in the indent on the blank the hole is made quite central. Punch on a thick lump of lead.

an outline profile, while the lower or right half shows a section. Place the tee-piece in position, its height being such that when the spinning tool is placed on it the knob of the tool is at the centre of the lathe. Dip a rag in oil or in a mixture of oil and tallow, and rub the rotating blank. Hold the tool firmly against the steel peg on the tee-piece, and with the left hand hold a strip of hardwood, roughly wedge-shaped, against the back of the blank, pressing the

Fig. 1380.—Steel Punch.

handle of the spinning tool gently to the right. The metal will give way, the tool describing an arc of a circle with the steel peg as a centre, and the metal slipping between the knob of the tool and the hardwood wedge. When the arc is completed the flat blank will be brought to A (Fig. 1381). On repeating the operation the blank is brought to B (Fig. 1381). The ragged edge of the blank is removed with a diamond-pointed tool for brass or German silver, or with a very sharp wood-turning gouge for copper. Finish the spinning by bringing the metal down close to the chuck, and with the tool shown in Fig. 1360 press the metal firmly and smoothly against the wood, to which the blank will cling. Now remove the bung and steel rod. While the lathe is still running, oil the front face of the blank and apply the or German silver; copper must be spun as smooth as possible, and afterwards simply finished by emery cloth and Sheffield lime. To release the mount, set the lathe running, fix a peg in the tee-rest where required, and press the hardwood wedge

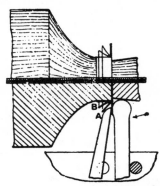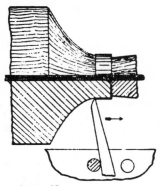

Figs. 1381 and 1382.—Spinning Pepper Caster Mount.

same tool, placing it at the lowest edge and moving it towards the centre, pressing hard against the chuck, thus obtaining a sharp angle and finishing the metal.

TURNING SPUN METAL.

Now with the diamond-pointed tool turn out a little disc of metal, leaving an opening of the same size as the opening of the bottle. In turning, draw the tee-rest a little way from the work and take the tool shown in Fig. 1366 or Fig. 1367; rest the fingers of the left hand with their inside upwards on the rest; make three fingers a rest for the tool. With the right hand on the handle approach the cutting edge of the turning tool to the metal surface and remove a very fine shaving. The way in which the tool is held insures great delicacy of touch; by holding the cutting edge obliquely against the metal surface, the beginner will soon find that he can remove an excessively fine, continuous shaving, which leaves the metal shining like a mirror; a little polishing with No. 0 emery cloth, and with Sheffield lime and blue oil, finishes the mount. As all turning of spun metal is done in the same way, the process will not require further description, but, of course, this turning applies only to brass

against the edge of the mount and from the headstock (see Fig. 1382).

PEPPER CASTER TOP.

The caster top is made in three pieces: the cylindrical part, the flat cover, and the ornamental button in the centre. The cylindrical part is made after the same fashion as the mount (as shown in Fig. 1383).

ANNEALING METAL DURING SPINNING.

To prevent the metal getting hard and unyielding, and possibly bursting before

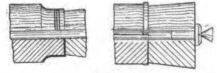

Figs. 1383 and 1384.—Spinning Pepper Caster Top.

it is spun home, stop when half-way, remove the metal and anneal it. Thus, for copper, put it on a clear fire or expose it to a blowpipe flame to heat it to a bright red; then throw it in pickle—that is, in a 5 per cent. solution of sulphuric acid, which will dissolve the oxide and leave the metal, after swilling and scouring, a fine dull pink. In the case of brass or German silver, previous to annealing, hammer out the tension. Take the hard metal, place

Fig. 1385.—Spinning Pepper Caster Top.

it on a beak-iron or on the horn of an anvil, and with a tinman's mallet cut to shape, deal it a series of blows which indent it all over; now anneal as above, but let the metal cool before putting it in pickle. Then, when dry, finish the work. When the metal is home, with the point of a diamond-pointed tool near the edge turn three equidistant lines, which serve as ornaments to break the monotony of the surface. With the same tool turn out the bottom, after removing the bung and screw, to complete the ring-shaped part. If the chuck is turned the least bit taper there will be no difficulty in getting the metal to leave it.

Spinning Pepper Caster Top.

As the flat top has no hole in its centre, the contrivance shown in Fig. 1384 is used.

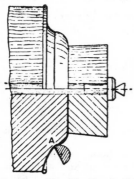

Fig. 1386.—Tray Fixed in Chuck with Bung.

In a bung with a central hole a steel plug fits tightly; the plug has a conical hole for the point of the tailstock. When the

chuck is turned to shape, place the **bung** as in Fig. 1384 and turn it a shade **smaller** than the chuck; place the blank centrally between chuck and bung and screw up **the**

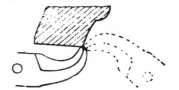

Fig. 1387.—Bending Edge of Tray.

tailstock. A spot of oil on the point will make the whole run quite smoothly. The narrow edge is spun half-way down, turned true, and brought home. Remove the bung and proceed with the front, the same as in Fig. 1381, to get a flat and smooth surface; turn three equidistant rings and a small central hole, say $\frac{1}{8}$ in. in diameter, to receive the turned knob. The small holes are drilled later on. Both the top and the ring are turned and polished before being taken off the chuck. To join these two, proceed as in Fig. 1385; the top is placed in a flat recess on a chuck, the ring put inside, and a bung similar to the one in Fig. 1384 presses the two firmly together. Turn the narrow rim of the top to a knife-edge, and with the tool shown in Fig. 1361 bring it down to the ring; if this is well done, it is hardly possible to see the joint. Turn the knob from hard-drawn brass rod, fix it in the central hole of the top, and either rivet it or fix it with a pellet of soft solder with the aid of a blowpipe. To drill the holes, put the caster top on a piece of wood and press it against a small twist-drill running in the lathe. The number of holes and their arrange-

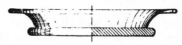

Fig. 1388.—Metal Tray with Wooden Bottom.

ment is, of course, entirely a matter for individual decision. The caster top is now ready, and can be plated or left as it is, according to taste. In the foregoing, the easiest operation has been described first, but in practice it is usual to start with the flat top, then to form the ring, and lastly the mount, because after each part has been finished the chuck can be turned down for the next shape and the whole spun on the same piece of wood without removing it from the headstock.

SPINNING METAL TRAY.

A small tray, like those used for whist-markers, can be made either wholly of metal or with a wooden bottom. Fig. 1386 shows the first, half in elevation, half in

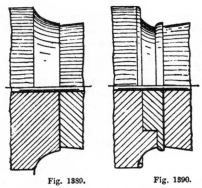

Fig. 1389.　　　　　Fig. 1390.

Fig. 1389.—First Operation in Spinning Tray.
Fig. 1390.—Finishing Spinning of Tray.

section. It must be held on the chuck by a bung as shown, and should be annealed once. To bring the metal home in the sharp corner A (Fig. 1386), use a tool in the manner shown. When the metal is turned true, grip the edge with the nose of the pliers, well oiled; while the metal slips round, move the pliers steadily through a quarter of a circle, when the edge assumes the shape in Fig. 1387. To finish the bead, use the tool shown by Fig. 1361, or the plane face of Fig. 1363, finishing round, not flat. For finishing inside, turn a recess in a piece of wood after the manner of Fig. 1385, then turn and polish.

METAL TRAY WITH WOODEN BOTTOM.

For a tray with a wooden bottom, choose mahogany or walnut $\frac{3}{16}$ in. thick and either French polished on both sides or with one side covered with green baize. Fig. 1388 shows the tray in elevation and

section. First spin the metal as in Fig. 1389. For narrowing (Fig. 1390), the chuck is made in two parts which fit together, and by spinning the metal home to the surface of the chuck the front piece of the

Fig. 1391.—Metal Egg-cup.

chuck is encased in the metal. The diamond-pointed tool applied at the edge turns off a round blank which, when unscrewing, will come off, together with the front part of the chuck. The bead is formed before removing the bottom blank, as in the previous case. To fit the wooden bottom, put the polished wood between a chuck turned flat and covered with wash-leather, and a bung treated similarly; bring up the tailstock, and the wood can be turned with a sharp turning chisel. Place the metal ring on the chuck as in Fig. 1390, with the front part removed; put the wood in place, and the bung with wash-leather to it; bring up the tailstock, and the metal can be fitted tightly over the wood. Of course, if the tray is to be

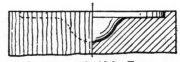

Fig. 1392.—Mould for Egg-cup.

plated, that must be done before mounting the wood. Polished copper might be left bright from Sheffield lime, but would then have to be lacquered, or coated with transparent celluloid varnish.

SPINNING METAL EGG-CUP.

The egg-cup (Fig. 1391) is essentially a miniature goblet made of two parts, the

bowl and the foot. A stand for three
or more cups can also be spun. The bowl
being rather deep is not spun up directly
on the chuck. A flattish piece of beech
(Fig. 1392) is fitted on the lathe, and a recess is turned in it about ⅛ in. deep, and

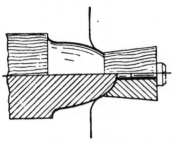

Fig. 1393.—Egg-cup Bowl on Chuck.

just large enough to hold the blank. Now
in the chuck turn a hollow with rounded
edges of about one-third the depth of the
bowl of the egg-cup and rather larger.
Put the tee-rest in position, press the
blank in the recess, and, while the lathe
is running rapidly, place the knob of the
tool shown in Fig. 1360 in the centre of the
blank; give light pressure and move the
knob upwards till the wood of the chuck
is felt. Return with the tool to the centre
and continue thus till the wood of the
chuck is felt all over; then release the
blank. Next make a chuck for finishing
the bowl. Select a piece of sound, well-
seasoned hornbeam, fix it and turn it to
shape to a template of thin zinc. Fig.
1393 shows the chuck carrying the blank
and also a bung, turned hollow where it
presses against the blank and fitted with
the steel centre used with the dead centre.
Before screwing up the dead centre, set
the lathe going, and if the blank runs
"drunken," true it by holding a tool or
a piece of wood against its face and pressing gently. Screw up the centre, lubricate the blank, and spin as previously
described. When the metal goes hard,
release, anneal, pickle, and finish spinning.
The edge is rounded into a bead and the
bowl finished by turning and polishing.
The outside can be finished on the chuck
used for spinning the bowl; for the inside,
chuck a piece of beechwood and turn a

hollow to hold half the bowl lightly. To
assist in soldering on the foot, turn a small
circle at the bottom of the bowl.

SPINNING FOOT OF EGG-CUP.

As the foot of the egg-cup is small, use
the small chuck-holder (Fig. 1370) either
with boxwood or brass for the finishing
chuck. Figs. 1394 and 1395 show stages of
the process, and as the metal will require
perhaps two softenings it is well not to
turn the chuck to its proper profile till
after the last annealing. The foot is nar-
rowed into a neck, which can be formed
in either of two ways. After the last annealing, turn the chuck with the profile
A B C (Fig. 1395) and spin home the metal
between B and C that is to form the staffs
and bead, but leave the part between A
and B untouched. Next turn the chuck to
the profile in Fig. 1396 and put the metal
on again; the bottom part now fits as before, but an annular space is left between
the chuck and the wood. Now apply the
round face of the tool shown in Fig. 1363
underneath the work as if about to prise
the whole upwards. Start from B (Fig.
1395) and move towards A, taking small
feeds each time; on part of the chuck the
metal will be pressed home. Turn out the
small bottom and thin to a knife-edge to
make a neat joint with the bowl; then
turn and polish it.

SOLDERING FOOT TO BOWL OF EGG-CUP.

To solder the foot to the bowl, tie the
two together with wire, the circle turned
on the bowl serving as a guide; drop

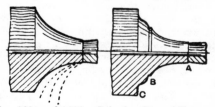

Figs. 1394 and 1395.—Spinning Base of Egg-cup.

inside a little soldering liquid and a pellet
of soft solder, and apply the blowpipe till
it flows. Any solder sweating out is easily
removed by placing the egg-cup on the
bowl chuck, steadying by a bung and tail-

stock, and applying an outside turning tool, emery, and Sheffield lime.

For the second method consult Fig. 1397. The loose front part of the chuck should be of brass, and joins the fixed part at the narrowest point. When the small bottom has been turned out the loose part of the chuck remains ready for another foot. For all the above-named articles use metal of a thickness of No. 25 or No. 26 s.w.g., except for the bowl of the egg-cup, which ought to be one number thicker.

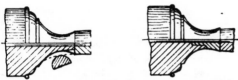

Figs. 1396 and 1397.—Two Methods of Finishing Egg-cup Base.

CHUCK FOR SPINNING SPILL-CUP.

A cup to hold pipe-lighter spills may have a slightly concave bottom, be 5 in. in height, and 2 in. in diameter at the bottom, increasing to 3 in. at the top. Turn up a piece of dry beechwood to the above dimensions and smooth it off nicely. Fix a piece of wood 1 in. thick tightly to the end of the chuck by means of the back-centre, and turn the outside slightly convex, so as to fit in the concave bottom of the chuck. Then reverse the piece, forming a permanent centre at the same time, putting a spot or two of oil in the centre to relieve friction, and turn off, continuing the taper in line with the body of the chuck; try this with square or straight-edge, and correct any inequality which may appear.

SPINNING SPILL-CUP.

To introduce the sheet of metal, which has been cut to circular form of some 11 in. or 12 in. in diameter, first remove the end piece of wood, placing the sheet between it and the chuck bottom, and with a piece of wood between the rest and the edge of the sheet make it central. This operation

should, at first, be done while the lathe is at rest. The more experienced operator may do it while the lathe is running; but a nasty accident may occur if the sheet of metal flies out of the lathe. When the truth of the metal has been ascertained, set the lathe at full speed, and apply a chisel to the edge to make true, and gradually increase the pressure, by

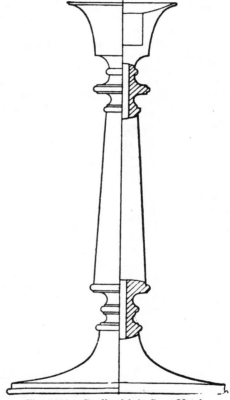

Fig. 1398.—Candlestick in Spun Metal.

means of the back-centre, on the bottom of the spill-cup which is being spun This pressure will form the slight hollowness requisite for the cup to stand firmly. Then place the rest so that it will just escape the contact of the metal, put two pegs in consecutive holes, and, with a semicircular movement of the curved end of the round burnisher above the centre, endeavour to make the metal lie on the body of the chuck. After this has been done, move the rest so that it is parallel with the chuck, and with the spade tool, rounded side uppermost, proceed to flatten the metal to the mould below the centre. If this is done quickly, the metal, which has become hot with friction, can be manipulated as easily as if it were paper. In holding the burnisher, use it as a gouge would be used in roughing out, and with exactly the same movement—that is, above the centre, and with a circular to-and-fro movement, gradually pressing down the metal as the burnisher is shifted peg by peg. In actual practice, the spinner makes a sweep from end to end of the article in a succession of to-and-fro movements, but this is difficult for a beginner. The cup now having received its form, cut

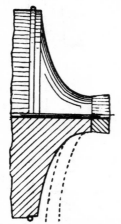

Fig. 1399.—Spinning Candlestick Foot.

off the excess of metal from the top, and polish with soft leather and whiting, and finish with another leather and rouge.

MAKING CANDLESTICK.

The candlestick (Fig. 1398) can conveniently be divided into foot, stem, vase, and two necks; the foot, the stem, and the vase are spun, the two necks partly spun, the bigger parts being cast in brass and turned. The foot and the vase, with its accessories, are spun out of flat blanks, and the stem is cut from the sheet, bent round a mandrel, and hard soldered. It is

best to mark the outlines with a needle point from a zinc template laid on the sheet metal. Cut the metal with the shears and bend it round a conical mandrel with the aid of a tinman's mallet, taking

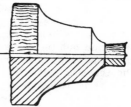

Fig. 1400.—Spinning Candlestick Vase.

care that the two edges of the joint are on the same level.

HARD SOLDERING CANDLESTICK STEM.

To make a perfect joint to the tube, fix a flat smooth file between bits of sheet lead in a vice, slip the slightly opened joint over the file, and draw it a few times forwards and backwards while gently pressing the edges against the file teeth. This also roughens the surfaces to hold the solder better. Anneal so as to remove all grease and dirt which may accumulate whilst working, and after pickling, tie the edges well together. Apply hard silver solder inside the tube, so that a slight ridge, which can be filed away, will appear on the outside. Solder applied outside is apt to show as a line, which is very unsightly. To finish the stem it must be either hammered very gently and accurately on a

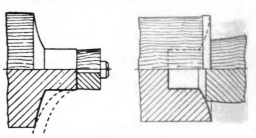

Figs. 1401 and 1402.—Spinning Lining of Candlestick Vase.

mandrel with a wooden mallet, or else put on a turned mandrel rotating in the lathe and spun by the tool shown in Fig. 1364, and finished in the usual way by turning and polishing.

SPINNING CANDLESTICK FOOT.

The spinning of the foot is illustrated, with two intermediate stages, by Fig. 1399.

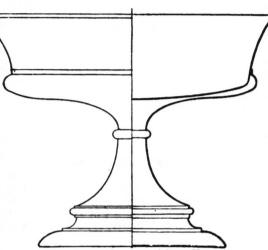

Fig. 1403.—Ornamental Spun Vase.

Move the spinning tool not only from the centre to the circumference, but also in the opposite direction, so as to force some of the metal towards the centre, otherwise the metal is apt to become too thin in the narrow neck of the foot, which not only weakens it but also makes it liable to break during the spinning process. When the work is finished, remove the bung and screw, and turn out the small bottom to a circular opening in which the small neck can be fitted tightly and soldered.

SPINNING CANDLESTICK VASE.

The vase consists of two parts, the vase proper and the lining. The first can be spun with the aid of the screw, but as the lining must have no hole in it, a bung should be held in position by the tailstock. Furthermore, as the rim of the lining is laid over the rim of the vase, it will require two chucks. Fig. 1400 shows the spinning of the vase, and Figs. 1401 and 1402 show stages in the spinning of the lining. The vase must have a small bottom turned to

fit the brass neck, and is joined to the lining as in Fig. 1385, using the chuck shown in Fig. 1401 to fix the lining and pressing the vase home with the tailstock and a bung turned to fit the bottom part of the vase. The two necks must be turned so that each has a cylindrical part at each end, as shown in Fig. 1398.

LOADING AND FINISHING CANDLESTICK.

It is advisable to bore the necks out with, at smallest, a ¼-in. drill, because when finished the candlestick ought to be filled with plaster-of-Paris, pitch, or resin mixed with a little dry sand. While the filling is liquid, drop an iron wire right through ; this filling gives strength to the candlestick and prevents it being top-heavy. The candlestick must be suspended bottom upwards, so that only the proper quantity of liquid is poured in, and kept level. When it is set, a piece of green baize is glued on and trimmed, when the candlestick is ready. No. 22 or 23 s.w.g. metal should be used for the spun portions of this job.

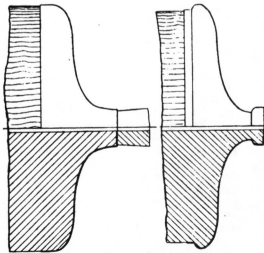

Figs. 1404 and 1405.—Spinning Base of Vase.

SPINNING ORNAMENTAL VASE.

The next piece of work to be described is an ornamental vase (Fig. 1403), and it is assumed that only one article is to be made; for more than one, complex chucks would be necessary. The vase is made in three pieces, of which the bowl may be spun similarly to the bowl of the egg-cup —that is, half deepened, half spun up. It must be turned and polished inside and outside, as far as it is not covered. The foot is in two pieces, made in a similar way. In Fig. 1404 the metal is shown spun home on a chuck; anneal the metal and turn the chuck to shape, put on metal, and bring it well home to Fig. 1405. The metal cannot be removed from the chuck without destroying the latter. When liberated from the chuck the upper part of the foot should grip the bottom of the bowl, on which it is fixed by the aid of a few pellets of tinman's solder. The lower part of the foot is also made to fit the chuck tightly,

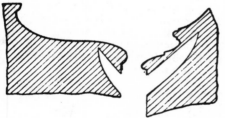

Fig. 1406.—Cutting Away Chuck from Vase Base.

and the wood destroyed by turning part away. Fig. 1406 shows this; the ring-shaped piece of wood which remains inside the metal is easily dug out with a joiner's chisel or sharp pocket-knife. It will be seen that the two parts of the foot differ at the point where they are to be joined; Fig. 1407 shows the reason. When the upper part has been spun home and finished, the small bottom is turned out as usual, and the wood filling that part is turned away. The round head on the bottom part of the foot just fits inside the upper part, and if pressed home by a suitable bung and the tailstock, the metal can be folded round to make the two pieces one.

SECTION CHUCKS.

Sometimes the shape of the neck of a vessel is such that once spun it could not be got off the chuck or mould. In this and similar cases the chuck is made in two portions, the top and the bottom. The bottom portion is bored with a round hole. The top portion has a projecting piece

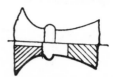

Fig. 1407.—Jointing Base of Vase.

turned to fit accurately the recess in the lower portion, and then the mould itself is turned to shape. The blank is then slipped on in the usual way, secured by a piece of wood as has been described, spun to the form of the block, and then cut through top and bottom ready for taking off. In this case the top portion of the mould comes away with the spun article; it is easily knocked out, and the process is repeated until the series is completed.

SECTION CHUCK FOR SPINNING JUG.

A section chuck in wood, suitable for spinning a jug, say, of silver, may be made as follows:—Fix a piece of hornbeam of

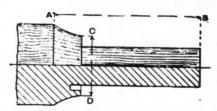

Fig. 1408.—Part of Section Chuck for Jug.

the requisite size on the mandrel and turn it to the shape of Fig. 1408; A B is the height of the jug, C D the diameter at its narrowest point, and A C the profile of its upper part. The diameter of the long cylindrical part C B should be as large as possible without weakening the chuck. Next join a number of wedge-shaped pieces of hornbeam, as shown in Fig. 1409; one of the wedges, marked 1, should be so shaped that its broadest part turns away from the outside, while the opposite is the case with the other wedges. The joints must be perfect, and are best finished on their joining surfaces with a toothed plane, being so glued together that a piece of brown

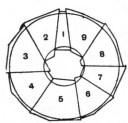

Fig. 1409.—End of Section Chuck for Jug.

paper is inserted between each pair of wooden surfaces. Join 1, 2, 3, 4, and 5 together; next 6, 7, 8, and 9. It will now be seen that if the free surfaces of 1 and 5, and 6 and 9, are lying in one plane, the last joining will be fairly easy to accomplish. The better plan is to make a drawing, plan down the shape of the wedges, and work accordingly. When all are joined and dry, chuck the roughly cylindrical piece; bore it out, and turn a ring on one end which will fit nicely in the annular recess shown at D (Fig. 1408), the cylindrical part C B fitting tightly in the hole bored without forcing the wedges from one another. When this is accomplished, the chuck can be finished to template as Fig. 1410.

Using the Section Chuck.

Now separate the wedges, first marking them with lead pencil so as to secure their

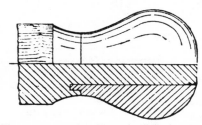

Fig. 1410.—Section Chuck for Spinning Jug.

proper positions. Remove the loose part of the chuck, insert a thin knife blade in

any of the glued joints, and tap gently with a mallet on the back of the knife. The wedges, owing to the brown paper inserted between them, can easily be separated; and when they are placed on the fixed part of the chuck in their proper rotation, they will appear like one entire piece. When the metal has been spun home and is removed from the lathe, it is evident that all the wedges are inside the bowl of the jug; but when this is released from the fixed part of the chuck, piece 1 (Fig. 1409) can be pushed towards the centre and drops out, the other pieces following. Take care that none of the

Fig. 1411.—Spun Kettle.

wedges is of larger transverse dimensions than will permit of its passing easily through the narrowest part of the jug's neck; a drawing of the sections should be made before joining them together.

Spinning Kettle.

The designs about to be given are suitable for tea-kettles that can be placed on a trivet or suspended on brackets. The kettle shown by Fig. 1411 is made up of a body of three pieces, A, B, and C, besides the lid, which drops into place without being hinged. The bottom part C is spun in one piece, fixed on the chuck with a bung and dead centre, and ought to be at least No. 20 s.w.g. thick. The bead is finished as in the lower part of Fig. 1412. The upper part of the body may be made in the same way, but many workers would probably prefer to make a truncated cone

Fig. 1412.—Producing Bead on Kettle Body.

of sheet-metal brazed together as shown by the dotted lines in Fig. 1413. Then turn a large cup-shaped chuck, and " lift " the metal till it has assumed the shape indicated in Fig. 1412. Polish these parts before they are put together. If the kettle is not to be electro-plated it must be tinned inside, which is most profitably done at this stage.

Joining Top and Bottom of Kettle Body.

To join the two parts together after tinning, proceed as shown by Figs. 1412, 1414, and 1415 ; Fig. 1412 shows the bottom part ready to be put into the top part ; for Fig. 1414 the parts are forced together with a bung and dead centre ; for Fig. 1415 they are spun together. To make the joint watertight, go over it inside the kettle with a soldering iron and soft solder, any solder showing on the outside being thoroughly cleaned away and polished.

Rim, Lid, and Handle of Spun Kettle.

The rim and lid offer no difficulty ; the rim is, of course, soldered on. The handle

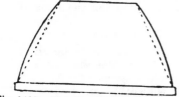

Fig. 1413.—Forming Upper Part of Kettle.

is in three sections ; a central part made of a bad conductor of heat, such as ivory, ebony, or china, and two equal uprights of cast-metal or tubes of sheet-metal. The

three are generally joined together with an iron rod which goes through the centre of the middle part ; this is fixed with screws in holes tapped in the two uprights.

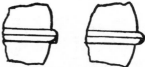

Figs. 1414 and 1415.—Producing Bead on Kettle Body.

The complete handle is fitted to the body of the kettle as shown in Fig. 1411, and either fixed by soldering or with short studs of brass rod threaded and tapped into the uprights and held firmly on the kettle by nuts inside.

Another Design of Kettle.

The kettle shown in Fig. 1416 would look well if electro-plated. The three parts of the body are joined by broad bands of fancy wire, and can be spun from flat blanks. For the largest band, cut a sufficient length and join it by hard solder to make a ring ; true it by hammering on a beck iron with a wooden mallet. Fix in a chuck and turn two recesses inside, as

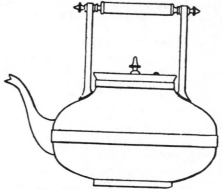

Fig. 1416.—Another Spun Kettle.

shown in Fig. 1417, which also shows how the wire joins the parts together ; they drop into the two recesses, and are soldered inside. The rim is joined in a similar manner, but, although the fancy wire ought to be of the same pattern as the large ring, it should not be more than two-thirds of its width. The lid is assumed to

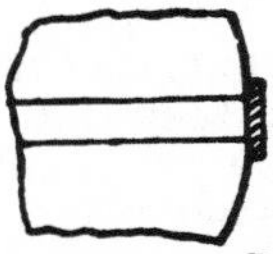

Fig. 1417.—Section of iron Band on Kettle.

be flat and hinged. In the handle the iron rod goes through the uprights, and is kept in place by the two nuts outside the uprights.

MAKING KETTLE SPOUT.

To make a neat-looking, well-fitting spout requires considerable skill. It is often advantageous to stamp the spout in two halves, remove the surplus metal with a piercing saw, file flat, and hard solder the two halves together; this secures uniformity. Another way of making a spout is illustrated by Figs. 1418 to 1420. Cut sheet metal (Fig. 1418) to template, the shape varying according to that of the

ness to the finished article, and by hammering with light and properly shaped hammers on the stakes, shown in Figs. 1421 and 1422, it becomes still more

Fig. 1420—Making Kettle Spout.

shapely. These stakes are fixed in a hole in the work-bench or in the jaws of a big bench vice, Fig. 1421 being used for the convex part, and Fig. 1422 for the concavo-convex.

FINISHING KETTLE SPOUT.

Further to finish the spout, it is filled with molten lead, and, while held in the left hand, hammered with very light hammers, planishing out all irregularities till it feels smooth. To bend the straight part, the spout is placed on a beak-iron, and by striking a few smart blows with a wooden mallet on the end, as shown in Fig. 1420, the final swan-neck shape is obtained. It is gone over slightly with a smooth file

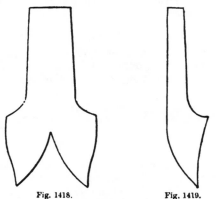

Fig. 1418.   Fig. 1419.
Fig. 1418.—Pattern for Kettle Spout.
Fig. 1419.—View of Kettle Spout Unfinished.

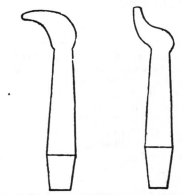

Figs. 1421 and 1422.—Stakes for Shaping Kettle Spout.

finished spout; bend the straight part into a tube over a mandrel and make the curved edges meet by bending, hammering gently, and coaxing till they are in contact as in Fig. 1419. Tie this up and solder with silver solder. The spout has a rough likeness

before being sent to the polisher to be bobbed after the lead has been melted out. If the template is true very little fitting should be necessary; though, at all cost, the spout must fit close to the body to make a tight joint.

FIXING KETTLE SPOUT.

To fix the spout, hold it in position on the kettle and mark all round it with a fine scriber. Fasten a drill in the lathe and drill a series of holes so close together that they overlap inside the mark made on the kettle, when a blank will fall out, leaving a ragged hole; cut and file this raggedness away so that an annular space

Fig. 1423.—Knurls.

⅛ in. wide is left inside the scribed mark. Moisten the spout with soldering liquid, put a few pellets of soft solder on the inside edge, and, with a blowpipe, fix these without making them run—that is, prevent their falling from their places when the spout is tied on. Secure this now very firmly with tying wire, moisten the joint with a drop of soldering liquid, and move a thin, keen blowpipe flame round the joint till the spout is firmly fixed. Then clean, and polish with a calico mop and Sheffield lime.

BUILT-UP ARTICLES OF SPUN METAL.

Occasionally, an article may have to be built up in two or more parts. Thus, a certain shape of tea service is spun in two parts, or, if the lid is counted, in three parts, the belly being spun from the flat sheet, while the neck is cut from a template, joined at the seam, and then spun. The belly is rounded up in the lathe from the nearly flat bottom to the straight side at right angles; this is so for the convenience of the fluter, who otherwise could not get the finished article into the press without serious damage; but the partly shaped belly, with its nearly straight sides and rounded bottom, gives a general shape which will most nearly fit his dies. Therefore, when the article has passed through the fluter's hands, it goes back to the spinner, who "rolls" in the top and shapes it for the addition of the neck, which,

when soldered to the belly, forms a graceful and well-balanced body.

OVAL SPINNING.

Of oval spinning there is little to be said, for the same rules apply to oval as to circular work; but in oval spinning it is specially important to maintain the burnisher always in a central position. The lathe is somewhat different, a double-oval lathe being sometimes used—that is, a lathe having a movable oval back-centre, so that oval work may be chucked and spun the same as circular work. Some oval spinners dispense with the double-oval lathe, and use, instead, simple but ingenious chucks.

SPINNING SPHERICAL VESSELS.

Vessels intended to be perfectly spherical are executed only in halves, and the bodies of those teapots which are of a globular form, or which exhibit concentric or circular swells, are also made by the spinner. In the performance of this work it is rather by great manual dexterity than by means of intricate machinery that the workman produces such rapid transformations of the tractable substance on which he operates.

KNURLS.

Knurling is a method of ornamenting circular projections, such as the raised rim of any fancy article, the head of a

Fig. 1424.—Holder for Knurls.

set-screw, small handles, etc. A knurl is a steel wheel, shown by Fig. 1423, revolving in a forked holder (Fig. 1424), and having on its periphery a pattern, sunk or raised. Fig. 1425 (p. 497) represents a selection of the exceedingly numerous designs in common employment, for each of which a separate wheel, though not a separate holder, is required. In making knurls in the lathe, shaped strips of hard steel (master patterns) are moved up and down, while being applied with pressure

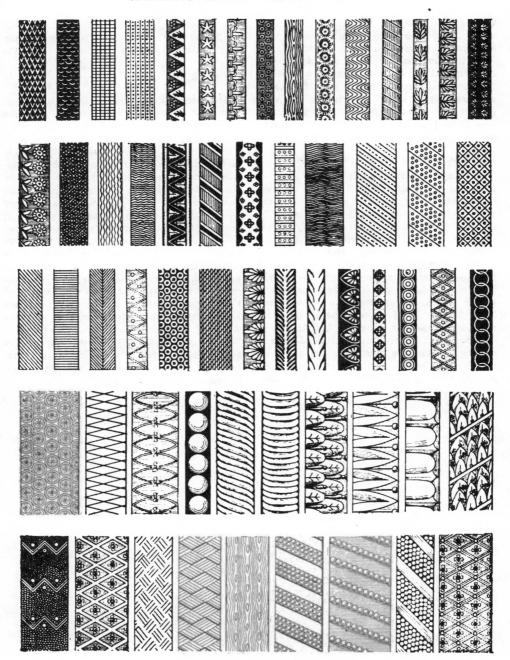

Fig. 1425.—Designs Produced by Knurling.

to the periphery of the knurl blanks or discs until the design is satisfactorily transferred; this work need not be gone into further, for but few craftsmen will care to make their own knurls, which can be obtained already made in such extensive variety. Knurling wheels are from $\frac{1}{2}$ in. to $\frac{3}{4}$ in. in diameter, and generally less than $\frac{1}{4}$ in. in width, but occasionally very much wider.

ACTION OF KNURLING TOOLS.

Holtzapffel gives the following instructions on the use of the knurling tool on soft metal. It is held against the revolving work, with which it is placed in forcible contact by pressure, and by lowering the handle; it imprints its pattern partly by compression and partly by nibbling away or abrading the sides of the indentations it forms on the work. The compression exerted first causes a series of slight indentations somewhat like the teeth of a cog-wheel, but the teeth or ornament upon the knurling tool not being of a form admitting of purely rolling contact, a scraping or abrasive action then begins; this secondary action is enhanced by the difference in the angular velocities at which the work and tool respectively travel, the one being generally of much larger diameter than the other.

USE OF KNURLING TOOLS.

In using the knurling tool the pressure is at first given moderately, and then it is gradually increased until the perfect pattern appears. Then the tool is withdrawn, as its further use would unduly reduce the diameter upon which it operates, and, if continued, blur or disfigure the pattern. Similar tools are used upon iron and hard brass; upon the latter they will endure a considerable amount of work, but they are quickly worn out upon iron or steel; separate tools are therefore reserved for iron and for brass, as a worn tool makes but little progress even upon the softer metal. The knurling tool is sometimes applied to wood, much of the German toy turnery from the Black Forest being very effectively ornamented with it; also, it is employed in England for ornamenting porcelain and terra cotta; the wheel is then called a "runner," is 1 in. or $1\frac{1}{2}$ in. in diameter, and is made of very hard wood.

TOOLS FOR MEASURING AND TESTING METALWORK.

TOOLS DESCRIBED IN THIS CHAPTER.

THE tools used by metalworkers are, roughly, of two kinds—those used for shaping, principally by removing material, and those used for marking, measuring, and testing. It is with the second of these classes that this chapter deals, the tools described being surface plates, straight-edges, squares,

SURFACE PLATES AND LINING-OFF TABLES.

Surface plates and lining-off tables are made in various forms and sizes. The lining-off table is used by the metalworker for the same general purposes as the joint board is used by the woodworker. Work is marked out on the lining-off table, and small work is built up on it, or is taken to it for checking the accuracy or otherwise of

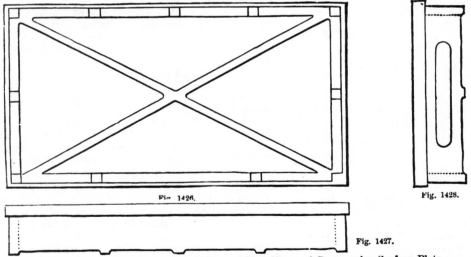

Fig. 1426.

Fig. 1428.

Fig. 1427.

Figs. 1426 to 1428.—Underneath, Side and End Views of Rectangular Surface Plate.

bevels, callipers (both inside and outside), dividers, compass callipers or moffs (probably a workshop corruption of hermaphrodite), trammels, telescope gauges, depth gauges, v-blocks and scribing-blocks.

lines and centres. The only legitimate use of the surface plate is the testing of faces and edges to which it is desired to impart the highest possible degree of accuracy. The functions of the two being distinct from one another, they are prepared by different methods. In their general form they are similar, consisting in each instance of a broad plated surface, well stiffened with vertical ribs underneath. But while the lining-off table is simply planed over, and so left, the surface plate is planed, filed, and scraped; the former is approximately true only, but the latter is absolutely so. The general truth of the broad area alone of the former is of importance; in the latter every minute localised section of area is in the same absolute

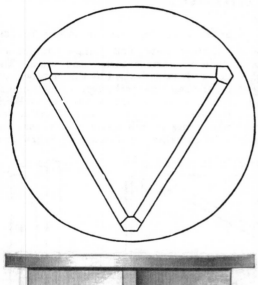

Figs. 1429 and 1430.—Underneath and Side Views of Circular Surface Plate.

plane with the rest. The former is easily and quickly prepared, the latter involves the labour of many days, perhaps of weeks. In making a surface plate, unless the aid of a true surface plate already in existence is available, it is necessary to perfect accuracy that three surface plates be made.

Shapes of Surface Plates.

For all work of large and moderate size, surface plates are made of rectangular and oblong forms (see Figs. 1426 to 1428): for very small work they are often circular

(Figs. 1429 and 1430). In each case, patterns have to be made in such a manner that the face of the plate is cast downwards. It is best for this purpose that the stiffening ribs be only dowelled upon the plate, and they will be thinner upon the top edge than at the bottom. If the plate is large, diagonal stiffening ribs (Fig. 1426) will be required in addition to those which run round the edges. If the plate is small, say 15 in. or 18 in. long, hand-holes should be cast in the two end ribs to lift the table by. If large, say 2 ft. or more in length, handles of wrought-iron are screwed in, as at B (Fig. 1431); but such plates are only used in big workshops. The design shown in Figs. 1426 to 1428 will answer very well for plates up to about 24 in. long. The illustrations are to scale, so that they will answer for a plate of any smaller size, down to, say, 8 in. by 6 in., which is about the smallest that it is worth while to make. The oblong plates are the most useful. The rectangular surface is handier than the circular, and the edges afford a convenient base for the trying of the square across. But many small circular plates are made, and they are commonly of the form shown by Fig. 1429, the three legs enabling them to stand steadily on an uneven bench. All surface plates, whatever their shape, should stand preferably on three legs only.

Making Surface Plates.

Having received the castings, grind over the surface and edges to see if they are sound before spending further labour upon them. Then if they can be planed over in a planer or shaper, so much the better. If not, work on them with a coarse file and straight-edge until the surface of each is reduced to a fair level condition. If the surface plate is small, the file may follow immediately after grinding. If it is of moderate or large size, and in any case if its surface is uneven and rough, it must be chipped before the file is employed. The chipping should be begun with a cross-cut chisel about ¼ in. wide, cross-hatching the surface with a series of shallow grooves. Test the truth of the grooves from edge to edge with a straight-edge, and afterwards remove the metal between the grooves with an ordinary broad-faced chipping chisel. When chipping near the edges, do not cut outwards towards the edges for fear of breaking them, but strike inwards. Test the general level from time to time with a straight-edge, and get it as accurate as possible with the chisel, because it is quicker and easier to chip than to file. It will also be found that by cross-hatching it is much easier to get below the hard skin, and a truer surface can be obtained in less time than by using the broad chisel in the first place.

TRUING SURFACE PLATES.

In the first rough stage of filing, when the chisel marks are being obliterated and the surface brought to a general level free from winding and curvature, special precautions are not necessary except testing with straight-edges. When the finer files are brought into use, they will have to be applied more precisely, and red-lead paste will have to be smeared over the straight-edges, and transferred thence to the surface of the plate. The plates are tested by mutual contact when the superfine file and the scrape come to be used. Each plate is made as true as possible, quite independently of the others, and then each is tested by the others. The plates, to all appearance true when tested singly by the straight-edge, will be proved untrue when brought together in succession with red lead on their faces; only a small portion of the colour will be transferred from one to the other. From the higher-coloured portions, material must be removed with the point of the fine-cut file, and this alone will occupy considerable time. Later, when the colour becomes more equally distributed, it will not be easy to localise the action of the file, and then the scrape has to be employed, this allowing of the most minute localisation of action. The only limit to the use of the scrape is the patience of the operator, as the number of minute points of contact between the three plates constantly increases, and the action of the scrape is more and more localised, and the red-lead mixture is made thinner

and thinner until it is a mere film of coloured oil, and presently even this will be superfluous. It is better in the latest stages to trust simply to the contact of surfaces, the polish produced by the rubbing of the metallic surfaces upon one another being sufficient to indicate the higher points. At the last stage the surface of the plates will be covered with these points, which impart that lustrous appearance common to all scraped work.

THE SURFACE OF A SURFACE PLATE.

So intimate is the contact of surfaces in plates which have been prepared in the above way, that the upper one is capable of lifting the lower merely by molecular attraction. In surface plates not prepared

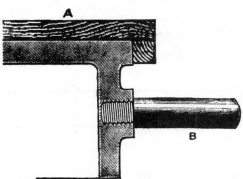

Fig. 1431.—Handle and Cover for Surface Plate.

with this high degree of accuracy, there is nevertheless a very sensible force required to pull them asunder. When the plates are put together, the upper one floats for a while on an interposed film of air, and when this is squeezed out it requires a good pull to separate the plates, unless the top one is slid sideways off the lower one.

CARE OF SURFACE PLATES.

So much pains having been taken to make them true, plates should be treated with care and consideration. Work should not be tumbled about or hammered upon their surfaces. If, for the sake of convenience, lining out is done upon them, care should be taken not to scratch or otherwise impair their smoothness. They should be wiped clean with waste or with a wiper after use, and when not in use should be protected with a wooden cover as shown at A (Fig. 1431). The cover has a plain ledge screwed on round the edge.

PRINCIPLES OF TRUING SURFACE PLATES, STRAIGHT-EDGES, ETC.

Straight-edges, surface plates, and squares are distinguished from other kinds of tools in this respect, that their accuracy can be best assured by constructing them without reference to any pre-existing standards; they are trued by a laborious process of trial and error. The fundamental principle is simple enough in itself, but the labour involved is both tedious and minute. The principle may be stated thus :—If three surfaces are mutually and interchangeably coincident, then each of those surfaces must be a true plane.

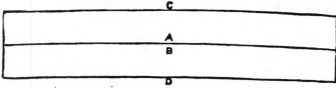

Fig. 1432.—Coincidence of Two Parallel Strips with Curved Edges.

Not less than three surfaces will suffice. Assume a straight-edge with its edges planed parallel, and as true in the linear direction as they can be made by testing upon a plain surface; the test involves scribing a line along one edge, and turning the straight-edge over and placing its opposite edge beside the same line, when it may appear to coincide. But now make a second straight-edge, the precise duplicate of the first, and then try edges to edges, and it will most certainly be found that the light is visible between some portions, while others are in contact. There will be some amount of rounding and hollowing of the edges, and although the two edges may be altered and modified until there is thought to be coincidence between them when tested in succession by changing end for end, there will not be even then absolute accuracy, although the result

would be sufficiently good for much ordinary work. The inaccuracy is proved by making a third straight-edge, the counterpart of the first two, and it will be seen on trial that the three will not be mutually interchangeable. It is clear that if the edges of straight-edges Nos. 1 and 2 are coincident, it may happen that both are slightly inaccurate in opposite directions—that is, convex and concave. But trying the third against both No. 1 and No. 2, it will indicate whether any inaccuracy exists; because it is impossible that it should coincide with two other edges unless those edges are true. In this manner inaccuracy may be detected and gradually reduced to an infinitesimal amount. But the same results may be obtained by making two parallel straight-edges, and using these edges as a check upon each other. Thus, making two straight-edges alike in width, with edges at precise right angles with their faces, calling their edges A and B, C and D (Fig. 1432), they may, when A and B are brought edge to edge, have the relationship, shown in the illustration, where the curvature of the edges is exaggerated purposely. The trial of these edges alone, therefore, would not indicate anything wrong, because, to all appearance, they coincide. But the edges, when reversed, would have the appearance of Fig. 1433 (also exaggerated). Obviously, the concavity of one pair of edges and the convexity of the other have now to be reduced in exactly equal proportions until either edge coincides with the two edges of the other strip.

USING SURFACE PLATE.

To surface a piece of metalwork which has been planed or filed approximately flat, mix a little red ochre with oil, and smear a little of this on the top of the surface plate, applying only just sufficient to colour the plate uniformly. Then place the work on the plate, and with a firm, even, downward pressure, draw the work across the face of plate. Remove it, and then, if the work has been machined or filed true, it will show on its surface a slight, uniform colouring. If, on the other hand, it has not been machined true, the colouring will show only in places on the face of the work. Those places that show the colour are high, and must be eased

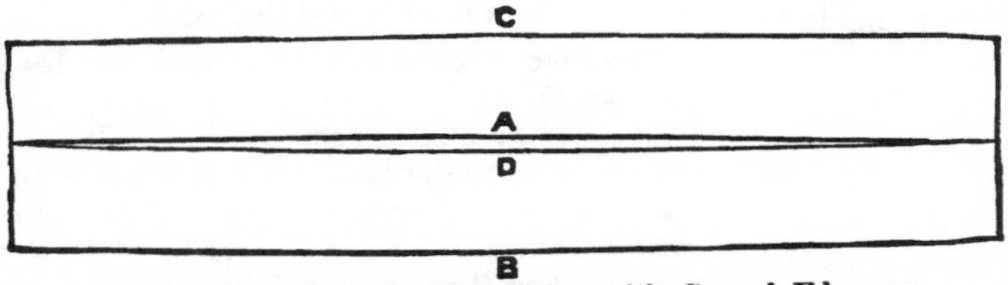

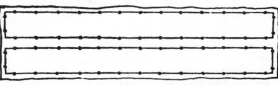

Fig. 1433.—Reversed Parallel Strips with Curved Edges no longer Coincident.

with a smooth file or by other means. Then it is tested again, and further correction is made until the surface of the work is satisfactory.

MAKING STEEL STRAIGHT-EDGE.

Straight-edges are made of steel of moderate hardness, temper, and elasticity. A piece of steel thinned down by hammering may be used, but it is better to use a strip properly prepared to the correct thickness. Large straight-edges are of wrought-iron, "master" straight-edges are of cast-iron stiffened and ribbed, whilst the ordinary tools from 1 ft. to 2 ft. long are of steel plate from $\frac{1}{16}$ in. to $\frac{3}{16}$ in. thick. If the steel plate available is not of the proper width, roughly line it out. Centre-pop the outline upon the steel as a guide for the first stage of cutting out (see Fig. 1434). Then, if the metal is thin, it may be cut round roughly with a cold chisel, but if it is as much as $\frac{1}{8}$-in. or $\frac{3}{16}$-in. thick, drill a number of $\frac{1}{8}$-in. or $\frac{3}{16}$-in. holes close together alongside the outline. If the steel is too hard for drilling, it is too hard for a straight-edge, and it must be tempered, and then annealed to give it equal elasticity throughout. The edges are now filed or ground roughly to width and length. By the time this cutting off and roughing out are done, it is certain that the plate, even if true originally, will no longer be so. Therefore, the sides must at once be brought into linear accuracy, and be

made free from winding and unevenness. Probably they may have to be hammered more or less upon a block of iron, or of hard wood set end grain upwards. The use of a surface plate is desirable at this stage for the purpose of testing the accuracy of the faces. Failing that, a true lathe-bed will answer the purpose. If the plate has been hammered, the hammer-marks must be obliterated by draw-filing; test the surface, meanwhile, by bedding on the lathe-bed or surface plate. During filing, the strip of steel will be laid upon a block of wood, and held steadily between nails driven in around the edges, the wood being clamped in the vice jaws. The straight-edges can now be filed singly, or both at once; if the latter, they are united with a particle or two of solder.

FINISHING STRAIGHT-EDGE.

It will assist greatly if a light cut can be taken down the edges in a planing machine; the next best thing is to use a surface plate or lathe-bed as a guide in

Fig. 1434.—Steel Straight-edge Lined Out and Centre-popped.

giving the initial edges to the strips. If the plate or bed is accurate, and the strips are brought into parallelism with callipers, there will be little left to be done afterwards in the way of alteration and correction. But the difficulty of truing up strips is very much increased with every increase in length and in thickness. Metal straight-edges 2 ft. long and $\frac{1}{8}$ in. thick give proportionately much more trouble than strips 1 ft. long and $\frac{1}{8}$ in. thick. It is difficult to hold the file so that the edge shall be at perfect right angles with the faces, and bevelling at any one part involves that the

strip has to be reduced in width for the
entire length. The strips are checked
together in the later stages by laying them
side by side and edge to edge on a surface
known to be true, and by checking the contact of the edges in the vertical as well as
in the longitudinal direction. During the
later stage of fitting the scrape is made to

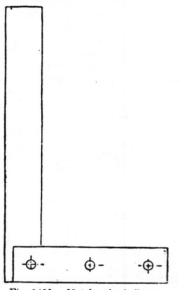

Fig. 1435.—Metalworker's Square.

remove very minute quantities of material.; and as a guide in localising the action
of the file and scrape, a very thin paste of
red lead and oil is smeared along the edge
by which the straight-edge is being corrected. At the final stages this must be
very thin ; in fact, wiping the edge with an
oily wiper will suffice.

METALWORKER'S SQUARE OR TRY-SQUARE.

The construction of the square requires
more care and skill than that of almost
any other tool. Fig. 1435 shows a square
with a blade, or upright part, about 6 in.
long, and the stock, or horizontal part,
about 3¼ in. long. Saw-blade should be
used for the blade, and it should be at least
$\frac{1}{16}$ in. thick by. say, ¾ in. wide. This should
be scribed to the square or   shape, and
then filed to the lines with some care,

though great accuracy need not be attempted. A rough square is thus produced
about $\frac{1}{16}$ in. thick. The sides of the horizontal portion of the square should next
be surfaced, and pieces of steel prepared
for the stock.

STOCK OF STEEL SQUARE.

The pieces for the stock must be at least
⅛ in. thick, and may be 3¼ in. long by ¾ in.
wide. After filing each side fairly flat and
square with the adjacent sides, surface one
flat side of each piece so that it will bed
against one side of the square already surfaced. The three pieces forming the stock
of the square are then held in position by
small cramps, and, say, three $\frac{1}{16}$-in. holes
marked and drilled, one being at the
centre and each of the others near an end
of the stock. A good plan is to use five
rivets, one at the centre of the blade and
one at each end, these being staggered
across the centre line so as to close the ex-
tremities of the blade quite tight. The
pieces may then be taken apart, the burrs
on them removed, and the outer ends of
the holes countersunk as in calliper washers. Steel pins being fitted to the holes,
the parts can be riveted together ; a horizontal section of the stock after drilling
is shown by Fig. 1436. If a piece of suitable steel cannot be obtained of sufficient
width, the square can be made in four
pieces, a short piece, with the two ⅛-in.
pieces already described, being used for
the stock, and one long piece forming the
blade. A horizontal section of the stock

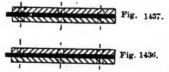

Fig. 1437.

Fig. 1436.

Figs. 1436 and 1437.—Sections of Stocks of Squares.

of such a square is shown by Fig. 1437, and
it is of great importance that the blade
and central part of the stock be of the
same thickness, while the rivet that passes
through the blade should be as near as
possible to the centre of the surface of
contact between the two outer pieces of
the stock and the lower end of the blade.

Another method of making the square would be to use a stock of the full width, and in its end to saw-cut a slot of such a width that the jaws make a tight fit with the end of the blade, which may then be

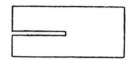

Fig. 1438.—Slotted Surface Plate.

riveted in place. The method first described is, however, the best, when all things are considered.

SURFACING STEEL SQUARE.

The square being riveted up, the hammered ends of the pins should be filed down flush with the sides of the stock. Each face of the square should then be filed fairly flat and square with the adjacent sides, and the top edge, or face of the stock, should then be surfaced. In this work the ordinary surface-plate is next to useless, as the ends of the stock beside the blade cannot be worked. A surface-plate to Fig. 1438 should therefore be prepared, the slot cut along the centre being just sufficiently wide to admit the blade of the square. The top edge can then be surfaced, the work being done, of course, with the greatest skill at the disposal of the workman. From the surfaced top, the two flat faces forming the sides of the stock and the end of the blade can be finished off, a square known to be true being used in the work. When one side has been prepared, the truth of the other can be tested by small callipers, the two sides forming parallel planes, so that the stock is of uniform thickness. The bottom face of the stock can then be similarly prepared, the work being tested as before with squares and callipers. It is of the greatest importance that each pair of faces be tried until they are truly parallel with each other, and that adjacent faces be square. It will probably be better, especially if the worker is not highly skilled, to finish the inner edge of the blade directly after sur-

facing the top edge of the stock. To test the squareness of these edges a true square may be employed, but in finishing, a surface should be obtained with one edge perfectly flat and square with the face. The stock of the square can then be placed against this edge as shown to the left of Fig. 1439, and a line scribed on the surface along the edge of the blade to be tested. The square should then be turned over as shown to the right of Fig. 1439, and a similar line scribed close to the one first drawn. The work must be tried until the two lines thus drawn are quite parallel with one another. The long outer edge of the blade can then be surfaced and squared in the same way. The top of the blade can then be squared off, any slight want of squareness between the faces and the edges of the blade being neglected. It is necessary to repeat that the whole of the work must be done with the greatest skill at the disposal of the operator. The tool thus made is often called a try-square to distinguish it from the set-square of triangular form so largely used in drawing offices.

SET BEVEL.

For the set bevel (Figs. 1440 and 1441) the blade may be of $\frac{1}{16}$-in. steel. For the other part, two pieces of $\frac{1}{8}$-in. mild steel have a smaller piece riveted between the ends. A set-screw has a portion of its head filed away on each side, leaving a projection that works in the slot, and prevents the screw turning with the wing nut.

Fig. 1439.—Testing Truth of Square.

CALLIPERS.

Callipers may be made from any mild sheet steel. Formerly, before such steel was easily obtainable, they were often marked out on saw-blade, and holes were drilled to the lines, but the cheapness of mild steel obviates this. Cast steel would, of course, be a good material, being stiffer than mild steel, and the tools could be forged in it; but its use is not necessary, as many turners can "feel" better with callipers having unhardened, rather than hardened, ends. Fig. 1442 shows a pair of outside callipers, Fig. 1443 a pair of inside tools, and Fig. 1444 an edge view of the top of either kind.

MAKING OUTSIDE CALLIPERS.

In making a pair of outside callipers to take, say, 6 in. easily, as shown one-third size by Fig. 1442, select a piece of steel about ⅛ in. thick by 1 in. wide and 10¼ in. or so long. It would be better to use wider steel, say 1⅛ in., if obtainable, so that the washers might be 1 in. in diame-

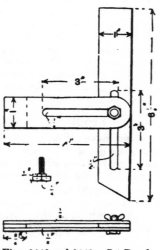

Figs. 1440 and 1441.—Set Bevel.

ter. However, with 1-in. steel, scribe circles at the ends ⅝ in. diameter, as shown by Fig. 1445, connecting the two circles by a diagonal line as shown. Along this line the material should be parted, a flat chisel being used for this purpose, and cuts being made on both sides of the flat piece of steel. In this way two pieces more or less triangular in shape are produced. The centres of the circles at the ends should be centre-punched, and the top half of at least one circle dot-marked, so that the curve is clearly seen. The corners may then be cut off, as shown dotted in Fig. 1445. A ¹⁄₁₆-in.

hole should be drilled in the centre, the usual plan being to hold a suitable drill in a drill chuck fastened in the running head of the lathe, the back centre at the poppet head being removed, and the steel being held against the face of the cylinder while the latter is forced against the revolving drill by the hand-wheel. The flat sides of the legs should then be levelled with a smooth file, and a pin turned to fit the holes, the burr on the latter being removed by the end of a half-round scraper. With this pin rivet the two legs firmly together, taking care that the dot-marked curve is outside.

SHAPING CALLIPER LEGS.

The legs thus riveted can be filed to the lines, going a little below so as not to leave too much for finishing. The edges of the legs should be filed smooth, the ends being finished as shown in Fig. 1442. Next the legs should be heated and bent to shape over the nose of the anvil, a chalked board being marked in pencil to the curve desired, and acting during the bending process as a template or gauge. The edges are then filed quite smooth, and finished with emery paper. The legs are next taken apart, one of the rivet heads being filed off flush with the face of the legs, the opposite face being then held over a collar having a hole rather larger than the remaining head of the rivet. A dot punch or a long centre punch applied to the filed end of the rivet can then be used with a hammer to drive the pin through the legs into the hole in the collar.

FINISHING CALLIPER LEGS.

The flat faces of the legs should next be finished by smooth filing, followed by fine emery-cloth, first dry and then with oil; the emery-cloth is, of course, wrapped round the flat file. For this purpose each leg should be held on a block of wood fastened in the vice, and the leg may be secured either by brads driven in the wood, in which case a straight sweep may be made with the file right across the leg, or by clamps made of hoop iron and jammed down to the leg by wood screws, one on each side of the clamp. By the second plan only the part of the leg clear of the clamp can be dealt with by the file, and consequently the work must be done in several sections. Especial care must be taken with the drilled ends of the legs, which must be quite flat, and which should

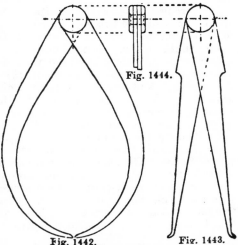

Fig. 1444.

Fig. 1442.
Fig. 1443.
Fig. 1442.—Outside Callipers. Fig. 1443.—Inside Callipers. Fig. 1444.—Edge View of Top of Callipers.

therefore, for good work, be surfaced. A surface-plate being obtained, its face should be covered with a light layer of reddle—red-lead mixed with oil to a thin consistency. The part to be surfaced is then held flat on the plate and given a circular swirling motion. In this way its high parts take up some of the reddle from the surface-plate, and the portions thus marked should be removed by a flat scraper.

LAPPING RIVET HOLES TO SIZE.

Thus, by constant trials and scrapings, the working surfaces of the calliper joint are made perfectly flat, and a good joint results. To attain this end the holes in the legs and in the washers, when turned, may also be lapped out to one size, an iron rod, slightly smaller in diameter than the bore of the holes, being centred in the lathe and driven in the usual way. This rod is grooved and emery powder is placed in the grooves. The lathe being revolved,

the hole is ground out to size, plenty of oil being run on the emery.

MAKING CALLIPER WASHERS.

In making the washers they are first marked out in flat mild steel, chipped and filed to the lines and drilled, the holes corresponding in size with those in the legs. They are then forced on a turned mandrel, which is driven between the lathe centres. The edge of each washer is thus turned to size by a front tool and the faces surfaced by a knife tool. The cuts must, of course, be very light, the front tool being very pointed. The outer face of each washer may be rounded on the edge, as shown in Fig. 1444, the other face being either perfectly flat or slightly concave. The former is better, but in no case should the face be rounding. The thickness of each washer should be rather more than $\frac{1}{8}$ in. The washers should next be countersunk to the extent shown by Fig. 1444, being clipped on a bench in any convenient manner for this purpose; a rose-bit and a belly-brace will do the work. Lapping, as before described, is the next process.

MAKING CALLIPER RIVET.

The steel pin or rivet should be turned in the lathe, and its diameter should be such that, when turned, it is a fairly tight fit in the holes. The pin should be cut off something like $\frac{1}{16}$ in. or $\frac{1}{4}$ in. longer than the distance between the outer faces of the

Fig. 1445.—Setting out Callipers.

washers when in place; the whole may be riveted together.

SURFACING OR GRINDING CALLIPER JOINTS.

Before riveting, however, the working faces should be thoroughly cleaned and a thin layer of the best oil rubbed on them. At one time, instead of surfacing the joints it was usual to grind them. A flat true surface on the lathe-bed was selected, usually at the side of the lathe-head over which the slide-rest or carriage never passed, and over it was sprinkled a little

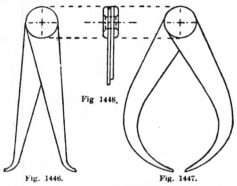

Fig 1448.

Fig. 1446.   Fig. 1447.

Fig. 1446.—Small Inside Callipers.
Fig. 1447.—Small Outside Callipers.
Fig. 1448.—Edge View of Top of Callipers.

powdered emery moistened with oil. The piece to be faced was held quite flat over this and given a circular motion, much as in testing with the surface-plate. The emery then ground off the high places, but, if the material were not uniform in quality, would attack soft parts preferentially. A good working face could in this way be obtained, the chief objection to the method, apart from the greater ease and certainty of the process of scraping, being the liability of the emery to work into the metal and then to work out when the tool was in use, thus spoiling the joint.

Riveting Calliper Legs Together.

The riveting should be done on an anvil or on something equally rigid, and it is of the greatest importance that the ends of the pin be spread to fill the countersink. If in cutting off the pin it is left too long, the joint should be lightly riveted so that the whole is fairly rigid. Then file off the excess at the ends of the pin and rivet the whole together. The ends may next be filed off flush with the faces of the washers, the work being finished carefully with a

smooth file. If the outer faces of the washers have been left slightly rounding, it will be easy to file off the rivet ends without at the same time spoiling the turned face of the washer with file marks.

Finishing Outside Callipers.

The circular ends of the callipers can now be finished, the two legs being moved so that their top parts lie one over the other. Both legs can then be finished together, a smooth file being used to bring down the circular ends to the diameter of the washers, and the edges then being draw-filed. To do this, place the file across the edges of the calliper and grasp its ends, then, instead of pushing the file, draw it along the edges and finish in the same way with fine emery-cloth wound round the file. The ends of the pin can be polished by wrapping round the circular ends a piece of emery-cloth well moistened with oil. This should be placed between two blocks of soft wood and the whole fastened loosely but securely in the jaws of

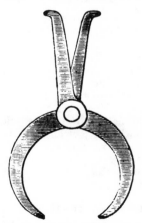

Fig. 1449.—Combination Outside and Inside Callipers.

the vice. The thin ends of the legs are next taken, and the tool is given a circular motion by hand, the pin forming the axis. When this polishing is finished the face of the washers should be perfectly smooth, and it should be impossible to detect the presence of the rivet, which in Fig. 1442 is consequently not shown.

Making Inside Callipers.

A pair of inside callipers to match the outside callipers just described is shown by Fig. 1443. The construction of the two kinds is much the same, but, of course, the points only of the insides are bent or turned round. The length of the steel to be used should be about 9 in., the width

Fig. 1450.—Circumferential Callipers.

and thickness being as before. In Fig. 1443, the legs are shown with circular nicks, the top parts being parallel for about 1¾ in., and then tapering down to the points from small rounds. The legs, however, may taper direct from the washer, but this is a matter of taste.

Small and Combination Callipers.

Figs. 1446 to 1448 show, half size, pairs of small inside and outside callipers, the former being usual for measuring up to a diameter of 2¼ in. The steel required should be at least ¾ in. wide by ₁⁄₁₆ in. thick by 5¼ in. long for the outside callipers, and 4½ in. long for the inside tools. The washers may be 1⅛ in. in diameter by ³⁄₃₂ in thick, the central hole, afterwards countersunk, being ¼ in. in diameter. Combination outside and inside callipers are shown by Fig. 1449.

The Use of Callipers.

Callipers are used for determining the diameters of cylindrical surfaces, as shafts, studs, and other objects, and also of circular and other recesses. The callipers, the beam-gauge, and the fixed-gauge are instruments that measure diameters by touching on opposite sides of a cylinder. This measurement is accomplished entirely by feeling, and this sense of feeling for mechanical purposes must be educated;

no uneducated man can possibly feel the touch of the two jaws of callipers or gauge on the two sides of a shaft, or stud, twice alike. If he uses the callipers, or any yielding gauge, he will barely touch the points at one time, and spring them over at another time. Even if the gauge is a rigid one, as a beam-gauge or a fixed-gauge, he is liable to "bear down" at one time, and "hold up" at another time. Even when the touch of the mechanic is educated so as to calliper the same at any and all times, his touch will be no guide for the feeling of another; and where one would calliper right, another would calliper small, and another large. Then again, callipers touch a cylinder at two opposite points only, and if the cylinder is not exactly circular in cross section, the measurement may be quite incorrect.

Circumferential Callipering.

What is claimed to be a more perfect method than using ordinary callipers for determining the proportionate sizes of

Fig. 1451.—Using Circumferential Callipers.

cylinders as compared one with another, involves the use of a metallic tape, which, of course, is applied over the entire circumference instead of at two points only. It is evident that if the measurement of a cylinder can be got from its entire circum-

ference, it has three times as much prob-
ability of being correct as if taken from
the diameter, as the circumference is three
times greater than the diameter. Fig.
1450 is intended to represent a flattened
wire, which is preferably of soft copper,
with its ends doubled over and secured with
string. The wire may be of any convenient length—one of 12 in. or 14 in. may be

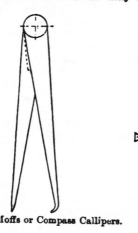

Fig. 1452.—Moffs or Compass Callipers.

adapted to studs and shafts from $2\frac{1}{2}$ in.
down to $\frac{1}{2}$ in. Across the top surface of
the wire, at any convenient point, is drawn
a line with a scratch awl, and when the
right diameter of the piece to be turned
is found, another corresponding line is
drawn, which coinciding, as in Fig. 1451,
determines the true size. Fig. 1451 shows
the method of using the "wire measurer,"
or "flexible callipers."

MOFFS OR COMPASS CALLIPERS.

A pair of moffs or compass callipers is
illustrated by Fig. 1452. For this the steel
should be about $\frac{7}{8}$ in. wide, the thickness
being about $\frac{3}{32}$ in. and the length (say)
$7\frac{1}{2}$ in. The washers should be at least $\frac{3}{4}$ in.
in diameter by $\frac{3}{32}$ in. or $\frac{1}{8}$ in. thick. It will
be seen that one leg of this tool is bent
round like an inside calliper leg, the other
leg being ground taper to a point at the
lower end after being hardened. The
dotted lines give the appearance of the
tool if the legs are left parallel at the top
for a distance of about $1\frac{1}{2}$ in.

MAKING DIVIDERS.

A pair of steel points or dividers may be
made similarly, but in this case both the
legs are hardened and ground to points
at the ends. A large pair of dividers is
shown in side elevation by Fig. 1453, and
in front elevation by Fig. 1454. This tool
cannot be made by hand with ease, and for

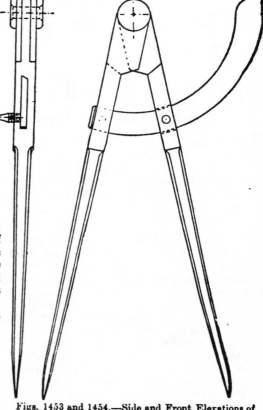

Figs. 1453 and 1454.—Side and Front Elevations of
Dividers.

this reason a milling machine, or, failing
that, a shaping machine, or even a slotting
machine, will be found handy. One leg,
the left in Fig. 1454, has a forked end into
which the top end of the other leg fits,
both legs being slotted to receive a curved
arm or sector. The legs may be worked
down from cast steel $\frac{3}{4}$ in. wide by $\frac{3}{8}$ in. or
$\frac{5}{16}$ in. thick, each leg being 9 in. long. In

making this tool, first face the outsides of each leg, and then, say, the sides presented to view by Fig. 1454. Then, in a vice in the machine, clamp this pair of sides together so that they touch one another, and shape the undercut part, this, of course, being previously scribed out and dot-punched. One end should then be slotted to Figs. 1455 and 1456, the other being shaped to Figs. 1457 and 1458. These ends being fitted together, they can be clamped in place, marked out, and the hole for the pin drilled through the pair.

SECTOR ARM OF DIVIDERS.

The slots for the sector arm should next be marked out carefully, and may then be cut in a slot-drilling machine, or, with

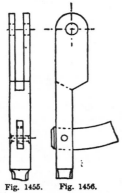

Figs. 1455 and 1456.—Slotted Leg of Dividers.

great care, in a shaping machine, provided that holes of a diameter equal to the width of the slot are drilled at the two ends of the slot. The centre of the pin or rivet at the top of the tool is the centre from which the arcs forming the sides of the arm are struck. The arm should be turned to diameter in the lathe, and one end is then fitted tight in the slot in the left leg and riveted there as shown, the slot in the right leg being sufficiently wide to allow the arm to move quite freely in it. One end of the arm is rounded as shown at the top of Fig. 1454, the other end being bevelled, as will be seen on reference to Figs. 1454, 1456, and 1459. Before this, however, the lower parts of the two legs may be worked down to the drawings, the corners being

chamfered as shown. A hole can next be drilled and tapped for the small set-screw that pinches the arm and holds the legs to any given distance ; this screw, shown

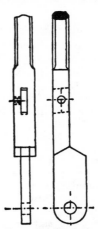

Fig. 1457.   Fig. 1458.

Figs. 1457 and 1458.—Corresponding End of Other Leg of Dividers.

separate by Fig. 1460, and to be seen also in Fig. 1453, being turned, threaded, and filed to shape, the end being well chamfered to prevent burring in use. Then, when the washers have been turned, the pin has been riveted in place, and the ends are filed up, the points may be hardened and ground, and the tool will be complete. Dimensions or details are not given, but as the drawings are half size, they can be scaled where necessary.

ALTERNATIVE METHOD OF MAKING DIVIDERS.

Instead of making the whole of the tool in steel as here directed, it would be quite possible to have the whole of the upper

Fig. 1459.    Fig. 1460.

Fig. 1459.—Bevelled End of Sector Arm.
Fig. 1460.—Set screw for Dividers.

part in iron, the lower portion of the legs only being of steel, welded on so that the points may be hardened. The work of fitting the slots, etc., would then be lightened considerably. This is often done for dividers, especially with those that are made for sale, and also in the case of callipers.

CIRCULAR SCRIBER.

A circular scriber handy for marking circles for bolt holes on cylinder covers, etc., is shown by Figs. 1461 and 1462. The body is steel $\frac{3}{32}$ in. thick. One end of each $\frac{3}{8}$-in. steel pin B is turned down to a shoulder and riveted to the plate, to project $\frac{1}{4}$ in. to $\frac{1}{2}$ in. In the centre at C a $\frac{1}{4}$-in. hole is drilled. Mild steel, $\frac{3}{8}$ in. diameter, is turned at one end to a shoulder and riveted to the plate, the hole being countersunk. At $\frac{1}{4}$ in. from the plate a $\frac{1}{8}$-in. hole is drilled through the piece, the left-hand end is drilled and tapped for a $\frac{3}{16}$-in. steel set-screw, with a milled head. This grips the scriber when tightened. In use, the pins B are pressed against the side of, say, a cylinder cover, and the point of the scriber is set to mark a circle of the required diameter. The scriber is then rotated round the circumference of the work,

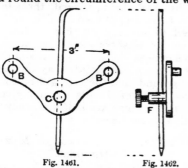

Fig. 1461.   Fig. 1462.

Figs. 1461 and 1462.—Circular Scriber.

when a little chalk rubbed on the surface will show the line clearly.

MAKING TRAMMELS.

The trammel heads illustrated in front and side elevation by Figs. 1463 and 1464 will fit a rough wooden bar $1\frac{1}{2}$ in. by $\frac{1}{2}$ in. thick. One of them carries a bracket for auxiliary points of $\frac{3}{8}$-in. round iron, Fig. 1465 showing a horizontal section through this bracket. The pattern for the head may be made from a solid block of

mahogany, with recesses for the spaces in the sides and with the core prints A B and C D (Fig. 1463) nailed on. These core prints include the projecting lips at the bottom. The bosses on the ends are best

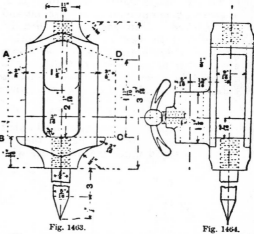

Fig. 1463.   Fig. 1464.

Figs. 1463 and 1464.—Front and Side Views of Trammel Head.

turned in the lathe, and should be made to a diameter equal to the thickness of the body of the trammel. Another way is to cut out a centre-piece to the shape of the body, including the prints, making it the thickness of the core, and then gluing on side pieces to the shape of the body without the prints and lips, and having the spaces cut out. These pieces should bring the body up to the proper thickness. This pattern is made the same on both sides. The auxiliary bracket, to the sizes given, has its

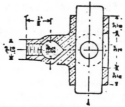

Fig. 1465.—Horizontal Section of Trammel Head.

bottom cut to fit into the recess on one side of the body pattern, to which it is attached by a long, thin, central brass screw. The bracket can then be taken off after a mould has been made, as one head only has a bracket. The core-box can be made either of wood or of sheet brass bent up square. The patterns do not need much draw, as they are all thin. The castings should be filed parallel inside; then

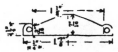

Fig. 1466.—Pad for Trammel Head.

centre the bosses in line with the centre of the cored hole and with the centre of the sides. Put the casting between centres in the lathe, drive a bit of flat wood into the cored hole to act as a catch, and rough-turn the bosses. Next chuck between the running centre and a cone-plate, and bore a ⅜-in. tapping hole in each end, tapping each hole as it is bored, and steadying the tap in the back-centre. Do the holes for the pinching screws first, and chamfer each hole slightly so that it will run true on the centres afterwards. The hole in the bracket can now be marked and bored to $\frac{11}{32}$-in. diameter, parallel to the centre line of the head. Soften a piece of ⅜-in. round tool steel 8¼ in. long, divide it into two equal bits, square the ends, and screw them, for ⅜ in. up, a tight fit for the heads. Grip the rod in the vice, put a bar through the head, and force it hard on the screw. Put this between the centres and see whether it runs true; if not, set the rod near the neck till it does run true. Then finish off the bosses and turn the steel, tapering it to $\frac{5}{16}$ in. diameter, and pointing it as shown. The

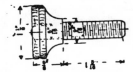

Fig. 1467.—Trammel Head Pinching Screw.

hole in the auxiliary bracket can now be filed with a ¼-in. square file to a **V** of 90° (see Fig. 1465). Try the hole by putting in a bit of ⅜-in. straight rod, measuring to see whether it is parallel to the

centre of head; if it is not, it can be corrected now. The **V** gives the rod a better grip. Drill and tap the ¼-in. set-screw hole.

COMPLETING THE TRAMMELS.

The bodies of the trammels can now be finished off, the slot at B C being square to the centre of the head. The steel points can then be hardened and tempered to a reddish purple. File up the pads (Fig. 1466) $\frac{9}{16}$ in., to be an easy fit sideways in the heads, and mark and drill two $\frac{3}{32}$ in. holes. When the pads are in their place, two bits of brass wire are driven in tight to project ⅛ in. on each side. Make the screw heads of brass and the threaded part of iron to the dimensions given in Fig. 1467. The threads should be a good fit in the trammel heads. The screws are made of iron, as brass is easily bent. The set-screw for the bracket is of steel, turned and screwed ¼-in. with a knob head ⅜ in. in diameter. Drill

Fig. 1468.—Simple Telescope Gauge.

a ⅛-in. hole through the head, flatten the end of a bit of ⅛-in. steel wire, drive it into the hole, flatten the other end, and then set down both ends as shown. Blacken the whole thing in the fire, and harden the point of the screw and put into its place. Mount the heads on a bar of hard wood—teak, mahogany, oak, or ash for preference—and the trammels are complete.

SIMPLE TELESCOPE GAUGE.

A tool often employed in obtaining the distance between opposite flat faces, say between bosses on the two side frames of a machine, is illustrated by Fig. 1468. It is called a telescope gauge, and is used in much the same way as inside callipers. As a further instance of its use, the telescope gauge might be employed in facing to length the two ends of a long roller fixed on a shaft that projects at both ends. The roller and shaft having been turned to size as regards diameter, the telescope gauge would be set to length and one face of the roller finished off. The other end is then faced and the gauge placed on top of the roller. A straight-edge is then held flat against the faced

cylinders, etc.; it can be fitted with three or more wires of various lengths, and will take diameters from 8 in. up to 30 in., with short and long wires respectively. Fig. 1469 is an elevation, Fig. 1470 a section.

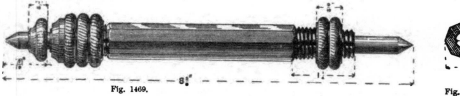

Fig. 1469.

Fig. 1469.—Improved Telescope Gauge.

Fig. 1471.

Fig. 1471.—Section of Telescope Gauge Body.

end of the roller and one end of the telescope placed so as to butt against it; the gauge resting parallel with the axis of the roller spindle, its other end will just be touched by a straight-edge held against the other faced end when the roller is turned true to length. The gauge consists essentially of a steel tube, one end of which is closed in any convenient way and tapered down to a point that is faced off. The opposite end may bear a collar with a tapped hole carrying a set-screw. In the tube slides easily a steel circular rod, with one end ground down taper and faced; this rod can be fixed in any required position by the set- or thumb-screw. Obviously, the tube must be quite stiff, so that it will not sag or bend in gauging long distances, and for this reason the sliding rod should not stand out far. Fig. 1468 shows, one-third full size, a telescope gauge that can be used to measure distances between 6½ in. and 11 in. or 12 in.

Fig. 1471 a cross-section of the body, and Fig. 1472 an end view. To use the gauge, run out the loose pointer to about the proper size, and tighten its milled nut. The final adjustment is made with the threaded pointer, which is secured in position by a milled nut bearing on the conical portion of the split end. Before sawing down the thread to split this end, drive in a piece of hard wood; the saw will not then jerk so much. The shank of the gauge is made of mild steel tube, 7½ in. long by 1⅜ in. in diameter; part of an old gun-barrel will do. If it is desired to make the body from the solid, reduce the length of the material to 5 in., drill a hole just under ⅞ in., and put the rose bit through; then mount the piece between the centres of a lathe, skim over the portion to be filed octagonal (Fig. 1471), and shoulder down each end for the gas threads. It will be noticed in Fig. 1469 that at the right-hand end the thread is tapered; this may be done with the stocks

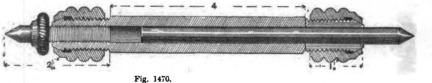

Fig. 1470.

Fig. 1470.—Section of Telescope Gauge.

Fig 1472.

Fig. 1472.—End View of Gauge.

IMPROVED TELESCOPE GAUGE.

Figs. 1469 to 1472 represent an improved form of gauge for taking the diameter of

and dies, beginning at the base and gradu-ally adjusting till the point is reached. The ring nut should not be more than half-way up when it tightens the wire. There is less work in this form than in the section shown by Fig. 1470. The pointers should be hardened at the tips and the tool polished. If a small gauge is required, reduce by one half the sizes given.

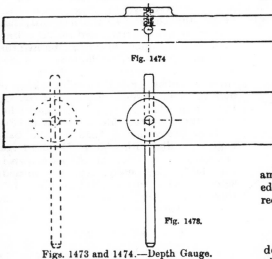

Fig. 1474

Fig. 1473.

Figs. 1473 and 1474.—Depth Gauge.

DEPTH GAUGE.

Figs. 1473 and 1474 show a form of depth gauge in plan and elevation respectively. A piece of steel, cast preferred, about 4 in. long by $\frac{7}{8}$ in. wide by about $\frac{3}{8}$ in. thick, should be shaped perfectly flat on one side and two edges, adjacent faces being square. A plain central hole, about $\frac{1}{8}$ in. in diameter, should then be drilled parallel with, and about $\frac{3}{8}$ in. distant from, the faced side. The end can then be filed off square and the piece fastened flat on the face-plate or on any convenient chuck in the lathe, the centre line of the drilled hole passing across the centre of the lathe. The front face is then turned down so that the tool has a thickness of about $\frac{3}{8}$ in.; the boss, which may be central, but which for convenience in special cases may be placed towards one end, as shown dotted in Fig. 1473 stands up about $\frac{1}{8}$ in., thus making the greater thickness of the tool equal to $\frac{1}{2}$ in. This boss is drilled and tapped for a set-screw, not shown in Figs. 1473 and

1474, and as its use is simply to provide a longer bearing for the screw, it may be omitted, the gauge then being a plain rectangular bar. Into the plain hole drilled across the gauge a circular steel rod, about 4 in. long, is fitted so that it can slide easily, and the tool is then complete.

USE OF DEPTH GAUGE.

The above gauge can be used for much the same purposes as a very short telescope gauge, and a good example is the measuring of the depth to which a hole in a boss may be supposed to be recessed. One edge of the tool is placed in the faced boss and the rod is entered in the hole, being slid along until it just touches the face of the recess; the amount by which the rod stands from the faced edge of the gauge is then the depth of the recessing.

ANOTHER FORM OF DEPTH GAUGE.

The elevation and sectional plan of a depth gauge with two positions for the sliding wire and screws are represented

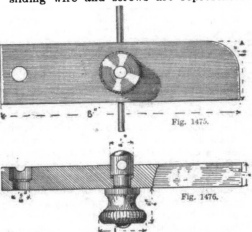

Fig. 1475.

Fig. 1476.

Figs. 1475 and 1476.—Another Depth Gauge.

by Figs. 1475 and 1476. The end position is useful for getting in corners, etc. When the holes are drilled for the clamping screw, with a square and scriber mark the position of the groove for the depthing wire ; then start the groove with a three-cornered file and finish with a rat-tail file, true the edges, and polish the sides and ends.

COMBINATION GAUGE.

Fig. 1477 shows a combination tool, comprising a centre gauge, square, and hexagon gauge ; it may be made from steel

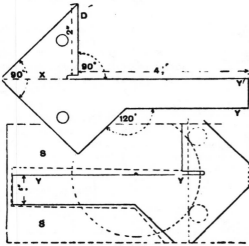

Figs. 1477 and 1478.—Combination Gauge and Method of Setting It Out.

plate $\frac{3}{32}$ in. thick. The method of marking out the gauge is shown in Fig. 1478. Mark it off, allowing quite $\frac{1}{8}$ in. all round for truing up, and drill small holes at the angles to facilitate cutting out with the hammer and chisel ; level the plate if it has buckled, trim off the rough edges, and draw-file on the flats, smoothing with No. 1 emery cloth. It is usual to discolour polished steel plate with lemon juice or vitriol so that the lines of the scriber may show up readily. Carefully mark off the finishing sizes, and on a line at right angles to the centre line x, and at equal distance on each side of it, drill two $\frac{1}{16}$-in. holes, countersinking them on the top side. Turn the pins (Fig. 1479) parallel and both $\frac{11}{32}$ in., each shoulder being slightly hollowed to ensure its closing to the plate properly when the end is riveted over. Leave a little on each pin to avoid damage in the riveting ; it can be cut off afterwards. In

truing up this tool, as the pins are a fixture, the adjustment must be done on the blade. Take a 10-in. or 12-in. square, scribe a line on a marking-off slab s (Fig. 1478), place the gauge against the slab, and file the inside of the blade square with the line. Next try the centre gauge on a piece of turned stuff, a 5-in. plug gauge being best. Rub chalk over the flat end, hold the centre gauge firmly, and scribe a line ; then from the opposite side scribe another line, this probably overlapping the first one ; the true centre will be the mean of these two lines. File the blade, taking off half the lap, frequently referring to the lines on the slab and plug gauge, and when this edge is correct file the edge D (Fig. 1477) square with it. Then file the back edge y parallel with edge y[1], file the hexagon gauge to an angle of 120° as shown in Fig. 1477, and, finally, true the outside square. The dotted lines in Fig. 1478 represent in an exaggerated form the blade y y out of truth.

THE USE OF V-BLOCKS.

A V-block is shown in front elevation by Fig. 1480, and in plan by Fig. 1481. These blocks are generally used when finding the centre of turned work, and although a single block may be employed for short jobs, a couple of blocks are desirable for use with long shafts, spindles, etc. The blocks are therefore generally made in pairs, and the ends of the work to be marked out rest in the V's. The blocks are placed on a surface-plate, and the scriber of a scribing block is set by

Fig. 1479.—Pins of Combination Gauge.

the eye so that it can mark a horizontal line slightly above the centre of the job. The latter is then rotated in the V's for half a revolution and another line scribed parallel with the first one, and slightly removed from it as the scriber is set above the centre. The work is now turned through a quarter revolution, or one right angle, and, without altering the position of the scriber, another horizontal line is scribed. The work is again turned through half a revolution, and a line parallel to the third line is scribed. In this way a small square is scribed at the middle of the end of the job, and its centre can be marked by a dot- or centre-

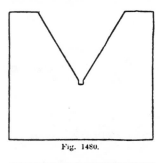

Fig. 1480.

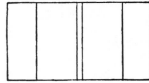

Fig. 1481.

Figs. 1480 and 1481.—V-block.

punch. The same process being applied at the other end, the job is centred. These blocks should not be used to mark off rough work, which is apt to score the V's.

MAKING V-BLOCK.

In making the blocks, choose a rectangular piece of iron that will work down to the dimensions required. It is, of course, quite possible to file the blocks by hand, but this is seldom or never done, a shaping machine usually being employed. Therefore, first shape the two flat sides of each block perfectly flat and parallel, and then clamp the two blocks tightly together in the vice of the machine and shape the base, and, after shifting the blocks, the two edges. The top can also be shaped across, a fair amount of care being taken to get all opposite faces parallel and all adjacent faces perpendicular. With the base perfectly parallel with the stroke and feed of

the machine, the V's (see Fig. 1480) are roughed out, the two blocks being clamped together. To finish the V's, the head of the machine is set over to the necessary angle and the finishing tool is run into a groove previously cut in the centre of the block with a parting tool. It is of the greatest importance that the two sloping sides should be equally inclined to the vertical, and to test this when the two blocks are taken from the machine, turn one round so that what was its right-hand sloping side is near the left-hand slope of the other block. If the two blocks are now placed

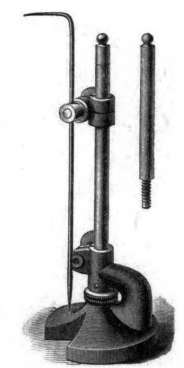

Fig. 1482.—Scriber Block and Extension Piece.

on a surface-plate with the two bottom grooves in line, the planes of the sloping sides of the two V's should coincide. If the two slopes are not equally inclined, a ridge may be seen by the eye or will be felt when passing the finger in a horizontal direction over the two V's. Dimensions for the blocks are not given, as much depends on the nature of the work to be dealt with ; generally, two pairs of blocks will be found useful for marking out—a large pair and a smaller pair.

SCRIBING BLOCKS.

These tools have been repeatedly mentioned already, and their use in a number of typical jobs has yet to be referred

the extension provided to lengthen the spindle when work higher than 12 in. is to be scribed.   The universal scribing block (Fig. 1483) has a heavy grooved base adapted for use against circular work. The spindle passes through a rotating head, jointed to a rocking bracket pivoted in the base   The bracket is adjusted by a knurled screw in one end acting against

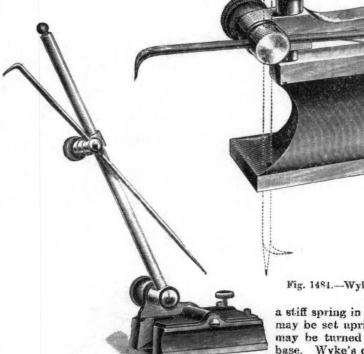

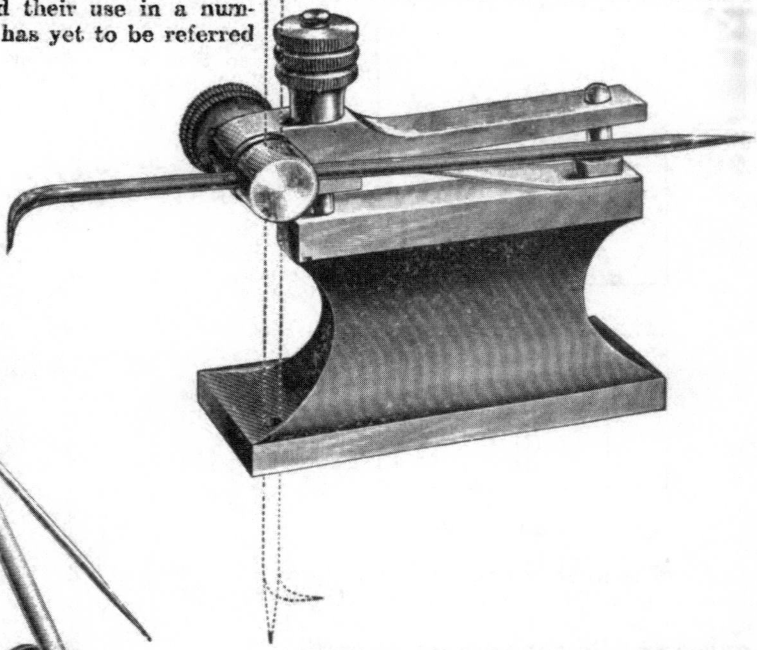

Fig. 1484.—Wyke Combination Scribing Block.

a stiff spring in the other ; and the spindle may be set upright or at any angle, or it may be turned so as to work under the base.   Wyke's combination scribing block (Fig.1484) is fitted with a fine adjustment, a hole in the base allowing of depthing as shown.

Fig. 1483.—Universal Scribing Block.

to later in this work. Fig. 1482 shows a scribing block in which the sleeve and scriber clip, when loosened for adjustment, are both held by a light spring friction ; they both are clamped by a knurled nut. For fine adjustment, the spindle in the base is raised or lowered by a knurled nut, and all backlash is taken up by a spiral spring in the base. Fig. 1482 also shows

MAKING 9-IN. SCRIBING BLOCK.

The scribing block illustrated by Fig. 1485 is 9 in. high from the bottom of the base to the top of the pillar, without the knob. The base and pillar may be forged solid, preferably of mild steel, but if this cannot be procured solid, a round blank of wrought iron, about $3\frac{1}{4}$ in. in diameter and 1 in. thick, would serve for the bottom if turned to shape. A $\frac{7}{8}$-in. hole should be drilled in the centre and tapped $\frac{1}{2}$ in. The pillar, of $\frac{5}{8}$-in. round mild steel, should be turned down to $\frac{1}{2}$ in. diameter about 1 in. up and screwed tight into the base and slightly riveted over at the bottom to prevent its slacking back. The bottom should not be finished flat all over but should be hollowed out about $\frac{1}{16}$ in. deep; leaving about $\frac{1}{16}$ in. round the outside to form a flat surface to rest on the surface plate. To pass the scribing point to form a depth gauge when necessary, a $\frac{3}{8}$-in. hole should be drilled through the base in a straight line with the scribing point when

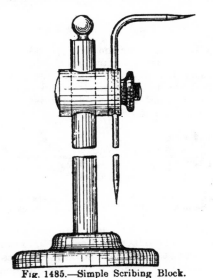

Fig. 1485.—Simple Scribing Block.

the latter is perpendicular. The knob shown on top of the pillar is screwed in $\frac{3}{16}$ in. in diameter by $\frac{1}{2}$ in. deep; this knob can then be screwed out and replaced by an extension pillar should circumstances require it; the pillar is $\frac{1}{2}$ in. in diameter, and should be turned after it is screwed in the base.

CLIP, OR CRAMPING SHELL AND PIN, OF SCRIBING BLOCK.

The cramping shell shown in section by Fig. 1486 consists of a piece of mild steel 1 in. in diameter and 1 in. long bored $\frac{3}{4}$ in.,

$\frac{7}{8}$ in. up, with a flat-bottomed drill, and then a $\frac{3}{8}$-in. hole is drilled through the remaining $\frac{1}{8}$ in. thickness. The pin (Fig. 1487) should be turned to fit this shell and rounded on the large end; the reduced

Fig. 1486.—Cramping Shell.

Fig. 1487.—Cramping Pin.

portion should be screwed $\frac{3}{8}$ in. in diameter by about $\frac{3}{4}$ in. long. When turned and screwed this pin should be inserted in the shell and nipped by the milled nut shown by Fig. 1488; a $\frac{1}{2}$-in. hole should then be drilled completely through both shell and pin to allow the pillar to pass through, and the pin should then be put back in the lathe and about $\frac{1}{32}$ in. turned off the end nearest the screwed end, this allowing both shell and pin to have a tight grip on the pillar when the milled nut is screwed tight up against the washers, shown by Fig. 1489. These washers have each a half-round hole filed across one side, thus gripping the scriber tight just before the faces of the washers are nipped. It will be seen that when the milled nut is slackened it will release both the pillar and scribing point; the latter can then be moved in either direction vertically or in a complete circle, and is made out of $\frac{1}{8}$-in. steel wire and about 10 in. long, bent over at one end about $1\frac{1}{2}$ in.

ANOTHER SCRIBING BLOCK.

The scribing block shown by Fig. 1490 is made from a rod of mild steel 10 in. long and $\frac{1}{2}$ in. in diameter. This is turned down

Fig. 1488.—Nut for Cramping Pin.

Fig. 1489.—Washer for Cramping Pin.

to $\frac{7}{16}$ in., finished smooth, and quite parallel throughout its length. The top is finished off as at A (Fig. 1490), and the other end, for rather more than 1 in., is turned and threaded ⅜ in. A collar, B, is then screwed on tightly. The bottom disc or iron or gunmetal C, with the bottom dished out, is tapped to suit the post. Some may prefer to have the disc turned up of mild steel. The hole in the sliding block (Fig. 1491) should be a sliding fit on the post, the saw cut meeting the large hole ; this will enable the block to grip the rod and scribing point when the nut (Fig. 1492) is tightened. The steel pin (Fig. 1493) should be turned, drilled, and

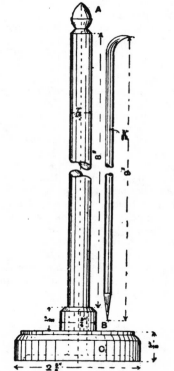

Fig. 1490.—Pillar, etc., of Scribing Block.

threaded to fit the nut. Two washers (Fig. 1494) are required, one being grooved diameterwise for the scriber to bed into ; they are placed one at each side of the block (Fig. 1491). The ends of the scriber should be hardened and tempered.

CLIP FOR SCRIBING BLOCK.

Figs. 1495 and 1496 show a form of scribing block clip greatly in favour a few years ago, simply because turning, rather than fitting, was principally required. Fig. 1495 shows the clip complete in elevation. It consists essentially of three pieces—the clip A, the square washer B, and the handle C. The clip is shown in plan by Fig. 1496,

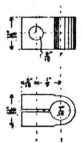

Fig. 1491.—Sliding Block.

and in end elevation by Fig. 1497. It may be made from square steel, drilled with a twist drill at one end to fit the post, this end, the left in Figs. 1495 and 1496, being rounded off to suit. The sides of these holes having been faced on a mandrel in

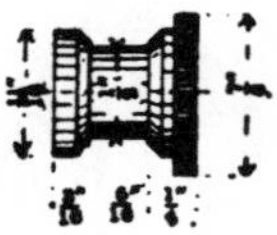

Fig. 1492.—Nut of Scribing Block.

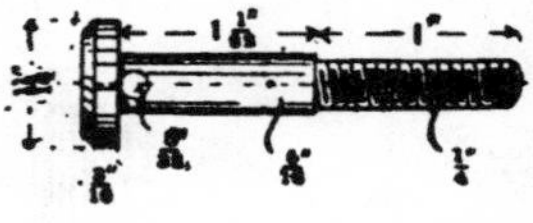

Fig. 1493.—Scribing Block Pin.

the lathe, these edges can be placed on any true surface, and centre lines scribed across at the ends. Or the piece can be placed on a mandrel, and, the rounded end being centred, the ends of the mandrel resting in V-blocks, the point of a knife

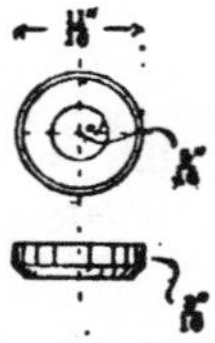

Fig. 1494.—Washer.

tool is set to the mark, the tool withdrawn by the bottom slide only, the piece turned round, and the point of the tool moved up to mark the other end. The top slide must not be moved in these operations. Of course, the ends should have been prepared for scribing previously by filing and chalking. This method will ensure that

the shank at F is for ease in chasing the thread. G (Fig. 1497) shows the slots left by the turning back at D (Fig. 1496). Figs. 1498 and 1499 are plan and end ele-

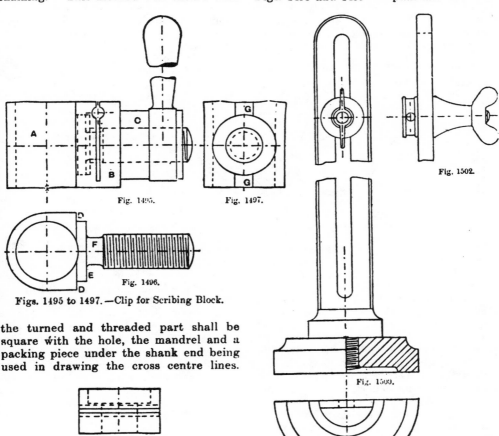

Fig. 1495.

Fig. 1497.

Fig. 1496.

Figs. 1495 to 1497.—Clip for Scribing Block.

the turned and threaded part shall be square with the hole, the mandrel and a packing piece under the shank end being used in drawing the cross centre lines.

Fig. 1498.

Fig. 1499.

Figs. 1498 and 1499.—Washer for Scribing Block Clip.

The rest of the work does not present difficulty, but the face at D (Fig. 1496) should be turned back just past the hole, a collar being formed at E. The turning down of

Fig. 1502.

Fig. 1500.

Fig. 1501.

Figs. 1500 and 1501.—Elevation and Half-plan of Slotted Scribing Block. Fig. 1502.—Side Elevation of Top of Sliding Block.

vation of the washer. The outer surface of this corresponds to the shape of the large end of the clip, and a recessed hole is bored in it, the larger end fitting on the collar E (Fig. 1496), while the small hole slides over the threaded end of the clip. The washer is slit down the centre nearly but not quite to the bottom, a hole for the scriber having previously been drilled across as shown at the top of Fig. 1495. The handle c (Fig. 1495) is threaded to fit the screwed end of the clip. The cross section of the boss and of the handle itself is circular. The washer also may be circular instead of rectangular, and will then work easier on the post.

SCRIBING BLOCK WITH SLOTTED POST.

A scribing block with a slotted post is depicted by Figs. 1500 and 1501. The block may be made throughout in mild steel, and, at any rate, this material is recommended for the base, which, for a post about 12 in. high, should have an outside diameter of 3 in. or 3$\frac{1}{4}$ in. It may be $\frac{3}{4}$ in. thick. The central hole as shown is tapped, the end of the post being suitably threaded. This hole may, however, be left plain, the end of the post fitting it fairly tightly. In both cases the hole should be slightly countersunk, the end of the screw or plain part, as may be, being riveted over as shown in Fig. 1500. In making the base the hole is first drilled and tapped and the block is then turned on a screwed mandrel. This is a mandrel that carries at one end a faced collar with a short length of screw-thread adjacent to it. On this the circular base is screwed, the thread on the mandrel being recessed so that the faces of mandrel collar and base may jamb together. In this way the bottom surface of the base-block may be faced, but not right across, the central portion being recessed as shown to the right of Fig. 1500.

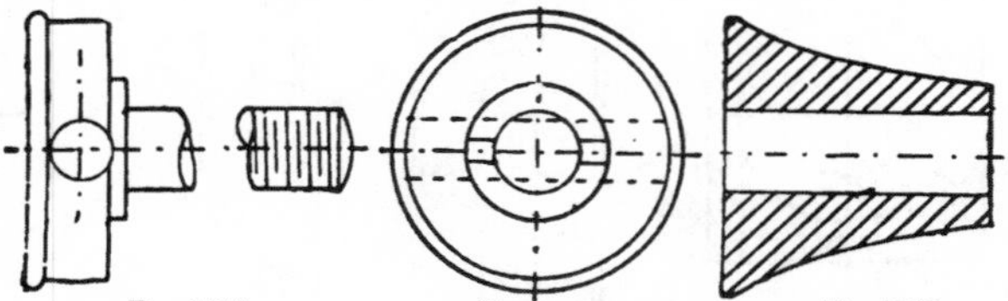

Fig. 1503. Fig. 1504. Fig. 1505.

Figs. 1503 and 1504.—Central Pin of Scriber-holder.
Fig. 1505.—Scriber-holder Distance-piece.

SLOTTED POST OF SCRIBING BLOCK.

The post may be turned down, if necessary, from 1$\frac{3}{8}$-in. material, the collar towards one end being 1$\frac{1}{2}$ in. in diameter. For the top part of its length the post may be turned to 1 in. or 1$\frac{1}{8}$ in. in diameter, this being rounded into the 1$\frac{1}{2}$ in. diameter as shown. The small end may then be threaded and the collar faced. After this the top length of the post should be slotted or shaped so that it is almost rectangular in section, 1$\frac{1}{8}$ in. by $\frac{5}{16}$ in., the long ends being rounded. This is shown by the half plan (Fig. 1501). The central slot, about $\frac{5}{16}$ in. wide, may then be machined.

OTHER PARTS OF SCRIBING BLOCK.

The scriber-holding device is shown in front and side elevation by Figs. 1500 and 1502 respectively. The device consists of a central pin. shown in detail by Figs. 1503 and 1504, a washer, a distance block (see Figs. 1502 and 1505), and a thumb- or fly-nut. The head of the pin should have a larger diameter, of $\frac{7}{8}$ in., the adjacent diameter being slightly smaller and drilled across as shown, the hole allowing the scriber to pass through easily. The inner face of the head should be turned back so that in plan (Fig. 1504) a couple of small slots are formed, a small boss, say $\frac{7}{16}$ in. diameter, being made before reaching the shank of the pin, about $\frac{1}{4}$ in. in diameter and 1$\frac{1}{2}$ in. long. On the boss thus formed a thin washer fits easily (see Fig. 1502). and between its face and the opposite side of the hole the scriber is jammed tight.

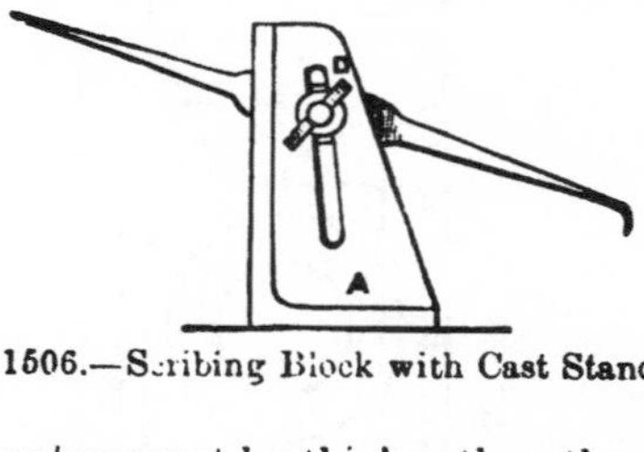

Fig. 1506.—Scribing Block with Cast Standard.

The washer must be thicker than the width of the boss or the scriber will not be jammed when the nut is screwed up. The distance piece, shown in longitudinal section by Fig. 1505, is bored $\frac{1}{4}$ in. full so as to pass easily over the shank of the central pin. The whole is fitted together as shown in Figs. 1500 and 1502.

Scribing Block with Cast Standard.

Another and quite different type of scribing block is shown by Fig. 1506. It possesses the merit of great steadiness, but involves some pattern work and casting. There are three parts—the base A, Figs. 1506 and 1507, in cast-iron or gunmetal, the scriber, Fig. 1508, the screw and wing-nut, D. The pattern of the base (Fig. 1507) is like its casting, and the slot can be cast out by using a print and making a core-box, making it a little narrower than the screw to allow of cleaning out with a file. When cast, plane or file the three faces, a, b, c, at right angles with one another; and smooth over the other parts with the file for the sake of good appearance. The scriber (Fig. 1508) is made of

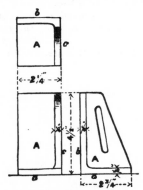

Fig. 1507.—Cast Standard of Scribing Block.

steel, first roughed to outline on the anvil, then filed flat and parallel about the central portion, upon which the parallel lines forming the slot and the bounding metal will be marked and filed. The tapering ends may be turned for a portion of the distance out from the centre, but it is not necessary, as a little rough grinding, followed by filing, will bring them to a good shape. The screw is provided with a turned head and a square shoulder, to fit the slot. The wing-nut D is cut from iron, steel, or gun-metal.

Scribing Block with Two Short Scribers.

The scribing block shown by Fig. 1509 has a circular base, and a slotted upright

cast solid; the slot made in the upright receives the squared end of the clamping screw (shown enlarged by Fig. 1510) allowing an up and down motion of about 2 in.;

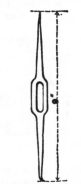

Fig. 1508.—Scriber.

there are two scribers, one straight and the other hooked, the one not wanted for use dropping out of the way when the milled head is slackened. This pattern of scribing block forms, if made in gun-metal, a fairly good specimen of workmanship, if got up with mouldings, etc., and the needles made from flat steel; but for utility it is not to be recommended. It may be somewhat improved by slotting the needles, and so allowing the length of the projecting part to be altered to suit requirements. The base may be turned up true between the lathe centres, making a centre-punch dot on the top of the upright, and one in the

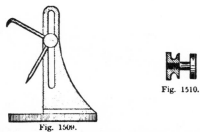

Fig. 1509.— Scribing Block with Two Short Scribers.
Fig. 1510.—Clamping Screw.

centre of the base. The upright itself, and the upper surface of the base, have to be filed up. A brass or steel screw, with a milled head, serves for the clamping screw.

SIMPLEST FORM OF SCRIBING BLOCK.

Fig. 1511 shows the simplest form of efficient scribing block that it is possible to make. This block will bear any amount of rough usage. To make it, get a piece of

Fig. 1511.—Simplest Scribing Block.

hardwood such as close-grained mahogany, measuring 3 in. by 2 in. by 1½ in., and square all the surfaces. File out the scribing pointer, rather pear-shaped as shown, of sheet steel, and make a chamfered hole in the centre of the large end to take the head of a thick wood screw; round off all the edges so that when screwed on to the block it will move freely. The point must be nicely shaped, and then hardened and tempered. For a block of the size above named, the screw hole should be ½ in. from the side to the left, and 1 in. from the upper end. All four sides can be used for the base, and thus, with the pointer standing out at right angles to the perpendicular side of the block, the following heights will be obtained, with the different sides as bases:—With left, ½ in.; with top, 1 in.; with back, 1½ in.; and with bottom, 2 in.; and, allowing only ¼ in. of angular movement for the pointer, all heights, from ¼ in. to 2¼ in., can be got. The pointer is adjusted by giving it a knock on the edge of a bench or other convenient place. Such a block should always be used for rough work so as to save the more elaborate tool. It costs hardly anything, and can be made by the merest tyro.

BUILDING A 4½-IN. CENTRE LATHE.

Though a screw-cutting and self-acting sliding and surfacing lathe is often desirable, the work of fitting it is too great for many craftsmen, whereas most could fit a plain lathe, and many workmen would not require anything else. The lathe to be described is practically identical with that shown by Fig. 1307, p. 457, except that the screw-cutting and surfacing attachments there shown are here absent. The lathe without headstocks, etc., is shown in front elevation by Fig. 1512, in plan by Fig. 1513, and in side elevation by Fig. 1514. It is of 4½-in. centres, with a 4 ft. 6 in. bed; the length of the bed can be increased at extra expense. It has a **T**-rest and a hand-operated slide-rest. Everything is in metal. The weight is 5 cwt. The bed is of the usual English gap type. The fast and loose heads fit closely between the bed shears, and the headstock spindle runs in reverse cones of steel. It is back geared, with an eccentric throw out. The driving is with a cord over four-speed **V**'d pulleys. There is an extra groove on the driving pulley on the crank for slow speeds and specially heavy cutting. The two-throw crank axle pivoted in centres is driven by chains from the treadle. Power driving from overhead can be substituted for the treadle motion. The centres have Morse tapers, the screws are square threaded, and the whole design is good. One advantage is that the heads and **T**-rest, the slide-rest, the bed, the driving-wheel, and crank-axle can each be purchased separately, either in the rough, or tooled ready to go together. The headstock and slide-rest can also be obtained completely fitted, ready to be put into place.

DIMENSIONS OF LATHE.

For general work the choice may lie between the 4-in. and 5-in. centre lathes. The usual practice is to make the standard length of bed in feet equal to the centres in inches, so that a 5-in. lathe would have a 5-ft. bed. It is better sometimes to exceed the standard length, even though this increases expense; thus a 5-in. lathe may have a 6-ft. bed. This will often save the trouble of extending the shears temporarily for the turning of extra-long work.

LATHE BED.

It will be well to build up the lathe on its own bed, doing as much work as possible to the heads after they have been fitted upon the bed, and utilising the treadle drive for certain details. In case anyone should intend to tackle the heavy work of the bed, the following will be of service: The bed, being long in proportion to its section, is very liable to spring if bolted carelessly. It should be held on the table of the planing machine by clamps pressed against the ends, rather than by bolts pulling down on the edges of the cross-bars. The clamps may be placed at the ends with little risk, but they are not necessary if end chocks are used. First plane the facings for the standards on the under-side, and then if these are laid down on the parallel packing blocks on the table, the bed will remain true after it is bolted to the standards. If the bed is filed to save expense, the edges should be lined out all may be reversed. Finally, the inner and outer edges should be scribed, the bed being laid upon its side, and those filed.

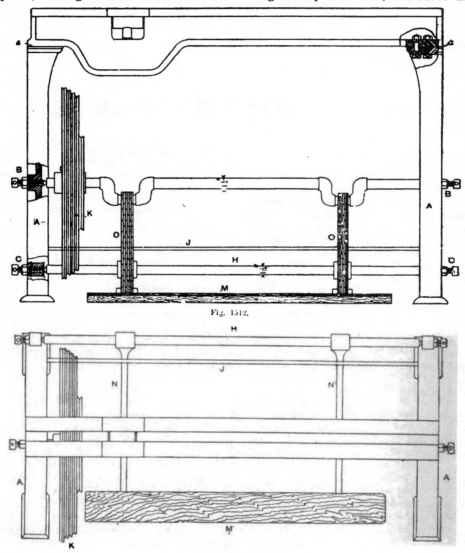

Fig. 1512.

Fig. 1513.

Figs. 1512 and 1513.—Front Elevation and Plan of Lathe without Headstocks, Rest, etc.

round with a scribing block, as a guide for working over the face. After the face is done, the facings for the standards should be lined and filed parallel. Or the method

The bed surfaces of a plain lathe are not so large as for a screw-cutting lathe, and there are no **V**'d edges. The skin should be removed first either by light chipping or by grinding. The truth is tested with a long straight-edge, with winding strips, and with a surface-plate.

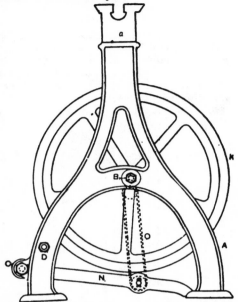

Fig. 1514.—Side Elevation of Lathe Standard, etc.

GAP AND GAP-PIECE.

Figs. 1515 and 1516 illustrate the gap-piece A, which is held in position with a bolt B, tapped into the boss, and having a collar which bears against the plated portion in the bed beneath. There are bearings on the shoulders at a, and the clearance at b allows the bolt to keep the bridge on its seatings. The fitting at c should be close. The head of the bolt is brought down well below the gap, so that it can be easily turned with an ordinary spanner; if it is shorter, a box spanner might be necessary. The fitting of this bridge-piece is done before the bed is planed.

LATHE STANDARDS.

Next attach the bed to the standards, and complete all the underwork. Having the bed planed, prepare the standards A (Figs. 1512 to 1514) for attachment to the bed. The top ends a may be planed or filed; filing will not take long. In squaring, either by shaping or filing, the test should be taken, not with a try-square set on the rough edges of the castings, but from a long straight-edge running the whole length of the standard. The holes

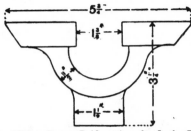

Fig. 1515.—Gap or Bridge-piece for Lathe Bed.

for the treadle and crank pivots and the stretcher rod should be drilled, either through the two frames at once, or through one first and thence marked on to the other. In each case set the two standards carefully in the horizontal position, one over the other, and see especially that the two top planed faces correspond when tested with a straight-edge. The holes for the crank centres at B (Figs. 1512 and 1513) and for the treadle at C are ⅝ in. in diameter, tapped to Whitworth threads;

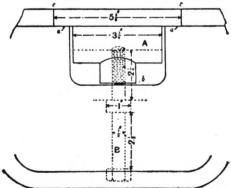

Fig. 1516.—Gap or Bridge-piece for Lathe Bed.

holes D, for the long stretcher bolt, should be ½ in. clearing.

TURNING CRANK-AXLE.

Turning the crank-axle E (Fig. 1512) requires centre-plates (Fig. 1517 and 1518) for turning the pins, with stretchers of wood. The turning is done in two operations; first, for the axle length and the end for the pulley, packing blocks in the dips taking the thrust of the centres; and secondly, the crank is centred in the holes A (Fig. 1517 and 1519) with thrust packings of wood, while the crank pins are turned to the centres B.

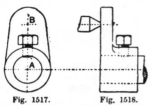

Figs. 1517.   Fig. 1518.

Figs. 1517 and 1518.—Centre Plate for Turning Crank.

PIVOTING CRANK-AXLE.

The pivoting of the crank-axle is shown in Fig. 1519; compare this with the left hand of Fig. 1512. E is the end of the crank-shaft, steel plugged, and F the pivot tapped into the standards A and secured with a check nut G. The three general types of bearings in use for crank-axles are pivots as in Fig. 1519 and roller- and ball-bearings. The last-named system is the best, but involves more work. The pivot bearing is simple, but is subject to wear,

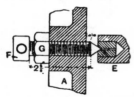

Fig. 1519.—Pivot for Crank and Treadle Shaft.

and requires frequent readjustment, while the pivots and their centres must be hardened.

TREADLE-SHAFT.

The centres c for the treadle pivot shaft (Figs. 1512 and 1514) are of the same size and shape as those for the crank-shaft (Fig. 1519). The treadle-shaft H (Figs. 1512 and 1513) is a bar of steel 1 $\frac{3}{16}$ in. in diameter.

The stretcher or stay J (Figs. 1512 and 1513) is a $\frac{1}{2}$-in. black bolt, screwed at each end to about 1 in. inwards.

DRIVING-WHEEL.

The four-speed driving-wheel K (Figs. 1512, 1513, and 1514) is a troublesome job when attempted outside the engineers' shop; it weighs 84 lb. For those who may desire to undertake the task, an enlarged and dimensioned section is given in Fig. 1520. Do not let the boring tool run, or

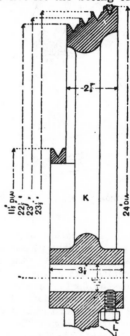

Fig. 1520.—Section of Lathe Driving Wheel.

the rim will wobble on the crank-axle. To turn it after fitting to the crank-axle would be rather troublesome, but would ensure true running. The grooves are for $\frac{3}{16}$-in. cord, and correspond in section with those on the cone pulley in the headstock. To secure the wheel, a pointed set-screw L (Fig. 1520) enters a slight countersink in the axle, or bears directly on a flat.

TREADLE-BOARD.

The treadle-board M (Figs. 1512 and 1513) is attached to two cast-iron levers N, bored and secured to the treadle-bar H with set-screws b (Figs. 1521 and 1522). Each lever has an opening (see Figs. 1521 and 1522) for a roller or treadle runner, over which the treadle chains o (Figs. 1512 and 1514)

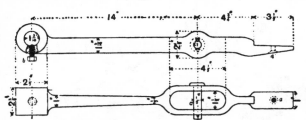

Figs. 1521 and 1522.—Lathe Treadle Levers.

run. The iron rollers (Fig. 1523) are turned and run on the turned pins c (Figs. 1521 and 1522). The pins may be riveted into the sides of the levers or fitted as Shown. The treadle-board is attached with wood screws put into countersunk holes a (Figs. 1521 and 1522) in the treadle.

Putting Together the Lathe.

At this stage begin to put the parts together. Bolt together the bed and standards with bolts, shown in Fig. 1524. Fig. 1525 shows the positions of these holes in the standards, and from these the corresponding holes may be marked to the beds. Then the standards may be drilled, and the bolts inserted and tightened. This gives a solid basis for the rest of the work. It will be well to bolt the bed to its foundation over substantial joists, or directly on

Fig. 1523.
Fig. 1524.
Fig. 1523.—Chain Roller.
Fig. 1524.—Bolt for Securing Lathe Bed.

a stiff floor. When doing this, note must be taken of the truth or otherwise of the top faces of the bed by means of a spirit level, and thin packings of paper, card, wood, or sheet metal should be inserted under the feet where required, to level the bed as accurately as possible. When

bolted down finally, the bed and standards should be a perfectly rigid whole, to which the details can be fitted. The crank-axle can then be inserted, and the pivots screwed up to it and held with the check

Fig. 1525.—Diagram showing Bolt Holes on Lathe Standards.

nuts. The treadle-bar comes next, being pivoted similarly to the crank-axle. The chains may be slipped into place, the rollers dropped into the bight of the chains, and their pins riveted. The stretcher bolt should be tightened, leaving all ready to commence the work upon the head and rests.

Fitting Lathe Headstocks.

The fitting of the headstock and the sliding poppet of this lathe can be conveniently done on the bed. Another way would be to bore the heads, slide them over a man-

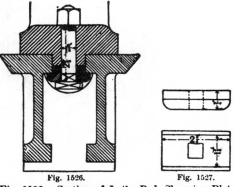

Fig. 1526.
Fig. 1527.
Fig. 1526.—Section of Lathe Bed, Clamping Plate, etc. Fig. 1527.—Clamping Plate.

drel to keep them in alignment, bolt them to the table of a machine, and plane the feet. It is better to plane or file the feet first and bore subsequently, and this would also be more easily accomplished in most cases. Filing is not difficult, because the bed affords a basis for chucking and filing. Take care not to file the tongues that fit between the shears too narrow, as this and by a scraper. A very little red-lead in oil smeared thinly over the bed, faces, and internal edges will indicate the degrees of accuracy of fitting. The heads will be bolted to the bed while being bored, the

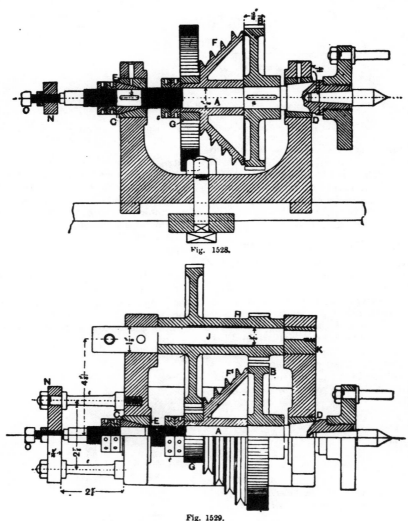

Fig. 1528.

Fig. 1529.

Figs. 1528 and 1529.—Sections of Lathe Headstock.

error cannot be rectified easily. The head-stock, like the poppet, fits closely within the shears instead of allowing cross adjustment by screws. The final fitting of the heads must be made by a fine flat file same bolts and plates being utilised as for permanent service. The cross section (Fig. 1526) is taken through the bed, bolt, and clamping plate. The square neck of the bolt fits a square hole cast in the plate.

The nut is of extra depth. The view of the plate lengthwise and in plan is given in Fig. 1527. The bolt shown in Fig. 1526 is for the fast head; that for the poppet is of the same diameter and shape, but is 2⅜ in. long between the head and the nut, instead of 2⅜ in., and the nut is ¼ in. thicker.

Boring Lathe Headstocks.

As stated on p. 186, the castings of the heads are supplied either solid or cored to suit purchasers. Where a stout drilling machine is available, it is better to drill through the solid metal, at least in the fast head. Most workers will find it more convenient to have the poppet cored, though the fast head should be solid; the cored hole must have a diameter of ¾ in. to admit a ⅝-in. boring-bar. For boring, two traverses are necessary, a roughing and a finishing cut; and the whole operation of boring the two headstocks is explained in detail with illustrations on pp. 186 to 188.

Headstock Mandrel

The completion of the fast headstock shown in Figs. 1528 to 1530, will now be considered. The mandrel A is of sound steel; wrought iron and double shear steel are often used, but a hard crucible steel is found very convenient, as there is then no necessity to harden the necks. If wrought iron is used, the necks must be case-hardened with yellow prussiate of potash, or steel necks must be welded on, when distortion is liable to occur, and subsequent grinding is necessary. The mandrel must be finished with a broad-nosed tool, the necks being polished with very fine emery charged on gripping laps. The threads for the check nuts at the rear (twenty-five to the inch) are cut from change wheels while the spindle is between the centres. The mandrel nose must have a clean-cut thread to 1 in. Whitworth standard. To bore the hole for the centre, the mandrel is chucked with the nose in a cone plate, and the slide-rest is set at an angle of 2½° with the centre line.

Cutting Mandrel Key-ways.

Two key-ways must be cut in the mandrel, one at a (Fig. 1528) for the wheel B, and another at b to take the hinder cone c. These can be cut, in the absence of a machine, with a narrow chisel and file; or, better, in the lathe by mounting a cutter in a chuck, fixing the mandrel on the cross slide of the rest, and traversing it past the revolving cutter.

Headstock Cone Bushes.

The cone bushes D and E in the head are often of steel, and should then be hardened and ground; but a hard quality crucible steel can be used without hardening; or

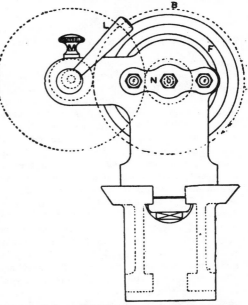

Fig. 1530.—End Elevation of Lathe Headstock.

hard gun-metal or phosphor-bronze, materials which are very durable, may be used; they require no treatment after turning, except scraping, or slight hand adjustment, by grinding the spindle in the bush with a little fine emery. The greatest care must here be exercised, because, if they are inaccurately adjusted, there is no satisfactory method of correction. The remainder of the work on the headstock is easy by comparison.

HEADSTOCK PULLEY.

The cone pulley F (Figs. 1528 to 1530) is cast from a pattern having allowances for turning. The grooves for the cords are turned from the solid in the manner explained on p. 528 in connection with the driving pulley on the crank-axle. The

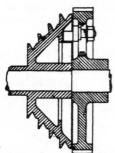

Fig. 1531 —Clamping Headstock Wheel and Pulley.

diameters at the bottoms are $3\frac{1}{4}$ in., $4\frac{1}{16}$ in., $4\frac{13}{16}$ in., and $5\frac{3}{4}$ in. The pulley should be turned inside for perfect balancing at high speeds. But here enters the question of locking to the wheel B. In some small lathes the wheel and cone are locked simply by friction at the bosses and rims by the tightening of the check nuts c at the rear; this is not so satisfactory a means as a bolt that slides in a lug cast either in the cone or on a plate screwed to the cone. Fig. 1531 illustrates a sliding bolt arrangement. A ring is cast within the cone and slotted, and a slot is cast in the change wheel. By sliding the bolt inwards the two are disengaged, and by sliding it outwards the two are locked together. The boss on the rear of the cone is turned to receive the back gear pinion G (Figs. 1528 to 1530). The pulley is retained endwise by the circular check nuts c (Fig. 1529), each of which has half a dozen tommy holes.

LATHE BACK-GEAR.

The back gears are cut; they can be purchased, or blanks may be cast and turned. There are twenty teeth of 10 diametral pitch in the pinions, and sixty teeth in the wheels. The wheel on the spindle is keyed, and the pinion is forced over the boss on the cone. The wheels on

the eccentric spindle H (Fig. 1529) at the back of the head are cast in one with the sleeve. By an alternative method (Fig. 1532) the wheels are made separately and driven over the sleeve. As gear cutting is rather costly, it may be worth while, if a lathe with overhead and division plate is available, to make or purchase two cutters and cut the teeth in the lathe. The back gear is thrown in and out with an eccentric spindle J (Fig. 1529) turned to fit the bore of the sleeve H. The eccentric is turned larger than the spindle, as shown at the left hand. At the right hand it is turned down to enter an eccentric bush K, so that the spindle can be inserted endwise. The bush is retained in position with a grub screw. The eccentric spindle is thrown over by the handle L (Fig. 1530), which is screwed into it. The pin M locks the spindle in the required position. The end thrust of the main spindle is taken against the bridge N (Figs. 1528 to 1530) of cast or wrought iron; this is secured to the rear of the headstock with pillar bolts e tapped into the head with $\frac{1}{2}$-in. Whitworth threads. Through the centre of the bridge a $\frac{3}{8}$-in. steel screw O is tapped with a fine thread; this is fitted with a lock nut, and hardened at the point to bear against the spindle end.

POPPET MANDREL.

The mandrel A (Fig. 1533) of the loose poppet is made from solid steel, turned

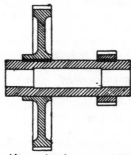

Fig. 1532.—Alternative Arrangement of Back Gear.

and bored in the manner commonly adopted on lathes of English design. Sometimes a piece of thick tubing is taken, which is very well if of sufficient thickness;

but light tube is of no use, because a bush has to be welded in to receive the bore for the point centre. If difficulty is experienced in boring a steel or wrought-iron spindle the mandrel may be cast and cored in hard gun-metal; this would wear well, but is out of harmony with one's ideas of lathe work. The mandrel should be turned outside first, and then bored. Both operations should be done by charging a common hinged up lap with fine emery, and traversing it along lightly a few times. It will be well also lightly to grind out the bore of the poppet casting, with a lap running between lathe centres, the poppet being traversed along it by hand. Whenever the spindle is tried in, the emery should be wiped off and the spindle and bore lubricated with clean oil. Without this precaution results will be deceptive. The mandrel is next chucked between the point centre at the headstock end and a cone plate at the end opposite, and bored to receive the ½-in. screw, allowing a little clearance, as shown. This may be done with a drill, followed by a reamer. The rear end is counter-bored with a common boring tool to receive the steel bush B, which is threaded for the screw. The bush is bored and turned to a driving fit, and is prevented from revolving by a grub screw (see Fig. 1533). The thread should be cut in a screw-cutting lathe, after the insertion of the bush, and should be of square section, as shown. Failing this, a **V** thread may be cut with a tap, and if the screw makes a good fit therein there will be a good deal of wear in it. Next the mandrel is reversed in the lathe, and, still using the cone plate, the taper is bored at the front end for the nose. This taper is the same as that in the nozzle of the fast headstock, and is bored in the same way. The spindle is prevented from turning by a groove *a* on its under side, in which a nib on the locking bolt c (Fig. 1534) fits. The groove can be made with a rotary cutter driven by the lathe, while the spindle is traversed past it in the slide-rest.

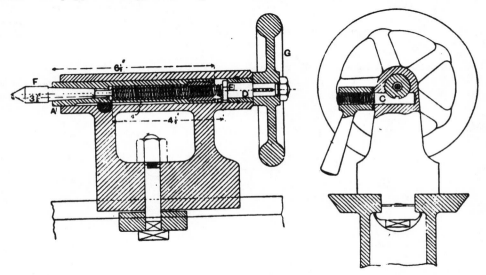

Fig. 1533.—Vertical Section of Lathe Poppet.   Fig. 1534.—End Elevation and Section of Lathe Poppet.

POPPET PUSHING SCREW.

There are several ways of retaining the pushing screw endwise; Fig. 1533 shows a common one. A bush D is turned to fit the rear of the bored end of the barrel, and is held in place with a couple of pins, in holes drilled transversely. An equally frequent practice is to screw a collar or boss at the rear end of the barrel, to receive the thrust of the collar on the pushing screw. The screw E is turned with a collar to fit against the inner face of the collar D, and the boss of the hand-wheel retains it on the other side. At the front end a portion is turned parallel beyond the screw to push out the point centre F. The latter is made of temper steel, turned and ground to fit the bore, the point only being hardened. The hand-wheel G is of cast-iron, made from a pattern; it is retained in place against a shoulder by means of a nut, and drives by a feather key. The spindle is locked up by a bolt C (Fig. 1534) having a feather entering the groove in the spindle. A hole is drilled in the casting, encroaching on the bore, and the bolt is filed to fit the curve of the spindle and its key groove. The tail is screwed to receive a long nut which tightens it with a handle. The thread is ½-in. Whitworth.

through a hole drilled near the end. The eccentric is a forging, having an eye at one end drilled to fit over the spindle, and screw at the other end to a length to suit the washer. The tightening boss for the shank of the T-rest is drilled and tapped to take a square-headed bolt, or one with a round head D, drilled for a tommy. Figs. 1538 and 1539 show the rest for hand-turning, a plain casting, which is simply cleaned over on the top edge with a file.

T-REST AND BORING R-REST.

Figs. 1540 to 1543 show a T-rest and a boring rest, the first of which may be either a casting or a forging. The latter should be forged, because considerable strain is put upon it.

SLIDE-REST.

Slide-rest construction has already been so completely described (see pp. 426 to 436)

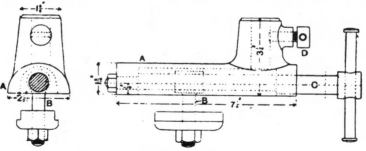

Figs. 1535 and 1536.—Hand-rest Socket, etc.

HAND-REST SOCKET, ETC.

The hand-rest socket, etc., shown in Figs. 1535 and 1536, is tightened with an eccentric which is more rapid in action than a screw. The foot A is cored out in an arched form to leave clearance for the eccentric B. The metal at the ends is bored to receive the eccentric spindle C; this can be done in the lathe or in a drilling machine after the foot has been planed or filed. Then this face will be blocked up and bolted either to the cross slide or the rest of the lathe, or to the face of an angle-plate under the drilling machine, and the holes are drilled and reamered. The steel spindle (Fig. 1537) is turned to fit, and is operated by a lever passed

that it is not necessary to treat the slide-rest of this lathe in tedious detail. It is hand operated. Its base has a tongue fitting closely between the shears of the bed. The upper part swivels for taper turning and boring. In Figs. 1544 to 1547, A is the saddle, B the cross slide, C the longitudinal and swivelling slide, D the top or tool slide, and E the tool clamp. These are all in cast iron, tooled everywhere for precise fitting and for good appearance. The tooling necessary will offer difficulties to those who have no shaping or planing machine. Filing, by reason of the large extent of the surfaces, is a tedious job. It will be better to pay for planing, but this need not include more than one cut, which, though showing the tool marks, will yet give true surfaces on the whole for final finishing with the file and scraper.

FITTING SLIDE-REST TO LATHE BED.

The base A is fitted to the bed with the tongue *a* (Figs. 1545 and 1546), and is

Fig. 1537.—Eccentric Spindle for Tightening Hand-rest.

clamped to the bed through the lugs *b* at the sides of A, which extend over the ends of the tongue. The clamping bolts *c* are tapped into the clamping plate F, within the bed. The fitting of the tongue in relation to the saddle or base A is of the first importance. The two must be exactly at right angles, otherwise work cannot be faced accurately. The casting is therefore first planed or shaped, and the tongue fitted by file and scraper between the bed shears, red-lead being used to effect final corrections. If too much is taken off the tongue, a slack fitting results; this cannot be corrected, and the casting must be thrown away. This fitting therefore must be attended to before that of the transverse edges of the saddle. When this is fitted, some means of taking up future wear would be desirable, but is not easily

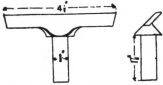

Figs 1538 and 1539.—T-rest.

practicable. Besides, if the tongue fits exactly, its broad area can stand much wear, and if refitting becomes necessary, a strip can be riveted on.

TESTING TRUTH OF SLIDE-REST.

At the next stage, bolt the saddle down to the bed near the headstock, and check the accuracy of the guiding edges by a scribing point or short bit of wire fastened to a face-plate on the mandrel nose, and brought into contact with the transverse

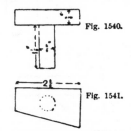

Figs. 1540 and 1541.—T-rest.

edges on opposite sides. This will be more accurate than testing with a square by the inner edges of the bed shears. The truth of the top faces of the saddle can be tried with a straight-edge laid over them—the longer the better—parallelism being tested by means of internal callipers between the bed face and the under side of the straight-edge. A good deal of correction will be required by file and scraper to ensure the accuracy of the surfaces of the top and edges both lineally and in relation to the faces and edges of the bed; for unless the base is correct the work on the upper parts will be as good as thrown away.

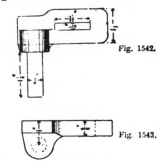

Figs. 1542 and 1543.—Boring Rest.

CROSS-TRAVERSE SLIDE.

The cross-traverse slide B not only has to to be fitted to the base A, so that it will slide with equal freedom in any part of its traverse, but the top face has to be turned and the annular groove for the **T**-headed bolt has to be central with the slide. The best course is to fit the slide first, making

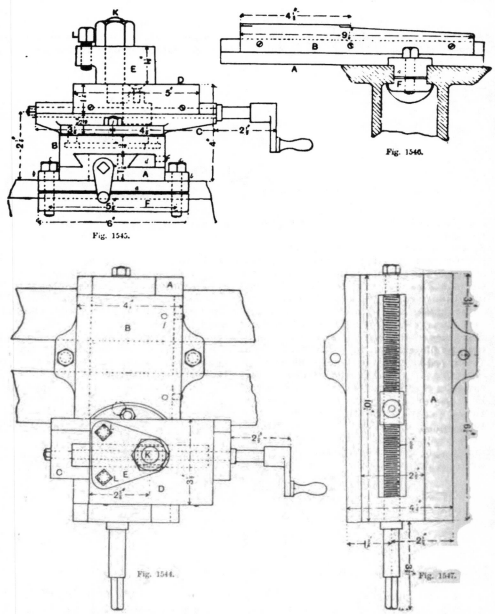

Fig. 1545.

Fig. 1546.

Fig. 1544.

Fig. 1547.

Figs. 1544 and 1545.—Plan and Elevation of Slide-rest. Fig. 1546.—Cross Slide, etc., and Section of Lathe Bed. Fig. 1547.—Saddle, etc, of Slide-rest.

B fit A by scraping the surfaces already planed. The loose strip d (Fig. 1545) is of cast- or wrought-iron, and fitted in the manner shown. Slack due to wear is taken up by three setting-up screws e. If these are slightly countersunk into the strip, screws need not be inserted at right angles to prevent end movement. The positions of these screws, if inserted, are seen at f (Fig. 1544). After the cross-slide is made a good fit to the saddle, a centre line should be scribed from the ends and across the top. This can be done by laying a parallel straight-edge across the face-plate, and scribing from that with compass calliper to the face of the slide. By this centre line the slide is chucked for turning the circular face and boring the annular T groove for the swivel slide. The outer edges are also scribed off for neat finishing, since, in the course of fitting B to A, the outer edges may overlap a trifle. The slide B will be packed up on the face-plate with parallel metal packing, so that the top can be faced true with the bottom. The edge will be skimmed over, and the T groove turned and a shallow pin turned to enter the recess on the bottom of the swivel slide. The truth of these parts will, of course, depend on the care with which the slide B has been chucked and packed on the face-plate of the lathe. This slide is operated by a screw and nut. These are seen in Fig. 1547, the gun-metal nut being shown by Fig. 1548. The traverse screw is properly square-threaded, but one with V threads, cut with dies, will do fairly well if difficulty is experienced in cutting a square thread in a lathe. The screw must be perfectly straight, or the slide will move easily and tightly alternately. The holes in the ends of A must be central with the tapped hole in the nut. The latter should be fitted and screwed to the slide B before the hole for the thread is drilled. Then the centres can be scribed on the nut face, and on the ends of A. Or the centres can be scribed at the ends of A only, the nut being brought near one end, and drilled through from A without marking off. The

nut can then have its thread cut in a screw-cutting lathe, or, if a V thread is used, with a tap.

SWIVEL SLIDE

In the swivel slide c the faces and V'd edges are finished first, and the slide is then chucked by its top face for turning the circular foot and the edges and the recess. Holes in the edges of the foot receive the bolts j (Fig. 1545), which move round in the annular groove for setting the slide at any angle. The edge of the foot is graduated into degrees, fifty on each side of the zero line, which is marked in the bottom slide. Unless these divisions are accurate, they will be useless.

COMPLETING SLIDE-REST.

On c is fitted slide D (Fig. 1549), with its loose strip g. The remarks made in reference to the fitting of B apply also in the

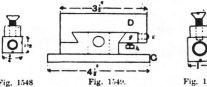

Fig. 1548.   Fig. 1549.   Fig. 1550.
Fig.1548.—Nut for Cross Traverse Slide. Fig.1549.—Top Slide. Fig 1550.—Nut for Top Slide.

main to this. The base strip must have a screw h inserted from beneath to retain it against the face of the slide. The grub screws i set it up and take up wear. The traverse screw is fitted similarly to that in the lower slide, and its nut is shown in Fig. 1550. The clamping piece E (Figs. 1544 and 1545) is slipped over a bolt K, which is tapped in D. The screws L are tapped in it to pinch the tool. This fitting is handier for a small lathe than a couple of clamping plates on springs would be. All the non-working faces of the rest may here be polished with emery-cloth for appearance. The lathe with headstocks, hand-rests, and slide-rest is now complete, and full instructions are given in earlier pages for making such needful accessories as face-plates and chucks.

GOLD AND SILVER WORKING.

This chapter will afford information on the practice of gold and silver working, and will in particular discuss the making

the end off as in Fig. 1551 ; make the strip perfectly flat and parallel, and bevel the edges with the file, so that when the plate is turned up to form the chenier (Figs. 1552 and 1553) its edges will abut as in Fig. 1554,

Fig. 1551.—Strip of Flat Metal for making Chenier.

of simple brooches, ear-rings, and other ear ornaments, and scarf pins and similar articles. Processes and examples will be copiously illustrated so as to make them intelligible to any persons taking up gold and silver working for the first time.

and not be as shown in Fig. 1555. This is not of much importance in small chenier, but is very necessary in large chenier, in order to avoid any sign of bad workmanship, such as filling up with solder. The flat strip is made into chenier with the aid of a hammer and bending block.

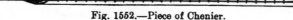

Fig. 1552.—Piece of Chenier.

MAKING CHENIER FOR BROOCH TONGUES.

The materials for making brooch tongues are " chenier " (tubular metal) and wire. The term " chenier " is derived from the French *charnière*, a hinge or joint. The

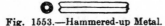

Fig. 1553.—Hammered-up Metal.

chenier is made from flat plate (gold, silver or German silver) of a thickness suitable to the size of tube required ; the thickness ranges from size 37, new standard gauge, for small brooches, to size 27, for very large brooches. The metal, having been flattened, is cut to the length and width required, as in Fig. 1551. The plate should be three times as wide as the diameter of the chenier. Cut

THE BENDING BLOCK.

The bending block is usually of hardwood or horn for small work, and of bronze, brass, or iron for heavier work. It has a number of grooves varying in size (see Figs. 1556 to 1558), and is held in the vice, or by the grooved out bottom of a sparrow-hawk block ; the last-named is a

Fig. 1554. Fig. 1555.

Figs. 1554 and 1555.—Joints Rightly and Wrongly Made.

small anvil, held between the knees, having two horns, one of which is conical and the other of the shape of a pyramid.

HAMMERING UP CHENIER.

The successive stages in the formation of chenier are illustrated by Figs. 1556 to 1558. The metal must first have been annealed. The edges are got up before the centre, and the pane of the hammer

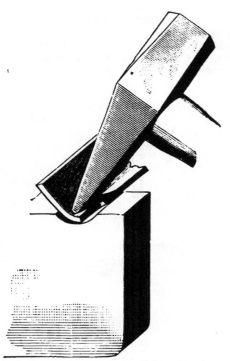

Fig. 1556.—First Stage in Making Chenier.

must be nicely rounded and not sharp at the edge, where it might cause awkward dents and bruises. A wrong way of making chenier is shown by Figs. 1559 and 1560, in which it is easy to see that the outside edges will be reduced considerably before the draw-plate is able to get the tube round on the outside, and even then the hole will be pear-shaped. No amount of clearing the hole out with a reamer or joint-brooch will correct it, and, unless the tubing is properly made, it will be impossible to get a smooth working joint. Chenier drawn over a mandrel will be dealt with later.

DRAWING CHENIER THROUGH DRAW-PLATE.

In order to finish the chenier and make it ready for soldering on, it now only requires to be drawn through a draw-plate to get it to the requisite size. During this process the point may prove to be too weak to stand the strain of drawing down, in which case it will have to be strengthened by soldering it up, or by introducing a piece of wire and soldering that as well. Generally, however, for the common run of brooch joints these expedients are unnecessary

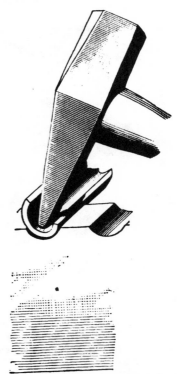

Fig. 1557.—Second Stage in Making Chenier

ANNEALING AND PICKLING CHENIER.

The drawn chenier is annealed to clean it from any grease it may have picked up in its passage through the draw-plate, and put in the pickle, consisting of one part of sulphuric acid to thirty parts of warm water. When the metal is clean, remove it from the pickle, rinse it well in clean water, and dry it. If the metal is rather hard, it would be as well to tie several pieces of iron binding-wire round it, or even to bind it with wire from end to end, in order to

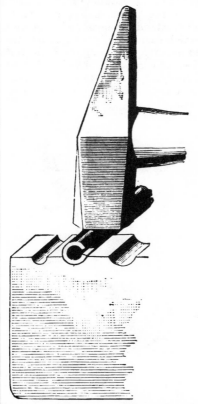

Fig. 1558.—Third Stage in Making Chenier.

keep the seam closed up. Score the chenier across the join with a saw or three-square file, as in Fig. 1552, and it will then be ready for use.

MAKING BROOCH TONGUE.

The wire out of which the tongue is fashioned should be drawn down to the proper size, and should not have been annealed for the last six to twelve holes of the draw-plate, so as to leave it rather hard. Making the tongue consists merely of soldering a piece of wire on a piece of tube. Cut off a piece of chenier and place it joint upwards on a flat charcoal or pumice-stone

block, grooved to retain it. Place the wire so that its end crosses the chenier at right angles (see Fig. 1561) and rests upon the joint. The parts in contact should be scraped clean, and the part of the wire to be soldered should be very slightly flattened with the hammer. Solder by means of the blowpipe flame (see pp. 227 to 233), taking care not to make the wire red-hot right along its length. Prevent the solder filling up the hole of the chenier by running a single horsehair through it. The piece of chenier must be soldered right to the ends.

HARDENING BROOCH TONGUE.

The tongue is finished by boiling out or pickling (a process already described) to

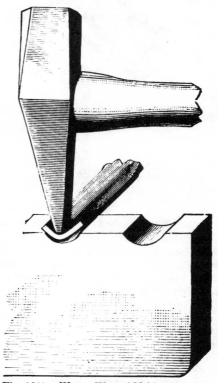

Fig. 1559.—Wrong Way of Making Chenier.

remove the charred borax, etc. Then it has to be hardened, fitted to the joint, filed up, and given the proper shape ready for the polisher. Work from the boiling-out or pickle pan must always be rinsed in clean water. With regard to the hardening, if

Fig. 1560.—Wrong Way of Making Chenier.

the tongue be of 9-carat gold, merely hammering it round and round on the flat iron, small anvil, sparrow-hawk, or beak iron will be sufficient; but gold of a better quality, or German silver, must be twisted (see Fig. 1562) before it is hammered, otherwise it will not be hard enough. Fig. 1562

Fig. 1561.—Soldering Chenier to Wire.

shows it being twisted only near where it has been soldered; the part that is held in the vice is purposely drawn down hard, in order to save the trouble of doing the work twice.

COMPLETING BROOCH TONGUE.

The tongue is fitted into the joint in so simple a manner that description is unnecessary; but care is needed to get it to fit well. As to filing up, it should be given a good shape in section; generally, it is either round or oval, though occasionally it is triangular or bayonet shape. The points should be filed to the shape shown by Figs. 1563 to 1569, and not to any of the shapes shown by Figs. 1570 to 1572. Figs. 1570 and 1571 are obviously weak, and Fig. 1572 is clumsy as well as ugly. The end must be filed to a point, and not left partly done, as is sometimes the case. With reference to the shape or curve (if any) that a

Fig. 1562.—Hardening Gold by Twisting.

brooch tongue should take, Figs. 1563 to 1569 (p. 542) afford sufficient guidance, it being remembered that a brooch tongue pierces like a bayonet or sword, not like a cork-screw; it does not wriggle through cloth or other dress material. Fig. 1566 shows a tongue intended to prevent a top-heavy brooch from tilting forward. Figs. 1573 and 1574 show two brooch pins that differ from those previously given. Fig. 1573 is simply the wire turned round and soldered, and this is used with a form of so-called Etruscan joint, shown on p. 543. The pin shown by Fig. 1574 is made with

and should be suitable to the size of the brooch and in keeping with the character of the ornament. The most common form is that where the fly-up plate is soldered to the joint so that the tongue obtains its

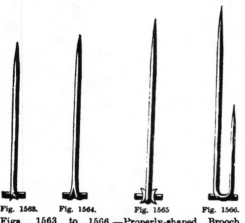

Fig. 1568. Fig. 1564. Fig. 1565. Fig. 1566.
Figs. 1563 to 1566.—Properly-shaped Brooch Tongues.

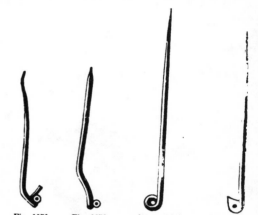

Fig. 1571. Fig. 1572. Fig. 1573. Fig. 1574.
Figs. 1571 and 1572. — Wrongly-shaped Brooch Tongues. Figs. 1573 and 1574.—Wire Turned and Soldered.

a flat plate, which is afterwards drilled and filed to shape. The hardening of the tongue is of the greatest importance; because on the power of the tongue to act as a spring always pressing against the top of the catch, much of the safety of the brooch depends.

spring from it (see Figs. 1563 to 1566). Then there are those without the fly-up plate, for which the tongues shown by Figs. 1567 and 1569 are used. Finally, there is the form of joint where the tongue is removable at pleasure.

MAKING BROOCH JOINT.

To make a brooch joint, take a piece of chenier of suitable length and thickness from the centre of which there has been cut away a portion less than one-third of the length, and only half-way through (see Fig. 1575). Then tie it on a plate of fair thickness (Figs. 1576 and 1577), not less

Fig. 1567. Fig. 1568. Fig. 1569. Fig. 1570.
Figs. 1567 to 1569. — Properly-shaped Brooch Tongues. Fig. 1570.—Wrongly-shaped Brooch Tongue.

JOINTS OF BROOCHES.

The joints of brooches should be made strong enough for a fair amount of wear,

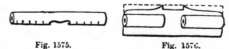

Fig. 1575. Fig. 1576.
Fig. 1575.—Chenier cut for Making Joint. Fig. 1576 —Chenier Soldered on Ready for Brooch Tongue.

than No. 30 gauge, and solder it with the joining or soldering seam in actual contact with the plate. Then remove entirely the piece of chenier at that part previously cut, so that the tongue can fit in between the two knuckles or side pieces of the joint; a door hinge gives the idea. The other way of making a joint—by soldering each knuckle on separately—is but rarely

THE JOINT PIN.

The joint pin itself should be of hard wire (but not of hard steel), and should taper gradually, if at all, and be finished with a smooth file; it should bind on the

Figs. 1578.   Fig. 1579.

Figs. 1578 and 1579.—Joint Knuckles Supported.

Fig. 1577.

Fig. 1577.—Brooch Tongue Joint.

used for single brooches, and more properly belongs to the joints of lockets and boxes. It is preferable to raise the chenier slightly from the back of the brooch, as in Figs. 1576 to 1579, though some workers put it close down on the brooch. In Fig. 1577, the dotted lines show the maximum angle to which the tongue should be able to move. If the fly-up plate has been made too weak, the side knuckles of the joint must be supported by grains as in

two outside knuckles so as not to rotate when the tongue moves; it should also be tight enough to keep the tongue perfectly steady, instead of wobbling up and down, and from side to side. Although in practice the joint pin is done last of all, yet it is convenient now to complete the description of it. The joint pin, if properly made and fitted, should stay in its place by its own friction, but slightly riveting the ends of the pin adds to the security, and if

Fig. 1580.   Fig. 1581.   Fig. 1582.

Fig. 1580.—Joints without Fly up Plate. Figs. 1581 and 1582.—French Barley-corn Joint.

Figs. 1585.   Fig. 1586.   Fig. 1587.   Fig. 1588.

Figs. 1585 and 1586.—Ball Joint. Figs. 1587 and 1588.—Badly formed Narrow Joint.

Fig. 1578, or by a bar as in Fig. 1579; but a skilled workman will not need either. Next file up the fly-up plate—something like Fig. 1576 or Fig. 1578 will do, or any other shape, providing that it is not liable to catch in lace that a lady might wish to wear with it. When the joint is soldered

carefully done will not damage either brooch or joint. The joint pin, however, must fit properly, and riveting should not be adopted to compensate for bad workmanship.

JOINT WITHOUT FLY-UP PLATE

If the joint is not to have a fly-up plate (see Figs. 1580 to 1582), prepare the chenier

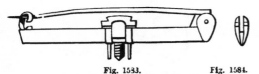

Fig. 1583.   Fig. 1584.

Figs. 1583 and 1584.—Ball Joint.

Fig. 1589.   Fig. 1590.

Figs. 1589 and 1590.—Etruscan Joints.

in its place and the tongue fitted, the hole should be regulated with a reamer or joint-broach from the top end.

in the same way as before—that is, with part cut away from the centre—and proceed to solder it on the back of the brooch, or on a plate such as is shown in Figs. 1580 to 1582; then cut away the remaining half of the centre as before, and fit the tongue. The tongues for this class of joint have to be made like Figs. 1567 to 1569 in order to obtain the necessary spring, which in the previous case came from the fly-up plate. Figs. 1582 to 1590 show some different forms of these joints. The Barleycorn (Figs. 1581 and 1582), is made from thick chenier, and is filed to shape ; it is mostly used on French jewellery. The ball-joint (Figs. 1583 to 1586) is the form usually

Fig. 1591.—Plan of Antique Brooch Joint.

found in the fittings for diamond work. It is made from two pieces of thick gold, like Figs. 1583 and 1584 ; or, instead, a grain of gold is melted to form a globule, and the centre is cut out, as in Figs. 1585 and 1586, to take the tongue—formed with a flat plate instead of chenier—and the whole is drilled through together to take the joint pin. Either way is an improvement on the method illustrated by Figs. 1587 and 1588, in which the knuckles are weak ; but Fig. 1583 is decidedly the better of the two, inasmuch as each knuckle can be well and securely soldered to the side of the brooch mount. Fig. 1589 shows a joint whose knuckles are made out of a coil of rings soldered together. This, which is sometimes called Etruscan, is undoubtedly one of the best forms of joint obtainable ; there is no soldering seam to become afterwards opened, and this method is even

Fig. 1592.—Side View of Antique Brooch Joint.

stronger, it is thought, than the seamless chenier that has been introduced, for that might split if there were a flaw in the gold, whereas any flaws show themselves in drawing down the wire, and cause it to

break long before the wire is reduced to the small size required for this work.

ETRUSCAN OR ROMAN BROOCH JOINTS.

Fig. 1590 shows a form of joint used in the modern Etruscan or Roman gold work ; it is made from one piece of wire, as the diagram clearly shows. A tongue for this

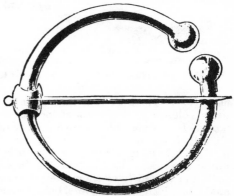

Fig. 1593.—Celtic Brooch Fastening.

is illustrated by Fig. 1573. Generally a piece of tube takes the place of the joint pin, and it is chamfered or burred over, so there is no fear of dropping the tongue out. This is a joint of the narrow and strong class.

ANTIQUE BROOCH JOINT.

The coiled wire replacing the ordinary joint in a safety-pin brooch does not vary much in pattern, except that occasionally there are three turns instead of two. Figs.

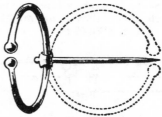

Fig. 1594.—Celtic Brooch Fastening.

1591 and 1592 show such a coiled wire on an ancient brooch now in the British Museum. The method of obtaining the spring is ingenious, and there is but small risk of the tongue breaking off short. In this case and in that of the brooch shown by Figs. 1593 to 1595 there is no attempt to

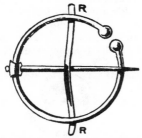

Fig. 1595.—Method of Securing Celtic Brooch.

hide the means of fastening, the aim being to turn a necessary part of the construction into an ornament. With regard to the means of securing the brooch shown by Figs. 1593 to 1595, there is no catch, the article of clothing itself preventing the brooch from falling. First put the tongue through the shawl, etc., as in Fig. 1594; then place the point between the ends shown by dotted lines, and bring it on to the top or front of the rim, as in Figs. 1593 and 1595, and slide it round. The shawl now acts in the way shown by the rod R (Fig. 1595). The hold is quite secure, but this method is only suitable for large brooches that have to fasten shawls and plaids, and similar heavy articles.

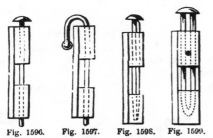

Figs. 1596 to 1598.—Brooch Joints with Removable Pins. Fig. 1599.—Brooch Joint with Partly-removable Pin.

JOINT WITH REMOVABLE TONGUE.

A joint that allows of the temporary removal of the tongue is often required in valuable ornaments that have to do duty for brooch, pendant, or locket, bracelet-

centre, hair-pin, etc., at the pleasure of the owner. The simplest way is to make the joint pin to take in and out ; just solder

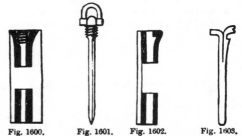

Fig. 1600. Fig. 1601. Fig. 1602. Fig. 1603.

Figs. 1600 and 1601.—Screw Joint and Pin. Figs. 1602 and 1603.—Spring Joint and Pin.

a top on like Fig. 1596 or Fig. 1597, and remember that all these easily fitting pins are to be inserted from the top end of the joint. Fig. 1598 is a split joint pin, made with two half-round pieces of wire soldered together at the two ends ; this can be made to spring apart just a little, and so give some sort of a hold. Fig. 1599 is much the same, but it has a peg inserted and soldered to the knuckle ; this allows of only the partial withdrawal of the joint pin, and prevents its loss. It is a way much used to fasten Indian bracelets, for which it is more suitable, it being rather clumsy on a

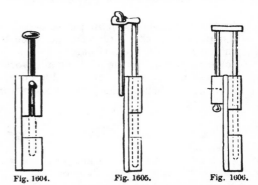

Fig. 1604. Fig. 1605. Fig. 1606.

Fig. 1604.—Joint with Partly-removable Pin. Fig. 1605.—Joint acting on Guiding-bar. Fig. 1606.—Joint with Bar acting through Tube.

brooch. Fig. 1600 has the top knuckle enlarged, and a plate or ring soldered in to take a screw. The joint pin (Fig. 1601) is made generally with a fall-down handle, but that will have to be governed by circumstances ; it is so made that it can be unscrewed with the fingers. In the joint illustrated by Figs. 1602 and 1603, there is a spring in place of a screw. These entirely withdrawing and removable joint

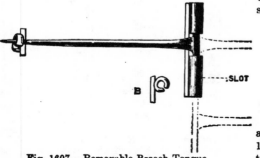

Fig. 1607.—Removable Brooch Tongue.

pins are always getting mislaid or lost, and a method of retaining the joint pin in its place is desirable. Fig. 1604 shows a slot cut in the top knuckle, in which a peg can work up and down, and to the peg the joint pin is soldered ; it should not be made to withdraw further than to allow of the removal of the tongue. The joint shown

Fig. 1608.—Brooch Tongue.

by Fig. 1605 has the joint pin soldered to a cap, and the cap is pierced and fitted to run on a parallel guiding-bar, or the bar is soldered parallel to the joint pin and works through a tube T (Fig. 1606), soldered underneath or behind the joint or wherever

Fig. 1609. Fig. 1610. Fig. 1611.

Figs. 1609 and 1610.—Correct Brooch Catches. Fig. 1611.—Brooch Catch Badly Shaped.

is most suitable for the particular job (see p. 548 for full particulars of a fall-down catch which should be used with these removable tongues). Thus far only the withdrawal of the joint pin to allow of the removal of the tongue has been considered, and for that purpose all has been worked

from the top knuckle, so that if by accident or wear the joint pin becomes loose, it will not drop down and disengage the tongue. In the next example (Fig. 1607), the bottom knuckle is made of service, it having a slot cut right through it in the position indicated by B. The joint pin is soldered to the tongue, which is flattened

Fig. 1612. Brooch Catch and Sections of Tongue.

and broadened out (see Figs. 1607 and 1608) ; it is placed in position sliding along the slot, as indicated by the dotted lines ; then when the tongue arrives in between the knuckles, it is turned down towards the catch in the usual way. This is the most general joint of its kind ; but the soldered connection of the joint pin and tongue must be sound and yet leave the round joint pin quite clear where it joins the tongue. The slot must be just wide enough to allow the joint pin to come out sideways.

Fig. 1613. Fig 1614.

Figs. 1613 and 1614.—Brooch Catches with Springs.

BROOCH CATCHES.

A brooch catch can be defined as a kind of hook, so shaped that it will retain the tongue in its position, and yet allow of its withdrawal easily. The forms of catches

Fig. 1615. Fig. 1616. Fig. 1617.

Figs. 1615 to 1617.—Safety Catches.

vary considerably, but the principle remains the same. They are all hooks of one kind or another. To begin with the simplest and most general forms, such as Figs. 1609 and 1610, it will be seen that the shape is a nice O-scroll ; careless working will give the shape shown by Fig. 1611. In making a good scroll, it is the very ends of the wire or metal that have to be bent first, and when these are properly curved it is easy enough to get the other part into shape. There is no actual need for the curls to be wider apart than just to let the tongue pass easily in and out. Fig. 1612 illustrates this, and shows the tongue in section at several places. Both of the ends are thinned down to improve the shape, and, at the top end, to faciliate the tongue's entry and exit.

Fig. 1618. Fig. 1619.
Figs. 1618 and 1619.—Safety Catches.

SPRING CATCHES.

Spring catches ought really to be called safety catches; but custom decrees that " safety " is the word to be used when " protecting " catches are what is actually meant. In the catches shown by Figs. 1613 and 1614, there is a spring acting against the lower part of the top curl of the catch,

Fig. 1620. Fig. 1621. Fig. 1622. Fig. 1623.
Fig. 1620.—Safety Catch. Fig. 1621.—Metal for Making Safety Catch shown by Fig. 1620. Figs. 1622 and 1623.—Wire Safety Catch.

which is prolonged as much as possible, in order not to let the tongue drop down below it, and also to give an unmistakable lead when it is desired to unfasten the brooch. Of the two, the one shown by Fig. 1613 is the better, because the spring is a separate piece of gold, and

Fig. 1624. Fig. 1625. Fig. 1626. Fig. 1627.
Figs. 1624 to 1627.—Wire Safety Catches.

therefore can be made of any quality and hammered as much as is desired; if it is soldered as in Fig. 1614 there is the chance of breaking it off, besides the greater difficulty of

getting it hard enough to act as a spring. The separate spring in Fig. 1613 is fixed in its place by a pin or screw passing through it; or it can be pewter-soldered in place, though the use of pewter solder as a general thing is not desirable. Safety catches —catches that protect the point of the

Fig. 1628. Fig. 1629.
Figs. 1628 and 1629.—Safety-pin Brooch.

tongue—are made as in Figs. 1615 to 1627. All of these are a means of guarding the point of the tongue, and this is the case also with every one of those shown by Figs. 1628 to 1632. It may be said that Fig. 1621 is the shape of metal from which the catch illustrated by Fig. 1620 is made. Catches

Fig. 1630. Fig. 1631. Fig. 1632.
Figs. 1630 to 1632.—Safety-pin Brooch Catches.

shown by Figs. 1619, 1620 and 1625 are used for narrow bar brooches, the width being represented by the right-hand part of Fig. 1619 and by Figs. 1621 and 1626. The ancient Etruscan brooch (Fig. 1633) has a large protective catch, and such work was,

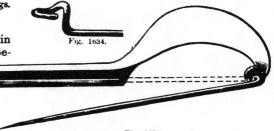

Fig. 1633.

Fig. 1633.—Ancient Etruscan Brooch. Fig. 1634.—Bad Form of Safety-pin Brooch Catch.

as a rule, beautifully ornamented with fine wire and grains of gold. Fig. 1628 is a typical example of a safety-pin brooch, which may have two or three turns. It will be understood that Fig. 1629 is another view of the construction shown by Fig.

1628 ; this is the neatest style of all those illustrated. The catch with five turns (Fig. 1631) is the safety-pin or non-scratching pin

Fig. 1635.   Fig. 1636.
Figs. 1635 and 1636.—Fall-down or Hinged Catch.

of surgical practice ; and it is this use that may account for the fact that it is known as both pin and brooch. A very bad example is shown by Fig. 1634 ; this is a shape of catch to be avoided.

vent the catch falling outwards a stop is formed, either by the catch itself (Fig. 1636) or by soldering a grain of gold to the side of the joint, which does quite as well. The method shown, however, keeps the work

Fig. 1638.—Lyre-shape or Clip Catch.

more compact, for the grain of gold would only be used when the catch is soldered on top of the joint. This is the catch to be used with any of the removable tongues already referred to.

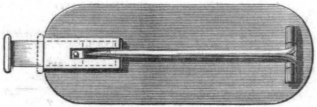

Fig. 1637.—Patent Sliding Brooch Catch.

FALL-DOWN BROOCH CATCH.

Figs. 1635 and 1636 show the most usual way of arranging a fall-down catch. The catch is simply soldered on to a joint in

SLIDING CATCH.

Fig. 1637 shows a slide arrangement which catches the end of the tongue. It is shown open, with the tongue in posi-

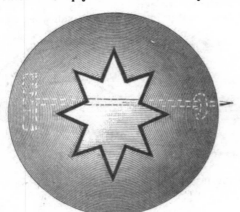

Fig. 1639.—Circular Brooch.

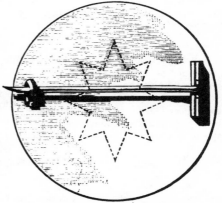

Fig. 1640.—Back of Circular Brooch.

such a manner that it can move about the quarter of a circle—that is, from its proper upright position to one where it lies down close to the back of the brooch. To pre-

tion, the dotted lines indicating its position when closed. This was introduced in 1848 by the firm of Unite, silversmiths, of Birmingham.

CLIP CATCH.

The lyre-shaped catch (Fig. 1638) is a clip catch that can be occasionally used to fasten scarf-brooches and ear-ring hooks.

Fig. 1641.—Edge View of Circular Brooch.

This shape of catch is easier to adjust for steadying the brooch than the **C** shape, but, of course, does not hold so well. The two horns or ends act as springs to some slight degree, and retain the wire—clipped in between them—steady, but hardly secure.

Fig. 1642.—Double-pointed Brooch Tongue.

BROOCH MOUNTS.

Directing attention to Fig. 1639, the front view of a circular brooch, it may be said that the perpendicular lines on the front of the brooch (be they either ornament or stones) depend on the position of the tongue, which must be so placed that it is parallel with the true horizontal line, or straight across the brooch, as in Figs. 1639 and 1640, but a little above the middle line, else the brooch will topple forward. The

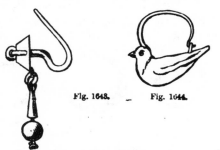

Fig. 1643.—Ancient Roman Ear-ring.
Fig. 1644.—Greek Bird Ear-ring.

star shown in the front view (Fig. 1639) is to suggest the horizontal and perpendicular lines ; the correct position of tongue—straight across and above the middle line—

is shown by dotted lines in Fig. 1639, and by full lines in Figs. 1640 and 1641. Fig. 1639 is the front view, showing by dotted lines the position of the various mounts as

Fig. 1645.—Modern English Ear-ring.

they would appear if seen through the brooch. Fig. 1640 is the back view, giving the positions of the mounts ; note that the opening of the catch is downward. Fig. 1641 is an edge view, to show parallel position of tongue when fastened. To get the tongue right, the joint and catch have to be placed properly. It is not difficult to see where the centre of the joint should be —it is in the same line as the tongue— and there is no difficulty with the catch either, if it is remembered that the natural spring of the tongue will force it to rest against the highest part of the inside of the top curl. This point being fixed, solder the highest point in line with the centre of the joint. Solder the joints and catch parallel, thus I I and not thus / \. Besides obtaining

Fig. 1646.—Modern Continental Ear-ring.

the horizontal position of the tongue in reference to the front, get a proper distance between it and the back of the brooch (see Fig. 1641). This distance varies with the size of the brooch, and with the material that it will have to fasten ; there need be but little space when a silk tie or lace is to be secured, while something like ¼ in. may be wanted for a heavy shawl brooch. The side view of the brooch (Fig. 1641) indicates the position that should be tried

Fig. 1647.   Fig. 1648.

Fig. 1647.—Greek Ear-ring with Spring Fastening.
Fig. 1648.—Greek Ear-ring with Coiled Spring.

for: it is, as can be seen, quite parallel with the back all along. As to the length of the tongue, the ordinary tongue projects a little beyond the catch, say ⅛ in. to ¼ in. If it is too short, it will not adjust easily; and if it is too long it will scratch the wearer of the brooch. Turn up the point a little way from the brooch, in order that the point may rest against the dress, and so obtain a little protection. The length of the tongue in a protecting or safety catch decides itself. The catch, it will be noticed, is placed with its opening downwards. Fig. 1642 indicates a suitable

Fig. 1649.   Fig. 1650.

Fig. 1649.—Indian Ear-stud.
Fig. 1650.—Russian Ear-ring.

way of using the double tongue, for if a narrow bar brooch has a heavy or high front, then it will look towards the ground, and show its edge instead of its front,

unless there are some such means as this to keep it up.

EAR-RINGS AND THEIR MODE OF ATTACHMENT.

In England, ear ornaments are attached to the lobe only, but in some parts of India

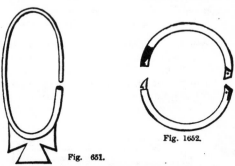

Fig. 651.

Fig. 1651.—Ancient Sardinian Ear-ring.
Fig. 1652.—Ear-wire or Sleeper.

the ear is pierced in the upper part as well, so that several pairs of ear-rings can be worn at one time; sometimes in addition, the ear is slit and lengthened by wearing heavy solid ear-rings. Another kind of ornament may be worn on pierced ears—the ear-stud; this rests on the ear, whereas the ear-ring hangs below it. Ear-rings for unpierced ears will be dealt with later.

SOME DESIGNS OF EAR-RINGS.

Fig. 1643 illustrates an ancient Roman ear-ring, with wire fashioned in the shape

Fig. 1653.   Fig. 1654.   Fig. 1655.

Fig. 1653 to 1655.—Ear-wires.

of what is now known as the "German hook." Fig. 1644 shows the Greek bird-shaped ear-ring, not now worn. Fig. 1645 illustrates a modern English ear-ring having the same type of fastening. Fig. 1646 depicts an ear-ring produced in Normandy, Germany (Bremen), Italy, and Spain. Each country stamps its nationality on the ornament by means of a slight variation of design, but in all cases the fastening is the same as in Fig. 1644. The ornament shown by Fig. 1647 is not much followed now in

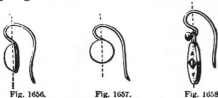

Fig. 1656.   Fig. 1657.   Fig. 1658.
Figs. 1656 to 1658.—Ear-ring Wires.

Europe, though it is in use in India at the present time ; it is, as will be seen, in the form of a monster's head, with a tail formed by a coil of tapering rings. This gives a fairly secure and practical fastening. Fig. 1648 also depicts an Eastern ear-ring which possesses the merit of security. The most general method of fastening Indian ear-rings or ear-studs (see Fig. 1649) is by means of a screw which is made by coiling double wire round a mandrel ; this mandrel is a piece of hard wire. A piece of wire is soldered inside a tube to form the female screw, and another piece is soldered on a peg to produce the male screw. Such screws are used in England for bracelet fastenings, being too large for ear-rings. Figs. 1650 and 1651 are interesting because they show types of ear-rings which hang in a manner contrary to custom ; they show their edges and parts of their backs to anyone facing the wearer.

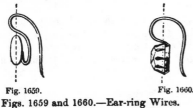

Fig. 1659.   Fig. 1660.
Figs. 1659 and 1660.—Ear-ring Wires.

EAR-WIRES OR SLEEPERS.

Ear-wires or sleepers are the simplest form of ear-rings now in use. They are usually made of plain round or faceted —that is, lapped—wire. Fig. 1652 shows the way they are made, their actual size

being indicated by Fig. 1653. The hardness of the gold itself acts as the spring to keep the catch in its place, and generally these ornaments are well and skilfully made, although sold at a low price. The ear-wire shown by Fig. 1654 is of the same

Fig. 1661.—Ear-ring with Correctly-shaped Wire.

kind, but has a more ornamental shape, this being a serpent with its tail in its mouth. This also gives certain mechanical advantages ; for instance, the head is broader and deeper, therefore stronger, and better suited to take the catch ; the catch is rightly formed from the tapering tail, and this tail may be just as thin as desired, for it has to pass through the ear. Whereas usually the joint is the weakest part of these " sleepers," in this case the body thickens the part where the joint is, and gives greater strength. Taken as a whole, Fig. 1654 shows thought, and is an excellent example of a high degree of mechanical efficiency combined with ornamental effect, and all the advantages of the ductility of the gold are obtained.

Fig. 1662.   Fig. 1663.
Fig. 1662.— Position for Soldering Ear-ring Wire
Fig. 1663.—Wrongly-shaped Ear-ring Wire.

FASHIONING FINE GOLD EAR-RINGS.

Fine gold ear-rings of an antique pattern are fashioned thus—a plain piece of wire is taken and its two ends are tapered, a loop being formed with one of them. The other end is passed through the ear, then through the loop, and coiled round itself, as shown in Fig. 1655. Only fine gold can be employed in this way, but it would be foolish to attempt to make a pair of jointed sleepers in such soft metal.

EAR-RING HOOKS.

The hooks of the ear-rings shown by Figs. 1656 to 1662 are plain pieces of bent wire. On the shape of the wire depends the position of the top of the ear-ring, and sometimes its safety as well. It will be noticed that nearly all the ear-ring-tops (not ear-studs) worn with and without pendants hang as in Figs. 1663 and 1664; that is, they look downwards, and consequently show much of their edges. Now, ear-rings are generally designed with the idea of showing their fronts, not their edges; and, consequently, the fronts are the most ornamental parts; but owing to the wrong shape of the hook, the edges and not the

Fig. 1664.—Ear-ring with Wrongly-shaped Wire.

fronts are the more easily seen when the ear-rings are in use. The defect is easily remedied. All that is necessary is to make the highest part of the hook in line with the centre or heaviest part of the ear-ring top. The dotted lines in Fig. 1663 show what is meant. To test ear-rings to see if they hang properly, hang them on a needle file or a piece of wire, and it will at once be seen if they are defective. Even if a pendant is hung from the ear-top, there is not much improvement, for it will hang as in Fig. 1664, instead of as in Fig. 1661, as it should do. All this points to the necessity of hooks resembling Figs. 1656 to 1662. It does not matter whether the hook swings like Fig. 1658 or is a fixed one. Fig. 1659 is one of the forms of hook that allow the pendant to be removed at pleasure. Hooks like Figs. 1665 and 1666 should never be made, as they turn upside down, as shown

by the dotted lines, in which position the ear-ring will drop out. They can be made all right by soldering a grain, shot, or ring to them, to prevent their turning over. There is another tendency to guard

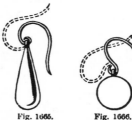

Fig. 1665.   Fig. 1666.   Fig. 1667.

Figs. 1665 and 1666.—Ear-ring Wires without Stops. Fig. 1667.—Ear-ring Wire with Ungraceful Bend.

against, that of the wire to get fixed in a reversed position. In soldering the wires on the work, take care to attach each one in the direction it has to take—that is, forward, as in Fig. 1662; then the curve of the wire will be prettier than that produced by bending it sharply forward. This is only a little matter, but it gives style to the work. Fig. 1667 shows the way that is not recommended. It will be seen that the wire does not look as though it belonged to the top, but appears to stand away, and it has an ugly bend where the wire starts curving forward.

HOOKS FOR EAR-RINGS.

The same principles that apply to brooches must be borne in mind, at any rate as far as they concern the soundness and strength of joint, the position of the catch, and the spring of the tongue—in this case the ear-wire. Ear-rings with catches are sometimes made without a joint either to hook or catch, but though they are

Fig. 1668.   Fig. 1669

Figs. 1668 and 1669.—Fixed Ear-ring Wires and Catches.

simple to look at it is difficult to make them act properly. The ear-rings shown by Figs. 1668 to 1671 are all different in shape and arrangement, but all act upon the same principle, that is, the wire itself gives sufficient spring to hold itself in the catch when it is placed there, and to spring away far enough on being released to allow the ear

Fig. 1670.   Fig. 1671.   Fig. 1672.

Figs. 1670 and 1671.—Fixed Ear-ring Wires and Catches. Fig. 1672.—Fixed Ear-ring Hook with Fall-down Catch.

to pass easily between it and the catch. For the latter purpose something over ⅛ in. is wanted, and to get this distance is no easy matter. The dotted lines show how far the hook should spring back. That shown by Fig. 1668, being very short, is the most difficult of the four, and after much hammering and twisting the hook generally has to be bent each time of removal. The others, having longer wires, give better results, and can be made to act fairly well. With regard to Fig. 1671, it may be pointed out that with one piece of wire is ingeniously formed—first, an ear-ring hook; second, a catch to retain it; third, a loop to swing a collet; and fourth, another loop for a pendant or pearl to hang on.

FALL-DOWN CATCHES FOR EAR-RINGS.

If time is not a consideration a fall-down catch or jointed hook is much to be preferred. With fall-down catches there is an improvement, inasmuch as the great strain necessary to obtain so much spring is taken away from the wire, and the necessary space for the ear to pass is produced by simply turning the catch down. Fig. 1672 indicates how it is turned down out of the

Fig. 1673.   Fig. 1674.

Figs. 1673 and 1674.—Ear-ring Catches.

way, and the dotted lines show the position ready for fastening the hook. The usual **O**-shape catch is the one most often

employed, and on a pair of ear-rings the catches should be placed with their openings on different sides, as in Figs. 1673 and 1674, the reason for this being that ladies find it easier to fix them from outwards

Fig. 1675.   Fig. 1676.

Figs. 1675 and 1676.—Fall-down Loop Catches.

towards the head. Occasionally another form of catch is used; this is shown in side and front views by Figs. 1675 and 1676. This loop-catch is made with a piece of tubing, which is soldered on to the ear-ring, and with a piece of wire bent up in the form shown, this wire loop moving as indicated by the dotted lines. This form of catch answers if the length of the ear-wire is properly adjusted. The small grain of gold shown soldered on the top is simply to afford a better hold for the finger nail. The ear-wire is made to spring a little, just sufficient to hold in the catch.

JOINTS FOR EAR-RING HOOKS.

Hooks soldered on a joint are in general use, and one or two points that have to be attended to may be noted:—First there must be enough action in the joint to allow of sufficient opening at the catch; the joint must have strength combined with a workmanlike neatness of fitting, and all is to be so arranged that there is something to prevent the wire coming too far back; just

Fig. 1677.   Fig. 1678.

Figs. 1677 and 1678.—Ear-ring Hooks with Chenier Joint.

as a fly-up plate in a brooch is used as a fixed point from which to obtain resistance, so there must be in a brooch a corresponding fixed point for the ear-wire to spring from. The simplest way is shown in Fig. 1677, and it will be noticed that this ear-wire and all the others are bent much in the form of those previously noted, but

Fig. 1679.   Fig. 1680.   Fig. 1681.

Figs. 1679 and 1680.—Ear-ring with Ball Joint.
Fig. 1681.—Ear-ring with Joint made from Rings.

with this slight difference—that it is just far enough away on the upper part to allow of sufficient opening at the other end, near the catch, when the wire is brought forward. In the grain-collet ear-ring (Fig. 1678) the joint is let into the collet; this is for neatness and strength. If the very top of the ear-ring were of such small size as to reduce the length of the joint too much for a fair amount of strength, then another form—most likely Figs. 1679 and 1680, or some adaptation of it—would be necessary. It is, as will be seen, a ball joint, made from three flat plates; it is narrow, strong, neat and compact. So far nothing has been said of ear-rings formed from two ornaments or two settings. If it is necessary to swing a collet, make a mount to carry the catch, something like the lower part of Figs. 1669 and 1670. In these, although more irregularly arranged, are found

Fig. 1682.   Fig. 1683.   Fig. 1684.

Figs. 1682 and 1683.—Good Forms of Joint Hooks.
Fig. 1684.—Bad form of Joint Hook.

all the same parts that are in a brooch, but modified to suit the altered conditions. Yet another joint for an ear-ring hook is shown by Fig. 1681. It is formed with rings, a collet or bead being in front. The ear-wire is loose in the ring soldered on the

drop or pendant until it is fastened by the catch, when the two rings jamb, and give the wire the necessary resistance.

Fig. 1685.   Fig. 1686.

Figs. 1685 and 1686.—Joint Hooks with Stops.

JOINT HOOKS.

Joint hooks, such as are shown by Figs. 1682 and 1683, were formerly used in England to a much greater extent than they are now; in fact, most of the jewellery made more than fifty years ago will be found to have this class of fastening, and even at the present time the French use it a great deal. The hooks in French work are of a proper shape, and have the fastening formed into a well shaped lip piece, which acts as a guide to lead the wire into the hole, whose edge retains the hook in its place (see Fig. 1683). The hook is governed by the same rules as the ear-wire, and must be brought well forward if the work is to hang properly. It is because the hooks are generally bought ready made, and are put on the work just as bought, that English-made collet ear-rings so often hang down like Fig. 1684, instead of as in Figs. 1682 and 1683. Some fastenings of this class have the hook and its catch-piece straight up and down, as in

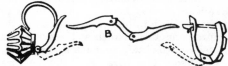

Fig. 1687.   Fig. 1688.   Fig. 1689.

Figs. 1687 and 1688.—French Spring Fastenings.
Fig. 1689.—French Jointed Ear-wire.

Figs. 1685 and 1686, and yet there is nothing wrong with them when in the ear, for the tilting forward is prevented by some arrangement or stop. These stops are generally small, as illustrated in Figs. 1685 and 1686; but now and then a wire mount is found, devised to keep the ear-top in its place. The arrangement is large enough to rest against the firm cartilage of the ear. Stop pieces are necessary under some conditions; but otherwise they should not be employed. Similar to the foregoing, but acting differently, are those sketched in

Fig. 1690.   Fig. 1691.
Figs. 1690 and 1691.—Ear-ring for Left Ear.

Figs. 1687 and 1688. Fig. 1687 is an ear-ring with a patent spring ear-ring joint, Fig. 1688 showing the joint as sold ready for soldering on. In Fig. 1687 the limits of the spring action are shown by the dotted lines. The principle on which this is made seems

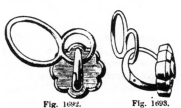

Fig. 1692.   Fig. 1693.
Figs. 1692 and 1693.—Ear-ring for Right Ear.

a good one, for the hook is a fixture, being hard-soldered on, and thus lessening the chance of losing the ornament even if the spring failed to act. The movable or spring piece has no strain on it, and that is of course an advantage in favour of

Fig. 1694.   Fig. 1695.   Fig. 1696.
Fig. 1694.—Ear-stud Split Tube. Fig. 1695—Plug Fastening. Fig. 1696.—Screwed Ear-stud.

security. Another way of fixing an ear-top is shown in Fig. 1689. It fastens in this way:—First the straight piece that passes through the ear is soldered on in a slightly downward direction, and is flattened and notched as illustrated. Then the jointed piece is pierced with an oval hole to allow

the notched end to pass through until it clips down, and thus it is secured. Figs. 1690 and 1691 show an ear-ring jointed for attachment to the left ear, whilst Figs. 1692 and 1693 show one for the right ear.

Fig. 1697. Fig. 1698. Fig 1699. Fig. 1700. Fig. 1701. Fig. 1702.
Figs. 1697 to 1702.—Nuts for Ear-studs.

EAR-STUDS.

Ear-studs are ornaments which rest on the lobe of the ear, ear-rings having been defined as the ornaments that hang below it. The piece of tube or wire attached to the ear-stud and actually resting in the thickness of the lobe should be straight, and have the ornament fixed on its end, just as a picture nail may have a brass

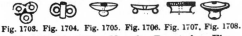

Fig. 1703. Fig. 1704. Fig. 1705. Fig. 1706. Fig. 1707. Fig. 1708.
Figs. 1703 to 1707.—Nuts for Ear-studs. Fig. 1708.—Spring Fastening.

head fixed on. This straight piece is the one feature distinguishing ear-studs from ear-rings. It passes through the thickness of the lobe of the ear, for it is not a hook to be used as an ordinary ear-ring hook is used, as will be seen later. Generally, ear-studs are retained on the ear by one of three methods—first, by pegs, as in Figs. 1694 and 1695; secondly, by screws and

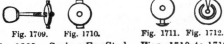

Fig. 1709.   Fig. 1710.   Fig. 1711. Fig. 1712.
Fig. 1709.—Spring Ear-Stud. Figs. 1710 to 1712.—Backs of Spring Ear-studs.

nuts, as in Figs. 1696 to 1707; thirdly, by a spring piece to clasp the wire, like Fig. 1708; or to fasten on the end of it, like Figs. 1709 to 1711. The mode of fastening by a peg or plug can be judged from Figs. 1694 and 1695. A thin unsoldered, or split tube for the front piece passes through the ear from front to back, and when in position the back plate (Fig. 1695), which has a peg soldered on it, is inserted, and, if properly made, holds well. The split chenier is made of rather thin (say No. 38, new standard gauge) tough gold, and the peg fits rather tight. Ear-studs fastened with screws are now most often made, and they are shaped as in Figs. 1696 to 1707. The other way

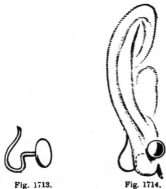

Fig. 1713.   Fig. 1714.

Figs. 1713 and 1714.—Correct Forms of German Hooks.

of screwing the back into the chenier seems to have been superseded entirely; in that method the two pieces resemble Figs. 1694 and 1695, but the chenier is soldered and tapped to take the peg, which would also be given a screw thread so as to screw in. Fig. 1696 shows that the wire which passes through the ear is screwed at the end, and on this screwed part a nut is made to run. Figs. 1697 to 1707 show designs for the nuts, the best of which have a good thickness for the screw; for instance, those shown by Figs. 1704 to 1707 are steady-running and wear well. Spring pieces (see Fig. 1709), that clip the bulb-like end of the wire are not satisfactory; a wire that will give a good grip to the spring is necessarily

Fig. 1715.   Fig. 1716.

Figs. 1715 and 1716.—Incorrect Forms of German Hooks.

clumsy, whereas a wire that will pass through the ear nicely is too thin to give a good hold. Spring pieces act by releasing the wire when the thumbpiece is pressed,

fastening being automatic when the beaded end of the wire is placed in the hole made to receive it. Fig. 1708 shows a fastener introduced from America. For this the wire must simply be straight and smooth.

Fig. 1717.   Fig. 1718.   Fig. 1719.

Fig. 1717.—Simplest Ear-ring for Unpierced Ears. Figs. 1718 and 1719.—Ear-rings Fastened from the Front and Back respectively.

The fastener is made so that it will slide easily up the wire, and will remain and hold just where it is placed; it comes away easily when the bottom bead is held. Fig. 1712 shows an attempt to use a grip on a plain wire.

GERMAN HOOK.

The German hook (Fig. 1713) is one of the best fastening arrangements for an ear-stud. It has no loose nut or spring to get lost, and is formed from a simple piece of wire, usually bent as shown in Fig. 1713 and worn as shown in Fig. 1714. To make this properly, the wire must not be very soft, and the four following points must be remembered: The part that rests in the ear must be straight, and at right angles with the front of the stud; this part must be

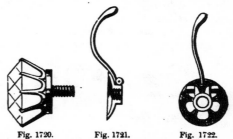

Fig. 1720.   Fig. 1721.   Fig. 1722.

Fig. 1720.—Ear-ring Front with Screw. Figs. 1721 and 1722.—Ear-ring Socket and Wires.

just a shade longer than the average thickness of an ear; the wire must be curved to permit of easy insertion and withdrawal; and the very end of the wire must be in line with the front of the bottom loop, so as to steady itself against the back of the ear. Figs. 1715 and 1716 show wrong forms of this book. For instance, Fig. 1715 is turned wrong way about, and if worn as an

Fig. 1723.   Fig. 1724.

Figs. 1723 and 1724.—Ear-ring Back Wires.

ordinary ear-wire, the front would hang face downwards; if worn in the proper way, the end is much too far away and will not touch the ear, as it should do. Fig. 1716 is a bad ear-ring hook, possessing none of the advantages of the German hook; it is, however, much used.

Ear-rings for Unpierced Ears.

In such ear-rings, the wire is divided at the top, the two ends nipping the ear between them. The blunted or enlarged ends give a gentle pressure, which is regulated by a screw arrangement. These fastenings can be pulled off, but are safe enough to carry the weight of the ear-rings. A screw is preferable to a spring, as with the latter there is the difficulty of keeping it at a proper pressure, for if it is too great it will hurt the wearer, and if too little, then the ear-ring drops off. For the purpose of removal or attachment, it is common for one end of the divided wire to be made movable

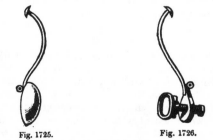

Fig. 1725.   Fig. 1726.

Figs. 1725 and 1726.—Front and Back Wire and Mount of Ear-ring for Unpierced Ears.

either by a spring or a screw. The simplest arrangement is given in Fig. 1717, in which the dotted lines explain its action clearly. It is a **U**-shape wire, to one end of which is attached a nut, in which a screw

works; the end of the screw is blunted and is made to approach or recede from the other end of the **U**-shape wire by simply rotating the screw. This is but a form of ear-ring wire, and to it an ornament has to be attached to make it into a proper ear-ring. In Figs. 1718 and 1719 there is a greater resemblance to the ordinary wire. To elucidate Figs. 1718 and 1719, the front of the ear-ring shown by the first of these figures is illustrated by Fig. 1720; this front acts as a handle to the screw fixed to it; Fig. 1721 shows section of the socket for the screw to work in, and shows where the wire and part of the joint are soldered; Fig. 1722 is a back view of this, showing

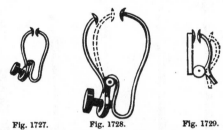

Fig. 1727.   Fig. 1728.   Fig. 1729.

Figs. 1727 and 1728.—Wire Mount for Attaching Drop. Fig. 1729.—Spring Ear-ring Fastening.

position of socket-joint and wire, and also showing how it can be pierced when the stone is to be set "à jour" or transparent; Figs. 1723 and 1724 show the back wire, with the other part of joint to hinge to the socket. This is the piece that actually nips the ear, for the screw acting on its lower end will regulate the position of the upper end. To show the construction of the ear-ring represented by Fig. 1719, its front with joint and ear-wire soldered in position is shown by Fig. 1725, whilst Fig. 1726 shows the back wire and mount which hinges on to the front. This carries the screw which governs the distance between the ends at the top, and Fig. 1726 also shows a small collar or stop soldered on to prevent loss of screw. Fig. 1727 is a wire mount for attaching a drop, and Fig. 1728 is the same enlarged and shown open. This acts on the same principle as the two others. In all these one part of the wire is fixed to the front of the ear-top, while the back part of the wire is jointed to it in such a way that the two open sufficiently to let the ear pass ; a screw, acting on the lower part of the jointed wire, causes the upper end to approach the end of the fixed front wire, and it is in between these ends that the ear is nipped. Fig. 1729 illustrates a spring fastening, in which the spring is contained in the hinge. The limit of its action is shown in the diagram. This more nearly approaches an ear-stud in appearance, for the front straight bar is intended to have an ornament fixed on it, and is made of hollow wire for that purpose.

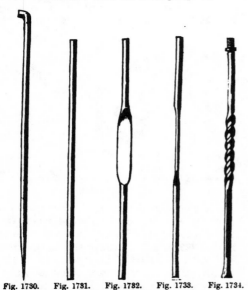

Fig. 1730.   Fig. 1731.   Fig. 1732.   Fig. 1733.   Fig. 1734.

Fig. 1730.—French Scarf Pin-stem. Fig. 1731.— Plain Wire for making Pin-stem. Figs. 1732 and 1733.—Wire after Hammering. Fig. 1734.— Twisted Wire with Collar Soldered on.

SCARF PINS

The term "scarf pin" applies to a pin or stem set with a jewel or with an ornament of precious metal, intended for a tie or cravat. The pin-stem must be pointed, for it has to be inserted in the tie, and it must be strong enough to bear forcing through several thicknesses of material. On the other hand, if it is thicker than is necessary, it makes a large hole, and possibly spoils the tie. With reference to the pin-head, suppose (as an extreme case) that this is to be formed with a hollow bead of delicate wirework. Now, if a piece of filigree work such as this were attached to a thick or clumsily made stem, it would be squeezed up by the finger long before the stem could be forced through the tie. The stems must be of greater or less thickness according to the strength or fragility of the head of the pin, or the use to which it is to be put. In some cases it is advisable to use a steel needle in place of a gold stem. To give the stem a secure hold in the tie, the French make the stem of great length (nearly 3 in.), it being a plain pin of hard gold, nicely pointed. English workers generally groove or twist the pin-stem, and occasionally add spikes, nuts, or some other further means of security. The French pin-stem is shown in Fig. 1730, the English ordinary stem being illustrated by Figs. 1731 to 1736, the first showing the plain wire before treatment ; the second and third the wire after hammering ; the fourth the wire after twisting it and soldering on the collar, and the fifth the finished stem. The sixth figure (Fig. 1736) shows a stem with raised thread, made by coiling a thin wire round the stem and soldering it in place. Old scarf-pins sometimes had a second pin attached to the first by means of a chain (see Fig. 1737). Fig. 1738 depicts a highly ornamental pin or brooch known as the comet, and this was once in high favour for short frills or cravats. It is now seldom or never worn.

FRENCH PIN-STEMS.

The simplest pin-stem of all is that illustrated by Fig. 1730. This is used in all French work, and is a quite plain piece of pointed wire, made in 18 ct. gold (alloyed all copper), which has necessarily to be drawn down very hard ; that is, annealing is dispensed with when drawing the wire through the last twenty or more holes in the draw-plate. Even that may be insufficient, and it is sometimes necessary to twist and hammer it, and not anneal it at all. The length is about 3 in. for smooth stems, and the thickness is No. 17, Birmingham wire gauge. The pin should be nicely pointed. Plain stems occasionally form part of the light pins used by ladies for fastening lace, the length and strength of stem being made suitable, of course, to the use they are to be put to, and to the size of the ornament they carry.

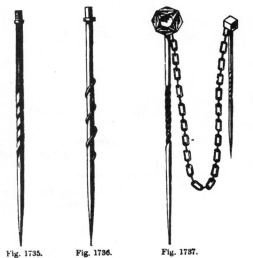

Fig. 1735.   Fig. 1736.   Fig. 1737.

Fig. 1735.—Finished Pin-stem. Fig. 1736.—Pin-stem with Raised Thread. Fig. 1737.—Scarf-pin with Chain and Smaller Pin.

They will therefore be several sizes thinner than the stems of ordinary tie- or scarf-pins.

TWISTED PIN-STEMS.

The average length of these is 2½ in., and there are two ways of making them; one is by filing the groove, the other by hammering and twisting the stem. The first is somewhat objectionable because the bottom of the groove is likely to be not round, but sharp—a continuous nick, as it were. A sharp nick is a place of great weakness. The second method is preferred; the gold wire (9 ct., usually) has to be drawn to a proportionate size—say, 17 B.W.G.; then it is cut into lengths of about 2½ in., straightening it in the draw-plate before cutting it, of course. The piece of wire (Fig. 1731) has now to be annealed precisely in the centre, where the twist comes, and then flattened out with the hammer,

as in Figs. 1732 and 1733, and not with an unequal thickness as in Fig. 1739. Again anneal it, and proceed to twist it until it resembles Fig. 1734; do this with a couple of hand vices, or in the same way as a brooch tongue is twisted, as already explained. Any soldering that is necessary should be done now, before the stem is hardened. The stem may have to be soldered to the head, or a collar may be required; if it is, then let it be a small one —not anything like the one shown in Fig. 1740. It is now soft all over, except at the point, which is not to be annealed, and a good deal of trouble will be saved if it has not been annealed, but kept hard. If through inattention or carelessness it has become soft, hammer it first one way and then the other, so as to harden it; but do not get it "double"—that is, split—if it can be avoided. If all is right, then hold the stem on a sparrowhawk, small anvil, or smooth iron, and proceed to reduce it to the same diameter all along. The blows of the hammer should be many but light, for

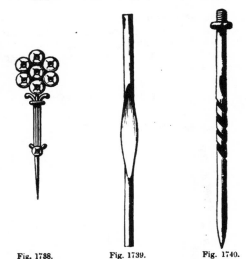

Fig. 1738.—Comet Brooch. Fig. 1739.—Pin-stem Flattened Incorrectly. Fig. 1740.—Bad Example of Twisted Stem.

it is necessary to obtain a gradual and thorough hardening, and to that end the stem should be rotated with the finger and thumb, so that the hammer blows will strike successive parts ; thus the round section of the wire will be brought back again. This also straightens the stem. Now proceed to file it up, first roughing the points

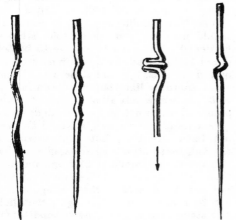

Fig. 1741.　　Fig. 1742.　　Fig. 1743.　　Fig. 1744.

Fig. 1741.—Spiral Stem for Lace Pin.　Fig. 1742.— Zigzag Stem for Lace Pin.　Fig. 1743.—Starting the Spiral Stem.　Fig. 1744.—Cork-screw Pin-stem.

down, and then with a smooth file going over all of it from collar to point ; the result should be something like Fig. 1735. The point, be it noted, is of the same shape as the French one.

BADLY MADE PIN-STEMS.

Fig. 1740, a very bad example, may be compared with Fig. 1735. The pin which it

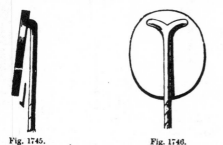

Fig. 1745.　　　　　　Fig. 1746.

Figs. 1745 and 1746.—Attachment of Pin-head to Stem.

depicts is hammered too thin in the centre of the twist, the wire itself is too thick, the twist is too long and too near the point,

and is of course irregular, and finishes off with a kind of gash ; also the point is too blunt. The collar at the top is too large and its size is out of proportion to the

Fig. 1747.　　　　Fig. 1748.

Figs. 1747 and 1748.—Pin-head Badly Fitted to Stem.

work it has to do—namely, to act as a stop against which to tighten the screw.

POLISHING PIN-STEMS.

The pin-stem is finished by polishing. Take a bundle of threads, and charge one

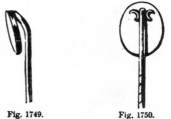

Fig. 1749.　　　　　　　Fig. 1750.

Fig. 1749.—Pin-head Badly Fitted to Stem.　Fig. 1750.—Pin-head Soldered Direct to Stem.

of them with crocus and oil ; coil it round in the groove and well rub from one end

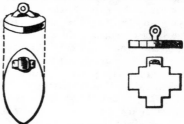

Fig. 1751.　　　　　　　Fig. 1752.

Fig. 1751 and 1752.—Screw Sockets or Stirrups on Pin-heads.

to the other ; after that, use Water-of-Ayr stone and buffs ; do not take the work to the lathe and polish off all the corners with a wheel brush; the edge of the groove should be sharp and square, both for the sake of appearance and to keep the pin steady in the scarf.

Fig. 1753.   Fig. 1754.   Fig. 1755.

Figs. 1753 to 1755.—Pin-head Screw Sockets.

LACE PIN-STEMS.

The crinkled stem is an approved form for lace pins, where a sharp-edged groove might cause damage. It was formerly used for gentlemen's scarves as well. Fig. 1741 shows a crinkled stem for a lace-pin. A slightly varied form (Fig. 1742) is simply made by twisting the middle of the stem wire round a small mandrel until it is of the shape shown by Fig. 1743; then after removing the mandrel the wire is pulled out again, until some slightly waving form only is left.

CORKSCREW PIN-STEM.

The stem illustrated by Fig. 1744 is intended to make its way corkscrew fashion into the scarf. To make one with several turns, the easiest way is to take a large steel "wood" screw, and coil the wire in between the threads; this will give regularity of spiral, and do away with nearly all the plier work.

Fig. 1756.   Fig. 1757.

Fig. 1756.—Horseshoe Pin-head with Screw Socket.
Fig. 1757.—Shepherd's Crook Pin-stem with Swung Ornament.

SCREW AND OTHER SOCKETS FOR PIN-STEM.

With regard to the means of attaching the stem to the ornament there are three possible methods. The simplest is by soldering it direct on to the pin-head in the French fashion; another method is to make a female or socket mount on the pin-head, and in this to screw the stem; thirdly, in-

Fig. 1758.   Fig. 1759.   Fig. 1760.

Figs. 1758 to 1760.—Head and Stems with Spikes.

stead of screwing the stem in, it can be soft-soldered into the socket. It must be remembered that the ornament has to be shown to the best advantage. To do that, it should look well out to the front, and just a little upwards. It should remain steady in that position. In Figs. 1745 and 1746 the front of the pin-head makes a good angle with the stem, and therefore the ornament is shown properly; the space between pin and front is about sufficient to let the scarf pass easily, and no more. The scarf must be allowed to travel unimpeded as near to the top as possible; it should never be stopped by a peg as in Fig. 1747, or by having the "stirrup" as low down as in Fig. 1748. The pin-head must not be tilted as in Fig. 1749. Figs. 1745 and 1746 and Figs. 1750 to 1756 show a few of the different ways in which the stem is attached to the work; they are various and general forms of (stirrup) sockets, which

Fig. 1761.   Fig. 1762.   Fig. 1763.

Figs. 1761 to 1763.—Butterfly Nuts.

are employed when the stem is to be either screwed or soldered in. The centre of all of these can be made from chenier, from a grain of gold melted on a piece of charcoal in which a small round depression has been made with a doming-punch, or from a coil of small rings soldered together. Two ways not yet alluded to are illustrated by Figs. 1737 and 1757. The first is where the stem is soldered straight into the ornament, which will be either globular or egg-

Fig. 1764. Fig. 1765. Fig. 1766
Fig. 1764.—Butterfly Nut. Fig. 1765.—Bead Nut.
Fig. 1766.—Disc Nut.

shape in form, such as a pearl or a lapis, gold or coral bead. The other (Fig. 1757) is adopted when the ornament (perhaps a coin or pearl pendant) is to swing, this style being sometimes called the "shepherd's crook."

OTHER METHODS OF SECURING ORNAMENTS TO PIN-STEMS.

In the system illustrated by Figs. 1758 to 1760, a spike is or spikes are so placed

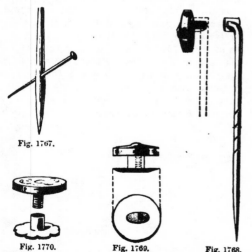

Fig. 1767.

Fig. 1770. Fig. 1769. Fig. 1768.
Fig. 1767.—Pierced Stem with Pin. Fig. 1768.—Interchangeable Pin and Stud. Figs. 1769 and 1770.—Studs.

as to catch in the tie and prevent the pin riding up. Fig. 1758 shows the spike

soldered to the stem. Now, as this cannot be screwed in the socket, it must be soft-soldered in so as to make both the part that fits the socket and the socket itself either square or oval; this, of course, prevents

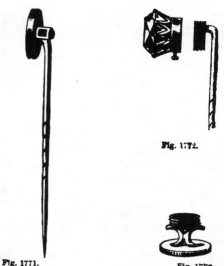

Fig. 1772.

Fig. 1771. Fig. 1773.
Figs. 1771 to 1773.—Interchangeable Pins and Studs.

the stem from turning. Fig. 1759 shows a double spike arrangement, which is part of the "stirrup," consequently part of the head of the pin. Great care must be exercised in getting the points hammered hard after the work is coloured, this colouring being the very last operation of all. The points themselves should be polished bright. Another way of obtaining extra security is to form the end of the stem into

Fig. 1774. Fig. 1775.
Figs. 1774 and 1775.—Plain Band Scarf Slides.

a screw, and make a nut to fit on it. Any shape of nut will do, provided it is large enough for the wearer to handle conveniently; but one of butterfly shape (Figs.

1761 to 1764) will lie flat, and is more comfortable to wear than one resembling Fig. 1765 or Fig. 1766. This screw and nut arrangement is the safest, but is liable to

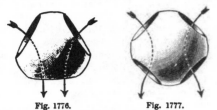

Fig. 1776. Fig. 1777.

Figs. 1776 and 1777.—Scarf Rings or Slides.

catch in the silk of the tie. Still another method is to have a perforated stem, as in Fig. 1767, where a common pin is intended to be pushed through the scarf and the stem.

COMBINED PIN AND STUD.

It is usual for a combined scarf pin and stud to have the screw right in the centre. Owing to this, the pin cannot set properly in the scarf. When the screw is put above the centre, it tends to keep the pin steady, and when used as a stud no inconvenience is found if the pillar of the stud be made oval, the screw, of course, fitting in one end (not the centre) of the oval pillar. Fig. 1768 shows this, the stud alone being illustrated by Fig. 1769; the ordinary method is shown by Figs. 1770 and 1771. Still another way is shown in Fig. 1772. Here the screws are much larger in diameter; and as but two or three turns can

Fig. 1778. Fig. 1779.

Figs. 1778 and 1779.—Skeleton Scarf Slides.

be used, it is well, where possible, to add a small screw, s, which will pass through both mounts, and so prevent the small amount of unscrewing which might

result in the entire loss of the jewel. The corresponding part when the ornament is for use as a stud is illustrated by Fig. 1773.

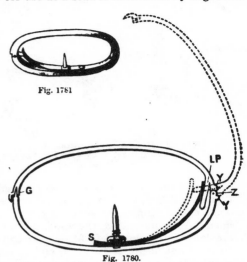

Fig. 1781

Fig. 1780.

Fig. 1780.—Jointed Scarf Ring with Spring and Lever Piece. Fig. 1781—Jointed Scarf Ring with Sunk Joint.

SCARF RINGS OR SLIDES.

The simplest form of scarf ring or slide is one that merely slides over the scarf, and is retained in its position by the two folds of the scarf, which together fit the ring rather tight. The simplest form of scarf ring is shown by Fig. 1774, which is an oval plain gold band, about 1 in. long and

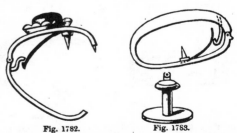

Fig. 1782. Fig. 1783.

Figs. 1782 and 1783.—Other forms of Jointed Scarf Rings.

½ in. wide, the depth being about ¼ in. to ⅜ in. Three other simple ones are shown by Figs. 1775 to 1777; these are all made from hollow, round beads, pierced by round holes in the various ways shown, the direction the scarf takes being indicated by the arrows. Of these, Fig. 1776 is the most usual, and can be worn just as well the

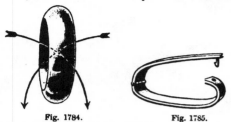

Fig. 1784.   Fig. 1785.

Fig. 1784.—Front View of Vertical Scarf Ring.
Fig. 1785.—Jointed Scarf Ring with Peg Snapping
into Hole.

other way up. In the ring shown by Figs. 1778 and 1779 spaces or passages for the scarf are arranged by attaching a wire mount. As illustrated, the upper parts of the loops are intended to fit over or hang on the front of the collar stud, and consequently to keep the ornament always well up to the top. This skeleton pattern seems the best to use when the front consists of a large oval stone—say, a cameo—for then the stone and its setting need only form the front; while, if a band like Fig. 1774 is used for such a shaped stone, it will probably project on each side of the setting, and will not always look as if it belongs to the ornament, but more or less of a makeshift. A spike is often added, and this, while penetrating the scarf, keeps the ring in its place; but with the one shown by Figs. 1778 and 1779 there is no need of such a spike, because the wire

Fig. 1786.   Fig. 1787.

Figs. 1786 and 1787.—Another Jointed Ring on
Same Principle.

mount is made to fit over the top part of the collar stud, and, generally, spikes cannot be used with any advantage unless there is a joint (as in Figs. 1780 to 1784) which allows the ring to reach its proper

place on the scarf before it is closed, for it is by the action of closing the scarf ring that the spike is made to penetrate. The first departure from the simple, rigid form is by making the band of two pieces and jointing them together, as in Figs. 1785 to 1791. Three (Figs. 1785, 1786 and 1788) are held fastened by a peg on one piece, snapping into a hole or over the bar on the other piece. To either of these a spike, such as that shown in Fig. 1781, can be added. The fourth (Figs. 1790 and 1791) has three hinges, and the position, when closed, is shown in dotted lines. The closed portion here is retained, because the front of the slide—which is somewhat of the shape of a bow—has the ends made to spring slightly inwards, and in that way it acts as a spring to force the inside plate

Fig. 1788.   Fig. 1789.

Figs. 1788 and 1789.—Jointed Scarf Ring which
Snaps over Bar.

and spike forwards against the two parts of the scarf that are passed through it. The spike and plate are on a loose piece of joint. For security, as well as for convenience in adjusting, the greater number of scarf rings are now made as in Figs. 1780 to 1784. These all have springs so fitted that their force is exerted to close the scarf ring, and keep it closed. In spring scarf slides, the lever peg must be soundly and firmly attached to one half, and so placed and shaped that the spring is able to exert sufficient force on it to keep the two parts closed up. The spring itself must be of a suitable shape and strength, and so placed that it will act on the peg as it should do, and yet allow of sufficient opening of the scarf ring to easily insert the scarf. These lever pegs take different shapes, as illustrated, and there are also several forms of springs (see Figs. 1792 to 1794).

OTHER CONSIDERATIONS ON SCARF PINS.

The large ellipse in Fig. 1780 represents a side view of a scarf ring, jointed at one end so as to allow an opening of rather less than a right angle, as indicated by the dotted lines. If the scarf slide has to open

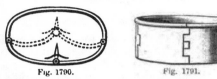

Fig. 1790.   Fig. 1791.

Figs. 1790 and 1791.—Three-joint Scarf Ring.

wider than those illustrated, a longer spring will be required, otherwise it will slip under the lever peg, and prevent the closing of the ring until the joint pin is taken out and the spring readjusted. One other advantage of this restricted opening is that pressure is always on the spring, so it does not require to be fastened in its place with a screw, but will remain perfectly steady and act perfectly well if it is passed under a bar and against a stop, as in Figs. 1781, 1792 and 1793. Points to be noted are:—The strength of the hinge; the firm attachment and a good form of the lever peg; the strength and firm position of the spring at one end, so as to allow the other end always to act steadily on the peg in the one direction. The spring must be made of rather tough gold; polish it bright, and also polish that part of the lever peg against which it acts, in order to reduce the friction as much as possible.

QUALITIES OF GOLD: THE TERM "CARAT" DEFINED.

Gold not containing a trace of base metal is known as "fine gold," but of course this is too soft and too expensive for ordinary use, and so it is alloyed with silver and copper to harden and to cheapen it. For the purpose of estimation, the mass of gold alloy is divided into 24 imaginary equal parts, termed "carats," and the alloy is known by the number of carats of fine gold it contains. (It is a mistake to suppose that the term "carat" applied to

gold has, in the United Kingdom, any reference to the weight, for it refers to relative purity only; however, diamond merchants use a weight so named.) Fine gold is 24 carat, that is, it does not contain alloy. Twenty-two carat gold contains $\frac{2}{24}$ of base metal; 18-carat gold contains 6 parts of base metal and 18 parts of fine gold, and so on. There are five standards of alloyed gold in general use, 22 carat, 18 carat, 15 carat, 12 carat, and 9 carat, and as far as hall-marking is concerned intermediate grades are not recognised. In testing at the assay offices, if gold fails by ever so little of being 22 carat, it is hall-marked as 18 carat, except in Dublin, where a 20 carat standard is recognised. Gold failing to come up to the 18-carat standard, even though it may contain more than 17 carats of fine gold, is stamped 15 carat, and so on down the scale.

COLOURING GOLD.

The object of colouring alloyed gold is to give it the appearance of fine gold, and the principle of the process is easily understood, it being but the dissolving out of the base metals from the surface of the alloy to a very slight depth, leaving a skin of fine (pure) porous gold. The simplest method of colouring gold jewellery

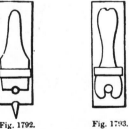

Fig. 1792.   Fig. 1793.   Fig. 1794.

Figs. 1792 to 1794.—Springs for Jointed Scarf Rings

is to bring it to a uniform heat, allow to cool (and thus become annealed), and then boil until bright in a pickle of 8 oz. of rainwater and 1 oz. of sulphuric acid. Gold alloys of not less quality than 15 carat may be made to assume the colour of fine gold by carefully boiling them in a mixture of saltpetre (nitrate of potash) 15 oz., table salt 7 oz., alum 7 oz., and spirit of salts

1 oz. The work must be previously annealed and boiled out in aquafortis pickle, and wired with platinum wire. It must only be exposed to the colouring mixture for five minutes at a time, and well rinsed in boiling water between each operation. For 18-carat gold alloys, the colouring mixture may consist of 1 oz. more of each of the above ingredients, omitting entirely the spirit of salts, and making the other powders into a paste with hot water. In all cases it is advisable to thin the colouring mixture with hot water as the process of colouring progresses, so as to avoid overdoing the work.

Another Method of Colouring Gold.

Another method is to anneal the gold as before, boil it in a pickle of nitric acid and water, again anneal, and dip in the following colouring mixture. Two parts (by weight) of saltpetre and 1 part of table salt are heated in their dry state in a colouring pot or blacklead crucible; when hot, make into a paste with hot water, boil, add 1½ parts of muriatic acid, and stir well. Use at boiling point; leave the gold in the solution for not more than 90 seconds, as the solution removes more or less of the gold. On taking the gold from the colouring solution, rinse it in a pickle, dip it in hot water, and dry in hot sawdust; the gold will be spotted if not thoroughly dried. This method may be used with gold ranging between 12 and 20 carats fine, the best results being obtained with 15-carat gold.

Melting Gold.

Gold, either pure or alloyed, is melted in plumbago crucibles, either Morgan or Doulton. Before using the crucible, well rub the inside with flour charcoal to smooth down any irregularity in order that the gold may not be held in and lost. To anneal the crucible, place it in a furnace whose fire has nearly gone out, or else over a fire in a furnace, and weigh out the gold and other metals if such are being used. When copper is added, this must be placed at the bottom of the crucible and covered with a layer of charcoal to prevent undue oxidation. When the metal is well melted, stir with a clean lumbago or smooth iron stirrer; the metal is then ready for pouring. If only gold, which may contain slight impurities, is being used, a small quantity of sal-ammoniac should be added; this will purify and toughen the gold so that it may be well worked. The heat required to melt gold satisfactorily can be learned only by experience.

Alloying Gold.

In melting the metals for 18-carat gold, use a plumbago crucible lined with finely powdered charcoal, and put the copper in first, then add the silver and gold. When the mixture is at the point of fusion, throw on its surface about two tablespoonfuls of finely powdered vegetable charcoal and finely powdered best sal-ammoniac intimately mixed. Use no other flux. When completely fused, stir the whole with the point of a red-hot iron rod, bring to the proper fluid condition for pouring, and hold a strip of wood to the mouth of the crucible to keep back loose flux whilst pouring the metal into the mould.

Causes of Brittle Gold.

The following are some of the causes of brittle gold. (a) Oxidation of copper and absorption of the copper oxide by the molten metal. (b) A pasty condition of the molten metal at the moment it is poured into the mould or ingot. (c) The mould may be too hot or too cold at the time of pouring the metal. (d) Absorption of some impurity from the flux. (e) Some impurity in the added copper or silver. The impurities in added metals may be arsenic, phosphorus, iron, or nickel in the copper, and lead or zinc in the silver. Impurities in the flux may be grit and iron in the sal-ammoniac, and free mercury in the corrosive sublimate. Annealing will often remove the brittleness of the gold.

Melting Silver on Open Fire.

Silver can be melted in a small fireclay crucible on an ordinary open fire, equal quantities of finely powdered charcoal and sal-ammoniac being used as the flux. Make up a large bright coal fire in an open grate, and when the fire is quite clear break a hollow space in the centre. In this space place the crucible, and allow it to get red-hot ; then put in the silver, and draw some of the hot coals closely around and over it. Blow the fire with the bellows until the crucible gets white hot, when the silver will melt, the fusing point being at 1873° F. (1022·7° C.). Then add the flux to clear the surface from scum. Again make the crucible hot, and quickly pour the contents into an iron ingot mould previously made scalding hot. One or two ounces of silver may be melted at a time in this way. The flux may be stirred with a pointed rod of iron previously made red hot.

MOULD FOR SILVER CASTING.

The silver casters in the neighbourhood of Birmingham use for their moulds a special sand known as Moxley loam, which is obtained from Moxley, near Bilston, Staffordshire. The face of the mould is dressed with best flour charcoal and pea-flour as usual. Plaster-of-Paris is useless for making moulds for silver casting, since the heat of the molten metal is sure to perish the plaster and cause it to crack all over. If the article is only small, making the moulds with Parian cement might be tried, but this method cannot be guaranteed.

RESTORING LUSTRE OF BLACKENED SILVER.

Silver merely oxidised by exposure to the atmosphere, and not by repeated cleaning, is restored simply by brushing with a clean tooth-brush and a little carbonate of soda. The best way to restore the original lustrous whiteness of silver goods which has been lost or impaired by exposure to sulphurous atmospheres or by frequent or careless cleaning, is first to anneal and then to pickle the silver, the latter portion of the process resembling the colouring of gold alloys already described.

ANNEALING SILVER GOODS.

The annealing for the purpose explained in the previous paragraph may be done in a charcoal fire or in the flame of a gas or oil blowpipe ; the heat destroys all organic matter adhering to the surface of the arti-cle, at the same time oxidising on the surface the base metals with which the silver is alloyed. The annealing requires some care and attention, or the workmanship of the piece will be destroyed. If the silver has been soft-soldered previously, it is unfit to be annealed, as the heat employed for this would melt the solder. It is necessary to remove all stones, steel, or any material not silver or liable to be injured by the heat, and it is also advisable to remove pins, tongues, or other steel work from brooches, etc. Both over-heating and under-heating must be prevented ; if the article is overheated, the silver is liable to melt ; and if under-heated, the adhering organic matter is not effectually destroyed, and the surface not sufficiently oxidised. In order to obtain the required degree of heat, and not to run a risk of either under- or over-heating, the article is held with a pair of pincers very close over the flame of the lamp so as to be entirely covered with soot, and is then exposed to the blast of a flame by means of a blowpipe until the soot burns or disappears.

PICKLING SILVER.

When the article is cool, it is immersed in a boiling solution of from 1 part to 5 parts of sulphuric acid in about 20 parts of water. The quantity of the water depends upon the quality of the silver ; the coarser this is, the stronger is the solution. The solution dissolves the extracted deposit of oxide and leaves a coating of fine silver on the surface. Good sterling silver will be whitened almost in an instant, common silver will take a minute or even longer ; if the articles are left too long in the solution, they turn an unseemly greyish colour, and the process has to be repeated. Common silver has to be treated repeatedly in this manner before the desired whiteness is obtained, and in some cases will even have to be silvered by electro-plating. As soon as the article in the acid turns white it is transferred quickly to lukewarm water. The articles are then dried in sawdust, kept in an iron vessel near the stove or in any warm place. Any places on the article desired to look bright are burnished with a steel burnisher.

MATERIALS, ETC., FOR TESTING GOLD.

It is by comparing the way in which nitric acid acts on certain known qualities of gold with the way it acts on the article to be tested, that the quality of the latter is ascertained. Necessary appliances are (a) a touchstone (usually a piece of Lydian stone, which is a black variety of jasper): this has its surface smooth and partly polished, but is not bright; or a piece of Wedgwood black ware does very well. (b) A series of "needles," or pieces of gold, to be used as standards of comparison; for general use small pieces of gold wire of 9, 12, 15, and 18 carat should be sufficient. (c) Nitric acid kept in a special bottle having a long pointed stopper, by means of which a small drop of the acid can be removed from the bottle without dropping any or without any coming in contact with the fingers, for it destroys the skin where it touches. The ordinary commercial nitric acid will do. Aquafortis is another name for the same chemical.

TESTING GOLD WITH TOUCHSTONE.

In testing gold, first get a rubbing on to the touchstone from one or two places which fairly represent the whole article. If possible, get on the touchstone a clean streak of gold, ½ in. long and ⅛ in. wide, and by the side of this take rubbings from the standard pieces, then put a narrow streak of nitric acid across the whole. Now note which of the standards is acted on in the same way as is the rubbing from the article. The effects may vary from no change at all with good qualities to complete destruction of the rubbing in low qualities. On removing the surplus acid by means of tissue paper or a rag, the effect can be better judged. A piece of work may be very thickly plated, or even gilt, in which case the rubbing taken must go through the gold into the metal; and, on the other side, if the soldering seam alone be tested, the gold will show poorly. Therefore judgment is necessary in selecting the place where the test is to be applied; and even then it is better to file away a portion of the surface as well, and apply the stone or acid direct to it. After some practice, when one knows the effect that acid has on the different qualities, it is customary to do without the touchstone in most cases; but it is well to use it, at any rate, until one is quite used to the process roughly indicated above. Following up the manner in which gold alloys are tested, comparison should be made in the same way with the effects of nitric acid on silver, brass, copper, etc.; for although brass will boil green, so also will some very common qualities of silver and gold. The capability of judging qualities, after all, is largely a matter of experience. To people used to handling gold the very appearance of the article is often sufficient to denote its quality.

TESTING COLOURED GOLD.

Coloured gold articles do not always need the application of acid; for if, by scraping the surface, there is found to be a granular brown layer, then the quality of the article can be judged very nearly by the extent of the change that has taken place below and on the surface while the article was undergoing the process of colouring. In colouring, the gold alloy is dipped in an acid pickle which dissolves out the base metal on the surface, leaving what is in fact a case of pure or fine gold enclosing the alloy. If the article is coloured, is must be more than 12 carats fine, and is generally 15 carat.

ROLLED GOLD AND FILLED GOLD.

"Rolled gold," "gold filled," and "gold cased," therefore, mean that the article so described has a hard covering of gold of an appreciable thickness, and anyone selling gilt goods under the above descriptions can be proceeded against for fraud. As a substitute for solid gold, the most common device is to make the articles of a base metal—generally brass—and electro-gild them. This process, in the case of articles subject to hard wear, such as watch cases, chains, brooches, etc., is of little use. The coating of gold is soft and thin, and very soon wears off on the most exposed parts. A better substitute for gold is made by coating brass or other hard alloy by mechanical means with a thin layer of hard gold. There are several methods of doing this. Some American watch cases, and some English cases, made by a Birmingham firm, are known as "filled gold." The result is a case made of hard brass, of which all the surfaces, outside and inside, are covered with a fairly thick plate of gold, calculated to wear almost a lifetime. The gold on these cases is so thick as to bear engraving without cutting through into the base metal, and when such cases, weighing 2 oz. or 3 oz., are sent to the refiners to be melted, they frequently produce 15s., showing the actual value of the gold covering. Rolled gold is mostly of German origin, and is made by brazing a plate of gold on a thicker plate of brass, and rolling it out thin into sheet, from which the articles are then manufactured. Rolled gold jewellery is, therefore, the same as a "gold-filled" watch case, and consists of hard brass, mechanically covered with a layer of hard gold. The gold covering may obviously be of any quality or thickness. The best is equal to American gold-filled cases. The commonest made is still greatly superior to gilt goods. The cheap jewellery seen in fancy shop windows is mostly of this kind in its commonest form, while in Germany the best rolled gold bears an official stamp, guaranteeing the quality of the gold covering and its thickness. Government assay offices in Great Britain do not recognise "rolled" or "filled" gold.

GOLD CASING.

Gold casing is older than either gold filling or gold rolling, and the method has been practised in England for at least a century. It consists in covering the completed article with a thin gold shell and uniting the two by soft solder. In this case, also, the gold covering is thick and hard, can be engraved, and has a considerable value when the articles have to be consigned to the melting pot. The articles most commonly in use that are gold cased are pencil cases and pocket pens, the bows of watch cases, etc. Much old jewellery, brooches, bracelets, etc., are found to be gold cased after having been in wear the greater part of a century; their present owners often believe them to be of solid gold, and are undeceived only when the articles are taken in exchange for more modern jewellery, and have to be melted.

MAKING A SKELETON CLOCK.

MECHANISM OF SKELETON CLOCK.

SKELETON clocks are good timekeepers, and are very interesting because they afford opportunities of examining the mechanism at work. A simple eight-day skeleton clock, standing 18 in. high, may be made for about 25s., complete with marble stand and glass shade. Fig. 1795 is reproduced from a photo of a complete clock made exactly according to the instructions given in this chapter, and from the materials to be described. Its mechanism is extremely simple, consisting of a barrel containing the mainspring ; a fusee and gut line or steel chain, to equalise the uneven pull of the mainspring and enable the clock to keep equable time throughout the week ; a centre wheel carrying the hands ; a third wheel to convey the power to the escape wheel, which in its turn gives impulse to the pendulum (the true time-measurer) through the medium of the pallets and crutch. This completes the clock, and it may be summed up thus—a main-spring, four wheels, and a pendulum. As before observed, the pendulum is the time-keeper ; the rest of the mechanism merely serves to record its vibrations upon a dial and to supply to it the necessary power to keep it vibrating. The timekeeping quali-ties of a clock mainly depend upon the " arc of vibration " of the pendulum being kept uniform. Obviously this can be at-tained only by delivering to it impulses absolutely equal in force and at equal in-tervals. It is the duty of the mainspring to do this through the train wheels and escapement.

MOTIVE POWER OF CLOCK.

The motive power will be first consid-ered. A weight is the best, being invari-able in its pull throughout the entire fall. A weight in a small house clock being out of the question, a steel mainspring has to be used.

THE FUSEE.

Naturally, the force exerted by a coiled-up spring in uncoiling itself is far from being uniform. When coiled up closely it exerts very much more force than when partially uncoiled, and the diminution in force is not regular. To counteract this variation, the fusee is introduced. This is a coned pulley with a spiral groove cut upon it, and so arranged that when the mainspring is fully wound up, and there-fore at its strongest, the line pulls upon the smallest part of the cone, acting, as it were, upon a short lever ; while, as the spring uncoils and winds up the line or chain upon its barrel, it pulls, upon a con-stantly increasing diameter, a lever that becomes longer and longer until the clock is run down. The curve and proportion of the cone are suited to the diminution in force of an average mainspring, and though in practice it never absolutely equalises the power, yet it is near enough for the purposes of ordinary timekeeping, and is the best substitute for a weight.

THE ESCAPEMENT.

The centre and third wheels are to con-vey the power to the escapement, and pro-vided they are cut fairly well and the pinions are moderately good, they will not seriously affect the uniformity of the effort of the spring. Next to the motive power, the escapement claims attention. The one escapement capable of delivering invariable impulses to the pendulum is Lord Grimthorpe's "gravity escapement," used in turret clocks and in some very fine regulators. This, however, would be quite unsuitable for such a clock as the one to

little use in so small a clock as this. The temperature error of such a clock is probably caused much more by the varying elasticity of the mainspring and the unequal resistance to motion of the train and escapement, caused by the fluidity or otherwise of the oil than by the actual variation in length of the pendulum. Therefore, the material of which the pendulum is constructed is not a matter of

Fig. 1795.—Skeleton Clock and Glass Shade.

be described, and recourse is had to the well-known recoil or anchor escapement, illustrated in Fig. 1795. Its action is simple, its construction easy, and its performance very fair indeed, and quite up to ordinary standards of timekeeping.

THE PENDULUM.

Of pendulums there is a variety; all are liable to a temperature error, except those truly compensated, and such would be of

much importance, and for the sake of general appearance brass is usually selected. Doubtless a wood rod pendulum would be a slight improvement, but it does not look so nice nor is it so easy of construction or so strong.

MATERIALS FOR CLOCK.

The materials from which it is advisable to construct the clock are not bought in the form of rough sheet brass and steel rod but in such a form and condition that, by means of a few simple tools, they can be finished and put together. Thus, the barrel had better be bought finished; the fusee

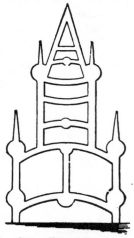

Fig. 1796.—Frameplate Casting.

with the body turned and the spiral groove cut; the wheels with the teeth cut and almost ready to mount on their pinions; the pinions, in the form of pinion wire,

Fig. 1797.　　　　　　　　Fig. 1798.

Fig. 1797.—Foot of Skeleton Clock. Fig. 1798.— Back-cock or Pendulum-cock Casting.

having the leaves already formed; the pendulum finished and ready to put in the clock; and the plates, etc., in the form of clean brass castings.

Fig. 1799.　　　　　　　　Fig. 1800.

Fig. 1799.—Minute-cock Casting. Fig. 1800.— Barrel Ratchet-click Casting.

PARTS OF THE CLOCK.

Most of the parts are shown in the rough condition in which they are generally bought by Figs. 1796 to 1815. Figs. 1796 to 1800 show the castings of the frame-

plate, foot, back-cock or pendulum-cock, minute-cock, and barrel-click, respectively. The finished pendulum is represented by Fig. 1801, and the main wheel, centre

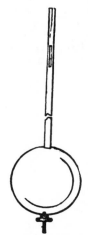

Fig. 1801.—Pendulum.

wheel, third wheel and 'scape or escapement wheel by Figs. 1802 to 1805. The pallet is illustrated by Fig. 1806. Three motion wheels are shown by Figs. 1807 to 1809, the finished barrel by Figs. 1810 and 1811, the rough fusee by Fig. 1812, a piece of pinion wire by Fig. 1813, a barrel ratchet by Fig. 1814, and lastly the fusee

Fig. 1802.—Main Wheel.

ratchet by Fig. 1815. The sizes and numbers of wheels, etc., are here appended:— Barrel—2¼ in. diameter and 1⅜ in. high. Fusee—⅝ in. diameter small end, 1¼ in. large end; length of body, 1⅜ in. Centre wheel—1⅞ in. diameter; eighty-four teeth;

pinion, eight leaves ; size, $\frac{2}{3}$. Third wheel —$1\frac{3}{4}$ in. diameter ; seventy-eight teeth ; pinion, seven leaves ; size, $\frac{7}{}$. 'Scape wheel — $1\frac{3}{8}$ in. diameter ; thirty-three

Fig. 1803.—Centre Wheel.

teeth ; pinion, seven leaves ; size, $\frac{7}{}$. Minute wheels—1 in. (bare) diameter ; forty teeth ; minute pinion, six leaves ; size, same as 7. Hour wheel—$1\frac{3}{4}$ in. (full) diameter ; seventy-two teeth. Frames—10

Fig. 1804.—Third Wheel.

in. high and 5 in. wide. Pendulum—$6\frac{1}{2}$ in. long from centre of bob to point of suspension.

FRAME OF SKELETON CLOCK.

The pattern of the frame plate (Fig. 1796) is fairly plain, but nevertheless looks well when made up. It gives the minimum of

Fig. 1805.   Fig. 1806.
Fig. 1805.—'Scape Wheel. Fig. 1806.—Pallet.

trouble in finishing, no files of out-of-the-way shapes being required. First of all, the pair of plates is cleaned up and the

frame is made. First square up the edges all round, inside and out, and then file up the faces ; for this, lay them flat upon the work-bench, and to prevent them moving

Figs. 1807 and 1808.—Motion Wheels.

sink three screws into the bench, leaving only $\frac{1}{8}$ in. projecting. The heads will thus be below the level of the brass and out of the way. File them up with a new file about 1 in. broad and 8 in. or 10 in. long. Do not trouble about smoothing them off,

Fig. 1809.—Large Motion Wheels.

but get them clean and flat. The smoothing process comes on at a later stage, when the clock is nearly completed. Clamp the plates together carefully, and drill four $\frac{3}{16}$-in. holes right through for the four pillars.

FRAME PILLARS.

For an exceptionally rigid and good-looking frame, six pillars can be made ; but four are ample for strength. At clock material shops, soft brass castings are sold

Fig. 1810.   Fig. 1811.
Figs. 1810 and 1811.—Barrel of Skeleton Clock.

for pillars, but they are awkward to turn, and, on the whole, brass rod is to be preferred ; therefore, take a foot length or so of ⅜-in. brass rod and centre it in the lathe, and proceed to turn four pillars like Fig. 1816. If a screw-cutting lathe is available, cut the threads ; but, if not, a good screw-plate or dies will do. The pillars are to be screwed into the back plate, and the front plate is dropped on to the shoulders, the

Fig. 1812.—Fusee in the Rough.

ends of the pillars being drilled through and pins inserted. The shoulders of the pillars must be turned with care, and should be a shade hollow or undercut, rather than round. The length from plate to plate inside the frame is 1⅞ in.

Frame Feet.

To complete the frame, the four feet can be turned and finished. Fig. 1797 shows one of the castings. They are best chucked by being soft-soldered on to a piece of sheet brass and clamped on to the face-plate. It is a good plan to have a plate chuck like a face-plate but with no holes or slots in it. It is made of sheet brass ¼ in. thick, and is 3 in. in diameter. It is used solely for soft-soldering work for turning, and is extremely useful. A series of concentric circles turned upon its face facilitates centering. Each foot can be finished at one chucking. Turn them to the

Fig. 1813.--Piece of Pinion Wire.

shape shown in Fig. 1817, and drill a ⅛-in. hole down the centre of each, right through. This can be done with ease as they run in the lathe. Do not detach them until they have been well polished, as there

will not be another opportunity of holding them in the lathe without much trouble Fig. 1818 shows how the feet are attached to the frame. A hole is drilled and tapped in the bottom of the plate and a piece of steel wire screwed in. The wire passes through the foot and the clock stand, and is screwed up with a nut underneath.

Fig. 1814.—Barrel Ratchet.

Fig. 1815.—Fusee Ratchet.

Barrel and Barrel Arbor.

Figs. 1810 and 1811 show a finished barrel, with a hook in it ready for the main-spring. The arbor will require turning to the shape shown in Fig. 1819, one end—the front—being squared to fit the barrel ratchet (Fig. 1814) tightly. Be sure that the right end is squared ; this can be ascertained by the hook, it being remembered that to wind up the spring the arbor is turned to the right, or the same way as a screw is turned. The pivots can be smooth-filed in the lathe and polished as they run with a very fine emery-stick (emery-paper on wood). If the square projects ¼ in. beyond the ratchet it will be long enough. Clamp the plates together again and drill the pivot holes for the barrel. Open them out to fit with a good

F g. 1816.—Clock Pillar.

broach or reamer. Drill them as low down as can be done with safety, so as to get as much freedom as possible between barrel and fusee. Remember that all pinions and arbors throughout the clock must have a little endshake ; that is, they must not be quite 1⅞ in. from shoulder to shoulder, but just a little less—enough to ensure perfect freedom under all conditions.

BARREL RATCHET CLICK.

The barrel ratchet click (Fig. 1800) is merely filed up and screwed to the front plate in such a position as to engage the ratchet properly (see Fig. 1795).

Fig. 1817.—Clock Foot.

FUSEE.

Now take the fusee (Fig 1812), of which the spiral groove is already cut. Put it in the lathe and pivot it like the barrel, making the front pivot come $\frac{1}{16}$ in. in from the end of the fusee. The small end comes in front. Make a square for the winding key, $\frac{3}{8}$ in. long and the least trifle tapered. The fusee pivots, being important ones, must be well polished. After smooth-filing them in the lathe, get a piece of flat steel or gun-metal about 8 in. long, $\frac{3}{16}$ in. wide, and $\frac{1}{8}$ in. thick. File it perfectly true and square on the edges. This is for a polisher. Charge it with oilstone dust and

Fig. 1818.—Fixing Clock Foot to Frame.

oil and proceed to smooth the pivots in the lathe, using it like a file. When smooth, clean them off thoroughly, polisher as well, and charge again with red stuff and oil and bring up a brilliant polish.

FUSEE RATCHET.

Take the fusee ratchet (Fig. 1815) and open out the centre hole to go on the back

end of the fusee arbor right up to the brass body. Drill two holes through it, as in Fig. 1820, right into the body of the fusee. Then take off the ratchet and tap the holes in the fusee and screw in two brass pins tightly, right home, leaving them projecting just the thickness of the ratchet. Then open out the two holes in the ratchet so

Fig. 1819.—Barrel Arbor.

that it will push tightly on to the pins. See that the ratchet is on the right way about, remembering that the fusee turns to the right. Fig. 1820 shows the ratchet put on the right way.

THE MAIN WHEEL.

Take the main wheel (Fig. 1816). The recess cut in it is for the fusee ratchet to lie in. Therefore, open out the centre hole of the ratchet to go easily over the centre boss of the main wheel, and allow the ratchet to lie completely in the recess. The centre hole of the main wheel must now be enlarged to pass over the fusee arbor easily, but without shake. It would not do to simply reamer it out, as the hole may not be, and usually is not, truly in the centre of the wheel. Screw a piece of deal board to the face-plate of the lathe and proceed to turn out a shallow recess

Fig. 1820.—Fixing Fusee Ratchet.

into which the main wheel can be pressed tightly. This will hold it quite true by the points of the teeth, and firm enough for the centre hole to be turned true and opened out to size. The main wheel can then be pushed on to the fusee arbor, and should come right up to the brass body of the latter, completely covering and hiding the ratchet, which should be between them, pinned on to the fusee and lying in the recess of the main wheel. If the ratchet is a little too thick to allow the main wheel to come right up to the fusee body, thin it by filing.

Click and Spring.

The click and spring, shown in position by Fig. 1821, must now be made. The steel click with a hole through it works on a fixed steel stud, screwed into the main wheel from the inside and riveted over on the outside and filed off smooth. The click spring is of brass. To make it, take a straight or curved strip a little less than $\frac{1}{16}$ in. thick and file it up; afterwards hammer the thin part well to harden it and make it elastic, and then bend it to shape. If it is filed up to the curve of the recess in the main wheel, one small brass rivet will be sufficient to fix it.

Fig. 1821 —Fusee Clickwork.

Key for Fusee Arbor.

The fusee can be put together now and the " key " made. This key is a small disc of brass, $\frac{1}{16}$ in. thick and $\frac{1}{4}$ in. diameter, which goes tightly on to the fusee arbor against the main wheel. It is held in position by a pin passing through the arbor and lying in a groove filed in the key. The groove is to ensure that the key does not turn upon the arbor.

Completing Fusee.

English clocks with fusees almost always have stopwork to prevent winding too far. Upon the small end of the fusee a projection will be found; this is left for the purpose of catching the " stop " when the gut or chain is all upon the fusee. But in a skeleton clock, where everything is visible,

there is no necessity for stopwork, and its omission will get rid of a troublesome piece of work and be no disadvantage to the finished clock. Therefore, file off the projection referred to and turn off the small

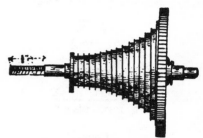

Fig. 1822.—Finished Fusee

end of the fusee clean and smooth on the edge, and no time and space will be wasted in describing stopwork, which can be dispensed with. The finished fusee is shown in Fig. 1822. Do not drill its pivot-holes yet, but make the entire train first, or difficulties will be experienced in the spacing of the wheels.

Centre Wheel and Pinion.

The centre wheel and pinion will now engage attention. Fig. 1803 shows the wheel, and Fig. 1813 a piece of pinion wire from which the pinion must be turned. Pinion wire is of steel drawn to the shape of a pinion through a special draw-plate.

Fig. 1823.—Finished Centre Wheel.

The leaves, therefore, extend throughout its entire length, and it is necessary to strip them from the parts where they are not wanted.

Making Clock Pinions.

To turn clock pinions, a pair of fine female centres (see p. 398) are required.

The best lathe for the purpose is a watch lathe, the next best being a pair of turns and bow. But with care they can be turned in an ordinary 3-in. centre lathe, with a pair of fine female centres, but the

Fig. 1824.—Pinion Wire Centred for Turning.

running centre must be absolutely true. Fig. 1823 shows the centre wheel and pinion finished. The total length is just under $3\frac{1}{4}$ in. ; therefore, take the length of pinion wire of eight leaves and cut off a piece $3\frac{1}{4}$ in. long by filing all round till nearly through. Do not bend it in breaking. Hold it in the hand and centre it with a file, and try it in the lathe for truth. Draw the centres by further filing until it runs true ; it will then look like Fig. 1824. Put it in the lathe, and with the hand-rest and a sharp graver turn it like Fig. 1825. See especially, before turning, that the two parts A and B, in Fig. 1825, run true ; the rest of the pinion does not matter. Screw it in the vice and file off the leaves where they are not wanted—just down to the solid centre, and no further. It must now be hardened. Hold it by the thin end, and heat the part A (Fig. 1825) as far as C to a full red, and plunge it in water. Then take the other end and heat it for rather more than 1 in., and harden as before. Clean it off with emery-cloth and temper it to a blue colour. Both ends of the pinion will now be hard, and the centre of the arbor soft. It can now be put in the lathe again and tested for truth. The pinion body A (Fig. 1825) must run quite true, so draw the centre till it does so. The portion B is wanted for the front pivot, and must also run true. Turn it to the shape

Fig. 1825.—Pinion Partly Turned.

shown in Fig. 1826, which is a section through it. A is the seating for the wheel, and must be cut square—a very little deeper than the thickness of the centre wheel. The outside face is undercut, as

shown, leaving the points of the leaves standing up for riveting over. The centre wheel must run as close to the back-plate as possible—less than $\frac{1}{16}$ in. in clearance will do. The pinion body must extend $\frac{3}{8}$ in.

Fig. 1826.—Section of Completed Pinion.

from the centre wheel. The front pivot B (Fig. 1826) should be a little longer than the thickness of the plate, and a second shoulder should be turned as shown. The arbor should extend $\frac{7}{8}$ in. through and beyond the front plate, and be turned slightly taper.

TURNING PINIONS.

For turning tempered steel a very sharp graver is required, which must be constantly re-sharpened on an oilstone. If turning be tried with a dull edge, the surface of the steel will become burnished like glass, and nothing but a file will cut it. Those who are unaccustomed to turning clock pinions and pivots will find the file a useful tool. The leaves can be just nicked in with a graver, and a file then put on, with the safe edge against the pinion face or pivot shoulder, as the case may be. By this means the metal can be reduced until the solid body is reached. The work will then be out of truth on account of the tendency of the file to follow the steel. A little careful turning with a graver will correct this, and when the work is true

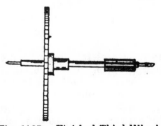

Fig. 1827.—Finished Third Wheel.

again, proceed with a very fine-cut safe-edge file, and with it reduce the pinion face, arbor, or pivot to the final dimensions. All this filing is done with the pinion revolving in the lathe between centres, and will be found an easy and rapid way, besides being safer than turning with a graver in inexperienced hands. The one portion that must be finished with the graver is the wheel seat, as it must be absolutely true. The pivots are to be

Fig. 1828.—Section of Wheel Collet.

nicely smoothed and polished, as described in connection with the fusee, but a narrower polisher must be used.

SMOOTHING PINION LEAVES.

The pinion leaves will be found to be rather rough, and will require filing up with a smooth, thin file. This done, smooth them out with some knife-powder emery on a piece of wood. After cleaning this off thoroughly, polish them with red-stuff or crocus and oil on a piece of wood. It is a rather tedious process, but pays in the end; the better the pinions are polished, the easier will the clock go.

PREPARING CENTRE WHEEL.

The centre wheel should now be filed out—that is, "crossed out"—between the arms. Its centre hole should be turned true by chucking in wood as before described, and opened out by broaching to go on to the shoulder turned to receive it on the pinion body. File it smooth and flat on both sides, and finish with a very fine emery-stick. It will then be ready for mounting. It should press on tightly down to its shoulder on the pinion, and then the points of the leaves, which were left standing up, must be riveted over into the brass with a small flat-faced punch and a light hammer. When put in the lathe, the wheel should run quite true. With the graver, just take a finishing cut over the riveted pinion face to clean it off and give a good appearance.

THIRD WHEEL.

The third wheel and 'scape-wheel pinions have seven leaves, and are made from the same length of wire. The third wheel is illustrated by Fig. 1804, Fig. 1827 showing it mounted upon its pinion. The length of pinion wire can be cut off and centred in the lathe as before described, and the leaves filed off where not wanted. Harden each end separately as before, and true it up. The third wheel is not mounted upon the pinion body like the centre wheel, but upon a brass collet soldered on to the arbor. To make the collet, drill up a piece of $\frac{1}{4}$-in. or $\frac{3}{8}$-in. brass rod, and solder it on the steel arbor where required, afterwards washing it well in hot water to remove the acid. Turn it to the shape shown in Fig. 1828. Turn out the centre of the wheel true, and open it out to go on the collet tightly; then smooth the wheel and trim up the arms with a file, and rivet it on the collet with a small punch. If care be taken in turning a good seating for the wheel, and it fits well, it will go on quite true.

'SCAPE WHEEL.

The 'scape wheel (Fig. 1805, p. 573) can then be taken in hand. Fig. 1829 shows it completed. Turn the pinion like the last, and mount the wheel on a brass collet near the centre of the arbor. Before mounting the 'scape wheel, cross it out carefully—that is, file out the arms and spaces neatly: also carefully smooth the teeth and point them up. See that it is mounted the right way about. A reference to the escapement diagram (Fig. 1832, p. 580) will decide this point.

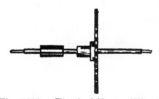

Fig 1829.—Finished 'Scape Wheel.

" PITCHING " TRAIN OF WHEELS IN FRAME.

The train of wheels being now completed, they can be "pitched" in the frame. Clamp the plates together accurately first, and mark carefully the position of the front centre wheel pivot in the exact centre of the frame (see Fig. 1830). In Fig. 1830 the circle A represents the barrel, B the fusee, C the centre wheel, D the third wheel, E the 'scape wheel, F the cannon pinion, G the minute wheel, and H the hour wheel. Also mark the position of the

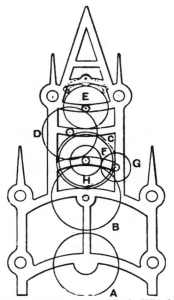

F g. 1830.—Arrangement of Train Wheels, etc.

'scape pivot hole on its bar in the exact centre. The positions of the fusee and third wheel must now be measured from these as fixed points.

ACTION OF CLOCK WHEEL TEETH.

Before going further, it will be as well briefly to consider the question of tooth action. The teeth of clock wheels consist of straight radial lines up to the pitch circle, beyond which they are curved and terminate in a point. Where the straight portion ends and the curve begins is the pitch line or pitch circle, and this is the true diameter of the wheel. Similarly the leaves of a pinion consist of radial lines up to the pitch circle, and beyond that point are rounded off by a semicircle. In a wheel tooth the curved part does the driving, while in a pinion it is the straight

portion which takes the impulse. In a properly pitched depth, the pitch circles of the wheel and pinion must roll upon one another, as it were, as in Fig. 1831. The pitch circles A A are shown just in contact. It is therefore evident that in measuring wheels for the purpose of pitching the depth, only the pitch circles need be measured.

THE "PITCHING" DESCRIBED.

Take the fusee first. With a fine-pointed pair of dividers, measure the pitch circle of the main wheel. Draw a fine straight line on a sheet of note-paper and transfer to this line the full diameter of the pitch circle of the main wheel. Then, in the same way, measure the diameter of the pitch circle of the centre pinion. Add this distance to that already laid down. With the dividers, by a system of trial and error, find the exact half of the total distance. This then represents the distance of the fusee pivot holes from the centre holes. With the dividers, score it off upon the plate, and mark it with a fine centre-punch. To find the position for the third pivot holes, measure the centre wheel and third pinion, and, taking the half as before, score it off upon the plate, making a small arc of a circle just upon the bar. Measure the third wheel and the 'scape pinion, and from the 'scape pivot holes as a centre mark another arc upon the bar, crossing the former one. The point of intersection is the position for the third wheel pivot holes. If the third pivot

Fig. 1831.—Action of Wheel and Pinion.

holes come too near the lower edge of the bar, the position of the centre holes will have to be altered by placing them a little higher, and re-marking the depth; also raise the position of the fusee holes. Having marked them all, drill them with a

small drill, afterwards opening them out
to fit their respective pivots. The pivots
should be quite easy in the holes, and show
no signs of sticking ; but, at the same time,
there must not be perceptible shake. In
opening the pivot holes with a broach,

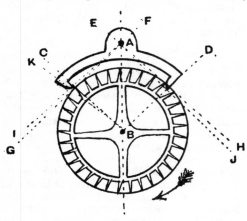

Fig. 1832.—Clock Escapement.

care must be taken to keep them quite
upright. The train can now be put in, and
the depths tried. The teeth of each wheel
must have a little shake between the leaves
of the pinion in any position. If it has
not, the depth is too close, and one of the
holes must be altered; if it has much
shake, they must be brought a little closer
together. Provided the measurements
are made carefully, and the drills started
exactly in the centre-punch marks, the
depths should require little or no alteration. What little is required can generally be effected by bending the bars of
the plate a trifle. The actual amount required to alter a depth is so slight that no
distortion will be perceptible in the frame.
Where bending is not possible, the holes
will have to be bushed. This is done by
opening them out by broaching and drawing them in the desired direction by pressing on one side of the broach, or filing with
a fine rat-tail file. They then have a piece
of stopping wire riveted in. Stopping wire
is brass rod with a central hole through it.
When this is filed off clean on both sides,
and opened out to fit the pivot, it is hardly
noticeable.

THE RECOIL ESCAPEMENT.

Fig. 1832 shows the construction of the
escapement. The pallets are filed up from
the small steel forging shown in Fig. 1806
(p. 573). The escapement recommended is
a " recoil," but with very slight recoil and
thin pallets, approaching very nearly to
that form of escapement known as the
" half-dead " Graham. Fig. 1832 is drawn
exactly to scale and full size, and measurements of depth, etc., can therefore be
taken from it. Those who would like to
set it out for themselves should follow
these instructions: Draw a vertical line
first, A B (Fig. 1832), and from B as centre
strike a circle of the diameter of the 'scape
wheel. On another piece of paper take an
impression of the 'scape-wheel teeth by
pressing the points into the paper with the
fingers. Transfer these points to the circle
just described, taking care that one
point comes exactly on the vertical line
A B. Then draw in the teeth, or at least
as many of them as are required—say, five
on each side of A B. On the left-hand side
count four spaces from A B, and through
the point of the tooth draw B C. On the
right-hand side count four and a half
spaces, and, exactly midway between two
teeth, draw B D. Then draw the tangents

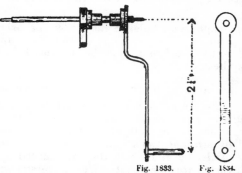

Fig. 1833.—Pallets and Crutch.   Fig. 1834.—Back
                View of Crutch.

G F and H E, just touching the circle of the
'scape wheel at the point of the fourth
tooth from the centre line. Where the
tangents cross A B at A is the pallet staff
centre. About 4° of impulse are required ;
so draw A I at an angle of 4° from A G.

Draw A J at an angle of 4° from A H. Draw
B K at an angular distance from B C of half
a tooth-space.

PALLETS.

The pallets can now be drawn in. The
lines B C and B D (Fig. 1832) represent the
straight backs of the pallets. The left
hand or entering pallet just touches the
point of the fourth tooth from the centre.
Where A I crosses B K marks the point on
which the next tooth will drop; therefore,
from this point to the end of the pallet face
is the impulse, and backwards from this
point will be the recoil face. It is advisable to give very little recoil, so draw the
pallet face to the form shown, the curve
passing through the point of intersection
of A I and B K. Similarly the other pallet
can be drawn in. It intersects the circle
of the 'scape wheel by the amount of the
impulse, 4°, and its point therefore comes
at the intersection of B D and A J. The
point on which the tooth drops is shown
by the point of the tooth itself, which is
shown as having just dropped. So draw
the curved face through this point like the
other pallet. A pair of pallets set out as
above will have equal impulses and light
recoil, and will prove a great improvement
on the usual haphazard pair found in most
clocks of this kind, in which the exact impulse and amount of recoil are largely
left to chance.

PALLET STAFF AND CRUTCH.

Fig. 1833 shows the pallet staff and crutch.
The latter is shown separately by Fig. 1834.
The pallets are mounted on a brass collet
on the staff just like the 'scape wheel.
The crutch is riveted on to another collet
near to the back end. The usual practice
of carrying the back pallet staff pivot
through to the pendulum cock has been
departed from, as it will be found easier
and more straightforward to make the
pivot hole in the plate and crank the
crutch as in many American clocks. The
crutch itself (see Fig. 1834) is made of brass
$\frac{1}{16}$ in. thick, and is riveted on to its collet
on the pallet staff, and has the impulse pin
screwed and riveted in the lower end.

This pin must be nicely smoothed and polished, as must also be the inside of the
slot in the pendulum rod, in which it must
work quite easily, but with scarcely perceptible play.

CORRECTING AND FINISHING PALLETS.

The pallets should be filed up as smooth
as possible, leaving a little too much metal
in them. They should then be hardened
and not tempered. When the pivot holes
are drilled, the depth can be tried and the
pallets trimmed up and polished with
emery-sticks until the depth is correct. Of
course, any great inaccuracies in the depth
must be corrected by drawing and bushing
the pivot holes, or bending the bars of
the frame-plate. The 'scape-wheel teeth

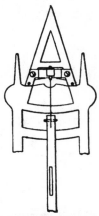

Fig. 1835.—Back Cock and Pendulum Suspension.

should have a little "drop" on to each
pallet face, and the amount of drop should
be equal on each pallet. If the drop on
the entrance or engaging pallet is more
than on the exit pallet, it shows that the
points of the pallets are too close together,
and to remedy it the flat back of the entrance pallet must be cut back a little to
increase the drop on the exit pallet until
both are equal. When this is done, should
the drop be excessive, deepen the depth
a trifle. If the drop is excessive on the
exit pallet only, the points of the pallets
are too far apart, and the flat back of the
exit pallet must be cut back. Spare no
pains to get a good escapement, and on

no account leave the pallet faces in a rough condition, but let them be well polished.

Back Cock or Pendulum Cock.

The back cock, or pendulum cock, shown by Fig. 1798, p. 572, must be filed up and screwed on to the back of the back plate, being shaped up at the sides as shown in Fig. 1835. The under side of the projecting piece must come on a level with the pallet staff pivots.

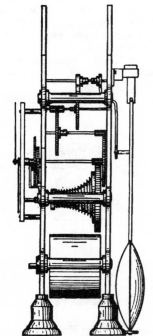

Fig. 1836. — Side Elevation of Clock Movement.

Pendulum.

The pendulum (Fig. 1801, p. 572) is supplied finished, except for the pinning-on of the suspension spring; this is bought ready to pin on. Slit the top of the pendulum rod with a fret-saw and pin the spring through with a brass pin. The projection on the back cock is slit with a wider saw, cut to admit the top of the suspension spring. A pin through the latter rests upon the top of the cock. The whole arrangement is shown in Fig. 1835. The pendulum

should be left of such a length as just to swing clear of the stand upon which the clock will be placed. Fig. 1836 shows the arrangement of the train and pendulum

Motion Wheels.

The motion work consists of three wheels (see Figs. 1807 to 1809, p. 573) whose duty

Fig. 1837.—Cannon Pinion. Fig. 1838.—Minute Wheel and Pinion.

is to impart the motion of the clock to the hands. Fig. 1830, p. 579, shows the position of the motion work seen from the front, and Fig. 1836 shows it in elevation. There are two small wheels exactly alike —one is for the cannon pinion, and the other for the minute wheel. The larger wheel is the hour wheel.

Cannon Pinion.

The cannon pinion is shown in Fig. 1837. It fits tightly upon the tapered part of the centre-wheel arbor, and goes right down to a slight shoulder just clear of the front plate. Its "pipe" can be made of a piece of clock stopping-wire, supplied with the set of materials. Put this wire

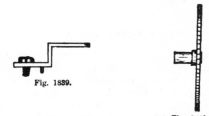

Fig. 1839.

Fig. 1840.

Fig. 1839.—Minute Wheel Cock with Screw and Steady Pin. Fig. 1840.—Hour Wheel and Pipe.

between male centres in the lathe and turn the pipe down to size and slightly taper; its diameter should be less than $\frac{1}{8}$ in. At the lower end a good square shoulder is cut to form the wheel-seating, and the wheel is riveted on. The centre hole is then opened out with a broach to the correct size, and the end squared up to take the socket of the minute hand. The centre of the pipe, for about ¼ in. or so, is filed on both sides through to the centre hole. The sides left are then squeezed in a trifle to form a kind of spring, and hold the cannon pinion tightly on to the centre-wheel arbor at all times.·

Fig. 1841.—Fancy Dial Plate.

MINUTE WHEEL.

The minute wheel (Fig. 1838) is similar to the cannon pinion. Its pinion has six leaves, the wheel being riveted direct on to the pinion like the centre wheel. The minute wheel cock (Fig. 1799, p. 572) is fixed to the front plate with a screw and steady pin, as in Fig. 1839. In pitching the depth of the minute wheel into the cannon pinion, make it as deep as possible so as to avoid shake; but it must not be so deep as to bind. If the pitching is done correctly the hour wheel depth into the minute pinion will be quite right.

HOUR WHEEL.

The hour wheel pipe must be made from a piece of ¼-in. round brass rod and drilled up. If a jaw-chuck, or other means of holding this size of rod, is available, fix a short piece of rod in the lathe and drill it up as it runs, turn it to size and shape, and finish it before cutting it off. Fig. 1840 shows the shape to which it must be turned. The total length of the pipe is ¾ in. Before turning it, however, take the hour hand and cut its pipe or socket down

to half its former length, and open the hole a little; then turn the hour wheel pipe to fit it tightly. The hour wheel must be riveted on to its pipe like the cannon pinion. When finished, the hour wheel should revolve freely upon the cannon pinion, but without side shake.

PUTTING ON HOUR AND MINUTE HANDS.

Push the hour hand on tightly; it should project a little beyond the end of the pipe. Then push on the minute hand; its socket should go inside the hour hand pipe without touching it. Make a small washer to go on outside the minute hand on the centre arbor, and drill the arbor through with a very small drill and pin it on. When the hands are pinned on properly, the hour wheel should still be free to revolve and have a little endshake under the minute hand. The minute wheel also must be quite free and have endshake.

DIAL OF SKELETON CLOCK.

The dial and dial feet, when made, will inside. Fig. 1795 (p. 571) shows a plain the sheet of brass ⅟₁₆ in. thick. It should be 5 in. outside diameter and 2¾ in. or 3 in. inside. Fig. 1795 (p. 571) shows a plain circle; but if desired it can be turned out inside and filed outside to the shape shown in Fig. 1841. Whichever shape is chosen, when done it must be filed flat on the front and made perfectly smooth, every scratch being got out. Lay it upon the front plate, making sure that it is quite central, and mark near its edge the position of the feet. One should come on the centre bar of the plate, just below the six o'clock, and the two others up by eleven and one o'clock, near the top pillars of the frame. At these

Fig. 1842.—Dial Foot.

points drill small holes in the dial and countersink them neatly on the outside. Place the dial again on the plate and mark through the holes the position of the feet upon the front plate. Drill holes ⅛ in. in diameter at each place and tap them.

FEET OF DIAL.

Turn three dial feet from ¼-in. brass rod, like Fig. 1842, and thread their lower ends to screw into the front plate. The tops are left quite flat, and have a small hole drilled and tapped in each to receive a small chamfer-headed screw to fix the dial on. These screws should be polished on the heads and blued for appearance. The dial, when finished, should be sent to a specialist to be silvered and painted. Every part of the clock will now be made, and it only requires finishing off and putting together.

PUTTING MAINSPRING IN BARREL.

The mainspring must be put in the barrel. To do this, start at the outer end and hook it on the hook in the barrel.

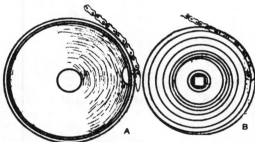

Fig. 1843.—Methods of Hooking Chain in Barrel and Fusee.

With a firm wrist, coil it in gradually towards the centre, forcing each coil down until it is all in. This job requires a firm and strong wrist.

DRIVING CHAIN, OR GUT LINE.

The worker must decide whether he will have a steel chain or a gut line. The steel chain is the better, but costs about 4s. or so, whereas a gut line costs only about 8d., best quality, and answers well until it breaks. If a chain is used, make the hook hole close to the opening in the barrel cover, and ⅛ in. from the edge of the barrel, as at A (Fig. 1843). Cut a slot in the last coil on the fusee (the large end) and put a pin through, as at B (Fig. 1843). If a gut line is used, drill three holes in the

barrel and fix the line as in Fig. 1844. Burn the end to prevent it pulling through. To do this well, cut the end off close and heat a piece of brass very hot (not red hot) and apply it to the end of the gut. This will sear it and harden it, spreading it out like the head of a French nail. At the fusee end a hole is drilled straight down into the body, and another large hole in the end to meet it. The gut is passed through and a knot tied, searing it as before. The knot lies in the large hole drilled in the fusee end underneath the ratchet, where it is quite out of the way.

FINISHING PARTS OF CLOCK.

The plates must be carefullly smoothed by filing with a fine file. The edges can be burnished with an oval burnisher. The flats must be smoothed off with fine emery-sticks and polished with rotten-stone on a rag, and afterwards fine pumice-powder and water on a rag or pad. All the small pivot holes should previously have been cupped out on the outside to form "oil-sinks" to prevent the oil spreading over the plates. The barrel can be polished in the same way as the plates. Pillars, feet, etc., should all be well polished. Screw-heads should be filed smooth, and burnished or glued. All the arbors should be run in the lathe once more and polished from end to end with red stuff or rouge on a bell-metal or brass polisher. The barrel square and fusee-winding square should be both filed smooth and burnished.

PUTTING CLOCK TOGETHER.

The clock should now be put together. Take the barrel first and apply some clock oil to its pivot holes where it runs upon the arbor. Oil the coils of the spring freely. If a gut line is used, fasten it in as before described. The fusee should be oiled where the main wheel rubs against it. The click should be oiled at its pivot and where the click spring touches it, but do not oil its point or the ratchet teeth. Oil the fusee arbor where the main wheel turns upon it, and the main wheel where the circular key presses against it. The fusee and barrel can then be put upon the back plate in their places. Also the centre, third, and 'scape wheels and pallets. Pin on the front plate and oil all the front pivot holes. Before going any further, see that every wheel has endshake between the plates. If all is right, proceed to put on the chain or line.

Putting on Chain or Gut Line.

If a line, it will be already fastened at each end to the barrel and fusee respectively. Place a key upon the barrel square and wind the line up upon the barrel in a regular spiral, and when all is on and the line pulled tight, put on the barrel ratchet and click and " set up " the mainspring about half a turn or so, after which screw the click up tightly. Then place a key on the winding square, and carefully wind the line on to the fusee. The line must be most carefully guided straight during this process, and on no account be allowed to drag sideways, or it will cut against the edges of the fusee groove. When it has been once wound on right it will always run straight afterwards. The chain is put on in much the same manner. First hook it on to the barrel and wind it all up, then hook the other end on to the fusee. Set up the mainspring and wind up as before, using great caution, as the effect of letting a chain run over the edges of the groove is most disastrous to the latter.

Completing the Clock.

The movement, when placed upright, should now " trip " through the escapement without hitch. Put on the motion work, dial and hands, and give a little oil to the minute-wheel pivots. Apply oil to the pallets and all the back pivots; put

a little on the impulse pin which works in the slot in the pendulum-rod; just grease it—that is sufficient. Let down the pendulum to its lowest point and proceed to time the clock. Gradually raise it until it goes fast enough; then the projecting part of the screw, except about $\frac{1}{4}$ in. or so, can be cut off. When to time, it will just swing clear of the stand.

Clock Stand and Shade.

The stand and shade now only remain to be considered. An ebonised and velvet-covered wood stand, made of 1 in. wood,

Fig. 1844.—Method of Fastening Gut to Barrel.

does very well. It should have four short feet near its edges, and a groove round the top edge to take the glass shade. The join is generally covered by a piece of chenille round the bottom, of a colour to match the velvet on the stand. A marble stand can be used, but this is troublesome to drill for the feet, and, being cemented together by plaster-of-Paris or other cement, is likely to come apart; besides which, it is heavy and awkward to handle. The shade will be about 13 in. high, 9 in. wide, and 5 in. from back to front.

BUILDING A SMALL HORIZONTAL STEAM ENGINE.

SIZE AND POWER OF ENGINE.

A ¼-HORSE-POWER engine is illustrated, one-fourth full size, in side elevation by Fig. 1845, the pump being in section. The cylinder (see Fig. 1846) is bored 2 in., and the stroke is 4½ in. Allowing a maximum boiler pressure of about 50 lb. per square inch, the average pressure in the cylinder may reasonably be taken at 26 lb. per square inch. Now, find the area of the piston, square the diameter in inches, and multiply by ·7854. The diameter is 2 in., and the area will be $2 \times 2 \times ·7854 = 3·14$, or say ·3 sq. in. Thus, the average total pressure on the piston is $3 \times 26 = 78$ lb., and if the engine is regulated by the governor to run at 150 revolutions a minute, the piston will travel $2 \times 4½ \times 150 \div 12 = 106$ ft. (say) in one minute. Therefore, 78 lb. × 106 ft. = 8268 foot-pounds of work will be done per minute, which is about ¼-horse-power.

CASTINGS, ETC., FOR THE ENGINE.

The following list includes all the castings and materials required to make the engine alone. In cast-iron:—Bedplate, cylinder, top cover, bottom cover, piston, piston bottom, steam chest, steam chest cover, cross-head, fly-wheel, cross-head guides (2), angle bearings (2), caps (2), eccentric, eccentric half-strap (rod end), eccentric half-strap (outer end). In gun-metal:—Gland for cylinder cover, piston rings, piston-rod nut, gland for steam chest, slide-valve, valve-rod nuts (4), bearing brasses (bottom) (2), bearing brasses (top) (2), connecting-rod brasses (small) (4), and connecting-rod brasses (large) (2). A connecting-rod, valve-rod, and an eccentric-rod will be required in malleable cast-iron, and a crank shaft in wrought-iron. There will also be needed five Stauffer lubricators, smallest size, and one ½-in. brass screw stop-valve with nipples to fit. The following is a list of castings required for the feed pump. In gun-metal: Pump body, pump cover, pump gland, pump union nut, ¼-in. spindle valve, ⁵⁄₁₆-in. spindle valve, pump plunger, and liner. Also a pump eccentric rod in malleable cast-iron, and an eccentric with two half-straps in cast-iron. The governor requires two balls and a bracket in cast-iron, and a governor sleeve or body, pivot pin, driving pulley, throttle valve, valve case, and throttle valve case gland in gun-metal. Other materials needed include 4 in. of ¼-in. steel tube ⅛-in. bore, and one pair of 1-in. brass mitre wheels.

TOOLS FOR MAKING ENGINE.

The tools required are:—A 4½-in. centre metal-turning lathe, with slide-rest; if the fly-wheel is to be bored, the lathe must have a gap capable of taking in a diameter of 16 in., and if it is not a self-acting lathe the cylinder had better be ordered ready bored. A universal chuck 3 in. or 4 in. in diameter is desirable; a face-plate chuck 8 in. to 10 in. in diameter is necessary, and must have dogs and bolts. Other requirements include an angle-plate chuck for the face-plate; a vice; a set of flat and round files, and some cold chisels; screwing tackle, including a screw plate and taps up to ¼ in., and a stock and dies for brass gas-threads; a set of straight flute or Morse twist drills up to ¼ in. with drill chuck; a surface-plate, scribing block, squares and callipers.

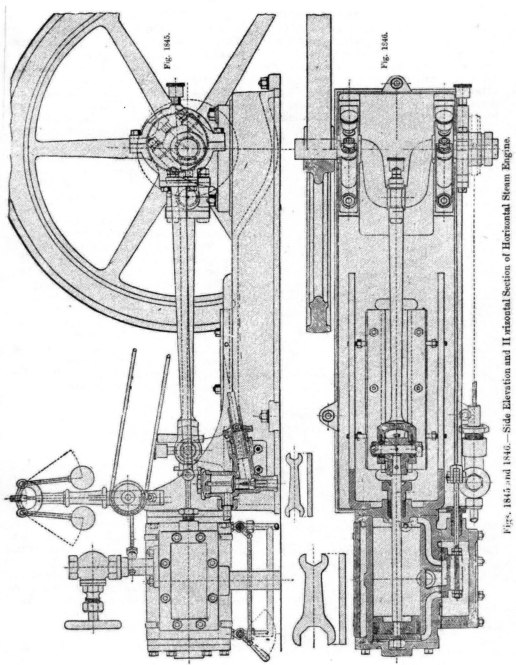

Fig. 1845.

Fig. 1846.

Figs. 1845 and 1846.—Side Elevation and Horizontal Section of Horizontal Steam Engine.

CYLINDER OF STEAM ENGINE.

Figs. 1845 and 1846 show that the cylinder overhangs the bed-plate, which is cast with a round flange at the back end, against which the front cover of the cylinder is bolted: thus the cylinder, being attached by one end only, is free to expand lengthways when heated by the steam. Fig. 1847 shows the back end of the engine.

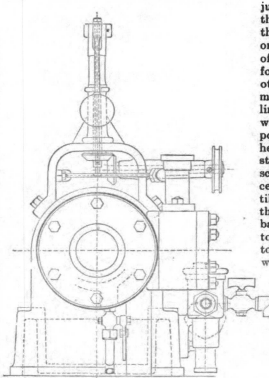

Fig. 1847.—End Elevation of Engine.

The cylinder should be bored with a bar, which will require a slide or self-acting lathe.

MAKING CYLINDER BORING-BAR.

To make the boring-bar, get a 14-in. length of 1¼-in. round iron, and after drilling the centres, finish the ends and turn the bar up true. It will now be about 1¼ in. in diameter. At its middle, round the bar with a point tool scratch two fine lines ⅛ in. apart; between these lines and through the middle of the bar a hole or slot ⅞ in. long and ³⁄₁₆ in. wide must be cut to take the cutter and wedge. To do this, first mark the position of the slot on both sides, the marks being exactly opposite, so as to ensure that the cutters shall pass through the middle. Wrap a slip of paper round the bar, and cut it so that it will just meet; folding this in half gives half the distance round the bar. Take ¹⁄₁₆ in. in the dividers, and mark that distance off on one of the lines marked on the middle of the bar; measure half round with the folded paper, mark two other spots on the other side of the bar, and indent these marks with a centre-punch. To draw four lines from each of these four points parallel with the bar and with each other, use a point tool turned over to lie on its side and held in the slide-rest, the bar being held still by advancing the back-centre by its screw so as to jam the bar between the centres. The bar would be turned round till the dots of the centre-punch come to the point of the tool; the screw of the back-centre would then be forced a little, to prevent the bar from turning; and the tool, moved forward to touch and then sideways by the slide screws, will mark the

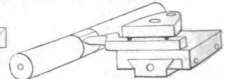

Fig. 1848.— Marking Boring Bar wi.h Slide-rest Tool.

lines, as shown in Fig. 1848. Next, with a ³⁄₁₆-in. drill, work half-way through, first from one side and then from the other, arranging the holes so as almost to cut into one another. A narrow cross-cut chisel will remove the metal between the drill holes, and the slot can then be finished by filing, cutting away just up to the lines at both sides, and keeping the sides and ends of the hole flat, and not rounded.

CUTTER AND WEDGE OF BORING-BAR.

Next obtain for the cutter and wedge two pieces of tool steel ½ in wide, ¹⁄₁₆ lb full thick, and 2¼ in. long. File these up, flat-ways first, to exactly ₁³₆ in. in thickness, callipering them to ensure their being the same thickness throughout, and trying them in the slotted hole; when they go in about ⅛ in., file them no thinner, but fit the slot to them, until the wedge and cutter can be pressed in with the fingers. This is done because the hole is almost sure to be wider at the mouth than in the middle. Now square up one edge of each of the pieces of steel, round the corners a little, and use them to finish the ends of the slot, carefully filing till the steel pieces bed down solidly on each end. Fig. 1849 is a section through the bar, showing the cutter and wedge in position; and Fig. 1850 is an end view. The cutter may now have the shallow notch filed in one edge, this notch fitting small flats filed across the mouths of the hole on each side of the bar. The cutter having been well fitted, so that it requires pressure to force it into place, its back may now be filed across and tapered to suit the wedge, the taper being, say, ₁³₆ in. bare. Having thus finished filing up the cutter, file up the edges of the wedge, making sure that it fits the whole length of the hole; then shorten its ends, put the cutter in position, drive in the wedge firmly, put the bar in the lathe, and turn turn up the ends of the cutter till it measures 2 in. long; take also a cut down the front or cutting edges of the cutter, to ensure those edges being true. The cutter must now be removed from the boring-bar, but before doing so, be careful to mark both it and the bar. Then " back off " the cutting edges *a* and *b* (Figs. 1849 and 1850) with the file, the cutting angle being about 80°. The edges must slant in contrary directions, and in doing this " backing off " be careful to leave just a trace of the tool marks left by turning, so as to ensure that the edges shall cut equally. Next back off the guiding or scraping edges *c* and *d*, removing rather less in this case than in the case of the cutting edges *a b* (Figs. 1849 and 1850), and the cutter will then be ready for hardening.

HARDENING AND TEMPERING BORING-BAR CUTTER.

Any fire will do for hardening the cutter, provided there be a good amount of red-hot coals; twist a piece of stiff iron wire round the cutter, leaving a foot or so of the wire for a handle; have about a quart or so of water ready (this may be slightly warm), and when the cutter is red all over, plunge it endways into the water and leave it a minute. Take it out and try it with a file, when, if quite hard, it is taken to the grindstone and the edge slightly ground; it does not require to be " let

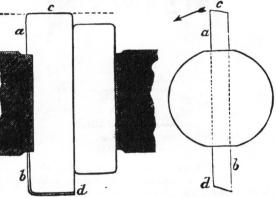

Figs. 1849 and 1850.—Section and End Elevation of Boring Bar.

down " or tempered, because the edge is quite strong enough without it, and, thus treated, the cutter will go right through the cylinder without showing sign of wear. The boring-bar and wedge will be useful for other work, as many cutters may be made for it of different lengths, so as to bore holes from 1½ in. to 2½ in. in diameter.

BORING STEAM ENGINE CYLINDER.

Secure the cylinder casting on the saddle in such a way that its centre line may be true with that of the lathe centres. Figs. 1851 and 1852 show one way of doing this. The upper part of the slide-rest is removed, and in its place is bolted a flat iron plate A, ¼ in. to ½ in. thick; two pairs of wooden clamps are prepared to fit the cylinder, and bored through for the four ¼-in. holding-down bolts, these being tapped into the square plate at their lower ends, and fitted with nuts and washers at the top. Cut away the lower clamps by degrees till the cylinder comes level at the exact height of the centres; it can be taken out of twist by slacking the two holding-down bolts,

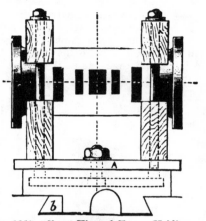

Fig. 1851.—Front View of Clamps Holding Cylinder.

and turning round the square plate, which must have two holes in it to pass the holding-down bolts which secure it on the slide, as well as the four tapped holes. The side on which the plate is bolted would be fixed by tightening the screws of the gib *b* (Fig. 1851). Put the cutter into the slot in the bar, and fix it with the wedge so that it presents its two cutting edges towards the hole; put on the requisite change-wheels to give a feed of about seventy to eighty threads to the inch, and bore with the back gear in, and with the belt at a slow speed. This cut should be carried right through without stopping, and it will take perhaps about half an hour. The cylinder casting will soon get rather warm under the cutting, and will expand slightly in size; if the cutting were stopped for long, so that the metal began to cool, there would be produced a slight groove or inequality in the bore quite perceptible to the finger. Therefore, just before beginning to cut, slightly warm the casting, which could be done equally all round by holding inside it

a red-hot poker till it felt warm to the hand; then replace the bar, and begin to bore at once. The cutter can take a cut of $\frac{1}{16}$ in. all round, and one cut should be enough to finish the bore, leaving a fine surface bright and smooth, and perfectly parallel. If it were found to be necessary to take another cut, the cylinder would be

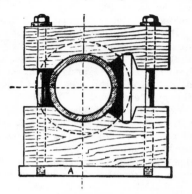

Fig. 1852.—End View of Clamps Holding Cylinder.

left as it is whilst another cutter was made a little larger than the first, to bore, say, 2⅛ in.

EXAMINING CYLINDER BORE.

As soon as the cylinder is removed from the lathe, rub it out carefully with an oily rag, to clean it and to preserve it from rust, and examine the bore on the side of the portways; it is possible that the boring may expose in the casting a blow-hole, making a communication between the inside of the cylinder and the port. If the blow-hole is a bad one that cannot be stopped by screwing in a little plug with red-lead, the casting should be returned for another.

OTHER METHODS OF BORING CYLINDER.

Should the lathe not be self-acting, then if the slide-rest has a traverse of at least 5¼ in., and the mandrel of the lathe is strong and not too short, the cylinder might be mounted on the face-plate and bored with a strong inside tool about 1 in. square. The cylinder, however, will stand overneck a good deal, and so will the tool.

The hole would be made slightly larger at the mouth, and then carefully corrected until a solid gauge, made something like the piston and rod, will fit perfectly all the way down. Yet another way would be to use a couple of flat drills, but both these plans would produce inferior work, and would render necessary the use of a grinder to finish the bore true and smooth.

CONING OUT ENDS OF CYLINDER BORE.

The next operation is to turn the flanges of the cylinder true with the bore, and to cone out the bore slightly at each end for ½ in., so that the packing rings of the piston may pass beyond the surface on which they rub; otherwise, they might wear a slight depression in working, and leave a ridge at each end of the bore, which would make it impossible to get them out. This precaution is always taken in similar cases; as, for instance, in the slide-valve and the slipper-guide. Chuck the cylinder true with the bore, so that a flange and part of the bore shall be accessible to the tools, Fig. 1853 showing how this can be done. Turn to a diameter of 2 in. a piece of round iron M, about a foot long, and reduce it carefully till it drives on the cylinder to within about 1 in. of one end; now turn up one flange, and also enlarge the bore by slightly coning it out at that end. Then drive off the cylinder, and drive it on the other way, so as to treat the other end in the same manner. A pin P is driven or screwed into the mandrel to catch the arm of the chuck D, if there is no 2-in. carrier. The cylinder must not be driven on very hard, or it may burst.

CYLINDER COVERS.

Taking in hand now the cylinder covers, begin with the bottom cover. An American universal chuck will hold it by the outer edge, or it may be driven into a recess in a piece of hardwood turned out to receive it. Face up the inside surface of the cover flat and true, and make the shallow projection fit exactly into the back end of the cylinder, trying this on as it is fixed in the chuck; then with a point tool strike a fine circle 3 in. in diameter, to indicate the centres of the six screw holes.

Mark these with the centre-punch, noting that, whilst six holes equally divide the circle into sixths, one pair should be placed farther apart by ⅜ in., to avoid clashing with the ports. Next, taking the top, or lid, which has the stuffing-box cast on it, turn up the inside in the same way, fitting it to the front end of the cylinder, and, after striking the 3-in. circle for the screw holes, at the same chucking bore a ⅜-in. hole through the centre of the boss for the piston-rod, to ensure that the piston-rod shall be central with the bore. Next drill the six holes in the two covers with a drill ³⁄₁₆ in. full. To turn the other side of the covers and their rims true with the part already finished, a recess must be made in a hardwood chuck B (Fig. 1854), into which the cylinder fitting on each lid can exactly fit. When making this recess, try each cover C in, and begin with whichever has the smallest fitting, because then the recess can be slightly enlarged for the other. Having fitted one cover into the chuck of hardwood (which is screwed to the front of a face-plate A), clamp it firmly with two or three plates D by its extreme edge, so as to allow the tools to face up almost all the front surface. In the case of the front cover, the shallow fitting on

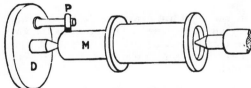

Fig. 1853.—Cylinder Centred for Turning.

the outside will be formed, and the stuffing-box will be bored, and screwed about 14 threads to the inch, with the screwing lathe; or a ⅜-in. gas-thread tap may be used, followed by a chaser, by those who do not possess a self-acting lathe.

TURNING EDGES OF CYLINDER COVERS TO SUIT FLANGES.

Finish both covers, and polish with emery-cloth, and, before removing either from the chuck, turn up and polish the edges to exactly the size of the flanges on the cylinder. To do this, secure the cover to the chuck by two or three wood screws, passed into the wood chuck through the screw holes in the covers; this method of holding will also allow of finishing the edge of the face of the covers. To complete the cover, take the brass casting for the gland, bore it with a ¾-in. hole while chucked in the lathe, mount it on a small mandrel and turn and screw it in the screwing lathe,

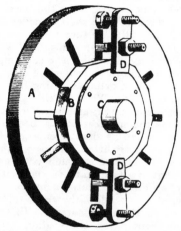

Fig. 1854.—Cylinder Cover Chucked for Turning.

finishing with a chaser to fit the internal thread in the cover. Or the thread might be marked out by the ¾-in. gas stock and dies, when the chaser would be used to finish.

Making Top of Gland Eight-sided.

The top of the gland is made eight-sided, so that it may be grasped by a spanner. The best way to do this would be by means of a division plate on the lathe pulley, but, failing this, a template or pattern may be made by drawing the required octagon on tin or sheet metal, and after cutting away the inside and filing to the lines, the template can be used to test the sides of the gland as they are gradually filed to shape. The same template will serve for filing up the pump gland. Another equally good method would be to file up a little octagon in tin, and use that to file out the spanner;

the jaws of this largest spanner being thus correctly shaped, it can itself be used to test the octagons on the piston-rod and pump glands.

Filing and Scraping Piston Rings.

The piston and piston-rod are shown in detail by Fig. 1846. The casting for the brass packing rings should be chucked by the outside and bored. As it measures 1⅝ in. inside when compressed, it should be bored out $\frac{1}{8}$ in. larger than this. Drive it on a mandrel, and turn the outside to $2\frac{1}{8}$ in. in diameter; turn the edges and part it in two, so that each ring is ¼ in. wide. Now take the rings to the vice, the jaws of which must be protected with lead clams—that is, sheet lead about ⅛ in. thick, bent over the jaws to prevent them marking the work. A perfectly even surface will be wanted for filing and scraping the edges of these rings quite flat without destroying their parallelism; the scraped surface of the lathe-bed might do, but a surface-plate, say 3 in. by 5 in., will be better. If of this size, the plate will do to true the bed-plate facing on which the slipper-guide works, as it will go between the flanges. Get some Venetian red, and mix with machine oil in a flat tin box, and smear a very little of this "marking" on the face-plate, equally all over it; then when the rings are rubbed on it the parts which touch the surface can be seen. To hold the rings while being filed or scraped, take one of them and, placing it on a suitable piece of wood, draw round it with a pencil, and hammer a few small nails into the wood on the line, so that the ring can be placed between the nails, which prevent its slipping while it is being filed. The heads of the nails should be knocked down a little lower than the upper edge of the ring, so that they may not interfere with the file. Callipers are sure to spring slightly, and as the rings must be perfectly parallel, make a rigid calliper by filing a notch in sheet iron to act as a gauge of width. File up and scrape one edge of each of the rings till it bears well all over on the face-plate; then file the notch for the gauge, trying it on the rings all round till it will almost go on at the smallest place; then bring up the other surface of each ring true, so that the gauge will go on equally tight all round. The rings must be handled carefully, as they are very easily bruised.

Sawing Piston Rings.

Screw a ring gently in the vice in the lead clams, one edge projecting about ¼ in., and with a hack-saw cut it across obliquely, because then the edges of the ring are not likely to wear grooves in the cylinder. The rings must be cut twice, so as to leave an oblique gap ¼ in. wide; this is necessary to get the rings into the cylinder.

Piston Cover, etc.

The cover of the piston is next chucked. Face one side, and bore a ⁵⁄₁₆-in. hole through the middle. Chuck the body of the piston, hollow side out; true that side, and bore a ⁵⁄₁₆-in. hole through it. Place aside these two parts, and take a piece of ½-in. mild steel 8½ in. long; centre this, and turn it down to ⅜ in. full. Turn down one end to fit tightly into the body of the piston; cut a ⁵⁄₁₆-in. Whitworth thread on the end of this part, and make the brass nut for it. Now drive this end of the piston-rod into the body of the piston and put on the nut, screwing it hard up to secure the piston body on the rod; put all between the centres, and turn up the piston body. Then scrape up the inner surface of the piston cover, remove the piston-rod nut, put the piston cover in its place, and screw on the nut tightly. Now try the brass packing rings in the groove; if they are too tight, turn a little off the side of the groove which is part of the piston; if too loose, take off the nut and piston cover, and turn away a little from the face of the piston against which the cover fits. The width of the groove must be such that when the rings are in position and the nut is screwed up hard, the rings will be held tightly enough to stand turning in the lathe.

Turning Piston to Size.

Now turn the outside of the rings again, because when they are compressed into the cylinder they will not be exactly round. First turn the outside diameter of the piston body and cover to fit into the cylinder, and when trying it in the cylinder, be careful to wipe the turning dust off and have the bore of the cylinder quite clean. It is better to pass the piston-rod through the stuffing-box in the cylinder cover when trying the fit, because that will keep the piston from getting askew in the cylinder; therefore, at this stage it will be well to finish the piston-rod to size. Use the brass gland as a gauge, and, putting in a very sharp tool, with a light cut reduce the piston-rod till the gland will almost go over it; then take a worn, smooth file, and use that on the rod as it turns in the lathe till the gland can be pushed over it with some difficulty. Now take a smooth file and

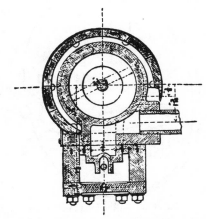

Fig. 1855.—Cross-section of Cylinder through Steam Chest.

"draw-file" the rod, rubbing lengthways from end to end, whilst the lathe is at rest, so as to lay the grain of the scratches longways of the rod, and remove all circular scratches, which would rapidly wear out the packing, besides increasing the friction. The gland should now slide freely over the rod from end to end, and in trying the fit of the piston in the cylinder the cover would be put on, and the piston-rod, with piston on it, would be passed in from the back end. The piston-rod, passing through the cover, would keep the piston true, so that the fit could be observed.

FITTING PISTON RINGS.

When the piston has been adjusted so that it will pass freely up and down the cylinder, take off the nut and cover, and put the piston body just inside the cylinder, the rod passing, as before, through the stuffing-box in the front cover. Now take one of the packing rings, and, compressing it slightly with the fingers, press it into the coned-out mouth of the cylinder and over the body of the piston ; then add the other ring, the piston cover, and nut, but do not screw up the nut. On pressing,

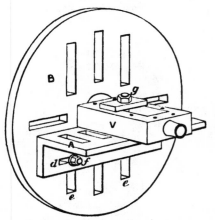

Fig. 1856.—Chucking Valve-box.

or gently driving, the piston forward, the rings will gradually close till they come to the parallel part of the cylinder. Try drawing the piston up and down, tighten the nut, and take out the piston and rod. Now bright spots will show where the rings rubbed hard on the cylinder. Put the piston as it is in the lathe, and take a light cut over the rings ; then put the piston in the cylinder, loosen the nut, and pull the piston up and down ; fix the nut ; take out the piston and examine it. The bright marks should now show all round the rings. If, however, the piston rubs hard in the cylinder, turn a little more off the rings

till there is only enough spring left in them to hold up tight without causing too much friction. One thing more remains to be done to the piston. The surface on the piston body against which the rings rest must now be got up with the scraper, so much metal being taken off that the rings shall be able to move, even when the nut is tightened on the piston cover ; yet the rings must touch both sides of the groove. The shape of the piston body makes it impossible to use the surface-plate to try this last annular surface, therefore rub the edge of one of the rings with the marking, and rub the ring on the piston surface. Further, take care that, when putting in the rings, the two cuts do not come together. The cut place on one ring should be at right angles with the other ; then the steam cannot pass, and the rings will press equally all round the inside of the cylinder. The piston, being now finished, should be oiled and put away.

CYLINDER VALVE-FACE, VALVE-BOX, ETC.

Next file up the valve-face of the cylinder, keeping it parallel with the bore of the cylinder by means of callipers, measuring from inside the bore to the valve-face, or placing this face on the surface-plate, and trying with a square whether the flanges of the cylinder stand square with the plate. When they do, finish the face by filing and scraping to the face-plate. File up also the valve-box edges that make the steam-tight joint, keeping the sides parallel by trying all round with callipers ; then scrape and finish these surfaces. File up the cover or lid on its flat sides, scraping the inner surface to make a steam-tight joint. Place the valve-box on the valve-face in position, and the cover on that, and observe whether the outside edges correspond : if not, they must be made to do so by the file. It is not, however, advisable to file up the outside of the box, but only to make the edges come fairly even. The inside of the box should be filed out square and true, so that its walls may form a guide to the slide-valve (see Fig. 1855).

Squaring Cylinder Ports.

As cast, the ports in the cylinder will not have their edges perfectly straight and square. Therefore, on a piece of stiff card the size of the valve-face of the cylinder, draw very accurately, with a hard, sharp pencil, the exact shape and position of the three ports from the dimensions given in Figs. 1846 and 1855. Cut out the ports in the card, place it on the valve-face, so as most nearly to correspond with the three holes cast in it, scribe the holes on the iron, and file the edges of the ports to these marks. To fasten the parts together, eight long $\frac{3}{16}$-in. studs are required with Whitworth threads, and these may be made from iron wire; the eight nuts should be of steel, and should be hardened. The washers are not necessary, though they look well. First mark out the positions of the eight holes on both sides of the valve-box, squaring across carefully from one side to the other; or, better still, take the card again, mark the holes on that, and then, after cutting out the holes, place the card on the four surfaces and mark them with the scriber, taking care to keep the same side of the card always to the positions of the same studs: thus, the card might be placed first on the valve-face, and, after marking there the eight holes, then on the inside of the valve-face, then on the outer face of it, and lastly, on the cover. It is not an easy matter to get these holes exactly in line, and great care must be exercised. If the box will not go on at the first trial, use a small rat-tail file in the holes through the sides of the box, filing away towards whichever side of the hole binds against the stud.

Drilling Valve-box, Cylinder Cover, etc.

Having marked, by means of the card template, all the eight holes as accurately as possible on the four surfaces, drill the valve-box first from both sides with a $\frac{1}{8}$-in. drill, to meet in the middle. The drill should revolve in the lathe, and the work be pressed forward by the boring-flange.

Watch the drill as it starts, to see that it does not run to one side; if the holes do not meet correctly, rub them outside with a rat-tail file, used all round equally, so as not to alter their position. Now send a $\frac{3}{32}$-in. drill straight through the holes from one side; $\frac{3}{32}$-in. being the tapping size for $\frac{3}{16}$ in., therefore it is the size to be used. Choose two holes far apart on the cylinder face, and drill and tap these first. Take their distance apart in the compasses, and try whether it corresponds with the distance apart of the corresponding holes in the valve-box. If it does, drill and tap them, keeping the tap upright; screw in the long studs; enlarge the two corresponding holes in the valve-box to $\frac{3}{16}$ in. full, and put it on over these two studs. Leaving the cover for the present, screw on two nuts, and so fix the

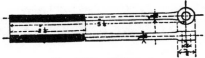

Fig. 1857.—Valve-rod.

valve-box firmly up to the cylinder-face. Now take the cylinder as it is to the lathe, pass the $\frac{3}{32}$-in. drill through the six remaining holes, and drill into the cylinder-face; by this means the drill will be so guided that it will be in correct position. Take off the valve-chest, enlarge the six $\frac{3}{32}$-in. holes to $\frac{3}{16}$ in., tap the six untapped holes in the cylinder-face, screw in the six remaining studs, and try on the valve-chest. Bend the long studs a little if they are not perfectly upright, and the holes in the valve-box may be made, say, $\frac{7}{32}$ in. in diameter, allowing $\frac{1}{32}$ in. for clearance, errors, etc. The holes in the cover may be drilled $\frac{3}{32}$ in. first, and then the cover should be tried on the valve-box, when it will be seen whether the holes all come opposite; if any appear to be to one side, correct them with a rat-tail file, and then bore them all $\frac{7}{32}$ in. Mark the cover and the valve-box with the centre-punch, so that they may be put together always the same way; take all apart, except the studs, and slightly countersink the holes on both sides of the valve-box and cover, so as to remove any rough edge that may have been raised by the drilling, trying the surfaces on the surface-plate again to make sure they are still true.

TURNING STUFFING-BOX.

The stuffing-box should next be turned. To chuck this, an angle-plate A (Fig. 1856) is required; B is the face-plate, with parallel slots, e being useful in connection with the long slot d in the angle-plate. Two bolts f pass through these slots, and thus the angle-plate can be moved to any position, up or down. The valve-box v is secured on the angle-plate by the bolt and oblong washer g; and then the stuffing-box can be centred by moving the angle-plate on the face-plate, and fixed by the bolts f. The two surfaces of the angle-plate must be strictly at right angles. The valve-box being fixed, $\frac{5}{8}$ in. from the inner face and $\frac{1}{2}$ in. from the other, drill a $\frac{3}{16}$-in.

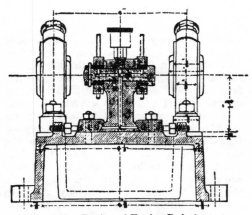

Fig. 1858 —Section of Engine Bed-plate, etc.

hole for the valve-rod; enlarge this with a $\frac{3}{8}$-in. drill for the gland to within $\frac{3}{16}$ in. of passing through; true out the hole with a slide-rest tool, and enlarge to $\frac{7}{8}$ in. Then tap it in the lathe, or use a $\frac{1}{2}$-in. gas-thread tap. The outside of the stuffing-box should be turned, and then the valve-box is finished.

SURFACING SLIDE-VALVE.

Next file and scrape up the flat face of the slide-valve, and then file the side edges square with the face and parallel to each other, making the valve $1\frac{1}{2}$ in. wide, so that it will fit between the sides of the valve-box (see Fig. 1855). Now, using square and scriber, draw four lines across the face, and file to these, so as to make the flat edges $\frac{1}{2}$ in. wide, and the hollow space $1\frac{1}{8}$ in. wide (see Fig. 1846).

SLIDE-VALVE ROD.

The valve rod (Fig. 1857) might be turned down from a piece of $\frac{7}{16}$-in. steel bar $5\frac{1}{4}$ in. long, but a piece of malleable cast-iron will make the rod. It must be turned down to $\frac{3}{16}$ in. and draw-filed, like the piston-rod. A backstay will be required to support the rod against the lathe tool, and a $\frac{3}{16}$-in. thread should be cut on the end of the rod for $2\frac{1}{4}$ in. The four little nut castings would be chucked in a three-jaw chuck, which will hold them true by alternate sides. Then each nut should be drilled and tapped, then chucked on a suitable mandrel and turned true between the lathe centres. File the sides of the valve-rod boss parallel, leaving a thickness of $\frac{1}{4}$ in., and drill a $\frac{1}{4}$-in. hole for the pin at right angles with the rod. The nuts on the rod hold the valve between them, and are of brass, because otherwise they might rust fast to the rod. Further, they are in pairs, because then they can be locked together, turning one nut hard against its fellow; besides this, it is possible to adjust the position of the valve.

SECURING CYLINDER COVERS.

At this stage it may be well to fit the screws into the covers of the cylinder, and so complete that part of the engine. The screw-holes have already been bored in the two covers. Then place the bottom cover in position on the cylinder, taking care to put the two holes which are farthest apart opposite the port side, pass the scriber point through the holes in the cover, and mark their positions on the flange of the cylinder. Drill these $\frac{5}{32}$ in., and tap them with a $\frac{3}{16}$-in. thread. Make six $\frac{3}{16}$-in. screws of steel, $\frac{1}{2}$ in. long in the body, to secure this cover, and harden their heads; in addition, make six screws, $\frac{7}{8}$ in. long, for the top cover, as these have to go through three thicknesses of metal into the bed-plate. Before filing up the heads of these screws, the spanner may be made first to act as a gauge. The screws would be made of soft steel that will not harden of itself, but the heads may be heated in a blowpipe flame, rolled in powdered prussiate of potash, again heated full red, and plunged in water, when they will have a hard skin without being brittle and liable to twist off. Only the heads require to be hardened, and they need not be cleaned up bright after the hardening, but should be well oiled, as the blackened appearance looks well, and forms a contrast to the

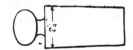

Fig. 1859.—Wooden Mandrel.

bright work. The bottom cover can now be put in. The top cover is secured in a different way, the screws going through the cylinder flange first, then through the cover, and finally screwing into the bed-plate, although it is not quite orthodox to secure two joints with one set of screws, because one cannot be separated without the other. Use the top cover, as before directed, to mark the screw-holes on the cylinder flange, but drill these holes $\frac{3}{16}$ in. full.

Truing Bed-plate Casting Surfaces.

The bed-plate casting has four surfaces to be trued: first, the end surface against which the cylinder is bolted; second, the large facing in the middle on which the slipper-guide works; third and fourth, the two smaller facings on which the crankshaft bearings are bolted. The last three are all in one plane, and, though not necessary, it will be most convenient to have

them planed. It is essential that the surface against which the cylinder beds should be at right angles to the other three, or the slipper-guide cannot work without bending the piston-rod; further, the rod must point fairly down the middle of the bed. Again, the recess into which the cylinder cover fits must place the cylinder at the correct height, namely, $1\frac{3}{4}$ in., as dimensioned in Fig. 1858. Probably it would be well to mount the bed on the saddle of a lathe, so that the line of the lathe centres coincides with the centre line of the cylinder; then a boring-bar may be used both to face the end surface and to bore out the hole in it. However, another way will do very well. If the three level surfaces are not planed, these should be filed up fairly true. First of all, place a spirit-level on the three surfaces, and pack up with slips of wood, tin, etc., under the bed-plate as it lies on the bench, until the three surfaces come as level as possible; now file them up, keeping them in one level plane by trying them alternately with level and surface-plate. The bed-plate had better be screwed down on the bench by wood screws passing through the three lugs. The three flat surfaces being filed or planed, turn to the end surface, the hole in which must fit the cylinder cover. The diameter of the hole is to be $2\frac{1}{4}$ in., and the height of its centre above the plane surfaces is $1\frac{3}{4}$ in.; therefore, the height of its bottom edge above these surfaces will be $\frac{5}{8}$ in. Place a scribing-block on the second surface, and set its point $\frac{5}{8}$ in. high. Now with a half-round file treat the bottom of the hole till the application of the scribing-block shows it to be $\frac{5}{8}$ in. above the level surfaces. Then turn a short conical mandrel (Fig. 1859) of hardwood, making it small enough to go into the hole at the smaller end, and at the larger end make it the size the hole ought to be. It may be 2 in. to 3 in. long. Rub it all over with chalk, and push it gently in the hole; take it out and file where the chalk appears; continue thus till the cone will go nearly up to the large end of the hole, touching also the bottom of the hole, when the cylinder cover should be tested. By this means the cover is fitted without looseness at the proper height.

TEMPLATE FOR USE IN FILING.

The template shown in Fig. 1860 is turned in hardwood; with it, it is possible to file up the first surface at the end of the

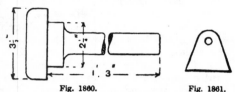

Fig. 1860.   Fig. 1861.

Fig. 1860.—Wooden Template. Fig. 1861.--Sheet-metal Template.

bed-plate. The neck, 2¼ in. in diameter, should fit the hole without shake; the shoulder is turned exactly square and flat, and the long shaft, reduced for lightness, may be 1 in. in diameter, and about 1 ft. 3 in. long, and should be quite parallel. Chalk the shoulder, put the pin through the hole in the bed-plate, rub the shoulder against the end by turning the wood in the hole, and note the direction of the long pin vertically and horizontally; take out the template, and file the end surface accordingly. Measure carefully from underneath the pin to the facing for the slipper-guide at both ends to see whether the pin is quite level, and from the inner edges of the crank-shaft facings to the sides of the end of the pin, to see whether it points fairly down the middle of the bed. The surface-plate may be applied to finish the end surface; it need not be scraped, however, but left from a smooth file. Next apply the cylinder cover and cylinder, adjusting for the screw-holes while the valve-face is vertical; remove the cylinder without disturbing the cover, scribe through the holes to the back of the bed, drill and tap these holes, and screw the cylinder to the bed.

SLIPPER-GUIDE AND GUIDE-PLATES.

The slipper-guide, or cross-head, and the guide-plates follow for treatment. First finish the facing on the bed-plate, scraping it true with the help of the surface-plate.

Take next the guide-plates between which the slipper moves, file these top and bottom, bringing them to the same thickness all over, scrape up their lower surfaces and draw-file their upper faces, and square their outer edges and draw-file them. For the chamfered edges (see Fig. 1858), make a gauge by filing a notch in sheet metal with a smooth triangular file; use this notch to try the angle of the chamfer, filing to the gauge, and keeping the pieces parallel. This chamfered edge must be brought to a true surface by means of the surface-plate and the scraper. Drill three holes for screws in each piece in the position shown in Fig. 1846, and make and fit the four screws seen in the same view, which are used to push the V's into touch with the slipper, but do not bore in the bed the holes by means of which they are held down till the slipper is made. Now take the slipper itself, and file up first the sole on which it slides, taking off just enough to make it 1¾ in. high to the centre of the bosses.

CROSS-HEAD AND PIN.

The upper part of the slipper-guide is the cross-head, the pin which goes through it, and is grasped by the small end of the connecting-rod, being the cross-head pin: the whole forms a very important part of the engine, and requires careful workmanship. The centre lines of the holes in the cross-head, for the piston-rod and for the

Fig. 1862.--Setting Angle-plate on Face-plate.

cross-head pin, must be right angles, and in a plane parallel to and 1¾ in. from that of the sole of the slipper. Make a little template, to Fig. 1861, of sheet iron, boring the hole ⅜ in. to fit the piston-rod, and making the distance from the centre of this hole to the base 1¾ in. Thread this template on the piston-rod, and try whether it

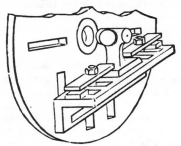

Fig. 1863.—Chucking Cross-head.

will touch the guide surface with equal pressure whether the rod be pushed in or drawn out. This will show whether the cylinder is really true with the guide facing. Now put the angle-plate on the face-plate chuck, drive a piece of wood into the hole in the lathe mandrel, and turn it down to ⅜ in. Then put the template (Fig. 1861) on this, and bring up the angle-plate till it touches the base of the template, and fix it there (see Fig. 1862).

BORING CROSS-HEAD.

Now remove the template and the wood centre, and fix the cross-head on the angle-plate, the dogs holding down the sole, and the bosses for the cross-head pin being centred to run true. Fig. 1863 makes this plain. Now face up the boss and bore it through, when the hole will be, of necessity, true with the sole. To drill this hole straight, a half-round, or D-bit, would do, but nothing else should be trusted; if, therefore, the workman does not possess a ⅜-in. D-bit (one can be made as described on p. 178) let him bore with a ₁₆⁄-in. drill, and true out the hole with a fixed tool in the slide-rest, leaving it rather less than ⅜ in., since it must be cleaned out with a fluted reamer to fit the pin. Turn a little rod about 4 in. long; this fits into the hole, and is quite parallel its whole length.

PISTON-ROD BOSS, ETC.

Then twist the cross-head one-quarter turn on the sole, and fix it so that the boss

for the piston-rod runs true; thread the rod through the hole just bored, and measure very carefully from the ends of this rod to the face-plate chuck, altering the setting of the cross-head until these measurements show by being equal that the hole is parallel with the face of the chuck. Now face up the boss for the piston-rod, turning the outside as far as possible; then bore with a ₁₆⁄-in. bit into the first hole, as seen in Fig. 1846, and cone out this hole to ⅜ in. with the fixed tool in the slide-rest, using a very sharp and rather pointed tool, and setting the cutting point exactly at the height of the centres. Then put the piston and its rod in the lathe once more, and turn the end of the rod to fit the hole, to the dimensions given in Fig. 1846. Drive the cross-head on the rod, drill a ⅛-in. hole right through for a pin, and put a slightly taper reamer in to enlarge the upper end of the hole.

SECURING PISTON-ROD IN CROSS-HEAD.

Usually a flat key would secure the piston-rod in the cross-head, but it would be difficult to fit, and as the piston-rod is of ample strength, it will not pull in two even though half its cross-section is cut away by the round pin. Moreover, the pin may be of steel, hardened and spring-tempered; it will not then shear in two. Also it is quite easy to put a little "draft" on the pin, so as to make it tend to draw the rod into the cross-head; all that is required, after reaming out the hole and fitting the pin, is to put a small rat-tail

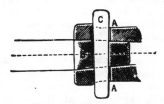

Fig. 1864.—Fastening Piston-rod in Cross-head.

file first into the hole in the piston-rod, and file away a very little at B (Fig. 1864), and then into the cross-head, and file away a little at A.

TESTING CROSS-HEAD, GUIDE-BARS, ETC.

Having put the cross-head on the rod, driven in the pin securely, and marked the rod so that it may be put in always the same way, rub some marking on the guide surface, and try the working of the cross-head by pulling it up and down whilst the

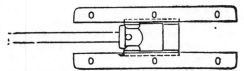

Fig. 1865.—Testing Cross-head Strips.

piston and cylinder are in position. Notice whether the sole of the slipper bears all over the guide surface, or whether the front or back edge is lifted ; if necessary, take the cross-head off the rod and correct the sole with the scraper. Now file up the slanting edges of the slipper parallel with the piston-rod and with each other, and to the same angle as the guide-bars. Test the work as it proceeds by placing the cross-head and two guide-bars in position upon a true surface, as in Fig. 1865, and observing whether the guide-bars and piston-rod are all three parallel. The piston should be taken off the rod, or it will be in the way. When satisfactory, finish filing up the cross-head, elongate the holes in the guide-bars, as shown in Fig. 1865, place the bars in position on the guide surface, mark through the holes in the guide-bars on to the bed to give the positions of the six holding-down screws, drill and tap these holes, and put in the screws to hold down the guides and the four other screws for advancing them sideways into contact.

CORRECTING CHAMFER OF SLIPPER-GUIDE.

Slide the cross-head and piston-rod up and down, the piston being removed and the rod passing through the stuffing-box. Watch whether the chamfered edges of the slipper are in perfect contact with the guide-bars, and, if necessary, correct them till they are. The red marking would be rubbed on the guide surface and inside the guide-bars, and the cross-head would be

rubbed up and down and then pulled out, the piston-rod coming out of the stuffing-box. Thus the chamfer of the slipper can be quickly tried, examined, and corrected till the marking shows the chamfered edges and sole of the slipper touching all over. Then put on the piston, and test once more.

THE FUNCTION OF THE SLIPPER-GUIDE.

The particular form of guide here described is suitable for an engine which will run chiefly, if not entirely, in one direction. As the eccentric precedes the crank by a little more than a right angle, the engine wheel, as seen in Fig. 1845, will turn round in the same direction as the hands of a clock. In Fig. 1866, A B represents a piston-rod, B C a connecting-rod, C F a crank, and C D E a crank circle. Now, if the piston be driving the crank forward from C towards D, were there no guide at B for the cross-head the tendency would be to bend the piston-rod downwards. It is this tendency which is resisted by the broad sole of the slipper-guide. But let the motion of the crank be continued past D to E, and let the connecting-rod take up the position of the dotted line B E ; now the piston-rod is pulling the crank, yet the pressure at B is still downwards, and not upwards, as might have been supposed. If, however, the direction in which the engine runs were to be reversed, then the

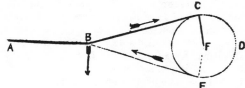

Fig. 1866.—Diagram Showing Pressure on Slipper-guide.

pressure on the guide would be upward, and the guides would require to be modified.

CROSS-HEAD PIN.

The cross-head pin (see Fig. 1858, p. 596) is of mild steel. Take a piece $\frac{5}{8}$ in. in diameter and about $2\frac{1}{4}$ in. long, and bore lengthwise right through the centre a $\frac{1}{32}$-in.

hole ; this is for the oil or grease. Countersink both ends of this hole to fit the lathe centres The hole must afterwards be plugged up at both ends ; or it might be bored not quite through, as shown at Fig. 1858 ; at any rate, the piece must be a little too long at first, so as to allow for removing the countersunk part and squaring up the ends. Then, with a ⅜-in. taper fluted reamer, clean out the hole in the cross-head so that the pin can be driven up to within ¹⁄₃₂ in. of the head. While it is in that position, bore the hole for the ⅛-in. set-screw opposite the piston-rod, letting the point of the drill cut a little way into the cross-head. Take out the pin from the cross-head, tap this hole, and fit the set-screw ; then drive in the pin again, and fix it with the set-screw.

BORING OIL CHANNELS.

The oil channels, as shown in Fig. 1858, have now to be bored ; first bore with a ³⁄₃₂-in. drill from the centre of the top of the cross-head right down through the cross-pin and out at the sole ; this will supply oil to the sole of the slipper. Then bore across the bottom of the hole from one side chamfer to the other ; bore also the two short holes from the top of the cross-head pin to the central hole, which will enable the lubricant to reach the two bearings at the forked end of the connecting-rod. The single lubricator (see Fig. 1858) screwed into the top of the vertical hole supplies lubrication to all these five surfaces.

TURNING AND BORING FLYWHEEL.

Fig. 1867 shows a convenient method by which the flywheel may be mounted on a 10-in. face-plate chuck for boring the ⅞-in. hole for the crank-shaft. The wheel is 16 in. in diameter, and if the gap in a 5-in. lathe were 3 in. deep, the wheel could be mounted as shown, the hole bored, and slotted for the key, and the boss turned. Possibly the rim might also be turned, but it would be with difficulty. It is not necessary to have it turned, though it improves the appearance of the engine. This rim, like the planing of the bed, might well

be left to an engineer or to whoever supplies the castings. In Fig. 1867, A is the face-plate ; three of the arms are grasped by clamps B of iron, and under each arm beneath these clamps are three pieces of hard wood which hold the boss of the wheel off the face-plate. When the wheel has been adjusted to run true and the clamp bolts are fixed, the boss would be turned and the hole bored, first with a small drill, and then with larger ones, till it is of nearly the right size, when it would be finished with a fixed tool in the slide-rest, using a sharp tool and a very light

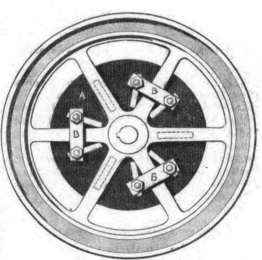

Fig. 1867.—Chucking Flywheel.

cut, and trying in the hole a piece of ⅞-in. turned bar, to make quite sure it is turned parallel.

SLOTTING OUT THE KEY-WAY.

The key-way would be most easily made by slotting in the lathe—that is, by fixing a slotting tool on the slide-rest, holding the work still, and moving the tool into the hole by working the saddle along the bed ; or simply mark the position of the key-way by lines drawn inside the hole with the point of the tool which was used in the slide-rest for finishing the hole. To do this the work would be held still whilst the lines are marked by moving the saddle along the bed. This method would ensure that the key-way should be marked out parallel with the hole of the wheel, and then the cutting could be done with the chisel and file. The key-way is not sunk in the

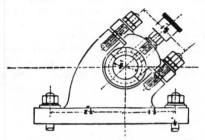

Fig. 1868.—Side View of Crank-shaft Bearing.

shaft, as is usual, but bears simply upon a flat filed on the shaft, and therefore it should be rather larger than usual, since it is essential that it should fit well, and hold the wheel to the shaft without the slighest possibility of turning on it, for this would produce a knock at every stroke; and any slight looseness there might be at first would continually increase. Therefore make the key-way $\frac{3}{8}$ in. wide and $\frac{3}{16}$ in. deep.

ANGULAR CRANK-SHAFT BEARINGS.

The angular crank-shaft bearings are shown in Fig. 1858, p. 596, and in Figs. 1868 to 1870. These bearings are arranged so as to require as little filing as possible. The brasses are round and therefore can be turned and their seats can be bored. Lest, however, they should turn in their seats, the upper brass of each pair has a pin cast on it, which fits into a hole in the cap, and both prevents all danger of turning and conducts the lubricant down to the journal, without allowing any of it to escape between the cap and brass, as may happen in the usual form of construction. There are two castings of iron for each of the two bearings, and two of gun-metal. Begin with the two upper brasses, which must be chucked for turning the round pins. File each of the four brasses so as to bring up their two edges flat and smooth, and make

them semicircular. Then centre the round pins on the two upper brasses; centre-punch and drill them. Next in the lathe, face up the end of a block of wood, about 2 in. in diameter; place the half bearing on this, and bring up the back centre to enter the centre hole in the pin, advancing the point till the little casting is firmly held. The pin to be turned is so small that this pressure may be sufficient; but if not, two nails a (Fig. 1871) can be driven into the end wood, to ensure that the brass shall turn with the wood. Having turned the pins on both the upper brasses, tin the edges of the two pairs of brasses, and put them together edge to edge, in pairs; heat till the solder melts and they are united. They can now be smoothed out inside with a file till the inside of each pair is fairly smooth, and then they are ready to be chucked by the hole whilst they are turned to 1 in. outside the body (Fig. 1870) and on the inner sides of the flanges. However, first take up the two bearings and their caps. File up the soles of the bearings so that they will stand upright and square with the plate; file also the sides of the bearings, bringing them to $\frac{1}{4}$ in. in thickness.

CAPS FOR THE BEARINGS.

Now fit the caps in their openings; these must measure 1 in., or the bearings brasses cannot be put in. Fit the $\frac{1}{4}$-in. screws that hold down the caps, provide nuts, and bolt down; file up the sides of the caps

Fig. 1869.

Fig. 1870.

Fig. 1869.—Plan of Bearing Cup.
Fig. 1870.—Section of Upper Brass.

level with the sides of the bearings. The sides of the bearings are filed true first, because then the sides of the opening for the caps can be got square with the flat sides of the bearings. Now file the under sides of the caps, both under the ears or lugs, and under the edges of the projection, so that they will stand up square on these surfaces. Each cap can now be chucked by the oblong projection underneath, measuring 1 in. by $\frac{3}{4}$ in. ; bore the $\frac{7}{16}$-in. hole exactly in the centre of this projection. Chuck a piece of hardwood, turn

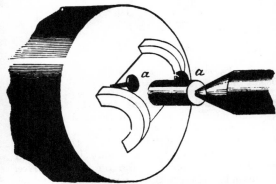

Fig. 1871.—Chucking Brass of Crank Bearing.

it about $2\frac{1}{2}$ in. in diameter, and face it up ; make a recess $\frac{1}{2}$ in. deep, and enlarge it gradually ; the projection of the cap can be pressed into it with the fingers ; this will centre it truly. Now pass two wood screws through the screw-holes in the ears to secure them to the wood chuck, and, starting with a graver, bore a $\frac{7}{16}$-in. hole, into which the pins on the upper brasses can fit easily.

CHUCKING BEARING BRASS.

Now make a chuck A (Fig. 1872) of hardwood ; turn down a pin B or short mandrel on which the brasses E can fit ; cut a screw C on the end, and fit a nut D so that the brasses may be pinched between the nut and the shoulder of A, and held without any tendency to burst asunder the soldered joint. A rather easier way would be to cut off the pin B where the screw begins, and, instead of a nut, have a piece of flat wood to receive the back centre point and transmit to the bearings the pressure of the back centre screw ; this is not so good a plan, because the point of the back centre will gradually indent the wood, and diminish the pressure on the bearing, which

may then turn round on the pin and require to be trued up again. The means shown at Fig. 1872 will not relax its hold, and, if the cutting is cautiously done without catching against the pin G, the necks H can be turned to exactly 1 in., the outsides of the flanges c to $1\frac{1}{4}$ in., and the inside edges b of the flanges can be turned true. In doing the last, the cap F would be tried on the turned pin G, and the flanges turned away till the cap will just go down between them when put over the pin. The caps and the tops of the pins of the brasses should be marked so that the caps may neither be exchanged nor turned half round, since, if the holes in the caps are not absolutely central, they will not fit both ways. The band, $\frac{7}{16}$ in. wide in the middle, cannot be turned on account of the pin, but can be very easily cut down level with the two side bands with a sharp chisel and files.

BORING CRANK BEARINGS.

Now returning to the bearings, fit on their caps, screwing them firmly down. Then take the face-plate and angle-plate used for boring the cross-head. If the angle-plate has not been moved, it will be in exactly the right position for boring the

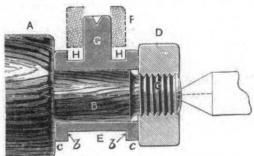

Fig. 1872 —Chucking Brass of Crank Bearing.

bearings, since their centres should be $1\frac{3}{4}$ in. high ; but if the angle-plate has been taken off, it can easily be replaced by means of the same template (Fig. 1861, p. 598). Bolt the bearings on the angle-plate one at a time, so that their sides shall be exactly parallel with the face of the plate, and standing perpendicular to the top of the angle-plate. To test the work, bring up a tool and let it touch the circular facing, on bearing and cap, as it turns, when it should touch equally, or nearly so, all round. Now bore out the seat for the brasses to 1 in., and take off

difficult part of the work, owing to the necessity for having two holes through the brasses at the two ends exactly parallel Begin by centering the connecting-rod and putting it in the lathe; the forked end will admit the lathe centre between the

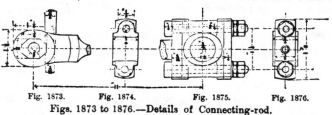

Fig. 1873.　　Fig. 1874.　　　Fig. 1875.　　　Fig. 1876.

Figs. 1873 to 1876.—Details of Connecting-rod.

the cap and try the brasses in. The flanges have been fitted to the sides of the cap, and the cap has been filed to the same thickness as the bearing; therefore the flanges of the brasses should fit the bearing equally well. They should require a little gentle rapping to drive them into their seat, and must not be loose. Having got one pair of brasses down solidly on their seat, put on the cap and nuts, screw them down, and bore out the brasses to $\frac{3}{4}$ in.; face up the front of the brasses, and round the corner at the mouth of the hole. The crank-shaft, where it passes through the bearings, is reduced from $\frac{7}{8}$ in. to $\frac{3}{4}$ in. in diameter; but shafts of this kind are always rounded at the corner, which leaves them much stronger. Therefore, round the corner of the hole in the brasses to a quarter circle of $\frac{1}{16}$-in. radius. When both bearings have been treated in the same way, re-chuck them in a reverse position so as to face up the other side of the brasses; turn them half round on the angle-plate, and adjust till the hole in the brasses runs true, and the face of the bearing is parallel with the face-plate. Then face the brasses and round the corner as before. Now heat the brasses till they come apart, mark them carefully and their bearings, drill the oil hole in the pin, and tap it for the lubricator, when the bearings may be temporarily laid aside.

CONNECTING-ROD.

The connecting-rod (shown in detail by Figs. 1873 to 1876) is perhaps the most

forks, and the carrier (Figs. 1877 and 1878 can be of the kind used to hold against a square. This carrier is simply formed of two pieces of flat iron A held together by two screws B; C is the connecting-rod end in section, and D is the pin of the driver chuck. Having then chucked the rod, turn the body, leaving it tapered from $\frac{9}{16}$ in. at the large end to $\frac{3}{8}$ in. at the small end, and polish it with emery-paper. Turn up the oblong flange at the large end, facing the end square, and scratching on the end surface two short arcs of a circle $1\frac{1}{4}$ in. in diameter, on which to mark the centres of the bolt-holes.

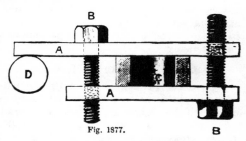

Fig. 1877.

Fig. 1878.

Figs. 1877 and 1878.—Connecting-rod Carrier.

FILING UP CONNECTING-ROD END FLANGE.

The only parts now remaining of the large end of the rod not got up bright are the sides of the end flange, and these must be filed up with reference to the small end of the rod. Place the rod B (Fig. 1879) on the surface-plate A so that it rests on the fork D at one end and on the flange at the other end. The edges c must be filed up square with the small end, but this can be done only if the depths of both the eyes

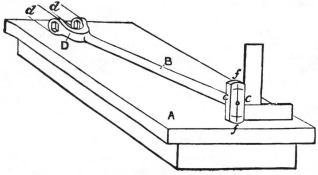

Fig. 1879.—Squaring Connecting-rod End.

d are equal. Having tried the surfaces c with the square under each eye, file small flats on which the rod may rest with the edges c vertical. Now file up these edges, making a flange a little more than ⅜ in. wide, so as to allow a little for after correction, but bringing both surfaces square and parallel. Then turn the rod over and file two other small surfaces on the other sides of the eyes, so as again to bring c vertical. Draw a line with the scriber point down the centre of the flange at f, to determine the centre of the bolt holes, and make a centre-punch mark at the intersection of this line with the arcs marked with the lathe tool.

CONNECTING-ROD GUIDE-SURFACES.

Looking now at Fig. 1880, two small flats, a and b, are shown on each eye square with the flange at the other end of the rod. Two other guide surfaces c are needed before the octagonal hole for the brasses can be touched, and these surfaces are obtained by standing the rod up on the surface-plate and filing till the rod stands vertical as tested by the square. The truth of these surfaces is very important.

FILING OUT OCTAGONAL HOLES.

The surfaces just mentioned can be used as guides for filing out the octagonal holes. The distance a to b is 1⅛ in., and the thickness of metal will be ¼ in. In order to mark the lines d at right angles to the length of the rod, put a piece of wood under the end, at D (Fig. 1879), so as to bring the rod horizontal and level with the plate; then these lines can be marked with a scriber by squaring up from the plate. Lines e, too, can be drawn at the same time with the scribing block. In filing out these holes a small square file will pass through both eyes at once, so that it must run straight, and the lines d and e will be marked on the outsides of both eyes. Then, by taking ¼ in. in the callipers, and making that dimension the thickness of the metal inside each of the little guide surfaces, the holes will be correct. Lines at f can be drawn with the scriber by means of an ordinary square, resting on the face-plate, whilst the rod is packed up level with wood.

CONNECTING-ROD BRASSES.

Now take the four little castings provided for the brasses of this end, and, rubbing their edges on a file, tin and join

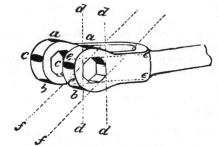

Fig. 1880.—Guide Surfaces on Connecting-rod End.

them in pairs with solder. File each pair so as to true one side from which to work; then begin with two opposite sides of the octagon — say those which fit against *e* (Fig. 1880)—square these with the first face, parallel with each other, leaving them just so far apart that the brass can almost be pushed into the hole sideways (see Fig. 1881). The brasses being only $\frac{1}{16}$ in. thick, and the top and bottom sides of the octagon a little thicker, it is possible to try the two first sides of the brasses in that way; but as the octagon is not a true one, but somewhat elongated, the other six sides being shorter than $\frac{1}{16}$ in., careful measurement must be made until the brasses begin to drive in a little way, when the shape of the hole in the rod will get marked on them, and they can be filed

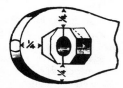

Fig. 1881.—Trying Brass in Connecting-rod End.

accordingly to fit tightly, so as to require a little gentle driving.

ENLARGING HOLES IN CONNECTING-ROD BRASSES.

Having fitted the two pairs of brasses into the forked end of the connecting-rod, enlarge the hole to the size for the cross-head pin by means of the $\frac{3}{8}$-in. fluted reamer. Begin by passing through the holes a small round file, using it very carefully so as not to get the hole askew; enlarge gradually, stopping continually to measure with callipers from the hole to the three test surfaces. As soon as the $\frac{3}{8}$-in. reamer will enter, use it to round the hole and measure again, then correct with a file if needful.

TESTING BEARINGS FOR CROSS-HEAD PIN.

It is very important that both sides of the fork shall bear equally on the cross head pin. To ensure this, test still further by turning a little iron mandrel A (Fig. 1882), about 4 in. long, to the same taper as the reamer, so that it will fit into the

holes for the pin; then turn each end to, say, $\frac{1}{4}$ in.; pass this mandrel into the two holes and stand the connecting-rod up vertically on the face-plate. Now measure from the plate to the small diameter ($\frac{1}{4}$ in.) at each end of the mandrel to test whether the mandrel is perfectly parallel with the plate and therefore square with the length of the rod. If not, ease the hole inside with the file, and round it again with the reamer. The reamer must finally be worked in to make the hole about the same size as that in the cross-head; the pin should, however, fit more easily in the brasses than in the cross-head, where it should be rather tight, and further secured with the set-screw.

COTTARS FOR TIGHTENING BRASSES.

The brasses then being driven into their places in each side of the fork, and the hole for the pin being finished, cut the two little slots for the cottars or wedges which tighten the brasses. The cottars cut into the brasses to keep them in place, and prevent the fork end of the connecting-rod slipping sideways. This, however, is already prevented by the cross-head fitting between the fork; still, it may serve to keep the brasses in place when the rod is taken off. These little holes must be cut by drilling from both sides and finished with small files. The cottars should be of steel, and may be hardened and spring tempered.

BUSHING FORK END OF CONNECTING-ROD.

It would be easier, and perhaps as efficient, to insert hard steel bushes or ferrules in the fork ends, and to harden the cross-head pin instead of fitting the brasses and cottars. There is then no means of taking up wear; but the movement of this joint is very small, and, if both bushes and pin are hard, the wear will be almost inappreciable; moreover, it is always possible to make a new pin and to insert new hard bushes. To fit the fork end this way, file the octagonal holes to a round shape in connection with a $\frac{3}{8}$-in. fluted reamer, using the same precautions to ensure the hole being square with the rod. The cross-head pin would be made of cast steel, and hardened; the steel bushes would be bored slightly small, turned on a mandrel, hardened, driven tightly into place, and then ground out by working an iron or a brass rod with emery and oil through both bushes while in position in the rod end, the rod being turned to exactly the same taper as the cross-head pin.

CONNECTING-ROD BOLT HOLES.

For the large end of the connecting-rod, unite the two brasses with solder as before described, and file up the end surfaces parallel with each other; mark the positions for the bolt-holes, and bore them,

be of mild steel, and the nuts should be hardened; the holes they fit in should be cleaned out with a ¼-in taper reamer. The bolts look best if they have the heads inside and the nuts outside, but this necessitates the reamer being put in from the rod side, which is not very convenient. However, the reamer may be put in from the cap end, and the bolts may also go in that way if preferred; the body of the bolts should be turned to fit well into the holes.

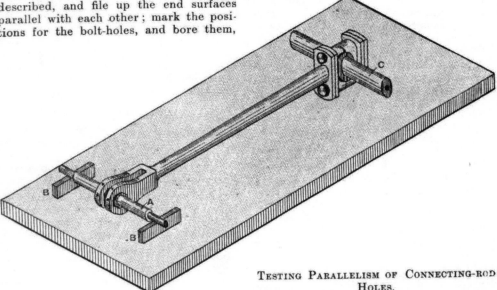

Fig. 1882.—Adjusting Brasses for Parallelism.

TESTING PARALLELISM OF CONNECTING-ROD HOLES.

and make the oblong cap of steel and bore similar holes in it. By means of these bolt-holes, the brasses, with the cap, can be secured to the angle-plate on the face-plate of the lathe, whilst the holes in the brasses are bored out and faced, and the corners rounded. Next bore the holes for the bolts in the end of the rod; one of these holes may be bored first and the bolt fitted, but before the other is bored, ensure that it shall be so placed that the holes through the brasses at both ends of the rod are parallel. The bolts to secure the large end brasses should

To ensure the holes at both ends of the rod being parallel, fit two turned rods into the holes, using them as winding strips, the cap and brasses of the large end being secured by the one bolt first fitted. A small mandrel A (Fig. 1882) has already been fitted to the small end of the rod, which may be supported by means of the two small strips B, which are of equal width; the other end of the rod rests on the face-plate. Now, when the larger mandrel C is fitted into the brasses at the large end, it will be easy to see whether it is parallel with the plate; if not, the brasses can be slightly twisted on the one bolt till it is so.

CENTERING CONNECTING-ROD CAP.

When the brasses have been correctly adjusted, the bolt may be firmly secured and the ¼-in. drill put through the cap and brasses to bore the remaining hole in the rod; and when this has been cleaned out with the reamer, the second bolt may be fitted. A centre at the top of the cap by which to turn the whole rod can be found by resting the rod on two similar V's, or angles, standing on a level surface, on which the rod can be turned round (see Fig. 1883). Lines are scribed across the top of the cap at A, the rod being turned partly round to make the marks, but not moved endways in the V's. These V-pieces may be of wood, nailed together and cut out both at once; but a pair of cast-iron blocks, carefully made, would be very useful things for centering rods for the lathe, etc. The centre being marked with the centre-punch, the rod may be put in the lathe and brought true with the square centre, and the rounded edges of the head of the rod, brasses, and cap may be turned up and finished. It only remains to drill a hole for the lubricant, starting from the centre hole just found, through the cap and brass; enlarge this hole and tap with ¼-in. thread in the cap, and it is ready for the lubricator. Then the sides of the flanges can be filed up, and the large end of the rod is finished.

FILING UP FORK END OF CONNECTING-ROD.

The fork end must now be filed up; the four flat surfaces through which the holes for the cross-head pin pass must be accurately at right angles to this hole, or the joint cannot work. To make sure of this, use the cross-head pin as a guide. The head of the pin abuts against one of the sides of the fork; file that side flat first, and test it by inserting the pin with a little red marking under the shoulder; do not be satisfied till the red marking shows equally all round the hole; then the other three surfaces can be filed parallel with it, testing with callipers. Though, however, the first surface may be square with the hole, before filing the second or left-hand side surface to it, place the rod on the V's with the finished side up, and, with the scriber, measure from the plate up to that surface, then, after turning the rod round, measure to the second surface to try how much must come off to bring the rod central. See also whether, when both the first two surfaces on the connecting-rod are parallel, square with the hole, and equidistant from the centre-line of the rod, the fork will measure 1⅜ in. from side to side. When all these conditions have been complied with and the two outer surfaces have been finished, the two inner ones, which embrace the cross-head guide, may be filed parallel with them, and tested with the callipers to exactly ⅟₁₆ in. thick. Now it will only remain to file up the top and bottom of the fork and round the end, take out the three pairs of brasses, and separate them where soldered.

COMPLETING CROSSHEAD.

To complete the crosshead, file up the sides, or cheeks, of the upper part true with the hole, and bring it to 1 in. wide (consult Fig. 1858, p. 596), so that it may fit between the sides of the fork of the connecting-rod. The little mandrel A (see Fig. 1882, p. 607) is turned to the taper of the reamer used for cleaning out the hole for the crosshead pin, and will therefore fit into that hole and project at both sides. Now, if on that mandrel a collar about 1 in. in diameter is fitted so that, while on the mandrel, it can be brought up to touch one of the sides, it will form a guide for the filing just as the head of the crosshead pin did for the fork of the connecting-rod. Therefore in a piece of round iron or brass about 1 in. in diameter and, say, 1 in. long, bore a ¼-in. hole, and try it on the small end of the mandrel, enlarging it, if necessary, with the reamer till it will touch the face of the crosshead. Next take it off, drive out the mandrel, and drive the collar on it; put it thus in the lathe and square up both sides of it; knock it off the mandrel, replace this in the crosshead, and put the little piece on the projecting end, when it should be used to test the side of the crosshead and bring it true with the hole. The ther side may be made parallel by means of the callipers, or the collar may be opened a little with the reamer, so that it can be pushed farther up the mandrel, and so be used on the other side of the cross-head. The remainder of the crosshead may be filed up to look well, and the oil-hole at the top enlarged and tapped for the lubricator; then the crosshead may be oiled and placed aside.

CRANK-SHAFT.

For the crank-shaft (Fig. 1884) a round bar, 1 in. in diameter, is bent, and the flattened ends are turned up at A to assist in the turning, and are afterwards cut off.

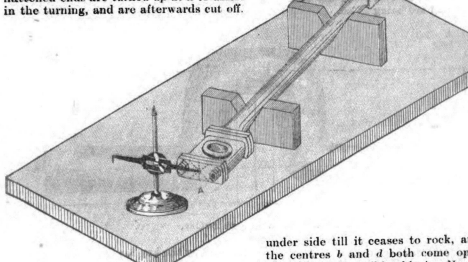

Fig. 1883.—Centering Cap of Connecting-rod.

The outer surfaces (a b and c d) may be roughly squared up first, and then the centres (b and d) found by placing the crank on the V-blocks, and marked with the centre-punch. Drill and countersink in the lathe, and turn the crank-shaft, the arms A acting as carriers. Next, a point-tool may be used to strike a circle on the ends A, at a and c, just 2¼ in. from the centre of the shaft. When the arc is struck at c the tool will be moved side-ways and the crank turned end for end, when the tool will be moved again to strike an arc at a without otherwise dis-turbing it, so as to ensure the radius being

exactly the same at each end. The radius of 2¼ in. must not be exceeded, or it will be necessary to reduce the thickness of the piston or to take other measures to prevent its knocking against one or both ends of the cylinder. The centres at a and c must be absolutely in the same plane as the centres b and d. To ensure this, try the forging on the surface-plate, and file the under side till it ceases to rock, and till the centres b and d both come opposite the point of the scribing block. Now, with the point so set, scribe across the arcs previously marked at a and c; the points will then be, all of them, equidistant from the surface-plate, and therefore will all lie in one plane. Countersink and drill the centres at a and c, and put the crank in the lathe to turn the main part of the shaft. Though the surfaces a b and c d are prob-ably fairly square and true, it will be well, first, to turn them true in the lathe. The dimensions in Fig. 1884 are for the forging, and give the size of the work before it is turned.

TURNING CRANK-SHAFT.

In turning up the ends of the shaft and the arms A, bring the extreme length to

8½ in., working to the dimensions given in Fig. 1846 (p. 587). When this is done, the end surfaces may be used to measure from, and will help to get the necks for the two bearings and for the crank-pin correctly placed. Now proceed to turn the shaft on the centres b and d. If the crank were weaker it would be necessary to fit a piece of hard wood between the throws, or arms, of the crank; but this will hardly be required in the present case, except in square centering. Turn the main part at each end to ⅞ in. full, leaving a little to come off for finishing, when the arms A have been cut off and when the eccentric and flywheel are being fitted.

before described. The crank-pin being thus finished, the arms A may be cut off with the hack-saw, the ends of the shaft turned, and (if desired) the arms or throws of the crank can be filed up and polished or painted.

Securing Flywheel on Crank-shaft.

The flywheel is supposed to be already finished, and to have a key-way cut in it ⅜ in. wide and ₃⁄₁₆ in. deep. The key for this and for the eccentric should be of steel, but the latter need not be more than ₁⁄₁₆ in. wide and ⅛ in. thick. It would, of course, be more orthodox to sink a groove in the shaft instead of simply filing a flat,

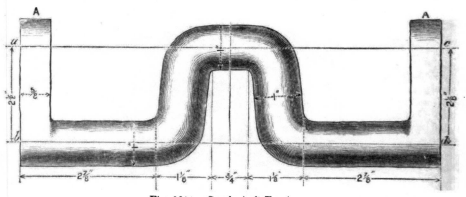

Fig. 1884.—Crank-shaft Forging.

Turning Necks for Bearings.

Turn the necks for the bearings, taking out the brasses and trying one of each pair on the neck in which it is to turn, as the fit must be good, both as to diameter and as to the rounding of the corners. It is comparatively easy to make a fit with one of the brasses as a gauge. Get the necks as smooth and round as possible, and well polished; then rub them over with marking, and rub the brasses round on them, to see whether they bear evenly; a little scraping may be necessary inside the brasses to make them fit perfectly. Having finished the main bearing necks, centre at a and c, and turn the neck of the crank-pin in a similar way, trying the brasses of the connecting-rod on it as

but making the key for the flywheel rather wider than usual causes the flywheel to be held firmly if well fitted. The keys are about ₃⁄₃₂ in. thinner at the point, so as to cause them to tighten up like a wedge, the bottom of the key-ways being carefully filed to the same slope so that the key may bear equally hard all along its length. The flywheel being bored, finish one end of the crank-shaft, and make it fit accurately into the wheel; it may then be laid aside till the eccentric is finished.

Eccentric Strap.

The eccentric and its strap and rod are shown in Figs. 1885 to 1889. The eccentric for the feed-pump is so arranged that it may be easily added if a feed-pump is required. It is shown, with its method of attachment, in Figs. 1889 to 1891, and is simply bored out to fit the boss of the slide-valve eccentric, and is fixed to it with two screws. Taking in hand the eccentric strap first, file up the ears for the screws. These four surfaces should be trued on

means of two dogs, in a similar way to that illustrated by Fig. 1854 (p. 592), the dogs clasping only about ⅛ in. of the ears, so that almost the whole of the front surface of the strap can be turned true. Bring the strap to ½ in. thick, and turn

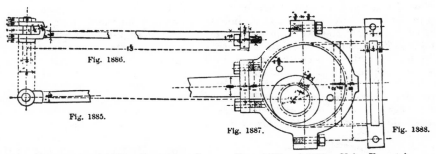

Fig. 1886.

Fig. 1885.

Figs. 1885 and 1886.—Eccentric-rod.

Fig. 1887.

Fig. 1888.

Figs. 1887 and 1888.—Valve Eccentric.

the surface - plate, testing with a square whether the half-strap stands vertical on the plate; the ears are ⅜ in. thick. Therefore set the point of the scribing block to ⅜ in. high, and, holding the half-straps in one hand, scribe across the ears of both, on each side, a line ⅜ in. above each of the four finished surfaces, so as to mark the thickness of the ears; then file to these marks so as to get a flat surface for the heads of the screws. Next set the scriber point to 1⅜ in. and mark off the large flat surface on the one half-strap and the small one on the other. Tin the ears and solder the two halves of

out the interior; scratch a fine line round as a guide for filing the outside, and take the strap from the lathe to finish with the file. Then chuck it again the reverse way to smooth and polish the side, melt the solder, and separate the two halves.

TURNING THE ECCENTRIC.

Now grasp the boss by the eccentric itself in a universal chuck, bore the hole ⅞ in., and cut or mark the key-way; fit it on a mandrel, put it between the centres, and finish the outside of the boss and the

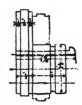

Fig. 1889.—Valve Eccentric Sheave.

the strap together; then fit the screws and screw them firmly in. Now file one side of the strap thus united perfectly flat; screw a piece of hardwood on the face-plate chuck, and turn it true and flat; place the flat side of the strap on this and fix it by

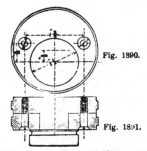

Fig. 1890.

Fig. 1891.

Fig. 1890.—Pump Eccentric Sheave.
Fig. 1891.—Section of Eccentrics.

sides of the eccentric, bringing it to ½ in. wide. To turn the outside of the eccentric a die chuck may be used; but, failing that, employ the face-plate A (Fig. 1892), with

a flat piece of hardwood B on it, truly
faced up. Prepare also another piece C
of hardwood, 1⅛ in. or more thick, and
about 5 in. by 4 in., with the edges
squared ; put two long screws through the
corners and screw it on the middle of the
first piece, sinking the heads of the screws
below the surface, so that it can be faced
up in the lathe. Now bore a 1½-in. hole in
the centre of this piece, 1 in. or more deep,
and into it press the boss of the eccentric.
The line a b of the eccentric which joins
its two centres must be placed parallel
with two sides, c d and e f, of the squared-
up piece. Now the two screws holding C
to B may be taken out, and the piece C
moved down ½ in. along the lines c d and

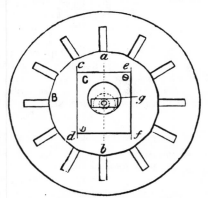

Fig. 1892.—Chucking Eccentric.

e f, and refixed. Thus the piece C acts
like the slide of an eccentric chuck. Lest
the eccentric should move while being
turned, an iron clamp g may be placed
across the hole and secured by a screw
passing into the wood. The half of the
eccentric strap will, of course, be used to
test the work, and the eccentric should be
made to fit, both on the ½-in. middle band
and on the smaller ⅛-in. bands on each side
of it, so that the eccentric may bear on the
whole ½-in. width of the strap. When the
straps and brasses of the connecting-rod,
etc., wear, they must be taken apart, and
a little filed or scraped off the ears, or off
the brasses where they touch, until the fit
is again satisfactory. If this is not done
the engine will knock on each dead centre

—that is, there will be a thump every time
the piston changes its direction of motion.

ECCENTRIC-ROD AND VALVE-ROD.

For the eccentric-rod, also the valve-rod
(Fig. 1857, p. 595), castings are provided.
The valve-rod must be turned up, and
draw-filed like the piston-rod, to avoid
wearing out the stuffing ; but, as it is only
$\frac{3}{16}$ in. in diameter, it will require some kind
of a back-stay to support it while being
turned. When the shaft is finished to size,
the thread may be cut with stocks and dies,
but it must not be too high up, so that
the threaded part will be drawn into the
stuffing-box. Then the nuts may be
threaded in a three-jaw universal chuck,
and trued up in position on the rod. For
the joint, file up first the part that belongs to the valve-rod. It should be ¼ in.
thick ; and the flat surfaces on each side
must be parallel with the centre line of the
rod, and equally distant from it. These
conditions can be ensured simply by pressing one of the surfaces down on the surface-
plate, when it will be easy to see whether
the length is parallel with the plate or not.
Then file up one side of the joint on the
eccentric-rod, taking care that it is parallel
with the length, and out of winding with
the flange at the large end. File the
three other surfaces of the joint to this
first one, bore both parts of the joint with
a ¼-in. drill, put them together, clean out
the hole with a ¼-in. reamer, and turn the
pin of cast steel to fit. Then harden
both parts of the joint with prussiate of
potash, harden the pin, and grind it in with
emery powder. Clean off the emery, and
the joint is finished. No directions appear
to be needed for fitting the flange at the
larger end to the eccentric shaft ; and the
eccentric-rod itself may be got up bright,
draw-filed, and finished with emery-cloth,
or it may be simply left rough and painted.

FITTING TOGETHER ENGINE PARTS.

It will be well at this stage to fit together
the parts thus far finished. The cylinder
being bolted to the bed-plate, the piston
and rod being in place, and the cross-
head and guide-plates fixed, put the crank-

shaft into the two bearings, and put the connecting-rod on to the crank-pin ; after laying these parts on the bed-plate, bring together the fork end of the connecting-rod and the crosshead, passing the pin through, and securing the joint. Push the piston to the bottom of its stroke ; place the crank horizontal, and mark on the bed the position occupied by the soles of the two crank-shaft bearings. Now turn the crank half round, pulling out the piston to the end of its stroke. Set the crank again horizontal, and mark again the position of the soles of the bearings on the bed-plate. This position should be the same as before, but if the bearings have moved a little in the direction of the length of the bed, place them in an intermediate position, so as to split the difference ; scribe through the bolt-holes in the feet of the bearings on to the bed, drill and tap these holes, fix the four studs, and bolt the bearings down on the bed. The brasses of the large end of the connecting-rod having been clasped on, the crank-pin should suffice to place the crank-shaft fairly at right angles with the centre line of the engine. Any error that might arise would be eliminated by splitting the difference between the two positions. The studs which secure the bearings should fit fairly well in the holes, so as to act partly as steady-pins. Now file a flat on the crank-shaft, and make the key to secure the flywheel. Fit this carefully, and drive it in firmly, and then turn the engine round by the wheel. It is very important that it move evenly. If it sticks or moves stiffly at any point, find what is wrong by taking off one thing after another. For instance, disconnect the crosshead joint, and try the three brasses on the crank-shaft alone. Suppose there is still a tightness, loosen the brasses one by one, then tighten one at a time, turning the wheel every now and then.

BEARINGS WORKING STIFFLY.

Aim to have the bearings all equally tight, so that the shaft can revolve freely without much friction, and yet without being loose enough to knock. However, there is sure to be a slight stiffness in a new engine, but this will wear off, though it could be prevented by grinding in the joints. A shaft may be ground in with emery if, like the mandrel of a lathe, both it and the collar, or bearing, are left quite hard, because emery powder cannot get embedded in hard steel. A joint or bearing of iron or brass would be ruined if ground together with emery, because the emery, once in, could never be got out again. A joint of brass or iron can be ground together with the grit from the grindstone trough, if, like the plug of a tap, it is moved only occasionally ; but when the shaft has to turn continuously, it is better to trust to scraping out the brasses where they seem to bind, and to keep all abrasive powders far away.

ADJUSTING SLIDE-VALVE.

Having now, thus far, put together or erected the main parts of the engine, proceed to adjust the slide-valve. Pass the

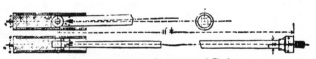

Fig. 1893.—Pump Plunger and Rod.

valve-box over the studs, and, leaving off the cover for the present, put in the slide-valve, and pass the valve-rod in through its stuffing-box ; add two nuts, the slide-valve, and two more nuts, locking these in pairs on each side of the valve. Now connect the eccentric-rod with the valve-rod and with the eccentric, this being in its place on the shaft, but not keyed—the keyway being cut in it, and fitted with a temporary key bearing on the round shaft, on which the flat has not yet been filed. Place the piston in the position shown in Fig. 1846 (p. 587), the crank being horizontal, and the piston at the end of its stroke. Now turn the eccentric round on the shaft till it points upwards, and away from the cylinder, in the position of Fig. 1887 (p. 611), making an angle of about 120° with the crank. The valve should now be just beginning to uncover the port leading to the bottom of the cylinder, as it is seen in Fig. 1846. If this is not the position of the valve, that will be owing to one of two causes: either the angle of the eccentric is not correct, or the length of the connection between the eccentric and the edge of the valve is not exact. Turn the eccentric on the shaft till the valve uncovers the edge of the port, so that the corner of a visiting card can be inserted. That distance is called the lead of the valve. Fix the eccentric by the temporary wedge, and turn the crank just half-way round till it stands level, whilst the piston is at the opposite end of the stroke. Now, the valve, to occupy the proper position, should be uncovering the other port by the same distance, when the lead would be equalised; this, however, is not likely to be the case till the length of the eccentric-rod, etc., has been adjusted. Suppose the valve, instead of having only just opened the port, has opened it by half its width—that is, it has moved too far by $\frac{1}{8}$ in. This would prove that the eccentric- and valve-rods were too long by half that distance. Therefore shorten the effective length of the valve-rod by moving the nuts so as to diminish the travel of the valve by $\frac{1}{16}$ in., so that, without moving the other parts, the valve shall be open but one-quarter of the width of the port. Now loosen the wedge of the eccentric, and turn it round on the shaft till the valve is open by the thickness of a card, fix the eccentric by the wedge, turn the crank round to the first position, and observe the valve, when the lead should be about right. Enough has now been said to show how the lead can be equalised, and the proper angular position for the eccentric found. When this has been determined, mark on the shaft the position of the keyway of the eccentric, and take off the eccentric to file the flat on the shaft for the permanent key. When the key is being finally fitted, test the lead again, to make sure that the flat is filed just right; it is easy to adjust with a flat on the shaft, a thing that cannot be done on the shaft, but not with a sunk keyway. This is a recommendation of the method here advised, and may compensate for the departure from usual practice.

FEED-PUMP.

It would be possible to stop here, simply connecting the steam-pipe with the boiler, and conducting the exhaust steam away by another pipe. The engine would, however, be deficient in two respects: it would have to be regulated as to its speed by hand, and it would not possess a pump for supplying to its boiler the feed-water necessary to make up for what is used. Of course, if steam is obtainable from the boiler of another engine, or if an injects is employed to supply the boiler with water, no feed-pump would be necessary. As 1 cub. in. of water would give about 500 cub. in. of steam at 37 lb. pressure above the atmosphere, the pump, being single acting, must have a capacity of at least $\frac{1}{500}$ that of the cylinder. Suppose the steam cut off at $\frac{1}{4}$ of the stroke, then a length of 6 in. of the cylinder is filled with steam for the double stroke, or one complete revolution; add another inch for leakage and call it 7 in. Now, the bore 2 in. has an area of 3·14 sq. in., and 7 in. × 3·14 = 22 cub. in., which is the quantity of steam expended for each revolution. The pump has a diameter of $\frac{1}{2}$ in. and a stroke of $\frac{3}{4}$ in., its area being ·2 sq. in.; then ·2 × ·75 in. = ·15 cub. in., which is the quantity of water thrown at each revolution. No pump will throw its full capacity but allowing two-thirds of this, the quantity of water delivered by the feed-pump per revolution will be ·1 sub. in., or $\frac{1}{220}$ the capacity of the cylinder. The actual amount thrown by the feed-pump, therefore, will be greater than is theoretically required.

RETURN COCK, ETC.

It is convenient to be able to regulate the amount of water thrown by the pump so as to balance as nearly as possible the quantity extracted from the boiler by evaporation. This is done in a very simple way by means of a "return cock," which is either fitted as a branch to the delivery pipe or screwed into that part of the valve box of the pump from which the delivery pipe strings. Fig. 1847 (p. 588) shows such a cock. It serves a double purpose ; acting as a pet-cock, it will release any air that may have accumulated, and by gradually adjusting the cock it can be made to let out and allow to return to the supply tank any water forced by the pump and not required by the boiler. It is also convenient to have a pump a little larger than the size absolutely needed, because when about to stop work—at dinner-time, for instance—it is possible, by filling up the boiler, both to ensure its safety from burning, and also to lower the pressure by the introduction of a considerable quantity of feed-water. Some engineers, too, prefer to have both the pump and the injector, as they are sure then to have one available ; also the injector can be used when the engine is standing still, provided there is steam in the boiler ; thus it could be used to fill up the boiler during the stoppage for meals.

PUMP ECCENTRIC, PLUNGER, AND ROD.

The pump appears in section, plan, and end view respectively in Figs. 1845, 1846, and 1847, whilst Figs. 1890 and 1891, p. 611, show the eccentric. The plunger and rod are illustrated in Fig. 1893. A second pair of straps will be required. Then the eccentric (Fig. 1890) would be undertaken ; the two holes for the screws may be drilled first of all, and through these holes wood screws may be passed into c (Fig. 1892, p. 612), which can then be adjusted upon B, so as to bring true first the hole to be bored so as to fit the boss of the main eccentric, and then the outer rim to fit the strap. Also, if the screw-holes are countersunk and the heads of the two screws do not project, the face of the eccentric can be turned ; if not, a narrow band only can be turned round the edge, and the remainder might be finished by the file ; then the other side would be either turned up flat or filed. Notice, however, that the eccentricity is in this case only $\frac{3}{8}$ in., and not $\frac{1}{2}$ in. as in the valve eccentric. About the plunger and rod there is but little to say. The small end of the rod with the hole for the pin should be case-hardened, and the pin itself must be of hard steel, driven firmly into the eye of the brass plunger. The plunger is made in two pieces, for which castings are supplied, and the two parts are to be soldered together when the joint has been fitted, and cannot then be separated without unsoldering again ; this, however, will most likely never be required.

PUMP BODY.

The casting for the body of the pump has a flange, by which it may be chucked on the angle-plate on the face-plate. This flange must first be brought flat and vertical, so as to be parallel with the centre line, in such a way that, when standing on the face-plate, the scriber point will come to the centre of all four holes—namely, the suction, the delivery, the top of the valve-box, and the barrel. To ascertain these centres, plug them up temporarily with wood, file off the wood level, and find the centre with the dividers. There should have been a facing on the bed against which the pump flange would have bolted, but none was made because the pump is not always needed, and the flange can be well enough fitted by rubbing the bed with red marking and trying the flange on it. First, however, the flange must be upright ; also it must be brought into a plane parallel with that in which the centre lines of the pump lie ; then it can be clamped on the angle-plate, and each of the three centre lines can be brought perfectly true with the centre of revolution one after the other.

PUMP VALVE-BOX.

Begin with the longest part—the valve-box. Set this true in the lathe by adjusting the flange on the angle-plate, and clamp it firmly ; then bore out the inside to $\frac{1}{8}$ in., $\frac{1}{2}$ in., and $\frac{7}{8}$ in., taking care to bring the valve-seats perfectly true and smooth ; also cut the thread for the cap that closes the top, and turn the mouth of the hole true. The outside as far as the delivery branch might be turned, but that would involve getting up the whole of the outside bright, and it may as well be painted.

Conical Valves.

Take off the face-plate chuck without disturbing the setting of the pump body, and turn the two small conical valves (see Fig. 1845). These valves are made from two little castings. Taking the larger one first, centre it and bore the little hole for the reduced spindle of the lower valve; countersink this hole slightly, put the carrier on the small end, and put the casting between the centres. Or it might be grasped at the small end by the universal chuck, and supported at the other end by the back centre point in the drilled hole. Turn it, bringing the three flat wings to fit the bored part of the valve-box, and the part that fits the seating to exactly the same cone, rubbing red marking on the turned seat, and trying the valve in till it fits. Then see that the largest part of the valve does not exceed $\frac{5}{8}$ in. in diameter, so that there is room for the water to pass. The smaller valve is turned in a similar way, the top of its spindle entering the hole in the upper valve to steady it, but not fitting that hole closely. The small valve should be capable of lifting $\frac{1}{16}$ in., an amount which can be regulated by holding down the upper valve and lifting the lower one with wire. Lift up the suction, and file off the top of the spindle till the lift is correct. Then on the lathe put the face-plate with the pump body on it, and make a handle for the valves by boring a hole in the end of a piece of wood, into which can be driven the small end of the spindle; then, while the pump is going half round and back in the lathe (for it must not be driven continuously), grind the valves to their seats.

Grinding in Valves.

Grind in the valves with some of the grit from the grindstone trough, or, better still, with a piece of oilstone crushed up with the hammer on any hard surface. The great danger to be guarded against in grinding-in valves, or plugs, is getting the surfaces scored in rings and rough. This may arise through not keeping the grinding material distributed. Press very lightly

and withdraw constantly, to pass the finger over the surface and re-distribute the powder, and continue till the smooth dulness on the plug of a gas-tap appears. The grinding powder may be mixed with water.

Other Details of Pump.

The two valves being ground to their seats so that air cannot be sucked through, turn the pump body round on the angleplate and set the barrel true to bore it out to fit the plunger, and cut the thread for the gland; then turn it round again, and screw and face up the delivery branch. The pump is to be strongly bolted to the bed, and the studs should fit well into the holes in the flange, so that it cannot move. An indiarubber pipe is to be stretched over the bottom of the valve-box for the suction. The delivery-pipe has a small ring brazed to it after the union-nut is put over it; the nut holds the pipe by this ring, and clasps it up to the face of the delivery branch. Between the two faces would be interposed a ring of indiarubber to make the joint good. The return-cock already referred to, as seen in Fig. 1847, is supposed to be screwed into the valve-box, but it need not be placed there; it might be fixed as a branch downwards from the delivery-pipe.

The Action of the Pump.

Many pumps fail, or give trouble, on account of the presence of air inside. When the plunger is drawn up this air expands, but no water rises into the barrel; when the plunger comes back the air contracts again. In this pump the delivery-valve is on the highest point of the inside, so that the air may be got rid of first. The valves are arranged to have a lift of $\frac{1}{16}$ in. When a valve of this sort lifts a distance equal to a quarter of its diameter, then the annular space opened under its edge is equal to the area of the valve. By this rule the lift of the smallest valve should be $\frac{1}{64}$ in. more than $\frac{1}{16}$ in., but it is well to allow a small lift. In the case of the upper valve, which must of necessity be of larger diameter, there is sure to be sufficient passage, and the lift might be rather less than

₁⁄₈ in., which will diminish the shock and noise at each closure. Water confined in a pump or pipe is so incompressible that it must be considered almost as a solid ; were the plunger of the pump to meet in the middle of its stroke a solid piece of heavy metal, there would be a thump and a shock ; just such a shock may be expected in a force-pump, when the plunger, having filled up the empty space left in the body of the pump after the up-stroke, meets the column of water in the feed-pipe, and drives it a few inches farther on. To prevent this shock, a cushion of air is provided in an air vessel, and, instead of mounting a separate air-vessel on the top of the valve-box, the valve-box itself is prolonged upwards a little way, so as to imprison above the delivery branch a small quantity of air, which, by its elasticity, acts as a buffer.

THE ENGINE GOVERNOR.

The details of the governor are best seen in Fig. 1894. The only difficulty connected with this form of construction is the tubular spindle, 3½ in. long, ¼ in. diameter, with a ⅛-in. hole through it, but steel tubing of this size can be obtained. Begin by bolting the side of the main bracket on the angle-plate, and after boring it through with a ¼-in. drill, in the top turn both the recess which holds the oil and the rounded bead on the edge. Now twist the casting on the angle-plate and bore the long boss, also with a ¼-in. drill, and face the end. Take the brass casting with the two arms ; hold it in a universal chuck, and bore a ¼-in. hole. Now put the steel tube between the centres of the lathe to get it perfectly straight by bending ; then try it in the standard or body, and reduce it by turning or simply by filing till it will drive firmly in, when it will not need the little round pin, seen at the top, for securing the two together. Now turn the outside of the body, and then reduce the lower half of the tube till it will turn easily in the bracket, carefully squaring up, and smoothing the bottom surfaces of the brass body, which turns on the top of the standard. The throttle-valve is shown by Fig. 1895.

GOVERNOR MITRE-WHEEL.

The mitre-wheel underneath will be the next to fix. The under-side of the hole in the bracket would be first filed up square for the wheel to work against, and then the wheel may be fixed by the pin, as shown. This pin cannot go through the middle, because of the ⅛-in. rod which

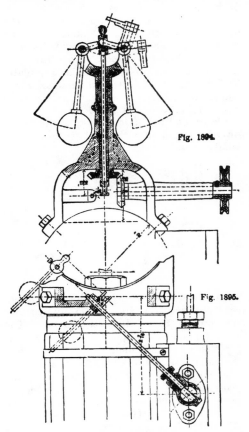

Fig. 1894.—Engine Governor.
Fig. 1895.—Arrangement of Throttle-valve.

slides inside the tube. The wheel may be keyed if it is easier, or it might be fixed with a pinching-screw.

GOVERNOR BALLS.

The governor balls are 1 in. in diameter. For them, fix a bit of sheet-metal into a wood chuck, and bore a 1-in. hole in it; then cut it in two to serve as templates in turning up the balls. Drive the balls into a recess in a hardwood chuck till very slightly more than half the diameter projects; turn the projecting part to an exact half-sphere, using the template to test it; then bore a hole in the middle of the ball, and tap it ready to screw to the arm. When both the balls have thus been treated, alter the chuck so that they can be fixed in the opposite way up to the shoulder left by the turning, the half-sphere now being contained within the chuck. Now turn the projecting portion true to the template. Through the back of the wood chuck there should be a small hole by which to drive out the balls, and a little screw may be made on any bit of steel held in the lathe, on which to mount the balls to polish the whole of the outside.

Fig. 1896.—Governor Arm Casting.

Fig. 1897.—Top of Governor.

ARMS OF GOVERNOR.

Take the arms on which the balls hang, centre them, and turn them up, cutting the screws to fit into the balls. The shoulder against which the balls screw is a collar ¼ in. in diameter, and that is just the thickness of the joint at the other end. File the flat of the joint parallel with the arms, placing them down sideways on the face-plate to try the joint, and file out the slot which forms the joint in the arms of the body of the governor. The file can be passed through both arms (see Fig. 1896), so that it is easy to file out these two joints and make a perfect fit.

OTHER DETAILS OF GOVERNOR.

The arms of the governor carry small levers at an angle of 120° with the longer part of the arms, and the short arms or levers meet in the centre, in the slot of the brass piece which forms the top of the governors, and are halved together in that slot. One of these small levers is projected above Fig. 1894, and the top piece with slot is shown at Fig. 1897. This little piece is pinned on to a ⅛-in. steel wire, which passes down the centre of the governor and revolves with it, but can be raised up or down ⅜ in., according as the balls are hanging down or flying out. On the lower end of this steel wire is pinned another piece of brass, in which is turned a groove that enters the forked end of a small lever, clearly seen in Fig. 1897, this lever being also shown projected in Fig. 1894. It is secured to a piece of ⅛-in. brass wire, which passes straight through the gland of the throttle-valve. The lever might have on it an arm (dotted in Fig. 1895), with an adjustable weight for regulating within small limits the speed at which the governors would control the engine. But if the engine is required simply for driving a lathe or similar tool in a small workshop, a high degree of regularity will not be required.

THROTTLE-VALVE.

The wire which controls the throttle-valve is made of brass, because, like the pump and its plunger, it would be liable to rust if it were of iron, and that would cause unnecessary friction in the stuffing-box. This stuffing-box is clearly seen in section in Fig. 1895. It must be carefully packed and screwed up lightly, just to prevent the escape of steam. It would be very easy to screw it so tight that the governor could not rise at all, and would have no control whatever over the engine. The throttle-valve and its case, shown separately at Fig. 1898, may be held in a universal chuck while the oval flange is faced; then the holes for the bolt might be made through this flange, and then it could be clamped by the flange on a face-plate, the hole inside bored out true and smooth, the screw-thread cut (gas thread) to fit the stop-valve, the shoulder faced, and the edge turned bright.

STOP-VALVE AND NIPPLE.

It is supposed that a stop-valve would be bought, and not made. A ⅜-in. stop-valve is large enough for this engine, and ¾-in. is the size for the steam pipe ; but ½-in. would be better for the exhaust. That being the case, it will be simpler to have both steam and exhaust ½-in. bore, and then the stop-valve must be ½-in. too. There will be required a ½-in. "nipple" to connect the stop-valve with the standard, since both these have female threads. A "nipple" is a short ferrule of pipe with threads on the outside; it would be screwed firmly into the standard, and the stop-valve would screw on that.

ARRANGING GOVERNOR AND THROTTLE-VALVE.

The throttle-valve case, or standard, may be mounted on the angle-plate chuck to bore and tap the stuffing-box. The casting for the throttle-valve itself appears at Fig. 1899. It may be centred and turned till it will fit into the bored hole in the standard. The hole for the brass wire which carries and controls it may, perhaps, best be bored by placing the valve in position in the case or standard, and then passing the drill in through the stuffing-box and gland. Next, the governor standard would be fixed and screwed down in position, and the hole for the other end of the brass wire bored through the arm of the bracket, when the wire would be put through that arm and the stuffing-box of the throttle-valve standard, while the latter is placed in position on the valve-box. Thus arranged, the holes through the oval flange could be correctly marked on the steam-chest. The stems, cast on the throttle-valve to aid in turning, may now be cut off, and the hole through the valve and wire drilled, and the little brass pin, seen in Fig. 1898, inserted. All that is required now to complete the connection between the governor and the throttle-valve is to make the little lever that embraces the grooved piece at the bottom of the small rod which passes down the centre of the governor. Now fit the shaft in the long boss, bore the small pulley which receives the band, drive it on at one end, and turn it up in position ; then bore and fix the little mitre-wheel at the other end,

adjusting it to gear to the correct depth and fixing it with a key or a pin driven right through.

EFFECT OF GOVERNOR.

It must not be supposed that the throttle-valve will stop the steam without any leakage ; it could not do so fitted in this way. The stop-valve will do it ; and the throttle-valve will prevent nearly all passage of steam, and when closed by the governor there should not be sufficient leakage to work the engine. The parts of the governor are so proportioned as to

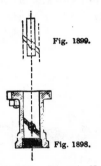

Fig. 1899.

Fig. 1898.

Figs. 1898 and 1899.—Throttle Valve.

allow the engine to run at about 100 revolutions a minute. The pulley on the horizontal driving shaft is the same size as the groove in the boss of the eccentric, so that both engine and governor would run at the same rate.

ACCELERATING SPEED OF ENGINE.

If it were desired to make the engine run twice as fast, that could be done by replacing the pulley on the horizontal shaft by one of twice the diameter, so that the governor would still turn at the same speed, while the engine went twice as fast. It would increase the power of the governor if it were allowed to run at 200 revolutions whilst the engine made 100. To do this, make a brass or hardwood pulley 3 in. in diameter, and key it on the boss of the eccentric, over the present groove ; then a spring or weight would have to be so arranged as to bring down the arms of the governor when making

200 revolutions, to their normal, or midway, position—the one they now occupy when revolving at 100 revolutions. A weight could be arranged as shown dotted in Fig. 1895 (p. 617), and springs might be applied as shown at Figs. 1900 and 1901. It is not proposed to show how to calculate the spring or the weight, because it is a very easy thing to test the governor by spinning it, counting the revolutions, and then making the pulleys to the correct size.

DRAWING GOVERNOR BALLS TOGETHER.

In Fig. 1901 it is supposed that small eyes are screwed into the centre of the

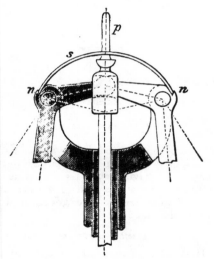

Fig. 1900.—Governor with Bow Spring.

balls on both sides, so that two springs can be fixed to draw the balls together. A neater plan is shown in Fig. 1900; here two notches *n* are filed in the knuckle of the governor arms, and into them enter the ends of the bow spring *s*. These ends are kept in place, sideways, by the slot in which the arms move, whilst the middle of the bow is pierced with a small hole, and passes over a small pin *p* screwed into the lower part of the little ball, the upper part of which has been cut away.

ADJUSTING GOVERNOR.

Supposing it be decided to use a spring, a flat one of this kind would be very easy to make and apply; the pulley on the horizontal shaft of the governor could be driven off a groove on a chuck in the lathe, then the governor could be spun till the arms took up their midway position, when the revolutions would be counted; if the rate were evidently too great, the spring would be taken off and weakened by grinding it a little thinner. When the arms of the governor come to the mean position at about 250 revolutions, the adjustment would be correct. It would be well also to ascertain what speed is required to raise the arms to their highest point and close the throttle-valve, and, again, at what speed the arms begin to rise. Thus, 240 revolutions might raise the arms, 250 bring them half-way up, and 260 raise them to the highest point; then, if the engine is speeded to run at half the revolutions of the governor, it may vary in speed from 120 to 130 revolutions a minute, but not more. It might be impossible to count the revolutions of the governor, but it would be easy to count the lathe mandrel, and the groove on the chuck could be made twice or three times the size of the pulley on the governor shaft.

LUBRICATORS FOR ENGINE.

The lubricators are not of the usual kind. However, those who may wish to use the common kind of oil cup can of course do so; but Stauffer's lubricators are better. The ordinary oil cup, moreover, requires to be set in a vertical position, which these do not. The lubricators shown in Fig. 1902 are the smallest size of Stauffer's patent lubricator; they are filled with a semi-solid grease. The lubricator consists of two parts: a cap, which is filled with the grease, is screwed on the other part till the grease is forced down the small hole into the bearing. As the shaft revolves it distributes the grease, which soon begins to ooze out of the sides of the brasses; the bearing is then ready for work, and will require no further attention for a day or so, or even, perhaps, for a week. When a fresh supply of grease is required, it is only necessary to give the cap about half a turn with the fingers, so as to force out a little more grease into the bearing. Of course, when the cap has been screwed quite down it must be re-filled. This form of lubricator can be used in any position: it may be horizontal, as on the eccentric-

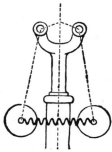

Fig. 1901.—Governor with Two Springs.

and connecting-rod; it may force grease into several rubbing surfaces at once, as that on the top of the crosshead; it may be placed upside down, or it may revolve with a boss of a loose pulley. No lubricator has been provided for the slide-valve and piston, because when working at moderate pressures the water contained by the steam lubricates sufficiently. A simple way to lubricate both slide-valve and piston at once is to arrange an oil cup so that it may deliver into the steam-pipe; by thus lubricating the steam-pipe, the oil is carried to all the rubbing surfaces inside.

STEAM EXHAUST PIPE.

The exhaust-pipe is supposed to be simply screwed into the cylinder casting; it bends downwards to carry off the water, and may be led down towards a drain and a hole made at the lowest point there to allow the water to escape, after which it may turn upwards to carry the bulk of the exhaust steam into the air. Cylinder cocks carry off the condensed steam from the cylinder; they are opened on first starting the engine, when the cold metal of the cylinder condenses a good deal of steam. The two cocks are both moved by one handle. These taps cannot really be considered necessary on such a small engine, especially with a descending exhaust.

CATCHING DRIPS OR LEAKAGE.

All drips or leakage about the engine should be prevented as far as possible. Engines are often seen working without the slightest whiff of steam appearing even at the stuffing-boxes. A little tin tray may be hung by a wire under each stuffing-box gland, so as to catch any drops that may fall; the pump gland is one especially liable to drip and make a mess.

LAGGING OF CYLINDER.

The lagging or wood casing surrounding the cylinder appears in section in Fig. 1846 (p. 587). This is useful in two ways: first, it prevents loss of heat from the cylinder, and consequent loss of power; and secondly, it prevents the engine from unduly heating the apartment in which it works. Also, it looks well if neatly fitted, with the wood polished, and the brass bands kept bright. The cylinder would first be clothed with felt or, say, part of an old blanket; then add the wood lagging in strips of mahogany; and then secure the brass hoops, as is shown in Fig. 1855 (p. 593). Otherwise, instead of

Fig. 1902.—Section of Stauffer's Lubricator.

the hoops there might be a piece of thin sheet metal bent round the wood and painted.

MAKING STEAM JOINTS.

Making the steam joints is a very simple operation. Get a little red-lead powder and some paste white-lead, and mix a little of the former into some of the latter just to stiffen it; work it well together, and rub this on the threads of the pipes before screwing up. It is usual to spread some of this same cement between the flat surface joints, such as those of the cylinder cover, valve-chest, etc., but a piece of pink blotting-paper has been used with great success. It was cut out with holes for the bolts, and when cut round the edge of the joint, looked as if it were a red-lead joint, with the advantage that it could be taken apart often, whilst none of it squeezed out of the joint into the inside of the cylinder. It would be an advantage to oil the blotting-paper before screwing up the joint

ENGINE BELT.

The flywheel being 16 in. in diameter, a flat belt round this wheel should transmit all the power the engine is capable of exerting. It is well to arrange for the belt to run upwards, because then it may relieve the bearings of most of the weight of the flywheel; and it is better, if the belt must be inclined upwards, to arrange it to incline forwards from the engine—not backwards, towards, and over the engine. In the first case, the pull of the belt would tend to cause wear on the middle of the upper brass of the crank-shaft bearing. which wear can be taken up by tightening the cap of the bearing; but in the second case the wear would come on the sides of the brasses, where they are divided, and here the wear cannot be taken up.

CONCLUDING REMARKS.

There are three lugs provided on the bed-plate, by means of which the engine can be bolted down to a foundation. This may be a frame of wood made so as to raise the engine high enough to allow the flywheel to run clear of the floor; or it might be a stone, the foundation bolts being fastened in with lead. When not at work, the engine should be carefully covered with a cloth or piece of green baize, to keep it from dust and damp, remembering that these, and above all emery-dust. must not be allowed to get on to the sliding guide or into the bearings.

MAKING A $\frac{1}{4}$-H.P. VERTICAL STEAM ENGINE.

THE NECESSARY TOOLS.

THE vertical engine to be described in this chapter can be constructed with the aid of a $3\frac{1}{2}$-in. back-geared lathe, having an ordinary slide-rest, with a travel of about 4 in. on each slide. Other tools required are stocks and dies cutting $\frac{1}{8}$-in., $\frac{3}{16}$-in., and $\frac{1}{4}$-in. threads; a screw-plate cutting $\frac{1}{16}$-in. to $\frac{1}{8}$-in. threads; a surface-plate, about 4 in. by 6 in.; and a few drills, files, etc.

ENGINE CYLINDER.

The engine is shown complete in front elevation, side elevation, and plan by Figs. 1903, 1904, and 1905 respectively. The cut-off occurs at five-eighths of the stroke, so that with a boiler pressure of 60 lb. per sq. in., and a speed of 300 revolutions per minute, the brake-power of the engine may be about $\frac{1}{4}$-horse-power. Begin by cramping the cylinder (Figs. 1906 to 1908) true on the face-plate of the lathe, using a flange and placing packing between it and the face-plate, so that the tool may take a clear cut. Bore out to within $\frac{1}{64}$ in. of the finished diameter, then finish to $1\frac{1}{2}$ in. diameter. If a sharp tool is used, polishing should not be needed. Next, on a mandrel turn each end to reduce the length to $3\frac{1}{16}$ in., leaving the flanges of equal thickness. Then take a cut down the inside of each flange $\frac{7}{32}$ in. from each end of the cylinder, until a diameter of $2\frac{9}{16}$ in. is reached; this ledge forms a support for tin lagging. Then turn the flanges to $2\frac{13}{16}$ in. diameter and polish on the outside diameter.

STEAM AND EXHAUST PORTS.

The port face must now be filed and surfaced true with the bore of the cylinder, measuring $1\frac{5}{16}$ in. from the centre of the cylinder. Now mark off the steam and exhaust ports, the former being $\frac{1}{8}$ in. by $\frac{5}{8}$ in., and the latter $\frac{3}{16}$ in. by $\frac{5}{8}$ in., with bars between $\frac{1}{8}$ in. wide. The steam ports must be drilled and clipped or cut with a traverse drill to a depth of $\frac{7}{16}$ in. from the steam-port face, and then two holes $\frac{7}{16}$ in. diameter are drilled from each end of the cylinder to meet the ports. A small nick, $\frac{7}{16}$ in. wide by $\frac{5}{32}$ in. deep, must be cut in each end of the cylinder where the steam enters. Before cutting the exhaust port, drill a hole $\frac{3}{16}$ in. diameter and $\frac{5}{16}$ in. from the steam-port face, parallel to it, and at right angles to the bore of the cylinder, right through from side to side. Enlarge each end of this hole to $\frac{11}{32}$ in. diameter for a depth of $\frac{1}{4}$ in., and tap it $\frac{1}{8}$ in. gas thread, to carry the nipples to which the exhaust pipes are attached. Then drill and chip the exhaust port to a depth sufficient to communicate with the $\frac{3}{16}$-in. hole previously drilled.

EXHAUST NIPPLES.

The exhaust nipples may be turned from soft steel bar $\frac{5}{16}$ in. diameter, and should be drilled first and then turned on a mandrel to the dimensions given in Fig. 1906. A $\frac{1}{8}$-in. gas thread must be cut on one end, and three holes drilled and tapped, $\frac{1}{8}$ in. diameter, in the end carrying the flange, the pitch circle being $\frac{5}{8}$ in. diameter. A little red-lead may be smeared on the thread, and the nipples then screwed into place. Eight stud holes, in both cylinder flanges, should be drilled and tapped $\frac{1}{8}$ in. diameter, and the studs should be of soft steel wire $\frac{1}{8}$ in. diameter, screwed to the dimensions given in Fig. 1907. The studs for the top flange project $\frac{13}{32}$ in., those for the bottom flange project $\frac{1}{4}$ in. from the cylinder faces. Four $\frac{3}{16}$-in. studs, $1\frac{1}{4}$ in. long, with nuts, should now be fitted into holes in the port face, as shown by Figs. 1906, 1907 and 1908.

NUTS AND SCREWS.

All nuts and screws may be made from soft steel rod to the following sizes :—One-eighth-inch nuts, width across flats $\frac{1}{4}$ in., depth $\frac{1}{8}$ in. ; $\frac{3}{16}$-in. nuts, width across flats $\frac{5}{16}$ in., depth $\frac{3}{16}$ in. ; $\frac{1}{4}$-in. nuts, width across flats $\frac{3}{8}$ in., depth $\frac{1}{4}$ in.

STEAM-CHEST AND STUFFING-BOX.

The steam-chest casting should be filed or faced in the lathe until the thickness is $\frac{3}{4}$ in., taking care that an equal amount is taken off each side, and that the projection, which forms the stuffing-box, is kept central. The inside should be dressed out to measure $1\frac{3}{8}$ in. by $\frac{3}{16}$ in. ; then cramp it on the face-plate, and drill four holes $\frac{3}{16}$ in. diameter at centres corresponding to those of the studs screwed in the port face. Now mark out the stuffing-box, central both ways, and cramp the steam-chest on an angle-plate fastened to the face-plate, so that the centre of the stuffing-box runs true ; then drill a recessed hole shown in Fig. 1907. Two holes (see Fig. 1908) $\frac{3}{32}$ in. diameter and $\frac{3}{16}$ in. deep, must now be drilled and tapped on each side of the stuffing-box for gland studs, $\frac{1}{2}$ in. long. The nuts for these should be $\frac{3}{16}$ in. across the flat by $\frac{3}{32}$ in. thick.

LAGGING THE CYLINDER.

A piece of stout tin must now be cut to bend round the cylinder, overlapping at the ends ; this is fastened in place by two screws, $\frac{3}{32}$ in. diameter by $\frac{1}{4}$ in. long (see D D, Figs. 1906 and 1908), placed $\frac{1}{16}$ in. from the

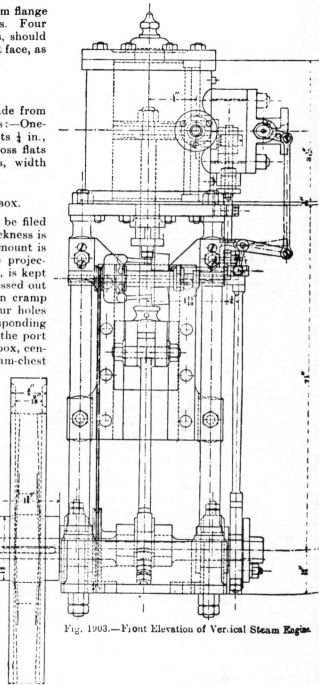

Fig. 1903.—Front Elevation of Vertical Steam Engine.

cylinder end. Make a paper template to fit round the cylinder, having pieces cut out where the exhaust nipples come, and mark off the tin from this. The space between the cylinder and the tin lagging may be filled in with layer of felt.

Bottom Cylinder Cover.

The bottom cylinder cover (Figs. 1909 and 1910) should be filed to dimensions; then cramp it true on the face-plate, with the projection which is to form the stuffing-box on the outside, and drill a recessed hole as shown. Then mount the cover on a mandrel turned to fit both holes, the smaller diameter ($1\frac{1}{8}$ in.) passing through the cover a distance of about 1 in. Take a light cut with a left-hand tool down the stuffing-box side as far as possible without cutting into the projecting metal, the small surface left being removed afterwards with a file. Next, with a right-hand facing tool, face down the opposite side until $\frac{1}{4}$ in. thick to a diameter of $2\frac{13}{16}$ in.; then draw the tool away $\frac{1}{16}$ in. and face down until a diameter of $1\frac{1}{2}$ in. is reached. The latter forms a spigot to fit the cylinder. Draw the tool away another $\frac{1}{16}$ in., and face down to the mandrel, the thickness of metal at this point being $\frac{3}{8}$ in. It may now be taken off the mandrel, and a portion of the spigot $\frac{7}{16}$ in. wide must be cut down flush with the face of the cover on which the cylinder rests, and opposite to where the steam

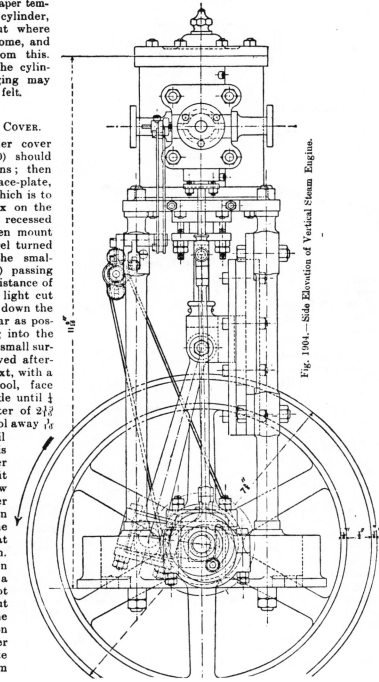

Fig. 1904.—Side Elevation of Vertical Steam Engine.

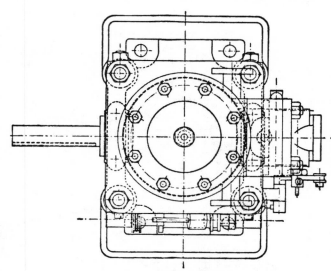

Fig. 1905—Plan of Engine.

enters. Two $\frac{1}{8}$-in. studs must now be screwed in on each side of the stuffing-box as shown. The brass guide for the valve spindle is made to the dimensions in Figs. 1909 and 1910 and bolted as shown. Drill a $\frac{5}{32}$-in. hole through the guide with its centre $1\frac{1}{8}$ in. from the centre of the cover.

DRILLING HOLES IN CYLINDER COVER.

The four holes at the corner of the cover may now be drilled from a sheet-iron template $\frac{1}{8}$ in. thick, which will be used again for the corresponding holes in the engine bed. With a sharp scriber mark two centre lines across the plate at right angles to each other, as it is by these lines that the template is set in position. A hole $1\frac{1}{2}$ in. diameter should be cut in the centre to allow the plate to lie flat against the stuffing-box side of the cover. One of the centre lines should come in line with the centre of the valve spindle guide, and be

parallel to the sides of the cover. Now cramp this tight to the cover, and bore the corner holes through the cover. The drill may be revolved by the lathe mandrel, and the work fed by the loose headstock. Next mark out and drill the eight holes $\frac{1}{8}$ in. diameter through which the studs in the bottom flange of the cylinder pass.

BRASS GLAND.

The brass gland (Figs. 1911 and 1912, p. 628) is now turned. First drill the hole through the centre, then mount the casting on a mandrel, and face the flange and turn the body. Drill two $\frac{1}{8}$-in. holes, $\frac{7}{8}$-in.

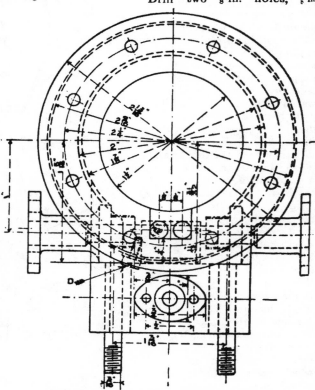

Fig. 1906.—End Elevation of Engine Cylinder.

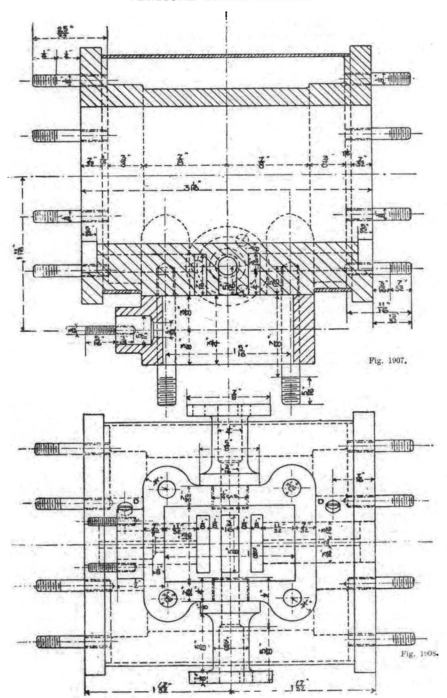

Fig. 1907.

Fig. 1908.

centres, through the flange, which should
be filed up oval, measuring 1¼ in. by 1⅛ in.
across the extremes.   Place the gland in

TOP CYLINDER COVER.

The top cylinder cover (Figs. 1913 and
1914) may be treated in a similar manner

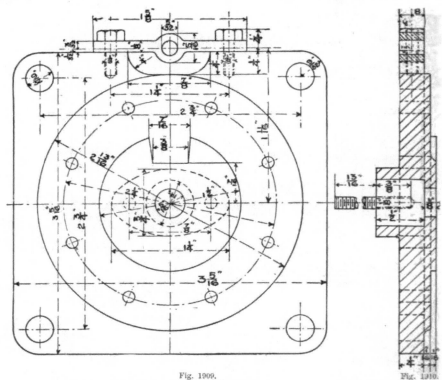

Fig. 1909.

Fig. 1910.

Figs. 1909 and 1910.—Plan and Section of Bottom Cylinder Cover.

position, and, after inserting a sheet of
brown paper smeared with a very little

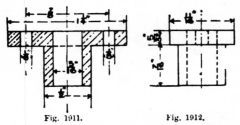

Fig. 1911.   Fig. 1912.

Figs. 1911 and 1912.—Section and Elevation of
Piston-rod Gland.

red-lead between the cylinder flange and
cover, bolt up tight by means of the nuts.

to the one just completed. Next turn up
the screwed plug from steel rod to the
dimensions given in Fig. 1914. This should
fit well in the screwed hole, as it must
be taken out now and then for the ad-
mission of oil to the cylinder. Drill eight
holes ⅛ in. diameter to correspond with the
studs in the cylinder flange, and place the
cover in position with the nuts on.

BRASS SLIDE VALVE.

The brass slide valve is made as shown
in Figs. 1915, 1916 and 1917, the exhaust
recess being ⅜ in. long by ⅝ in. wide and
⅛ in. deep. Two lugs projecting from the
back of the valve carry a piece of brass
which is a good working fit, with a hole

$\frac{3}{32}$ in. diameter drilled through, and a saw cut from the side into the hole. This is

valve face must be surfaced dead true. A brass gland for the stuffing-box (Figs. 1918

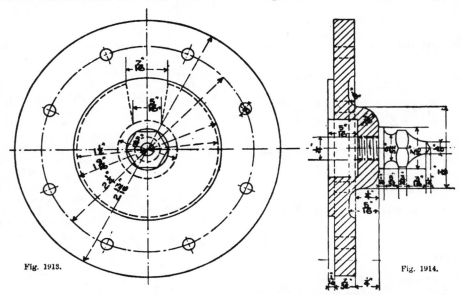

Fig. 1913.

Fig. 1914.

Figs. 1913 and 1914.—Plan and Section of Bottom Cylinder Cover.

fastened to the valve by a set-screw. The holes in the lugs are made $\frac{3}{32}$ in. by $\frac{3}{16}$ in. to allow the valve to keep close to the port

and 1919) should be treated as described for Figs. 1911 and 1912.

SLIDE VALVE ROD.

The valve rod (Figs. 1920 and 1921) must be turned from soft steel wire $\frac{3}{16}$ in. diameter; one end must be turned down to $\frac{1}{8}$ in. diameter for a length of $\frac{7}{32}$ in. for screwing into the crosshead. This may be made from soft steel rod $\frac{3}{8}$ in. square; file down a piece $1\frac{9}{32}$ in. long until reduced to $\frac{3}{16}$ in. by $\frac{11}{32}$ in., and then cut out the centre part. Now drill a $\frac{1}{8}$-in. tapping hole right through into the nick, and mount it on a

Fig. 1916.

Fig. 1917.

Fig. 1915.

Figs. 1915 to 1917.—Plan, Vertical Section, and Side View of Slide Valve.

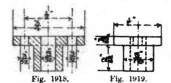

Fig. 1918.

Fig. 1919.

Figs. 1918 and 1919.—Section and Elevation of Valve-rod Gland.

face as wear takes place, the valve-rod remaining in the same position. The

mandrel, and turn it to the dimensions given in Fig. 1921. Another hole, $\frac{1}{32}$ in.

diameter, must now be drilled at right angles to the first one, passing through both jaws at the crosshead. Through this a steel pin, reduced at one end to take a $\frac{1}{8}$-in. nut, is passed to carry one end of the eccentric rod; the crosshead may now be

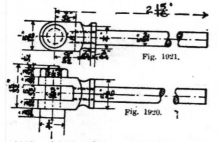

Fig. 1921.

Fig. 1920.

Figs. 1920 and 1921.—Plan and Side View of Valve Rod.

screwed on the rod and riveted to prevent it coming loose.

STEAM-CHEST COVER AND THROTTLE VALVE.

The steam chest cover and throttle valve (Figs. 1922 to 1925) may now be taken in hand. Cramp the cover on the face-plate by the flange, and drill the centre recessed hole; mount it on a mandrel, and face both sides to the dimensions given. Next turn a piece of brass, $\frac{3}{8}$ in. diameter and $\frac{1}{8}$ in. long, and drive it into the hole in the cover flush with the inside face. Then place the cover face downwards upon an angle-plate fastened to the face-plate, and drill a hole $\frac{1}{16}$ in. diameter and $\frac{3}{8}$ in. from the face of the cover, to a depth of 1 in. from the side of the cover; enlarge this to $\frac{3}{16}$-in. tapping, for a depth of $\frac{1}{8}$ in. from the edge, and tap this. Take the cover off the angle-plate and drive out the plug, which will now have a $\frac{1}{16}$-in. hole running across its centre. A flat piece must now be sawn out $\frac{1}{8}$ in. thick at an angle of 45° to the length, and with the $\frac{1}{16}$-in. hole running through the centre; this should be filed to the dimensions given in Fig. 1925,

and a spindle, $\frac{1}{16}$ in. diameter by $1\frac{1}{4}$ in. long, should be fitted into the hole in the valve. When this spindle is in position, drill a small hole through the enlarged part of the valve, cutting a little of the metal away on one side of the spindle. Fit a small pin in this, and, when the valve is in position, this will prevent the spindle turning without it. A screwed gland (E, Figs. 1922 and 1923) may be turned from $\frac{7}{16}$-in. brass rod. Hold the rod in a chuck, and bore a $\frac{1}{16}$-in. hole up the centre to a depth of $\frac{1}{2}$ in.; now turn down the outside to $\frac{3}{16}$ in. diameter for a distance of $\frac{1}{4}$ in. from the end, and cut a $\frac{3}{16}$-in. thread upon it; turn the remainder down to $\frac{3}{8}$ in. diameter, and file a hexagon upon it, then chamfer the edges, and cut off to a total length of $\frac{3}{8}$ in. Four $\frac{1}{16}$-in. holes must now be drilled in the cover, and three $\frac{1}{8}$-in. holes drilled and tapped in the

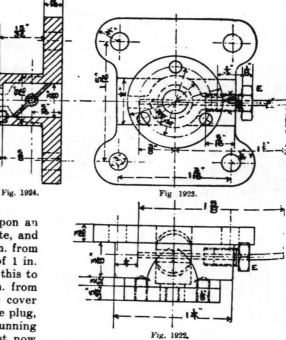

Fig. 1924.

Fig 1923.

Fig. 1922.

Figs. 1922 to 1924.—Plan, Front View, and Vertical Section of Steam-chest Cover and Throttle Valve.

flange on the outside, as shown by Figs 1923 and 1924.

PISTON.

The piston, piston-rod, crosshead, and slipper are shown in Figs. 1926 to 1929. Drill the hole in the centre of the cast-iron piston and turn it on the mandrel to $1\frac{1}{2}$ in.

Fig. 1925.—Elevation of Throttle Valve.

diameter by $\frac{3}{8}$ in. thick, recessing the sides as shown in Figs. 1926 and 1927. Then turn the groove and the ring. The latter is made from $\frac{1}{4}$-in. brass. This should be cramped upon the face-plate and bored to $1\frac{7}{8}$ in. diameter; then take another cut

PISTON-ROD.

The steel piston-rod (Figs. 1926 and 1927) is turned and the ends are threaded as shown. The $\frac{1}{4}$-in. nut, after having been screwed tight against the piston, is secured by a pin $\frac{1}{16}$ in. diameter, passing through the rod and nut.

CROSSHEAD.

The crosshead may be cut from soft steel rod $1\frac{1}{4}$ in. square by $1\frac{3}{8}$ in. long. File one side and one end square and smooth and cramp it on an angle-plate, and drill a $\frac{1}{4}$-in. hole right through, $\frac{7}{8}$ in. each way from one corner (see Fig. 1927). Next turn the block half round, so that the hole just

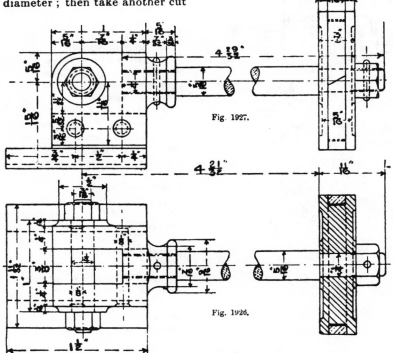

Fig. 1927.

Fig. 1926.

Figs. 1926 and 1927.—Piston, Piston-rod, Cross-head and Slipper.

through the brass, $1\frac{5}{8}$ in. diameter, leaving a complete ring. Face this on a mandrel to $\frac{3}{16}$ in. wide, and, with a saw, cut through it at an angle of about 60°. Then spring it into place on the piston.

drilled crosses the lathe centres, and, $\frac{7}{16}$ in. from the top of the block and central the other way, drill and tap a $\frac{1}{4}$-in. hole for the piston-rod. Cramp the block on the face-plate and drill a hole $\frac{7}{16}$ in. diameter through the metal where the bottom of the jaw will come. Now with a saw make cuts from the end of the block to each side of the hole just drilled to form the jaw. This must be finished with a file to the dimensions given in Figs. 1927 and 1928. A piece on the underside must also now be cut away. Now drive a mandrel in the $\frac{1}{4}$-in. hole and face each side equal until 1 in. thick ; then face down to leave bosses projecting $\frac{1}{16}$ in. on each side and $\frac{1}{2}$ in. diameter. Next mount on a screwed mandrel, passing through the hole tapped for the piston-rod, and turn down until a collar

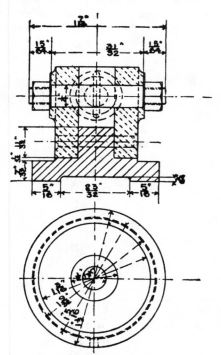

Figs. 1928 and 1929.—Crosshead and Slipper.

is formed, as shown. File the bottom of the block and screw in the piston-rod, fastening it by a $\frac{1}{16}$-in. pin.

CAST-IRON SLIPPER.

The slipper is of cast-iron, with a projection which fits between the cheeks on the underside of the crosshead. When this is in place, two holes, for $\frac{1}{8}$-in. rivets,

are drilled from side to side ; these holes should be countersunk on each side, and riveted over flush with the sides of the crosshead. The wearing surfaces of the slipper must be surfaced, and a portion underneath recessed (Figs. 1928 and 1929). After the steel crosshead pin is turned and threaded, it must be hardened and polished again, the dies being run over the threads. The cylinder may now be fitted up complete and the parts wiped with an oil rag to guard against rust.

ENGINE BED.

The cast-iron bed (Figs. 1930 to 1932) should be trued on the bottom with a rough file. Cramp it on the face-plate, and take a cut down the faces on the top face to a total depth of $1\frac{1}{4}$ in. The pockets for the brasses must next be filed out ; these are $\frac{1}{8}$ in. deep and $\frac{5}{16}$ in. wide. Next drill and tap two $\frac{1}{8}$-in. holes for each pocket, and screw studs into these as shown. Two nuts should be fitted for each stud, the top one, forming a lock nut, being chamfered on both sides. The template made for the bottom cylinder cover will now be used for drilling the holes for the columns. Bolt this on the bed by the studs just mentioned, taking care that one of the centre lines on the plate lies exactly central with the pockets, the edges of the template coming flush with the outsides. Drill four holes $\frac{9}{32}$-in. in diameter through the bed, using the template as a guide. Next remove the template, and recess the holes to $\frac{3}{8}$ in. diameter for a depth of $\frac{3}{16}$ in. Drill four $\frac{1}{4}$-in. holes through the lugs for the purpose of bolting the engine to its foundation when completed. The brasses (Figs. 1933 and 1934) should be fitted into the pockets, red-lead and a scraper being used.

CAST-IRON CAPS FOR BRASSES.

The cast-iron caps (Figs. 1935 to 1937) may be filed, and holes $\frac{1}{8}$-in. in diameter drilled. Place the brasses, with the caps and nuts, in position, a centre line being marked on the outside of each pair of brasses crossing the joint at right angles

With dividers scribe a circle $\frac{7}{16}$ in. diameter, taking as centre the crossing point of the two centre lines. The bed may now

set in line with the centres of the lathe, and also set true by the circle scribed on the brass. With a $\frac{3}{8}$-in. drill cut through

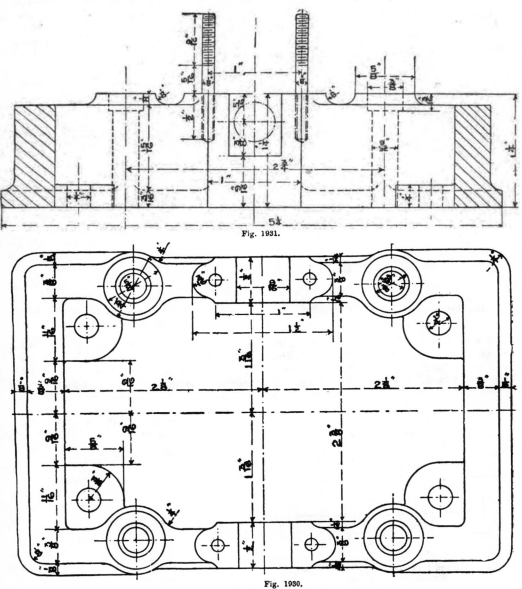

Fig. 1931.

Fig. 1930.

Figs. 1930 and 1931.—Plan and Section of Engine Bed.

be bolted to an angle-plate fixed on the face-plate of the lathe, the brasses being

one brass, and, with a small boring tool, finish the hole to $\frac{7}{16}$ in. diameter; then reverse the bed, and proceed with the other brass in a similar manner.

CRANK-SHAFT.

The steel or iron crank-shaft, governor pulley, and catch-plate come next (see Figs. 1938 to 1940). Place the crank-shaft between the centres of the lathe, and cut

and place one at each end of the crank-shaft, fixing it with a $\frac{1}{2}$-in. set-screw. These blocks must come in line with the webs of the crank, as they form dummy centres by means of which the crank-pin is finished true with the shaft. The space between the webs may be cut out with a parting tool to a width of $1\frac{7}{8}$ in. A hole $\frac{1}{8}$ in.

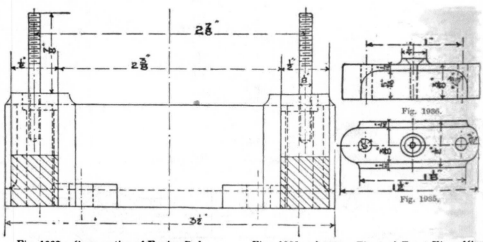

Fig. 1932.—Cross-section of Engine Bed.

Figs. 1935 and 1936.—Plan and Front View of Cap for Brasses.

it to a length of $6\frac{1}{8}$ in. ; then drill a centre at each end to be left in. Turn each end of the shaft to $\frac{9}{16}$ in. diameter for a distance of $\frac{3}{4}$ in. from each end. Two pieces of metal must now be found, not less than

diameter is drilled through the webs and pin, and a steel rivet driven through and riveted over at each end. When the crank-pin is finished, the end-plates may be taken off and the shaft turned to the dimensions given. Fig. 1940 shows the key-way, which may be cut either with a file or with a traverse drill.

GOVERNOR PULLEY.

The cast-iron governor pulley should be polished bright all over, and must be a

Fig. 1933.

Fig. 1934.

Figs. 1933 and 1934.—Front and Side Views of Brasses for Crank-shaft.

Fig. 1937.—Side View of Cap for Brasses.

$\frac{5}{8}$ in. by $1\frac{1}{4}$ in. by $2\frac{1}{4}$ in. ; two centres must be made on each of these $1\frac{5}{8}$ in. apart. Now drill a $\frac{9}{16}$-in. hole through each block,

driving fit on the crank-shaft, a radius of $\frac{1}{8}$ in. being turned on the hole to allow the boss to lie close to the web.

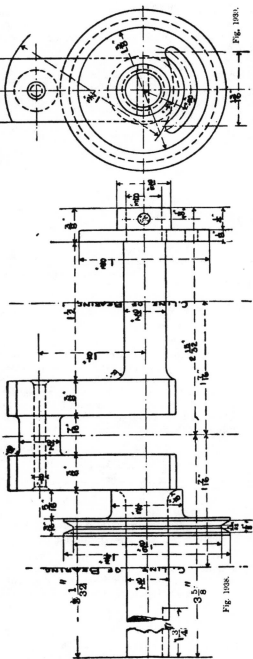

CATCH-PLATE FOR REVERSING.

The catch-plate, by means of which the engine is reversed, is turned of steel. A slot ⅛ in. wide should be cut at a radius of ½ in. from the centre, the centres of the ends being 1⅜ in. apart, as shown in Fig. 1939. This may be marked on the plate first, and then several ⅛-in. holes drilled; and the remaining metal chipped and filed out. The plate must be secured to the crank-shaft with a taper pin in the position shown.

CONNECTING-ROD AND BRASSES.

The connecting-rod and brasses are shown at Figs. 1941 to 1945. The rod may be made from steel bar 1⅜ in. by ⅜ in. by 5⅛ in. long. Dress both ends of the bar square, then mark out a hole at each end and 4¼ in. apart, one ⅜ in. diameter, the other ⅝ in. diameter; the centre of the

Fig. 1940.—Section of Engine Crank-shaft showing Keyway.

smaller hole should be 1⁄16 in. from the end of the bar. Drill these holes to the sizes given, and mark out two 1⁄16-in. holes, for the bolts which hold the brasses, at the big end of the connecting-rod. Drill these holes to a depth of 1⅛ in. from the end of the bar. To ensure these holes being true, cramp the bar on the top slide, and feed the bar towards the drill revolving in the lathe-head. The bar should be centred at both ends, and as much metal turned off as possible, leaving enough to clean up to dimensions, the remainder being filed off. A strip 3⁄16 in. thick must be cut off the wide end of the bar, after which this end may be filed until it is 1⁄16 in. from the centre of the large hole. Two saw-cuts must now be taken from the end of the bar to each side of the ⅝-in. hole; the jaw may then be filed out, and the back brass (Figs. 1943 and 1944) fitted in. The latter is made from a casting, the body of the brass being cast right round to enable it to be turned on a mandrel. Drill a ¼-in. hole through it, and, on a mandrel, turn down between the flanges to ⅜ in. diameter, the width

brasses in place may now be filed from the strip cut off the big end. This should be held in place by four nuts, two on each bolt (see Fig. 1942).

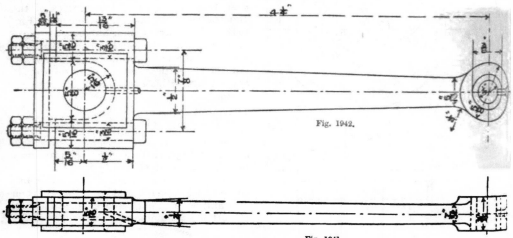

Fig. 1941.

Figs. 1941 and 1942.—Plan and Elevation of Connecting-rod.

between being ₁⁵₆ in. Next face the outside. The brass may now be cut, and one half fitted into the big end of the connecting-rod. The bottom brass must now be filed from the casting, and fitted to the dimensions given in Figs. 1943 and 1944. Both brasses must now be taken from the connecting-rod; after they have been cramped together, one flange of each brass may be fastened to the face-plate of the

CONNECTING-ROD BUSH.

The bush for the small end may be turned from hard steel ₁⁷₆ in. diameter: first bore the hole, and, on a mandrel, turn the outside to fit tight in the hole bored in the connecting-rod. Harden this bush and polish it again, and drive it in place, using wood to protect it from the hammer face.

Fig. 1945.—Bolt for Connecting-rod.

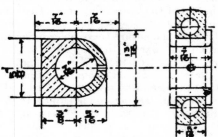

Figs. 1943 and 1944.—Connecting-rod Brasses.

lathe and bored to ₁⁷₆ in. diameter. They may then be put back in place. The bolts (Fig. 1945), turned from ₁⁷₆-in. soft steel rod, must be a driving fit in the holes in the big end. The strap which keeps the

ECCENTRIC STRAP.

Figs. 1946 and 1947 show the eccentric strap and rod. The casting for the strap is solid to ensure soundness. Cramp this on the face-plate and bore a hole about ¾ in. diameter through; then, on a mandrel, face each side to the required thickness. The holes for the ₅₃₂-in. bolts may next be marked off and drilled, the strap being meanwhile cramped on an angle-plate. The strap should next be sawn in halves and filed roughly to shape on the outside, that the bolts and nuts may be put in place. Then cramp the strap on the face-plate, setting it true by the centres of the bolts one way, and by the division the other. Then bore it to $1\frac{3}{16}$ in., and make a recess $\frac{1}{16}$ in. wide in the centre to $1\frac{1}{4}$ in. diameter. Then cramp the strap on the angle-plate, so that the neck into which the eccentric rod screws may be turned, drilled, and tapped. Now polish the outside of the strap. The nuts on the bolts must not be more than $\frac{1}{4}$ in. across the angles.

ECCENTRIC ROD.

The eccentric rod (Figs. 1946 and 1947) may be turned to dimensions from soft steel rod $\frac{3}{8}$ in. diameter by $5\frac{3}{8}$ in. long, the ends being threaded as shown. The knuckle-piece at the top of the rod may be filed from steel rod, $\frac{1}{2}$ in. by $\frac{5}{16}$ in., to

of the metal; then, on a mandrel, face each side until reduced to $\frac{5}{16}$ in. thick. Next turn the outside diameter to $1\frac{1}{4}$ in., and then reduce each side, for a distance of $\frac{1}{8}$ in. from the edge, to $1\frac{3}{16}$ in. diameter, leaving a projection $\frac{1}{16}$ in. wide to fit the

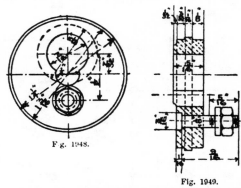

Fig. 1948.

Fig. 1949.

Figs. 1948 and 1949.—Front View and Section of Eccentric Sheaf.

recess in the eccentric strap. Now mark out two holes, both on the same centre line, one being $\frac{7}{16}$ in. diameter and $\frac{7}{32}$ in.

Fig. 1947.

Fig. 1946.

Figs. 1946 and 1947.—Plan and Elevation of Eccentric Strap and Rod.

the dimensions given in Figs. 1946 and 1947. Figs. 1948 and 1949 show the steel eccentric sheaf. For this a punching, $\frac{3}{8}$ in. thick by $1\frac{3}{8}$ in. diameter, will do very well. First drill a $\frac{1}{4}$-in. hole through the centre

from the centre of the sheaf, and the other $\frac{1}{8}$ in. diameter and $\frac{1}{2}$ in. from the centre of the hole just marked out. Cramp the sheaf on the face-plate, and bore both these holes, the smaller one being opened

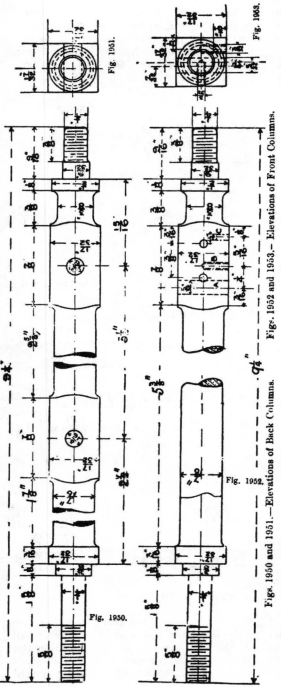

Fig. 1951.

Fig. 1953.

Figs. 1952 and 1953.— Elevations of Front Columns.

Fig. 1952.

Figs. 1950 and 1951.— Elevations of Back Columns.

Fig. 1950.

to $\frac{1}{16}$ in. diameter to a depth of $\frac{3}{32}$ in. to take the head of the bolt, as shown in Fig. 1949. Next, with a mandrel through the larger hole, face $\frac{1}{32}$ in. off the sheaf (from the side on which the head of the bolt will come) to $\frac{5}{8}$ in. in diameter. The

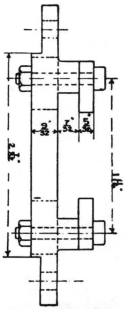

Fig. 1954.—Side View of Slide Bracket.

bolt and nut may now be made, a small washer, $\frac{5}{16}$ in. diameter by $\frac{1}{16}$ in. thick being placed at the back of the nut, as this tightens on to the catch-plate.

SLIDE BRACKET COLUMNS.

Two columns, which carry the slide bracket, are shown by Figs. 1950 and 1951. Procure some soft steel bar, $\frac{15}{16}$ in. square and cut off two lengths $9\frac{1}{4}$ in. long ; centre these, and turn them to the dimensions given in Fig. 1950. Now reduce the two squares left to $\frac{17}{32}$ in. each way, taking an equal amount off each side. Next mark off and drill two $\frac{3}{16}$-in. holes in each column, to the centres given in Fig. 1950 for the bolts that hold the slide bracket. The two front columns (Figs. 1952 and 195' may now be turned from two similar bars

Two holes, $\frac{3}{32}$ in. diameter and $\frac{9}{32}$ in. deep, must be drilled on the centre line of each column to the centres given in Fig. 1952, and tapped for the screws which fasten the governor spindle bracket. The holes marked A, B, C (Fig. 1952) must be drilled in one column only. Use three $\frac{1}{4}$-in. nuts for each of the four columns, one, $\frac{3}{16}$ in.

which the slipper works being surfaced. Eight $\frac{1}{8}$-in. holes may be drilled to the centres given in Fig. 1955, and also through the strips made from steel bar $\frac{3}{16}$ in. by $\frac{3}{8}$ in. wide. These strips should be cramped in position, and both bracket and strip drilled together, so that the holes may be in line. Eight $\frac{1}{8}$-in. bolts may now

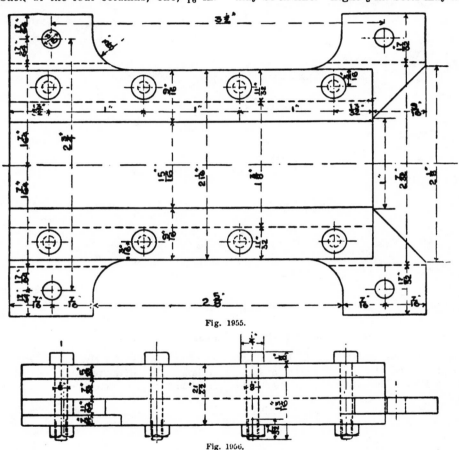

Fig. 1955.

Fig. 1956.

Figs. 1955 and 1956.—Plan and Front View of Slide Bracket.

thick, being used as a lock nut at the bottom.

THE SLIDE BRACKET.

The slide bracket may now be filed up from an iron casting to the dimensions given in Figs. 1954 to 1956, the face on

be turned ; these, if made a tight fit in the strips, may have round heads. Four $\frac{3}{16}$-in. holes must be drilled to correspond with the holes in the two back columns (Fig. 1950). Four bolts and nuts may now be made, $\frac{3}{16}$ in. diameter and $1\frac{1}{8}$ in. long, with hexagon heads, for bolting the bracket to the columns.

GOVERNOR FOR ENGINE.

A plan and elevation of the governor fixed in position on the columns, with (Figs. 1961 and 1962) use steel bar $\frac{1}{8}$ in. by $\frac{1}{2}$ in. by $1\frac{1}{8}$ in. long, two notches being cut, into which the governor arms are

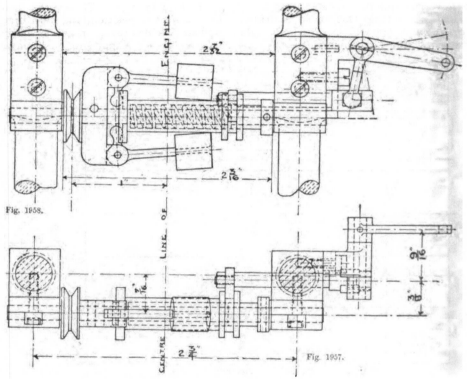

Figs. 1957 and 1958.—Plan and Elevation of Engine Governor.

levers attached, are shown in Figs. 1957 and 1958, and the detail parts in Figs. 1959 to 1983. The spindle (Figs. 1959 and fixed by means of a rivet $\frac{1}{16}$ in. diameter. A hole $\frac{5}{32}$ in. diameter may be drilled at right angles to the rivet holes through the

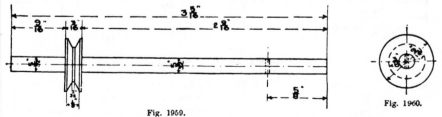

Fig. 1959.

Figs. 1959 and 1960.—Front and Side Elevations of Governor Spindle.

1960) may be turned from soft steel rod $1\frac{1}{6}$ in. diameter, the pulley being turned solid with the shaft. For the crosshead centre of the crosshead. This must be fixed on the spindle against the pulley by a $\frac{3}{32}$-in. pin, which cuts slightly into the spindle. The arms may be filed from sheet steel $\frac{1}{8}$ in. thick to the dimensions given in Figs. 1963 and 1964, the ends being passed through the balls and riveted over ; these may now be riveted in position, the

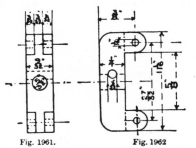

Fig. 1961.   Fig. 1962

Figs. 1961 and 1962.—Front and Side Elevations of Governor Crosshead.

arms being free to move in an outward direction. A small collar (Figs. 1965 and 1966) should be pushed along the spindle until the centre of the $\frac{1}{16}$-in. hole is $\frac{3}{8}$ in.

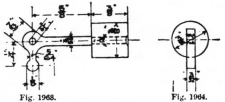

Fig. 1963.   Fig. 1964.

Figs. 1963 and 1964.—Front and Side Elevations of Governor Arms and Balls.

from the end. The hole must now be drilled through the spindle, and a small taper pin fitted, thus securing the collar in place. Next, from a piece of rod $\frac{3}{8}$ in.

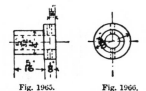

Fig. 1965.   Fig. 1966.

Figs. 1965 and 1966.—Front and Side Elevations of Collar for Governor Spindle.

diameter by $1\frac{3}{8}$ in. long, drill and turn on a mandrel the steel sleeve (Figs. 1967 and 1968). A groove $\frac{3}{32}$ in. wide by $\frac{3}{64}$ in. deep should be cut across the end. Remove the

small collar from the spindle, and push the sleeve towards the crosshead until the governor arms fit in the groove. A fairly stiff spiral spring must now be placed in the space between the sleeve and spindle,

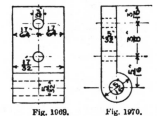

Fig. 1967.   Fig. 1968.

Figs. 1967 and 1968.—Front and Side Elevations of Sleeve for Governor Spindle.

the collar being replaced, and the taper pin driven in as shown in Fig. 1958.

GOVERNOR BEARINGS AND MOTION WORK.

The two cast-iron bearings (Figs. 1969 and 1970) must be filed up bright to dimen

Fig. 1969.   Fig. 1970.

Figs. 1969 and 1970.—Front and Side Elevations of Bearing Brackets for Governors.

sions, and holes drilled for the screws, which fasten these to the columns. Four $\frac{3}{32}$-in. cheese-headed screws will be re-

Fig. 1971.   Fig. 1972.

Figs. 1971 and 1972.—Front Elevation and Section of Motion Shaft from Governors.

quired, $\frac{3}{8}$ in. long, with heads $\frac{3}{16}$ in. by $\frac{1}{8}$ in. deep ; holes $\frac{3}{32}$ in. diameter must also be drilled to form bearings for the governor spindle. The spindle and fork by which the motion is transmitted from the sleeve on the governor spindle to the levers on the outside of the column are shown in Figs. 1971 to 1973. The fork is cut from sheet brass $\frac{1}{8}$ in. thick with a sharp chisel, filed to lines previously scribed on the brass, the small hole being drilled and tapped to $\frac{1}{32}$ in. diameter. The steel spindle must be turned to the dimensions given in Figs. 1971 and 1972, a notch $\frac{3}{16}$ in. long being filed on each side. Fig. 1972 shows how the metal is to be cut away. In these notches works the lower end of the lever (Figs. 1974 and 1975). The brass fork should be screwed on and secured by a nut $\frac{3}{32}$ in. across the flats and $\frac{3}{32}$ in. thick.

MOTION LEVER.

The lever (Figs. 1974 and 1975) may be made from steel rod $\frac{5}{16}$ in. by $\frac{3}{8}$ in. by $\frac{7}{8}$ in. long. Drill a $\frac{1}{8}$-in. hole through $\frac{3}{32}$ in. from

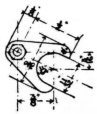

Fig. 1973.—Side Elevation of Motion Shaft from Governors.

the end, and entering from the $\frac{5}{16}$ in. side of the metal. Then mount this on a mandrel, and remove as much metal as possible in the lathe, turning up the boss and facing the sides; it must be finished to dimensions with a file, and is fastened on the shaft by a pin $\frac{1}{32}$ in. diameter, passing through the boss and spindle. The latter must not be drilled until the lever is in place. Figs. 1976 and 1977 show a lever and the spindle just mentioned. This lever may be made from steel rod $\frac{5}{16}$ in. square and $1\frac{1}{8}$ in. long, and may be drilled and turned to dimensions. Both levers may now be fixed on the shaft, the distance between the inside of the bosses being $\frac{3}{8}$ in.

BEARING BRACKETS FOR MOTION SPINDLE

The casting for the iron bracket (Figs 1978 and 1979) may now be finished. First drill the $\frac{1}{8}$-in. hole in which the spindle works; then, on a mandrel, face each side and turn the bosses. Then drill the $\frac{1}{32}$-in

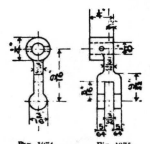

Fig. 1974.   Fig. 1975.

Figs. 1974 and 1975.—Motion Levers and Spindle

hole, to correspond to that marked B in Fig. 1952 (p. 638), the bracket being attached to the column by a screw $\frac{1}{32}$ in. diameter by $\frac{3}{16}$ in. long. Next, from the back of the bracket, on the centre line of the bearing, drill a $\frac{1}{16}$-in. hole, and fit a steady peg, as shown. A semicircular recess is filed near the bottom of the bracket (Fig. 1979) to clear the spindle shown in Fig. 1971. Two links (Figs. 1980 and 1981) must be cut from sheet steel $\frac{1}{16}$ in. thick, and filed, drilled, and tapped as shown. Two $\frac{3}{32}$-in. screws should be turned, $\frac{3}{16}$ in. long over all, being reduced

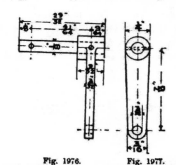

Fig. 1976.   Fig. 1977.

Figs. 1976 and 1977.—Motion Levers and Spindle

to $\frac{1}{16}$ in. diameter for a distance of $\frac{1}{16}$ in. from the end, and a thread cut upon it: the heads should be $\frac{3}{32}$ in. diameter by $\frac{3}{32}$ in. deep.

THROTTLE VALVE LEVER.

The throttle valve lever shown in Figs. 982 and 1983 is made from steel rod $\frac{3}{16}$ in. square, and must be finished with a file, as a mandrel in the hole in the boss would be too small. A set-screw, $\frac{1}{16}$ in. diameter and $\frac{1}{8}$ in. long, fixes the lever is position on the valve spindle.

FLYWHEEL.

The cast-iron fly-wheel is shown in Figs. 1903 and 1904 (pp. 624 and 625), which also show the engine complete. The flywheel must be turned on the outside, the sides of the rim being recessed and faced, and a hole $\frac{7}{16}$ in. diameter bored through the centre. It may be necessary to raise the centres of the lathe to deal with the flywheel. A square key-way being rather difficult to file, it will be easier to plug the hole with soft steel rod, and to drill a hole

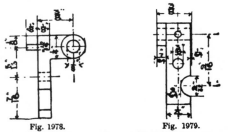

Fig. 1978. Fig. 1979.

Figs. 1978 and 1979.—Front and Side Elevations of Bearing Brackets for Motion Spindle.

$\frac{5}{16}$ in. diameter, with the centre running between the inside of the boss and the outside of the plug, right through. Now drive out the plug, and a groove $\frac{1}{8}$ in. wide and $\frac{1}{16}$ in. deep will be left ; a key must be fitted, half of which lies in the key-way, cut in the crank-shaft, while the top side must be rounded off to fit the groove in the wheel.

ERECTING THE ENGINE.

To erect the model, begin from the engine bed. The crank-shaft should be placed in the bearings, and the brasses and caps put on. Now take off the catch-plate at the end of the crank-shaft, and slip on the eccentric sheaf, having first

taken the nut and washer off the bolt which runs through it. Next replace the catch-plate, allowing the bolt to come through the slot, and screw on the washer and nut. The two back columns should now be put in place, and the nuts on the bottom of each tightened up. Now bolt the slide

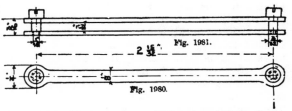

Fig. 1981.

Fig. 1980.

Figs. 1980 and 1981.—Plan and Elevation of Motion Links.

bracket to the columns, after which the front columns may be fastened in place from the under side. The joint of each cylinder cover being made, and also that between the steam-chest and steam port face, the cylinder, with piston, piston rod, gland, crosshead, and slipper, and the valve, valve rod, and gland, may be bolted on the columns, the slipper and crosshead being passed down between the guide bracket and strips. The connecting-rod should next be coupled to the crosshead pin at one end, and to the crank pin at the other. Now bolt the strap round the eccentric sheaf, and couple the other end

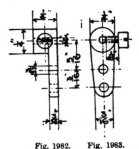

Fig. 1982. Fig. 1983.

Figs. 1982 and 1983.—Front and Side Elevations of Throttle Valve Lever.

to the crosshead of the valve rod. The valve must now be set : first bring the crank to the top centre, then unscrew the bolt holding the eccentric sheaf to the catch-plate, and push round to the end of the slot towards the front of the engine ; now set the valve on the spindle, so that the top port is open $\frac{1}{64}$ in. Next turn the crank in the direction of the arrow (Fig. 1904, p. 625) until it reaches the bottom centre ; there should now be an equal amount of opening to the bottom port. If this is less than $\frac{1}{64}$ in., the nut on the catch-plate must be unscrewed and the bolt pushed a little further from the crank ; if, on the other hand, it is found that the lead is more than $\frac{1}{64}$ in., the slot in the catch-plate must be lengthened to allow the sheaf to move a little nearer the crank. By throwing the bolt round to the other end of the slot, the engine is reversed. The steam-chest cover may now be bolted on, and the small lever (Figs. 1982 and 1983) placed on the throttle valve spindle. The shaft and fork (Figs. 1971 and 1973) can now be placed in position, and the bracket and levers (Figs. 1974 to 1979) fastened to the column. Now couple the

two levers shown in Figs. 1977 and 1983 by the links (Figs. 1980 and 1981). The governors, with the bearings, may now be fixed on the front of the columns. The motion should be adjusted so that the throttle valve is closed when the governor balls are fully opened. These may be driven by catgut or whipcord from the pulley on the crank shaft. The flywheel may now be keyed on.

FINISHING THE ENGINE.

To finish the engine, give it a coat of enamel of a good drying nature, which will not blister with the heat. The lagging round the cylinder, and the steam-chest and cover, should be coloured green, as also the engine bed ; the flywheel may be red. The engine should be mounted on a wooden foundation, and, when the enamel has had time to dry thoroughly, the glands should be packed, and, after well oiling, the engine may be tried under steam.

BOILER BUILDING.

Model Horizontal Marine Boiler.

Three types of boiler will be described in this chapter. The first is a model horizontal marine boiler, for which a suitable oil burner will also be shown. The second is a small power vertical boiler, and the third is another vertical boiler, but of 8 h.p. A start will be made with the small horizontal boiler suitable for a model steamship. Fig 1984 shows a longitudinal section the back end of the boiler as much heat as possible. The fuel to be used by the two lamps for a boiler of this design is benzoline under pressure. The temperature obtained is between 2,000° and 2,200° F., the flame being expelled from the nozzle of the lamp with some force, and it is thus not advisable to interrupt the course of this flame by using a bridge or other form of baffle, which might make the flame "flash back," causing the lamp to

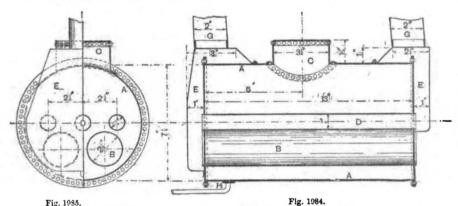

Fig. 1985.   Fig. 1984.

Figs. 1984 and 1985.—Sections of Model Horizontal Marine Boiler.

of the boiler, Fig. 1985 a half-sectional view of the back or after end, and Fig. 986 a full elevation of the front or fore end. These three illustrations are one-sixth full size. The boiler has two furnace tubes, both of which extend from the fore end to the after end without interruption. As the boiler is to be heated by means of a blow-lamp, it is advisable that the flame should not meet with any obstacle for the full length of the furnace tube, in order to ensure better combustion and to give explode. By using two funnels, there is an easier exit for the hot air caused by the forced draught; only a part of the hot air will go up the after funnel (see Fig. 1987), whilst a like quantity will return along the smoke tubes and up the funnel leading from the smoke box at the fore end. The shell A and furnace tubes B (Figs. 1984 and 1985) are of solid drawn brass, or preferably of No. 16 B.W.G. copper tube. The furnace tubes, however, may be a little thinner—say No. 17 B.W.G., or ˙056 in. ; while the shell, end, and dome c are ˙064 in. in thickness. The processes of brazing and flanging must be thoroughly performed, as the boiler is intended to carry a pressure of 45 lb. per sq. in. The capacity of the boiler should be four or five times as much as the water it evaporates per hour, and the steam space should be at least ten times as large as the consumption of steam at each stroke. The three illustrations of the boiler (Figs. 1984 to 1986) are, as will be seen, dimensioned in the main parts. Any other dimensions may easily be got by measuring the illustrations, which, as has been said, are drawn to the scale of 2 in. to the foot, or one-sixth full size.

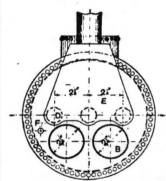

Fig. 1986.— Front Elevation of Model Horizontal Marine Boiler.

END PLATE FLANGES, ETC.

The extreme length of the boiler, measuring from outside of flanges before the end plates are riveted on, is 13 in., and the diameter inside is 7½ in. As the end plates are to be riveted to the ends of the boiler, allowance must be made for the two flanges when obtaining the material for the shell. The rivets should be ⅛ in. in diameter ; therefore, the width of the flange will require to be a little more than three times the diameter of the rivet ; 1⁷⁄₈ in. must be allowed at each end for flanging, or altogether a tube 13⅞ in. long by 7½ in. inside diameter is required to form this part of the boiler shell.

FURNACE TUBES AND SMOKE TUBES

The furnace tubes will be 13½ in. long by 2⅛ in. inside diameter. This extra length is allowed so that the ends may stand " proud " of the plate in order that they may be slightly expanded in their place and afterwards brazed with silver solder to withstand the high pressure. This a ˙ applies to the smoke or return tubes ɒ which are to be ⅞ in. inside diameter.

THE DOME.

The dome c (Fig. 1984) is to fit accurately on the top of the boiler. It is of 3¼-in. diameter inside, and when finished measures 1½ in. from the top of the boiler. This is not the length of the tubes before the dome is made from it, but when flanged and fitted to its place. Therefore it will be necessary to grip a 4-in. length of tube in the vice, and to file with a half-round file until it fits the top of the boiler, slipping it temporarily upon a wooden mandrel, in order that it may not be flattened whilst gripped in the vice. Flange the dome in the same manner as the ends of the boiler, allowing the same breadth of flange. The holes for the rivets may then be pitched off and drilled, and the dome placed in position as shown in Fig. 1984. Having ascertained that the centre line of the dome is in line with the fore and aft centre line of the boiler, mark off two holes on each side of the dome for drilling the rivet holes in the boiler These holes must be marked out with a steel scriber through the holes already drilled in the dome. The rivets in every part of this boiler should be ⅛ in. in diameter, and they should be pitched at a distance of ⅜ in. from centre to centre of rivet. Whilst the drilling is proceeding round the flange of the dome, bolts and nuts may be used to secure the dome in its place temporarily ; they also help to get the holes in accurate relation to each other, an important factor in boiler-making. The holes for the cover of the dome may be formed either before it is placed in position for drilling the holes in the flange, or after, as it is obvious that the dome itself is not to be riveted on and razed before its cover is made a permanent fixture.

BUILDING THE MODEL HORIZONTAL BOILER.

The boiler is built up as follows:—After he shell of the boiler has been flanged, cut he hole for the dome; next build up the lome, and rivet and solder it with silver older firmly to the boiler shell, using a lowpipe. Next, fix on either of the ioiler ends, taking care that the centre ine, running from side to side of the boiler .nd through both furnace tubes, stands .t right angles with the vertical centre line unning down through the dome. Both urnace tubes and smoke tubes may be .ntered into their respective places in the late or boiler end thus fixed. The opposite boiler end may then be placed in

the water-line of the boiler, and fixed by means of $\frac{1}{8}$-in. studs and nuts above (see Fig. 1984). The holes for these studs, after the latter have been screwed into the boiler, should be caulked to render them steam-tight, and if found not so when the boiler is tested under steam, they should be sweated in with silver solder, using the blow-pipe.

BOILER MOUNTINGS.

The boiler-mountings — including the force-pump for feeding (which must be worked by hand when fresh water is required for the boiler), safety-valve, water-gauge, and steam-tap or stop-valve—are bought more cheaply than they can be made. The spring safety-valve should blow off at a pressure of 45 lb. per sq. in. It is, of course, only approximately shown

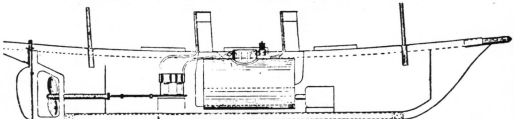

Fig. 1987.—Section of Model Steamship, showing Boiler.

position and riveted up. Great care must be exercised in getting the last end plate of the boiler in position; some difficulty will probably be experienced in entering all five tubes into their holes in the plates. The furnace and smoke tubes may now be tapped through the two boiler ends until their ends stand "proud" to an equal distance, after which they may be expanded and brazed in. This latter operation is not, of course, to be done until it is ascertained that all rivets are sound and the ends are sweated up. The two outside combustion chambers, made of ordinary tin-plate for lightness, are shown at E (Figs. 1984 to 1986), and require little explanation further than that given in the drawings. The combustion chamber, properly speaking, is at the back end of the boiler, whilst the chamber at the front end is the smoke-box; both are soldered below

in Fig. 1987, which is a longitudinal section of a model steamer fitted with the boiler; it may lie in any convenient position on the dome, and, for neatness of appearance, should blow off its waste steam into the forward waste steam pipe. The boiler should have a feed-check or back-pressure valve at F (Fig. 1986), and the pump should be fitted with suction and delivery-valve. The pipe below the suction-valve has open connection with the water outside. The funnels G (Fig. 1984) should be 2 in. in diameter, and made of hard rolled sheet brass ·02 in. thick; they may be either round or oval. The bottom blow-off cock should be soldered to the ship's side, and the pipe H shown at the bottom of the boiler led to it. All tubing for steam, exhaust, and blow, as well as for feed and waste-steam pipes, should be of $\frac{1}{16}$ in. bore.

PETROLEUM RESERVOIR FOR BURNER.

Oil burners for heating boilers are usually fed with oil under pressure obtained by compressing air on top of the oil in the reservoir. For a marine boiler, the most suitable shape for the reservoir would be a cylinder resting horizontally in a cradle. The general arrangement of reservoir and fittings, with the burner in position, is shown by Fig. 1988. The reservoir may be of tinplate, copper, or brass, and should be strong enough to withstand a pressure of 60 lb. per square inch. If the first-named metal is used, 6 **X** plate will be sufficiently stout; if copper or brass is used, metal weighing 1½ lb. to the square foot would answer the purpose. A

round the two seams. The air pump and supply tube are next securely soldered in position.

THE BURNER.

The burner is made preferably of ½-in. brass tube. It is formed by joining four pieces of curved tube together, each tube being joined at right angles with its neighbours. The top ends are mitred to form a conical top, and the lower ends are joined to the threaded cap attached to the supply tube. Fig. 1989 shows the supply tubes κ in position, and Fig. 1990 the pair of tubes J which convey the gas to the nipple L. The top of the threaded cap H (Fig. 1990) has five holes drilled in it as shown by Fig. 1991. Inside the cap a brass disc

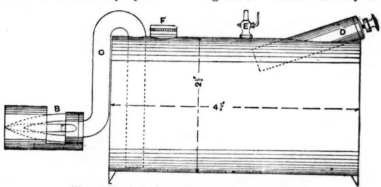

Fig. 1988.— Oil Burner for Model Marine Boiler.

reservoir of the size shown would hold rather more than half a pint of oil, and would, without replenishing, supply the burner for from two to three and a half hours, according to the pressure at which the burner was fed. The pattern for this reservoir would be a rectangle 7½ in. by 4¾ in., with holes cut for the pump D, screw cap F, blow-off cock E, and supply tube C. Turn the cylinder and groove the seam, and then well float it with solder along the inside at the seam; the filler cap and blow-off cock are placed in position and strongly soldered from the inside. Throw off a small edge at each end of the cylinder, take up an edge on the two circles forming the ends, and pane down this edge over that on the body, and then solder

(Fig. 1992), with two holes drilled, is brazed, the position of the holes being directly below either pair of opposite holes in the cap. The nipple I (Fig. 1990) is drilled so that No. 29 wire will just pass through the hole o; the nipple may be either screwed or brazed in the cap. Fit the pair of tubes J together so that the top ends mitre, and the lower ends pass through opposite holes in the cap; next file a semicircular opening on each side of the tube as shown. Bend the tubes κ (Fig. 1989) and file them at the top ends so that they fit the tubes J. The lower ends of the tubes κ pass through the top of the cap and also through the holes in the inner disc. Bind the four tubes securely in position with wire, braze them together at the top end, and then at the top of the cap. Next braze the ends of the tubes K to the inner disc. The collar G (Fig. 1990) is brazed to the end of the supply tube. Fig. 1993 shows a combined cylinder and starting trough, which should be made of

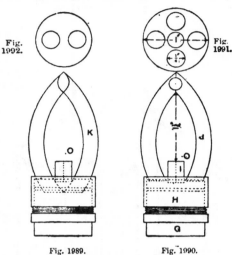

Fig. 1989.—Vaporising Tubes of Burner. Fig. 1990.—Supply Tubes of Burner. Fig. 1991.—Top of Burner Cap. Fig. 1992.—Inner Plate of Burner.

stout sheet brass, or may be cut from a length of brass tube, the lower part being removed as shown. A brass collar plate is next cut, half the shape being shown by the hatched part N P (Fig. 1993). A second plate is also cut of a shape found by producing the line P across the complete circle. This piece is brazed in the position shown by the dotted line at Q, the complete collar plate being brazed to the part of the cylinder touching the ground line, so that the point P on each side of the disc is flush with the sides of the cut part of the cylinder. The space between the two plates forms a trough for the methylated spirit when the cylinder is placed horizontally.

MOUNTING THE BURNER.

When mounting the burner for use, place a thin asbestos washer on the collar G (Fig. 1990), then place the circular opening at N (Fig. 1993) in the end of the cylinder

plate over the collar thread so that it butts against the shoulder, with the trough part of the cylinder in the position shown in Fig. 1988; then place a second asbestos washer in position and screw the burner in tightly, thus securing the cylinder also in position.

MANAGEMENT OF BURNER.

When using the lamp, nearly fill the reservoir with paraffin oil through the cap F (Fig. 1988); screw this up tightly, and then fill the starting trough with methylated spirit; light this, and when it is nearly burnt out, give one or two strokes with the pump. The extra pressure of air will force the oil through the supply tube to the burner. The oil would then pass through the hot tubes K (Fig. 1989), be converted to the gaseous state, and pass in that condition down the tubes J (Fig. 1989) into the top of the cap H, and thence issue at the outlet O. The jet of gas would then strike the top parts of the tubes and become ignited, the heat generated by the

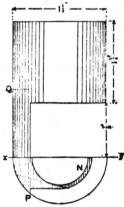

Fig. 1993.—Burner Cylinder and Starting Trough.

continued combustion of the gas keeping the tubes hot enough to vaporise the oil as fast as it is supplied to them. The cylinder (Fig. 1993) combines the flame and causes it to shoot forth in a compact body. When increased heat is required the pressure at which the oil is supplied to the burner must be increased by pumping more air into the reservoir. To decrease the supply of oil, or to extinguish the flame, let out air through the blow-off cock.

A ¼-HORSE-POWER VERTICAL BOILER.

The usual method of making a small power boiler is attended with many difficulties, requiring, as it does, tools and appliances not usually found outside a regular boiler-shop. The method about to be described does away, however, with all flanging of plates, is quite within the scope of the average mechanic, and at the same time makes a neat and thoroughly reliable and safe boiler. The measurements to be given are for a boiler powerful enough to drive either of the ¼-h.p. engines dealt with in previous chapters; but the method of construction is applicable to boilers from one horse-power down to mere models. The necessary materials are copper body, tubes, and brass rings. Fig. 1994 is a vertical section of the boiler, showing water space, brass rings, internal tubes, fire-box, etc. Fig. 1995 is a horizontal section on line A B, immediately above the fire-box crown, showing the water space round the fire-box, arrangement of seven tubes, division of rivets, etc. Fig. 1996 is an enlarged view of one inner tube end, fixed in crown plate with nuts on each side. Fig. 1997 is a full-size section of ring. In the section (Fig. 1994) B B show the tube forming the fire-box; C, firebox crown; D, steam-space crown; E E E, three of the inner tubes; F F, brass rings; G G, bottom brass rings; I, smoke-box crown; J, brass ring at base of chimney; and K, chimney. Every part must be made and fitted carefully and accurately, as no slipshod work can possibly be tolerated in a boiler that has to be subjected to a test pressure of 100 lb. per sq. in., and to work afterwards at a pressure of from 50 lb. to 60 lb. per sq. in. First of all make a full-size drawing, increasing or diminishing the dimensions as desired, but keeping the same proportions.

FIRE-BOX TUBE AND SHELL.

The fire-box tube B B and the shell are usually made from copper plates, the edges

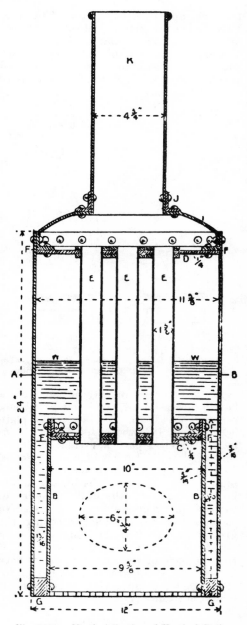

Fig. 1994.—Vertical Section of Vertical Boiler.

being joined by dovetailing them into each other, and then brazing; or the edges may be overlapped and riveted.

BOILER CROWN PLATES.

The crown plates C D are of copper. They must be fitted to drive in tightly into their respective tubes. The edges of these plates should be turned in a back-geared

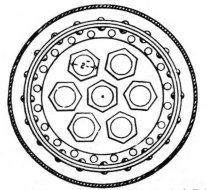

Fig. 1995.—Horizontal Section of Vertical Boiler.

6-in. lathe to be accurate. In each plate seven holes are cut for the tubes; find the exact centre of each plate, draw a 5½ in. circle, and with the same radius divide the circles into six equal parts; draw a 1⅛-in. circle for each tube; divide round the edge

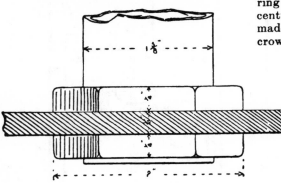

Fig. 1996.—Nut Fastening of Internal Tube of Boiler.

of each plate for the rivets (see Fig. 1995), and make the holes ¼ in. in diameter.

BRASS RINGS.

The brass rings F F are cast from patterns, and turned on their face and outer edge to exactly the size of the plates C D.

To bore them for rivets to fix to plates, bore two holes, one on each side of the ring, and bolt or rivet the plate to the ring flush all round; then bore the other holes in the ring through those already in the plate. Now rivet the rings to their plates all round. The bottom ring G is made to fill exactly the difference between the shell and fire-box tubes. For the furnace door opening, an oval ring set to the curvature of the boiler has to be made; it is the same thickness as ring G, and riveted all round.

BOILER TUBES.

The length of the seven tubes may be found by placing temporarily the fire-box and upper crown in their places; they should project through the plates ¼ in., to receive the outer nuts. About ⅞ in. of the tubes, both ends, are cut with fine screws for the nuts, which are brass castings to finish ¼ in. thick and 2 in. across the flats.

SMOKE-BOX CROWN.

The smoke-box crown is of copper, ¹⁄₁₆ in thick, hammered to shape, and fitted inside the shell, its edge resting on the edge of ring D. The chimney is erected on the centre of this crown, an angle ring being made for the chimney. The smoke-box crown, with chimney, is held sufficiently in

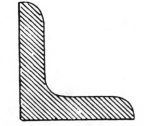

Fig. 1997.—Section of Angle Ring.

place by the fittings, like the safety-valve, which passes through the crown to the plate D. A brass ring, semicircular in section, is fitted on top of the chimney.

ERECTING THE BOILER.

The two rings C D have been riveted to their plates, also the seven tubes screwed, and nuts fitted; now coat the screwed parts of the tubes with a mixture of red-lead and oil, place a nut on one end plate c, which is the fire-box crown; coat seven other nuts on flats, and screw up tight on under side of plate c. Clean off

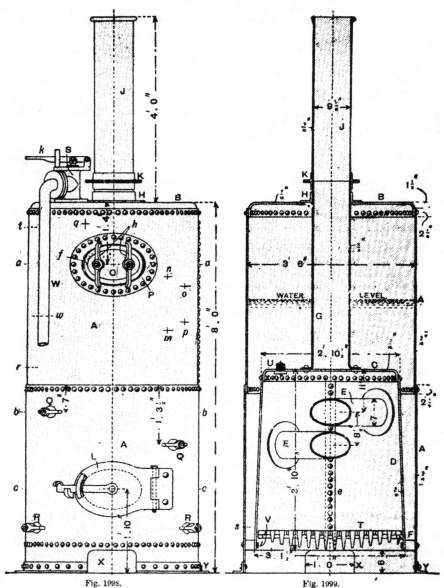

Fig. 1998.

Fig. 1999.

Figs. 1998 and 1999.—Elevation and Vertical Section of 8-h.p. Boiler.

of each, allowing ½ in. to project through, coat flats of nuts, and place tubes in the superfluous lead; place seven nuts on the upper ends of the tubes, turn them down the tubes till ½ in. projects, and see that they are all exactly on the same level, and that the tubes measure the proper

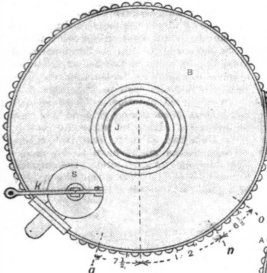

Fig. 2000.—Plan of Boiler.

length between C and D. Again coat the nuts and place the plate D in position on the tube; screw up the outer nuts tight, as before. Having cut the oval for the furnace door through both shell and fire-box, fit the oval ring by filing to shape. This ring is 1 in. broad and ½ in. thick. The proper way is to make a pattern after the hole is cut in the fire-box, and get the ring cast from it, and it should be fitted accurately to the fire-box tube. The ring G is ½ in. thick, which represents the water space around the fire-box. Having fitted the oval at the furnace door, lay it aside until the fire-box is riveted to the crown C; then rivet the oval ring all round. Place the fire-box, with tubes and crown D in position in the shell, insert ring G, and bore and rivet all round top and bottom. All rivets for body and fire-box are ¼ in. thick, with snap-heads, and the hammered ends are finished with a snap punch. The rivets at base of chimney are ₁₆ in. diameter. The rivets in all parts of the body are 1 in.

apart, centre to centre, and all joints before riveting are coated with red-lead, as before mentioned. The boiler should have a cast-iron base (not shown in the illustrations); on the top of this base, and fitting inside the fire-box, is a circular grating, on which the fire is placed. The fuel is charcoal or gas cinders. In Fig. 1994, w indicates the proper level of water boiler when charged for use.

AN 8-HORSE-POWER VERTICAL BOILER.

The boiler illustrated by Figs. 1998 and 1999 on a scale of ⅛ in. to the foot would be termed commercially an 8-horse-power boiler. Its shell, 3 ft. 6 in. diameter by

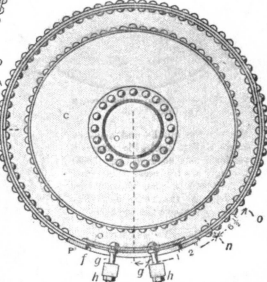

Fig. 2001.—Section of Boiler through Manhole an Uptake.

8 ft. high, and fire-box are made wholly of steel, as most boilers are now constructed. Certain fittings are made of wrought iron or steel, the choice being quite a matter of indifference. Important dimensions are given in the illustrations; minor ones may be measured by scale. Fig. 1998 is an external elevation of the boiler, with the positions and centres of its mountings indicated. Fig. 1999 is a vertical section, showing the shell, fire-box, uptake, etc.

Figs. 2000 to 2003 are on a scale of ¾ in. to
the foot. Fig. 2000 is a plan of the complete boiler. Fig. 2001 is a sectional plan

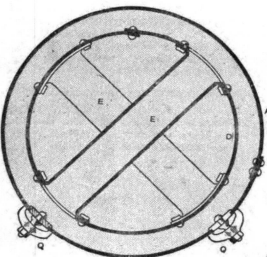

Fig. 2002.—Section of Boiler through
Top Cross Tubes.

through the uptake and manhole
on *a—a* (Fig. 1998). Fig. 2002 is a
sectional plan on *b—b* (Fig. 1998)
through the top cross tube ; and
Fig. 2003 is a plan through the
fire-hole in the plane *c—c* (Fig.
1998), just above the fire bars.

SHELL PLATES AND
SHELL CROWN.

The shell plates are $\frac{5}{16}$ in. thick,
single riveted circumferentially,
but with double riveting vertically. The shell crown B, ⅜ in.
thick, is also single riveted, as, too,
is the crown C of the fire-box, $\frac{7}{16}$ in.
thick, to its shell D. Fig. 2004 shows the
riveting of the shell crown enlarged, and
also serves to illustrate the fire-box crown.
Note the large radius given to the flanging
of the crown plates ; this is essential, to
provide elasticity to the plates, and so
lessen the risk of grooving. The ⅛-in or
$\frac{1}{16}$-in. bevel given to the edges is for caulking purposes. Fig. 2005 shows the zig-zag
riveting of the vertical shell seams.

RIVETED SEAMS.

Fig. 2006 illustrates the breaking of the
seams of the top and bottom rings of the
shell plates. This is highly important,
since many boilers have exploded in consequence of having the seams in line. The
amount by which they break joint is not
of importance, so long as the horizontal
seam that covers the vertical ones has an
interval occupied by its seam rivets. The
four rivets clear, seen in Fig. 2006, will
represent good practice. There is only
one vertical seam in each ring, as it is easy
to get steel plates of sufficient length.
With iron, two at least, and even three,
vertical seams might have to be made.
The thinning and spreading of the corners,
shown by *d d* (Fig. 2006) permit the making
of a good joint where three thicknesses of

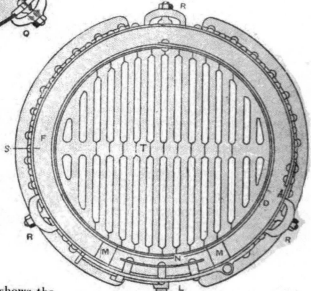

Fig. 2003.—Section of Boiler through Fire-hole
and Door.

plate cross. Some makers plane the bevel
instead of thinning down at the corners.

FIRE-BOX SHELL AND CROWN.

Returning to Fig. 1999, the fire-box shell
D is made $\frac{1}{8}$ in. thicker than the outer

shell A because of the extra wear and tear to which it is subjected. The crown c is $\frac{1}{16}$ in. thicker than the shell. Single riveting is used for the vertical seam e, lap-jointed like the shell. Fire-boxes of over

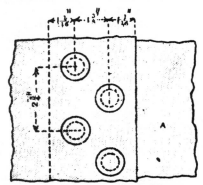

Fig. 2004.—Riveting of Crown Plates.

3 ft. diameter are double riveted, but still lap-jointed. Butt straps are liable to cause overheating at the joint, and welded joints, though the best on the theory of a perfect weld, are not regarded favourably in practice. The single-riveted lap-joint is safe because the fire-box is short and well stayed with two cross tubes.

HORIZONTAL CROSS TUBES.

The horizontal cross tubes E E seen in Fig. 2002 are each made by welding a plate into a tube and flanging the ends to a good radius subsequently. This requires much skill and care to secure perfect welds

Fig. 2005.—Riveting of Vertical Seams.

and exact fitting inside the fire-box. Many makers incline the tubes in the box to improve the circulation, but, as the inclination is slight, the trouble of fitting

without special appliances is hardly repaid. The riveting of these tubes to the shell is shown in detail in Fig. 2007; the rivet tails are conical.

UNITING SHELL TO FIRE-BOX AND UPTAKE.

The fire-box is attached to the shell through the medium of a foundation ring F, seen in Figs. 1999 and 2003, and enlarged in Fig. 2008. This is of steel or wrought iron of $2\frac{1}{4}$-in. square section, and it is riveted closely because of the tendency to leak. At the top in this boiler the two

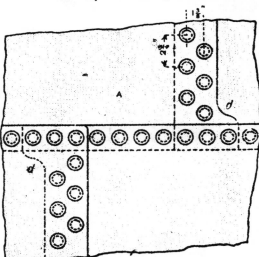

Fig. 2006.—Breaking Joints of Vertical Seams.

shells are united by the uptake G only. This is sufficient in boilers under 4 ft. diameter, but, above that dimension, vertical screwed rod stays are carried from the fire-box to the shell crowns. In boilers of 4 ft. 6 in. diameter and more, the fire-box and shell are further united with horizontal screwed stays passing through the water space. The uptake G is welded like the cross tubes, and the bottom end flanged with a good radius, as in Fig. 1999. The pitch of riveting is similar to that of the cross tubes (Fig. 2007). The upper end is fitted to the shell crown with an angle ring H, which is riveted both to the crown and to the uptake. The uptake is prolonged to receive a light angle ring, to which another ring K, encircling the chimney J, is bolted. In many cases the shell crown B is flanged upwards with a good radius to encircle the bottom of the up-

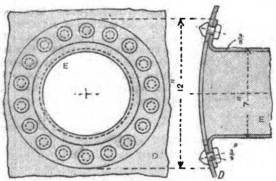

Fig. 2007.—Riveting of Cross Tubes.

take. Other methods also are in use. In connecting the two crowns by the uptake, the endeavour is to make an elastic union, so that the expansions due to rise in temperature shall not produce grooving of the crown plates. With the large radii imparted to the flanged crowns, and the radius at the flange of the uptake, sufficient elasticity is assured in the example illustrated. In many boilers the bottom of

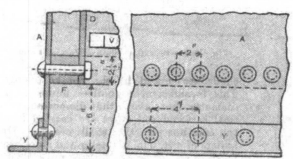

Fig. 2008.—Riveting of Foundation Tubes.

the fire-box shell D is flanged outwards to meet the outer shell A, and either the foundation ring F is omitted or a thinner one is used. But then sediment is more apt to collect than in the design shown ; and the

flanging, unless done with the greatest care, is liable to distress and to thin down the plate

UNITING SHELLS AT FIRE-DOOR.

Another point of connection is made between the two shells at the fire-door L

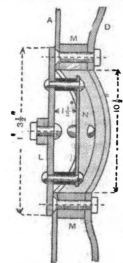

Fig. 2009.—Vertical Section through Fire-hole and Door.

(Fig. 1998), shown enlarged in vertical section in Fig. 2009. An elliptical ring M of steel or wrought iron, $2\frac{1}{4}$ in. square in cross-section, forms the fire-door opening, and the shell plates are riveted through this, the fire-box sheet D being dished outwards to meet it. The rivets are countersunk on the front to permit the door L to lie close to the boiler plate. In many boilers the fire-box is dished out to a greater extent and the ring M is thinner than in Fig. 2009. The fire-door L is hinged and latched, as seen in Fig. 2009. A baffle plate N (Fig. 2009), riveted to it, protects it from becoming buckled by the heat. The actual boiler is complete with the riveting-up and caulking of the various seams and joints.

BOILER MANHOLE DOOR.

The various fittings following after the fire-door will now be mentioned. Fitting into the elliptical hole above the water level is the manhole door o, shown in Fig. 1998, and in enlarged section in Figs. 2010 and 2011—Fig. 2010 showing a horizontal and Fig. 2011 a vertical section through door and shell. A strengthening ring P (Fig. 1998) is riveted round the hole to compensate for the metal cut away. The door o is a plate bent to the curvature of the shell, to which is riveted a light angle ring

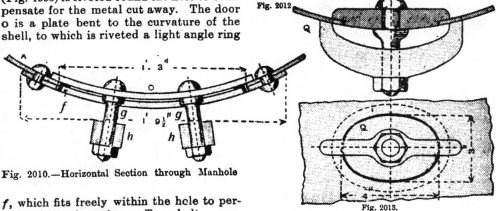

Fig. 2010.—Horizontal Section through Manhole

Figs. 2012 and 2013.—Section and Elevation of Mudhole and Door.

f, which fits freely within the hole to permit of its insertion. Two bolts g g, riveted to the plate, pass through bridges h h of wrought iron or steel, which are pulled against the ring encircling the manhole by screwing up the bolts. The joint is made steamtight with red-lead.

Fig. 2011.—Vertical Section through Manhole.

MUDHOLES AND DOORS.

Opposite one end of each cross tube there are mudholes and doors Q. These

are shown in Figs. 1998 and 2002, and in enlarged detail in Figs. 2012 and 2013. The holes are made by punching in the shell plate previous to bending and riveting. The doors are castings, shouldered into

the holes, and each is held in place by a bolt pulling against a wrought-iron or steel bridge, the joint being made steamtight with gasket and red-lead. Near the bottom of the boiler three similar mudholes and doors R R R are ranged equidistant in a circle. These are seen in Figs. 1998 and 2003, while Figs. 2012 and 2013 will serve as detail drawings. Through these holes the sediment from the bottom of the boiler is raked out, as that from the cross tubes is raked out through Q Q.

BOILER SAFETY-VALVE.

The safety-valve and its casing s, seen in Figs. 1998 and 2000, and in detail in Figs. 2014 and 2015, is of cast-iron, bolted with studs to the boiler crown. The valve j and its seating are of gun-metal, and the lever k receives the screw of a spring balance through a hole, the nut being tightened on the flat boss at the free end. A boss must be tapped into the shell at t (Fig. 1998) to receive the pin which connects to the barrel of the spring balance. A branch cast with the casing s receives a flange and pipe w, through which steam is taken from the boiler to the engine, the stop-valve being located about *w*.

FIRE-BARS.

The fire-bars T, of cast or wrought-iron, are seen in Figs. 1999 and 2003, and one of the central bars is shown in elevation in Fig. 2016. At the sides, two or three bars are cast in one. They rest upon a ring V of wrought-iron, about 1¼ in. square in section, seen in Figs. 1999 and 2008. The ring is supported upon three pins arranged

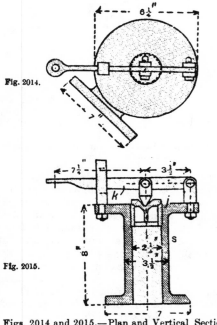

Fig. 2014.

Fig. 2015.

Figs. 2014 and 2015.—Plan and Vertical Section of Safety Valve.

equidistant, and tapped into the metal of the fire-box.

FUSIBLE PLUG.

Two, or even three, fusible plugs, like U (Fig. 1999) should be used on the fire-box crown. The plug is shown enlarged in Fig. 2017, and is a brass casting, screwed into the crown, and having the interior filled with lead, or an alloy of lead and tin,

which will melt if the crown becomes red-hot.

WATER GAUGES.

The positions of the water gauges and other mountings are indicated in Figs 1998, 2000 and 2001, but will be best under-

Fig. 2016.—Fire Bar.

stood by reference to the normal water level in Fig. 1999. The water-gauge centres *m n* (Fig. 1998) are so arranged that, with the water at normal height, the gauge will have about 5 in. or 6 in. height of water. The pet- or try-cocks *o p* are so placed that the lower one should, when opened, always blow off water, and the upper one always steam. If the lower one blows steam, the water has sunk too low. The position of the gauge and cocks in the horizontal direction is not important, but they should be near the front, as shown in Figs. 2000 and 2001.

PRESSURE GAUGE, ETC.

The pressure gauge *q* should also be at or near the front, and high up where the steam is driest. A combined feed and check-valve is placed at about the position *r* shown in Figs. 1998 and 2001, and a blow-off cock near the bottom *e* (Figs. 1999 and 2003). The openings x x, below the boiler at back and front, are used for raking out the ashes. The angle-iron ring Y, around the bottom of the shell, may be placed

Fig. 2017.—Section of Fusible Plug.

either internally or externally as shown. The boiler when new will work safely at 80 lb. per sq. in., but should be tested before with water to about 140 lb. per sq. in.

BUILDING A PETROL MOTOR.

INTRODUCTION.

THE petrol motor or engine to be described here is specially designed for use on a cycle, its main feature being simplicity. It is of the light class, for the engine, with a 14-lb. flywheel, should weigh about 25 lb., which may be reduced by using a lighter flywheel. Thus it will be specially adapted for fitting on to an ordinary roadster bicycle, whose front forks have been strengthened with tubular stays. The light class of motor has been selected in preference to the heavy type (which requires a specially built frame), as it will probably be more generally acceptable, being adaptable to any strongly built cycle, and thus enabling anyone to try the pleasures of motor cycling for a very moderate outlay. There is a growing tendency at present to increase the weight, and consequently the engine has to be heavy and of greater power, engines of $2\frac{1}{4}$ to $2\frac{3}{4}$ horse-power being common, with the weight of the whole machine complete from 100 lb. to 170 lb. The engine to be described will develop $1\frac{1}{2}$ horse-power, or, if very well made, should approach $1\frac{3}{4}$ horse-power, and this should give a speed of twenty-five to thirty miles an hour on the level, and should climb any ordinary hill without pedalling. The position chosen for the engine is well up on the bottom tube, in a vertical position, thus enabling a long belt to be used with a sound grip of the motor pulley. Although the motor is light and simple in construction, nothing has been sacrificed to efficiency, strength, or bearing surface. The main shaft bearing, for example, is $3\frac{5}{16}$ in. long by $\frac{3}{4}$ in. in diameter, giving long life and steadiness in running. To make the engine from the rough castings a screw-cutting lathe of $4\frac{1}{2}$-in. or 5-in. centre is essential, with the ordinary small tools of an engineer's workshop.

SPECIFICATION OF MOTOR.

One and half horse-power air-cooled vertical engine. (For those who fancy an inclined position for the engine, the lug which fastens it to the frame may be suitably fitted on the pattern.) Cylinder bore, $2\frac{1}{2}$ in. by $2\frac{1}{2}$-in. stroke ; outside flywheel ; aluminium crank case ; high tension electric ignition ; spray carburetter of simple design ; drive is by a **V**-belt on to the back wheel.

PATTERN FOR CRANK CASE.

Start on the crank case (Figs. 2018 and 2019). The pattern for this should be made with due allowance for shrinkage and machining. A core-box will be required, of rather a complicated pattern, to core the large interior, with depressions in the back of the case as shown for the two-to-one gear, the hole through for the main bearing bush, and the upper opening connecting the cylinder. The pattern may be made to leave its own core in the back of the case for the exhaust valve cam chamber.

MACHINING CRANK CASE.

To machine it, hold the case by the main bearing extension in a three- or four-jaw chuck, and bore out the main bearing hole and the depression for the small gear wheel to the dimensions given in Fig. 2018. Face the edge of the case, and true up its inside edge $\frac{1}{4}$ in. deep. If a suitable chuck is not available, the hole may be drilled halfway from each side, and the operations of truing up the face and inside edge and recessing may be done on a mandrel between the

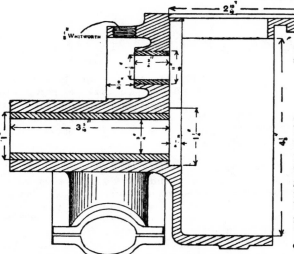

Fig. 2018.—Section of Petrol Motor Crank Case.

centres: but this will not make such a true job of it as if all the operations were done at one chucking. The hole for the main shaft bushing is 1 in. in diameter. The recess for the small 16-tooth pinion is $\frac{1}{4}$ in. deep and $1\frac{1}{4}$ in. in diameter. The inside back of the case will not require machining, except at the two recessed parts for the 16-tooth pinion and the 32-tooth wheel. The 1-in. hole must be quite parallel, and if a 1-in. reamer is available, bore the hole $\frac{1}{64}$ in. under size and reamer out by hand in the vice.

HOLE FOR CAM SHAFT BUSHING.

Next bore out the hole for the exhaust cam shaft bushing and the recess for the 32-tooth wheel; this should be $1\frac{1}{8}$ in. in diameter and the recess $\frac{1}{4}$ in. deep by $2\frac{1}{4}$ in. in diameter. If the chuck is large enough, the casting may be set over till the hole runs true to perform this operation. If not, it may be done on an angle-plate bolted to the face-plate, the boring of this

hole and the recessing being left until the top flange for the cylinder has been machined. Whichever way it is done, it must first be marked out with its centre $1\frac{1}{2}$ in. full from the centre of the main shaft hole. To do this, plug it with a piece of hardwood, mark off the $1\frac{1}{2}$ in., and scribe a $\frac{1}{4}$ in. circle from this centre, and put centre-punch dots round this circle as a guide for setting and boring.

MACHINING TOP FLANGE, ETC.

To machine the top flange, fasten the casting to the angle-plate with a long bolt through the main-shaft hole, the trued-up face being on the plate. A piece of paper should be placed between this face and the angle-plate to avoid getting it dented or bruised, as this face and the trued-up inside edge must be kept perfect to ensure an oil-tight fit for the cover (Figs. 2020 and 2021). With the casting true, true up the face of the flange and the outside edge to

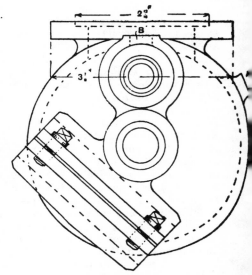

Fig. 2019.—Plan of Petrol Motor Crank Case.

$3\frac{7}{8}$ in. in diameter, and turn out the recess for the cylinder edge, $2\frac{13}{16}$ in. in diameter by a bare $\frac{1}{8}$ in. deep. Now bolt the casting to the angle-plate on the trued-up flange face, and machine up the back of the case. The cam shaft hole must be $\frac{3}{4}$ in. through, and the face for the cover (Figs. 2022 and 2023) for the exhaust cam chamber is $\frac{9}{16}$ in. from the edge of the camshaft bearing.

ADVANCE SPARKING APPARATUS.

If the advance sparking apparatus is to be purchased, it should be bought before the crank case is machined, some being small enough to work in the space here specified, while some are larger, and in this case a small recess must be turned in the top of the main bearing part. The cover (Figs. 2022 and 2023) will then not require the piece cut out of the bottom, but will be quite circular. At this chucking, the face of the main shaft bearing may be trued up, so that its length from inside the face to outside is $3\frac{5}{16}$ in.

FITTING PARTS OF CRANK CASE TOGETHER.

The clip holding the motor to the frame tube may next be bored to suit the tube. Roughly file up the faces of the two parts till they fit squarely together. Then mark off and drill for four $\frac{1}{4}$-in. or $\frac{5}{16}$-in. screws to hold the parts together, the holes being clearance in the bottom half and tapping in the top half. Bolt the two halves together, and bore them to size, either on the angle-plate in the lathe or on the drilling machine. To ensure a smooth, true hole, a bit or reamer should be put through to finish. Remove the screws and file $\frac{1}{32}$ in. off each to give the necessary clearance to allow the clip to grip the tube. The pins may, if thought necessary, be long enough to come through and take a locknut. In tapping aluminium, use paraffin as a lubricant, and do not allow much swarf to accumulate in the flutes of the tap, or the thread will tear up. Drill and tap a hole at A (Fig. 2019) to suit the oiling arrangement. For injecting oil with an oil can a $\frac{1}{4}$-in. Whitworth thread, with a plain screwed plug, will do ; but for an oil pump worked from the saddle a larger hole, to suit the particular connection

used, will be necessary. Another hole must be drilled and tapped in the bottom of the case for letting out the waste oil, and this may be fitted with a $\frac{1}{4}$-in. screw plug ; but a small waste-oil tap is handier

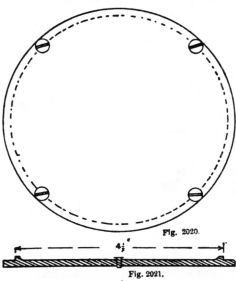

Fig. 2020.

Fig. 2021.

Figs. 2020 and 2021.—Crank Case Cover.

and gives a better appearance. The $\frac{1}{4}$-in. Whitworth hole B (Fig. 2019), to receive the exhaust-valve lift-rod, must be left till after the cylinder is fitted.

CRANK-CASE COVER.

This may now be turned and fitted to the case. A $\frac{3}{4}$-in. lug is left on the back by which to chuck it and to form the air outlet. Only the edge and the part which fits in and against the case will require machining. It must be made an oil-tight fit by being turned nearly to size and being then ground in with powdered pumicestone. It should be left about $\frac{1}{16}$ in. larger in diameter than the case, and the edge should be milled to facilitate holding whilst grinding-in and for removal. The cover is secured to the case by four $\frac{1}{4}$-in. screws, as shown in Fig. 2020. The cover must be chucked again, back outwards, and the air-hole in the centre formed. A flange should be left round this, as shown in Fig. 2021,

so that the oil splashed over the cover may
run round the hole and down again to the
bottom of the case, instead of leaking out
on the outside. The hole may be $\frac{1}{16}$ in.
or larger, and the head of the flange $\frac{1}{4}$ in.
in diameter. A valve may be made of this
if thought desirable by fitting a $\frac{3}{16}$-in. cycle-
ball, but it is not necessary unless auto-
matic lubrication is fitted.

SMALL COVER.

This (see Figs. 2022 and 2023) may be
bored, turned on the edge to $1\frac{7}{8}$ in., and
faced on the inside. Bore the hole for the
bush $1\frac{1}{8}$ in., and just true up the projecting
edge. The piece is fastened to the case
by three $\frac{1}{8}$-in. screws with countersunk

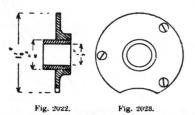

Fig. 2022.　　　Fig. 2023.

Figs. 2022 and 2023.—Cover for Exhaust Cam
Chamber.

heads, as these must not project beyond
the face of the cover.

BUSHES.

The main bush of phosphor bronze is
$3\frac{5}{16}$ in. long by 1 in. in diameter, with a
$\frac{3}{4}$-in. hole. Chuck the casting, bore the
hole $\frac{1}{64}$ in. under size, and finish with a
1-in. reamer. Knock the bush on a tree
mandrel, and turn it a tight fit for the case,
so that it requires driving in with a mallet.
A $\frac{1}{8}$-in. peg may be fitted half in the bush
and half in the case, $\frac{3}{8}$ in deep, to prevent
any possibility of the bush shifting. When
the bush is fitted, knock it on the mandrel
again, and face off the ends flush. Prepare the bushes for the exhaust cam shaft
in the same manner. The one in the case
is $\frac{3}{4}$ in. long by $1\frac{1}{8}$ in. in diameter, with a
$\frac{1}{2}$-in. hole. The cover bush is $1\frac{9}{32}$ in. by
$1\frac{1}{8}$ in., with a $\frac{1}{2}$-in. hole. Leave these holes
a shade under size, knock in the bushes,
screw the cover in place, and pass a $\frac{1}{2}$-in.

reamer through both whilst in position.
Be sure the main bush hole is at right
angles with the top flange face, the exhaust cam shaft hole parallel with the main
bush hole, and the hole in the tube-clip
at right angles with the main shaft hole.

CYLINDER CASTINGS.

Dress up the cylinder castings, taking
off all lumps and mould marks, level the
bottom part that fits on the crank case
square with the sides, and mark off. Plug
a piece of hard-wood across the mouth of
cylinder, taking care not to drive it in too
hard, and from the edges of the cored hole
find the centre. From this describe with
the dividers a circle $2\frac{1}{2}$ in. full in diameter,
and centre-dot lightly with about eight
dots at equal distances round the circle, as
a guide for boring. In a similar manner
plug the openings in the top of the cylinder for the inlet valve and sparking plug,
and mark off the two openings so that
their centres are exactly $2\frac{1}{16}$ in. apart. As
the finished sizes are $1\frac{9}{16}$ in. and $\frac{11}{16}$ in.
respectively, mark the circles about $\frac{1}{16}$ in.
larger, so that the guide marks will not be
obliterated in machining. Drill the sparking-plug hole $\frac{5}{8}$ in., to be ultimately tapped
out $\frac{11}{16}$ in. with seventeen threads to the
inch to suit the standard pattern De Dion
plugs. If a tap to suit this size and thread
is not to hand, the cylinder can be chucked
on the face-plate and screw cut to suit the
sparking plug, but it must not be threaded
till the cylinder is bored, as the plain $\frac{5}{8}$-in.
hole will be required for a bearing and
guide for the boring bar. Face off the
cylinder top to a thickness of $\frac{5}{8}$ in. (see
Fig. 2024). Fig. 2025 is a half plan of the
cylinder head.

BORING CYLINDER.

To bore the cylinder, bolt the casting
truly on the saddle of the lathe. It should
be held firmly in position by two stout iron
straps bent to the radius of the cylinder,
and the casting should be packed up to the
correct height of the centres. See that
everything is quite firm and the lathe properly adjusted before starting to bore, as
on the accuracy of the work on this part
depends in a great measure the efficiency

of the engine. Take three cuts at least through—four will be better—the finishing cut being a mere scrape. The finishing cut and the cut before it should be taken right through without a stop from start to finish, or a true bore will not be obtained.

BORING BAR.

The boring bar should be made from 1¼-in. or 1½-in. mild steel with one end turned down ⅝ in. to pass through the hole in the top of the cylinder far enough to allow the cutter to go to the top of the cylinder bore. This ⅝-in. part of the bar must fit the hole accurately, without shake from end to end. The bar, before being turned, should be drilled up each end and countersunk to the same angle as the lathe centres. Two cutters should be made, a roughing and a finishing cutter, the latter to be used on the finishing cut only, and to be dead to size—namely 2½ in. To make a perfect job the bore should be taken to about $\frac{1}{100}$ in. under size and reamered out by hand in the vice with a dead parallel 2½-in. reamer; but unless this tool is already in the possession of the worker, it will be an expensive tool to make or purchase, and would not be economical unless a number of cylinders are to be made. A substitute may be a copper or lead lap fed with flour emery and oil, but every particle of emery must be washed from the work with paraffin or petrol. If the work is to be lapped out the cylinder must be bored to within the merest shade of the finished size.

CYLINDER FLANGE, SHOULDER AND CHAMFER.

The flange, shoulder, and chamfer on the mouth of the cylinder may now be turned or cuttered. The work may be done at the same setting as the boring, by making cutters to fit the cutter bar, or it may be done on a mandrel between the lathe centres. To avoid making a mandrel the cutter bar may be used, a collar being turned to fit tightly on the ⅝-in. part of the

bar, and the outside may be turned up in its place to fit the bore of the cylinder. The small end of the bar can then be slipped through the ⅝-in. hole from the inside and driven with a carrier on this end. The flange should be turned up true on both sides and left ⅜ in. thick, and the shoulder should be turned $2\frac{13}{16}$ in. to fit tightly the recess in the crank chamber. This part should be the least shade taper,

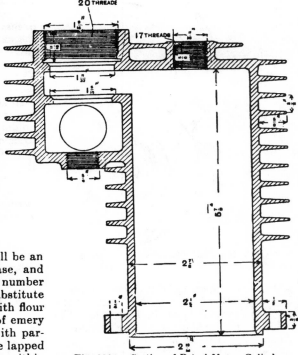

Fig. 2024.—Section of Petrol Motor Cylinder.

so that the screws which fasten it to the flange on the crank chamber will pull it up tight. The diameter of the flange should be exactly the same as that of the crank chamber flange. The mouth of the cylinder should be chamfered out as shown in Fig. 2024 in order to facilitate the insertion of the piston and rings.

BORING OUT VALVE CHAMBER.

The casting must now be chucked on the face-plate, head outwards, and fastened down with a bolt passing right through the sparking-plug hole, or held down by the flange with dogs or clamps. Get the dotted circle round the inlet valve opening quite true, and bore out and screw the hole for the exhaust-valve guide (Fig. 2026), $\frac{5}{8}$-in. Whitworth thread, and with a hook tool face the under side for the valve guide

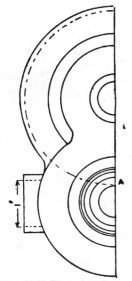

Fig. 2025.—Half Plan of Cylinder Head.

to bed truly against. Next bore the opening and seating for the exhaust valve to the sizes and angle shown in Fig. 2024. The sides of the exhaust chamber should be cleared up with the hook tool. The opening and seating for the inlet valve may now be machined, and the top part bored out and screwed, as shown in Fig. 2024. Face up the opening so that it is $1\frac{1}{8}$ in. from the top of the valve seating. The casting must not shift during these operations, as it is imperative that the exhaust-valve seating, the screwed $\frac{5}{8}$-in. hole, and the under part for the exhaust-valve guide be absolutely true with each other, or the exhaust-valve will never be a gas-tight fit. It is well to rough the parts first, and then finally go over the above-mentioned parts with a light finishing cut to make sure they are true. The seating and screwed part

for the inlet valve must also be dead true with each other.

EXHAUST PIPE OPENING.

The opening for exhaust pipe, shown in the half plan of the cylinder top (Fig. 2025), should now be bored or drilled out and tapped 1 in., with twenty-six threads to the inch. Take care not to burst this part in tapping, as it is rather light. It is well to drill it out rather full, so that the tap works freely, a full thread not being necessary, as there is only the weight of the exhaust silencer for it to support, and this has a long bearing.

CLEARANCE HOLES FOR THE HOLDING-DOWN PINS.

Next mark off the bottom of the cylinder flange for six $\frac{1}{4}$-in. clearance holes for the holding-down pins. Start the first hole to come at A (Fig. 2025), and mark off the other five equally from this. Get the holes the correct distance from the edge to allow the screw heads to clear the cylinder wall, as there is not much space. The best form of screw for the purpose has a square head with a circular collar underneath.

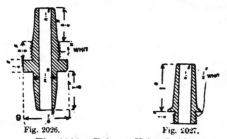

Fig. 2026.
Fig. 2027.
Fig. 2026.—Exhaust Valve Guide.
Fig. 2027.—Exhaust Push-rod Guide.

EXHAUST VALVE GUIDES.

The guide for the exhaust valve and the exhaust-valve push-rod guide are shown in Figs. 2026 and 2027 respectively. The guide shown in Fig. 2026 can be turned from $1\frac{3}{8}$ in. or 1-in. case-hardened mild steel, or from tool steel hardened and tempered, or a pattern can be made for this and for the push-rod guide, and phosphor-bronze castings obtained. Whichever method is adopted the machining will be the same.

Drill the ¼-in. hole right through, a shade under size, and reamer it out to size. Turn up a mandrel to fit the hole tightly, and finish up the outside between the lathe centres. The shoulder may be left round, and two or three ⅛-in. tommy holes drilled in it, or it may be filed up hexagon ; the latter is preferable, as it can then be

Fig. 2028.

Fig. 2029.

Figs. 2028 and 2029.—Exhaust Valve.

screwed up tighter than with a tommy wrench. The push-rod guide can be made in a similar manner, two flats being filed on the base by which to screw it up. To ensure the valve seating being true with the bore of the guide, a cutter bar may be made of ⅝-in. or ¾-in. mild steel with a leg turned down to fit the ¼-in. hole, a cutter being made to the size and shape of the valve opening and seating; this can be worked round with a lathe carrier by hand, and will make a true job.

EXHAUST VALVE.

This should be made to the dimensions given in Figs. 2028 and 2029. The head and stem are separate pieces screwed together, and the end is riveted over. The stem is a piece of ⅜-in. mild steel having at one end ⅟₁₆ in. of any convenient thread, the head being tapped to suit and countersunk. The stem is screwed in up to a shoulder

and riveted over. The valve should now be truly centred at each end and turned to the sizes given in Figs. 2028 and 2029. The part to rest on the seating should not be much more than ⅟₁₆ in. wide, and of an angle corresponding to that of the seating in the valve chamber. The projection and saw-cut on the head is for use when grinding the valve to its seating with a screwdriver, or, better, a screwdriver held in a brace. The valve grinding should be done with flour emery and oil, and may be finished off with powdered pumice and oil after all traces of the emery have been washed away. The hole in the tail end of the valve stem should be drilled $\frac{3}{32}$ in. and opened out to a slight taper. It is for the pin to hold the valve spring up to its work. It may be ⅟₁₆ in. from the end, but the exact position will depend on the length and strength of the spring used, and it will be best to leave this hole till the valve and spring are tied in their places.

INLET VALVE.

With the simple form of carburetter to be described later the inlet valve will be part of the carburetter, but should it be desired to fit any other form of spray or surface type carburetter, then the inlet valve will be required of the size and form of Fig. 2030. This should be as light as possible consistent with the work it has to do. The valve body (Figs. 2031 and 2032) is an iron casting, drilled ⅟₁₆ in. and

Fig. 2030.—Inlet Valve.

turned to size. The valve must be ground to its seating in the same way as the exhaust valve.

INLET VALVE SPRING.

The spring for this valve is much weaker than the exhaust-valve spring, as the valve is opened by the suction of the downward stroke of the piston. The end of the spring is passed through the hole drilled in the valve stem, the position of this hole being left till the spring is tried in its place. It is best to buy these valve springs, as they cost only a few pence and are then certainly of a suitable strength. The correct adjustment of the inlet-valve spring is a very important matter, and can only be arrived at by trial. If the spring is too strong the valve will not open sufficiently to admit a full charge of gas, and if too weak it will not close quickly enough, and will thus cause loss of compression, and possibly back-firing.

DRILLING CRANK-CHAMBER FLANGE.

The cylinder can now be fitted on the crank chamber, and the holes marked off to correspond with the six $\frac{1}{4}$-in. holes in the cylinder flange. Place the cylinder on the crank case with the exhaust-valve

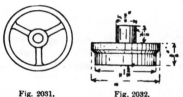

Fig. 2031.　　　　　Fig. 2032.
Figs. 2031 and 2032.—Body of Inlet Valve.

guide directly over the centre of the exhaust-cam chamber on the crank case. Mark off through the holes in the cylinder flange with a scriber, remove the cylinder, centre-dot the crank chamber flange carefully, drill with a bare $\frac{1}{32}$-in. drill, and tap out to suit pins $\frac{1}{4}$ in. by 20 threads, using paraffin to lubricate the taps. The holes must be tapped carefully, or the metal will tear up and spoil the thread. Remove the tap once or twice, and clear off the swarf to avoid tearing. Then replace the cylinder, fasten it down with the screws (three will be sufficient), and with a long drill made to fit the bore of the exhaust-valve guide, drill the hole in the cam chamber for the push-rod guide. This will ensure the push-rod being exactly in line with the valve stem. Tap it $\frac{1}{2}$-in. Whitworth to suit the guide, and screw it in place and test

for truth. The push-rod will be a short length of steel turned to fit the guide freely, one end, operated by the cam, being rounded and hardened. The length must be arranged so that the valve shuts down on its seating with $\frac{1}{32}$ in. between the rod and the valve stem.

THE PISTON.

The piston is shown in section by Fig. 2033 and in plan by Fig. 2034. The hole for the gudgeon pin is $1\frac{1}{4}$ in. from the front of the piston. The ring grooves are $\frac{1}{16}$ in. wide by $\frac{1}{8}$ in. deep, and are $\frac{1}{8}$ in. apart. The distance apart of the faces of the bosses for the gudgeon pin is $1\frac{9}{32}$ in.; this is $\frac{1}{32}$ in. more than the length of the small end of the connecting-rod, a small amount of play being necessary here to prevent the piston binding in the cylinder. Of course, the play must only be sideways, the fit of the pin in the connecting-rod bearing simply allowing it to work quite freely. The piston, and, in fact, all reciprocating parts, such as the connecting-rod, should be as light as is possible consistent with strength to attain a high speed with the least possible vibration.

WORKING UP THE PISTON CASTING.

Hold the piston casting in the chuck by the lug cast on the head, and turn up the outside parallel to a working fit in the cylinder. Take a very fine finishing cut with a freshly ground tool and with a slow feed. When smoothed off with a very fine smooth file, the piston should fit the cylinder so that if oiled it will sustain its own weight therein. True up the bottom edge, and turn a very narrow groove, $\frac{1}{32}$ in. deep, $\frac{1}{16}$ in. from the edge, to facilitate lubrication. As far as the lugs, the inside should be turned slightly taper as shown, the thinnest part being left $\frac{1}{16}$ in. full thick, and a rim should be left on the inside $\frac{1}{8}$ in. bare thick, this strengthening the edge somewhat. Face up the head to the lug by which the casting is held, leaving this part $\frac{1}{8}$ in. thick. Then with a fine-pointed tool mark a light line round the centre, as a guide for drilling the gudgeon-pin holes. Then carefully turn the grooves for piston rings. To get them all alike, make a tool,

similar to a parting tool, just $\frac{3}{16}$ in. wide
and well backed off on each side for clearance. Any burr that may have been
thrown up should be carefully smoothed
off, and the head parted off with a long
parting tool. Smooth off the burr, and
polish the head with several grades of

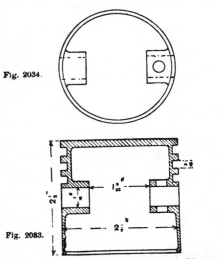

Fig. 2034.

Fig. 2033.

Figs. 2033 and 2034.—Petrol Motor Piston.

emery cloth. The more highly finished
the head of the cylinder the better, as the
burnt gases will then not so readily accumulate on it. If, for polishing, it is held
in the vice, great care must be taken not
to grip the thin edge, or it will get cracked.
Mark off on the centre line the positions of
the pin holes, which must be exactly opposite each other. Drill a shade under the
size, and then hand-reamer the holes. In
the absence of a reamer the holes should
be drilled with a twist drill to finish. The
best way to get these holes true is first to
drill halfway on the lathe centre with a
small drill, say of $\frac{3}{16}$ in. diameter.

FACING INSIDE FACES OF BOSSES.

The inside faces of the bosses must now
be tooled with a facing cutter, the cutter
bar fitting the hole without shake. The
cutter is fitted to the bar, turned up in
place, hardened, and let down to a light
brown. To use it, pass the leg of the bar

through the hole, and insert the cutter,
and hold the cutter bar in the chuck with
the back centre as support and feed for the
work. Drill a $\frac{1}{4}$-in. clearance hole right
through one boss, as shown in Fig. 2034,
to hold the pin in place. One hole in each
lug would make a more certain job. The
face of the hole should be faced for the
head of the pin.

GUDGEON PIN.

The gudgeon pin (Fig. 2035) should be
turned from $\frac{1}{2}$-in. tool steel; it should be
quite parallel, and such a fit as to require
tapping into place with a mallet. Smooth
it off to a high finish, knock it in place,
and drill through the hole in the boss $\frac{1}{4}$-in.
tapping size. Remove the pin and tap
$\frac{1}{4}$-in. Whitworth, smooth off the burr, and
harden, then polish and let down to a
brown shade. Perhaps the easiest way to
ensure the tapped hole coming exactly in
line with the clearance hole is to knock the
pin in place and drill to tapping size right
through, then remove the pin, and open
out the holes in the lug. When the pin is
in place the ends should be clear of the
face of the piston; for if level, or projecting in the slightest, the hard pin will mark
the cylinder when working. The screw to
keep the pin in place may have a cheese
head with a screwdriver slot or a square
head with a round collar under, similar to
those used to fasten the cylinder to the

Fig. 2035.—Gudgeon Pin for Petrol Motor
Piston.

crank case. The latter is better, as the pin
should be a very tight fit to avoid any
possibility of its coming loose when working, but a box key will be required to screw
it up with.

MAKING AND FITTING PISTON RINGS.

On the fit of the rings much depends, as
if they are not perfectly fitted, loss of compression, and consequently loss of power,
will result. There are several ways of
making these rings. A common method

in cheap motors is simply to bore up a cylindrical casting, turn the outside to finished size, part the rings off the required width, and cut them through; but this is bad practice, as there is no spring in the

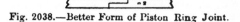

Fig. 2036.—Piston Ring Ends.

rings to keep them up to their work in the cylinder. A better method is to chuck the ring casting by the end lug, bore out inside $2\frac{1}{4}$ in., turn up outside to $2\frac{9}{16}$ in., and part off the required number of rings $\frac{3}{16}$ in. wide. Care should be taken to have the parting-off tool set quite square with the work, so that both sides of the rings are alike and at right angles to the face. A gauge should be made to test the width, so that the rings may be all exactly alike. Make four rings whilst on the job, as it is very probable that one will be broken in finishing or springing on the piston, and even if not it is well to have a spare ring. The rings should now be cut through. The more generally adopted method is to saw a piece out diagonally, as at Figs. 2036 and 2037, the piece removed measuring $\frac{1}{16}$ in. A preferable method is to make the joint as shown in Fig. 2038; this makes a sure gas-tight fit, and should the rings get so placed in the piston that all three joints come in line, there will be no loss of compression. To make the joint, mark off a line with the square at any point on the surface of the ring, and $\frac{1}{16}$ in. on each side of this line mark another line; scribe a line midway on the width through these three lines, and carefully file down to this line at each side. Before the rings are cut through, they should be tried in the piston grooves, and should go to the bottom without shake. If there is any variation in the width of the grooves, number the rings as fitted.

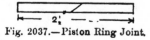

Fig. 2037.—Piston Ring Joint.

Should the rings require easing, place them on a flat board and lightly smooth the sides with a very fine smooth file, taking care to keep the file perfectly flat. Now when the rings are cut, press the joint

together and try the ring in the groove, and if it is too tight, ease a little off the insides of the joint. They should be without shake, but when the pressure is released the joint will spring open. With the rings

Fig. 2038.—Better Form of Piston Ring Joint.

all fitted and joints made perfectly, they must be sprung together and turned outside to $2\frac{1}{2}$ in. in diameter to fit the cylinder bore. To do this, a jig of some kind will be required to hold them. It may consist of a casting to be bolted to the face-plate with a circular projection to fit the inside of the rings when sprung in, and standing up $\frac{1}{32}$ in. less than the combined width of the rings to be turned. Now with a plate and a $\frac{3}{8}$-in. bolt through the centre they may all be held securely together. Before removing the rings from the jig, try them in the cylinder, and if they are a smooth working fit within it they can be removed and finally bored out to the finished size, $2\frac{11}{32}$ in.; this boring will be eccentric with the outside, as shown at Fig. 2039, the thick portion measuring $\frac{3}{32}$ in., and the thin part, $\frac{1}{16}$ in., being at the joint. The rings may be held so that all are bored at one operation. Make a band clip, about $\frac{1}{8}$ in. narrower than the combined width of the rings to be bored, of sheet steel about $\frac{1}{16}$ in. or $\frac{3}{32}$ in. thick; clip the rings with the joints tight together, and hold them in a jaw chuck. The rings can be set true

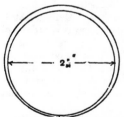

Fig. 2039.—Piston Ring Bored Eccentrically.

by the part of the ring beyond the surface of the clip. Bore with a fine-pointed sharp boring tool with a very light cut and feed. The rings are now finished, and may be sprung on over the head of the piston into place. This has to be done carefully, or a fractured ring will be the result. See that they do not stick in any part of the grooves; if they do, remove and carefully scrape the part to free it. With careful

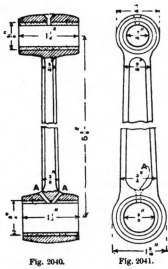

Fig. 2040.   Fig. 2041.

Figs. 2040 and 2041.—Section and Elevation of Connecting-rod.

grooving and fitting, the rings should just drop down the grooves by their own weight, but without side shake.

CONNECTING-ROD.

The connecting-rod (Figs. 2040 and 2041) can now be machined. Set the casting as true as possible, chuck the large end, and bore it out $\frac{3}{4}$ in. Chuck the small end, and bore it $\frac{5}{8}$ in. In chucking the small end, set it so that the two bores are parallel and with the centres $5\frac{3}{8}$ in. apart. It is usual in doing this work first to mark off the faces of the bosses on a level surface with a scribing block, and centre-dot the circles on these faces; but with the special set of castings designed for this motor, if the bosses are set true with the outsides, the bores will come exactly $5\frac{3}{8}$ in. apart without marking off. The phosphor-bronze bushes for the two ends may now be prepared. Chuck the large one, and bore and

reamer it out $\frac{5}{8}$ in. The small bush is finished $\frac{7}{16}$ in. in diameter. Knock each bush on a true mandrel of the proper size, and turn the large one $\frac{3}{4}$ in. full and the small one $\frac{5}{8}$ in. full. They should be a very tight fit for their respective bores, as they are to be shrunk in place. Heat the boss of the connecting-rod to about the heat of a hot soldering iron, and knock the bush in quickly with a mallet, or press it in between the vice jaws, and cool at once in cold water. If this is properly done, the bushes will never shift; but if thought desirable, or if the bushes are found to be not so tight a fit as was intended, a hole may be drilled half in the bush and half in the boss, and a small screw or peg may be driven in, and cut off flush. Now face off the sides on the mandrel in the lathe centres to the dimensions given in Fig. 2040, leaving no more projecting on one side than on the other. Two oil holes should be drilled in the large boss, as shown at A (Figs. 2040 and 2041), and well

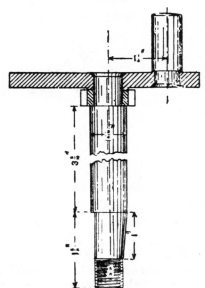

Fig. 2042.—Petrol Motor Shaft and Crank.

countersunk. To facilitate lubrication, file a small groove the whole length of the bushes on the side where the holes penetrate.

SETTING CONNECTING-ROD.

The connecting-rod will now require careful setting. The mandrels on which the bushes were turned should be inserted in the ends, and tested with the callipers to see whether the bores are parallel. The rod may be set cold if found out of truth. These mandrels should be made rather long, say 6 in. or 7 in., as, when used for

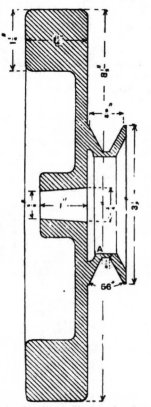

Fig. 2043.—Section of Flywheel and Pulley.

setting, the extra length shows up any irregularity better. The bores of the bushes should certainly be reamered out to obtain a smooth parallel hole, the reamer being finally put through after the oil holes and grooves have been drilled and filed.

THE MAIN AXLE.

This should be made from a piece of tough mild steel case-hardened, or of hard-

ened tool steel; the ends should be let down rather low, however, to avoid fracture, this treatment at the crank disc end extending to just beyond the pinion shoulder (see Fig. 2042), and at the flywheel end to just beyond the end of the keyway. Let these parts down to a blue colour, the remainder or centre portion being left brown. A tool steel axle will wear longer unless the mild steel is very carefully case-hardened. Cut off a piece of steel $\frac{7}{8}$ in. or $1\frac{7}{8}$ in. in diameter and $5\frac{11}{16}$ in. long. Centre, drill, and countersink each end to the same angle as the lathe centres, and turn the $\frac{3}{4}$-in. part to fit the $\frac{3}{4}$ in. reamered hole in the crank-case bush. A working fit without shake is needed. If the axle is to be ground after hardening, leave it large by $\frac{1}{64}$ in. bare for this. Grinding is decidedly preferable, as a perfectly true axle will result, but of course a good job can be made of it without grinding if a fit is made before it is hardened, and care is observed not to warp it in the hardening process. The end for the fly wheel lock nuts may now be turned to $\frac{3}{4}$ in for $\frac{3}{8}$ in. up, and then a line should be marked round the axle with a fine pointed tool, 1 in. farther up or $1\frac{3}{8}$ in. from the end. Now set the slide-rest to turn a smooth taper that will start at this line and finish exactly at the end of the $\frac{3}{8}$-in. part. Reverse the carrier and turn down the other end $\frac{3}{8}$ in. $\frac{3}{4}$ in. up.

SECURING CRANK DISC AND FLYWHEEL.

The method of fastening the crank disc must now be decided. The plan shown in Fig. 2042 is to make the axle end a driving fit through the small pinion and crank disc, countersink the outer side of the $\frac{3}{8}$-in. hole. and rivet the axle end up after a key or grub screw has been fitted right through the disc and pinion, half in the axle and half in the disc and pinion. If this is well done it will make a firm job, but for preference the hole in the disc and pinion should be tapped out $\frac{3}{8}$ in. by twenty-six threads, and the axle end screwed to suit. With a right-hand thread the working of the engine tends to screw it tighter, though if desirable a grub screw may be fitted, but then the pinion and disc must be screwed right home before drilling for the grub screw. The other end can now be screwed the same thread for the flywheel lock nuts. The keyway for the flywheel should be sunk 1 in. long by $\frac{3}{16}$ in. wide by $\frac{1}{8}$ in. deep, but if this cannot be cut by machine, a parallel keyway filed flat will be best.

THE CRANK DISC.

This should now be prepared. Find the centre of the axle boss, scribe a line centrally on the crank-pin end of the disc, and on this line mark off and centre-dot exactly $1\frac{1}{4}$ in. from the centre of the disc for the crank pin. Scribe a guide circle round these centre marks a shade larger than the holes to be drilled, and centre-dot these circles as a guide for drilling. If a true drilling machine is available, these holes may be drilled, using a $\frac{1}{8}$-in. or $\frac{7}{16}$-in. drill first, and enlarging to the finished sizes, which should be tapping sizes for $\frac{5}{8}$ in. by twenty-six threads, and $\frac{1}{2}$ in. by twenty-six threads, left hand, respectively, if the screwed method of holding is adopted. In any case, to facilitate assembling, the crank pin should be screwed to the disc. Then tap these two holes.

CRANK PIN.

For this, cut off a length of $1\frac{3}{8}$ in. of $\frac{3}{4}$-in. steel, and centre, drill, and countersink the axle. Regarding the material, the same remarks apply as for the main axle. Turn this up $\frac{5}{8}$ in. to fit the large end of the connecting-rod, round off an end, and turn down the other end to $\frac{1}{2}$ in. for $\frac{3}{8}$ in. up, and screw it to fit very tightly the tapped hole in the disc. This must be a left-hand thread. Provision must be made for screwing this home after the main axle and disc are in place in the crank case, by filing two flats on the outer end of the pin, or drilling a $\frac{1}{4}$-in. hole and drifting it out square for a key. The main axle and crank pin may now be hardened. Screw the pinion and crank disc on to the main axle the reverse way to the final position, that is, with the boss outwards. Face off this boss true in the lathe, so that the final combined thickness of the disc and boss is $\frac{3}{4}$ in., while at the same time the whole of the

boss side of the disc may be turned up and the other boss for the crank pin faced off. Remove the disc from the axle, screw the crank pin into the disc, and turn the outer side of the disc. If the edges of the disc are now filed up, these parts will be finished.

FLYWHEEL AND PULLEY.

These are in one casting (see Fig. 2043). Chuck it by bolting to the face-plate, or by a three- or four-jawed chuck on the inside of the rim. Bore out the hole to the taper on the axle, the fit being tested with red-lead and oil smeared on the axle end, which should mark the hole from end to end, and all round. If the lathe is true, the axle end may be finally ground in with fine emery and oil. The edge of the flywheel, the pulley side, and the pulley can be turned at this chucking, if the casting is held in a jaw-chuck, or it can be roughed at this chucking, and finished in its place

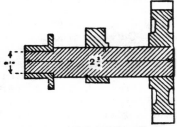

Fig. 2044.—Exhaust Cam Shaft.

on the main axle, after the keyway has been cut, the key fitted, and nuts are screwed home. The latter method will be more likely to give the truer job. This key must touch on both sides from end to end, and should have a slight taper, the thicker end being towards the pulley side. The flywheel is 8 in. in diameter, with a rim $1\frac{1}{4}$ in. by $1\frac{1}{4}$ in., but it may be lighter if desired. The heavier wheel, however, gives a steadier running motor. The pulley is $3\frac{1}{2}$ in. in diameter, and the belt groove turned 28 degrees from the perpendicular. The groove A, $\frac{3}{16}$ in. deep by $\frac{1}{8}$ in. wide, catches any oil which may leak out of the axle bearing, and thus prevents it working down the flywheel and splashing on to the rider. The groove for the belt is $\frac{9}{16}$ in.

wide at the top by $\frac{1}{8}$ in. at the bottom, and the rim of the flywheel will require drilling with $\frac{5}{8}$-in. or $\frac{1}{2}$-in. holes to balance the piston and connecting-rod.

THE GEAR WHEELS.

The two gear wheels to work the exhaust and ignition cams are phosphor-bronze

Fig. 2045.—
Exhaust Cam.

Fig. 2046.—
Ignition Cam.

castings having thirty-two and sixteen teeth respectively. The pinion is shown in section in position on the main shaft in Fig. 2042 (p. 669.) The larger one is shown in section on the exhaust cam shaft in Fig. 2044. If the teeth are cast they may be filed to shape after the wheels are bored and turned. The pinion may be tapped and screwed on its shaft and turned in position. Only just true the top of the teeth, face up the sides, and scribe a fine line on both sides as a guide for filing the bottoms of the teeth. The wheel may be fastened on the exhaust cam shaft by a well-fitted key, or screwed on with a $\frac{7}{16}$-in. left-hand thread with twenty-four or twenty-six threads.

EXHAUST CAM SHAFT.

This is $2\frac{5}{8}$ in. long by $\frac{1}{2}$ in. in diameter, and is of tool steel for preference, hardened and let down to brown, the end screwed for the wheel being let down to a blue colour. The other end may be turned down $\frac{3}{8}$ in. for $\frac{7}{16}$ in. up to take the ignition cam, which may be fixed by drilling a $\frac{1}{16}$-in. or $\frac{3}{32}$-in. hole right through both cam and shaft while a small split pin to suit is passed through ; or the cam and the end of the shaft may be threaded left-hand. The chief consideration in screwing the ignition cam is the difficulty in getting it right home to the shoulder on the shaft so that the cam is in the correct position for igniting ; for this, turn a shade off the back of the cam until on trial it is found to come

in the right position. Another method of fixing this cam is to make it with a $\frac{1}{4}$-in. square hole, filing the end of the shaft to suit, and allowing the end to project sufficiently to permit a small split pin to be inserted to prevent the cam working off. The exhaust cam is fixed by a plain flat key.

EXHAUST AND IGNITION CAMS.

These (see Figs. 2045 and 2046) should be turned from tool steel and filed or milled. The exact shape and size of the projection on the exhaust cam cannot be decided until the engine has been assembled and tested, which will be explained later in giving particulars of timing. The position of the exhaust cam on the shaft will be determined by slipping the shaft through its bearing in the crank case, with the thirty-two-tooth wheel screwed up tight. The cam must then come right up against the face of the bush so that there is no endshake to the shaft, but it must revolve freely. After the timing has been fixed, the cams should be hardened and let down to a brown colour. The ignition cam will require hardening only on the projection.

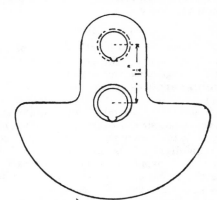

Fig. 2047.—Crank Casting

TIMING THE EXHAUST VALVE.

For this purpose partly assemble the motor parts. Connect the piston to the connecting rod by the gudgeon pin. Screw the thirtytwo-tooth wheel to the exhaust shaft, place this in position in the crank case, key on the exhaust cam, put on the cover of the exhaust cam chamber, and place the main shaft in position with the small pinion and crank disc attached. Now carefully slip the cylinder over the piston without injuring the rings. Fix the cylinder to the crank case by two screws, and slip the crank pin through the large end of the connecting rod and screw it into the crank disc. Insert the push rod into its guide, and put the exhaust valve with spring and cotter in place. Before going further, see that the exhaust valve stem is of such a length that there is $\frac{1}{32}$ in. play between its end and the top of the push rod when the valve is quite shut down on its seating. Now take a piece of $\frac{1}{4}$-in. or $\frac{5}{16}$-in. rod and, placing it through the sparking plug hole in the top of the cylinder, with the end resting on the top of the piston, mark off on the rod when the piston is at its highest and lowest positions. The wheels must now be in gear so that when the main shaft is turned forward (the way the engine will run) the exhaust valve will begin to open $\frac{3}{8}$ in. before the piston reaches its lowest point, and will shut exactly at the moment it reaches its highest point on the next upstroke. This can be seen by watching the marks on the rod passed through the top of the cylinder. When the right teeth in gear have been found, mark them with a centre punch. The exhaust cam must be filed to bring it into the correct position.

THE IGNITION.

The ignition cam may now be fitted on the end of the shaft so that the projection is about one quarter of a revolution in advance of the exhaust cam. The advance sparking gear should be put on first, and then the ignition cam. This cam should be so placed on the shaft that the spark will pass when the piston has $\frac{3}{4}$ in. to travel to complete its upward compression stroke for the earliest or most advanced sparking, and so that the spark will pass after the piston has descended $\frac{3}{4}$ in. for the latest or retarded sparking. The spark will pass at the plug points immediately on the break or coming-apart of the platinum points of the trembler blade and screw. On the frame of the advance sparking

91

apparatus and the edge of the exhaust cam chamber mark the "latest" and "earliest" positions, as guides when fitting the lever and rod to work from the top tube of the bicycle. The advance sparking apparatus best suited to this motor is the small one of Bassée and Michel. Current for the ignition is furnished by an accumulator or by dry primary cells, the former preferably. Fig. 2047 illustrates the crank casting.

COMBINED CARBURETTER AND INLET VALVE.

The effective little carburetter and inlet valve combined (see Fig. 2048) about to be explained is suitable for the petrol motor just described, or any other petrol

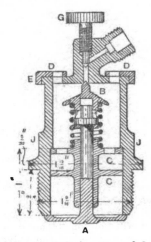

Fig. 2048.—Combined Carburetter and Inlet Valve.

motor of similar power. It is easily made, having few parts, is not liable to get out of order, and works well when the proper adjustment of air and petrol has been found by experiment. The fitting of the needle-valve and the conical sprayer calls for a certain amount of skill, as it is rather fine work and must be accurately done to give good results. A considerable saving in cost and labour is effected by the use of this carburetter, as the separate inlet valve and seating, and the ordinary union and connection necessary with the common form of carburetter, are done away with entirely.

PETROL FEED TO CARBURETTER.

It must be borne in mind that the petrol feed from the tank is by gravitation, therefore the tank must be so fitted on the bicycle that its bottom is at a higher level than the top of the carburetter. This, with the somewhat high position of the motor on the frame, will necessitate the tank being fitted to the front of the handlebars, with a flexible supply pipe from it to the carburetter. This pipe may be a length of flexible metal tubing, with screwed metal connections, or a piece of strong rubber tubing, such as that used for cycle-pump connections, in which case the making of the carburetter is still further simplified, as the inlet to the carburetter may be plain, turned to a size for the rubber tube to fit tight.

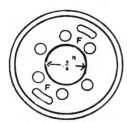

Fig. 2049.—Carburetter and Inlet Valve Cap.

ACTION OF CARBURETTER.

The action of the carburetter may be explained as follows: Upon the suction stroke of the piston forming a partial vacuum in the cylinder, the valve A (Fig. 2048) is drawn down, and with it the small needle-valve and the conical sprayer B, allowing the petrol to be sucked in on to the double baffle-plate C C, which is drilled with a number of holes to allow of the mixture passing through into the cylinder. At the same time air is drawn through the hole drilled in the top of the casing D, and this, mixing with the petrol vapour, forms the explosive mixture of gas. The holes in the top of the case D are covered by a plate E, which is also drilled with corresponding holes (see plan, Fig. 2049), and by means of the slots F sufficient rotary movement is allowable to permit of the plate wholly or partially covering the corresponding holes

in the top of the casing. This is to allow of the adjustment of the air inlet, by which the proper proportion of air to petrol is regulated. The milled screw G (Fig. 2048, is for regulating the supply of petrol. Fig. 2050 is an enlarged sectional view of the needle-valve and conical sprayer. One great advantage of this carburetter is that, owing to the heating of the carburetter by the cylinder head, petrol of any density can be used, as it is soon vaporised on coming into contact with the hot baffle-plates.

PARTS OF COMBINED DEVICE.

The whole of the parts, except the valve head and stem and the cross-pin H (Fig. 2050) and springs, are made from gunmetal or brass castings. Four patterns will be required—namely, the case (which will require a core-box), the cap (Fig. 2049), the valve-stem guide and double baffle-plate C, and the combined needle-valve and conical sprayer B. The milled screw G may be turned from a piece of $\frac{1}{2}$-in. brass rod. The $\frac{3}{16}$-in. hole in the baffle-plate piece and the $\frac{3}{16}$-in. hole in B should be drilled out from the solid, instead of being cored out. All the patterns should be made as illustrated, making due allowance for shrinkage and machining.

BODY, ETC., OF CARBURETTER.

Chuck the case casting in the jaw chuck, bore out to $1\frac{1}{16}$ in. diameter, and face the inside of the top and the bottom edge forming the valve seat; bore out $1\frac{1}{8}$ in. for 1 in. up, this leaving a $\frac{1}{16}$ in. square cornered shoulder as a rest for the baffle-plate. The outside, as far up as possible, may also be machined at this chucking. Turn the shoulder, just above the screwed part, $1\frac{13}{16}$ in. in diameter by $\frac{3}{32}$ in. wide at the extreme edge, and forming a rounded corner above just down to the edge of the hexagon part J. Turn the part to be screwed $1\frac{1}{16}$ in. diameter, and screw twenty threads to the inch. This size will fit the inlet of the motor described in this chapter. It may, of course, be any other size, to suit motors having a different screwing or size, within the limits of the casting. The edge below the thread is turned to suit the angle of the seating in the cylinder inlet, and must be a gas-tight fit when finished. Next chuck the air-inlet adjustment plate E, face the edge, and turn out the inside $1\frac{13}{32}$ in. by $\frac{1}{8}$ in. deep, and the central hole $\frac{3}{8}$ in. in diameter; mark a line $\frac{5}{32}$ in. from the edge of the $\frac{3}{8}$-in. hole as a guide for drilling holes. Chuck the conical sprayer casting and bore up the $\frac{3}{16}$-in. hole as far as the casting will allow without coming through. Turn up the outside $\frac{3}{8}$ in. diameter by $\frac{7}{8}$ in. long up to the extreme corner. The top part can be turned on a stud running true in the chuck; about five shallow grooves should be turned on the surface. Chuck the baffle-plate casting, drill up with a $\frac{3}{16}$-in. twist drill, knock it on a mandrel and turn up the outside all over so that it is a tight push-fit in the casing. The top portion should be $\frac{3}{8}$ in., or a good fit for the inside of the spring used. A line should be marked for the centre of the holes, eight in number and $\frac{3}{16}$ in. in diameter, to be drilled in each plate, the holes in the top plate coming midway between the holes in the bottom plate. The baffle-plate piece should be finally fixed by a small screw fitted through the outer casing.

PETROL INLET, ETC.

The hole in the top of the casing should now be drilled; this must be absolutely true with the valve stem hole in the baffle-plate piece. To do this, make a true drill by turning the shank to fit the $\frac{3}{16}$-in. hole in the baffle-plate piece, as a guide, with the end turned down small enough to form a $\frac{1}{16}$-in. drill; the $\frac{1}{16}$-in. hole should come right through, afterwards being opened out to $\frac{3}{16}$-in. tapping size, from the top end, for the required depth. The petrol inlet must be drilled to come through into the first hole just above where the point of the regulating screw takes its bearing, as shown. It is not advisable to drill the vertical hole too far down at first; it can be extended as required if not deep enough when the petrol inlet has been drilled. The enlargement of the petrol inlet will, of course, be made to suit the connection used, if this is of metal.

COMPLETING CARBURETTER CASING.

The baffle-plate piece should now be removed and the casing knocked on a true metal or hardwood stud, and the remainder of the outside turned up. The top edge must be a nice free fit to the inside of the plate already turned out. It should fit without shake, but should be capable of being easily turned in its place with the fingers. The hole for the regulating screw is next tapped, and the hexagon part J (Fig. 2048) filed up. The top of the cap should be marked off and drilled as in Fig. 2049. The holes as drawn are $\frac{3}{16}$ in., but if drilled $\frac{1}{4}$ in. they will be better, giving more range of air adjustment. In drilling, put the cap in

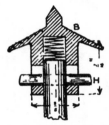

Fig. 2050.—Section of Sprayer and Needle-valve.

place on the casing and drill through the two at one operation, thus ensuring one set of holes being true with the others. The two radial slots should be marked off, drilled, and filed out to suit clearance for $\frac{1}{4}$-in. cheese-head screws; they should be of such a length as to allow of the holes in the casing being fully covered or opened by the top cap. Slip on the cap and mark off the top of the casing for the two holes for the screws, $\frac{1}{8}$-in. tapping size. Fit the screws in position and try the cap to see whether it works freely; if it does not, ease off where necessary.

INLET VALVE.

The inlet valve head and stem should now be made. The head may be of cast-iron or it may be a piece of sheet steel measuring $1\frac{3}{8}$ in. by $\frac{1}{16}$ in. in the rough. Drill centrally and tap $\frac{1}{16}$ in., countersinking one side. Take a piece of $\frac{1}{4}$-in. round steel $2\frac{1}{8}$ in. long, centre, and turn down one end

$\frac{3}{16}$ in. to a shoulder $\frac{3}{16}$ in. up ; screw to suit the stem, which should fit tightly and be riveted over on the countersunk side. File the riveted end level, and re-centre ; then turn up the valve to the section shown, leaving it $1\frac{3}{16}$ in. in diameter, and a bare $\frac{1}{8}$ in. thick at the edges, hollowing out the inside, as shown, for lightness, to within $\frac{1}{8}$ in. of the edge. On the outside, leave a projection about $\frac{3}{8}$ in. diameter by $\frac{1}{16}$ in. thick, and saw it across to form a slot for the screwdriver to facilitate grinding the valve to its seating. This sawing should

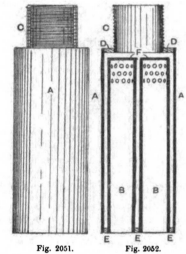

Fig. 2051. Fig. 2052.

Figs. 2051 and 2052.—Elevation and Section of Petrol Motor Silencer.

be the last operation of all. The finished length of the stem should be $1\frac{1}{8}$ in. from the inside edges to the top. The stem should be a free fit in its bearing, with the least possible shake, to ensure the valve shutting squarely on its seating.

NEEDLE-VALVE AND SPRAYER.

The needle-valve and sprayer B should also be a nice free fit on the end of the stem. Now get the exact distance from the bottom face of the casing to the point where the needle-valve takes its seating ; put the sprayer B on the end of the valve stem exactly to this measurement, and drill a $\frac{3}{32}$ in. hole right through the

end of sprayer and valve stem $\frac{1}{8}$ in. from the edge, and insert a pin $\frac{5}{8}$ in. long to fit tight. Insert the valve in the baffle-plate piece, slip on the sprayer B, knock in the pin, and push the whole into the casing up to the shoulder. If the sprayer has been fitted in the proper place, the needle-valve should come into contact with its seating $\frac{1}{64}$ in. before the inlet-valve. Put a little flour emery and oil on the needle-valve and on the face of the inlet valve, and grind the two together with a screw-driver inserted in the slotted head of the valve until they are a perfect fit. A little powdered pumicestone or bath-brick may be used instead of the flour emery. Remove the valve stem and sprayer, and with a small round file lengthen the hole in the sprayer so that it has about $\frac{1}{32}$ in. play both up and down the valve stem when the pin is inserted ; this must work very freely.

INLET VALVE SPRINGS.

A light coil spring will be required, made of spring wire about $\frac{1}{32}$ in. thick, and wound to $\frac{3}{8}$ in. diameter inside with coils fairly close together ; this should be a very free fit on the sprayer, and take a bearing under the ends of the cross-pin and not under the sprayer shoulders (see Fig. 2048), as the system of using one spring has not been found so satisfactory as using two springs. The second spring, made of very light wire, about half the strength of the first, will be a bare $\frac{3}{16}$ in. diameter outside, and fits into the space between the top of the valve stem and the bottom of the hole in the sprayer, as shown in Fig. 2050.

OTHER DETAILS OF COMBINED INLET VALVE AND CARBURETTER.

The inlet valve should not open less than $\frac{3}{16}$ in. or more than $\frac{1}{4}$ in. The correct strength of the springs can only be found after trial, by substituting stronger or weaker springs until the best results are obtained. The chamfered edge of the casing must be a gas-tight fit on its seat in the cylinder head ; therefore the length of the screwed part must be greater than that in the cylinder by at least $\frac{1}{16}$ in. ; or, if preferred, a gas-tight fit may be obtained by using a copper and asbestos washer between the shoulder of the casing and the top of the cylinder, in which case the size of the screwed part may be the same as in the cylinder, the thickness of the washer taking the chamfered edge off its seating. The latter method will give the least trouble in fitting. The whole of the outside parts may be polished or nickel-plated.

SILENCER FOR A PETROL MOTOR.

The silencer suitable for the motor has but few parts, and its construction is not difficult. The material used is weldless steel tubing brazed together; but in the absence of facilities for brazing, the parts may be screwed together, and if well fitted will give almost equal satisfaction. The following lengths of tubing should be cut. One piece of 1½-in. (No. 18 gauge), 4 in. long, for the outer case A (Figs. 2051 and 2052); two pieces of ⅜-in. (No. 18 or No. 20 gauge), 3¾ in. long, for the inner tube B (Figs. 2052 and 2053); and one piece of 1-in. (No. 16 gauge), 1 in. long, for c (Figs. 2051 and 2052). There will also be required a steel disc of about No. 16 gauge to fit inside the 1½-in. tubing as at D (Fig. 2052); another washer of the same diameter to fit inside the other end of the 1½-in. tube, as at E; and two washer blanks F to fit inside the ends of the ⅜-in. tubes.

SILENCER WASHERS.

Drill and tap out the top washer D 1 in. (twenty-six threads), and screw the short length of the 1-in. tube from end to end to fit it. Mark off the bottom washer E as shown in Fig. 2053, and drill two ⅜-in. holes to take the ⅜-in. tubes tightly (see Fig. 2052). Clean the ends of the ⅜-in. tubes inside at one end and outside at the other. Fit the small washers F in the top ends of the ⅜-in. tubes, and knock them in tightly flush with the ends. Braze in these washers, file up, and drill three rows of ³⁄₃₂-in. holes ¼ in. apart. Knock the two ⅜-in. tubes into the washer made to receive them, and then fit the whole to the end of the 1½-in. tube. See that the two ⅜-in. tubes are equally spaced at the other end,

and braze up the whole of the bottom. The top washer, with the 1-in. screwed tube in place, should now be fitted to its place and brazed up. Care must be taken not to get the thread filled up with brass and borax, or to unbraze the washers from the top ends of the ⅜-in. tubes, whilst making this joint. To prevent filling, paint the thread to within ⅛ in. of the joint with powdered blacklead and oil; and to avoid unbrazing the small washers, direct the flame to the end of the 1-in. tube first and work towards the joint, instead of blowing on to the joint straight away. The silencer can now be filed up, polished, and nickel-plated. It will only need screwing tightly into the tapped exhaust opening on the cylinder to be ready for use. No union or lock-nut is necessary, provided the threaded portion is a good tight fit in the casting. The length given is the most that space will permit to clear the bottom tube

Fig. 2053.—Underneath Plan of Silencer.

of the machine; but, if thought necessary, the diameter may be increased to 1¾ in. or even 2 in. for the outside chamber.

COMPLETING SILENCER.

Should it be decided to screw the parts together, a fine thread of about thirty to the inch must be used for the ends of the 1½-in. tube; and the washers should be screwed to fit very tightly, or the silencer may fall to pieces through the vibration. The two ⅜-in. tubes will be more reliable if of somewhat stouter gauge, say No. 16. It will also be necessary to peg the top washer after it has been screwed into place, or it will come away when the silencer is screwed to the engine. The short piece of 1-in. tube will require to be firmly riveted over for the same reason. If means of brazing are available, that method should be adopted, and will certainly make a more satisfactory job.

MAKING WATER MOTORS.

POWER OF A WATER MOTOR.

THIS chapter will describe the construction of three water motors. A cheap motor to drive a lathe or other small machine is often required, and when the pressure or head of water available is anything more than 30 lb. per sq. in. and can be obtained cheap, a water motor will be found a most convenient source of power. Unlike small gas engines, water motors are easily started; they are much cleaner, and, unlike small-power steam engines, require hardly any personal attention. A motor 18 in. in diameter, working at a pressure of 30 lb. per sq. in., will be capable of developing $\frac{1}{4}$ horse-power approximately; if the pressure is 60 lb., the yield will be $\frac{1}{2}$ horse-power; and again, if the pressure can be obtained at 90 lb., the horse-power of the motor will be $\frac{3}{4}$, or a little more. If a $\frac{1}{2}$-horse-power motor is desired, and a pressure of only 30 lb. per sq. in. is available, a larger motor will be necessary. The point to remember is that the pressure of the water available controls the diameter of the internal wheel. The supply pipe to the motor must be kept large enough, so that full pressure may be delivered at the rim of the internal wheel: sharp bends and insufficient area of pipe cause unnecessary friction of the water in passing through. In an instance where the pressure of the supply was 60 lb. per sq. in., the pressure delivered to the motor was 40 lb., or a clear loss of one-third of the pressure available; this was caused by the supply pipe being too small. Assuming that the pressure from the town main or other source is about 30 lb. per sq. in., begin by making a sketch of the motor.

The internal wheel, shown in Fig. 2054, must be 18 in. in diameter. The size of the spokes and the strength of the whole structure will, of course, vary considerably should the pressure reach anything like 90 lb. per sq. in.

RIM OF WATER MOTOR WHEEL.

The rim of the wheel is made of sheet iron or mild steel $\frac{1}{16}$ in. in thickness. Obtain a strip of sheet-iron $56\frac{1}{2}$ in. long by 4 in. broad and $\frac{1}{16}$ in. thick. This strip is to be bent round until it forms a hoop 18 in. diameter by 4 in. broad, the points to meet together, and be butt-jointed. The ends of the strip that forms the internal wheel should butt against each other, and the butt-plate which fits behind the joint inside of the rim should be 1 in. broad and as long as the wheel is broad, namely 4 in. It should also be dished out to fit the sweep of the rim inside, and the holes marked off for the rivets and drilled both on butt-plate and rim. Four rivets on each side of the joint will be sufficient to hold it firmly together, the rivets being $\frac{1}{8}$ in. diameter. Next rivet the butt strip to one side of the joint and leave it so until the boss or hub has been constructed with spokes complete.

WATER WHEEL BOSS.

For the boss, assuming that the shaft is to be made $\frac{7}{8}$ in. in diameter, obtain a piece of brass 2 in. long by $2\frac{1}{2}$ in. wide; through this drill a hole $\frac{7}{8}$ in. diameter, to be a good fit to the shaft, in order that it may be driven tight on to it. Cut a key-way through the hole, as shown in Fig. 2054, which represents the wheel only, with buckets complete. For the finished motor, see Figs. 2060 and 2061 (p. 681). The keyway should be ¼ in. broad by ⅛ in. deep.

SPOKES OF WATER WHEEL.

The boss must now be marked off to receive the spokes, which are to be screwed firmly into it. The holes in the boss are to be screwed ⅜-in. common thread; therefore the holes to be drilled will only be $\frac{5}{16}$ in. diameter and $\frac{7}{8}$ in. deep. The six spokes (Fig. 2055) are $\frac{7}{8}$ in. in diameter, and one end is screwed up ⅜-in. common

end of each spoke must reach the bottom of its hole in the boss. Now mark off accurately round the rim the holes to receive the spokes; make these of $\frac{5}{16}$ in. diameter, to be a good fit to the outside end of the spokes, which have been shouldered down. The holes are slightly countersunk, so that the projecting part of the spoke may be cupped round into the countersunk part.

BENDING RIM.

Bend the outside rim round until the

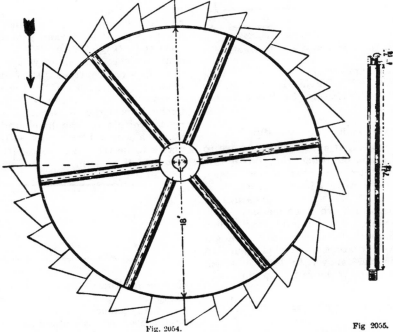

Fig. 2054.—Water Motor Wheel. Fig. 2055.—Water Wheel Spoke.

thread for a distance of $\frac{7}{16}$ in.; the other end of the spoke is to be shouldered down for a distance of ¼ in. to a diameter of $\frac{5}{16}$ in.; this end will protrude through the rim $\frac{3}{16}$ in. in order that it may be riveted to it. The spokes should be firmly screwed into position in their respective places round the boss, care being taken that they are screwed home, for if they are not, they will be hammered loose when attempting to rivet them finally; therefore the screwed

butt ends meet. This can be done by first bending it by hand round a bar of large diameter, and then working it into shape with light blows from a wooden mallet. When it has been turned to a true circular form, place the spokes in position and clasp up the loose end of the rim by means of small bolts and nuts to the portion of the butt-plate at the back of the joint: this should entirely enclose the boss and the six spokes.

RIVETING ENDS OF WATER WHEEL SPOKES.

The ends of the spokes may be riveted up in cuphead form. The wheel being rather large to handle, either grip each spoke separately in a vice, or put a temporary bar through the boss, grip the bar

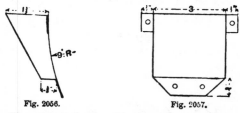

Fig. 2056.   Fig. 2057.

Figs. 2056 and 2057.— Bucket of Water Motor.

in the vice, and then rivet the six cupheads down. This having been done, the butt-joint in the rim may be firmly riveted up.

WATER MOTOR CUPS OR BUCKETS.

The twenty-eight cups or buckets are to be equally spaced round the rim of this internal wheel and riveted in their place, using cuphead rivets $\frac{1}{8}$ in. diameter. Figs. 2056 and 2057 are working drawings of these cups, which are to be made of $\frac{1}{32}$-in. charcoal plate or good mild steel; they are fitted to the rim so as to lie snugly against its curvature before being riveted to it. Fig. 2058 shows the form in which the buckets are to be cut from the sheet

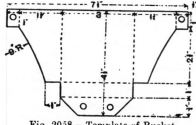

Fig. 2058.— Template of Bucket.

before the corners (which are shown by dotted lines) are bent, as seen in Figs. 2056 and 2057. Fig. 2057 is a view of the face of a bucket fitting on the wheel.

RIVETING CUPS TO WHEEL RIM.

The rivet holes for fixing the cups in position should be next drilled and the cups riveted in their place. The rivets must not be clenched up by using the hammer alone; if this is attempted, it will be found that after two or three cups have been fixed in place, further use of the hammer will only loosen the rivets that have been clenched. The tools needed for successful riveting are a draw-punch and a rivet-snap. The draw-punch is made from a piece of round steel of convenient length to handle, with the ends filed up square. In one end of this must be drilled a hole $\frac{3}{16}$ in. diameter and $\frac{1}{2}$ in. deep. The rivet-snap (Fig. 719, p. 252) is made of a similar piece of steel, in whose end there is a cup which represents the form of rivethead required. To use the tools, after inserting the rivet through both rim and bucket, place the draw-punch so that the rivet projects up the hole, and give the punch a smart blow; this will draw the plates to be riveted close together. Next put on the snap, without moving the

Fig. 2059.— Water Motor Stand.

rivet-head from where it is resting, and give the snap a quick blow, which will effectually clench the rivet and form the head. Two or three blows at the most are required for each rivet.

KEYING WATER WHEEL TO SHAFT.

Having completed the wheel, fix it firmly to the shaft by means of a sunken key. The key-way in the shaft should be 2 in. long by $\frac{1}{4}$ in. broad by $\frac{1}{8}$ in. deep, and may be cut out by means of a cross-cut chisel, and a piece of key-steel, 2 in. long and $\frac{1}{4}$ in. square, fitted to the recess. The key should be now lightly caulked in its place, after which the wheel can be driven on to the shaft.

STAND AND COVER FOR WATER MOTOR.

A neat stand, of trestle form, may now be constructed as shown in Fig. 2059. Its legs are to be made of flat bar-iron, ⅜ in. by 1½ in., whilst the horizontal supports for the plummer-blocks are to be made of light angle-iron, sufficiently broad to bolt the two ⅞-in. plummer-blocks to them. A skilled worker would construct the stand of angle-iron alone. The cover scarcely needs description, but it may be said that galvanised iron is the best material for it, and that it is built up in two parts, top and bottom, the form of which is shown in Figs. 2060 and 2061. The width of the casing, or cover, ought to be 5 in. inside,

plummer-block. The wheel should revolve freely by hand without side play, to avoid friction. A coat or two of good oxide paint should be applied inside and out. The supply pipe should not be less than 1 in. in diameter, and a screw valve should be used in preference to a cock. The waste pipe should be at least 1½ in. in diameter, in order that the water may be easily got rid of. The arrow shown in Fig. 2054 indicates the direction and position of the jet of water from the nozzle, the extremity of which should not be less than ⅛ in. in inside diameter. By following the instructions given above, a motor can be made without castings; however,

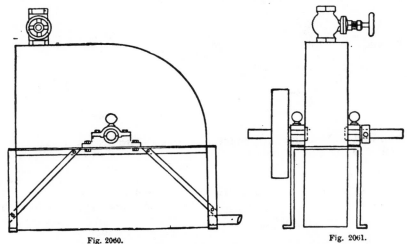

Fig. 2060.

Fig. 2061.

Figs. 2060 and 2061.—Side and End Views of Water Motor.

with ½-in. clearance all round the wheel, excepting at the corner where the supply valve is situated; this corner, for obvious reasons, is square.

COMPLETING WATER MOTOR.

To prevent lateral play, the internal wheel, before the top cover is put on, must be placed midway between the casing, there being ½ in. clearance on each side. On the outside of the plummer-block a stop-ring should be fixed against it by means of a set-screw; whilst to stop the lateral movement on the opposite side the belt-pulley should be keyed up against the

should the extra expense be immaterial, castings should be used, as they are, of course, neater and more convenient.

ANOTHER WATER MOTOR.

The motor about to be described develops ½ horse-power at 1,500 revolutions per minute on a fall of 30 ft., or ¾ horse-power at 2,000 revolutions per minute on a fall of 50 ft. The tools necessary for making it are a small screw-plate, taps, a pair of large scissors, a soldering bit, and a half-round file. The materials required are: one sheet of No. 10 zinc, 3 ft. by 2 ft.; one sheet of No. 14 zinc, 3 ft. by 2 ft.; brass

tubing, 2½ in. long and ¾-in. bore; four
pieces of brass, about $\frac{3}{32}$ in. thick by 1 in.
by 3 in.; 1 lb. of soft solder; one piece of
beech, ¾ in. by 1 ft. by 1 ft.; one piece of

covered with zinc. The hole in the centre
should be bored out for the tube, and the
arms are dovetailed in. The hole must be
perfectly square right through and the

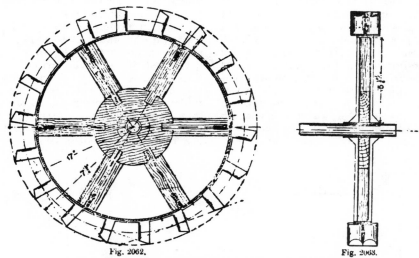

Figs. 2062 and 2063.—Elevation and Section of Water Wheel.

beech, 6 in. by 3 in. by 2 in.; and one piece
of bar iron, 6 in. by ¾ in. diameter.

WATER MOTOR DISC.

Figs. 2062 and 2063 show the construction

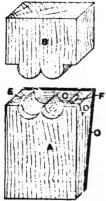

Fig. 2064.—Blocks for Forming Water Wheel
Buckets.

of the wheel. The disc on which the cups
or buckets are placed is built up of beech,

arms square across, or the wheel will not
run true. With compasses mark on the
stiff zinc two circles 12 in. diameter, with
a circle of ⅞ in. diameter at the centre of
each; cut out these exactly to the lines,
and they then ought to fit tight on the
brass tube. Punch suitable holes in the
discs, and through them screw the circular

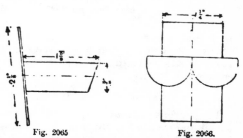

Figs. 2065 and 2066.—Water Wheel Buckets.

pieces on to each arm and to the centre
piece; file down the screw heads a little,
and drop solder over them. Then clean
the brass tube with emery cloth, and
solder the discs firmly to the tube.

Soldering Disc Rim.

Cut a strip from the 10-oz. zinc, $1\frac{1}{4}$ broad and long enough to meet round the

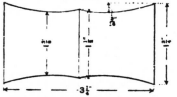

Fig. 2067.—Pattern for Bucket.

disc, and solder this in place, taking four pieces of beech, $\frac{7}{32}$ in. thick (the distance between the disc and the edge of the rim) and any size, to hold up the disc while the rim is being soldered. The tube is placed, say, in a hole in a box, the pieces placed

intervals, and solder all round. Eight zinc webs, as shown in Fig. 2063, are now soldered to the discs at equal intervals, four on each side.

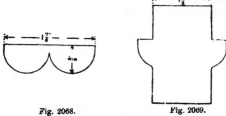

Fig. 2068. Fig. 2069.

Figs. 2068 and 2069.—Parts of Water Wheel Bucket.

Water Motor Cups or Buckets.

For the cups, blocks (Fig. 2064) made

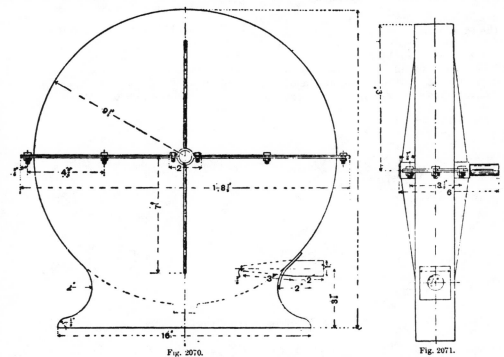

Fig. 2070.

Fig. 2071.

Figs. 2070 and 2071.—Side and End Elevations of Water Motor.

underneath, and the rim slipped over till it rests on the box. Then rub spirits round the seam, drop solder all round at 1-in.

from the 6-in. by 3-in. by 2-in. beech, cut in two, are necessary. The block A receives the zinc, which is punched into shape by the block B. In the block A one side is cut away at an angle of 73°; this is the angle at which the cup lies on the disc. The sawkerf D is at an angle of 80° to the end, and 1½ in. from the sloping face. A sheet of zinc or wood is held by three screws to the sloping side, zinc catches

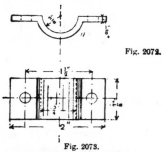

Fig. 2072.

Fig. 2073.

Figs. 2072 and 2073.—Bearings.

E and F revolving on screws and holding the curved zinc while it is being soldered. One of the eighteen cups soldered round the rim is shown by Figs. 2065 and 2066. Make three templates to the dimensions given in Figs. 2067 to 2069, and mark off eighteen pieces of zinc to each. Bend the

the centre by screwing up C (Fig. 2061), then solder; take the stamping out and clip all the edges on the outside. Solder these cups round the rim at equal distances.

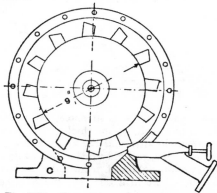

Fig. 2076.—Inner Elevation of Water Motor.

WATER MOTOR SHAFT AND CASING.

The shaft is iron, 6 in. long by ¾ in. diameter. Drill and tap two small holes through the brass tube and ½ in. into the shaft, and fit two screws. For the casing mark off a 19-in. circle on

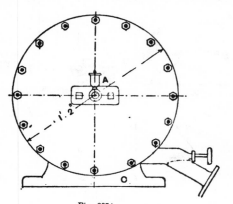

Fig. 2074.

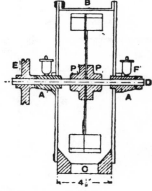

Fig. 2075.

Figs. 2074 and 2075.—Elevation and Section of Water Motor.

pieces like Fig. 2067 at the dotted line, place them in the block A, and punch them down with B (Fig. 2064). Take off the inside corners, holding each stamping by E and F. Next solder on the ends (Fig. 2068). Catch the pieces like Fig. 2068 in

the thick zinc with a ¾-in. circle in the centre, and cut it in halves through the centre. Cut also a strip of the No. 10 sheet 2½ in. broad, and long enough to stretch round one of these half circles, to which it should be soldered. Now mark and cut out two flanges for the cover and solder one to the top and one later on to the bottom. Six $\frac{1}{4}$-in. holes must be punched in those flanges for bolts, one at each end and two on each side. Mark off

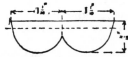

Fig. 2077.—Part of Bucket.

one more 19-in. circle with a $\frac{3}{4}$-in. hole, and draw a line through the centre, but do not cut the complete circle, as lugs will be required on the bottom, as in Fig. 2070. Now cut another strip $2\frac{1}{2}$ in. broad to go half round a 19-in. circle ; in the centre of this mark off a square $2\frac{1}{2}$ in., with a 1-in. square within it. Cut out the 1-in. one, and cut from its corners up to the corners of the $2\frac{1}{2}$-in. one ; then bend down and solder the corners ; the water escapes from this hole. Then make two strips to stretch from the circle just soldered on to the ends of the lugs ; these must be soldered on very firmly, and a flange has to be soldered to the open end as in the top half. The case complete is shown in Figs. 2070 and 2071.

COMPLETING WATER MOTOR.

The bearings (Figs. 2072 and 2073) are either cast or of sheet brass, and the bottom halves are tapped, as shown in Fig. 2070, and soldered in position firmly, with

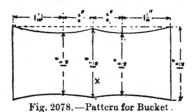

Fig. 2078.—Pattern for Bucket.

a web under them about 7 in. long and $\frac{3}{4}$ in. at the top. Mark off and cut out a 1-in. circle on one end, as in Fig. 2071, and in this fit the nozzle made from a piece of lead tube of 1-in. bore tapered down to a $\frac{3}{8}$-in. hole, by working round with a hammer till $5\frac{3}{4}$ in. long. This is for the supply,

and when the wheel is in place the motor will be complete.

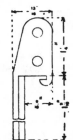

Fig. 2079.—Detail of Bucket.

MOTOR WITH TWELVE CUPS OR BUCKETS.

Figs. 2074 to 2076 illustrate, to the scale of $1\frac{1}{2}$ in. to the foot, the construction of a motor to be driven from the ordinary water supply, and suitable for driving a sewing machine or any light machinery. The capacity of the cups or buckets is sufficient for 20 gal. per minute, and with the nozzle to be described the power of the motor can be varied as required, with little loss of efficiency. The case is constructed of two metal discs of sufficient thickness to carry the two bearings A without vibration. The rim B (Fig. 2075) is of sheet metal of about No. 18 S.W.G., with two flanges soldered on as shown, and the back disc is riveted and soldered, and the front one bolted to the flange. The disc and rim of the case may be made or iron, brass, or zinc. The case is fixed to the pedestal C (Figs. 2074 to 2076), which may be cast in lead, with a central hole to take off the spent water. The shaft D is of $\frac{1}{2}$-in. mild steel ; E is the

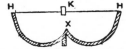

Fig. 2080.—Section of Bucket.

driving pulley, and F a collar. As the efficiency of a water motor depends on the construction of the buckets and nozzle, these will be more fully dealt with. The wheel, of sheet brass $\frac{1}{8}$ in. thick and 9 in. in diameter, should run true on the shaft, and is fixed between the two bosses P, which are secured to the shaft by nuts not illustrated.

THE CUPS OR BUCKETS.

Figs. 2077 to 2081 give details of the buckets, of which twelve will be required. Of sheet brass No. 18 s.w.g., cut twenty-four pieces to Fig. 2077 and twelve pieces to Fig. 2078 ; cut also twelve pieces of ½-in. sheet brass to Fig. 2079 to form the lugs by which the buckets are secured to the wheel. The buckets are built up by bending the pieces (Fig. 2078) to the shape shown in section at Fig. 2080 ; the bend at x should be quite sharp, a knife-edge if possible, as at this point the jet is split into two parts, which are thrown off at H, and with a well-designed bucket the water should have no energy left, but should fall

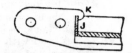

Fig. 2081.—Detail of Bucket.

away as dead water, the whole of the velocity being imparted to the wheel. Two end pieces (Fig. 2077) are soldered to the bent part of the buckets as shown at J (Fig. 2081), and the buckets are fixed to the lugs by slipping the catch K (Figs. 2080 and 2081) over the bucket and soldering the whole together. The buckets when finished are riveted to the wheel as shown in Fig. 2076.

NOZZLE OF WATER MOTOR.

The nozzle (Fig. 2082) may be cast in brass, or the body and flange may be dressed up from lead pipe, the nozzle B being of brass soldered to the body. To get a solid jet, the taper of the opening should be at an angle of 13½°. The sleeve c must be of brass tube tapped to suit the spindle D for regulating the flow of water,

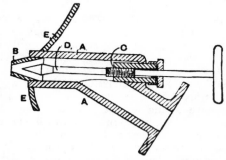

Fig. 2082.—Section of Water Motor Nozzle.

and must be fitted with a stuffing box similar in construction to a high-pressure screw-down water valve, to prevent leakage. A plate of ½-in. sheet brass, E, is cut to the width of the casing between the flanges, and to it the nozzle is soldered. This plate fits the rim and pedestal, and is fixed with small bolts and nuts in the position shown in Fig. 2076, great care being taken that the jet strikes each bucket in the centre. A shows the supply pipe.

BUILDING A DYNAMO AND ELECTRIC MOTOR.

INTRODUCTION.

THE machine to be described in this chapter is termed a 50-watt undertype dynamo and motor. It will either light four 10-volt 5-candle-power lamps, or work well as a motor when supplied with current from a battery. The construction of a small dynamo can well be recommended as a piece of mechanical work. It will be found that making a small dynamo to run at, say, 3,000 revolutions per minute, of good design and proportions, is both interesting and instructive. Such a dynamo, which may be driven from a lathe fly-wheel, hand-wheel, water motor, small gas or steam engine, or petrol motor, will prove a very good investment of time and labour. The illustrations are all reproduced to a scale of one-half full size. No knowledge of electricity is needed in building this machine. Patterns will have to be made and castings obtained. Assuming that the castings and other materials are ready, it will be convenient to consider the work of construction in three divisions—first, mechanical construction; second, insulation; third, winding on the wire and connecting up.

SHAFT BEARINGS.

Begin with the bearings A (Figs. 2083 and 2084); the holes for the shaft must be drilled. To do this, mark a centre at each end of the boss, measuring the position from the drawings. The holes should be drilled between the centres of a lathe with a drill rather less than ¼ in. diameter. Drill up a little way from one end and then reverse the bearing; drill up a little from the other end, reverse again, and so on until the holes meet at about half-way. The oil-cup bosses B (Fig. 2084) can now be drilled as shown, the small hole at the bottom, about $\frac{1}{16}$ in. diameter, being drilled through into the bearing hole to allow oil to pass. The holes for the bearing will be found to be somewhat rough and perhaps not exactly in line; therefore a reamer or rose-bit must be carefully passed through. Use a little oil as lubricant, and reamer through from one end completely—not first one side and then the other, as when drilling out. If the drilling has been carefully done with a drill which leaves about $\frac{1}{32}$ in. to come out, the finished hole will be quite smooth. Care must be taken not to spoil the surface when handling the bearings for other operations.

FIELD-MAGNET BEARINGS.

Now take the field-magnet casting C (Figs. 2084 and 2085). The bore should be cleaned out with a file to remove any lumps or irregularities, and if well cast will be very nearly a true circle. Those who make their own patterns and have a heavy lathe may prefer to leave sufficient metal to take a cut through; this of course makes the best job, but good results may be obtained without it if care is taken to secure a good casting. A dummy armature and shaft must now be made to enable the bearings to be fitted on in correct position. Take a piece of round iron or steel rod of any convenient diameter, such as ⅜ in., and just the length of the armature shaft. On this mount a piece of hardwood the length the armature core will be, and about 1¼ in. diameter. It should be tight on the rod, and in the same position as the armature core will be. Turn the wood to fit tightly in the bore of the field-magnet ; turn down the rod also at each end to fit the bearings. on the sides of the field-magnet ; if they are much out, the castings may be knocked

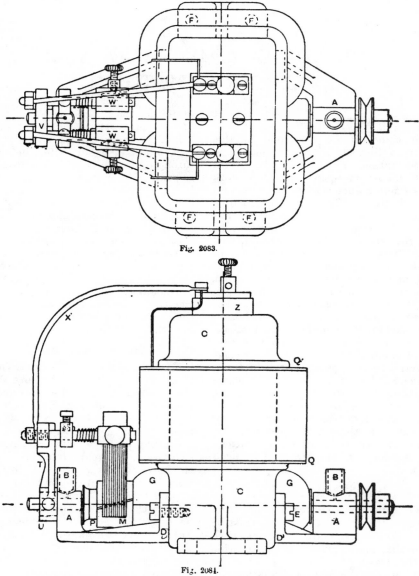

Fig. 2083.

Fig. 2084.

Figs. 2083 and 2084.—Plan and Front Elevation of Dynamo and Electric Motor.

Place the dummy tight in the bore, slip the bearings into place, and file the end lugs D (Fig. 2084) until they bed quite flat until they are somewhere near the mark, and then the remaining adjustment effected with the file. A little red ochre mixed with oil and smeared on the magnet will show where the lugs touch. It is important that the bearings bed properly, or they will not be in line when screwed up in place, and the shaft will run stiffly.

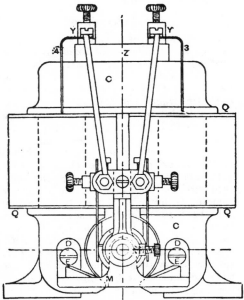

Fig. 2085.—End Elevation of Dynamo and Electric Motor.

In drilling the holes for the screws E, it is convenient to mark the positions according to the drawing and drill these first; then slip the bearings on the dummy shaft and mark off the holes for the screws in the field-magnet by drawing the point

circles. These holes may be about ⅜ in. deep, and should be tapped. Mark off the holes F (Fig. 2083) in the field-magnet feet for holding-down screws, the positions being taken from the drawing. The diameter of the holes may be ¼ in., or it may be made to suit the wood screws intended to be used.

ARMATURE SHAFT AND BEARINGS.

The armature G, 1½ in. diameter by 2¼ in. long, is now to be made. To make its shaft, centre and straighten a piece of steel 7¾ in. long; if this is made straight and true, the diameter can be $\frac{5}{16}$ in., so that the central portion need not be turned; but the best way is to have the diameter ⅜ in. and take a light cut all along, reducing to $\frac{5}{16}$ in. diameter. Now turn down a part at each end for the bearings. The length at the pulley end to be turned down is 1½ in., and at commutator end 2⅛ in. These parts should not be now turned to fit the brass brackets, but should be left a little large so that they can be fitted in after the core discs H (Fig. 2086 and 2087) are tightened up, the screwing up of the clamping nuts J having a tendency to spring the shaft out of truth. Now cut the screwed portions, about ⅞ in. at each end, as shown, for the clamping nuts. For this a fine thread is necessary (about twenty to the inch), and a screw-cutting lathe is the best tool to use; but if this is not available, the thread may be started with stock and dies, and finished with a chaser.

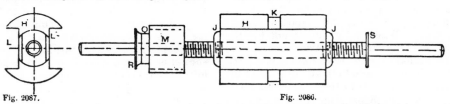

Fig. 2087.

Fig. 2086.

Figs. 2086 and 2087.—Front and End Elevation of Armature showing Commutator.

of a scriber round the holes in the bearings. Some chalk rubbed on the magnet where the bearings come will assist to show the line. Drill tapping holes for $\frac{3}{16}$-in. screws in the centres of the marked

ARMATURE CORE DISCS.

The core discs H must now be prepared. About $\frac{1}{50}$ in. thick is a common size, and of these about 120 will be required. See that the central holes go on the spindle easily, then take off any burrs with a file. This is important, as the screwing-up of the nuts would cause any uneven parts to bend the shaft. The circular groove K (Fig. 2086), $\frac{1}{4}$ in. long, in the middle of the core is to take the binding cord, and a few discs should be reduced in diameter and put on the shaft when half the other discs are on, so that the groove is produced without turning. It is a difficult job to turn out this groove if left till the core is complete. Before putting on the discs, one side of each should be painted with a thin coat of enamel or Brunswick black and left to dry. The clamping nuts J are circular pieces of brass $\frac{3}{4}$ in. diameter, with a $\frac{1}{16}$-in. hole drilled and tapped in the centre to fit the screwed shaft. The side which is to go next to the discs should be faced up on a screw mandrel, with the sharp edge on the outside rounded off as shown in the drawing. File two parallel flats to take the spanner. Now put the core discs on the shaft, the painted sides all facing the same way ; but before tightening up the nuts, see that the core is in the right position lengthways on the shaft, and that the channel L (Fig. 2087) for the insulated wire is straight ; then screw up as tight as possible, leaving the flats on the nuts flush with the channel ; a little oil between the face of the nut and the end disc is of assistance.

FITTING ARMATURE SHAFT.

Now try the shaft between centres. If it is out of truth, straighten it ; then finish the ends to fit the brackets. This is an important operation, and a good fit should be made. When fitting in do not file the shaft more than necessary, as it tends to make flats, or to leave the shaft elliptical. When the brackets are of a good fit, place the armature in the field-magnet and fix brackets in place. The armature should be central in the bore and turn freely with the fingers ; if it is stiff, ease the shaft bearings until it runs freely and without shake. An end movement of about $\frac{1}{16}$ in. may be permitted—in fact, is advisable. The core should just coincide lengthways with the field-magnet, and should not project from

one side more than the other, as the magnet will try to pull it into a central position. If in doing this it pulls it against a bearing, that bearing will run hot. If any of the core plates project, they can be levelled down with a file.

THE COMMUTATOR.

The commutator consists of a piece of brass tube M (Fig. 2086) fixed on a wood bush and carried on the shaft. The brass tube is divided into two equal portions by saw cuts parallel to the axis of the shaft (in practice it is usual to make the cuts a little slanting), one portion being for one end of the armature wire and the remaining portion for the other. The essential points are—the two portions must not be in metallic communication with each other except through the armature wire, and they must not be in metallic communication with the shaft. The tube may be a piece of thick seamless drawn tube, and should be a little over 1 in. outside diameter and $\frac{3}{4}$ in. long. The thickness is not of importance—say about $\frac{1}{8}$ in ; so as to allow turning the surface up from time to time as it becomes worn by the brushes ; the inside need not be bored unless a casting has been used.

BUSH FOR COMMUTATOR.

The hardwood insulating bush O is $\frac{7}{8}$ in. long, and large enough in diameter to fit the commutator. Drill the hole for the shaft with the drill used for the bearings, and reamer it out to fit tightly on the shaft. It is best to make a mandrel to turn the bush on, slightly smaller in diameter than the shaft, as the hole is apt to become enlarged in turning. The tube should be fixed in with a little cement, and left to set before proceeding. The bush must project $\frac{1}{8}$ in. at the end nearest the bearing.

SCREWING COMMUTATOR SEGMENTS TO BUSH.

Now mark off centres for two holes in the centre of the tube, and at diametrically opposite points, for the screws which hold the commutator segments to the bush.

These holes may be $\frac{1}{8}$-in. tapping for countersunk screws, drilled right through to the shaft. Countersink the tube so that the screw-heads will come nearly flush, just leaving enough projecting so that the hole will be quite filled up when the tube is turned up outside. These screws should be brass, and must screw in tightly, or when turning up the commutator they will come loose and cause trouble. It is very important that these screws do not come flush inside with the shaft-hole, so as to touch the shaft; they should be about $\frac{1}{16}$ in. short, as there must be no metallic connection between the tube and the shaft.

Completing the Commutator.

Now mark two centre lines on the tube from one end to the other, diametrically opposite one another, and midway between the screws just put in. These lines are to locate the positions for the cuts P (Fig. 2084) which separate the tube into equal parts. These cuts are made with a saw, about $\frac{1}{32}$ in. thick, and may be cut along the lines drawn, crossing exactly in the centre, as this causes the brushes to pass from one segment to the other gradually, and lessens the sparking and wear. The angle of slant is not important; about $\frac{1}{16}$ in. each side of the centre line will do. Cut right down to the bush and slightly into it; see that the slot is perfectly clear of cuttings and that the segments are separated from one another, then insert a thin strip of wood, or better still, mica, into the slot to keep the segments apart and prevent dust from getting into the slot. A little cement or shellac varnish on the wood before inserting will keep it in place. Now in each segment at the end next to the armature core, with a saw make a nick to take the ends of the armature wire, and just large enough to allow it to go in a good fit. The commutator is now complete, and may be gently driven on to the shaft, the slots being in line with the round part of core—that is, at right angles to the centre of channel L. The oil-throw R (Fig. 2086) can be of brass, and is driven tight on the shaft after the commutator is in place. The oil-throw S for the pulley end can be tapped to fit the screwed part of the shaft, or made to fit the plain part, and can be put in place after the armature is wound.

Rocker of Brush Gear.

In making the brush gear, begin with the rocker T (Fig. 2084). The rocker allows the brushes to be moved round the commutator in order to find the position which gives best results when working. Clip the rocker in a chuck, or fasten it on a face-plate, in order to bore out the hole in the boss U, which fits on to the boss on the bracket. Set it so that the arm is straight and square with the hole. The exact size of the hole is not important— $\frac{1}{2}$ in. diameter will do—but it should be parallel. Put it on a mandrel and face each side of the boss. Now put the bearing bracket on a mandrel, and turn down the projecting boss to fit the hole in the rocker. The small boss at the side of the rocker is to take a set-screw, as shown; this screw should fit well in the thread.

Crosspiece for Brush Pins.

A piece of hardwood is required for the crosspiece V (Fig. 2083) to carry the brush-pins. This should be filed up square to about $1\frac{1}{4}$ in. by $\frac{1}{2}$ in. by $\frac{1}{4}$ in. thick; the angle to receive it must be filed out square to the hole in boss, so that when the cross-piece is fixed in, the pins will come parallel to the shaft. Mark off a hole exactly in the centre of the crosspiece, and drill a clearing hole for a $\frac{1}{8}$-in. diameter countersunk screw. Now place the crosspiece in position in the angle of the rocker arm, so that it projects equally each side, and mark off the position of the hole on to the brass with a scriber; drill a tapping hole, and tap it to fit the screw. The crosspiece can now be fixed in place. Mark off centres for the brush-pins $\frac{3}{8}$ in. each side from the middle point, and on centre line of crosspiece drill and tap $\frac{3}{16}$-in. diameter holes.

Brush Holders.

The brush holders W (Fig. 2083) must be finished next. The castings should be filed up all over, and the ends squared up. Then mark off the centre at one end of the circular part of the casting, and drill a $\frac{3}{16}$-

in. diameter hole right through. The slots for the brushes can be cut with a saw, as illustrated, from one end. They should be $\frac{9}{16}$ in. long and fully $\frac{1}{8}$ in. wide, and parallel to the holes already drilled. Drill and tap the bosses for $\frac{1}{8}$-in. diameter clamping screws, as shown. These bosses may be turned up by putting a small mandrel in the screw-holes before they are tapped and holding the mandrel in a chuck.

BRUSH PINS.

The pins are made from brass rod. If straight $\frac{3}{16}$-in. rod is chosen to fit the holes in the brush holders, they will not require to be turned, but only polished. Screw one end of each pin for a length of $\frac{9}{16}$ in., and at the other end drill a small hole to take a pin, which prevents the brush-holder being forced off by the spring. The pins can now be screwed tightly in place in the cross-arm. Make two nuts to screw on the part which projects through to clamp the flexible wires x which carry the current to the terminals.

SPRING-TENSION ADJUSTMENT COLLARS.

Two collars, as shown, are required for adjusting the tension on the springs. They may be made from $\frac{3}{8}$-in. diameter brass rod, drilled to fit the pins, and each should be fitted with a set-screw to fix them in the required position. A small hole is drilled in the face of each collar at one side to take one end of the spring, and a similar hole should be drilled in each brush holder at the end to take the other end of the spring.

MAKING TENSION SPRINGS.

The tension springs are made of hard brass wire, about No. 24 B.W.G. To make the springs, take a piece of $\frac{1}{16}$-in. diameter rod, and hold it in a chuck so that $\frac{3}{4}$ in. or so projects. A small hole having been previously drilled at the outer end of the rod, the wire is pushed through this hole for about $\frac{1}{2}$ in., and then the lathe is rotated. Pull moderately on the wire, and allow it to coil along the rod until the chuck is reached; cut off the wire, and the spring is made, but requires the ends to be bent at right angles to fit the holes in

the brush-holder and collar. The lathe must be rotated backwards for one spring, and forwards for the other, so as to produce right- and left-coiled springs. Looking at the dynamo from the commutator end, the right-hand spring is a left-hand coil and the left-hand spring a right-hand coil.

TERMINAL BLOCKS.

The terminal blocks y (Fig. 2085) are very simple; they may be filed up all over and polished. Holes are drilled and countersunk to take wood screws, which hold them down on the terminal-board. A hole is drilled and tapped in each at the end next the commutator to take $\frac{1}{8}$-in. diameter cheese-head screws, which clamp the field-wires and brush-wires. Drill holes through the upright blocks to receive the outer-circuit wires with set-screws from the top; the sizes of these holes are not particular.

TERMINAL BOARD.

The terminal board z (Figs. 2084 and 2085) can be made from a piece of mahogany about $1\frac{1}{4}$ in. by 2 in. by $\frac{1}{4}$ in. thick, and polished or varnished according to taste. Holes are drilled in it to take two $\frac{3}{16}$-in. countersunk screws, which fix the board to the field-magnet. The holes in the magnet should be drilled and tapped last, and marked off for position from the terminal board.

DRIVING PULLEY.

The driving pulley can be made to take either a flat or a round belt of dimensions to suit the circumstances of driving. If the dynamo is to be driven from a foot lathe or hand wheel, a pulley about 1 in. diameter over all is a convenient size, with V-groove to take a $\frac{3}{16}$-in. diameter round belt. The width of the pulley should be about $\frac{3}{8}$ in., and it can be fitted with a taper fitting to the shaft or a parallel fitting with set-screw in the boss. The fit should be good, and the armature may be put between centres and the pulley finally touched up true in its place.

INSULATION OF DYNAMO.

The insulation can be undertaken when the mechanical construction is finished.

Take out the armature and take off the bearings. Examine the field-magnet where the wire is to be wound, and smooth off all corners and any rough places with a file, taking off all sharp edges and points which are liable to cut through the insulation. Wrap two layers of thick brown paper round each core, sticking them on with shellac varnish. Cut out four rectangular cardboard cheeks Q (Figs. 2084 and 2085) to fit the cores; the depth of the wire will be about ⅔ in. from the core, so the cheeks must be made to suit; they can be sprung into place on the cores if a slanting cut is made across one side and a piece of paper pasted over the cut to keep it together when the cheeks have been pushed to their places at the ends of the core. Look over the insulation carefully, and having seen that it is sound everywhere so that the wire cannot come into contact with the iron at any point, brush a good coat of shellac over the paper wrapping and cheeks, and leave it until quite dry. The armature must be treated in a similar way. Smooth all projecting points, edges, and corners along the channel where the wire is going to be wound; then cover the whole of the channel, the ends of the core and shaft, up to the oil-throw at the pulley end, and the commutator at the other end, with a layer of thick brown paper. Leave the paper projecting a little along the edges of the channel so as to be sure the insulation comes right up to the edges; it can be trimmed down after the wire is wound on. The edges at the ends of the channel can have an extra thickness of paper put on over the first covering, as the wire is liable to cut through at these points. Examine the insulation, give it a good coat of shellac if it is all right, and leave it to dry.

Winding Field-magnet.

The winding can now be put on. The field-magnet requires about 2 lb. of No. 22 s.w.g. single cotton-covered copper wire, which must be wound on by hand in the direction shown in Fig. 2088, layer by layer, keeping it as even as possible with a moderate tension; commence at the top of the magnet and finish at the bottom. The

exact number of layers does not matter, and the winding may be finished at the top if preferred; it does not matter greatly if the number of layers on each core is not quite the same, the essential point is that the winding shall be in the direction shown in the diagram and kept so throughout; try to put about 1 lb. of wire on each core. As each layer is finished it should be brushed all over with just sufficient shellac varnish to give the surface a good coat. The commencing ends of the wire, 3 and 4 (Fig. 2088), which reach from the core outwards, should be wrapped round with thin paper along the part which is

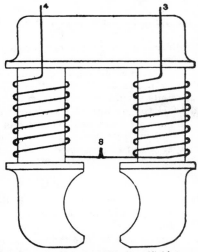

Fig. 2088.—Method of Winding Field-magnet.

buried in the end of the coil and varnished with shellac to make sure that the current goes first to the innermost layer and does not leak away to the other layers. The guiding principle is that the current must follow the wire throughout its length, and not be able to make a short cut across at any point; it is to ensure this that the wire is covered with cotton. If a bare or frayed place is found while winding, cover it with some thin paper.

Method of Winding Coils.

The most convenient way to wind the coils is to fix a strip of wood to the top of the magnet by the terminal board screws and then fasten the wood to a face-plate fixed on the lathe head, bringing each core in turn to the centre ; turn the face-plate round with the left hand, guiding on the wire with the right hand, assisted by the

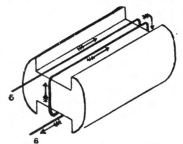

Fig. 2089.—Method of Winding Armature.

left where the wire requires passing between the cores. Completely wind one core first to the full depth of the cheeks, and then proceed with the other, joining up the ends when the final connections are made. If the weight of the overhanging core be counterbalanced the magnet will be rotated more conveniently. Finish the last layer with two or three coats of shellac varnish.

Winding Armature.

The armature requires about ¼ lb. of No. 20 s.w.g. double cotton-covered copper wire. Put the armature in the lathe between centres, with the commutator to the right hand. Begin from the right-hand end and lay the wire from the right, along the channel, to the left-hand end, then down the end and along the channel underneath to the right-hand end, and so on right across the channel for one layer, then back again for a second layer, and so on until the channel is quite full ; get on as much as possible. Each layer should have a coat of shellac varnish. If found more convenient, one side of the shaft may be filled up first and then the other side, the essential point being that the direction in which the wire is wound round the core must be the same right through (see Fig. 2089). It is very easy to reverse the direction of the winding when passing from one side of the shaft to the other and not

notice it, so care must be taken to avoid this mistake. The wire being all on, bind it down tight with some strong thin cord in the centre, the cord lying in the groove round the centre of core prepared to receive it ; make an even binding about ¼ in. long. The best way is to put the armature between centres, and turn the lathe slowly backwards whilst winding on the cord. This binding is to prevent the wires flying outwards by centrifugal force when the armature is rotating.

Commutator Connections, etc.

The beginning and end of the wire 5 and 6 (Fig. 2089) must now be connected to the commutator. One wire goes to each segment. The beginning wire 5 goes to the top segment 5 (Fig. 2090) and ending wire 6 to the other segment. The wires must be soldered into the nicks made for them ; it is well also to bind some cord round these wires to keep them in place. When all is finished give the wire and binding cord a good coat of shellac. The commutator can now be finally touched up in the lathe, taking a very light cut, to remove surplus solder and make it run quite true. If the pulley end oil-throw has not yet been put on, before doing so put a disc of thin card between it and the armature wire, to prevent any chance of bruising the insulation.

Putting Dynamo Together.

The dynamo can now be finally put together. The field wires are connected up

Fig. 2090.—Diagram showing relative Positions of Armature and Commutator.

to the terminal blocks (as shown in Figs. 2084 to 2086), and also joined together in centre at 8 (Fig. 2088) as shown. This last joint should be twisted and soldered over to make good contact. Connect the brush-holder pins to the terminal blocks, as shown, by flexible wires, so that the rocker may be easily moved about. To make these flexible wires, coil up some No. 20 insulated wire on a $\frac{1}{16}$-in. diameter rod, just as the springs were made, baring just enough of the wire at each end to make contact with the clamping screws.

MAKING THE BRUSHES.

The brushes may be made of sheet copper or copper wire; they should be flexible so as to make good contact with the commutator and brush holders. A good flexible brush may be made as follows: Take some copper wire, about No. 24 gauge, and, fixing one end in a vice, take hold of the other end with a pair of pliers and give a fair pull; this will stretch the wire a little and straighten it. Now cut off sufficient pieces each 2 in. long to make two brushes each $\frac{1}{2}$ in. wide, and solder the wires together at one end. It is a good plan to curve the brush a little where it touches the commutator, so that instead of touching at a point there is a surface of contact about $\frac{1}{8}$ in. broad all along the brush.

STARTING THE DYNAMO.

The dynamo is now complete, but requires its field - magnet excited to commence with; afterwards it will always excite itself. The right direction for running is in the same direction as the movement of the hands of a clock, the observer looking at the side of the pulley as if looking at a clock face. Now rotate the armature at a high speed; if it suddenly seems to work stiffly and sparks appear at the brushes, it has started itself all right and the field-magnet will not require any further assistance; but if these effects are not noticed, and the dynamo fails to light a 10-volt lamp, an electric battery will be required to give the field-magnet the necessary start. Put a piece of paper between one of the brushes and the commutator to prevent the current going through the armature. Now obtain an electric battery, the stronger the better, and connect the positive wire of the battery to the right-hand terminal (looking

from the commutator end), and the negative wire to the left-hand terminal. While the battery is thus connected gently tap the field-magnet with a hammer for half a minute : disconnect the battery, and remove the paper from under the brush. On driving the armature again the machine should work all right. The dynamo can be painted to suit taste.

SHELLAC VARNISH.

The shellac varnish used as directed in this chapter is made by dissolving good shellac, broken up small, in methylated spirit. Allow about three days for it to dissolve, or start the varnish before commencing to make the dynamo. Cork it up

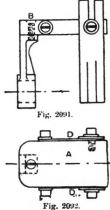

Fig. 2091.

Fig. 2092.

Figs. 2091 and 2092.—Elevation and Plan of Simple Brush Gear.

well and leave it until required. If it is too thick add more spirit. Paraffin wax melted and brushed on may be used instead, and is an excellent insulator, but is liable to come out as the wire warms up when the dynamo is working. It is a good plan to put the dynamo in a warm place, such as an oven, for a few hours when it is finished, to dry the spirit out of the coils ; but the temperature must not be very high or damage will be done to the insulation.

SIMPLE BRUSH GEAR.

A simple brush gear, as shown in Figs. 2091 and 2092, may be used instead of the one already described. It is not so good, but will work very well. The rocker arm is finished off as before, except that it fits on the bearing boss the other way round. A piece of hardwood (or better still vulcanised fibre) A, $1\frac{7}{8}$ in. by 1 in. by $\frac{3}{16}$-in. thick, is fitted on the arm as shown, being held in place by a $\frac{1}{8}$-in. countersunk screw B, passing down through the wood and tapped into the brass. The screws in the sides of the wood nearest the rocker are to take the flexible wires, and are tapped into the wood; a thin strip of brass D connects these screws to the brush screws, which are also tapped into the wood. The brushes are made from thin springy copper, and washers are used under the heads of all screws.

CONCLUDING REMARKS.

The output of this dynamo is about 5 ampères at 10 volts (50 watts) when the speed is 3,000 revolutions per minute; but it will give up to about 20 volts with less current if run at higher speeds. If obtainable, vulcanised fibre may be used instead of wood for the commutator, bush, etc. The magnet may be made with a joint through the top to allow of winding the coils in the lathe, if this joint is made to be in close contact all over the surface; but the extra trouble in winding the coils, as described, is nothing compared with extra trouble in construction if the joint is used, and it is advised that the magnet be made in one piece as described.

ELECTRO-PLATING.

SILVER-PLATING : INTRODUCTION.

THE art of silver-plating is an ancient one. Metalworkers in olden times learned how to overlay inferior metals with plates of silver, and, later, discovered how to make the silver adhere by soldering it to the metal beneath, being then able to use thinner plates of silver, which were, however, thick when compared with the thickest electro-silver-plating now done. To electro-plate with silver, the article must be immersed in a solution of silver, and electricity be passed through this.

THE PLATING SHOP.

The plating shop must be well lighted and ventilated, and kept at a temperature of about 60° F. Good light, preferably from high windows or skylight, is necessary for the examination of the work ; but direct sunlight should be excluded from the silver solutions. These should also be covered with canvas when not in use, to exclude dust. Ventilation is necessary, because all the exhalations of gas from the solutions are extremely poisonous and soon debilitate the strongest workmen when fresh air is absent. Not only is an even temperature conducive to regular working of the solutions, but the employees do better and more work when the shop is well ventilated and comfortably warmed.

SOLUTION VATS.

A vat to hold the solution may be of vitreous stoneware ; but this is liable to fracture, and is not easily made into large rectangular vessels. The next best is enamelled iron, which, being smooth, retards the creeping action of the cyanide salts, so that the sides of the vat are kept clean. In large plating establishments the vats are wood, lined with lead having burnt joints to prevent leakage and corrosion, then match-lined to prevent electric leakage. The ends are further strengthened with long iron bolts and plates, as shown in Fig. 2093. The sizes in general use vary from 3 ft. by 1 ft. 6 in. by

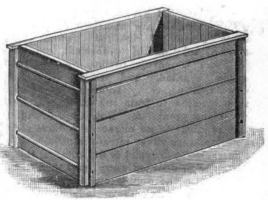

Fig. 2093.—Plating Vat.

1 ft. 6 in. depth, holding 30 gal., to 7 ft. by 3 ft. by 3 ft., holding 300 gal. Stout rods of copper or of brass, slightly longer than the vat, rest on the rims, and are furnished with screw connectors for attachment to the cable leading from the dynamo. Two of these rods are at the sides, and to these the anode plates are attached by hooks of stout copper, and one rod passes between the two others over the centre of the vat, from which the articles to be plated are slung by means of stout copper wire.

SUPPORTING PLATING VAT.

The plating vat, if large and heavy, should be supported by a strong foundation of masonry, and this should be so built as to ensure free access of air to the under side, thus preventing rot in the wood. Light vats may be supported on wooden trestles. If the floor can be coated with cement or asphalt it can be easily kept clean, and it is also advisable to have wooden grids on the floor by the sides of the vat, on which the workmen may stand.

MAKING SILVER-PLATING SOLUTION.

The best solution for silver-plating is the double cyanide of silver and potassium in distilled water. This salt may be made direct from pure silver plates or grains by first dissolving the metal in pure nitric acid diluted with distilled water, then evaporating all excess acid until silver nitrate crystals are obtained; then dissolve these in distilled water, and add a solution of potassium cyanide to form a curdy precipitate of the single cyanide, and finally dissolve this with a strong solution of potassium cyanide to form the double salt of silver and potassium. But, as the reduction of silver to its nitrate is tedious and noisome, pure silver nitrate is usually obtained from a druggist or drysalter, is then dissolved in distilled water, and then converted into the double cyanide of silver and potassium as above indicated. In doing this, to prevent waste of silver, care must be taken to avoid adding the cyanide solution in large quantities at a time. The precipitate must also be thoroughly stirred with a clean, smooth stick of wood after each addition of cyanide, and only enough of this added to throw down all the silver. When precipitation is complete, which is shown by all cloudiness disappearing from the liquid above the precipitate, the liquid must be carefully poured off and taken to the waste tub, and the precipitate of silver cyanide well washed by pouring clean water on it several times, so as to agitate it well;

then drain it as dry as possible. To this wet mass of silver curds add a strong solution of the best potassium cyanide in distilled water, until all the curds have been dissolved. The quantity of cyanide in the solution should have been ascertained by weighing the dry salt previous to its dissolution; then add one-fifth in excess to provide enough free cyanide of potassium to dissolve the silver anodes in working and thus maintain the strength of the plating solution. The whole concentrated silver solution must next be filtered through well-washed calico, and added to the distilled water previously placed in the vat intended to contain the silver-plating solution; it will then be ready for use.

AMOUNT OF SILVER IN SOLUTION.

Although good silver-plating may be turned out of solutions varying considerably in strength, it is of great importance to the plater to know how much silver is contained in the solution, and this may vary from 2 oz. to 6 oz. per gal. If grain or sheet silver is employed in preparing the solution, its strength may be easily ascertained; but a little calculation is necessary when silver nitrate is used, for in every 170 oz. of silver nitrate there are only 108 oz. of silver.

CARE OF PLATING SOLUTION.

The plating solution should be kept in good working order. Only distilled water should be allowed in it, and it should always be kept up to a certain mark and thus maintained at the same strength. No chemicals of any kind, except potassium cyanide (and this must be used cautiously) should be added to the solution. A certain quantity of free cyanide must always be present in the plating solution to dissolve the silver anodes at a rate equal to that of the silver deposited. Then the anodes are not coated with black slime, providing all other necessary conditions are fulfilled; but a slimy condition of anodes may be due to an insufficiency of silver in the solution or to an accumulation of dirt. Too much free cyanide in the solution will be shown by a coarse crystal-line condition of the anode surface and rapid erosion of the edges of the anode plates, which soon become ragged. In this solution the deposit is liable to become loose on copper, brass, and other metals readily soluble in cyanide solutions. In making silver solutions, only the best potassium of cyanide should be used; this is sold under the name of best grey potassium, 99 per cent., cyanide. Inferior cyanide may be employed for pickling solutions; but, because it contains a large percentage of uncombined potash salts, it should not be used in plating solutions, as these too soon become charged with an excess of potash salts, caused by a gradual withdrawal of cyanogen from the free potassium cyanide in solution. In process of time this excess of potash renders the old plating solution unfit for use, as shown by its muddy appearance and by the rough character of the silver deposited from it. If the muddiness of the plating solution is caused by dirt, filter the whole solution through good calico. This dirt may consist of dust from the workshop, carbon from the potassium cyanide, and finely divided silver from the anodes; it should therefore be saved and put in with the waste rinsing water to recover the silver. All water suspected to contain silver should be saved and evaporated in an enamelled iron vessel, the resulting salt being mixed with waste sawdust and sold to a refiner.

Anode Plates and Silver-plating.

The anode plates, which serve the double purpose of conveying the current into the solution and also keeping up its strength, must be of pure annealed silver, and their surfaces should always slightly exceed the surfaces of all the articles immersed at any one time. If they fall short of this for any length of time the solution will become impoverished, and, on the other hand, if their surface is excessive, the solution may become too rich. They should therefore be so slung as to be easily removed. Standard silver must not be employed for the anodes, it containing $7\frac{1}{2}$ per

cent. of copper. French silver coins contain even a higher percentage of copper; therefore coin silver must not be used at all, or the deposited silver will be hard and of a bad colour. Anode plates are usually suspended from the anode rods by hooks of pure silver wire, holes being punched in the upper edges of the plates for this purpose. The plates may be of any size and length suitable to the vat and work, but should not be less than $\frac{1}{8}$ in. in thickness. If kept moving whilst at work, the lower edges rapidly wear thin, and the plates should be reversed to wear both ends equally. This is also necessary if the plates remain stationary, as they are then liable to be cut through at the top of the solution. The anode rods may be kept in motion if they are made to rest on a steel frame moved to and fro by machinery. The frame should have broad steel rollers running on steel rails, and these must be connected to the positive pole of the dynamo. Steel is preferable to other metals because it is not readily corroded by potash salts or cyanide.

Plenty of Water Necessary.

As the silver salts will creep up the sides of the vat, and thus find their way to the floor, it is necessary to wash them back into the vat with a little distilled water applied with a stout brush almost every day. An abundance of clean water must be provided for rinsing, and the rinsing waters must be frequently changed. Failure in securing an adherent deposit of silver may often be traced to the use of dirty rinsing waters.

Source of Electric Current.

The primary batteries, such as the Smee, Daniell, and Bunsen, used in the early days of the art, have generally given way to specially designed dynamos. The advantages of these are many. The current is more constant, labour is lessened, and the cost of maintenance is much reduced. The current can also be regulated, and thus any class of work can be done from one machine. A plating dynamo gives a large current at a low pressure; the armature may be of the drum or ring pattern, and the machine is shunt wound. With a large machine, capable of giving, say, 200 ampères, large surfaces of work may be acted on in several vats simultaneously, and nickel-plating may proceed at the same time as silver-plating.

CARE OF PLATING DYNAMO.

The plating dynamo should be fixed in the plating shop where dust and damp are not likely to cause injury to its working parts. It should be kept thoroughly

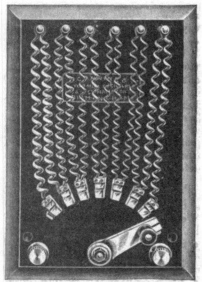

Fig. 2094.—Resistance Board.

clean. It must be bedded to a solid foundation to prevent vibration and bring its commutator and brushes to a convenient height for attention. The holders should be fitted with springs to ensure a light even pressure of the brushes on the commutator. If these press too hard, they will soon wear deep grooves in the commutator bars, and if the springs are too light, the brushes are liable to break contact occasionally, when the sparks will burn both brushes and commutator. Keep the tips of the brushes neatly trimmed square, and free all parts from copper dust before starting the machine. If the bars show signs of grooving, throw off

the brushes and work out the grooves with a flat stick covered with emery cloth, whilst the machine is running idle, then wipe the commutator with a rag smeared with vaseline. Always keep the machine clean and well oiled.

ELECTRICAL CONNECTIONS.

Connecting the dynamo with the vat are copper cables, and for these the following sizes may be used. For 10 ampères, seven strands of No. 18 s.w.g.; 20 ampères, seven strands of No. 16 s.w.g.; 30 ampères, nineteen strands of No. 18 s.w.g.; 50 ampères, nineteen strands of No. 17 s.w.g.; 75 ampères, nineteen strands of No. 15 s.w.g.; 100 ampères, nineteen strands of No. 14 s.w.g.; 150 ampères, thirty-seven strands of No. 15 s.w.g; and for 210 ampères, thirty-seven strands of No. 14 s.w.g. One of the cables should be in one length from dynamo to vat. The other cable should be in two or three lengths, the first from the dynamo to a resistance board near the vat; then a short length from the board to the ammeter; and then a short length from the ammeter to the vat. It matters very little which line is thus broken, but it is customary to have the line leading from the positive terminal of the dynamo to the anode system on the vat thus divided. The insulating covering must be stripped from the ends of the cables, the wires soldered together, then soldered into suitable connecting sockets.

RESISTANCE BOARD.

The resistance board is a slab of hard wood or slate, furnished with six or more brass studs on the upper edge, and eight or more brass contact pieces screwed to the board in the form of a semicircle on the lower third of the board. Lengths of stout copper wire, graduating in size, are fixed to some of the brass pieces and pass around the studs; then, German silver wires, offering ten or eleven times the resistance of the copper wires, are fixed to the remaining pieces and pass around the other studs, thus forming a zigzag line of wire of graduated resistance from one side of the board to the other, as shown in Fig. 2094. A brass lever is pivoted so as to sweep the semicircle of brass pieces and make the connection with each in turn. This central stud may be connected to a large binding screw at the bottom of the board, and one end of the cable is fixed to this screw. At the opposite end of the semicircle, one of the brass pieces is similarly connected to another section of the cable. This instrument enables the plater to control the current from the machine; when the switch lever rests on the contact piece connected direct to the cable, the circuit has no additional resistance, but when one of the lengths of copper wire is thrown into the circuit by moving the switch lever, resistance is added, and, when the lever rests on the last brass piece, all the resistances are thrown in. Thus the plater is enabled to silver small articles safely, even with current from a very large dynamo.

AMMETER.

The ammeter (Fig. 2095) measures the current, and is often fixed to the resistance board for ready reference. As the rate at which the silver is deposited depends on the current through the vat, and as the character of the deposit is greatly influenced by the rate of deposit, this instrument is of great importance, for by it the plater can determine how much silver is being deposited. The dial of an ordinary ammeter shows the current in ampères, 1 ampère being that current which will deposit 52 gr. of silver in one hour on $\frac{1}{10}$ sq. ft. of suitable surface, in a fit condition for polishing. Thus, in the special instrument made by T. Morris and Co., Birmingham, and illustrated in Fig. 2095, the lower scale shows also the weight of silver deposited per hour.

VOLTMETER.

The voltmeter is similar in external appearance to the ammeter, but the readings on its dial are in volts. From its readings the plater can adjust the brushes of a dynamo and regulate its speed so as to get the right voltage. Experience has proved that the silver in a silver-plating

solution may be separated from its salt and deposited in a good condition at as low a pressure as 2 volts; but this may be increased to 3 or even 4 volts without altering the condition of the deposit very much. However, when the pressure exceeds 4 volts, there is a tendency to a loose and powdery deposit, which gets more pronounced as the pressure is increased. A voltmeter is therefore useful, but, its coils being wound with fine wire of high resistance, it must be placed in a shunt bridging the two main lines, and furnished with a switch to cut the voltmeter out of

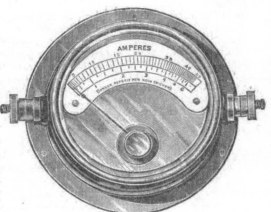

Fig. 2095.—Ammeter.

circuit after the dynamo has been adjusted to the required pressure (see the diagram of connections, Fig. 2096).

PREPARING ARTICLES FOR SILVER-PLATING.

The surfaces of articles to be silver-plated must be made perfectly clean. Not only must all rust and corrosion be removed, but also all trace of animal matter, even to the thin film left on by contact with an apparently clean hand. Unless this is done before the articles are immersed in the plating solution, the coat of silver is likely to strip and blister on all soiled spots, and thus spoil the whole work. Rust, verdigris, and other metal oxides can generally be loosened, and sometimes removed entirely, by immersing the corroded article in a pickle made of a diluted mineral acid, or a mixture of those acids, and then swilling in clean water. A mixture of 1 part sulphuric acid in 20 parts of water will loosen the oxides of copper and zinc on these metals and their alloys. The alloys of these metals may have many names, but are to be regarded by the plater as brass, and treated accordingly.

PICKLING CORRODED COPPER OR BRASS.

If the surface of the copper or brass article is deeply corroded and green, it may be necessary to use a stronger pickle, composed of 3 parts of sulphuric acid, 1¾ parts of nitric acid, and 4 parts of water. Rust on iron and steel may be loosened by immersion in a pickle composed of sulphuric acid 6 parts, muriatic acid 1 part, water 160 parts. The oxides of lead and tin may be loosened from these metals and their alloys (pewter, Britannia metal, soft solder, etc.) by immersion in a hot solution of caustic alkali, such as caustic potash or caustic soda.

REMOVING GREASE.

As it often happens that other metals are not only corroded, but also dirty with grease, oil, or other animal matter capable of resisting the action of acid pickles, it is advisable to swill them first in the hot alkali solution, then in hot water, and rinse in cold water before immersion in the acid pickle. The caustic alkali solution is made by dissolving ½ lb. of caustic soda, or caustic potash, or crude American potash, in each gallon of water contained in a wrought-iron tank.

RECEPTACLES FOR PICKLE AND ALKALIES.

The acid pickles should be mixed and contained in vessels made of vitrified stoneware. It should be understood that vitrified stoneware is the only suitable material for acid pickles, and plain wrought-iron for caustic alkalies. The mixed acids will undermine and dissolve the glaze from ordinary earthenware, and also the enamel from iron. Caustic alkalies also dissolve enamel, and extract the zinc from galvanised iron, and the tin from tinned vessels.

FURTHER TREATMENT OF PICKLED GOODS.

After all corrosion and dirt has been loosened, the surfaces of the articles to be silver-plated should be well brushed with a hard brush in water to get all loosened dirt out of crevices and pits. This may be done with an old scratch-brush made of wire, if preferred. An examination of the surface will then reveal numerous scratches, dents, and pits, some being due to corrosive action, and many to hard usage. These must all be removed, for the surface must be made like a new surface, all defects being made good before the silver is deposited upon it. Dents and misshapen parts can be put right only with the aid of hammers, pliers, etc. At one time all scratches and pits were also laboriously removed by hand, being first filed and scraped, rubbed with sand or emery, then with water-of-Ayr stone, and polished with fine abrading powders. Now the work is nearly all done on polishing lathes with revolving brushes, resulting not only in a great saving of time, but also in a higher finish. The ends of the lathe spindle should be threaded with a coarse taper screw which enters a hole in the boss of the brush or bob, and holds it firm whilst revolving. Larger lathes are furnished with flanged plates at one end in addition to the taper screws. These plates support calico mops, and grip the sides of emery wheels. Polishing lathes must be firmly bolted down on benches not liable to great vibration, and run at a speed of 1,400 revolutions a minute in polishing silver. For polishing steel or other hard metals they may be run at a higher speed, up to 2,500 revolutions per minute. The brushes used with these lathes differ in size and material according to the work required from them.

POLISHING BOBS.

In the preparation of articles to be plated, bobs and mops are employed. Bobs are of two kinds: one, formed of a disc of hardwood, having its edge coated with bull-neck leather, while the other is a solid disc of felt. Other varieties have felt or buff leather on their rims, whilst some are made of solid bull-neck, or of walrus leather. The sizes vary from 3 in. to 18 in. in diameter, and from ⅜ in. to 2 in. in thickness. To make a wooden bob, first select a piece of hard, well-seasoned wood and turn a true disc in a lathe. Then get a strip of the required covering material as wide as the wood disc, and long enough to go round and meet with butt edges. Next prepare some good glue and roll the edge of the disc in it whilst hot; then put on the leather or the felt, and secure this to the wood with long steel tacks, so driven in as to be easily pulled out when the glue is firm. At the end of

Mops.

Mops render the surface quite smooth, and are made chiefly from thin tough leather, such as basil leather and chamois leather, and from various grades of calico, the finest being the soft variety known as swansdown calico. Basil leather mops may be from 3 in. to 12 in. in diameter and from ½ in. to 4 in. in thickness. They are made of several discs of basil leather, cut and laid true on each other, between two smaller discs of thick leather, to form the bosses, which are then secured by long iron rivets passing through the whole mass of leather. These mops are very useful with

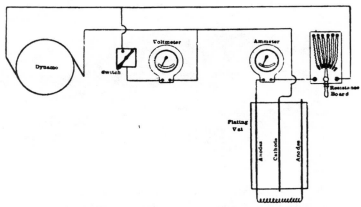

Fig. 2096.—Diagram of Electrical Connections for Plating.

twelve hours these tacks may be withdrawn, and wooden pegs dipped in glue driven in the holes; then true the whole in a lathe. The bob is then ready for the grinding material, which may be emery, sand, or tripoli, the material and grade being selected to suit the work in hand—emery for iron and steel, and tripoli for copper and brass. This is spread on paper in an even layer, and the rim of the bob is first rolled in the hot glue, then on the layer of powder until a sufficient thickness has been taken up, then it is set aside for another twelve hours to get firm. Solid leather or solid felt bobs are similarly coated. These bobs are employed to grind down rough surfaces and render them comparatively smooth.

tripoli compo in preparing surfaces of copper, brass, and other soft metals. Chamois leather mops are made in a similar manner, but are chiefly used to produce a high finish on jewellery. Sometimes these leather mops are stitched spirally to keep them more compact whilst revolving. Calico mops are made similarly. The thickest and coarsest calico is employed for mops used in preparing the work with tripoli compo, etc., and the finer grades for finishing the plated articles with rouge compo. The bleached varieties of calico are, as a rule, harsher in texture than the unbleached, and are used in making the mops for cutting out scratches, and are then followed by mops made of unbleached calico charged with rouge compo to give a higher finish. Calico mops run from 6 in. to 18 in. in diameter, and have from 50 to 105 folds to the inch of thickness according to the quality and thickness of calico employed. Swansdown calico mops run from 6 in. to 12 in. in diameter, and have from 24 to 60 folds in thickness, or about 24 folds to the inch.

CARE OF MOPS, ETC.

The polishing mops and materials used in finishing plated ware should be kept in boxes, to prevent gritty dust getting into them. Each kind of mop and polishing compo should have a separate box, and special attention should be paid to guard finishing mops and compos. from contamination with those of a lower and

height for the workman, who holds the article to be scoured in his left hand on the tray, dips his scouring brush (Fig. 2098) in water, shakes out the surplus, then dips it in the pumice powder, and brushes the article to and fro until every trace of oxide is removed. When it is rinsed in water, it should be quite clean with a uniformly dull surface all over it, which renders the surface suitable to take and hold the silver coat.

FINAL OPERATIONS BEFORE PLATING.

It must then be wired—that is, attached to the wire which will hold it in the plating solution—then rinsed in the mercury pickle to impart a thin coat of mercury, again rinsed in water and transferred to the

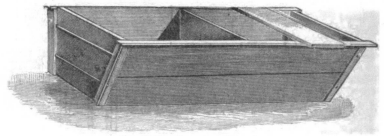

Fig. 2097.—Scouring Trough.

coarser grade. The same remarks apply to all chamois leather, rags, soap, and burnishers employed in finishing silver-plated articles. The burnishing of plated goods is described on pp. 145 to 148, and polishing is fully treated on pp. 131 to 139.

SWILLING AND SCOURING.

After the polishing, the articles will be found to be thinly coated with grease or oil. This is removed by swilling in the hot caustic alkali pickle, and rinsing in water. This process will probably coat copper, brass, and similar soft alloys with a film of black oxide which must be removed by scouring. For this the trough is of thick wood in two divisions, as shown in Fig. 2097. One division is filled with clean water for rinsing the scoured goods, and the other receives the waste scouring material. Over this is placed a shallow tray, and the whole is fixed at a convenient

plating bath, where deposition should commence at once.

TREATMENT OF SCRATCH-BRUSHES, ETC.

As new scratch-brushes and scouring-brushes may have traces of oil or grease on them, acquired in their manufacture, it is advisable to dip them for a moment or two in the potash solution, then rinse them in warm water before using them. New scratch-brushes should be first used on a piece of clean steel, to break down the sharp, hard tips of the wires, before employing them on silver; otherwise they may cause scratches, etc.

DRYING. SILVER-PLATED GOODS.

Boxwood furnishes the best sawdust for drying silver-plated goods, as it does not contain dye, acid, or resin liable to stain pure silver. But if the sawdust is heated in a vessel by direct contact with flame.

it will become charred, and in this condition will stain silver. The pan containing the sawdust should therefore always have a water or steam jacket.

THE PRELIMINARY COPPERING.

Iron, steel, and zinc goods, and, preferably, also those of lead and its alloys, such as pewter and Britannia metal, must be coated with copper after they have been scoured, before they can be made to take an adherent coat of silver. Silver may be made to adhere to lead alloys, with skilful treatment, without this coat, but the work is easier and better done when thus coated. The solution employed must be an alkaline one, as acid solutions, by their action on iron, steel, and zinc, undermine the deposit of copper and render it loose. The alkaline coppering solution is prepared by dissolving copper sulphate in hot rain-water in the proportion of 8 oz. of copper sulphate to 1 quart of water. When this has become cold enough, add first liquor ammonia to throw down the copper in the form of green mud, and then an extra quantity of ammonia to dissolve this mud and convert the whole into a bright blue liquid. To this must be added a sufficient quantity of potassium cyanide solution to take all the blue colour out and make the solution amber-tinted, or the colour of old ale. It should then be left exposed to the air for twelve hours, then filtered through calico, and diluted with clean rain-water until each original quart is made up to one gallon of solution. This solution may be worked hot or cold, but gives a fine clear deposit of copper when heated to 160° F. and worked with current from a plating dynamo. Anodes of pure copper must be employed.

"QUICKING."

When the articles of iron, steel, zinc, lead, pewter, and similar alloys, or those of brass with soft-soldered joints, are lightly coated with copper, they may be removed from the coppering bath to the alkaline mercury solution, and be there given a thin film of mercury, then rinsed and transferred to the silver-plating bath without delay. This process is named "quicking," or coating with quicksilver, and is done to secure the perfect adherence of the silver coating. The alkaline quicking solution is made by dissolving mercury slowly in dilute nitric acid, then adding enough strong solution of potassium cyanide to throw down the mercury in the form of black mud, and an extra quantity (with stirring) to dissolve this mud. Distilled water must be used solely in making this solution, and it should contain 1 oz. of mercury in each gallon. Some free cyanide must also be added to act on the copper and ensure a bright film of mercury. The acid solution of mercury proto-nitrate is employed and preferred by some platers. This is the first solution of mercury in nitric acid, largely diluted with distilled water. Its action is more speedy than that of the alkaline solution, but the film

Fig. 2098.—Scouring Brush.

of mercury is not so thin and uniform as that from the latter.

COPPER-PLATING.

The metals on which a coat of copper is deposited by electricity are lead and its alloys; tin and its alloys; iron, tinned iron; zinc; and steel. When articles made of these metals are to be silver-plated, nickel-plated, or gilded, it is always advisable and sometimes necessary previously to coat them with copper. This cannot be done in a copper sulphate solution, because that dissolves the metals. Of the various solutions the most successful one is described in the previous column, but it should have these proportions: Copper sulphate, 1 oz.; potassium cyanide, 3 oz.; liquid ammonia, 1 oz.; rain-water, 2 qts. Distilled water may be used instead of rain-water, but spring and river waters are not suitable because of the earthy matters held by them. The solution should be held in an enamelled iron vessel. If it is kept supplied with free cyanide and free ammonia it may be worked at from 6 to 8 volts ; but the deposit may be improved by heating the solution to from 150° F. to 170° F., and the vat may then be worked at from 4 to 6 volts. The best generator is a plating dynamo, the next a three-cell accumulator ; and among primary batteries the next best would be four ½-gal. Bunsen cells. Anode plates of pure copper must be employed ; these are connected by No. 16 s.w.g. copper wire to the positive pole of the generator. If the plates do not dissolve freely, but become encrusted with a green slime, a small quantity of potassium cyanide and of liquid ammonia should be added to the solution.

Preparing Articles for Copper-plating.

The surfaces of all articles to be copper-plated by this process must be cleaned and prepared. Iron and steel articles may be cleaned from rust by steeping and swilling in a pickle composed of 6 fluid oz. of sulphuric acid and ½ oz. of muriatic acid in each gallon of water. They must then be rinsed in clean water and immersed in a pickle composed of ½ lb. of American potash dissolved in each gallon of hot water. If the surfaces have been pitted, the corroded parts must be polished with emery held on a mop in a polishing lathe, after which the articles must be well swilled in the hot potash pickle to free them from oil and grease. All surfaces must be well polished before the copper is deposited, because the thin coat will not permit much polishing afterwards. Articles made of lead and tin, or their alloys, must be first scoured with sand and water, using a hard brush for the purpose, to free them from oxide ; then rinsed in the hot potash pickle ; again scoured with finer sand to polish them ; wired with short lengths of No. 24 s.w.g. soft copper wire ; again rinsed in the hot potash pickle, and transferred direct to the plating vat. The potash pickle will prevent rust forming on iron and steel articles, and will clear oxide from lead and tin and their alloys ; but it is advisable to transfer the articles quickly to the plating vat, and not to rinse them in water on the way. Zinc articles are cleansed in a similar manner ; but very

fine sand or finely powdered bath brick must be used in scouring. If articles are bright and free from rust and tarnish, only a light brushing with a vegetable fibre brush in the potash pickle will be necessary to prepare them.

The Actual Process of Copper-plating.

Each article must be attached to a short length of copper wire, which suspends it in the vat. Use No. 24 s.w.g. for small articles, and No. 18 s.w.g. for heavy ones. Each article should be held by the slinging wire during the final rinse, and the free end of this wire is bent over a brass rod on the plating vat, attached to the negative pole of the generator. Move each article to and fro with a rinsing movement when placing it in the vat, to remove any air bubbles on the surface. The current should be regulated by a resistance, usually a long length of German silver wire furnished with a switch. The resistance can also be increased by diminishing the surface of the anode exposed to the plating solution, and by placing the anode further from the article being plated. If the current is too strong, the deposited copper will be dark in colour and loose in character, and this will also happen if the solution contains too much copper. Movement of the articles whilst being plated will assist in securing a bright and smooth deposit. Some gas is given off from the articles whilst deposition is going on, but this should be regulated by adjusting the current. Only a few minutes is required for plating each article. The plated articles should be rinsed in plenty of clean water to free them from cyanide and copper salts. If the surface is to remain coppery, the article should be rinsed in hot water, placed at once in hot bran or hot sawdust, and moved about in it until quite dry and bright. Pure copper readily tarnishes in the air when damp, but may be brightened with a scratch-brush. If the surface is to be nickel-plated the articles must be rinsed and transferred at once to the nickel-plating vat. If a thicker deposit of copper is desired, use an electrotyping solution, after depositing a thin film of copper in the alkaline solution above mentioned. If the plated articles are to be gilded, get a very thin and bright deposit of copper, or brighten it with a scratch-brush; then rinse and transfer at once to the gilding vat. If they are to be silver-plated, coat with a thin film of mercury before placing them in the silver-plating solution. The solution is made by dissolving 1 oz. of mercury in very dilute nitric acid, say 1 part acid to 10 parts distilled water, then making it up to 1 gal. of solution with distilled water. Give a brisk swill in this, and then rinse in clean water.

COATING EARTHENWARE WITH COPPER.

So that copper can be electro-deposited on terra-cotta, earthenware, etc., the surface of these materials must be first rendered conductive to electricity. This is done by coating with blacklead, bronze powder, or some other finely divided metal. Blacklead is brushed into the pores of the material in a dry condition until the whole surface is evenly coated and well polished. Bronze powders are mixed with methylated spirit and applied in the form of a paste. If the surface is briskly brushed with a new brass-wire brush, it will become coated with brass and thus made conductive. A copper wire must then be tightly twisted around some part of the article and connected to the conductive surface by a liberal application of the powder. Thus prepared, the article is immersed in an electrotype solution, connected to a battery or dynamo, and copper deposited in the usual manner. Only a very thin coat must be applied if the pattern is to be retained or smoothness is desired. If the surface of a flat object is only covered, this coating may be afterwards peeled off; but if the object is surrounded with copper, as a vase or statue, the coat will be adherent.

ELECTRO-BRASSING SMALL IRON GOODS.

For brassing small bright iron articles by deposition, first a brassing bath will be needed, the following being suitable for cold working:—Carbonate of copper and carbonate of zinc recently prepared, of each 4 oz.; carbonate of soda in crystals, bisulphate of soda, and pure cyanide of potassium, of each 8 oz.; $\frac{1}{10}$ oz. of white arsenic; and about 2 gal. of water. Dissolve in 3 pints of water 5 oz. of copper and 5 oz. of crystallised sulphate of zinc, and add a solution of 14 oz. of carbonate of soda in 1 qt. of water. A greenish precipitate of carbonates of copper and zinc is formed; this must be well stirred and allowed to settle for several hours. Pour off the water, and replace by nearly 2 gal. of fresh water, in which are dissolved the bisulphite and carbonate. Dissolve in the remaining warm water the cyanide of potassium and the white arsenic, and pour this liquid into the other, which becomes rapidly decolorised and forms the brass bath. The vats for cold platings usually are gutta-percha-lined wooden tanks, the sides of which are also lined with brass sheets joined together, and thus connected with the carbon or copper of the battery. The articles to be brassed are suspended by copper or brass hooks to stout rods of the same metal, and thus connected with the last zinc of the battery. The intensity of the batteries is determined by the surface of the material to be brassed. The losses of the solution will need to be made up from time to time by the addition of copper and zinc salts dissolved in cyanide of potassium with the arsenic.

ELECTRO-GILDING.

Gold may easily be deposited in good condition on a large variety of metals and alloys from a solution of the double cyanide of gold and potassium. This may be made and kept warm in an enamelled iron saucepan over the flame of a gas-burner or of an oil lamp, which may be employed as the vat. One cell (popularly known as an electric battery, though this term implies the use of more than one cell) of any of the many varieties may be used, even one dry cell being suitable. It is only necessary, therefore, to get such a cell, and some gold solution heated to a temperature of 160° F. in a saucepan, to connect the trinket to be gilded by a length of No. 24 s.w.g. copper wire to the negative (zinc) pole of the cell, and a strip of pure gold by a similar length of wire to the positive (copper or carbon) pole of the battery, and then to suspend both trinket and gold in the hot solution for a few moments to coat the trinket with gold.

THE IMPOVERISHED SOLUTION.

By repeatedly heating the solution its solvent property is driven off, the gas having a noxious odour, cyanogen (prussic acid gas) being separated from the potassium cyanide, and leaving potash alone in the solution. As potash alone will not dissolve gold, more of this precious metal is withdrawn from the solution than is dissolved from the gold strip (anode) to make up the loss. As a consequence, the gilding becomes more and more poor in colour, because the strength of the solvent has diminished, and the quantity of gold in the solution has also decreased. Experienced electro-gilders add a small quantity of potassium cyanide occasionally to make up this loss, and also regulate the surface of gold anode to the work in hand so as to prevent impoverishment of the solution. The addition of cyanide to a gilding solution must be made carefully to avoid a great excess, as this will spoil the colour of the deposited gold, giving it a foxy-brown tint.

COLOUR OF DEPOSITED GOLD.

The colour of electro-deposited gold varies with the condition of the solution from which it is deposited, and the strength of the current employed. A solution containing only a few grains of gold to the gallon will yield a pale yellow deposit at a temperature of 140° F. when worked with a current from two Daniell cells in series, or at a pressure of two volts. By raising the temperature of the solution to 160° F., the colour of the deposit is also raised, and a still deeper tint is secured if the temperature is raised to 180° F. A similar result is obtained by increasing the voltage. A richer solution will more readily respond to these changes in temperature and to an increase in voltage; and a richer appearance is obtained from gilding done in gold solutions containing from 5 dwt. to 10 dwt. in the gallon than from poorer solutions. But the colour deepens in all solutions with the thickness of the deposit. For the first few moments it may be pale and bright, then it becomes deeper in tint and dull, this action increasing until the tone becomes a rich golden brown, which under favourable circumstances shows a fairly good deposit. This brown appearance disappears as the article is scratch-brushed, if the deposit is satisfactory. The same brownish tint appears too soon if the voltage and current density are too high, or if free cyanide is excessive.

OBTAINING THICK GOLD DEPOSIT.

If a thick deposit of gold is required, the articles must be frequently taken from the gilding bath and scratch-brushed to remove the brownish appearance, then rinsed and returned to the bath. If this is not done the gold will not adhere, but will simply surround the article in the form of brown mud, and this condition is soon observed when the solution is poor, and when the current is too strong.

COMPOSITION OF GILDING SOLUTION.

The gilding solution is composed of the double cyanide of gold and potassium in distilled water. The single cyanide of gold is a very light yellow powder, 223 grains containing 197 grains of gold. This is added to a strong solution of potassium cyanide and stirred until dissolved, then made up to the required strength with distilled water. The strength varies from 5 grains to 15 dwt. of gold in the gallon, the richer quality being used where large quantities of heavy gilding are done each day, and the poor solution for merely blushing the surfaces of trinkets with a gold tint. The solution may be made by dissolving gold direct into a heated solution of potassium cyanide, and passing an electric current through the solution from one strip of gold to another until sufficient of the metal has been acquired. This method is the most economical for small operations; but as the solution thus made contains an excess of potash, it is not to be recommended for large operations. The solution may also be made by the chemical method in a laboratory attached to the workshop if the necessary skill is available, but not otherwise, as unskilled attempts at making cyanide of gold usually result in much loss of gold. The gold salt is precipitated from its solution of gold terchloride by cautiously adding a weak solution of potassium cyanide; the resulting yellow powder is well washed, and then dissolved in a strong solution of potassium cyanide to form the gilding solution. But, as the signs of complete precipitation are not well defined, and as the single cyanide of gold is so easily dissolved in a very slight excess of potassium cyanide, there is danger of a great loss of gold in the making of the gilding solution by this method, which therefore is not recommended for general adoption.

Necessity of Using Fine Gold.

Pure gold (fine gold) must be employed in making gilding solutions and for the anodes. Coin gold, and standard or sterling gold, being alloys of gold with copper or silver, should not be employed. If silver is present in the gilding solution, the deposit will be of a pale tint, grading to a greenish yellow with each increase of silver. If copper is present in the solution, the deposit will have a ruddy tint, deepening to a reddish brown, and then to a dark bronze with an increase of copper. The effect of copper in deepening the tint of gold deposits has been used in fancy jewellery, slight additions of the inferior metal being made to the gilding bath from time to time, and small copper anodes employed to secure ruddy tints in the gilding; but solutions thus alloyed require more skill to work them successfully than those made of pure gold, and are always more or less uncertain in their action. The proportion of copper to gold is not determined by the proportions of these metals in the solutions and in the anodes, but by the selective power of the current working the solution. As this varies with the resistance of the circuit, which alters with each variation in temperature and change

of metal to be gilded, so does the proportion of the two metals, and the consequent colour of the deposit.

Gilders' Vats.

Although electro-gilding can be carried on with the simple apparatus first mentioned, this would be unsuitable in a regular workshop. The vats are generally of enamelled iron, mounted in iron frames furnished with iron legs and heated with Bunsen burners. In large establishments rectangular vats hold from 10 to 30 gal. of gold solution, and are fitted with steam jackets, as it is prejudicial to health to have the vats heated by gas jets, and with the use of the steam jackets the temperature of the solutions can be regulated much better.

Preparing Work for Electro-gilding.

The preparation of articles to be electro-gilded is much the same as for those to be coated with silver, with a few exceptions. It is not necessary first to coat bright iron and steel with copper, as gold may be firmly deposited in a hot solution of the double cyanide of gold and potassium. But it is advisable to copper-plate articles of zinc, lead, pewter, and other soft alloys before they are electro-gilded, because gold goes on them loosely. Copper, silver, brass, German silver, and similar hard metals with their alloys form suitable materials for articles to be gilded. Gilding metal, a kind of gunmetal, and the various kinds of brass which are sold under grand names for sham jewellery, all form excellent bases on which to deposit gold, taking a very fine polish in their preparation. The articles should always be highly polished before they are gilded, this greatly affecting the nature of the surface after gilding. One method of producing a matt or frosted appearance on gilded articles is to make the surface rough by means of frosting brushes, by the sand blast, or by the action of acid, before the article is gilded. It is not necessary to coat the surface with mercury before gold is deposited on it, except in the case when a very thick deposit is required on bare copper.

GILDING ALUMINIUM.

Articles made of aluminium cannot be gilded direct in a solution of the double cyanide of gold and potassium, because the alkali therein attacks and rapidly disintegrates aluminium. It should therefore be first coated with copper in a solution of copper sulphate, and then transferred to the gilding bath.

SLINGING WIRES.

Fine soft copper wire should be employed to connect the articles with the cathode rod of the gilding vat, and the lengths of wire thus used are named slinging wires. These may be inserted in any holes in the articles, or twined round any obscure projections, or formed into slings for the suspension of coins and medals. In gilding some patterns of long thin chains it is advisable to take the chain in loops, and insert the wire in several links, and if the chain is a bad conductor it will be necessary to twine it around the slinging wire. If this is not done, there will be patches of links imperfectly gilded, or left ungilded.

QUANTITY OF GOLD DEPOSITED.

Gold is deposited rapidly from ordinary electro-gilding solutions, and a sufficiently strong coat may be deposited in the course of a few minutes, the rate being about 37 grains an hour per ampère. Weigh the articles after they have been cleaned and after they have been gilded to ascertain the quantity of gold on them. When, therefore, by calculation of time and current, it has been estimated that enough gold is deposited, the article must be rinsed in clean water, scratch-brushed to remove the brown appearance, dried by rubbing in sawdust or otherwise, and then weighed. If there is not enough gold in the article, return it to the gilding bath. In gilding lockets, light brooches, and similar light trinkets, the solution will get inside and remain there during the process of scratch-brushing. This must be taken into consideration when weighing the goods, and care must be taken to have them dry. Some of the hollow ware is filled with a waxy composition which oozes out in the course of gilding, and this falsifies the calculations made to determine the deposit of gold.

CARE WITH IMITATION GEMS.

Bits of coloured glass, named " stones," are inserted in cheap brooches, rings, etc., under the names of rubies, diamonds, pearls, etc. In the commonest goods these stones are merely attached with gum or some soluble cement, which is dissolved in the solution, and thus the stones come off. An examination of the goods before gilding will soon detect these, and, if the " stones " are not held firmly in claws, they should be taken out by steeping the articles in hot water before they are prepared for gilding. They must be reset after the goods are gilded.

SCRATCH-BRUSHES FOR GILDED WORK.

The scratch-brushes used in brushing gilded work are made with very fine brass wire, some being crimped for extra elasticity combined with softness. Special shapes are required for such goods as rings and watch cases, so that the insides may be brushed and polished. Sometimes a good brushing with a scratch-brush is all the finish required for the insides of goods. When chains are heavily gilt, each link must be twisted around and brushed, whilst only a short length of chain is held between the fingers and thumbs of both hands.

FINISHING GILT GOODS.

Gilt articles are polished on soft mops made of swansdown, soft felt, and chamois leather, using finest rouge composition as the polishing material. The insides of rings are polished on felt fingers so tapered as to fit any size of ring. Chains are polished on broad-shaped bobs covered with soft leather on the convex sides. Thimbles and similar hollow ware have specially formed bobs made with wooden stocks of the required shape covered with fine soft felt. Contrasts in the various grades of finish are sometimes resorted to for effect. A frosted appearance is secured by using a coarse scratch-brush having long bunches sparsely set in the boss, and holding a stick to the revolving brush just before it strikes the gilded article. Raised parts are burnished with suitable burnishers made of steel, bloodstone, and agate. A final yellow blush is often imparted by a momentary dip in a new gilding solution, after which the article is rinsed in hot water and dried. Insides of mugs, spoons, salt cellars, etc., are gilded by means of a special arrangement in which the articles are connected to the cathode system ; then they are filled with gilding solution, and gold is deposited from a gold anode held by hand in the solution.

RECOVERING GOLD AND SILVER FROM PLATED WARE.

By one method, gold may be dissolved from gilt articles by swilling them in warm nitric acid and adding a few pinches of common salt as required until the gold has been removed. Silver may be dissolved in hot concentrated sulphuric acid by adding some saltpetre as required. Both gold and silver may be dissolved off articles by making these the anode in an old spent plating solution, using platinum as the cathode. The gold or silver is best recovered from these solutions by evaporating them first nearly to dryness, then adding litharge (plumbic oxide) to the nearly dry salt, and fusing the whole at a gentle heat, then pouring the molten amalgam into iron moulds. The lead ingots must then be rolled thin, and the gold parted in nitric acid. The silver is thrown down from the acid with common salt as silver chloride, which is then dried, mixed with an equal bulk of dried sodium carbonate, and fused in a large crucible. Metallic silver settles at the bottom of the crucible. The precious metals may also be parted from the lead by cupellation.

WIRE WORK.

Straightening Wire.

One of the earliest operations in the working of wire is straightening. Many different forms of straightening blocks are used, but all are alike in principle.

Fig. 2099.—Wire Straightening Block.

In Fig. 2099, for example, the wooden block is 3 in. or 4 in. deep, 4 in. wide, and about 15 in. long. In its top is inserted a number of iron pegs or nails, placed as shown in Fig. 2100. The wire to be straightened is then drawn through the pegs. The same block can be used for wires varying slightly in thickness by knocking the pegs a little inwards or outwards as may be necessary. When the thickness of the wires in use varies substantially, blocks of different sizes should

Fig. 2100.—Plan of Wire Straightening Block.

be employed. If the pegs are inclined towards each other as in Figs. 2101 and 2102, very irregular lengths of wire may be made fairly straight by being drawn

through the holes or arches formed. The block may be a fixture on the bench, or it may just be secured when in actual use by having a staple in one end (see Fig. 2103) and temporarily securing this by means of a hook swinging on a staple driven into

Fig. 2101. Fig. 2102.

Figs. 2101 and 2102.—Wire Straightening Blocks with Inclined Pegs.

the wall (see Fig. 2104) or into the bench (see Fig. 2105); it simplifies matters to screw a hook (Fig. 2106) directly into the wall.

The Swift.

A coil of wire, preparatory to being straightened or crimped, is placed upon a

Fig. 2103.—Staple in End of Block.

"swift" (Fig. 2107). In the bench is firmly inserted, in a perpendicular position, a stout iron or steel rod, upon which the swift revolves. It consists of two circular boards, one much smaller than the other, with canted or bevelled edges, connected to each other by three or more uprights, as shown. The vertical rod passes freely through a hole bored in each board. The larger boards rest upon the bench. The inside diameter of one coil of wire may happen to be less than

Fig. 2104.—Staple and Hook in Wall.

that of other coils, and it is to accommodate all dimensions that the swift is made tapering or cone-shaped. If the swift is properly made, it should be all that is required to assist in unwinding any particular coil. It is advisable that two swifts be used—one much smaller than the other

Fig. 2105.—Staple and Hook in Bench.

CRIMPING.

The crimped or corrugated wire now so extensively used in many branches of wire-working was invented and patented by the late Mr. Thomas Bellamy, of Birmingham. His invention consisted of

Fig. 2106.—Hook in Wall.

crimping, by the aid of machinery, wire intended to be placed together to form fences, garden borders, hen coops, fire guards, etc. The old method of making such things was that of placing the wires

over and under each other, and then securing them at their junctions by means of tying with pieces of wire (see Fig.

Fig. 2107.—Swift.

2108), this entailing a great amount of labour. In machine crimping, one end of a coil of "straight" wire is put into the

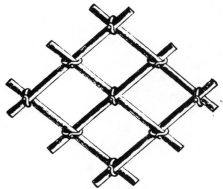

Fig. 2108.—Old Style Wire Trellis.

machine and drawn through, coming out at the other side crimped as shown in Fig. 2109. This is then cut up into the required lengths, which are put together across each other in the same way as the old method, and are held together by their

Fig. 2109.—Wire Crimped in Machine.

own union (see Fig. 2110), requiring no ties at the junctions, as did the former pattern. After being fitted into their

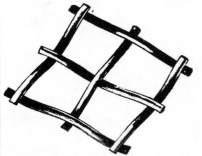

Fig. 2110.—Trellis Formation with Crimped Wire.

frames, they are as firm as, and have a better appearance than the straight wires.

PRINCIPLE OF THE CRIMPING MACHINE.

The salient feature of the Bellamy crimping machine consists in the use of pairs of cog wheels, fixed as in Figs. 2111 and 2112. A coil of wire is supported by the swift facing the edge of the machine, the worker traversing the length of the shop, pulling the remainder of the wire behind

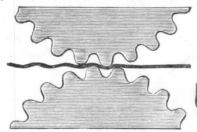

Fig. 2111.—Wire between Crimping Cogs.

him, which is finally cut up into the required lengths. The crimps are regulated according to the size of the wheels and the cogs, and the distance apart from each

other that they are fixed. If the crimps, however, are deeper or shallower than a certain limit, it will be impossible to place the wires together so that they may hold

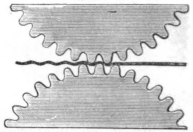

Fig. 2112.—Wire between Crimping Cogs.

themselves properly. The wheels are adjustable with regard to the distance between them.

Fig. 2113.—Light Crimping Machine.

LIGHT CRIMPING MACHINE.

A machine for crimping or corrugating wires of small gauges is shown by Fig. 2113. A thick standard is divided into two separate parts horizontally, as indicated by the cross line, and held together by a pair of screws. The screws penetrate sufficiently deep into the lower half of the standard to afford the requisite strength, and their heads project above the standards. On the front are two barrels, which receive the axles of the wheels. The wheels are removable from the axles. Through the ends of each axle, opposite properly fitted that the axles are made detachable from the barrels by removing the pin in the end of each.

HEAVY CRIMPING MACHINE.

The machine shown by Fig. 2114 is used for stouter gauges of wire and larger crimps. Here the axles are supported by side standards at each end, in which fit square blocks (see Fig. 2115) to receive

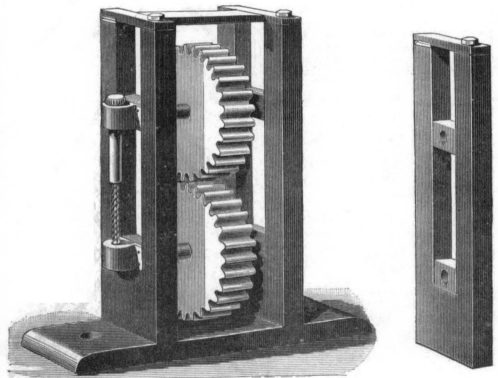

Fig. 2114.

Fig. 2115.

Fig. 2114.—Heavy Crimping Machine.   Fig. 2115.—Side Standard of Crimping Machine.

the wheels, is a removable pin to prevent the wheel and axle from being pulled out of its place accidentally. In working, the screws are turned until the wheels are brought sufficiently close to each other to receive the particular gauge of wire to be crimped. For various degrees of crimping, different pairs of equal wheels are used. It is to allow all wheels to be the axle ends. The square blocks slide up and down the standards, and carry with them the axles, wheels, etc., in order that a change of wheels can be made when necessary. At the top of each standard is a hinged cap; both caps are united by means of a cross-bar, and in front the caps are held firmly by means of a nut on a fixed screw. For closing the wheels together there is an attachment at each side, shown separately in Fig. 2116. On the outside of the top axle-block is a stout barrel, through which passes a movable screw-rod. In a similar manner, there is outside each lower block a barrel, into which is firmly embedded a horizontal screw, and through which passes the screw-rod already referred to.

PUTTING WIRES TOGETHER.

"Putting together" is the trade term for crossing the wires over and under one another. The crimped wires are cut into the required lengths by means of shears, preparatory to being "put together." First one wire is placed across the bench

Fig. 2116.—Adjusting Screw of Crimping Machine.

as at A (Fig. 2117). Another is then placed across it at one end, as B, crossing it at E. A third C (Fig. 2118), is fixed across in the same direction as, and exactly parallel to, the first A. Here care must be exercised in order that the end of the third wire shall be level with the ends of the first and second wires—that is to say, when the wires are all put together, the tops and bottoms should follow an imaginary straight line. The fourth wire D is next put across exactly parallel to B (Fig. 2119). This method is continued until sufficient wires are placed over and under each other to assume the appearance shown by Fig. 2120. This completes what is called the "start." The usual

practice is to place a weight of some sort upon the start. The "putting together" can then be continued from

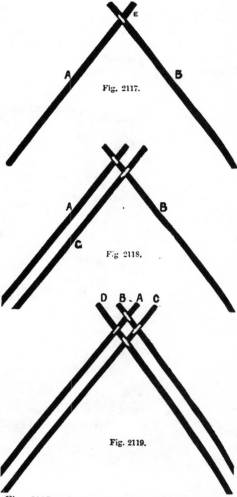

Fig. 2117.

Fig 2118.

Fig. 2119.

Figs. 2117 to 2119.—"Putting Together" Wires.

either end; there is no necessity whatever to work at both sides alternately when the start has been completed. By these means, dozens of yards of fencing can be made in a comparatively short space of time. It is well always to keep the start. Having completed the length required, undo from the remainder the wires which constitute the start, and lay them by for another time. It may appear to be a comparatively easy and simple matter to make the start, but it

bent over it by means of the pliers (Fig. 2123). When attaching the wires to a frame, temporarily secure them by turning over a wire here and there on each

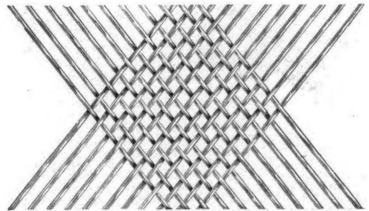

Fig. 2120.—" Start" produced by "Putting Together" Wires.

is really very difficult. Fig. 2121 shows a section of crimped wires when "put together," and also shows how each wire crosses over and under its companions. It is essential, when making the start, to allow the wires to project a distance beyond the points at which they cross

side. Then turn each wire up and partly over in consecutive order, and clip off, by means of the nippers, any superfluous wire. Finally, squeeze down each wire with the aid of the nippers or pliers. The use of the nippers (Fig. 2124) for this latter

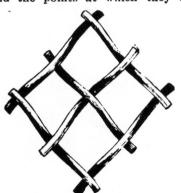

Fig. 2121.—Crimped Wires Put Together.

Fig. 2122.—Wire Frame.

each other E (Fig. 2117), so that there shall be sufficient length to enable the worker to turn over the wires, that is, when the put-together piece is attached to a frame, the parts of the wires which come in contact with the frame (Fig. 2122) should be

part of the work will be found to be the best for the larger sort, but the pliers (Fig. 2125) are to be preferred for the smaller descriptions. Fig. 2126 represents "squeezing down."

FLEXIBILITY OF CRIMPED WIRES.

When crimped wires are put together
they are capable of being stretched bodily

Fig. 2123.—Turning Over with Pliers.

to a great length, or, on the other hand,
of being closed together as in the case of
some window blinds. Fig. 2127 shows a
number of wires as " put together," and
Fig. 2128 the same wires when extended ;

Fig. 2124.—Nippers.

Fig. 2125.—Pliers.

while Fig. 2129 shows the same when
closed together.

" TWO-AND-TWO " ARRANGEMENT.

Sometimes the wires are fixed " two-

Fig. 2128.

Fig. 2126.—Squeezing Down with Nippers.

Fig. 2127.

Fig. 2127.—Wires as Put Together.   Fig. 2128.—
Wires Extended.

and-two," as in Fig. 2130.   To do this,
when the first and second wires are laid
across each other, another wire has to be

placed, in each case, on the outside of
them, parallel, as in Fig. 2119; the work

is then continued in a similar manner as in the single wire mesh, allowing two or more crimps to intervene alternately.

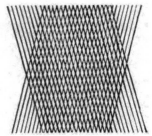

Fig. 2129.—Wires Closed Up.

SQUARE MESH.

Another mesh is the square one shown in Fig. 2131. The only difference between this and the diamond mesh consists in the

Fig. 2130.—Crimped Wire Mesh.

laying of the wires at right angles across one another, as shown in the progressive diagrams (Figs. 2132 to 2134).

Fig. 2131.—Crimped Wire Mesh.

USE OF GUIDING BOARDS.

Sometimes guiding boards are used to assist in putting together. Fig. 2135 shows

a section of guiding board for use with the diamond mesh. It is merely two boards, an inch or so in width, of any suitable length, and about ½ in. thick. They are

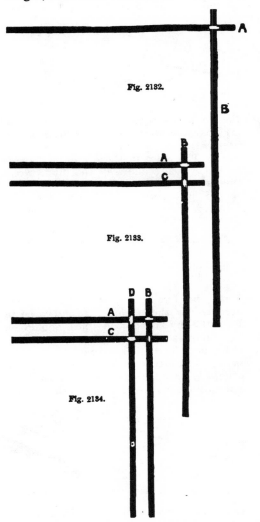

Fig. 2132.

Fig. 2133.

Fig. 2134.

Figs. 2132 to 2134.—Putting Together Square Mesh Trellis.

hinged together, and holes are bored through in the manner shown in Fig. 2136. They are secured with a hook and eye at each end. The wires are placed through this from the side A (Fig. 2135), either in the order already described—namely, across each other progressively—or else one set of parallel wires is fixed first, and then the cross wires are connected with them in the opposite direction.

Fig. 2135.—Wires in Guiding Board.

tion. For the square meshes, a rectangular board, as shown in Fig. 2137, could be used, also hinged, and with hooks at the ends. The holes in these guiding boards should measure one way the diameter of the wire; in the other direction they should just receive the crimps of the

Fig. 2136.—Guiding Board Open.

wires; by this is meant that the space occupied by the crimping, as indicated in Fig. 2138 by the dotted lines, should be the same as the width of the holes (Fig. 2137), in order that the wires may be held in their proper position, instead of the crimps lying upon their sides, as they are apt to do when being put together in the

Fig. 2137.—Another View of Guiding Board.

usual manner. Guiding boards, however, are not always considered advisable, as they are sometimes rather awkward to handle, and the work of putting together is done less quickly with their aid.

Lacing Wires with the Slide.

An extensively used mesh is that shown in Fig. 2139. The wires forming it are put together with the aid of a slide. This is a very simple affair, and a sketch of it is given in Fig. 2140. It is but a strip of metal or wood, with holes through it,

Fig. 2138.—Crimped Wire.

secured to the bench. In putting together a piece of work of the pattern shown in Fig. 2141, the long wires are placed through the slide, their front ends being allowed to project for a short distance, and to them, at this part, is temporarily fastened a cross wire, this being subsequently laced to the long wires.

Fig. 2139.—Laced Work.

The latter are then drawn partly through the slide, and cross wires laced at equidistant and, of course, parallel points to the first. When the first cross wire has been fastened, the work is almost the same as if the wires were in a frame, for the opposite ends to those to which the first cross wire is laced being in the "slide," they are there prevented from

Fig 2140.—Slide used in Making Mesh.

closing together, or otherwise becoming inconvenient to handle. It is usual to lace the alternate wires in opposite ways to the others: that is, the first, third, fifth, and so on, are laced from left to right, whilst the second, fourth, sixth, etc., are secured by the lacing wire travelling from right to left. Sometimes the middle one or two only, according to whether an odd or even number of cross wires is used, are laced in a reverse direction.

Fig. 2141.—Interlacing Crimped Wires.

HEXAGONAL MESH.

The hand method of working the hexagonal mesh is comparatively simple. The process is shown in Figs. 2142 to 2145. A number of hardwood or iron pegs, corresponding precisely to the number of clear holes required in the work, are inserted across the bench, as shown in Fig. 2146, in a straight line with one another, each peg being separated from the other on each side by about the space two or three thicknesses of the wire would occupy. Each peg (Fig. 2147) is hexagonal in shape, and is the same size as the mesh required to be made. An even number of wires of the proper gauge are laid, as in Fig. 2142, on the bench. One pair will

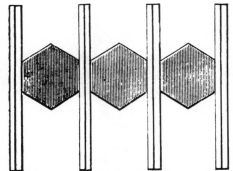

Fig. 2142.—Laying Out Wires for Hexagonal Mesh.

be on the outside of the end peg at one side of the bench, and another pair on the outside of the peg at the opposite extreme; while the remainder will be dis-

posed of in pairs between the rest of the pegs. A wire bound round the bench secures the wires temporarily if these are hooked on to it; or the wires

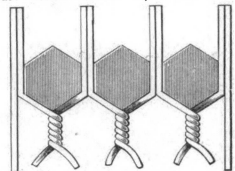

Fig. 2143.—First Stage of Twisting.

may be attached to small staples driven into the bench; or, again, the ends of the wires can be turned and driven into the bench. The inner wire at one side, and the one nearest to it of the adjacent pair, are twisted by the hands or pliers, according to the toughness of the work, until the twisted section is in length equal to one side of a hexagon (Fig. 2143); then the remaining single wire of the second pair is twisted in a similar manner

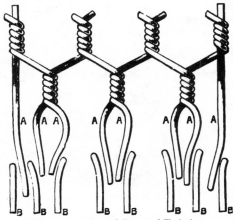

Fig. 2144.—Second Stage of Twisting.

in connection with the wire nearest to it of the third pair. So the operation is proceeded with until they assume the appearance shown in Fig. 2143. The two outside wires always remain in a straight line—that is, they do not follow a hexagonal shape, but are merely twisted round the others where necessary. The ends which

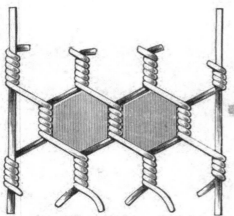

Fig. 2145.—Third Stage of Twisting.

have until now been secured to the bench must be twisted, and will appear as in Fig. 2144. At this stage the work is placed over the pegs, as in Fig. 2145, a half mesh remaining outside the pegs at each end. It will be found necessary to remove one of the pegs occasionally to permit this to be accomplished. When in the position last mentioned, the same process is followed as described above, a wire in each pair being twisted round the wire nearest to it in the pair close by. Every time they are twisted the same number of turns

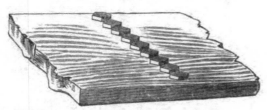

Fig. 2146.—Pegs for Forming Hexagonal Mesh.

must be given them to bring all twisted sections to the same length. By following these means, any length of netting or latticing can be made ; the finished portion being rolled as it increases (see Fig. 2148). It will be necessary to use a weight to

keep the work steady. When the ends of the wires are reached, they can be joined on to others in order to continue the length. Supposing the ends to be shown

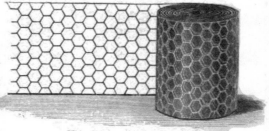

Fig. 2148.—Roll of Peg Latticing.

at A (Fig. 2144), the ends B can be continued with these. The joints are made by twisting B round A ; practice will soon show the correct length of the twist.

PEG LATTICING.

In the mesh illustrated by Fig. 2149 a single row of diamonds is introduced between every two, three, or more rows of hexagonals. To effect the difference, it is only necessary to allow the plain wires to cross one another, as in Fig. 2150, when a

Fig. 2147.—Peg for Making Wire Latticing.

row of hexagonals is finished, instead of placing them in a continuous straight line as is done in Fig. 2143. The wires, where they at first cross one another, are then twisted with two turns, again coming straight, as shown in Fig. 2144. Here they are proceeded with in the same manner as at the start of the work, each pair where they cross being twisted, as

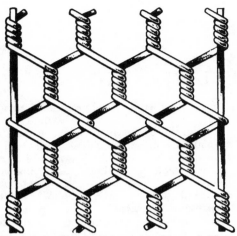

Fig. 2149.—Peg Latticing with Diamond Meshes Intervening.

in Fig. 2151. The mesh in which there are two rows of diamonds (see Fig. 2152) only

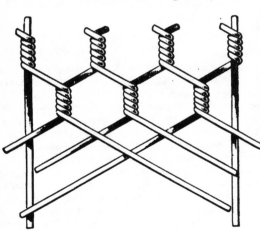

Fig. 2150.—Making Diamond Meshes in Peg Latticing.

differs from the others in that the wires are again crossed when one row of diamonds has been completed, and are

treated in exactly the same manner by being twisted twice, and then continued in a hexagonal mesh. There are other

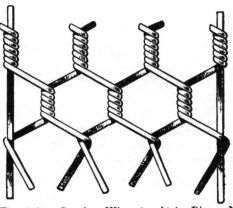

Fig. 2151.—Crossing Wires to obtain Diamond Meshes.

varieties of this mesh. Sometimes the twisted sides of the hexagon are lengthened to twice the distance occupied

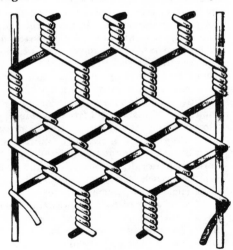

Fig. 2152.—Peg Latticing with Two Rows of Diamond Meshing.

by the plain sides; but those already described will convey an idea of their construction.

ELECTRIC BELL MAKING.

THREE TYPES OF ELECTRIC BELL.

ELECTRIC bells may be divided into three classes : (1) single - stroke bells ; (2) trembling - stroke bells ; (3) magneto - electric bells. Single-stroke bells are those in which the hammer is made to strike the gong once when the circuit is closed, and the number of strokes is controlled by the person closing the circuit. Trembling-stroke bells are those in which provision is made for automatically breaking the circuit after each stroke of the hammer on the gong. When the circuit is closed, the current passes through the magnet coils, as in the single-stroke bell, and thus causes the magnet to attract the armature on the hammer shaft. The movement of the armature breaks the circuit, and thus ceases to attract the magnet. The armature is then drawn back by a spring and closes the circuit, when it is again attracted by the magnet. This to-and-fro movement continues whilst the push button is pressed, and thus gives a rapidly vibrating or trembling movement to the hammer. This form is generally adopted for house electric bell work. Magneto-electric bells are constructed in a special manner to fit them for being worked by the pulsating current from a magneto-electric machine, the hammer vibrating between two gongs. These bells are sometimes used in hotels and large houses, and where signals are sent between two stations. They are frequently used in telephone circuits, but rarely in ordinary electric bell installations.

PARTS OF AN ELECTRIC BELL.

The electric gongs that are in general use are dome-shaped ; but the sheep bell, spiral steel wire, and church bell forms are also more or less commonly used. The common grade of English-made bells has the wood back of polished pine, the springs of German silver, and the contacts of silver ; whilst teak or walnut cases, boxwood bobbins, filled with No. 26 B.W.G. silk-covered copper wire, tempered steel springs, and platinum contact points, are used in the medium grades. In the best qualities the finish is better. A good bell has its working parts fixed to an iron frame, which is screwed to a seasoned teak or walnut back, and protected with a well-fitting cover of the same wood. The contact points are of platinum, and the wire of its coils is silk-covered.

COMMON QUALITY BELLS.

Bells of very common quality have stained deal backs and covers ; the steel gongs and working parts are fixed to the wood backs separately without iron frames ; the springs are of brass, without silver, aluminium, or platinum contact points ; the wire is cotton-covered, and the parts are often badly proportioned. In some bells an attempt at deception is made by soldering bits of German silver to the contact points, or merely putting a spot of solder alone on these parts. German silver and solder are both useless for contact points, as they soon corrode, and then the bell fails to ring. It is easy to test with a drop of nitric acid, which will turn green on German silver, and black on solder, but will not affect platinum. A drop of spirits of salts will dissolve aluminium ; it will, however, have no effect on platinum, silver, or German silver.

OTHER KINDS OF ELECTRIC BELLS.

In small houses, a cheap and pleasing system of electric bells may be established by employing a set of chime bells. These are sets of bell movements and gongs

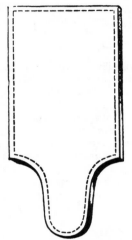

Fig. 2153.—Wood Base-board of Electric Bell.

mounted on one base and protected with one cover. There are from three to eight gongs in a set, of different tones, each indicating the room from which the bell has been rung. Church-pattern bells, so called because they resemble in form the bells of a church belfry, are constructed so as to have the working mechanism inside the gong, which forms a protective cover. Owing to their ornamental appearance, they may be fixed in any exposed position, as over doors, under archways, or mounted on brackets. The tone, although loud, is not dissonant or harsh, like that of ordinary bells. In addition to these varieties of electric bells, there are others adapted to special requirements, such as relay bells, employed on long outdoor lines; continuous-action bells for burglar alarm systems, which continue ringing even when the lines are cut; bells with indicator movements in their cases; heavily mounted gongs in cases water-proof and weather-proof, for underground and outdoor use; electric buzzers, or movements

without gongs, for sick-rooms and hospital wards; and electric trumpets, in which a vibrating diaphragm displaces the hammer and gong of the ordinary electric bell.

DEFECTS OF COMMON ELECTRIC BELLS.

An electric bell is simple in construction, and as the various parts can be bought ready made, no great mechanical skill is required in making it. Electric bells vary very much in quality as well as in type. A bell with badly wound magnets and leaky connections will exhaust the battery. If the iron of the magnets is hard or badly annealed, it will cause the armature to stick to the magnets on the first contact, and fail to ring the bell. Badly fitted screws or badly constructed contacts, or the improper fixture of the metal parts to the wooden base, may cause the bell to fail to act when required.

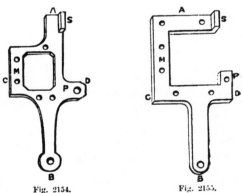

Fig. 2154.   Fig. 2155.

Figs. 2154 and 2155.—Electric Bell Frames.

GOOD AND BAD POINTS OF ELECTRIC BELLS.

The good and bad points of electric bells may here be noted as a guide in choosing. The wooden baseboard of a bell should be made of teak or mahogany, or some such wood not easily warped by changes in the moisture and temperature of the air. If the base is of metal, the contact pillar must be well insulated from the metal work with collars of ebonite above and below, and the connecting screws must be similarly insulated. The set screw to the contact breaker should have good threads, be well fitted, and be provided with a good lock-nut. If this part is defective, the armature spring will work out of contact under the jarring action to which it is subjected. When the bell is caused to ring, try to stop it by placing the forefinger lightly on the armature; if this has a tendency to stick to the magnet, and does not readily re-start itself, the bell should be

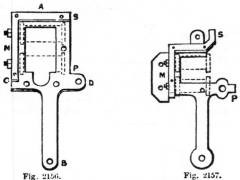

Fig. 2156.   Fig. 2157.

Figs. 2156 and 2157.—Electric Bell Frames.

rejected, as doubtless the iron of the magnet has not been annealed. A good electric bell magnet should attract iron filings when the current is passing through its coils, but should drop them the instant the current is interrupted. The contact screw should have a tip of platinum in contact with a small plate of platinum soldered or riveted to the armature spring. These parts must be of platinum, but German silver or aluminium is sometimes substituted, as has been said; tests (see p. 724) will soon show whether this is the case.

Baseboard and Frame of Electric Bell.

The foundation for an electric bell is a wooden baseboard. For an ordinary plain 3-in. bell, select a piece of sound well-seasoned teak or mahogany 8 in. by 4¼ in. by ⅞ in. thick. If the base has to carry a relay in addition to the bell, the wood should be 2 in. longer. Plane both sides smooth, then cut the wood to the form shown in Fig. 2153, and bevel the edges on

one face. Fill in the grain of the face and edges, and finish by French-polishing. A metal frame holds the magnet, armature, contact screw post and bell pillar, and prevents their being shifted from

Fig. 2158.—Magnet Core and Nut.

their positions relative to each other by changes of temperature and jarring motion in the bell itself. This frame may be of sheet iron, sheet brass, sheet copper, or even of stout sheet tin cut to the form shown in Fig. 2154; or it may be cast in brass, gun metal, or iron. The frame may have other forms and be equally useful; some of these are shown by Figs. 2155 to 2157. The reference letters in Figs. 2154 to 2157 correspond with each other. The dimensions are, from A to B 5¼ in., from c to D 3 in. The hole B is for the foot of the bell pillar; the hole P is for the foot of the contact screw pillar; to the lug s is fastened the armature spring; and M shows the position where the magnets are fastened. The coil bobbins are shown by dotted lines in Figs. 2156 and 2157. In this form the yokes of the magnet cores are made separate from the base plate, and fixed to it afterwards by studs or set screws.

Magnet Cores.

The magnets are the next parts to consider. These are made up in three parts: (a) the cores; (b) the bobbins or reels;

Fig. 2159.—Section of Core Screwed Internally.

(c) the wire. The magnet cores must be made of tough rod iron, preferably Swedish. If the iron rod selected is quite round and smooth the cores will not need to be turned, but the roundness and smoothness of the cores should be assured by turning if necessary, so that the cores and the bobbins of the coils may fit well, for it is most important that the wires which are wound on the bobbins should lie close to the magnet cores. The following

Fig. 2160.—Magnet Core.

table shows the sizes of the magnet parts for bells of several different sizes:—

PROPORTIONATE SIZES OF MAGNETS.

Diameter of Bell.	*Length of Cores.*	*Diameter of Cores.*	*Length of Bobbin.*	*Diameter of Bobbin.*	*Size of Wire, B. W. G.*
In.	In.	In.	In.	In.	
2½	2	5/16	1¾	¾	24
3	2¼	⅜	2	⅞	24
3½	2½	7/16	2¼	1	22
4	2¾	½	2½	1⅛	22
5	3¼	⅝	3	1⅜	18
6	3¾	¾	3½	1⅝	16
7	4¼	⅞	4	1⅞	16
8	4¾	1	4½	2⅛	14
9	5¼	1⅛	5	2⅜	14

The length given in the above table allows for the core ends being turned down and threaded to receive the nuts (Fig. 2158) which hold them in the yoke. Sometimes nuts are not used, but the cores are fixed to the yoke with screws entering the cores (Fig. 2159). If this plan is adopted the ends of the cores must be drilled and tapped to receive the screws (Fig. 2160), and it will not be necessary to cut the cores so long as when they are to be fastened with nuts. Fig. 2161 shows even a simpler way; in this the core is riveted into the

Fig. 2161.—Another Form of Magnet Core.

yoke. However good the iron may be, after the cores have been cut off it must be re-annealed to make it quite soft. Cores made of hard iron, or imperfectly annealed, will retain some magnetic influence over the armature after contact is broken and the current has ceased. Iron is usually annealed by heating to redness in a good fire, then covering fire and iron with hot ashes, and allowing all to cool gradually for some ten or twelve hours before disturbing the iron. After the cores are annealed, one end of each must be turned down to form a pin with shoulder to fit in the yoke, and the other ends filed level and smooth to form faces for the armature. If the cores are to be riveted to the yoke as illustrated in Fig. 2161, this riveting must next be done. If they are to be attached with nuts as in Fig. 2158, the pins must be screwed to fit the tapped nuts. If they are to be fastened with screws

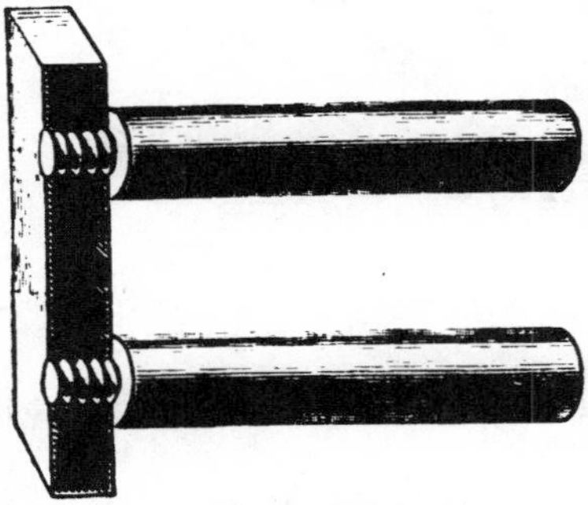

Fig. 2162.—Magnet Cores in Yoke.

as in Fig. 2159, the ends must be drilled and tapped. Yet another way of attaching the cores to the yoke is shown by Fig. 2162. This arrangement for fixing the cores in the yoke should be made before the bobbins are wound with wire.

Magnet Yoke.

The yoke, as shown in Fig. 2162, is the bar of metal to which the cores or legs of the magnet are attached. It should be made entirely of iron and the cores fixed to it; and the effect will be the same provided the two cores are connected by a strip of soft iron. A piece of angle iron of dimensions to suit the size of bobbin to be mounted on the cores will make a good yoke, and will also serve as a bracket.

Magnet Bobbins or Reels.

The bobbins for the coils of electric-bell magnets are usually turned of boxwood or ebonite, made as thin as possible consistent with strength. Special attention is paid to the thinness of the tubular part, the best effects being obtained with the insulated coil of wire as close to the core as it can be. This consideration will also show that the bobbins should fit the cores closely in every part. If the bobbins go loosely on the cores, fill the space by winding a slip of thin paper on the cores and fit the bobbins over this. Fig. 2163 shows

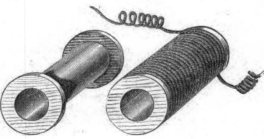

Fig. 2163.   Fig. 2164.

Fig. 2163.—Magnet Bobbin.   Fig. 2164 —Wound Magnet Bobbin.

the bobbin, and Fig. 2164 shows it when wound.

Winding Magnet Coils.

The wire for the coils of an electro-magnet should be of pure soft copper perfectly insulated. The wires in general use for electric-bell magnets are coated with green silk. This coating must be without defect in any part—that is, the copper wire must not have a bare place. Should a bare place be noticed whilst winding the wire on the bobbin, it should be carefully covered, because if two such bare spots come together on two different layers, the coil will be short-circuited, and a portion of the magnetic effect of the current be lost. The covered copper wire may be wound on the bobbins in a small lathe, but a little practice will be needed before the wire can be wound on regularly, like the cotton on a reel. The reel or spool containing the wire to be used is placed on a piece of iron wire where it is free to revolve, and that part of the wire which is to go on to the bobbin is held in one hand at a distance of about a foot from the bobbin. The bobbin being mounted between the lathe centres, the wire will go on in regular coils

and almost guide itself, if allowed to follow its course. A fold of white tissue paper between the layers of wire serves as a guide to wind the wire with exactitude. The end of the wire first placed on each bobbin must be brought out through a small hole bored in the end or laid in a small nick made in the edge of the end, and some 8 in. of it coiled spirally around a small rod to serve as an elastic connection between the two bobbins, or to connect the coils with any other part. The bobbins should be quite filled with wire, and the outer end secured from unwinding.

Putting Bobbins on Cores.

These outer ends should also be coiled spirally on a rod. The bobbins thus made may be slipped on the cores, and the two inner ends of the wires connected. To do this, bare the copper by stripping off ½ in. of the silk insulation, clean the bare copper by scraping, and twist the two ends tightly together. The joint may be soldered to ensure good contact, but this is not always done.

Armature of Electric Bell.

The armature must be made of iron, as soft and as well annealed as the cores of the magnet. The size of this armature

Fig. 2165.—Armature of Electric Bell.

(Fig. 2165) must be proportioned to the diameter of the magnet coils and their distance apart. It should be long enough to extend to the edges of the bobbins, and wide enough to cover the cores, and it should be thick enough to take the hammer shaft in one end. A size suited for the armature of a 4-in. bell is 2 in. × ⅝ in. × $\frac{1}{16}$ in. This piece of iron must be filed up flat and smooth; a hole is drilled in one end and tapped to take the screwed end of the hammer shaft; at the other end, in the positions shown by A B (Fig. 2165), two holes are drilled and tapped to receive two small iron set screws intended to hold the armature spring shown in Fig. 2166.

ARMATURE SPRING.

The armature spring may be made of spring brass, German silver, or steel. Its length and width are determined by the

CONTACT SCREW.

The upper part of the pillar (Fig. 2167) carries a brass contact screw which forms an electrical connection between the armature spring and the pillar. The screw is $\frac{1}{8}$ in. diameter and $\frac{3}{4}$ in. long, and has a

Fig. 2166.—Armature Spring.

dimensions of the armature, but it must be long enough to extend from the lug s to the pillar at P (Figs. 2154 to 2157). It should be stiff enough to bring the armature back to the contact screw sharply after the bell has been struck, but not so stiff as to require a high battery power to work it. Two holes are drilled at A B (Fig. 2166) to receive screws to hold it to the lug s (Fig. 2154), and two holes at C D to receive screws to attach it to the armature (Fig. 2165). At E another small hole should be drilled to receive a bit of No. 20 B.W.G. platinum wire, which, when riveted to the spring, forms the contact for the screw point.

CONTACT OR BREAK PILLAR.

The contact pillar and screw is shown with its accessories by Figs. 2167 and 2168 The pillar should be turned out of $\frac{1}{2}$-in. brass rod, the top part above the foot should be $\frac{3}{8}$ in. diameter, and the lower threaded part $\frac{1}{4}$ in. diameter, to receive the nut B (Fig. 2168), or to be screwed into the wood base ; either method is adopted. Where the pillar is secured to the frame by a nut beneath the base, a recess is cut beneath the base for the nut, the connecting wire is carried through a small hole into this recess, and the end secured between the nut B (Fig. 2168) and the thin brass collar C (Fig. 2168). The threaded part of this pillar will pass through the hole P (Fig. 2154, p. 725), in the metal frame, and must be insulated from the frame by the collar, which may be turned out of boxwood or ebonite to the shape shown by A (Fig. 2168).

milled head, as shown. A small hole in the tip of this screw is plugged with a platinum wire which projects to form contact with the platinum at E on the armature spring. Platinum is used because the electric spark which passes at the point when the bell is ringing has very little effect on this metal, whilst it will destroy most other metals. A hole is drilled through the pillar, about $\frac{3}{8}$ in. from the top, and tapped to receive the contact screw. The pillar should then be sawn across its diameter down to this hole.

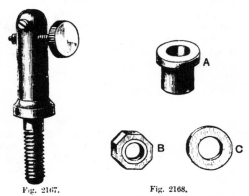

Fig. 2167.   Fig. 2168.

Fig. 2167.—Contact of Break Pillar.

Fig. 2168.—Insulating Collar, Brass Nut and Brass Collar.

Near the top of the pillar a small hole is drilled at right angles to the slit, and tapped to receive the small clamping screw shown in Fig. 2167. The contact screw, when nicely adjusted to ensure the best ringing action of the armature on the bell, is secured there by this clamping screw.

This clamping may be done by using lock nuts on the contact screw; but as these shake loose under the vibratory action of the armature, it is found best to secure the screw by means of the transverse screw (Fig. 2167), by which, also, the wear of the threads may be taken up.

ture to the magnet; the spring suitable in length and stiffness; and the hammer-shaft sufficiently long properly to ring the bell. The metal base plate, which should have countersunk holes for the heads of the screws used in fastening it to the wood base, may be fixed on first. Next fasten

Fig. 2169.—Armature and Spring.

HAMMER OF ELECTRIC BELL.

The hammer head of an electric bell is generally a small disc or ball of brass, secured to a shank made of hard iron or brass wire of about No. 11 or No. 12 B.W.G., as shown in Figs. 2169 and 2170. A hole drilled in the head is tapped to receive the screwed end of the shank, then screwed on tight. It is well to secure the head in its right position with solder in addition to the screw, to prevent the head shaking loose. The shank is secured to the armature by screwing in the same way as it is to the hammer head. The exact length must be found by measurement to ensure the head striking the bell in the most advantageous spot.

THE GONG.

The bell itself is generally nickel-plated, and distinguished by the name of gong. The pillar to support the gong may be about 2½ in. to 3 in. long, made of ⅜ in. rod iron, turned and screwed at the ends, as shown in Fig. 2171. The bottom part is screwed into the base and metal frame, or secured by a nut beneath the base. The top part passes through a hole in the centre of the gong, being secured by a brass milled head (Fig. 2172).

PUTTING BELL TOGETHER.

Supposing that all the necessary parts already mentioned have been got together, see that each part is proportionate. the metal frame to the base-board; the magnet coils and cores to the gong; the arma-

the yoke of the magnet to the base plate; then attach the cores to the yoke, and slip the wound bobbins on the cores. Strip off the silk covering from the inside ends of the two coils, clean the wires and twist them together as previously explained, and serve the outside ends of the coils in a similar manner. Attach one of these to the left-hand terminal, or binding screw (Fig. 2173) of the bell, and fix the other to the metal base plate or to the lug carrying the armature spring, in both cases using metal screws and making good connection between clean metals. This done, fasten the armature to its spring, and this to the lug on the base plate, and insert the hammer shaft in the ends of the armature. The contact pillar should now be

Fig. 2170.—Spring Hammer of Electric Bell.

fixed in its place, and in doing this see that the insulating collar A (Fig. 2168) entirely prevents the metal of the tang from touching the metal base plate. Next put in the gong pillar, and screw the gong in its proper place. Proceed to adjust the various parts of the movement. The armature spring must be bent so as just to touch the platinum tip of the screw on the contact post, when the hammer is about

¼ in. from the side of the gong ; the armature spring should be just free from contact when the hammer touches the gong.

ELECTRICAL CONNECTIONS ON BELL.

All connections may be made behind the base-board of the bell, to ensure safety from tampering, and for the sake of neatness. The wires are secured to the tangs

Fig. 2171.—Gong Pillar of Electric Bell.

of the various posts and terminals by brass nuts recessed in the base, and the wires led along in saw-kerfs made in the back of the base. These recesses and saw-kerfs should then be filled with paraffin wax.

TESTING ELECTRIC BELL.

All parts having been connected, test the bell by trying to ring it with current from a battery. Adjust the contact screw until the bell gives its best tone. If the armature taps the magnet cores as it vibrates, bend the spring a little outward so as to move it farther away from the cores. If it vibrates too freely, bring this part nearer the cores by bending the spring inward. It may be necessary to bend the hammer shaft to ensure it striking the gong properly. The battery power used in testing and adjusting the bell should be the same as that to be used in working it. A weak battery might just work the bell and a stronger battery not ring it so well, whilst it might ring well with a strong battery but fail altogether with a weaker one.

COVER FOR ELECTRIC BELL.

The working parts of an electric bell are very delicate, and liable to be injured by dust and damp, so they must be protected by a suitable cover. This is usually made from wood of the same kind as that employed in making the base-board of the bell. It is really a box with an open bottom, made of ½-in. wood, neatly put together

Fig. 2172.—Milled Head for Pillar.

with dovetailed joints or mitred corners strongly glued, and highly polished when finished. Holes or notches are cut to allow free working of the hammer shaft and lever of relay. The cover is secured to the base by two brass hooks screwed to the sides, and engaging in brass staples fixed

Fig. 2173.—Binding Screw.

in the base. The top of this cover is attached to its sides by brass screws, in addition to glue, to give it greater strength. Two holes are drilled in the base close to the outside terminals ; these are bushed with brass eyelets, and the bell is hung to a wall or a partition by means of screws passing through these, or by pins.

MAKING A MICROSCOPE AND A TELESCOPE.

A Stand Microscope.

THE instructions given in the first part of this chapter will enable a useful stand

used throughout, and as this can be purchased in almost any locality, it is assumed that this latter alternative is adopted.

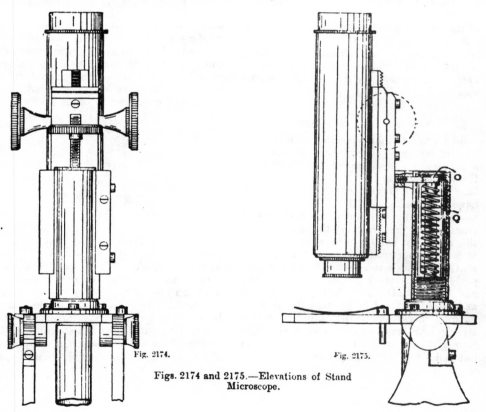

Fig. 2174.

Fig. 2175.

Figs. 2174 and 2175.—Elevations of Stand Microscope.

microscope (Figs. 2174 and 2175) to be constructed at a moderate outlay. If castings can be obtained, so much the better; but if not, sheet brass can be

Microscope Legs.

From a plate of brass $\frac{1}{4}$ in. thick cut the two legs (Fig. 2176), and, after filing them smooth on both sides, solder them together. The edges can then be filed up, holding them between lead clamps in the vice. Before separating them, drill a $\frac{3}{8}$-in. hole; $\frac{1}{2}$ in. from the centre of this

bent slip of thin, hard-rolled brass D, 2 in. long and $\frac{3}{16}$ in. wide, and tapered towards the end as shown. These pins fit easily into the holes B and C, and hold any object, or a slide, that may be placed

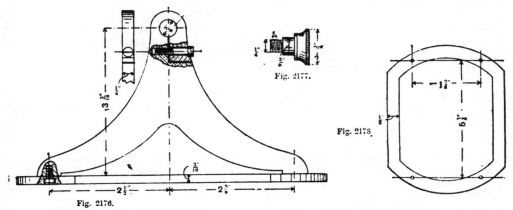

Fig. 2176.—Legs of Stand Microscope. Fig. 2177.—Screwed Pin. Fig. 2178.—Elliptical Plate.

hole, and at right angles to it, a smaller hole should be drilled and tapped with a $\frac{1}{8}$-in. thread, as shown in section by Fig. 2176. This hole should go nearly through. From the centre of the large hole, a saw-cut extends downwards about $\frac{3}{16}$ in., and is then continued to the edge of the plate as shown. The small hole is broached out to the saw-cut; the object of this is to allow it to grip the pin upon which the instrument turns when the screw is tightened. Two of these pins are made as shown in Fig. 2177. A plate shaped as shown by Fig. 2178 is next cut out. It is $\frac{3}{16}$ in. thick, and is screwed to the bottom of the two legs (see Fig. 2176), and provides a firm base for the microscope.

STAGE OF MICROSCOPE.

The stage is made from $\frac{3}{16}$-in. brass plate, and cut to the dimensions given in Figs. 2179 and 2180. In the centre of the square part a $\frac{3}{4}$-in. hole is drilled. Two $\frac{1}{8}$-in. holes are then drilled as indicated by B and C. Next turn the two pins (see Fig. 2181). A $\frac{1}{16}$-in. projection is left at the top, to which is riveted a

underneath them. At $2\frac{7}{8}$ in. from the left-hand end another hole is drilled and tapped with a $\frac{1}{4}$-in. fine thread; into this hole is screwed from underneath a short length of $\frac{5}{8}$-in. brass tube, one end of which is plugged with a piece of threaded brass rod (see Fig. 2182). Two semi-circular pieces of brass, $\frac{1}{4}$ in. thick, are

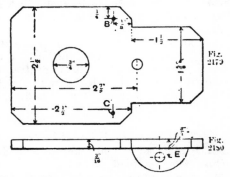

Figs. 2179 and 2180.—Microscope Stage.

then hard soldered to the plate; holes in the centres of these pieces (E, Fig. 2180) are drilled and tapped with a $\frac{1}{4}$-in. fine thread to receive the two pins, one of which is shown by Fig. 2177.

PILLAR TUBE AND FITTINGS.

Take a piece of brass tube (Figs. 2183 and 2184) $2\frac{1}{4}$ in. long and $\frac{3}{4}$ in. external diameter. A fine thread is chased about $\frac{1}{2}$ in. into each end (a pair of chasing tools fifty threads to the inch may be used), and at the top at F a slot $\frac{1}{8}$ in. wide and $\frac{3}{4}$ in. deep is cut. Turn a piece of brass

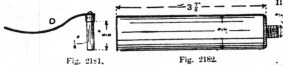

Fig. 2181.—Pin with Brass Clip.   Fig. 2182.— Mirror Tube.

rod or gun-metal to the size indicated at G (Fig. 2183); the small end is $\frac{1}{2}$ in. long, and threaded to fit the tube; a $\frac{1}{4}$-in. projection is left at the top. In the bottom of the large part three holes are drilled to take $\frac{1}{8}$-in. screws (see Fig. 2175). At the top of this tube a disc of brass is turned and threaded (see Figs. 2174 and 2183), and in the centre of this disc a $\frac{3}{16}$-in. hole is drilled and tapped with a fine thread, fifty turns to the inch; into this hole the fine adjustment screw H is inserted. This screw should be of steel, and pointed, the head being a disc of brass, with a milled edge. A plate $\frac{3}{16}$ in thick, the other dimensions of which are given in Fig. 2185, is now cut out, a slot being made at the top to correspond with the slot in the tube. The sides are then filed, to form a dovetail fitting. A circular groove is filed down the centre, and should correspond with the radius of the tube shown by Figs. 2183 and 2184, to which it is firmly sweated.

FINE ADJUSTMENT SLIDE.

The plate of the fine adjustment slide (Fig. 2186) is filed smooth and square, and a slip of brass, $\frac{1}{16}$ in. wide, $\frac{3}{16}$ in. thick, and $4\frac{3}{8}$ in. long, is then bevelled off to correspond with the gauge (Fig. 2187), and then sawn in half, one piece being sweated to the plate, the other half being screwed on (see J, Fig. 2186). The two holes in this loose slip should

then be elongated across the width, so that adjustment may be allowed for after wear. The method of fixing will be readily understood by referring to the illustrations; the slip is made to slide on the plate shown by Fig. 2185. A good, close fitting in all the slides is absolutely necessary, to which end grinding them with a little pumice-powder and oil, or fine oilstone powder, will help.

COARSE ADJUSTMENT SLIDES.

The top slide, which is $\frac{3}{16}$ in. thick, is illustrated by Fig. 2188. Its centre, which is filed out as shown in the plan, is $\frac{1}{8}$ in. deep, $\frac{9}{32}$ in. wide, and $2\frac{5}{8}$ in. long, and has a hole K cut through. This hole is $\frac{7}{32}$ in. wide, $\frac{1}{16}$ in. long, and is bevelled inwards. With a small round file a groove $\frac{1}{16}$ in. by $\frac{1}{8}$ in. is cut across the plate on the

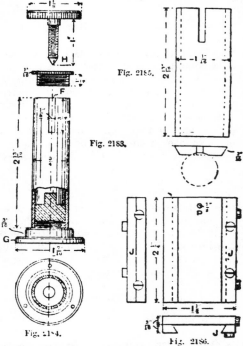

Fig. 2185.

Fig. 2183.

Figs. 2183 and 2184.—Pillar Tube. Figs. 2185 and 2186.—Fine Adjustment Slide Plates.

other side, as indicated by the dotted lines. This groove provides a bearing for the main tube adjustment (Fig. 2189), which is ground in. Two slips of brass, $\frac{9}{16}$ in. by $\frac{1}{8}$ in., are then filed to the gauge ; one piece is sweated on as shown at L (Fig. 2188), whilst the other is secured by two $\frac{1}{8}$-in. screws. The underneath part of this slip should be filed away about $\frac{1}{32}$ in., as indicated by the lines M M (Fig. 2188). The dovetail slide, being carefully fitted will allow the ends to spring on to the slide and take up any wear that may occur when the screws are tightened. The slide illustrated by Fig. 2190 is 4 in. long, and $\frac{3}{32}$ in. thick, and has sweated or screwed to its centre a piece of brass rack which is $\frac{9}{16}$ in. wide, $\frac{3}{32}$ in. thick, and has teeth of $\frac{1}{15}$ in. pitch.

formed in the same manner as is adopted when chasing a thread. Some care must be exercised, and then good results will be obtained.

MAKING ADJUSTMENT PINION.

The revolving tool can also be employed in making the pinion. To the centre of a 2-in. length of $\frac{1}{8}$-in. diameter steel rod sweat a disc of brass $\frac{9}{16}$ in. wide and $\frac{1}{4}$ in. diameter. Obtain a piece of hard wood 3 in. by 2 in. by $\frac{1}{2}$ in., and across its centre file a small groove. Cut a hole $\frac{3}{16}$ in. wide, $\frac{3}{8}$ in. long, and $\frac{1}{4}$ in. deep in the centre of this groove. Into the latter place the steel rod,

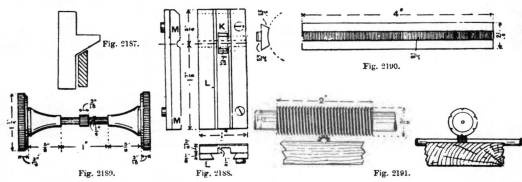

Fig. 2187.—Gauge for Bevelled Slides. Fig. 2188.—Coarse Adjustment Slide. Fig. 2189.—Main Tube Adjustment. Fig. 2190.—Rack Slide. Fig. 2191.—Method of Cutting Rack and Pinion.

CUTTING A RACK.

If this rack cannot be obtained, the following plan may be adopted. Turn up a piece of round cast steel $\frac{5}{8}$ in. diameter, 8 in. long, to the dimensions given in Fig. 2191, and on the large part cut a thread eighteen to the inch. With a triangular file make a few grooves, diagonally, across the threads to form cutting edges, and then harden and temper the steel. Solder a slip of brass, 4 in. by $\frac{1}{16}$ in. by $\frac{3}{32}$ in., to a plate 4 in. by 6 in. by $\frac{3}{16}$ in., and then allow the tool to revolve between the lathe centres, and adjust the tool-rest, upon which can be held the slip, previously made smooth and parallel. If it is kept in close contact with the quickly revolving tool, the teeth will soon be cut and the rack

with the brass disc in the hole ; then, by pressing the thumbs on the rod at each end of the pinion and holding the latter underneath and against the revolving tool and at right angles to the lathe centres, the tool will cut, and the pinion will work itself round. Reference to Fig. 2191 will simplify this description. Two discs are now turned to the dimensions given in Fig. 2189, and screwed on to the steel rod. Before screwing them home, a little solder put upon them will make a firm joint.

MAIN TUBE AND LENS FIXTURES.

The rack and slide are then sweated to the main tube (Fig. 2192), in the position shown by Figs. 2174 and 2175. This tube is $4\frac{7}{8}$ in. long, $1\frac{3}{16}$ in. outside diameter, and $\frac{1}{16}$ in. thick, and at the bottom a disc of brass is sweated, and turned down. The hole in its centre is $\frac{7}{8}$ in. diameter and chased with a fine thread. Into this the object lens is screwed. The one indicated in Fig. 2192 is $\frac{1}{2}$-in. focus. The top part is threaded to screw into the bottom of this tube, whilst the lower piece screws into the part marked N. The following is the method employed for fixing the lens. A hole is first drilled through the centre of the brass disc; in this case the hole is of $\frac{1}{10}$ in. diameter.

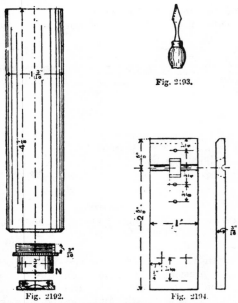

Fig. 2193.

Fig. 2192.—Main Tube of Microscope. Fig. 2193. —Burnishing Tool. Fig. 2194.—Fine Adjustment Brass Plate.

A circular recess for the lens is then turned out and in it is placed the plano-convex lens, taking care to have the convex side uppermost. A combined cutting and burnishing tool, $\frac{1}{8}$ in. in width, of the shape shown in Fig. 2193, will now be required. The work is made to revolve in the lathe, and as (with the cutting edge towards the operator) the point of the tool cuts a groove a little larger in diameter than the recess, the polished side will gradually burnish the edge of

the brass over towards the lens, thus holding the latter firmly in position. The brass should not fit too tightly over the lens, neither should the lathe run too fast, as the heat generated may cause the lens to crack.

FINE ADJUSTMENT BRASS PLATE AND OTHER DETAILS.

A brass plate is now cut to the size and shape shown in Fig. 2194, and in its centre a hole is cut, similar to the one at K (Fig. 2188). The hole should be bevelled inwards as indicated by the dotted lines in the side view given in Fig. 2194; a groove is also cut to correspond with the one in Fig. 2188, and six holes are drilled to take $\frac{1}{8}$-in. screws. The position of this plate will be seen by referring to Fig. 2175. The threaded steel pin O (Fig. 2175) is $\frac{1}{8}$ in. diameter, but the large part is $\frac{3}{8}$ in. diameter, the centre being countersunk. It is screwed into the top of the plate P (Fig. 2186), and works up and down the slot F (Fig. 2183). Q (Fig. 2175) is a stiff spiral spring which should be so adjusted that when in position it keeps the tubes and slides of the microscope at the top of the tube. The fine adjustment is made by screwing the milled head of the screw H (Fig. 2183). The point of the screw enters the countersink in the steel pin, and by screwing this either up or down, very fine adjustment is effected.

MICROSCOPE DRAW-TUBE.

Fig. 2195 illustrates the power tube, which is $4\frac{7}{8}$ in. long, $1\frac{1}{8}$ in. diameter, and $\frac{1}{16}$ in. thick. It should be put on a mandrel and a light cut taken off with a tool in the slide-rest until it fits easily into the tube (Fig. 2192). It should be so fitted that it can be easily raised by the thumb and finger, yet remain firmly in any position required. At $\frac{3}{8}$ in. from the top a $\frac{1}{16}$-in. milled collar is sweated; two stops, as shown at R (Fig. 2195), are pushed tightly into the tube. Turn down a piece of thin sheet brass until it enters a $\frac{1}{4}$-in. length of tube, to which it is sweated; fix this on a chuck, and turn the tube down until it fits rather tightly into the power tube. Turn a hole in the disc a little larger in diameter than the lowest power eye-piece. This stop is placed at the top of the tube, and just below the field lens of the eye-piece; the opening in the bottom stop is about ⅝ in. diameter.

MIRROR AND ITS SUPPORT.

The support for the mirror is shown by Figs. 2196 and 2197 and is now made. It is a short length of tube, ⅞ in. long and ¾ in. internal diameter, sawn through longitudinally, and made to slide rather tightly on the tube (Fig. 2182). A disc of brass, ⅜ in. diameter at the top, is then sweated on at the centre as shown. Take a slip of brass ⅜ in. wide, tapering to 1/16 in. at each end, and 1/16 in. thick, and at each end tap a 1/16-in. hole, into which two steel-pointed screws are inserted. These latter support the mirror, which has two small holes drilled diameter-wise, into which the steel-pointed screws enter, and whilst holding the mirror firmly in any position, allow it to be adjusted at any angle required. In the centre a hole is drilled, by means of which it is screwed to the projecting piece by a ⅛-in. screw and washer; it is then bent as shown. Fig. 2198 illustrates the mirror. A disc of ¼-in. brass is hollowed out, into which a circular mirror fits; the edge of the brass near the mirror is then burnished over, as previously described. On the other side of this disc fit a concave mirror of 3-in. or 3½-in. focus.

COMPLETING THE MICROSCOPE.

In screwing the fine adjustment tube and slides to the microscope stage, be sure that the central line of the main tube corresponds with the centre of the hole drilled in the plate. If they do agree, mark off with a fine steel scriber, and then drill and tap to take ⅛-in. brass screws. When all the parts have been constructed and the microscope has been satisfactorily fitted together, it may be taken to pieces and the various parts polished and lacquered. The inside of the draw-tube, eye-piece, and object-lens should then be blackened. To do this, mix a little ivory

95

black with gold size or thin varnish, and apply a coat; a dead black is necessary to prevent cross reflection of light.

A FOUR-DRAW TELESCOPE.

A small four-draw telescope is an instrument which requires some skill to make, but anyone who can use a lathe and chase a screw will find no difficulty. The telescope to be described is assumed to have a 1-in. object lens, but the instructions will apply to an instrument with any size of object lens or any length of focus. Brass tubing may be procured either stout or light. For a large telescope—such as for astronomical

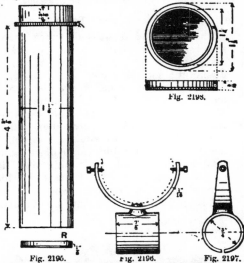

Fig. 2195.—Draw-tube. Figs. 2196 and 2197.—Mirror-holder. Fig. 2198.—Mirror for Microscope.

purposes—a stout tube is required, but for a light hand instrument, the less the weight the better. The tubing required is known as mandrel-drawn; it is thin, hard, and smooth on the inside, and may be had in all sizes from any well-known instrument maker. It is possible that small dents or pits from blows may be in the tubing. These must be removed, as they would appear very unsightly in the finished instrument. To remove them, lay the tube with the dents on a piece of hard level wood, and, with either a burnisher or a round piece of hard wood, rub the inside of the tube over the indentations, which will quickly bring up the surface level.

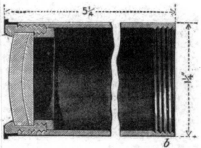

Fig. 2199.—Body Tube of Telescope.

MATERIALS FOR TELESCOPE.

The sizes and lengths of tubes needed are as follows: body, $5\frac{1}{4}$ in. by $1\frac{1}{4}$ in. diameter; first draw, $4\frac{3}{4}$ in. by $\frac{19}{20}$ in. diameter; second draw, $4\frac{3}{4}$ in. by $\frac{11}{20}$ in. diameter; tube to carry eye-piece, 5 in. by $\frac{3}{4}$ in. diameter; for the eye-piece, 2 in. of a size to slip into the last-mentioned tube, and 2 in. of a size smaller still for the erector lens. In addition, it will be necessary to get tubing for collars, which may be made out of tubing one size larger; a piece of sheet brass $\frac{1}{8}$ in. thick and about 6 in. square, from which to cut rings for the ends of tubes and collars.

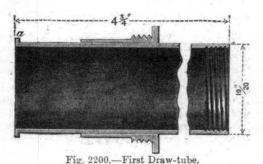

Fig. 2200.—First Draw-tube.

BODY AND DRAW-TUBES.

Chuck a piece of hard wood in the lathe, and turn it down to form a mandrel 4 in. long to receive the body tube (Fig. 2199). True off the ends of the tube at $5\frac{1}{4}$ in. long, and chase an internal thread at each end. The mandrel may now be turned down to receive the first draw-tube (Fig. 2200), which must be cut off $4\frac{3}{4}$ in. An internal thread must be chased on one end, as shown. The second draw-tube (Fig. 2201) must be treated in the same way. At one end of the eye-piece tube (Fig. 2202) a ring is soldered on, as at a; this can be cut off a thick tube—not mandrel tube. If it is not convenient to get a piece the exact diameter, then one a little too large can be made to fit by making a cut lengthwise in it. Place it on the tube, make the cut edges meet, and bind it in its place with a piece of wire. Run soldering fluid between the tube and ring, and then with a hot bit cause solder to flow into the joint until the ring and the tube are solid. Chuck the tube as the others, and cut it off to

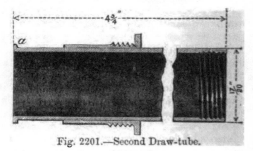

Fig. 2201.—Second Draw-tube.

5 in.; on the collar chase a thread, as shown at a, and chase an inside thread on the other end of the tube.

MAKING SLIDING COLLARS.

Three collars, as shown in Figs. 2200 to 2202, are now required. Fig. 2203 shows one in detail. Procure or make a tube, $1\frac{1}{2}$ in. long, to fit tightly the draw-tube on which it is to work. If a piece the exact size is not to hand, then, as before described, take a tube slightly larger, and cut a piece out and solder the joint. On one end of the collar a sweat a ring of thick tubing $\frac{1}{8}$ in. wide, as at b (Fig. 2203). Chuck it in the lathe, and cut off the end of the ring down to the collar, leaving the collar to project $\frac{1}{8}$ in. Out of sheet brass cut a ring c to fit the projecting end of the collar, and wide enough to project $\frac{1}{8}$ in. beyond b; solder it in place. If due care has been taken, it will be perfectly solid. Chuck the whole in the lathe, and turn down the collar as thin as possible consistent with strength. Chase a thread on b so as to screw into

it in the lathe and turn it up to the size mentioned, chamfering off the outside of the inner diameter as shown in Fig. 2199, and milling the outer edge. A thread must be chased on it, so as to screw it into the body tube; and on the

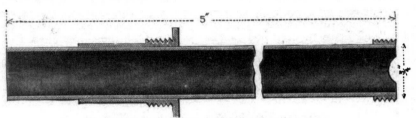

Fig. 2202.—Tube for carrying Erecting Eye-piece.

b (Fig. 2199), turn down c, and mill the edge. This milled edge must project beyond the body tube (Fig. 2199) about $\frac{1}{16}$ in. The other two collars must be made in the same way—that shown in Fig. 2201 must screw into Fig. 2200, and that shown in Fig. 2202 into Fig. 2201. At a (Figs. 2200 and 2201) solder a ring as shown, to prevent the tubes being drawn out. In each collar four slits ($d\ d$, Fig. 2203) are made on the slant. These slits, two each side, produce two free strips; a slight tap must be given to these, causing them to spring slightly inwards, so as to press more fully on the tube which slides in them. This prevents shake. The tubes can now be screwed together, and if the work has been properly done, the tube will be perfectly rigid.

CELL FOR OBJECT LENS.

This cell may be cast or built up, the latter plan being here adopted. It is assumed that the object lens is of 1 in. diameter and 10 in. focus. A ring must be made out of stout tubing of 1 in. internal diameter. Chuck it on the lathe, and turn down the ends true. Cut a ring from sheet brass with the inside diameter, say, $\frac{1}{32}$ in. smaller than the object lens, and the outer diameter $\frac{1}{4}$ in. larger than the body tube (Fig. 2199). This needs only to be done approximately at first. When soldered together, place

inside a thread must be chased to receive the collar to retain the object lens in position. Care must be taken that the inside of the cell is turned true, so that the lens beds perfectly. The stop collar shown in Fig. 2199 may be turned out of the solid, or may be built up like the cell into which it screws. The edge must be milled. The cell, when made, should hold the lens without any shake, and yet without pinching it, so as not to set up a strain.

TELESCOPE EYE-PIECE.

For this will be required a field lens $\frac{7}{16}$ in. diameter and, say, 2 in. focus; eye-

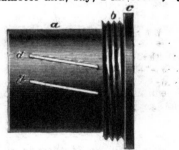

Fig. 2203.—Sliding Collar of Telescope.

lens $\frac{7}{16}$ in., and of a focal length one-third that of the field lens. Add the combined focal lengths together, and divide by 2; this will be the distance which must separate them. They must be plano-convex, with their plano sides towards the eye. The novice when purchasing them, should say what they are for, and ask the distance they are to be separated. Procure a piece of tubing that will fit nicely into the tube (Fig. 2202); cut off, say, ¾ in., and turn out one end to the extent of $\frac{1}{8}$ in., so that the field lens will bed on it. The other end must be turned down, and a thread chased on it, as shown in Fig. 2204. Now place this cell in a chuck, with the end to receive the field lens at the right. Place the field lens in its position, keeping it steady with a finger of the left hand, and with a burnisher press the bezel a (Fig. 2204) over the lens, which will hold it securely. For the cell of the eye-piece c d turn away the metal on the outside to form a bed for the

required for the erector; the one nearest the object lens may be either plano-convex or double-convex; but if the latter, the greatest convexity must be nearest the object lens; it should be ½ in. diameter and 1½ in. focus. The other is a plano-convex, with the plano side nearest the eye; this may be ¾ in. diameter and 1½ in. focus. They must be separated one-half of their combined focus. After what has been said on cell-making, the diagram (Fig. 2205) will be sufficient to show how they are mounted. A stop with a small hole c must be placed as shown. The milled edge at a must project a little beyond the diameter of the

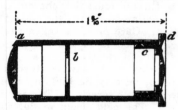

Fig. 2204.—Eye-piece.

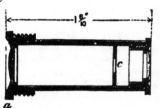

Fig. 2205.—Erector.

Fig. 2206.—Eye-piece Cap.

lens, as shown, and then pierce the centre, leaving a narrow flange. Now cut away the metal at an angle, as in Fig. 2204, leaving a bezel around the hole which has to be turned over on the lens, as in the other case. A thread has to be chased on the cell, as shown. A tube must now be cut of such a length that when the two cells are screwed in their places the lenses shall be separated the required distance. A stop must be placed, as at b (Fig. 2204), to cut off marginal rays.

THE ERECTOR.

If an astronomical telescope were being constructed, the work would now be practically complete; but the instrument, as it now stands, will invert the objects seen, and thus, for all terrestrial observation, it is almost useless. Two lenses are

tube (Fig. 2202), so that the collar shall prevent the tube from being pulled out.

COMPLETING THE TELESCOPE.

The eye-piece cap (Fig. 2206) is now made; this needs no special description. A thumb-piece must be taken out of tube (Fig. 2202), as shown at c, to facilitate the removal of the eye-piece when needed, the flange d (Fig. 2204) of the eye-piece being flush with the tube. It is usual to have a sunshade to draw over the object lens, with a cap to prevent damage. It this be desired, then a ring, say, 1 in. wide, must be soldered to the body tube, and a collar made to slide over the tube; the sunshade is screwed to this, and a cap or cover is fitted to the end. The insides of all the tubes must be made a dead black, and all the exposed parts of the tube burnished and lacquered.

INDEX.

(Illustrated subjects are denoted by asterisks.)